Fundamentals of Electric Propulsion

Fundamentals of Electric Propulsion

Second Edition

Dan M. Goebel, Ira Katz, and Ioannis G. Mikellides
Jet Propulsion Laboratory, California Institute of Technology, USA

Published by John Wiley & Sons, Inc., Hoboken, New Jersey.
Published simultaneously in Canada.

Library of Congress Cataloging-in-Publication Data

Names: Goebel, Dan M., 1954- author. | Katz, Ira, 1945- author. |
Mikellides, Ioannis G., author.
Title: Fundamentals of electric propulsion / Dan M Goebel, California
Institute of Technology, Pasadena, USA, Ira Katz, California Institute
of Technology, Pasadena, USA, Ioannis G Mikellides.
Description: Second edition. | Hoboken, New Jersey : Wiley, [2024] |
Includes index.
Identifiers: LCCN 2023023148 (print) | LCCN 2023023149 (ebook) | ISBN
9781394163212 (cloth) | ISBN 9781394163229 (adobe pdf) | ISBN
9781394163236 (epub)
Subjects: LCSH: Ion rockets. | Electric propulsion. | Space
vehicles–Electric propulsion systems.
Classification: LCC TL783.63 .G64 2024 (print) | LCC TL783.63 (ebook) |
DDC 629.47/55–dc23/eng/20230630
LC record available at https://lccn.loc.gov/2023023148
LC ebook record available at https://lccn.loc.gov/2023023149

Cover Design: Wiley
Cover Image: © Rendering of Gateway space station - Power and propulsion element (PPE) and habitation and logistics outpost (HALO) by NASA

Set in 9.5/12.5pt STIXTwoText by Straive, Pondicherry, India

Contents

Note from the Series Editor

This book is the latest contribution to the Jet Propulsion Laboratory (JPL) Space Science and Technology Series. This series is a companion series of the ongoing Deep Space Communications and Navigation Systems (DESCANSO) Series and includes disciplines beyond communications and navigation. DESCANSO is a Center of Excellence formed in 1998 by the National Aeronautics and Space Administration (NASA) at JPL, which is managed under contract by the California Institute of Technology.

The JPL Space Science and Technology series, authored by scientists and engineers with many years of experience in their fields, lays a foundation for innovation by sharing of state-of-the-art knowledge, fundamental principles and practices, and lessons learned in key technologies and science disciplines. We would like to thank the Interplanetary Network Directorate at JPL for their encouragement and support of this series.

Jon Hamkins, Editor-in-Chief
JPL Space Science and Technology Series
Jet Propulsion Laboratory
California Institute of Technology

Foreword

I am pleased to commend the Jet Propulsion Laboratory (JPL) Space Science and Technology Series and to congratulate and thank the authors for contributing their time to these publications. It is always difficult for busy scientists and engineers, who face the constant pressure of launch dates and deadlines, to find the time to tell others clearly and in detail how they solved important and difficult problems, so I applaud the authors of this series for the time and care they devoted to documenting their contributions to the adventure of space exploration. In writing these books, these authors are truly living up to JPL's core value of openness.

JPL has been NASA's primary center for robotic planetary and deep-space exploration since the Laboratory launched the nation's first satellite, Explorer 1, in 1958. In the years since this first success, JPL has sent spacecraft to each of the planets, studied our own planet at wavelengths from radar to visible, and observed the universe from radio to cosmic ray frequencies. Even more exciting missions are planned for the next decades in all these planetary and astronomical studies, and these future missions must be enabled by advanced technology that will be reported in this series. The JPL Deep Space Communications and Navigation book series captures the fundamentals and accomplishments of those two related disciplines, and this companion Science and Technology Series expands the scope of those earlier publications to include other space science, engineering, and technology fields in which JPL has made important contributions.

I look forward to seeing many important achievements captured in these books.

Laurie Leshin, Director
Jet Propulsion Laboratory
California Institute of Technology

Preface

Since the 2008 publication of the 1st edition of this book, the field of electric propulsion has exploded with thousands of electric thrusters now operating on Earth-orbiting satellites and ion and Hall thrusters having been successfully used for propulsion in deep-space science missions. The recent emergence of large constellations of communications satellites and a new generation of Earth observation satellites has greatly expanded the use of electric propulsion in space. In addition to these applications of electric propulsion technology, research and development has greatly expanded worldwide in universities, laboratories, and industry. Electric propulsion has arrived as a vibrant, growing field.

The 1st edition was intended to alleviate the community's dependence on empirical investigations and laboratory-based development programs for electric thruster invention and advances by presenting fundamental physics models to explain the behavior and scaling of ion and Hall thrusters. This required communicating a basic knowledge of plasma physics, cathodes, ion accelerators, electrical discharges, high voltage, gas dynamics, and plasma-surface interactions. As such, the 1st edition helped to educate a new generation of engineers and scientists interested in working in this field.

In the meantime, progress in the field and emerging applications have pushed researchers to continue to improve existing technologies and to develop new thrusters, which often requires complex multi-dimensional modeling to predict the plasma dynamics that drive the performance and life of these thrusters. The role of physics-based models and simulations in the design, development, and flight qualification of electric thrusters is now well accepted. The 2nd edition of this book retains much of the material of the 1st edition and captures progress in this field over the past 15 years especially with respect to hollow cathodes, modeling and simulation, and the discovery and implementation of magnetic shielding that has revolutionized Hall thrusters. We are fortunate to include Dr. Ioannis G. Mikellides as a coauthor of the 2nd edition to contribute his extensive expertise in these areas. The 2nd edition also includes new chapters on electromagnetic thrusters and future directions in electric thruster research and development. Work in this field is progressing rapidly, and we hope this new edition of the book will lead to further research and advances in our understanding of these surprisingly complex devices.

While this book encompasses a large body of literature in the area of electric thrusters, it is based largely on the research and development performed at the JPL. Therefore, this book should not be considered an all-inclusive treatise on the subject of electric thrusters or a review of their development history. Rather, this effort delves further into the fundamentals of some of the modern electric rockets that are finding increasingly more applications in an attempt to provide a better understanding of their principles.

Dan M. Goebel, Ira Katz and Ioannis G. Mikellides
March 2023

Acknowledgments

We are greatly indebted to our colleagues at the Jet Propulsion Laboratory (JPL), who collaborated in the research and development of these thrusters and provided valuable comments and material for this book. The research and development at JPL on electric thrusters is supported by NASA and the Jet Propulsion Laboratory, California Institute of Technology. The authors would also like to thank the developers and manufacturers of the flight thrusters described here for the performance data and photographs of their thrusters. We are especially grateful for the detailed review of the 1st edition by Prof. John Blandino, who identified corrections and points that needed better explanations. We also acknowledge the contributions, reviews, and new material in the 2nd edition by G. Becatti, R.W. Conversano, R.R. Hofer, Les Johnson, C. Joshi, M.R. LaPointe, P. Mikellides, J.E. Polk, and C.L. Sercel. Finally, we would like to thank the JPL Book Series Editor, J. Hamkins, for his support in the publication of the 2nd edition of this book.

Chapter 1

Introduction

Electric propulsion (EP) is an in-space propulsion technology aimed at achieving thrust with high exhaust velocities, which results in a reduction in the amount of on-board propellant required for a given space mission or space propulsion application compared to other conventional propulsion methods. Reduced propellant mass can significantly decrease the launch mass of a spacecraft or satellite, leading to lower costs from the use of smaller launch vehicles to deliver a desired mass into orbit or to a deep space target. Alternatively, the reduced propellant mass enabled using EP can be used to increase the delivered payload mass on a given launch vehicle.

In general, EP encompasses any propulsion technology in which electricity is used to increase the propellant exhaust velocity. There are many figures of merit for electric thrusters, but mission and application planners are primarily interested in thrust, propellant flow rate or specific impulse, and total efficiency in relating the performance of the thruster to the delivered mass and change in the spacecraft velocity (Δv) during thrusting periods. Although thrust is self-explanatory, specific impulse (Isp) is defined as the mean effective propellant exhaust velocity divided by the gravitational acceleration constant g, which results in the unusual units of seconds. The total efficiency is the jet power produced by the thrust beam divided by the electrical power into the system. Naturally, spacecraft designers are then concerned with providing the electrical power that the thruster requires to produce a given thrust, as well as required propellant flow, and in dissipating the thermal power that the thruster generates as waste heat.

In this book, the fundamentals of EP are presented. There is an emphasis on ion and Hall thrusters that have emerged as leading EP technologies in terms of performance (thrust, Isp, and efficiency) and use in space applications, but other thrusters such as electromagnetic thrusters are also described. Ion and Hall thrusters operate in the power range of hundreds of watts up to tens of kilowatts with an Isp of thousands of seconds to tens of thousands of seconds, and produce thrust levels of a fraction of a Newton to over a Newton. Ion and Hall thrusters generally use heavy inert gases such as xenon or krypton as propellants, but other propellant materials, such as cesium and mercury, have been investigated in the past. Xenon is generally preferable because it is not hazardous to handle and process, it does not condense on spacecraft components that are above cryogenic temperatures, its large mass compared to other inert gases generates higher thrust for a given input power, and it is easily stored at densities of about 1 g/cm^3 with low tank mass fractions. Krypton is also an emerging propellant that is cheaper than xenon and provides slightly higher Isp (due to its lower mass) than xenon for a given thruster electrical configuration. Electromagnetic thrusters can use nearly any propellant, but tend toward lower atomic mass to improve performance and Isp. However, this book will primarily focus on xenon as the propellant in ion and Hall thrusters, and performance with other propellants such as krypton can be examined using the information

Fundamentals of Electric Propulsion, Second Edition. Dan M. Goebel, Ira Katz, and Ioannis G. Mikellides.

provided here. The 2nd Edition of this book provides updates on the progress made in hollow cathode technology and ion and Hall thrusters since the 1st Edition, and includes new chapters on magnetically shielded Hall thrusters, electromagnetic thrusters, and future thruster concepts.

1.1 Electric Propulsion Background

A detailed history of EP up to the 1950s was published by Choueiri [1], and information on developments in EP since then can be found in reference books [2] and on various internet sites [3]. Briefly, EP was first conceived by Konstantin Tsiolkovsky [4] in Russia in 1903 and independently by Robert Goddard [5] in the United States in 1906 and Hermann Oberth in Germany [6] in 1929. Several EP concepts for some space applications were published in the literature by Shepherd and Cleaver in Britain in 1949. The first systematic analysis of EP systems was made by Ernst Stuhlinger [7] in his book *Ion Propulsion for Space Flight* published in 1964, and the physics of EP thrusters was first described comprehensively in the book by Robert Jahn [8] in 1968. The technology of early ion propulsion systems that used cesium and mercury propellants, along with the basics of low-thrust mission design and trajectory analysis, was published by George Brewer [9] in 1970. Since that time, the basics of EP and some thruster characteristics have been described in several chapters of textbooks published in the US on spacecraft propulsion [10–13]. An extensive presentation of the principles and working processes of several electric thrusters was published in 1989 in a book by S. Grishin and L. Leskov [14] (in Russian). A survey of the early flight history of ion propulsion projects in the US was published by J. Sovey et al. [15] in 1999.

Significant EP research programs were established in the 1960s at NASA's Glenn Research Center (GRC), Hughes Research Laboratories (HRL), NASA's Jet Propulsion Laboratory (JPL), and at various institutes in Russia to develop this technology for satellite station keeping and deep space prime propulsion applications. The first experimental electric thrusters were launched into orbit in the early 1960s [15] by the United States (US) and by Russia. The US demonstrated the first extended operation of ion thrusters in orbit with the Space Electric Rocket Test (SERT-II) mission [16] launched in 1970. The SERT-II mercury ion thrusters were also the first to use a hollow cathode "plasma-bridge neutralizer" to provide complete ion beam neutralization in space. Experimental test flights of ion thrusters and Hall thrusters continued from that time into the 1990s.

A detailed description of flight electric thrusters, with performance information and photos, is given in Chapter 12. Briefly, the first extensive application of EP was by Russia using Hall thrusters for station keeping on weather and communications satellites [17]. Since 1971, when the Soviets first flew a pair of SPT-60s on the Meteor satellite, over 250 SPT Hall thrusters have been operated on dozens of satellites to date [18]. Japan launched the first ion thruster system intended for north–south station keeping on the communications satellite "Engineering Test Satellite (ETS) VI" in 1995 [19]. However, a launch vehicle failure did not permit station keeping by this system, but the ion thrusters were successfully operated in space. The first commercial use of ion thrusters in the United States started in 1997 with the launch of a Hughes Xenon Ion Propulsion System (XIPS) [20], and the first NASA deep space mission using the NSTAR ion thruster was launched in 1998 on Deep Space-1 [21]. Since then, Hughes/Boeing launched their second generation 25-cm XIPS ion thruster system [22] in 2000 for station keeping applications on the high power 702 communications satellite [23]. The Hughes/Boeing 702 spacecraft is the first satellite to use electric thrusters for all propulsion applications (orbit raising, station keeping, momentum management, and attitude control).

The Japanese Space Agency (JAXA) successfully used the μ10 ion thrusters to provide the prime propulsion for the Hayabusa asteroid sample return mission [24] launched in 2003, and an upgraded version of this microwave ion thruster [25] was used for prime propulsion on Hayabusa-2 launched in 2014. The European Space Agency (ESA) used the Safran manufactured PPS-1350-G Hall thruster on its SMART-1 mission to the moon [26] also in 2003.The Russians have been steadily launching communications satellites with Hall thrusters aboard since 1971, and will continue to use these devices in the future for station keeping applications [18]. The first commercial use of Hall thrusters by a US spacecraft manufacturer was in 2004 on Space Systems Loral's (SSL) MBSAT, which used the Fakel SPT-100 [27]. ESA launched the QinetiQ/UK T5 ion thruster on the GOCE mission [28] in 2009. This was followed in 2010 with the launch of Aerojet's BPT-4000 Hall thruster [29] (now called the XR-5) on the AF/LMC AEHF satellite. ESA launched the QinetiQ manufactured T6 ion thruster [30] on the BepiColombo mission to Mercury in 2018. SSL launched its first spacecraft that used the 4.5 kW Fakel SPT-140 Hall thruster [31] in 2018 for orbit raising and station keeping. The first spacecraft that did not utilize any chemical propulsion (no kick-stage thruster for orbit insertion or chemical thrusters for orbit maintenance) was the Boeing's "all-electric" 702SP satellite [32] launched in 2015. NASA/GRC's NEXT ion thruster was flight demonstrated in 2021 on the DART mission [33]. Additional ion and Hall thruster launches are ongoing now for emerging LEO communications constellations, such as the thousands of SpaceX Starlink launches each with a Hall thruster used for orbit insertion and station keeping. This trend will continue going forward using various thrusters produced by commercial vendors, with Hall thrusters dominating the present market.

In the past 25 years, EP use in spacecrafts has grown steadily worldwide [34, 35], and advanced electric thrusters have emerged [36] for scientific missions and as a competitive alternative to chemical thrusters for station keeping and orbit raising applications in geosynchronous communication satellites. Rapid growth has occurred in the last 10 years in the use of ion thrusters and Hall thrusters in commercial communications satellites to reduce the propellant mass for station keeping and orbit insertion. The US, Europeans, and the Russians have now flown thousands of electric thrusters on communications satellites and will continue to launch more ion and Hall thrusters in the future. The use of these technologies for primary propulsion in deep space scientific applications has also been increasing over the past 15 years. There are many planned launches of new EP spacecraft, especially for the emerging LEO communications constellations and for challenging scientific missions, that use ion and Hall thrusters as the acceptance of the reliability and cost benefits of these systems grow.

1.2 Electric Thruster Types

Electric thrusters are generally described in terms of the acceleration method used to produce the thrust. These methods can be easily separated into three categories: electrothermal, electrostatic, and electromagnetic. Electrothermal thrusters are not discussed in this book, except for a brief description below, because the technology is very mature and covered in previous books [8], and the performance is relatively low compared to other types of electric thrusters. Common electric thruster types are:

1.2.1 Resistojet

Resistojets are electrothermal devices in which the propellant is heated by passing through a resistively heated chamber or over a resistively heated element before entering a downstream nozzle. The increase in exhaust velocity is due to the thermal heating of the propellant, which limits the Isp to low levels (<500 s).

1.2.2 Arcjet

An arcjet is also an electrothermal thruster that heats the propellant by passing it through a high current arc in line with the nozzle feed system. Although there is an electric discharge involved in the propellant path, plasma effects are insignificant in the increase of the exhaust velocity because the propellant is very weakly ionized. The Isp is limited by thermal heating to less than about 700s for easily stored propellants.

1.2.3 Electrospray/FEEP Thruster

These are two types of electrostatic EP devices that generate very low thrust (<1 mN). Electrospray thrusters extract ions or charged droplets from conductive liquids fed through small needles and accelerate them electrostatically with biased, aligned apertures to high energy. Field Emission Electric Propulsion (FEEP) thrusters transport liquid metals (typically indium or cesium) along needles, extracting ions from a "Taylor-cone" on the sharp tip by field emission processes. Owing to their very low thrust, these devices are used primarily for precision control of spacecraft position or attitude in space, but are now being pursued for CubeSat and small sat applications. Electrospray/FEEP thrusters are not discussed further in this edition because a book specifically on micropropulsion thrusters has previously been published [37].

1.2.4 Ion Thruster

Ion thrusters employ a variety of plasma generation techniques (DC discharge, rf discharge, microwave discharge, arcs, etc.) to ionize a large fraction of a gaseous propellant. These thrusters then utilize biased grids to electrostatically extract ions from the plasma and accelerate them to high velocity by voltages typically ranging from 1 to 2 kV, but which can even exceed 10 kV in laboratory thrusters. A hollow cathode positioned external to the thruster and grids provides electrons to neutralize the beam. Ion thrusters feature the highest efficiency (60 to >80%) and very high-specific impulse (2000 to over 10,000 s) compared to other thruster types.

1.2.5 Hall Thruster

This type of thruster utilizes an $\mathbf{E} \times \mathbf{B}$ cross-field discharge described by the Hall effect to generate the plasma. An electric field is established perpendicular to an applied radial magnetic field to electrostatically accelerate ions to high exhaust velocities. The transverse magnetic field causes the electrons to move in cycloidal orbits around an axisymmetric discharge channel (the "Hall current"), and inhibits electron motion axially toward the anode to create a resistivity in the plasma that supports the ion accelerating electric field. Although the force is transferred from the ions to the thruster through the magnetic field, these are classified here as electrostatic thrusters because the unmagnetized ions are accelerated electrostatically. Hall thruster efficiency and specific impulse is typically somewhat less than that achievable in ion thrusters, but the thrust at a given power is higher, the power density is higher, and the device is much simpler and requires fewer power supplies to operate. Recently, advanced Hall thrusters have begun utilizing a more complex magnetic field shape near the discharge channel exit called "Magnetic Shielding" that eliminated wall erosion and enabled Hall thrusters with lifetimes in excess of 50,000 h.

1.2.6 Magnetoplasmadynamic (MPD) Thruster

These electromagnetic devices use a very high current arc to ionize a significant fraction of the propellant, and then use electromagnetic forces (Lorentz $\mathbf{J} \times \mathbf{B}$ forces) in the plasma discharge to accelerate the charged propellant. Since both the current and the magnetic field are usually generated by the plasma discharge, MPD thrusters tend to operate at very high currents and high powers to generate sufficient force and thereby also generate high thrust compared to the other technologies described above. Since MPD thrusters tend to run at lower discharge voltages (<100 V) to avoid significant electrode erosion, they tend to use light atoms as propellant (H_2, Li and Ar) to achieve longer life at higher Isp.

1.2.7 Pulsed Plasma Thruster (PPT)

A PPT is an electromagnetic thruster that utilizes a pulsed discharge to ionize a fraction of a solid propellant ablated by a plasma arc, and then uses the electromagnetic Lorentz $\mathbf{J} \times \mathbf{B}$ force generated in the pulsed plasma to accelerate the ions to high exit velocity. PPTs that produce the order 1 mN of thrust at an Isp of over 1000 s have been flown at an average input power of about 100 W. The discharge pulse is short (tens of μsec) to avoid arc damage, and the repetition rate is typically used to determine the average thrust level.

1.2.8 Pulsed Inductive Thruster (PIT)

Like the PPT, the PIT is another electromagnetic thruster that operates in pulsed mode. However, its operational principle is distinctly different making it truly unique among all other mainstream electromagnetic thrusters. For example, the Lorentz force is produced inductively and, therefore, the PIT does not require electrodes. Also, in principle, it can operate with a wide range of gaseous propellants, although a single-shot operation has shown optimal performance with polyatomic molecule fuels. For example, a PIT operating with ammonia at 4.6 kJ per pulse demonstrated nearly constant efficiencies exceeding 50% at the range of specific impulse $4000\ \text{s} < \text{Isp} < 8000\ \text{s}$. Its propensity to work best with such polyatomic fuels renders the thruster as a potentially ideal candidate for a water-propellant operation.

Some of the operating parameters of chemical and electric thrusters, most with flight heritage, are summarized in Table 1-1.There are many other types of EP thrusters in development or merely conceived that are too numerous to be described here, some of which are described in Chapter 10. This book will focus on the fundamentals of electrostatic and electromagnetic thrusters as outlined in the next section.

Table 1-1 Typical parameters for selected thrusters [34, 35, 38].

Thruster	Specific impulse (sec)	Thrust	Input power (kW)	Efficiency range(%)	Propellant
Cold gas	25–75	0.1–100 N	–	–	Various
Solid chemical	250–304	10^7 N	–	–	Various
Liquid chemical (monopropellant)	150–235	1–500 N	–	–	N_2H_4 H_2O_2
Liquid chemical (bipropellant)	274–467	10^7 N	–	–	Various
Resistojet	100–300	0.5–6000 mN	0.1–1	65–90	N_2H_4 monoprop
Arcjet	130–600	50–6800 mN	0.9–2.2	25–45	N_2H_4 monoprop
Ion thruster	2500–4000	0.1–750 mN	0.4–4.5	40–83	Xe, Kr
Hall thruster	1500–3000	0.1–2000 mN	1.5–4.5	35–70	Xe, Kr
PPT	850–1200	0.05–10 mN	<0.2	3–30	Teflon
MPD	200–3200	0.1–2000 mN	<1	20–60	H_2, Li, Ar, Xe

1.3 Electrostatic Thrusters

1.3.1 Ion Thrusters

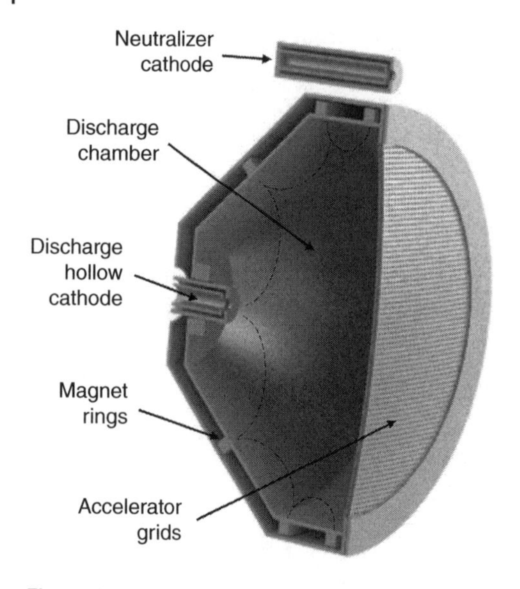

Figure 1-1 Ion thruster schematic showing grids, plasma generator and neutralizer cathode.

An ion thruster consists of basically three components: the plasma generator, accelerator grids, and the neutralizer cathode. Figure 1-1 shows a schematic cross section of an electron bombardment ion thruster that uses an electron discharge with a magnetic field produced by permanent magnets at the surface of the anode to generate the plasma. The discharge from the cathode to the anode generates the plasma in this thruster, and ions from the discharge chamber region flow to the grids and are accelerated to form the thrust beam. The plasma generator is at high positive voltage compared to the spacecraft or space plasma, and so is enclosed in a "plasma screen" biased near the spacecraft potential to eliminate electron collection from the space plasma to the positively biased surfaces. The neutralizer cathode is positioned outside the thruster and provides electrons at the same rate as the ions in the beam to avoid charge imbalance with the spacecraft.

A photograph of the NEXT ion thruster developed at NASA/GRC [39] is shown in Fig. 1-2. This thruster is capable of operating at 0.5–6.9 kW with a maximum Isp of 4000 s. Ion thrusters that use

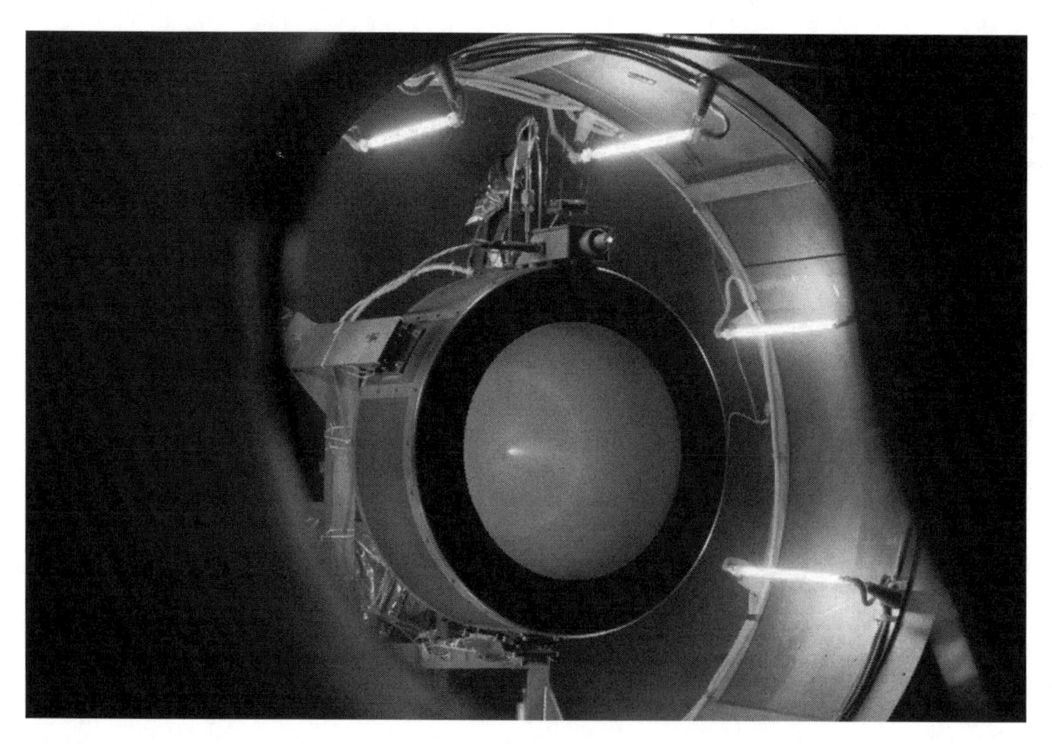

Figure 1-2 Photograph of the NEXT ion thruster during thermo-vac testing at JPL (*Source:* NASA/JPL, Public Domain).

alternative plasma generators, such as microwave or rf plasma generators, have the same basic geometry with the plasma generator enclosed in a plasma screen and coupled to a gridded ion accelerator with a neutralizer cathode. The performance of the thruster depends on the plasma generator efficiency and the ion accelerator design.

1.3.2 Hall Thrusters

A Hall thruster can also be thought of as consisting of basically three components: the cathode, the discharge region, and the magnetic field generator. Figure 1-3 shows a schematic cross-section of a Hall thruster. In this example, a cylindrical insulating channel encloses the discharge region. Magnetic coils induce a radial magnetic field between the center pole piece and the flux return path at the outside edge. The cathode of the discharge is an internal centrally mounted hollow cathode, although external cathodes mounted outside the thruster body are also used. The anode is a ring with some structure to evenly distribute the propellant gas in the annular channel, and is located at the base of the cylindrical discharge channel. Gas is fed into the discharge channel through the anode and dispersed into the channel. Electrons attempting to reach the anode encounter a transverse radial magnetic field, which reduces their mobility in the axial direction and inhibits their flow to the anode. The electrons tend to spiral around the thruster axis in the $\mathbf{E} \times \mathbf{B}$ direction and represent the Hall current from which the device derives its name. Ions generated by these electrons are accelerated by the electric field from the anode to the cathode potential plasma produced at the front of the thruster. Some fraction of the electrons emitted from the hollow cathode also leave the thruster with the ion beam to neutralize the exiting charge. The shape and material of the discharge region channel and the details of the magnetic field determine the performance of the thruster.

Figure 1-4 shows a photograph of a SPT-100 Hall that has extensive flight experience on Russian communications satellites [17]. This thruster, described in Chapter 12, has redundant externally mounted hollow cathodes and operates nominally [40] at a power of 1.35 kW and an Isp of 1600 s. The thruster includes a redundant hollow cathode to increase the reliability and features a lifetime in excess of 9000 h.The SPT-100 has also been flown on US commercial communications satellites [27].

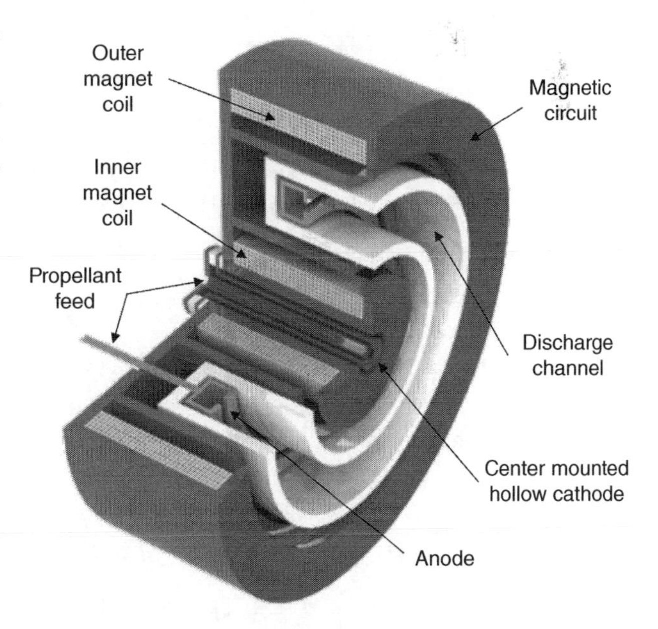

Figure 1-3 Illustration of a Hall thruster showing the annular discharge channel, magnetic circuit, anode and hollow cathode.

1.4 Electromagnetic Thrusters

There are many concepts for electromagnetic thrusters, but describing a couple of the more mature examples will illustrate the basic configuration. A more detailed description is contained in Chapter 9.

Figure 1-4 Photograph of the SPT-100 Hall thruster (*Source:* [27]/American Institute of Aeronautics and Astronautics, Inc.).

1.4.1 Magnetoplasmadynamic Thrusters

Magnetoplasmadynamic (MPD) thrusters [8] are high-power devices that produce thrust using the Lorentz electromagnetic force. The basic configuration consists of a centrally mounted cathode, a cylindrical or conical anode isolated from the center cathode, and for applied field thrusters magnetic field coils positioned outside the anode. Figure 1-5 shows an illustration of a typical MPD

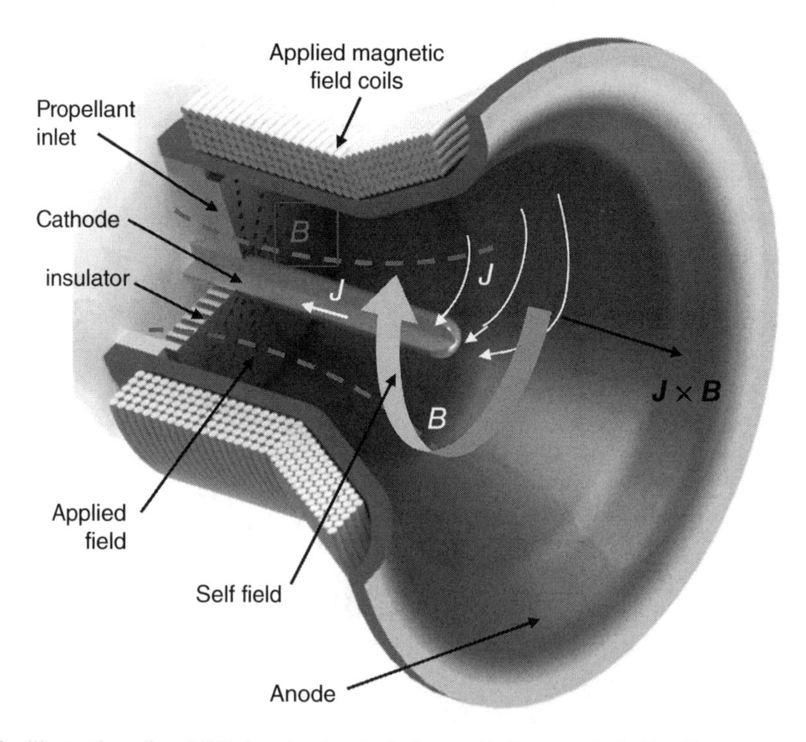

Figure 1-5 Illustration of an MPD thruster that includes applied magnetic field coils.

thruster design with these components. The cathode can consist of a refractory metal rod as shown in the figure, an assembly of refractory metal rods for more current capability, or a hollow cathode configuration, depending on the discharge current and cathode life desired. Current flowing in the cathode generates a "self-field" azimuthal magnetic field, and the radial component of the discharge current to the anode interacting with this azimuthal magnetic field generates axial thrust from the Lorentz force. This generation of transverse magnetic fields and plasma currents are inherent to all of the electromagnetic thruster concepts that use the Lorentz force to generate thrust.

Figure 1-6 Example of a laboratory self-field MPD thruster (*Source:* Photo credit: E. Choueiri, Princeton University).

In MPD thrusters with external coils, the "applied" magnetic field contributes to the Lorentz force and changes the plasma structure near the anode. There are also other forces on the plasma due to the complex structure of the magnetic fields and the current distributions, which are described in Chapter 9. MPD thrusters typically run in the multi-kiloamp, tens to hundreds of kilowatts power range, and scale well to over a megawatt in a compact, high-power package. An example of a laboratory self-field MPD thruster is shown in Fig. 1-6. There has been extensive research and development internationally on MPD thrusters, but to date only technology demonstration pulsed MPDs have been flown in space.

1.4.2 Pulsed Plasma Thrusters

The first electric thruster to provide spacecraft attitude control in space was an electromagnetic pulsed plasma thruster (PPT). This thruster concept uses an arc discharge across two electrodes that ablates and vaporizes solid propellant feed into a channel. The transverse plasma current in the arc and the induced magnetic field produces a $\mathbf{J} \times \mathbf{B}$ force that generates thrust out of the system. Figure 1-7 shows a simple schematic of a PPT thruster, and Fig. 1-8 shows a photograph of the PPT that flew on the Earth Observing (EO-1) Mission in 2000 [41]. PPTs are inherently pulsed devices to avoid damage to the electrodes and generate the desired amount of propellant into the system, and produce the order of 1 mN of thrust at an Isp of over 1000 s with an average input power of the order of 100 W. PPTs are now a commercial product and have flown recently in microsatellite applications.

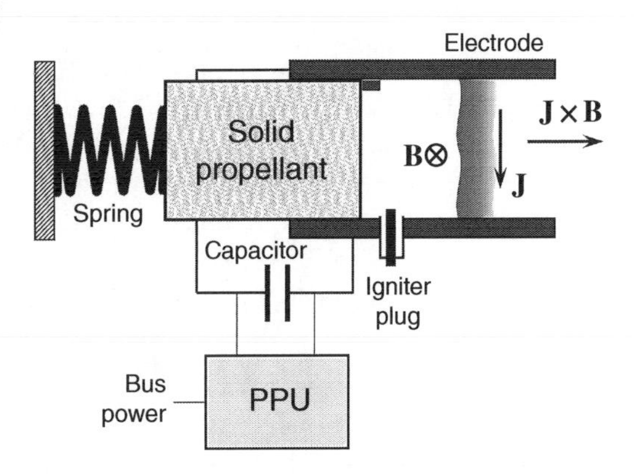

Figure 1-7 Simplified schematic of a PPT thruster.

1.4.3 Pulsed Inductive Thrusters

Finally, pulsed inductive thrusters (PIT) [42, 43] are electromagnetic accelerators in which energy is stored capacitively and discharged through an inductive coil. First, the propellant is transiently puffed onto the surface of the induction coil followed by the

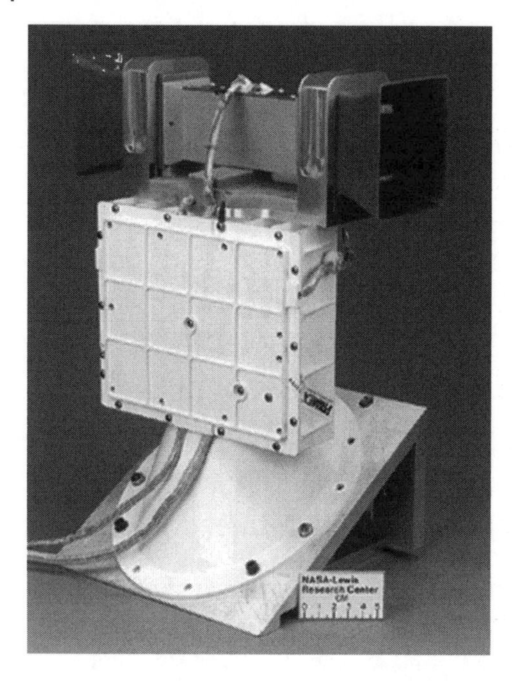

Figure 1-8 Photograph of the dual-direction PPT thruster and supporting electronics package flown on EO-1 (*Source:* NASA GRC, Public Domain).

release of the energy stored in the capacitors. The strong azimuthal electric field produced in this mannerbreaks down the propellant and establishes an azimuthal current that interacts with the rising radial magnetic field to accelerate the plasma along the thruster axis to high exhaust velocities. Such inductive acceleration, illustrated in Fig. 1-9, avoids the need for electrodes, the erosion of which has typically been the life-limiting process of traditional electromagnetic thrusters. Another unique benefit of the PIT is that it can potentially operate and perform well with any propellant, including water. Single-shot operation has demonstrated optimum performance with polyatomic molecule fuels such as ammonia which is plentiful and easily stored. Moreover, this thruster can maintain constant Isp and thrust efficiency over a wide range of input power by varying the pulse rate to maintain a constant discharge energy per pulse. PITs can also have a high energy per pulse, and by increasing the pulse rate they can operate at very high power levels producing high thrust from a single thruster. The first laboratory investigations of this accelerator were, in fact, performed in the late 1960s at Thompson Ramo Wooldridge Inc. (TRW) and continued later at Northrop Grumman Space Technology in the 1980s and 1990s [45]. It was not until the emergence of a renewed interest in high power EP in the early 2000s, however, that those largely

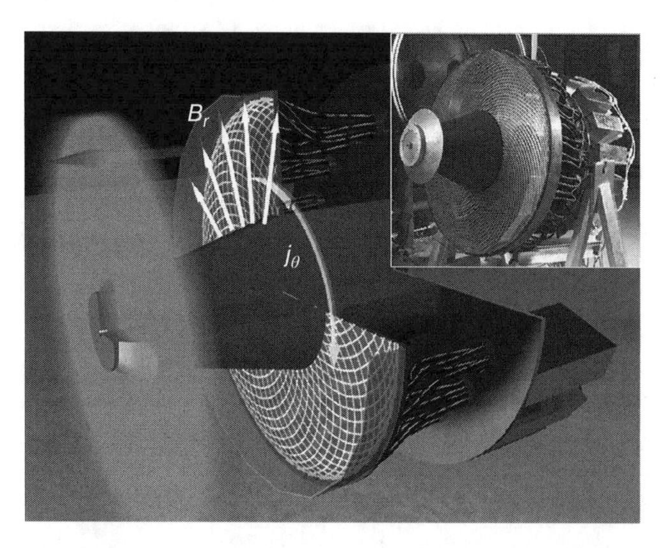

Figure 1-9 Schematic of the thruster illustrating the acceleration of a thin layer of ionized propellant by the interaction of the radial magnetic field and the induced azimuthal current, and a photograph (top right) of a PIT called the "Mark Va" developed at TRW Space & Technology Group (*Source:* [44]/AIP Publishing LLC).

empirical investigations were complemented by more in-depth analytical modeling and the first detailed 2-D numerical simulations [44, 46].

1.5 Beam/Plume Characteristics

The ion beam exiting the thruster is often called the thruster plume, and the characteristics of this plume are important in how the exhaust particles interact with the spacecraft. Figure 1-10 shows the generic characteristics of a thruster plume. First, the beam has an envelope and a distribution of the ion currents in that envelope. Second, the energetic ions in the beam can charge exchange with neutral gas coming from the thruster or the neutralizer, producing fast neutrals propagating in the beam direction and slow ions. These slow ions then move in the local electric fields associated with the exit of the acceleration region and the neutralizer plasma, and can backflow into the thruster or move radially to potentially bombard any spacecraft components in the vicinity. Third, energetic ions are often generated at large angles from the thrust axis either due to edge effects (fringe fields) in the acceleration optics of ion thrusters, large gradients in the edge of the acceleration region in Hall thrusters, or scattering of the beam ions with the background gas. Finally, the thruster evolves impurities associated with the wear of the thruster components. This can be caused by the sputtering of the grids in ion thrusters, the erosion of the ceramic channel in Hall thrusters, or the evolution of cathode materials or sputtering of other electrodes in the engines. This material can deposit on spacecraft surfaces, which can change surface properties such as emissivity, transparency, etc.

The plume from an electric thruster typically has a complex structure. For example, Fig. 1-11 shows an exploded view of a calculated three-dimensional plume from a three-grid ion thruster. In this case, the ion beam is shown as the extended plume and the molybdenum atom plume escaping through the third grid from sputter erosion of the center-accel grid is shown by the wider angular divergent dark plume and several beam lobes of high-angle particle fluxes. Likewise, Hall

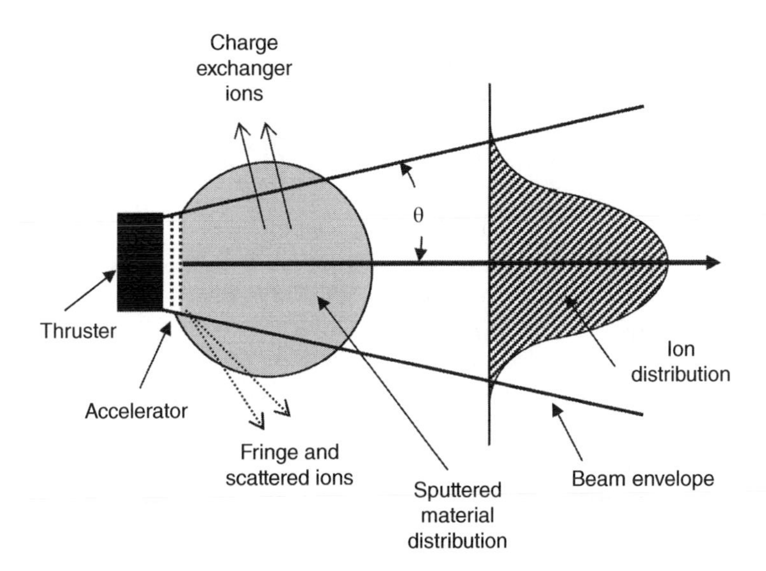

Figure 1-10 Generic thruster beam plume showing the ion distribution, sputtered material and "large angle" or charge exchange ions.

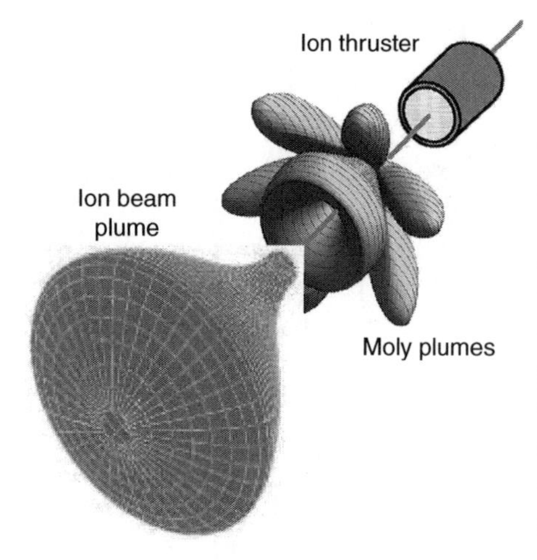

Figure 1-11 Example of a 3-D plot of an ion thruster plume (*Source:* [47]/Dr. Thomas LaFrance).

thruster plumes start from a conical discharge channel and merge downstream to form the thrust beam, and are characterized by a larger angular divergence of high-energy ions than typical from ion thrusters. Since the energetic ions in the thruster plume tend to sputter surfaces that they come into contact with, and metal or impurity atoms from the thruster tend to deposit and coat surfaces they come in contact with, the net interaction of these plumes with the spacecraft is very different from that encountered when using on-board chemical propulsion systems, and the spacecraft interactions must be examined with 3-D codes that include the spacecraft layout coupled to these types of thruster plume plots. Techniques and models for doing this are described in detail in Chapter 11.

References

1 E. Y. Choueiri, "A Critical History of Electric Propulsion: The First 50 Years (1906–1956)," *Journal of Propulsion and Power*, vol. 20, pp. 193–204, 2004.

2 R. G. Jahn and E. Y. Choueiri, "Electric Propulsion," *Encyclopedia of Physical Science and Technology*, 3rd edition, Vol. 5, New York: Academic Press, 2002.

3 https://en.wikipedia.org/wiki/Electrically_powered_spacecraft_propulsion.

4 K. E. Tsiolkovsky, "The Exploration of Cosmic Space by Means of Reaction Devices," *The Science Review* (in Russian), Gublit, pp. 12–46, 1903.

5 R. H. Goddard, *The Green Notebooks*, Vol. 1, The Dr. Robert H. Goddard Collection at Clark University Archives, Worcester: Clark University.

6 H. Oberth, *Wege zur Raumschiffahret (Ways to Spaceflight)* (in German), Munich-Berlin: R. Oldenbourg Verlag, 1929.

7 E. Stuhlinger, *Ion Propulsion for Space Flight*, New York: McGraw–Hill, 1964.

8 R. G. Jahn, *Physics of Electric Propulsion*, New York: McGraw-Hill, 1968.

9 G. R. Brewer, *Ion Propulsion Technology and Applications*, New York: Gordon and Breach, 1970.

10 H. R. Kaufman, "Technology of Electron-Bombardment Ion Thrusters," in *Advances in Electronics and Electron Physics*, vol. 36, edited by L. Marton, New York: Academic Press, pp. 265–373, 1974.

11 P. J. Turchi, "Electric Rocket Propulsion Systems," Chapter 9 in *Space Propulsion Analysis and Design*, edited by R. W. Humble, G. N. Henry, and W. J. Larson, New York: McGraw-Hill, pp.509–598, 1995.

12 J. R. Wertz and W. J. Larson, eds., *Space Mission Analysis and Design*, 3rd edition, New York: Springer, 1999.

13 G. P. Sutton and O. Biblarz, *Rocket Propulsion Elements*, New York: Wiley, pp. 660–710, 2001.

14 S. D. Grishin and L. V. Leskov, *Electrical Rocket Engines of Space Vehicles*, Moscow: Mashinostroyeniye Publishing House, 1989.

15 J. S. Sovey, V. K. Rawlin, and M. M. Patterson, "A Synopsis of on Propulsion Development Projects in the United States: SERT I to Deep Space 1," AIAA-99-2270, 35th AIAA Joint Propulsion Conference, Los Angeles, CA, June 20–24, 1999.

16 D. C. Byers, W. R. Kerslake and J. F. Staggs, "SERT II: Mission and Experiments," *Journal of Spacecraft and Rockets*, vol. 7, no. 1, pp. 4–6, 1970; doi:10.2514/3.29854

17 A. S. Boever, V. Kim, A. S. Koroteev, L. A. Latyshev, A. I. Morozov, G. A. Popov, Y. P. Rylov, and V. V. Zhurin, "State of the Works of Electrical Thrusters in the USSR," IEPC-91-003, 22nd International Electric Propulsion Conference, Viareggio, October 14–17, 1991.

18 V. Kim, "Electric Propulsion Activity in Russia," IEPC-2001-005, 27th International Electric Propulsion Conference, Pasadena, CA, October 14–19, 2001.

19 S. Shimada, K. Sato, and H. Takegahara, "20-mN Class Xenon Ion Thruster for ETS-VI," AIAA-1987-1029, 19th International Electric Propulsion Conference, Colorado Springs, May 11–13, 1987.

20 J. R. Beattie, "XIPS Keeps Satellites on Track," *The Industrial Physicist*, 1998.

21 J. R. Brophy, "NASA's Deep Space 1 Ion Engine," *The Review of Scientific Instruments*, vol. 73, pp. 1071–1078, 2002.

22 J. R. Beattie, J. N. Matossian, and R. R. Robson, "Status of Xenon Ion Propulsion Technology," *Journal of Propulsion and Power*, vol. 6, no. 2, pp. 145–150, 1990.

23 D. M. Goebel, M. Martinez-Lavin, T. A. Bond, and A. M. King, "Performance of XIPS Electric Propulsion in Station Keeping of the Boeing 702 Spacecraft," AIAA-2002-4348, 38th Joint Propulsion Conference, Indianapolis, July 7–10, 2002.

24 H. Kuninaka, K. Nishiyama, I. Funakai, K. Tetsuya, Y. Shimizu, and J. Kawaguchi, "Asteroid Rendezvous of Hayabusa Explorer Using Microwave Discharge Ion Engines," IEPC-2005-010, 29th International Electric Propulsion Conference, Princeton, October 31–November 4, 2005.

25 Y. Tani, R. Tsukizaki, D. Kodab, K. Nishiyama, and H. Kuninaka, "Performance Improvement of the μ10 Microwave Discharge Ion Thruster by Expansion of the Plasma Production Volume," *Acta Astronautica*, vol. 157, pp. 425–434, 2019; doi:10.1016/j.actaastro.2018.12.023

26 C. R. Koppel and D. Estublier, "The SMART-1 Hall Effect Thruster Around the Moon: In Flight Experience," IEPC-2005-119, 29th International Electric Propulsion Conference, Princeton, October 31–November 4, 2005.

27 D. J. Pidgeon, R. L. Corey, B. Sauer, and M. L. M. Day, "Two Years on-Orbit Performance of SPT-100 Electric Propulsion," AIAA-2006-5353, 24th AIAA International Communications Satellite Systems Conference, San Diego, June 11–14, 2006; doi:10.2514/6.2006-5353

28 N. Wallace and M. Fehringer, "The ESA GOCE Mission and the T5 Ion Propulsion Assembly," IEPC-2009-269, 31st International Electric Propulsion Conference, Ann Arbor, MI, September 20–24, 2009.

29 K. H. deGrys, B. Welander, J. Dimicco, S. Wenzel, and B. Kay, "4.5 kW hall thruster system qualification status. *41st Joint Propulsion Conference*, Tucson (10–13 July 2005).AIAA-2005-3682.

30 A. N. Grubisic, S. Clark, N. Wallace, C. Collingwood, and F. Guarducci, "Qualification of the T6 Ion Thruster for the BepiColombo Mission to the Planet Mercury," IEPC-2011-234, 32nd International Electric Propulsion Conference, Wiesbaden, September 11–15, 2011.

31 J. J. Delgado, R. L. Corey, V. M. Murashko, A. I. Koryakin, and S. Y. Pridanikov, "Qualification of the SPT-140 for Use on Western Spacecraft," AIAA-2014-3606, 50th AIAA Joint Propulsion Conference, Cleveland, July 28–30, 2014.

32 S. A. Feuerborn, D. A. Neary, and J. M. Pekins, "Finding a Way: Boeing's All Electric Propulsion Satellite," AIAA-2013-4126, 49th AIAA Joint Propulsion Conference, San Jose, July 14–17, 2013.

33 J. John, R. Thomas, A. HoskinJ. Fisher, J. Bontempo, and A. Birchenough, "NEXT-C Ion Propulsion System Operations on the DART Mission," IEPC-2022-281, 37th International Electric Propulsion Conference, Boston, June 19–23, 2022.

34 S. Mazouffre, "Electric Propulsion for Satellites and Spacecraft: Established Technologies and Novel Approaches," *Plasma Sources Science and Technology*, vol. 25, no. 3, 033002, 2016; doi:10.1088/0963-0252/25/3/033002

35 K. Holste, P. Dietz, S. Scharmann, K. Keil, T. Henning, D. Zschätzsch, M. Reitemeyer, B. Nauschütt, F. Kiefer, F. Kunze, J. Zorn, C. Heiliger, N. Joshi, U. Probst, R. Thüringer, C. Volkmar, D. Packan, S. Peterschmitt, K.-T. Brinkmann, H.-G. Zaunick, M. H. Thoma, M. Kretschmer, H. J. Leiter, S. Schippers, K. Hannemann and P. J. Klar, "Ion Thrusters for Electric Propulsion: Scientific Issues Developing a Niche Technology into a Game Changer Featured", *The Review of Scientific Instruments*, vol. 91, 061101, 2020; doi:10.1063/5.0010134

36 I. Levchenko, S. Xu, S. Mazouffre, D. Lev, D. Pedrini, D.M. Goebel, L. Garrigues, F. Taccogna, D. Pavarin and K. Bazaka"Perspectives, Frontiers and New Horizons for Plasma Based Space Electric Propulsion," *Physics of Plasmas*, vol. 27, 02061, 2020; doi:10.1063/1.5109141

37 M. Micci and A. Ketsdever, *Micropropulsion for Small Spacecraft*, American Institute of Aeronautics and Astronautics (AIAA), Progress in Astronautics & Aeronautics, 2000.

38 J. R. Wertz, D. F. Everett, and J. J. Puschell, eds., *Space Mission Engineering: The New SMAD*, Hawthorne: Microcosm Press, 2011.

39 M. Patterson, J. Foster, T. Haag, et al., "NEXT: NASA's Evolutionary Xenon Thruster," AIAA-2002-3832, 38th Joint Propulsion Conference, Indianapolis, July 7–10, 2002.

40 V. Kim, "Main Physical Features and Processes Determining the Performance of Stationary Plasma Thrusters," *Journal of Propulsion and Power*, vol. 14, pp. 736–743, 1998.

41 S. Benson, L. Arrington, W. Hoskins, and N. Meckel, "Development of a PPT for the EO-1 Spacecraft," AIAA-99-2276, 35th AIAA Joint Propulsion Conference, Los Angeles, June 20–24, 1999; doi:10.2514/6.1999-2276

42 R. H. Lovberg and C. L. Dailey, "Large Inductive Thruster Performance Measurement," *AIAA Journal*, vol. 20, no. 7, pp. 971–977, 1982; doi:10.2514/3.51155

43 K. A. Polzin, "Comprehensive Review of Planar Pulsed Inductive Plasma Thruster Research and Technology," *Journal of Propulsion and Power*, vol. 27, no. 3, 2011; doi:10.2514/1.B34188

44 P. G. Mikellides and C. Neilly, "Modeling and Performance Analysis of the Pulsed Inductive Thruster," *Journal of Propulsion and Power*, vol. 23, no. 1, pp. 51–58, 2007.

45 C. L. Dailey and R. H. Lovberg, "The PIT MkV Pulsed Inductive Thruster," NASA Contractor Report 191156, Prepared for Lewis Research Center Under Contract NAS1-19291, July 1993.

46 P. G. Mikellides and N. Ratnayake, "Modeling of the Pulsed Inductive Thruster Operating with Ammonia Propellant," *Journal of Propulsion and Power*, vol. 23, no. 4, pp. 854–862, 2007.

Chapter 2

Thruster Principles

Electric thrusters propel the spacecraft using the same basic principle as chemical rockets of accelerating mass and ejecting it from the vehicle. The ejected mass from electric thrusters, however, is primarily in the form of energetic charged particles. This changes the performance of the propulsion system compared to other types of thrusters and modifies the conventional way of calculating some of the thruster parameters such as specific impulse and efficiency. Electric thrusters provide higher exhaust velocities than available from gas jets or chemical rockets, which improves either the available change in vehicle velocity (called Δv or delta-v), or increases the delivered spacecraft and payload mass for a given Δv. Chemical rockets will generally have exhaust velocities of 3–4 km/s, whereas the exhaust velocity of electric thrusters can approach 10^2 km/s for heavy propellant such as xenon atoms, and 10^3 km/s for light propellants such as helium.

2.1 The Rocket Equation

The mass ejected to provide thrust to the spacecraft is the propellant, which is carried onboard the vehicle and expended during thrusting. From conservation of momentum, the ejected propellant mass times its velocity is equal to the spacecraft mass times its change in velocity. The "rocket equation" [1] that describes the relationship between the spacecraft velocity and the mass of the system is derived as follows. The force on a spacecraft, and thus the thrust on the vehicle, is equal to the mass of the spacecraft, M, times its change in velocity v:

$$\text{Force} = T = M\frac{dv}{dt}. \tag{2.1-1}$$

The thrust on the spacecraft is equal and opposite to time rate of change of the momentum of the propellant, which is the exhaust velocity of the propellant times the time rate of change of the propellant mass:

$$T = -\frac{d}{dt}\left(m_p v_{ex}\right) = -v_{ex}\frac{dm_p}{dt}, \tag{2.1-2}$$

where m_p is the propellant mass on the spacecraft and v_{ex} is the propellant exhaust velocity in the spacecraft frame of reference.

The total mass M of the spacecraft at any time is the delivered mass, m_d, plus the propellant mass:

$$M(t) = m_d + m_p. \tag{2.1-3}$$

Fundamentals of Electric Propulsion, Second Edition. Dan M. Goebel, Ira Katz, and Ioannis G. Mikellides.

The mass of the spacecraft changes because of consumption of the propellant, so the time rate of change of the total mass is

$$\frac{dM}{dt} = \frac{dm_p}{dt}. \tag{2.1-4}$$

Substituting Eq. 2.1-4 into Eq. 2.1-2 and equating with Eq. 2.1-1 gives

$$M\frac{dv}{dt} = -v_{\text{ex}}\frac{dM}{dt}, \tag{2.1-5}$$

which can be written as

$$dv = -v_{\text{ex}}\frac{dM}{M}. \tag{2.1-6}$$

For motion in a straight line, this equation is solved by integrating from the spacecraft initial velocity v_i to the final velocity v_f, during which the mass changes from its initial value $m_d + m_p$ to its final delivered mass m_d:

$$\int_{v_i}^{v_f} dv = -v_{\text{ex}} \int_{m_d + m_p}^{m_d} \frac{dM}{M} \tag{2.1-7}$$

The solution to Eq. 2.1-7 is

$$v_i - v_f = \Delta v = v_{\text{ex}} \ln\left(\frac{m_d + m_p}{m_d}\right). \tag{2.1-8}$$

The final mass of a spacecraft delivered after a given amount of propellant has been used to achieve the specified Δv is

$$m_d = \left(m_d + m_p\right)e^{-\Delta v/v_{\text{ex}}}. \tag{2.1-9}$$

The specific impulse, Isp, will be shown in Section 2.3 to be equal to the propellant exhaust velocity, v_{ex}, divided by the gravitational acceleration g. The change in velocity of the spacecraft is then

$$\Delta v = (\text{Isp} * g) \ln\left(\frac{m_d + m_p}{m_d}\right) = (\text{Isp} * g) \ln\left(\frac{M_i}{M_f}\right), \tag{2.1-10}$$

where M_i is the initial mass, M_f is the final mass, and g is the acceleration by gravity = 9.8067 m/s^2.

Equation 2.1-10 shows that for a given mission with a specified Δv and final delivered mass m_d, the initial spacecraft wet mass $(m_d + m_p)$ can be reduced by increasing the Isp of the propulsion system, which has implications on the launch vehicle size and cost. High delta-v missions are often enabled by electric propulsion because it offers much higher exhaust velocities and Isp than conventional chemical propulsion systems.

Equation 2.1-9 can be written in terms of the required propellant mass

$$m_p = m_d\left[e^{\Delta v/v_{\text{ex}}} - 1\right] = m_d\left[e^{\Delta v/(\text{Isp} * g)} - 1\right]. \tag{2.1-11}$$

The relationship between the amount of propellant required to perform a given mission and the propellant exhaust velocity (or the propulsion system Isp) indicates that the propellant mass increases exponentially with the delta-v required. Thrusters that provide a large propellant exhaust velocity compared to the mission Δv will have a propellant mass that is only a small fraction of the initial spacecraft wet mass.

The exhaust velocity of chemical rockets is limited by the energy contained in the chemical bonds of the propellant used; typical values are up to 4 km/s. Electric thrusters, however, separate the propellant from the energy source (which is now a power supply) and thus are not subject to the same limitations. Modern ion and Hall thrusters operating on xenon propellant have exhaust velocities in the range of 20–40 km/s and 10–30 km/s, respectively.

The dramatic benefits of the high exhaust velocities of electric thrusters are clearly seen from Eq. 2.1-11. For example, consider an asteroid rendezvous mission in which it is desired to deliver 500 kg of payload with a mission Δv of 5 km/s. A spacecraft propelled by a chemical engine with a 3 km/s exhaust velocity, corresponding to an Isp of 306 s, would require 2147 kg of propellant to accomplish the mission. In contrast, an ion thruster with a 30 km/s exhaust velocity, corresponding to an Isp of 3060 s, would accomplish the same mission using only 91 kg of propellant. High delta-v missions such as this are often enabled by electric propulsion, allowing either a significant reduction in the amount of required propellant that has to be launched, or the ability to increase the spacecraft dry mass for a given wet mass associated with a launch vehicle or mission requirement.

2.2 Force Transfer in Electric Thrusters

The propellant ionized in electric thrusters is accelerated by the application of electric and magnetic fields. However, the mechanism for generating the thrust and transferring the force from the ion motion to the thruster body, and thereby the spacecraft, is different for ion thrusters, Hall thrusters, and electromagnetic thrusters.

2.2.1 Ion Thrusters

In ion thrusters, ions are produced by a plasma source and accelerated electrostatically by the field applied between two (or more) grids, as illustrated in Figure 2-1. The voltage applied between the two grids creates a vacuum electric field between the grids of the applied voltage divided by the gap d. The ions represent additional charge q_i in the gap between the grids that modifies the electric field. Assuming infinitely large grids, the electric field distribution between the grids can be found from the one-dimensional Poisson's Equation:

$$\frac{dE(x)}{dx} = \frac{\rho(x)}{\varepsilon_o} = \frac{qn_i(x)}{\varepsilon_o} \tag{2.2-1}$$

where ε_o is the permittivity of free space, ρ is the ion charge density in the gap, q is the ion charge, n_i is the ion number density in the gap. Equation 2.2-1 can be integrated from the screen grid to the accel grid to give

$$E(x) = \frac{q}{\varepsilon_o}\int_0^x n_i(x')dx' + E_{\text{screen}} \tag{2.2-2}$$

where E_{screen} is the electric field at the screen grid. Assuming that the screen grid is a perfect conductor, its surface charge density, σ, is

$$\sigma = \varepsilon_o E_{\text{screen}} \tag{2.2-3}$$

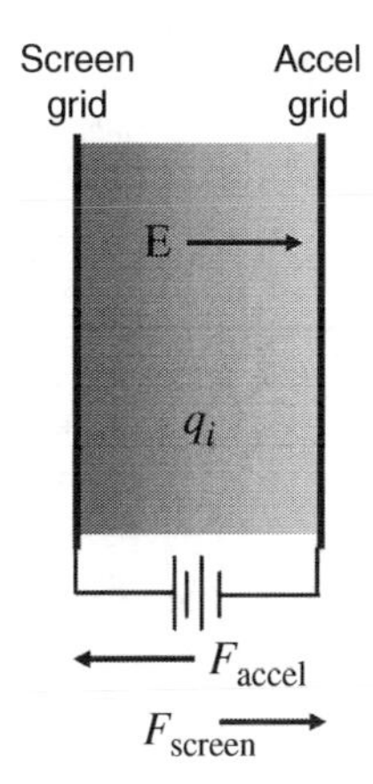

Figure 2-1 Schematic of ion thruster acceleration region.

The surface charge is an image charge and is attracted by the ion charge in gap. Since the field drops to zero inside the conductor, the screen grid experiences a force per unit area equal to the charge density times the average field (which is half the field on the outside of the conductor):

$$F'_{\text{screen}} = \sigma \frac{(E_{\text{screen}} + 0)}{2} = \frac{1}{2}\varepsilon_o E^2_{\text{screen}}, \tag{2.2-4}$$

where F'_{screen} is the force per unit area on the screen grid. Correspondingly, at the accelerator grid there is an electric field, E_{accel}, and a surface charge density equal to that on the screen grid but of the opposite sign. The accel grid feels a force per unit area in the opposite direction:

$$F'_{\text{accel}} = -\sigma \frac{(E_{\text{accel}} + 0)}{2} = -\frac{1}{2}\varepsilon_o E^2_{\text{accel}}. \tag{2.2-5}$$

The net thrust on the ion engine is the sum of the forces on the screen and accel grids,

$$T = \left(F'_{\text{screen}} + F'_{\text{accel}}\right) A = \frac{1}{2} A\varepsilon_o \left(E^2_{\text{screen}} - E^2_{\text{accel}}\right), \tag{2.2-6}$$

where T is the thrust in Newtons and A is the area of the grids. The force per unit area on the ions in the gap between the grids can be calculated using the fact that the force on an ion equals its charge times the local electric field, and by integrating that force across the gap:

$$F'_{\text{ions}} = q \int_0^d n_i(x) E(x) dx \tag{2.2-7}$$

Eliminating the ion density $n_i(x)$ using Eq. 2.2-1, the integral can be done directly:

$$F'_{\text{ions}} = \varepsilon_o \int_0^d \frac{dE(x)}{dx} E(x) dx = \frac{1}{2}\varepsilon_o \left(E^2_{\text{accel}} - E^2_{\text{screen}}\right) \tag{2.2-8}$$

The net force on the grids, which is the thrust, is equal and opposite to the electric field forces on the ions between the grids:

$$T = -F'_{\text{ions}} A = -\frac{1}{2} A\varepsilon_o \left(E^2_{\text{accel}} - E^2_{\text{screen}}\right). \tag{2.2-9}$$

Therefore, the thrust in an ion engine is generated by the applied electric field and transferred by the electrostatic force between the ions and the two grids.

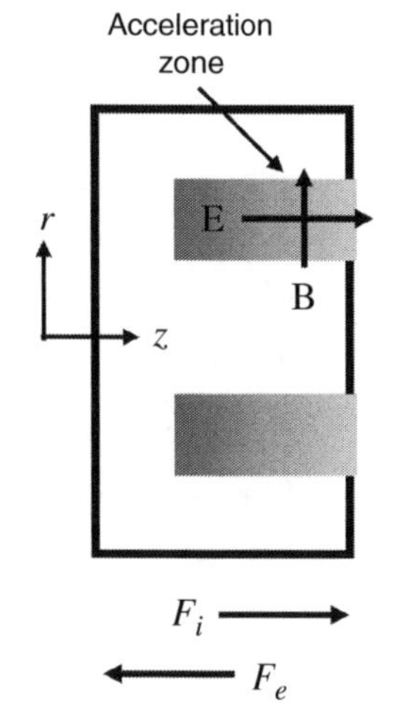

Figure 2-2 Cross section of a Hall thruster showing electric and magnetic fields.

2.2.2 Hall Thrusters

In Hall thrusters, the ions are generated in a plasma volume in the discharge channel and accelerated by an electric field in the plasma. The transverse magnetic field in the thruster increases the axial resistivity and the electric field in the plasma accelerates the ions to high velocity, and is responsible for the rotational Hall current that modifies the force transfer mechanism. Assume, for argument, that the Hall thruster plasma is locally quasi-neutral ($qn_i \approx qn_e$) in the acceleration region, where n_e is the electron plasma density, and that in the acceleration zone the electric and magnetic fields are uniform. The geometry is shown schematically in Figure 2-2.

The ions are essentially unmagnetized and feel the force of the local electric field, so the total force on the ions is

$$\mathbf{F}_{\text{ion}} = 2\pi \int\int qn_i \mathbf{E}\, rdr\, dz \tag{2.2-10}$$

The electrons in the plasma feel an $\mathbf{E} \times \mathbf{B}$ force and circulate in the system transverse to the electric and magnetic fields with a velocity

$$\mathbf{v}_e = \frac{\mathbf{E} \times \mathbf{B}}{B^2}. \tag{2.2-11}$$

The electrostatic force on the ions is the negative of the electrostatic force on the electrons because of their sign difference. The electrons are constrained to not move axially by the transverse magnetic field, so the force per unit area on the electrons (going to the left in Figure 2-2) is balanced by the Lorentz force:

$$\mathbf{F}_e = -2\pi \int\int qn_e \mathbf{E}\, rdrdz - 2\pi \int\int qn_e \mathbf{v_e} \times \mathbf{B}\, rdrdz = 0 \tag{2.2-12}$$

Using quasi-neutrality and the definition of the Hall current density $\mathbf{J}_{\text{Hall}} = -qn_e \boldsymbol{v}_e$ in Eq. 2.2-12, and substituting this result into Eq. 2.2-10, the force on the ions is then equal to the Lorentz force on the electrons

$$\mathbf{F}_{\text{i}} = 2\pi \int\int \mathbf{J}_{\text{Hall}} \times \mathbf{B}\, rdrdz. \tag{2.2-13}$$

Integrating Eq. 2.2-13 around the cylinder gives the total force on the ions

$$\mathbf{F}_i = \mathbf{J}_{\text{Hall}} \times \boldsymbol{B}. \tag{2.2-14}$$

By Newton's second law, the Hall current force on the magnets is equal and opposite to the Hall current force on the electrons and, therefore, is also equal and opposite to the force on the ions:

$$\mathbf{T} = \mathbf{J}_{\text{Hall}} \times \mathbf{B} = -\mathbf{F}_{\boldsymbol{i}}. \tag{2.2-15}$$

Therefore, the thrust force in Hall thrusters is transferred from the ions to the thruster body through the electromagnetic Lorentz force. These thrusters are sometimes called *electromagnetic thrusters* because the force is transferred through the magnetic field. However, since the ions are ***accelerated*** by the electrostatic field (like ion thrusters), we chose to call them *electrostatic thrusters*.

2.2.3 Electromagnetic Thrusters

Electromagnetic thrusters use the Lorentz force combined with internal pressure gradients to produce thrust. This can be shown by looking the momentum equation and assuming there is a quasi-neutral ($n_i \approx n_e = n$) plasma in the thruster composed of only electrons and singly-charged ions. We will neglect viscous effects that are usually small, and also neglect the momentum stored in the electron fluid because the electron mass is small compared to the ion mass. The fluid momentum equation for each species is given by

$$mn\frac{\mathrm{d}\mathbf{v}}{\mathrm{dt}} = mn\left[\frac{\partial \mathbf{v}}{\partial t} + (\mathbf{v} \cdot \nabla)\mathbf{v}\right] = qn(\mathbf{E} + \mathbf{v} \times \mathbf{B}) - \nabla \cdot \mathbf{p} - mn\nu(\mathbf{v} - \mathbf{v}_o), \tag{2.2-16}$$

where the convective derivation is shown, viscosity is neglected, and the Lorentz force term is:

$$\mathbf{F}_{\mathrm{L}} = q(\mathbf{E} + \mathbf{v} \times \mathbf{B}). \tag{2.2-17}$$

The momentum change from collisions can be written as

$$\mathbf{P}_{\mathbf{ie}} = -\mathbf{P}_{\mathbf{ei}} = mn\nu(\mathbf{v}_{\boldsymbol{i}} - \mathbf{v}_{\mathrm{e}}), \tag{2.2-18}$$

where $\mathbf{v_i} - \mathbf{v_e}$ represents the velocity difference between the electrons and ions, and conservation of momentum between the two species makes the collision terms equal and opposite. The 1-D equations of motion for ions and electrons are then

$$Mn_i \frac{d\mathbf{v}_{\boldsymbol{i}}}{dt} = en_i(\mathbf{E} + \mathbf{v}_{\boldsymbol{i}} \times \mathbf{B}) - \nabla p_i - \mathbf{P}_{\mathrm{ie}} \tag{2.2-19}$$

$$0 = -en_e(\mathbf{E} + \mathbf{v}_{\boldsymbol{e}} \times \mathbf{B}) - \nabla p_e + \mathbf{P}_{\mathrm{ei}} \tag{2.2-20}$$

Using quasi-neutrality, solving Eqs. 2.2-19 and 2.2-20 for the collision terms, and using Eq. 2.2-18 gives

$$Mn \frac{d\mathbf{v}_{\boldsymbol{i}}}{dt} = en_i(\mathbf{E} + \mathbf{v}_{\boldsymbol{i}} \times \mathbf{B}) - \nabla p_i - en_e(\mathbf{E} + \mathbf{v}_{\boldsymbol{e}} \times \mathbf{B}) - \nabla p_e \tag{2.2-21}$$

Defining the total pressure gradient and the ion density as

$$\nabla p = \nabla p_i + \nabla p_e, \quad \rho = Mn, \tag{2.2-22}$$

we write the fluid equation of motion as

$$\rho \frac{d\mathbf{v}_{\boldsymbol{i}}}{dt} = \mathrm{en}(\mathbf{v}_{\boldsymbol{i}} - \mathbf{v}_{\boldsymbol{e}}) \times \mathbf{B} - \nabla p = \mathbf{J} \times \mathbf{B} - \nabla p \tag{2.2-23}$$

where the difference in the ion and electron velocities has been expressed in terms of the net current density

$$\mathbf{J} = en(\mathbf{v}_{\boldsymbol{i}} - \mathbf{v}_{\boldsymbol{e}}). \tag{2.2-24}$$

Equation 2.2-23 reduces to the Lorentz force term, expressed as $\mathbf{J} \times \mathbf{B}$, and a pressure gradient term found in conventional rockets. The primary cause of the acceleration is the electric field that drives the current between cathode and anode. This discharge current generates the magnetic field, or interacts with the transverse component of any applied magnetic field, which results in the Lorentz force producing thrust. Electrons that are accelerated by the electric field also transfer momentum to neutrals and ions through collisions that result in heating, which also contributes (to a lesser extent) to the thrust.

2.3 Thrust

Thrust is the force transferred by the engine to the spacecraft. Since the spacecraft mass changes with time due to the propellant consumption, the thrust is given by the time rate of change of the momentum, which can be written as:

$$T = \frac{d}{dt}(m_p v_{\mathrm{ex}}) = \frac{dm_p}{dt} v_{\mathrm{ex}} = \dot{m}_p v_{\mathrm{ex}}, \tag{2.3-1}$$

where $\dot{m}_p$ is the propellant mass flow rate in kg/s. The propellant mass flow rate is

$$\dot{m}_p = QM, \tag{2.3-2}$$

where Q is the propellant particle flow rate (in particles per sec) and M is the particle mass.

The kinetic thrust power of the beam, called the jet power, is defined as

$$P_{\text{jet}} \equiv \frac{1}{2}\dot{m}_p v_{\text{ex}}^2. \tag{2.3-3}$$

Using Eq. 2.3-1, the jet power is then

$$P_{\text{jet}} = \frac{T^2}{2\dot{m}_p}. \tag{2.3-4}$$

This expression shows that techniques that increase the thrust without increases in the propellant flow rate will result in an increase in the jet power.

For ion and Hall thrusters, ions are accelerated to high exhaust velocity using an electrical power source. The velocity of the ions greatly exceeds that of any unionized propellant that may escape from the thruster, so the thrust in one direction T' can be described as

$$T' = \frac{dm_p}{dt} v_{\text{ex}} \approx \dot{m}_p v_i, \tag{2.3-5}$$

where $\dot{m}_i$ is the ion mass flow rate and v_i is the ion velocity. By conservation of energy, the ion exhaust velocity is given by

$$v_i = \sqrt{\frac{2qV_b}{M}}, \tag{2.3-6}$$

where V_b is the net voltage through which the ion was accelerated, q is the charge, and M is the ion mass. The mass flow rate of ions is related to the ion beam current I_b by

$$\dot{m}_i = \frac{I_b M}{q}. \tag{2.3-7}$$

Substituting Eqs. 2.3-6 and 2.3-7 into Eq. 2.3-5, the thrust for a singly charged propellant ($q = e$) is

$$T' = \sqrt{\frac{2M}{e}} I_b \sqrt{V_b} \text{ [Newtons]}. \tag{2.3-8}$$

Thrust is proportional to the beam current times the square root of the acceleration voltage. In the case of Hall thrusters, there is a spread in beam energies produced in the thruster, and V_b represents the effective or average beam voltage. If the propellant is xenon, $\sqrt{2M/e} = 1.65 \times 10^{-3}$, and the thrust is given by

$$T'(\text{xenon}) = 1.65\, I_b \sqrt{V_b} \text{ [mN]}, \tag{2.3-9}$$

where I_b is the beam current in Amperes and V_b is the beam voltage in Volts. If the propellant is krypton, $\sqrt{2M/e} = 1.32 \times 10^{-3}$, and the thrust is given by

$$T'(\text{krypton}) = 1.32\, I_b \sqrt{V_b} \text{ [mN]}, \tag{2.3-10}$$

which, of course, is slightly lower than that for xenon for the same beam current and voltage due to the lighter atom mass.

Equation 2.3-8 is the basic thrust equation that applies for a unidirectional, singly ionized, monoenergetic beam of ions. The assumption of a monoenergetic ion beam is generally valid for ion thrusters, but is only an approximation for the beam characteristics in Hall thrusters, which will be discussed in Chapter 7. This equation must be modified to account for the divergence of the ion beam and the presence of multiply charged ions commonly observed in electric thrusters.

This modification to Eq. 2.3-8 has been done in the ion thruster literature by introducing a correction factor γ in the thrust equation. The thrust correction factor is the product of the divergence and multiply charged species terms:

$$\gamma = \alpha F_t. \tag{2.3-11}$$

The correction to the thrust for beam divergence, F_t, is straightforward for a beam that diverges uniformly on exiting from the thruster. For a thruster with a constant ion current density profile accelerated by uniform electric fields, the correction to the force because of the effective thrust-vector angle is simply

$$F_t = \cos\theta, \tag{2.3-12}$$

where θ is the average half-angle divergence of the beam. If the thrust half angle is 10°, then $\cos\theta = 0.985$, which represents a 1.5% loss in thrust. If the plasma source is not uniform and/or the accelerator grids have curvature, then the thrust correction must be integrated over the beam and grid profiles. For cylindrical thrusters, the current-weighted correction factor is then

$$F_t = \frac{\int_0^r 2\pi J(r)\cos\theta(r)dr}{I_b}, \tag{2.3-13}$$

where I_b is the total beam current, $J(r)$ is the ion current density which is a function of the radius. The ion current density is usually determined from direct measurement of the current distribution in the plume by plasma probes. For a constant value of $J(r)$, Eq. 2.3-13 reduces to Eq. 2.3-12.

A more rigorous approach for finding the divergence correction factor is to calculate the momentum-weighted average divergence angle [2]. Assuming an axisymmetric beam and integrating over a hemispherical surface through the plume, the divergence correction factor is given by

$$F_t = \langle\cos\theta\rangle = \frac{2\pi R^2\int_0^{\frac{\pi}{2}}[\dot{m}(\theta)\overline{v}(\theta)/J(\theta]J(\theta)\cos\theta\sin\theta\,d\theta}{2\pi R^2\int_0^{\frac{\pi}{2}}[\dot{m}(\theta)\overline{v}(\theta)/J(\theta]J(\theta)\sin\theta\,d\theta} \cong \frac{2\pi R^2\int_0^{\pi/2}J(\theta)\cos\theta\sin\theta\,d\theta}{2\pi R^2\int_0^{\pi/2}J(\theta)\sin\theta\,d\theta} \tag{2.3-14}$$

where $\dot{m}(\theta)$ is the mass flux at the half-angular position θ in the beam and $J(\theta)$ is the ion current density which is a function of the angle in these coordinates. The momentum-weighted average angular divergence can be approximated as the charge-weighted average divergence in an axisymmetric plume [2], which enables probe current measurements in the plume to be used in evaluating Eq. 2.3-13.

The correction to the thrust for the presence of multiply charged ion species in the beam, α, is found by considering the force imparted by each species. If the beam contains both singly-charged and doubly-charged ions such that the total beam current is

$$I_b = I^+ + I^{++}, \tag{2.3-15}$$

where I^+ is the singly-charged ion current and I^{++} is the doubly-charged ion current, the total thrust for the multiple species, T_m, is the sum of the thrust from each species:

$$T_m = I^+\sqrt{\frac{2MV_b}{e}} + I^{++}\sqrt{\frac{MV_b}{e}} = I^+\sqrt{\frac{2MV_b}{e}}\left(1 + \frac{1}{\sqrt{2}}\frac{I^{++}}{I^+}\right). \tag{2.3-16}$$

where I^{++}/I^{+} is the fraction of double ion current in the beam. The thrust correction factor, α, for thrust in the presence of doubly-ionized atoms is

$$\alpha = \frac{T_m}{T'} = \frac{I^+ + \frac{1}{\sqrt{2}} I^{++}}{I^+ + I^{++}} = \frac{1 + 0.707 \frac{I^{++}}{I^+}}{I^+ + \frac{I^{++}}{I^+}}, \tag{2.3-17}$$

where T' is given by Eq. 2.3-8. A similar correction factor can be easily derived for even higher charged ions, although the number of these species are typically found to be relatively small in most ion and Hall thrusters.

The total corrected thrust is then given by

$$T = \gamma \dot{m} v_i = \gamma \sqrt{\frac{2M}{e}} I_b \sqrt{V_b}. \tag{2.3-18}$$

The total thrust for xenon can be simply written as

$$T_{Xe} = 1.65\, \gamma\, I_b \sqrt{V_b} [\text{mN}]. \tag{2.3-19}$$

For example, assuming an ion thruster with a 10° half angle beam divergence with a 10% doubles-to-singles ratio results in $\gamma = 0.958$. For a thruster producing 2 A of xenon ions at 1500 V, the thrust produced is 122.4 mN.

Likewise, the total thrust for krypton is given by

$$T_{Kr} = 1.32\, \gamma\, I_b \sqrt{V_b}\, [\text{mN}]. \tag{2.3-20}$$

For the same example of an ion thruster running at 2 A and 1500 V in krypton, assuming a 10° half-angle beam divergence with a 10% doubles-to-singles ratio ($\gamma = 0.958$), the thrust produced is 97.9 mN.

2.4 Specific Impulse

Specific impulse, termed Isp, is a measure of thrust efficiency and is defined as the ratio of the thrust to the rate of propellant consumption. Isp for constant thrust and propellant flow rate, is

$$\text{Isp} = \frac{T}{\dot{m}_p\, g}, \tag{2.4-1}$$

where g is the acceleration of gravity = 9.807 m/s^2. For a xenon thruster, the Isp can be expressed as

$$\text{Isp} = 1.037 \times 10^6 \frac{T[\text{N}]}{Q[\text{sccm}]} = 1.02 \times 10^2 \frac{T[\text{N}]}{Q[mg/\text{s}]}, \tag{2.4-2}$$

where Eq. 2.3-2 and the flow conversions in Appendix B have been used.

Using Eq. 2.3-1 for the thrust in Eq. 2.4-1, the Isp for any thruster is

$$\text{Isp} = \frac{v_i}{g}, \tag{2.4-3}$$

where v_{ex} is an effective exhaust velocity.

Defining the Isp in terms of the exhaust velocity relative to g is what gives rise to the unusual units of seconds for Isp. In electric thrusters, the thrust is due primarily to the ions. Using Eq. 2.3-5, the Isp is given by

$$\text{Isp} = \frac{v_i}{g} \frac{\dot{m}_i}{\dot{m}_p}, \tag{2.4-4}$$

where v_i is the exhaust velocity for unidirectional, monoenergetic ion exhaust.

The thruster mass utilization efficiency, which accounts for the ionized versus unionized propellant, is defined for singly-charged ions as

$$\eta_m = \frac{\dot{m}_i}{\dot{m}_p} = \frac{I_b}{e}\frac{M}{\dot{m}_p}. \tag{2.4-5}$$

In the event that the thruster produces a significant number of multiply charged ions, the expression for the propellant utilization efficiency must be redefined. For thrusters with both singly and doubly charged ions, the corrected mass utilization efficiency for multiple species is

$$\eta_{m^*} = \alpha_m \frac{I_b}{e}\frac{M}{\dot{m}_p}, \tag{2.4-6}$$

where α_m is a term that accounts for the fact that a doubly-charged ion in the beam current carries two charges but only one unit of mass. In a similar manner as the derivation of the thrust correction due to double ions, the mass utilization correction α_m is given by

$$\alpha_m = \frac{I^+ + \frac{1}{2}I^{++}}{I^+ + I^{++}} = \frac{1 + 0.5\frac{I^{++}}{I^+}}{I^+ + \frac{I^{++}}{I^+}} \tag{2.4-7}$$

For a small ratio of double-to-single ion content, α_m is essentially equal to one.

Substituting Eq. 2.3-18 for the thrust and Eq. 2.4-5 for the propellant utilization efficiency into Eq. 2.4-3 yields an expression for the Isp:

$$\text{Isp} = \frac{\gamma\eta_m}{g}\sqrt{\frac{2eV_b}{M}}, \tag{2.4-8}$$

where the propellant utilization efficiency for singly charged ions must be used because Eq. 2.3-18 defines the beam current this way, and again the effective beam voltage must be used for Hall thrusters. Using the values for g and e, the Isp for an arbitrary propellant is

$$\text{Isp} = 1.417 \times 10^3\,\gamma\eta_m \frac{\sqrt{V_b}}{\sqrt{M_a}}, \tag{2.4-9}$$

where V_b is the beam voltage in Volts and M_a is the ion mass in atomic mass units [1 AMU $= 1.6605 \times 10^{-27}$ kg]. For xenon, the atomic mass $M_a = 131.29$, and the Isp is given by

$$\text{Isp}_{Xe} = 123.6\,\gamma\eta_m\sqrt{V_b}. \tag{2.4-10}$$

Using our previous example of a 10° half-angle beam divergence and a 10% doubles-to-singles ratio with a 90% propellant utilization of xenon (in Eq. 2.4-5) at 1500 V, the Isp is (123.6)(0.958)(0.9) $\sqrt{1500} = 4127$ s.

For krypton, the atomic mass $M_a = 83.798$, and the Isp is given by

$$\text{Isp}_{\text{Kr}} = 154.8\,\gamma\eta_m\sqrt{V_b}. \tag{2.4-11}$$

Using our previous example of a 10° half angle beam divergence and a 10% doubles-to-singles ratio with a 90% propellant utilization of xenon (in Eq. 2.4-5) at 1500 V, the Isp is (154.8)(0.958)(0.9) $\sqrt{1500} = 5169$ s. Comparing with our example in the previous section for the same ion thruster example, the thrust decreased by 20% using krypton instead of xenon, but the Isp increased by 25%.

Isp is functionally equivalent to gas mileage in a car. Cars with high gas mileage typically don't provide much acceleration, just as thrusters with high Isp don't provide as much thrust for a given input electrical power. Of critical importance is the ratio of the thrust achieved to total power used, which depends on the electrical efficiency of the thruster (to be described in the next section).

2.5 Thruster Efficiency

The mass utilization efficiency, defined in Eq. 2.4-5, describes the fraction of the input propellant mass that is converted into ions and accelerated in the electric thruster. The electrical efficiency of the thruster is defined as the beam power P_b out of the thruster divided by the total input power P_T:

$$\eta_e = \frac{P_b}{P_T} = \frac{I_b V_b}{I_b V_b + P_o}, \tag{2.5-1}$$

where P_o represents the other power input to the thruster required to create the thrust beam. Other power will include the electrical cost of producing the ions, cathode heater or keeper power, grid currents in ion thrusters, etc.

The cost of producing the ions is described by an ion production efficiency term, sometimes called the discharge loss

$$\eta_d = \frac{\text{Power to produce the ions}}{\text{Current of ions produced}} = \frac{P_d}{I_b}, \tag{2.5-2}$$

where η_d has units of watts per ampere (W/A) or equivalently electron-volts per ion (eV/ion). Contrary to most efficiency terms, it is desirable to have η_d as small as possible since this represents a power loss. For example, if an ion thruster requires a 20 A, 25 V discharge to produce 2 A of ions in the beam, the discharge loss is then $20 * 25/2 = 250$ eV/ion.

The performance of a plasma generator is usually characterized by plotting the discharge loss versus the propellant utilization efficiency. An example of this is shown in Figure 2-3. At low propellant efficiencies, the neutral pressure in the thruster is high and the performance curves are relatively flat. As the propellant efficiency is increased, the neutral pressure in the thruster decreases, the electron temperature increases, and the loss mechanisms in the thruster become larger. Thrusters are normally operated near the knee of this curve such that high mass utilization efficiency is achieved without excessive discharge loss. Optimized thruster designs result in lower discharge losses and low loss at high propellant efficiency.

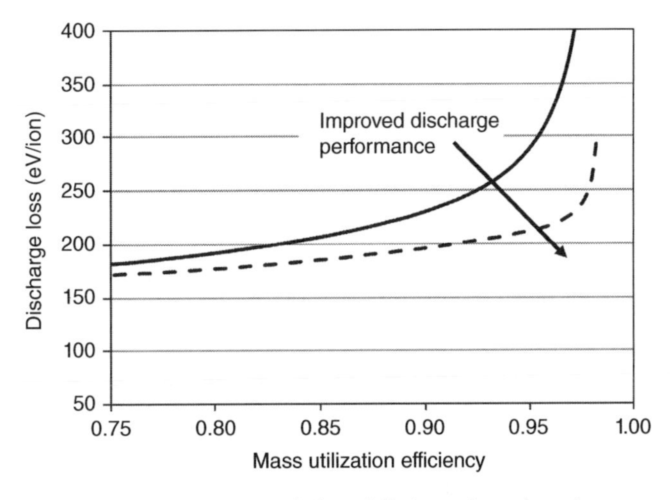

Figure 2-3 Ion thruster performance curves consisting of discharge loss plotted versus propellant utilization efficiency.

The total efficiency of an electrically powered thruster is defined as the jet power divided by the total electrical power into the thruster:

$$\eta_T = \frac{P_{\text{jet}}}{P_{\text{in}}}. \tag{2.5-3}$$

Using Eq. 2.3-4 for the jet power, the efficiency of any electric propulsion thruster is

$$\eta_T = \frac{T^2}{2\dot{m}_p P_{\text{in}}}. \tag{2.5-4}$$

Measurements made of the thruster's input electrical power, input mass flow rate, and thrust output (measured in the vacuum system by a thrust stand) during testing can be used to calculate the total efficiency of the thruster using Eq. 2.5-4. This is the preferred technique for determining efficiency of Hall thrusters because the beam parameters (current and velocity) are not known outright from measurements of electrical or gas flow parameters external to the vacuum system.

In ion thrusters, the beam is nearly monoenergetic, the exhaust velocity can be found from the net acceleration voltage applied to the thruster (using Eq. 2.3-6), and the beam current is measured by the high voltage power supply. This allows the total efficiency to be accurately calculated from the electrical and gas flow inputs to the thruster. Using Eq. 2.3-18 for the thrust, Eq. 2.3-6 for the exhaust velocity, and Eq. 2.4-5 for the propellant flow rate, the total efficiency in Eq. 2.5-4 can be written as

$$\eta_T = \frac{\gamma \eta_m T v_i}{2\dot{m}_i P_{\text{in}}} = \gamma^2 \eta_m \frac{I_b V_b}{P_{\text{in}}}. \tag{2.5-5}$$

The input power into the thruster, from Eq. 2.5-1, is

$$P_{\text{in}} = \frac{P_b}{\eta_e} = \frac{I_b V_b}{\eta_e}, \tag{2.5-6}$$

Substituting Eq. 2.5-6 into Eq. 2.5-5 gives:

$$\eta_T = \gamma^2 \eta_e \eta_m. \tag{2.5-7}$$

Measurements of the input propellant flow rate and electrical parameters (currents and voltages), and knowledge of the thrust correction factors from thruster plume measurements or code predictions, permit the total efficiency of ion thrusters to be calculated using Eq. 2.5-7. The electrical and mass utilization efficiencies are readily available for ion thrusters during testing from the external electrical signals and flow measurements. In Hall thrusters, however, the beam current and voltage are not directly proportional to the applied voltage and current, and so additional efficiency terms are normally included in the total efficiency expression to describe these effects. These terms will be described in Chapter 7.

Using our previous example of an ion thruster with 10° half-angle divergence, 10% double ion current, 90% mass utilization efficiency, and 250 eV/ion to produce a 2 A beam at 1500 V, the electrical efficiency is

$$\eta_e = \frac{2 * 1500}{2 * 1500 + 250 * 2} = 0.857,$$

and the total efficiency is

$$\eta_T = (0.958)^2\,(0.857)\,(0.9) = 0.708,$$

which indicates that the thruster converts 70.8% of the supplied electrical energy into useful kinetic energy imparted to the spacecraft.

Thrusters with high exhaust velocities, and thus high Isps, are desirable to maximize a mission payload mass. It was shown in Eq. 2.4-9 that to achieve high Isp, it is necessary to operate at a high ion acceleration voltage and high mass utilization efficiency. Reductions in ion mass also increase the Isp, but at the cost of thrust at the same power level. This is seen by examining the thrust to total input power ratio. The total power is just the beam power divided by the electrical efficiency, so the thrust to power ratio using Eq. 2.5-1 is

$$\frac{T}{P_T} = \frac{T\eta_e}{P_b}. \tag{2.5-8}$$

The beam power is the beam current times the beam voltage. Using Eq. 2.3-18 for the thrust and Eq. 2.4-8 to put this in terms of Isp, the thrust per unit input power is

$$\frac{T}{P_T} = \frac{2\gamma^2\eta_m\eta_e}{g\,\text{Isp}} = \frac{2}{g}\frac{\eta_T}{\text{Isp}}. \tag{2.5-9}$$

The thrust to power ratio ultimately is determined by the total efficiency and the Isp of the thruster. High thrust to power ratio is best obtained at lower Isp, which we will see later is more obtainable in Hall thrusters than ion thrusters.

Equation 2.5-9 can be rearranged to show how the total power is distributed in an electric thruster

$$P_T = \frac{T\,Isp\,g}{2\eta_T}. \tag{2.5-10}$$

This shows that for a given input power and total thruster efficiency, increasing the Isp reduces the thrust available from the electric engine. This trade of thrust for Isp at a constant input power can only be improved if higher efficiency thrusters are employed.

2.6 Power Dissipation

The power into a thruster that does not result in thrust must be dissipated primarily by radiating the unused power into space. If the thruster electrical efficiency is accurately known, the dissipated power is simply

$$P_{\text{dissipated}} = P_{in}(1-\eta_e). \tag{2.6-1}$$

If the electrical efficiency is not well known, alternative techniques can be used to determine the dissipated power. For example, in an ion thruster, the power in the beam is well known, and a simple difference between the total input power and the beam power represents the dissipated power. The various input powers can be measured externally to the thruster on the power supplies.

Using ion thrusters for this example, assuming the heaters have been turned off and the hollow cathodes are self-heating, the power into the ion thruster is given by the power into each component (described in Chapter 5):

$$P_{\text{in}} = I_bV_b + I_dV_d + I_{\text{ck}}V_{\text{ck}} + I_{\text{nk}}V_{\text{nk}} + I_{A1}(V_b + V_a) + I_{A2}V_a + I_{\text{DE1}}V_b + I_{\text{DE2}}V_G, \tag{2.6-2}$$

where the subscript "*b*" represents the beam current and voltage, "*d*" is the discharge current and voltage, "ck" is the cathode keeper current and voltage, "nk" is the neutralizer keeper current and voltage, "*A1*" represents beam ions incident on the accel grid, "*A2*" represents charge exchange ions at the accel grid potential, V_a, "I_{DE1}" represents the decel grid (if present) current from beam ions, and "I_{DE2}" represents the decel grid current from backstreaming ions from the beam plume. In reality, the accel and decel grid power are very small compared to the other power levels in the thruster.

The power that must be dissipated by the thruster is Eq. 2.6-2 minus the beam power:

$$P_{\text{dissipated}} = I_d V_d + I_{\text{ck}} V_{\text{ck}} + I_{\text{nk}} V_{\text{nk}} + I_{A1}(V_b + V_a) + I_{A2} V_a + I_{\text{DE1}} V_b + I_{\text{DE2}} V_G. \quad (2.6\text{-}3)$$

Using the same ion thruster example used previously in this chapter producing a 2 A beam at 1500 V as an example, Table 2-1 shows some example electrical parameters for a generic ion thruster. Assuming 10% of the grid currents are due to direct interception, using the table parameters in Eq. 2.5-12 gives a dissipated power of 528.3 W. Since the discharge power in this example is 500 W, the other power levels are relatively insignificant. However, the thruster will have to be of sufficient size to radiate all this power to space at a reasonable temperature that the materials and construction are designed to handle.

Unlike ion thrusters, power dissipation in Hall thrusters is not easily obtained from the external power supply readings. However, two techniques can be used to estimate the dissipated power. First, the dissipated power can be inferred from measurements of the thruster efficiency and the beam power (ion current and energy), which involves calculating the difference between the total beam power and the input electrical power. Another technique is to assume that the dissipated power is primarily the loss due to the electron current flowing from the exterior cathode through the thruster to the high voltage anode. If the fraction of the discharge current that becomes beam ions can be determined from the external diagnostics, then the difference between the discharge current and beam current times the discharge voltage is approximately the dissipated power. This technique neglects the ionization power and energy carried by the electrons in the beam and so is only a rough estimate. Hall thruster efficiency and performance useful in determining the power dissipation is described in detail in Chapter 7.

Table 2-1 Example ion thruster parameters used for the example power dissipation calculation in an ion thruster.

Parameter	Term	Units	Nominal value
Discharge voltage	V_d	Volts	25
Discharge current	I_d	Amps	20
Beam voltage	V_B	V	1500
Beam current	I_B	Amps	2
Discharge keeper voltage	V_{ck}	Volts	10
Discharge keeper current	I_{ck}	Amps	1
Neutralizer keeper voltage	V_{nk}	Volts	10
Neutralizer keeper current	I_{nk}	Amps	1
Accel grid current	I_A	mA	20
Accel grid voltage	V_A	Volts	250
Decel grid current	I_{DE}	mA	2
Coupling voltage	V_G	Volts	20

2.7 Neutral Densities and Ingestion

In electric propulsion thrusters, the propellant is injected as a neutral gas into a chamber or region where ionization takes place. Accurately knowing the flow rate of the propellant gas is important in determining the performance and efficiency of the thruster, and allows the operator to find the impact of finite pumping speed of test chambers on the thruster operation. The gas flow into the thruster, which is sometimes called the throughput, is often quoted in a number of different units. The most common units are standard cubic centimeters per minute (sccm) for ion thrusters and mg/s for Hall thrusters. Additional flow rate units include atoms per second, equivalent amperes, and torr-liter per second. Conversion factors between these systems of flow units are derived in Appendix B.

The neutral pressure in the thruster discharge chamber or in the vacuum system follows the standard gas law [3, 4]:

$$PV = NkT, \tag{2.7-1}$$

where P is the pressure in Pascals, V is the volume, N is the number of particles, k is Boltzmann's constant (1.38×10^{-23} W/s/K), and T is the temperature in Kelvin. Since there are 133.32 Pascals per Torr, the number density of the neutral gas is

$$n = \frac{P_T\,[\text{Torr}]\,133.32\,\{\text{Pascal/Torr}\}}{1.38 \times 10^{-23}\,[\text{J/°K}]*T[\text{°K}]} = 9.66 \times 10^{24}\frac{P_T}{T}\left[\frac{\text{particles}}{m^3}\right], \tag{2.7-2}$$

where P_T is the pressure in the vacuum system in Torr and T is the gas temperature in Kelvin. It should be noted that the pressure must be corrected for the gas type in whatever measurement system is used to obtain the actual pressure data. As an example, for a pressure of 10^{-6} Torr and a temperature of 290 °K, the density of gas atoms is 3.3×10^{16} per cubic meter.

The pressure in a vacuum system [5] in which a thruster is being tested is determined by the gas flow rate and the pumping speed

$$P = \frac{Q}{S}, \tag{2.7-3}$$

where Q is the total propellant throughput and S is the pumping speed. The most common units for pumping speed are liters per second, so utilizing a throughput in Torr-l/s directly provides the pressure in the vacuum system in Torr. The conversions of different flow units to Torr-l/s can be obtained from Appendix B.

The finite pressure in the test vacuum system causes a backflow of neutral gas into the thruster that may artificially improve the performance. This ingestion of facility gas by the thruster can be calculated if the pressure in the chamber is known by evaluating the flux of neutral gas from the chamber into the thruster ionization region. The equivalent flow into the thruster is then the injected flow Q plus the equivalent ingested flow. The ingested flow (in particles per second) is given by

$$Q_{\text{ingested}} = \frac{n\bar{c}}{4} A\,\eta_c, \tag{2.7-4}$$

where n is the neutral density in the chamber, $\bar{c}$ is the average gas thermal velocity, A is the total open area fraction of the thruster to the vacuum system, and η_c is a correction factor related to the conductance into the thruster from the vacuum system. The neutral gas density is given by Eq. 2.7-2, and the gas thermal velocity is given by

$$\bar{c} = \sqrt{\frac{8kT}{\pi M}}, \tag{2.7-5}$$

where M is the atom mass in kg. The conductance correction factor η_c is sometimes called the "Clausing Factor" [6] and describes the conductance reduction due to the finite axial length of the effective entrance aperture(s) to the thruster. This factor is generally negligible for Hall thrusters but appreciable for the apertured grids of ion thrusters. Owing to the large diameter-to-length ratio of the accelerator grid apertures in ion thrusters, the Clausing factor is usually calculated by Monte-Carlo gas flow codes. An example of a simple spreadsheet Monte-Carlo code for calculating the Clausing factor for ion thruster grids is given in Appendix G.

The ingested flow of gas from the finite pressure in the vacuum system is then

$$Q_{\text{ingested}} = \frac{133.2\,P}{4kT}\sqrt{\frac{8kT}{\pi M}}\frac{A\,\eta_c}{4.479 \times 10^{17}}\ [\text{sccm}]. \tag{2.7-6}$$

This expression for the ingested flow can be re-written as

$$Q_{\text{ingested}} = 7.82 \times 10^{8}\,\frac{P\,A\,\eta_c}{\sqrt{TM_a}}\ [\text{sccm}], \tag{2.7-7}$$

where P is the vacuum chamber pressure in Torr, T is the backflowing neutral gas temperature in °K, M_a is the gas mass in AMU, and A is the open area in m^2. The total flow rate into the thruster is then

$$Q_{\text{total}} = Q_{\text{injected}} + Q_{\text{ingested}}. \tag{2.7-8}$$

Problems

2.1 Assume that the ion charge density in a 1-D accelerator gap between two grids varies as $\rho = \rho_o x/d$, and that a voltage V_o is applied to the electrodes bounding the gap.

a) Find the potential and electric field as a function of position in the gap.
b) Find the force on each of the grids.
c) Find the total electrostatic force between the ions and the grids.

2.2 A mission under study desires to deliver a 800 kg payload through 8 km/s of Δv. The spacecraft has 3 kW of electric power available for propulsion. The mission planners want to understand the tradeoffs for different thrusters and operating conditions, and they want you to make plots of propellant mass and trip time required versus Isp for the following cases. Assume xenon is the propellant.

a) Ion thruster case: The ion thruster can run at full power from 1 kV to 2 kV. For all throttle conditions, assume the following parameters are constant: total efficiency of 55%, propellant utilization of 85%, beam divergence angle of 12°, and double-to-single ion current ratio of 10%.
b) Hall thruster case: The Hall thruster can run at full power from 300 V to 400 V. For all throttle conditions, assume the following parameters are constant: total efficiency of 45%, propellant utilization of 85%, beam divergence angle of 25°, and double-to-single ion current ratio of 15%.

2.3 Derive Eq. 2.4-7 for the mass utilization efficiency correction due to double ions.

2.4 Derive the thrust correction factor and the resulting thrust equation accounting for the presence of triply ionized atoms. Assuming 10% doubles, what is the error in the calculated thrust if 5% triples actually present have been neglected?

2.5 Mission planners have two candidate ion and Hall thrusters to place on a spacecraft and want to understand how they compare for thrust to power ratio and performance. The xenon ion thruster has a total power of 5 kW, a 1200 V, 3.75 A beam with 10% double ions, a total efficiency of 65%, and a mass utilization efficiency of 86%. The Hall thruster has a total power of 5 kW, a 300 V discharge voltage and a 12.5 A beam with 10% double ions, a total efficiency of 50%, and an input xenon gas flow of 19 mg/s.

a) What is the thrust to power ratio (usually expressed in mN/kW) for each thruster?
b) What is the Isp for each engine?
c) For a 1000 kg spacecraft, what is the fuel mass required to achieve a 5 km/s delta-v?
d) What is the trip time to expend all the fuel for each thruster type if the thrusters are on 90% of the time?

2.6 The thrust correction factor for multiply ionized species is based on the current of charges in the beam (see Eq. 2.3-15 for singles and doubles).

a) Derive an expression for the number of atoms of each ionized species in the beam for a given value of I^{++}/I^{+} and I^{+++}/I^{+}.
b) If $I^{++}/I^{+} = 10\%$ and $I^{+++}/I^{+} = 5\%$, what are the actual percentages of the number atoms of each species in the beam compared to the total beam current?

2.7 An ion thruster is being tested in a vacuum chamber with a measured xenon pressure during operation of 1×10^{-5} Torr at room temperature (300 °C). The thruster grids have a 25 cm grid diameter and a Clausing factor of 0.5.

a) If the thruster is producing a 3 A beam with 15% double ions with a total of 50 sccm of xenon gas flow into the thruster, what is the mass utilization efficiency neglecting gas ingestion?
b) What is the mass utilization efficiency including the effects of ingestion?

References

1 K. E. Tsiolkovsky, "The Exploration of Cosmic Space by Means of Reaction Devices," *The Science Review* (in Russian), Gublit, pp. 12–46, 1903.

2 D. L. Brown, C. W. Larson, B. E. Beal, and A. D. Gallimore, "Methodology and Historical Perspective of a Hall Thruster Efficiency Analysis," *Journal of Propulsion and Power*, vol. 25, no. 6, pp. 1163–1172, 2009.

3 J. M. Lafferty, *Foundations of Vacuum Science and Technology*, Wiley, 1998.

4 A. Ross, *Vacuum Technology*, Amsterdam: Elsevier, 1990.

5 G. Lewin, *Fundamentals of Vacuum Science and Technology*, McGraw-Hill, 1965.

6 P. Clausing, "The Flow of Highly Rarefied Gases Through Tubes of Arbitrary Length," *Journal of Vacuum Science and Technology*, vol. 8, pp. 636–646, 1971.

Chapter 3

Basic Plasma Physics

3.1 Introduction

Electric propulsion (EP) achieves high-specific impulse by the acceleration of charged particles to high velocity. The charged particles are produced by ionization of a propellant gas, which creates both ions and electrons and forms what is called a plasma. Plasma is then a collection of the various charged particles that are free to move in response to fields they generate or fields that are applied to the collection, and on the average is almost electrically neutral. This means that the ion and electron densities are nearly equal, $n_i \approx n_e$, a condition commonly termed "quasi-neutrality". This condition exists throughout the volume of the ionized gas except close to the boundaries, and the assumption of quasi-neutrality is valid whenever the spatial scale length of the plasma is much larger than the characteristic length over which charges or boundaries are electrostatically shielded, called the Debye length. The ions and electrons have distributions in energy usually characterized by a temperature T_i for ions and T_e for electrons, which are not necessarily or usually the same. In addition, different ion and electron species can exist in the plasma with different temperatures or different distributions in energy.

Plasmas in EP devices, even in individual parts of a thruster, can span orders of magnitude in plasma density, temperature, and ionization fraction. Therefore, models used to describe the plasma behavior and characteristics in the thrusters must be formed with assumptions that are valid in the regime being studied. Many of the plasma conditions and behaviors in thrusters can be modeled by fluid equations, and kinetic effects are only important in specific instances.

There are several textbooks that provide very comprehensive introductions to plasma physics [1–3] and the generation of ion beams [4]. This chapter is intended to provide the basic plasma physics necessary to understand the operation of electric thrusters. The units used throughout the book are based on the International System (SI). However, by convention we will occasionally revert to other metric units (such as A/cm^2, mg/s, etc.) commonly used in the literature describing these devices.

3.2 Maxwell's Equations

The electric and magnetic fields that exist in EP plasmas obey Maxwell's Equations formulated in a vacuum that contains charges and currents. Maxwell's equations for these conditions are:

$$\nabla \cdot \mathbf{E} = \frac{\rho}{\varepsilon_o} \tag{3.2-1}$$

Fundamentals of Electric Propulsion, Second Edition. Dan M. Goebel, Ira Katz, and Ioannis G. Mikellides.

$$\nabla \times \mathbf{E} = -\frac{\partial \mathbf{B}}{\partial t} \tag{3.2-2}$$

$$\nabla \cdot \mathbf{B} = 0 \tag{3.2-3}$$

$$\nabla \times \mathbf{B} = \mu_o \left(\mathbf{J} + \varepsilon_0 \frac{\partial \mathbf{E}}{\partial t} \right) \tag{3.2-4}$$

where ρ is the charge density in the plasma, $\mathbf{J}$ is the current density in the plasma, and ε_o and μ_o are the permittivity and permeability of free space, respectively. Note that ρ and $\mathbf{J}$ comprise all the charges and currents for all the particle species that are present in the plasma, including multiply-charged ions. The charge density is then

$$\rho = \sum_s q_s n_s = e(Zn_i - n_e) \tag{3.2-5}$$

where q_s is the charge state of species s, Z is the charge state, n_i is the ion number density, and n_e is the electron number density. Likewise, the current density is

$$J = \sum_s q_s n_s \upsilon_s = \mathrm{e}(Zn_i \upsilon_i - n_e \upsilon_e) \tag{3.2-6}$$

where υ_s is the velocity of the charge species, υ_i is the ion velocity, and υ_e is the electron velocity. For static magnetic fields ($\partial \mathbf{B}/\partial t = 0$), the electric field can be expressed as the gradient of the electric potential

$$\mathbf{E} = -\nabla \phi, \tag{3.2-7}$$

where the negative sign comes from the convention that the electric field always points in the direction of ion motion.

3.3 Single Particle Motions

The equation of motion for a charged particle with a velocity $\mathbf{v}$ in a magnetic field $\mathbf{B}$ is given by the Lorentz force equation:

$$\mathbf{F} = m\frac{d\mathbf{v}}{dt} = q(\mathbf{E} + \mathbf{v} \times \mathbf{B}). \tag{3.3-1}$$

Particle motion in a magnetic field in the $\hat{z}$ direction for the case of negligible electric field is found by evaluating Eq. 3.3-1:

$$\begin{aligned} m\frac{\partial \upsilon_x}{\partial t} &= qB\upsilon_y \\ m\frac{\partial \upsilon_y}{\partial t} &= -qB\upsilon_x \\ m\frac{\partial \upsilon_z}{\partial t} &= 0. \end{aligned} \tag{3.3-2}$$

Taking the time derivative of Eq. 3.3-2 and solving for the velocity in each direction gives

$$\begin{aligned}\frac{\partial^2 v_x}{\partial t^2} &= \frac{qB}{m}\frac{\partial v_y}{\partial t} = -\left(\frac{qB}{m}\right)^2 v_x \\ \frac{\partial^2 v_y}{\partial t^2} &= -\frac{qB}{m}\frac{\partial v_x}{\partial t} = -\left(\frac{qB}{m}\right)^2 v_y\end{aligned} \tag{3.3-3}$$

These equations describe a simple harmonic oscillator at the cyclotron frequency:

$$\omega_c = \frac{|q|B}{m}. \tag{3.3-4}$$

For electrons, this is called the electron cyclotron frequency.

The size of the particle orbit for finite particle energies can be found from the solution to the particle motion equations in the axial magnetic field. In this case, the solution to Eq. 3.3-3 is

$$v_{x,y} = v_{\perp} e^{i\omega_c t}. \tag{3.3-5}$$

The equation of motion in the y-direction in Eq. 3.3-2 can be re-written as

$$v_y = \frac{m}{qB}\frac{\partial v_x}{\partial t} = \frac{1}{\omega_c}\frac{\partial v_x}{\partial t}. \tag{3.3-6}$$

Utilizing Eq. 3.3-5, Eq. 3.3-6 becomes

$$v_y = \frac{1}{\omega_c}\frac{\partial v_x}{\partial t} = i v_{\perp} e^{i\omega t} = \frac{\partial y}{\partial t}. \tag{3.3-7}$$

Integrating this equation gives

$$y - y_o = \frac{v_{\perp}}{w_c} e^{i\omega_c t}. \tag{3.3-8}$$

Taking the real part of Eq. 3.3-8 gives

$$y - y_o = \frac{v_{\perp}}{\omega_c}\cos\omega_c t = r_L \cos\omega_c t, \tag{3.3-9}$$

where $r_L = v_{\perp}/\omega_c$ is defined as the Larmor radius. A similar analysis of the displacement in the $\hat{x}$ direction gives the same Larmor radius 90° out of phase with the $\hat{y}$-direction displacement, which with Eq. 3.3-9 describes the particle motion as a circular orbit around the field line at x_o and y_o with a radius given by r_L.

The Larmor radius arises from very simple physics. Consider a charged particle of mass m in a uniform magnetic field with a velocity in one direction, as illustrated in Fig. 3-1. The charge will feel a Lorentz force

$$\mathbf{F} = q\mathbf{v}_{\perp} \times \mathbf{B}. \tag{3.3-10}$$

Since the charged particle will move under this force in circular orbits in the $\mathbf{v}_{\perp} \times \mathbf{B}$ direction, it feels a corresponding centripetal force such that

$$\mathbf{F_c} = q\mathbf{v}_{\perp} \times \mathbf{B} = \frac{mv_{\perp}^2}{r}, \tag{3.3-11}$$

where r is the radius of the cycloidal motion in the magnetic field. Solving for the radius of the circle gives

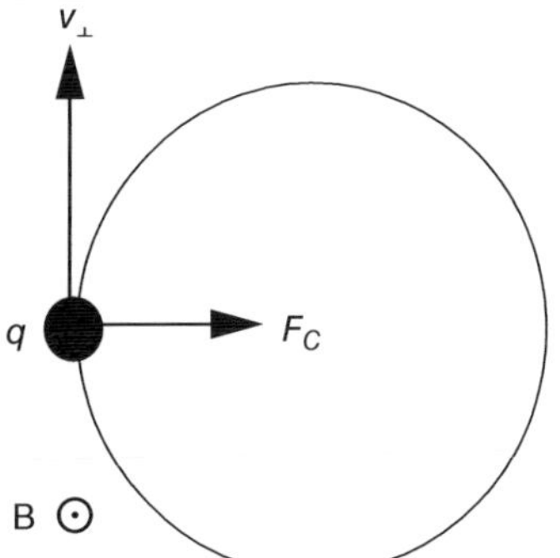

Figure 3-1 Positively charged particles moving in a uniform vertical magnetic field.

$$r = r_L = \frac{mv_\perp}{qB}, \tag{3.3-12}$$

which is the Larmor radius.

The Larmor radius can be written in a simple form to remember:

$$r_L = \frac{v_\perp}{\omega_c} = \frac{1}{B}\sqrt{\frac{2m\mathrm{V}_\perp}{e}}, \tag{3.3-13}$$

using $\frac{1}{2}m\mathrm{v}_\perp = eV_\perp$ for the singly-charged particle energy in the direction perpendicular to the magnetic field. The direction of particle gyration is always such that the induced magnetic field is opposite in direction to the applied field, which tends to reduce the applied field, an effect called diamagnetism. Any particle motion along the magnetic field is not affected by the field, but causes the particle motion to form a helix along the magnetic field direction with a radius given by the Larmor radius and a pitch given by the ratio of the perpendicular to parallel velocities.

Next consider the situation in Fig. 3-1, but with the addition of a finite electric field perpendicular to **B**. In this case, **E** is in some direction in the plane of the page. The equation of motion for the charged particle is given by Eq. 3.3-1. Considering the drift to be steady-state, the time derivative is equal to zero, and Eq. 3.3-1 becomes

$$\mathbf{E} = -\mathbf{v} \times \mathbf{B}. \tag{3.3-14}$$

Taking the cross product of both sides with **B** gives

$$\mathbf{E} \times \mathbf{B} = (-\mathbf{v} \times \mathbf{B}) \times \mathbf{B} = \mathbf{v}B^2 - \mathbf{B}(\mathbf{B} \cdot \mathbf{v}). \tag{3.3-15}$$

The dot product is in the direction perpendicular to B, so the last term in Eq. 3.3-15 is equal to zero. Solving for the transverse velocity of the particle gives

$$\mathbf{v} = \frac{\mathbf{E} \times \mathbf{B}}{B^2} \equiv \mathbf{v}_E, \tag{3.3-16}$$

which is the "**E** cross **B**" drift velocity. In this case, the drift is in the direction perpendicular to both **E** and **B**, and arises from the cycloidal electron motion in the magnetic field being accelerated in the direction of –**E** and decelerated in the direction of **E**. This elongates the orbit on one half cycle and shrinks the orbit on the opposite half cycle, which causes the net motion of the particle in the **E** × **B** direction. The units of the **E** × **B** velocity are

$$v_E = \frac{E[\mathrm{V/m}]}{B[\mathrm{tesla}]}\,[\mathrm{m/s}]. \tag{3.3-17}$$

Finally, consider the situation of a particle gyrating in a magnetic field that is changing in magnitude along the magnetic field direction $\hat{z}$. This is commonly found in EP thrusters relatively close to permanent magnets or electromagnetic pole-pieces that produce fields used to confine the electrons. Since the divergence of B is zero (Eq. 3.2-3), the magnetic field in cylindrical coordinates is described by

$$\frac{1}{r}\frac{\partial}{\partial r}(rB_r) + \frac{\partial B_z}{\partial z} = 0. \tag{3.3-18}$$

Assuming that the axial component of the field does not vary significantly with r and integrating yields the radial component of the magnetic field with respect to r:

$$B_r \approx -\frac{r}{2}\frac{\partial B_z}{\partial z}. \tag{3.3-19}$$

The Lorentz force on a charged particle has a component along $\hat{z}$ given by

$$F_z \approx -qv_\phi B_r, \tag{3.3-20}$$

where the azimuthal particle velocity averaged over a Larmor-radius ($r = r_L$) gyration is $v_\phi = -v_\perp$. The average force on the particle is then

$$\overline{F}_z \approx -\frac{1}{2}\frac{mv_\perp^2}{B}\frac{\partial B_z}{\partial z}. \tag{3.3-21}$$

The magnetic moment of the gyrating particle is defined as

$$\mu = \frac{1}{2}\frac{mv_\perp^2}{B}. \tag{3.3-22}$$

As the particle moves along the magnetic field lines into a stronger magnitude field regions, the parallel energy of the particle is converted into rotational energy and its Larmor radius increases. However, its magnetic moment remains invariant because the magnetic field does no work and the total kinetic energy of the particle is conserved. For a sufficiently large increase in the magnetic field, a situation can arise where the parallel velocity of the particle goes to zero and the Lorentz force reflects the particle from a "magnetic mirror". By conservation of energy, particles will be reflected from the magnetic mirror if their parallel velocity is less than

$$v_\parallel < v_\perp\sqrt{R_m - 1}, \tag{3.3-23}$$

where $v_\parallel$ is the parallel velocity and R_m is the mirror ratio given by $B_{\text{max}}/B_{\text{min}}$. This effect is used to provide confinement of energetic electrons in ion thruster discharge chambers.

There are a number of other particle drifts and motions possible that depend on gradients in the magnetic and electric fields, and also on time dependent or oscillating electric or magnetic fields. These are described in detail in plasma physics texts such as Chen [1], and although they certainly might occur in the EP devices considered here, they are typically not of critical importance to the thruster performance or behavior.

3.4 Particle Energies and Velocities

In electric thrusters, the charge particles may undergo a large number of collisions with each other, and in some cases with the other species (ions, electrons, and/or neutrals) in the plasma. It is therefore impractical to analyze the motion of each particle to obtain a macroscopic picture of the plasma processes that is useful for assessing the performance and life of these devices. Fortunately, in most cases it is not necessary to track individual particles to understand the plasma dynamics. The effect of collisions is to develop a distribution of the velocities for each species. On the average, and in the absence of other forces, each particle will then move with a speed that is solely a function of the macroscopic temperature and mass of that species. The charged particles in the thruster, therefore, can usually be described by different velocity distribution functions and the random motions calculated by taking the moments of those distributions.

Most of the charged particles in electric thrusters have a Maxwellian velocity distribution, which is the most probable distribution of velocities for a group of particles in thermal equilibrium. In one dimension, the Maxwellian velocity distribution function is:

$$f(v) = \left(\frac{m}{2\pi kT}\right)^{1/2} \exp\left(-\frac{mv^2}{2kT}\right), \tag{3.4-1}$$

where m is the mass of the particle, k is Boltzmann constant, and the width of the distribution is characterized by the temperature T. The average kinetic energy of a particle in the Maxwellian distribution in one dimension is

$$E_{\text{ave}} = \frac{\int_{-\infty}^{\infty} \frac{1}{2} m v^2 f(v) \mathrm{d}v}{\int_{-\infty}^{\infty} f(v) \mathrm{d}v} \tag{3.4-2}$$

By inserting in Eq. 3.4-1 and integrating by parts, the average energy per particle in each dimension is

$$E_{\text{ave}} = \frac{1}{2} kT. \tag{3.4-3}$$

If the distribution function is generalized into three dimensions, Eq. 3.4-1 becomes

$$f(u, v, w) = \left(\frac{m}{2\pi \mathrm{kT}}\right)^{3/2} \exp\left[-\frac{m}{2\mathrm{kT}}(u^2 + v^2 + w^2)\right], \tag{3.4-4}$$

where u, v and w represent the velocity components in the three coordinate axes. The average energy in three dimensions is found by inserting Eq. 3.4-4 in Eq. 3.4-2 and performing the triple integration to give

$$E_{\text{ave}} = \frac{3}{2} kT. \tag{3.4-5}$$

The density of the particles is found from

$$n = \iiint_{-\infty}^{\infty} n f(v) dv = \iiint_{-\infty}^{\infty} n \left(\frac{m}{2\pi kT}\right)^{3/2} \exp\left(-\frac{m(u^2 + v^2 + w^2)}{2kT}\right) du dv dw \tag{3.4-6}$$

The average speed of a particle in the Maxwellian distribution is

$$\bar{v} = \int_0^{\infty} v \left(\frac{m}{2\pi kT}\right)^{3/2} \exp\left(-\frac{v^2}{v_{\text{th}}^2}\right) 4\pi v^2 dv, \tag{3.4-7}$$

where v in Eq. 3.4-7 denotes the particle speed and v_{th} is defined as $(2kT/m)^{1/2}$. Integrating Eq. 3.4-7, the average speed per particle is

$$\bar{v} = \left(\frac{8kT}{\pi m}\right)^{1/2}. \tag{3.4-8}$$

The flux of particles in one dimension (say in the $\hat{z}$ direction) for a Maxwellian distribution of particle velocities is given by $n\langle v_z \rangle$. In this case, the average over the particle velocities is taken in the positive v_z direction because the flux is considered in only one direction. The particle flux (in one direction) is then

$$\Gamma_z = \int n v_z f(\mathbf{v}) d^3\mathbf{v}, \tag{3.4-9}$$

which can be evaluated by integrating the velocities in spherical coordinates with the velocity volume element given by

$$d^3 v = v^2 dv \, d\Omega = v^2 dv \sin\theta \, d\theta \, d\phi, \tag{3.4-10}$$

where the $d\Omega$ represents the element of the solid angle. If the incident velocity has a cosine distribution ($v_z = v\cos\theta$), the one-sided flux is

$$\Gamma_z = n\left(\frac{m}{2\pi kT}\right)^{3/2}\int_0^{2\pi} d\phi \int_0^{\pi/2} \sin\theta \, d\theta \int_0^{\infty} v\cos\theta \exp\left(-\frac{v^2}{v_{\text{th}}}\right) v^2 dv, \tag{3.4-11}$$

which gives

$$\Gamma_z = \frac{1}{4} n\bar{v} = \frac{1}{4} n\left(\frac{8kT}{\pi m}\right)^{1/2}. \tag{3.4-12}$$

Since the plasma electrons are very mobile and tend to make a large number of Coulomb collisions with each other, they can usually be characterized by a Maxwellian temperature T_e and have average energies and speeds well described by the equations derived in this section. The random electron flux inside the plasma is also well described by Eq. 3.4-12 if the electron temperature and density are known. The electrons tend to be relatively hot (compared to the ions and atoms) in electric thrusters because they are typically injected into the plasma or heated by external mechanisms to provide sufficient energy to produce ionization. In the presence of electric and magnetic fields in the plasma and at the boundaries, the electron motion will no longer be purely random, and the flux described by Eq. 3.4-12 must be modified as described below.

The ions in thrusters, on the other hand, are usually relatively cold in temperature (they may have high directed velocities after being accelerated, but they usually have relatively low random velocities and temperatures). This occurs because the ions are not well confined in the plasma generators for the reason that they must be extracted to form the thrust beam, and so leave the plasma after perhaps only a single pass. The ions are also not heated efficiently by the various mechanisms used to ionize the gas. Therefore, the plasmas in electric thrusters are usually characterized as having relatively cold ions and Maxwellian electrons with a high electron to ion temperature ratio ($T_e/T_i \approx 10$). As a result, the velocity of the ions in the plasma and the fluxes to the boundaries tend to be determined by the electric fields generated inside the plasma to conserve charge, and are different than the expressions derived here for the electron velocity and fluxes. This effect will be described in more detail in Section 3.6.

3.5 Plasma as a Fluid

The behavior of most of the plasma effects in electric thrusters can be described by simplified models in which the plasma is treated as a fluid of neutral particles and electrical charges with Maxwellian distribution functions, and only the interactions and motion of the fluid elements must be considered. Kinetic effects that consider the actual velocity distribution of each species are important in some instances, but will not be addressed here.

3.5.1 Momentum Conservation

In constructing a fluid approach to plasmas, there are three dominant forces on the charged particles in the plasma that transfer momentum that are considered here. First, charged particles react to electric and magnetic fields via the Lorentz force, which was given by Eq. 3.3-1:

$$\mathbf{F}_L = m\frac{d\mathbf{v}}{dt} = q(\mathbf{E} + \mathbf{v} \times \mathbf{B}). \tag{3.5-1}$$

Next, there is a pressure gradient force

$$\mathbf{F}_p = -\frac{\nabla \cdot \boldsymbol{p}}{n} = -\frac{\nabla(\mathbf{nkT})}{n}, \tag{3.5-2}$$

where the pressure is given by $P = nkT$ and should be written more rigorously as a stress tensor since it can, in general, be anisotropic. For plasmas with temperatures that are generally spatially constant, the force due to the pressure gradient is usually written simply as

$$\mathbf{F}_p = -kT\frac{\nabla \boldsymbol{n}}{n}. \tag{3.5-3}$$

Finally, collisions transfer momentum between the different charged particles, and also between the charged particles and the neutral gas. The force due to collisions is

$$\mathbf{F}_c = -m\sum_{a,b}\nu_{ab}(\mathbf{v}_a - \mathbf{v}_b) \tag{3.5-4}$$

where ν_{ab} is the collision frequency between species a and b.

Using these three force terms, the fluid momentum equation for each species is

$$mn\frac{d\mathbf{v}}{dt} = mn\left[\frac{\partial \mathbf{v}}{\partial t} + (\mathbf{v}\cdot\nabla)\mathbf{v}\right] = qn(\mathbf{E} + \mathbf{v}\times\mathbf{B}) - \nabla\cdot\mathbf{p} - mn\nu(\mathbf{v} - \mathbf{v}_o) \tag{3.5-5}$$

where the convective derivative has been written explicitly and the collision term must be summed over all collisions.

Utilizing conservation of momentum, it is possible to evaluate how the electron fluid behaves in the plasma. For example, in one dimension and in the absence of magnetic fields and collisions with other species, the fluid equation of motion for electrons can be written as

$$mn_e\left[\frac{\partial v_z}{\partial t} + (v\cdot\nabla)v_z\right] = qn_eE_z - \frac{\partial p}{\partial z}, \tag{3.5-6}$$

where v_z is the electron velocity in the z-direction and p represents the electron pressure term. Neglecting the convective derivative, assuming that the velocity is spatially uniform, and using Eq. 3.5-3 gives

$$m\frac{\partial v_z}{\partial t} = -eE_z - \frac{\mathrm{kT}_e}{n_e}\frac{\partial n_e}{\partial z}. \tag{3.5-7}$$

Assuming that the electrons have essentially no inertia (their mass is small and so they react infinitely fast to changes in potential), the left-hand side of Eq. 3.5-7 goes to zero and the net current in the system is also zero. Considering only electrons at a temperature T_e, and using Eq. 3.2-7 for the electric field, gives

$$qE_z = e\frac{\partial \phi}{\partial z} = \frac{\mathrm{kT}_e}{n_e}\frac{\partial n_e}{\partial z}. \tag{3.5-8}$$

Integrating this equation and solving for the electron density produces the "Boltzmann relationship" for electrons:

$$n_e = n_e(0)e^{(e\phi/kT_e)} \tag{3.5-9}$$

where ϕ is the potential relative to the potential ϕ_o at the location of $n_e(0)$.

Equation 3.5-9 is sometimes written as

$$n_e = n_e(0)e^{\left(\frac{\phi(z)-\phi_o}{T_{eV}}\right)} \tag{3.5-10}$$

where T_{eV} is the electron temperature in electron-volts and the exponent specifically shows the reference potential. This relationship simply states that the electrons will respond to electrostatic fields (potential changes) by varying their density to preserve the pressure in the system. This relationship is generally valid for motion along a magnetic field and tends to hold for motion across magnetic fields if the field is weak and the electron collisions are frequent.

3.5.2 Particle Conservation

Conservation of particles and/or charges in the plasma is described by the continuity equation:

$$\frac{\partial n}{\partial t} + \nabla \cdot n\mathbf{v} = \dot{n}_s, \tag{3.5-11}$$

where $\dot{n}_s$ represents the time dependent source or sink term for the species being considered. Continuity equations are sometimes called mass-conservation equations because they account for the sources and sinks of particles into and out of the plasma.

Utilizing continuity equations coupled with momentum conservation and with Maxwell's equations, it is possible to calculate the response rate and wave-like behavior of plasmas. For example, the rate at which a plasma responds to changes in potential is related to the *plasma frequency* of the electrons. Assume that there is no magnetic field in the plasma or that the electron motion is along the magnetic field in the z-direction. To simplify this derivation, also assume that the ions are fixed uniformly in space on the time scales of interest here due to their large mass, and there is no thermal motion of the particles ($T = 0$). Since the ions are fixed in this case, only the electron equation of motion is of interest:

$$mn_e\left[\frac{\partial v_z}{\partial t} + (v \cdot \nabla)v_z\right] = -en_eE_z, \tag{3.5-12}$$

and the electron equation of continuity is

$$\frac{\partial n_e}{\partial t} + \nabla \cdot (n_e\mathbf{v}) = 0 \tag{3.5-13}$$

The relationship between the electric field and the charge densities is given by Eq. 3.2-1, which can be written using Eq. 3.2-5 as

$$\nabla \cdot \mathbf{E} = \frac{\rho}{\varepsilon_o} = \frac{e}{\varepsilon_o}(n_i - n_e), \tag{3.5-14}$$

The wave-like behavior of this system is analyzed by linearizing using

$$\mathbf{E} = \mathbf{E}_o + \boldsymbol{E}_1 \tag{3.5-15}$$

$$\mathbf{v} = \mathbf{v}_o + \boldsymbol{v}_1 \tag{3.5-16}$$

$$n = n_o + n_1 \tag{3.5-17}$$

where $\mathbf{E}_o$, $\mathbf{v}_o$, and n_o are the equilibrium values of the electric field, electron velocity and electron density, and $\mathbf{E}_1$, $\boldsymbol{\upsilon}_1$, and n_1 are the perturbed values of these quantities. Since quasi-neutral plasma has been assumed, $\mathbf{E}_o = 0$, and the assumption of a uniform plasma with no temperature means that $\Delta n_o = \mathbf{v}_o = 0$. Likewise, the time derivatives of these equilibrium quantities are zero.

Linearizing Eq. 3.5-14 gives

$$\nabla \cdot \boldsymbol{E}_1 = -\frac{e}{\varepsilon_o} n_1. \tag{3.5-18}$$

Substituting Eqs. 3.5-15, 3.5-16, and 3.5-17 in Eq. 3.5-12 results in

$$\frac{d\upsilon_1}{dt} = -\frac{e}{m} E_1 \hat{z}, \tag{3.5-19}$$

where the linearized convective derivative has been neglected. Linearizing the continuity Eq. 3.5-13 gives

$$\frac{dn_1}{dt} = -n_o \nabla \cdot \boldsymbol{\upsilon}_1 \hat{z}, \tag{3.5-20}$$

where the quadratic terms, such as $n_1 \upsilon_1$, etc., have been neglected as small. In the linear regime, the oscillating quantities will behave sinusoidally:

$$\boldsymbol{E}_1 = E_1 e^{i(kz - \omega t)} \hat{z} \tag{3.5-21}$$

$$\boldsymbol{\upsilon}_1 = \upsilon_1 e^{i(kz - \omega t)} \hat{z} \tag{3.5-22}$$

$$n_1 = n_1 e^{i(kz - \omega t)}. \tag{3.5-23}$$

This means that the time derivates in the momentum and continuity equations can be replaced by *-iω,* and the gradient in Eq. 3.5-18 can be replaced by *ik* in the $\hat{z}$ direction. Combining Eqs. 3.5-18, 3.5-19, and 3.5-20, using the time and spatial derivatives of the oscillating quantities, and solving for the frequency of the oscillation gives

$$\omega_p = \left(\frac{n_e e^2}{\varepsilon_o m}\right)^{1/2}, \tag{3.5-24}$$

where ω_p is the electron plasma frequency. A useful numerical formula for the electron plasma frequency is

$$f_p = \frac{\omega_p}{2\pi} \approx 9\sqrt{n_e}, \tag{3.5-25}$$

where the plasma density is in m^{-3}. This frequency is one of the fundamental parameters of a plasma, and the inverse of this value is approximately the minimum time required for the plasma to react to changes in its boundaries or in the applied potentials. For example, if the plasma density is 10^{18} m^{-3}, the electron plasma frequency is 9 GHz, and the electron plasma will respond to perturbations in less than a nanosecond.

In a similar manner, if the ion temperature is assumed to be negligible and the gross response of the plasma is dominated by ion motions, the ion plasma frequency can be found to be

$$\Omega_p = \left(\frac{n_e e^2}{\varepsilon_o M}\right)^{1/2}. \tag{3.5-26}$$

This equation provides the approximate time scale in which ions move in the plasma. For our previous example for a 10^{18} m^{-3} plasma density composed of xenon ions, the ion plasma frequency is about 18 MHz and the ions will respond to first order in a fraction of a microsecond. However, the ions have inertia and respond at the ion acoustic velocity given by

$$v_a = \sqrt{\frac{\gamma_i kT_i + kT_e}{M}}, \tag{3.5-27}$$

where γ_i is the ratio of ion specific heats and is equal to one for isothermal ions. In the normal case for ion and Hall thrusters where $T_e \gg T_i$, the ion acoustic velocity is simply

$$v_a = \sqrt{\frac{kT_e}{M}}. \tag{3.5-28}$$

It should be noted that if finite-temperature electrons and ions had been included in the derivations above, the electron-plasma and ion-plasma oscillations would have produced waves that propagate with finite wavelengths in the plasma. Electron plasma waves and ion plasma waves (sometimes called ion acoustic waves) occur in most electric thruster plasmas with varying amplitudes and effects on the plasma behavior. The dispersion relationships for these waves, which describe the relationship between the frequency and the wavelength of the wave, are derived in detail in plasma textbooks such as Chen [1] and will not be re-derived here.

3.5.3 Energy Conservation

The general form of the energy equation for charged species "s" moving with velocity $\mathbf{v}_s$ in the presence of species "a" is given by

$$\begin{aligned} &\frac{\partial}{\partial t}\left(n_s m_s \frac{v_{s2}}{2} + \frac{3}{2} p_s\right) + \nabla \cdot \left(n_s m_s \frac{v_s^2}{2} + \frac{5}{2} p_s\right) \mathbf{v}_s + \nabla \cdot \boldsymbol{\theta}_s \\ &= q_s n_s \left(\mathbf{E} + \frac{\mathbf{R}_s}{q_s n_s}\right) \cdot \mathbf{v}_s + Q_s - \Psi_s \end{aligned} \tag{3.5-29}$$

For simplicity, Eq. 3.5-29 neglects viscous heating of the species. The divergence terms on the left-hand side represent the total energy flux, which includes the work done by the pressure, the macroscopic energy flux, and the transport of heat by conduction $\boldsymbol{\theta}_s = -\boldsymbol{\kappa}_s \nabla T_s$. The thermal conductivity of the species is denoted by $\boldsymbol{\kappa}_s$, which is given in SI units [5] by

$$\boldsymbol{\kappa}_s = 3.2 \frac{\tau_e n e^2 T_{eV}}{m}, \tag{3.5-30}$$

where T_{eV} in this equation is in electron volts (eV). The right-hand side of Eq. 3.5-29 accounts for the work done by other forces as well for the generation/loss of heat as a result of collisions with other particles. The term $\mathbf{R}_s$ represents the mean change in the momentum of particles "s" as a result of collisions with all other particles:

$$R_s \equiv \sum_n R_{sn} = -\sum_n n_s m_s \nu_{sn} (\mathbf{v}_s - \mathbf{v}_n) \tag{3.5-31}$$

The heat-exchange terms are Q_s, which is the heat generated/lost in the particles of species "s" as a result of elastic collisions with all other species, and Ψ_s, the energy loss by species "s" as a result of inelastic collision processes such as ionization and excitation.

It is often useful to eliminate the kinetic energy from Eq. 3.5-29 to obtain a more applicable form of the energy conservation law. The left-hand side of Eq. 3.5-29 is expanded as

$$\begin{aligned} & n_s m_s v_s \cdot \frac{D\mathbf{v}_s}{Dt} + \frac{m_s v_s^2}{2}\frac{Dn_s}{Dt} + n_s m_s \frac{v_s^2}{2}\nabla\cdot\mathbf{v}_s + \frac{3}{2}\frac{\partial p_s}{\partial t} + \nabla\cdot\left(\frac{5}{2}p_s\mathbf{v}_s + \boldsymbol{\theta}_s\right) . \\ & = q_s n_s \mathbf{E}\cdot\mathbf{v}_s + \mathbf{R}_s\cdot\mathbf{v}_s + Q_s - \Psi_s \end{aligned} \tag{3.5-32}$$

The continuity equation for the charged species is in the form

$$\frac{Dn_s}{Dt} = \frac{\partial n_s}{\partial t} + \mathbf{v}_s\cdot\nabla n_s = \dot{n} - n_s\nabla\cdot\mathbf{v}_s. \tag{3.5-33}$$

Combining these two equations with the momentum equation dotted with $\mathbf{v}_s$ gives

$$n_s m_s v_s \cdot \frac{D\mathbf{v}_s}{Dt} = n_s q_s \mathbf{v}_s\cdot\mathbf{E} - \mathbf{v}_s\cdot\nabla p_s + \mathbf{v}_s\cdot R_s - \dot{n} m_s v_s^2. \tag{3.5-34}$$

The energy equation can now be written as

$$\frac{3}{2}\frac{\partial p_s}{\partial t} + \nabla\cdot\left(\frac{5}{2}p_s\mathbf{v}_s + \boldsymbol{\theta}_s\right) - \mathbf{v}_s\cdot\nabla p_s = Q_s - \Psi_s - \dot{n}\frac{m_s v_s^2}{2}. \tag{3.5-35}$$

The heat-exchange terms for each species Q_s consists of "frictional" (denoted by superscript R) and "thermal" (denoted by superscript T) contributions:

$$\begin{aligned} Q_s &= Q_s^R + Q_s^T \\ Q_s^R &\equiv -\sum_n \mathbf{R}_{\mathrm{sn}}\cdot\mathbf{v}_s, \\ Q_s^T &\equiv -\sum_n n_s \frac{2m_s}{m_a}\nu_{\mathrm{sn}}\frac{3}{2}\left(\frac{\mathrm{kT}_s}{e} - \frac{\mathrm{kT}_n}{e}\right). \end{aligned} \tag{3.5-36}$$

In a partially ionized gas consisting of electrons, singly-charged ions, and neutrals of the same species, the frictional and thermal terms for the electrons take the form

$$\begin{aligned} Q_e^R &= -(\mathbf{R}_{\mathrm{ei}} + \mathbf{R}_{\mathrm{en}})\cdot\mathbf{v}_e = \left(\frac{\mathbf{R}_{\mathrm{ei}} + \mathbf{R}_{\mathrm{en}}}{\mathrm{e}n_e}\right)\cdot\mathbf{J}_e \\ Q_e^T &= -3n_e\frac{m}{M}\left[\nu_{\mathrm{ei}}\frac{k}{e}(T_e - T_i) + \nu_{\mathrm{en}}\frac{k}{e}(T_e - T_n)\right] \end{aligned}, \tag{3.5-37}$$

where as usual M denotes the mass of the heavy species and the temperature of the ions and neutrals is denoted by T_i and T_n, respectively. Using the steady-state electron momentum equation, in the absence of electron inertia, it is possible to write

$$Q_e^R = \left(\frac{\mathbf{R}_{\mathrm{ei}} + \mathbf{R}_{\mathrm{en}}}{en_e}\right)\cdot\mathbf{J}_e = \left(\mathbf{E} + \frac{\nabla p_e}{en_e}\right)\cdot\mathbf{J}_e. \tag{3.5-38}$$

Thus Eq. 3.5-35 for the electrons becomes

$$\begin{aligned} \frac{3}{2}\frac{\partial p_e}{\partial t} + \nabla\cdot\left(\frac{5}{2}p_e\mathbf{v}_e + \boldsymbol{\theta}_e\right) &= Q_e - \mathbf{J}_e\cdot\frac{\nabla p_e}{\mathrm{en}} - \dot{n}eU_i \\ &= \mathbf{E}\cdot\mathbf{J}_e - \dot{n}eU_i \end{aligned}, \tag{3.5-39}$$

where the inelastic term in expressed as $\Psi_e = \dot{n}eU_i$ to represent the electron energy loss due to ionization, with U_i (in volts) representing the first ionization potential of the atom. In Eq. 3.5-39, the $m_e v_e^2/2$ correction term has been neglected because usually in electric thrusters $eU_i >> m_e v_e^2/2$. If multiple ionization and/or excitation losses are significant, the inelastic terms in Eq. 3.5-39 must be augmented accordingly.

In electric thrusters it is common to assume a single temperature or distribution of temperatures for the heavy species without directly solving the energy equation(s). In some cases, however, such as in the plume of a hollow cathode for example, the ratio of T_e/T_i is important for determining the extent of Landau damping on possible electrostatic instabilities. The heavy species temperature is also important for determining the total pressure inside the cathode. Thus, separate energy equations must be solved directly. Assuming that the heavy species are slow-moving and the inelastic loss terms are negligible, Eq. 3.5-35 for ions takes the form

$$\frac{3}{2}\frac{\partial p_{\text{in}}}{\partial t} + \nabla \cdot \left(\frac{5}{2}p_{\text{in}}\mathbf{v}_{\text{in}} + \boldsymbol{\theta}_{\text{in}}\right) - \mathbf{v}_{\text{in}} \cdot \nabla p_{\text{in}} = Q_{\text{in}}, \tag{3.5-40}$$

where the subscript "in" represents ion-neutral collisions.

Finally, the total heat generated in partially ionized plasmas as a result of the (elastic) friction between the various species is given by

$$\sum_s Q_s^R = Q_e^R + Q_i^R + Q_n^R = -(\mathbf{R}_{\text{ei}} + \mathbf{R}_{\text{en}}) \cdot \mathbf{v}_e - (\mathbf{R}_{\text{ie}} + \mathbf{R}_{\text{in}}) \cdot \mathbf{v}_i - (\mathbf{R}_{\text{ne}} + \mathbf{R}_{\text{ni}}) \cdot \mathbf{v}_n \tag{3.5-41}$$

Since $\mathbf{R}_{\text{sa}} = -\mathbf{R}_{\text{as}}$ it is possible to write this as

$$\sum_s Q_s^R = -\mathbf{R}_{\text{ei}} \cdot (\mathbf{v}_e - \mathbf{v}_i) - \mathbf{R}_{\text{en}} \cdot (\mathbf{v}_e - \mathbf{v}_n) - \mathbf{R}_{\text{in}} \cdot (\mathbf{v}_i - \mathbf{v}_n) \tag{3.5-42}$$

The energy conservation equation(s) can be used with the momentum and continuity equations to provide a closed set of equations for analysis of plasma dynamics within the fluid approximations.

3.6 Diffusion in Partially Ionized Plasma

Diffusion is often very important in the particle transport in electric thruster plasmas. The presence of pressure gradients and collisions between different species of charged particles and between the charged particles and the neutrals produces diffusion of the plasma from high density regions to low density regions, both along and across magnetic field lines.

To evaluate diffusion driven particle motion in electric thruster plasmas, the equation of motion for any species can be written as

$$mn\frac{d\mathbf{v}}{dt} = qn(\mathbf{E} + \mathbf{v} \times \mathbf{B}) - \nabla \cdot \mathbf{p} - mn\nu(\mathbf{v} - \mathbf{v}_o), \tag{3.6-1}$$

where the terms in this equation have been previously defined and ν is the collision frequency between any two species in the plasma. To apply and solve this equation, it is first necessary to understand the collisional processes between the different species in the plasma that determines the applicable collision frequency.

3.6.1 Collisions

Charged particles in a plasma interact with each other primarily by Coulomb collisions, and also can collide with neutral atoms present in the plasma. These collisions are very important when describing diffusion, mobility, and resistivity in the plasma.

When a charged particle collides with a neutral atom, it can undergo an elastic or an inelastic collision. The probability that such a collision will occur can be expressed in terms of an effective cross-sectional area. Consider a thin slice of neutral gas with an area A and a thickness dx containing essentially stationary neutral gas atoms with a density n_a. Assume that the atoms are simple spheres of cross-sectional area σ. The number of atoms in the slice is given by $n_a A dx$. The fraction of the slice area that is occupied by the spheres is

$$\frac{n_a A \sigma dx}{A} = n_a \sigma dx. \tag{3.6-2}$$

If the incident flux of particles is Γ_o, then the flux that emerges without making a collision after passing through the slice is

$$\Gamma(x) = \Gamma_o(1 - n_a \sigma dx). \tag{3.6-3}$$

The change in the flux as the particles pass through the slice is

$$\frac{d\Gamma}{dx} = -\Gamma n_a \sigma. \tag{3.6-4}$$

The solution to Eq. 3.6-4 is

$$\Gamma = \Gamma_o e^{(-n_a \sigma x)} = \Gamma_o e^{(-x/\lambda)}, \tag{3.6-5}$$

where λ is defined as the mean free path for collisions and describes the distance in which the particle flux would decrease to $1/e$ of its initial value. The mean free path is given by

$$\lambda = \frac{1}{n_a \sigma}, \tag{3.6-6}$$

which represents the mean distance that a relatively fast moving particle, such as an electron or ion, will travel in a stationary density of neutral particles.

The mean time between collisions for this case is given by the mean free path divided by the charged particle velocity:

$$\tau = \frac{1}{n_a \sigma v}. \tag{3.6-7}$$

Averaging over all of the Maxwellian velocities of the charged particles, the collision frequency is then

$$\nu = \frac{1}{\tau} = n_a \sigma \bar{v}. \tag{3.6-8}$$

In the event that a relatively slowly moving particle, such as a neutral atom, is incident on a density of fast moving electrons, the mean free path for the neutral particle to experience a collision is given by

$$\lambda = \frac{v_n}{n_e \langle \sigma v_e \rangle}, \tag{3.6-9}$$

where υ_n is the neutral particle velocity and the reaction rate coefficient in the denominator is averaged over all the relevant collision cross sections. Eq. 3.6-9 can be used to describe the distance that a neutral gas atom travels in a plasma before ionization occurs, which is sometimes called the *penetration distance*.

Other collisions are also very important in electric thrusters. The presence of inelastic collisions between electrons and neutrals can result in either ionization or excitation of the neutral particle. The ion production rate per unit volume is given by

$$\frac{dn_i}{dt} = n_a n_e \langle \sigma_i \upsilon_e \rangle, \tag{3.6-10}$$

where σ_i is the ionization cross section, υ_e is the electron velocity, and the term in the brackets is the reaction rate coefficient, which is the ionization cross section averaged over the electron velocity distribution function.

Likewise, the production rate per unit volume of excited neutrals, n^*, is

$$\frac{dn^*}{dt} = \sum_j n_a n_e \langle \sigma_* \upsilon_e \rangle_j \tag{3.6-11}$$

where σ_* is the excitation cross section and the reaction rate coefficient is averaged over the electron distribution function and summed over all possible excited states j. A listing of the ionization and excitation cross sections for xenon and krypton are given in Appendix D, and the reaction rate coefficients for ionization and excitation averaged over a Maxwellian electron distribution are given in Appendix E.

Charge exchange [2, 6] in electric thrusters usually describes the resonant charge transfer between like atoms and ions in which essentially no kinetic energy is exchanged during the collision. Because this is a resonant process, it can occur at large distances and the charge exchange (CEX) cross section is very large [2]. For example, the charge exchange cross section for xenon is about 10^{-18} m^2[7], which is significantly larger than the ionization and excitation cross sections for this atom. Since the ions in the thruster are often energetic due to acceleration by the electric fields in the plasma or acceleration in ion thruster grid structures, charge exchange results in the production of energetic neutrals and relatively cold ions. Charge exchange collisions are often a dominant factor in the heating of cathode structures, the mobility and diffusion of ions in the thruster plasma, and the erosion of grid structures and thruster surfaces.

While the details of classical collision physics are interesting, they are well described in several other textbooks [1, 2, 5] and are not critically important to understanding electric thrusters. However, the various collision frequencies and cross sections are of interest for use in modeling the thruster discharge and performance.

The frequency of collisions between electron and neutrals is sometimes written [8] as

$$\nu_{\text{en}} = \sigma_{\text{en}}(T_e) n_a \sqrt{\frac{8kT_e}{\pi m}}, \tag{3.6-12}$$

where the effective electron-neutral scattering cross section $\sigma(T_e)$ for xenon can be found from a numerical fit to the electron-neutral scattering cross section averaged over a Maxwellian electron distribution [8]:

$$\sigma_{\text{en}}(T_e) = 6.6 \times 10^{-19} \left[\frac{\frac{T_{\text{eV}}}{4} - 0.1}{1 + \left(\frac{T_{\text{eV}}}{4}\right)^{1.6}} \right] [\text{m}^2], \tag{3.6-13}$$

where T_{ev} is in electron volts. The electron-ion collision frequency for Coulomb collisions [1] is given in SI units by

$$\nu_{ei} = 2.9 \times 10^{-12} \frac{Zn_e \ln\Lambda}{T_{eV}^{3/2}}, \tag{3.6-14}$$

where $\ln\Lambda$ is the Coulomb logarithm given in a familiar form [5] by

$$\ln\Lambda = 23 - \frac{1}{2} \ln\left(\frac{10^{-6} Zn_e}{T_{eV}^{3/2}}\right). \tag{3.6-15}$$

The electron–electron collision frequency [1] is given by

$$\nu_{ee} = 5 \times 10^{-12} \frac{n_e \ln\Lambda}{T_{eV}^{3/2}}, \tag{3.6-16}$$

While the values of the electron-ion and the electron–electron collision frequencies in Eqs. 3.6-14 and 3.6-16 are clearly comparable, the electron–electron thermalization time is much shorter than the electron-ion thermalization time due to the large mass difference between the electrons and ions reducing the energy transferred in each collision. This is a major reason that electrons thermalize rapidly into a population with Maxwellian distribution, but do not thermalize rapidly with the ions.

In addition, the ion-ion collision frequency [1] is given by

$$\nu_{ii} = Z^4 \left(\frac{m}{M}\right)^{1/2} \left(\frac{T_e}{T_i}\right)^{3/2} \nu_{ee}, \tag{3.6-17}$$

where Z is the ion charge number.

Collisions between like particles and between separate species tend to equilibrate the energy and distribution functions of the particles. This effect was analyzed in detail by Spitzer [9] in his classic book. In electric thrusters, there are several equilibration time constants of interest. First, the characteristic collision times between the different charged particles is just one over the average collision frequencies given above. Second, equilibration times between the species and between different populations of the same species were calculated by Spitzer. The time for a monoenergetic electron (sometimes called a *primary electron*) to equilibrate with the Maxwellian population of the plasma electrons is called the slowing time τ_s. Finally, the time for one Maxwellian population to equilibrate with another Maxwellian population is called the equilibration time τ_{eq}. Expressions for these equilibration times, and a comparison of the rates of equilibration by these two effects, are found in Appendix F.

Collisions of electrons with other species in the plasma lead to resistivity and provide a mechanism for heating. This mechanism is often called ohmic heating or joule heating. In steady state and neglecting electron inertia, the electron momentum equation, taking into account electron-ion collisions and electron-neutral collisions, is

$$0 = -en(\mathbf{E} + \mathbf{v}_e \times \mathbf{B}) - \nabla \cdot \mathbf{p}_e - mn[\nu_{ei}(\mathbf{v}_e - \mathbf{v}_i) + \nu_{en}(\mathbf{v}_e - \mathbf{v}_n)]. \tag{3.6-18}$$

The electron velocity is very large with respect to the neutral velocity, and Eq. 3.6-18 can be written as

$$0 = -en\left(\mathbf{E} + \frac{\nabla \mathbf{p}_e}{en}\right) - en\mathbf{v}_e \times \mathbf{B} - mn(\nu_{\text{ei}} + \upsilon_{\text{en}})\mathbf{v}_e + mn\nu_{\text{ei}}\mathbf{v}_i. \tag{3.6-19}$$

Since charged particle current density is given by $\mathbf{J} = qn\mathbf{v}$, Eq. 3.6-19 can be written as

$$\eta \mathbf{J}_e = \mathbf{E} + \frac{\nabla \mathbf{p} - \mathbf{J}_e \times \mathbf{B}}{en} - \eta_{\text{ei}} \mathbf{J}_i, \tag{3.6-20}$$

where $\mathbf{J}_e$ is the electron current density, $\mathbf{J}_i$ is the ion current density and η_{ei} is the plasma resistivity. Eq. 3.6-20 is commonly known as Ohm's Law for partially ionized plasmas and is a variant of the well-known generalized Ohm's Law, which usually expressed in terms of the total current density $\mathbf{J} = en(\mathbf{v}_i - \mathbf{v}_e)$ and the ion fluid velocity $\mathbf{v}_i$. If there are no collisions or net current in the plasma, this equation reduces to Eq. 3.5-7 which was used to derive the Boltzmann relationship for plasma electrons.

From Eq. 3.6-20, the total resistivity of a partiallyionized plasma is given by

$$\eta = \frac{m(\nu_{\text{ei}} + \nu_{\text{en}})}{e^2 n} = \frac{1}{\varepsilon_o \tau_e \omega_p^2}, \tag{3.6-21}$$

where the total collision time for electrons, accounting for both electron-ion and electron-neutral collisions, is given by

$$\tau_e = \frac{1}{\nu_{\text{ei}} + \nu_{\text{en}}}. \tag{3.6-22}$$

By neglecting the electron-neutral collision terms in Eq. 3.6-19, the well-known expression [1, 9] for the resistivity of a fully-ionized plasma is recovered:

$$\eta_{\text{ei}} = \frac{m\nu_{\text{ei}}}{e^2 n} = \frac{1}{\varepsilon_o \tau_{\text{ei}} \omega_p^2}. \tag{3.6-23}$$

In electric thrusters, the ion current in the plasma is typically much smaller than the electron current due to the large mass ratio, so the ion current term in Ohm's Law (Eq. 3.6-20) is sometimes neglected.

3.6.2 Diffusion and Mobility Without a Magnetic Field

The simplest case of diffusion in a plasma is found by neglecting the magnetic field and writing the equation of motion for any species as

$$mn\frac{d\mathbf{v}}{dt} = qn\mathbf{E} - \nabla \cdot \mathbf{p} - mn\nu(\mathbf{v} - \mathbf{v}_o), \tag{3.6-24}$$

where m is the species mass and the collision frequency is taken to be a constant. Assume that the velocity of the particle species of interest is large compared to the slow species ($\mathbf{v} \gg \mathbf{v}_O$), the plasma is isothermal ($\nabla p = kT\nabla n$), and the diffusion is steady state and occurring with a sufficiently high velocity that the convective derivative can be neglected. Eq. 3.6-24 can then be solved for the particle velocity:

$$\mathbf{v} = \frac{q}{m\nu}\mathbf{E} - \frac{kT}{m\nu}\frac{\nabla n}{n}. \tag{3.6-25}$$

The coefficients of the electric field and the density gradient terms in Eq. 3.6-25 are called the mobility:

$$\mu = \frac{|q|}{m\nu} \left[\mathrm{m^2/(V \cdot s)}\right] \tag{3.6-26}$$

and the diffusion coefficient:

$$D = \frac{kT}{m\nu} \left[\mathrm{m^2/s}\right]. \tag{3.6-27}$$

These terms are related by what is called the *Einstein relation*:

$$\mu = \frac{|q|D}{kT}. \tag{3.6-28}$$

3.6.2.1 Fick's Law and the Diffusion Equation

The flux of diffusing particles in the simple case of Eq. 3.6-25 is

$$\mathbf{\Gamma} = n\mathbf{v} = \mu n\mathbf{E} - D\nabla n. \tag{3.6-29}$$

A special case of this is called *Fick's law*, in which the flux of particles for either the electric field or the mobility term being zero is given by

$$\mathbf{\Gamma} = -D\nabla n. \tag{3.6-30}$$

The continuity equation (Eq. 3.5-11) without sink or source terms can be written as

$$\frac{\partial n}{\partial t} + \nabla \cdot \mathbf{\Gamma} = 0, \tag{3.6-31}$$

where $\mathbf{\Gamma}$ represents the flux of any species of interest. If the diffusion coefficient D is constant throughout the plasma, substituting Eq. 3.6-30 into Eq. 3.6-31 gives the well-known diffusion equation for a single species:

$$\frac{\partial n}{\partial t} - D\nabla^2 n = 0. \tag{3.6-32}$$

The solution to this equation can be obtained by separation of variables. The simplest example of this is for slab geometry of finite width, where the plasma density can be expressed as having separable spatial and temporal dependencies:

$$n(x,t) = X(x)\overline{T}(t). \tag{3.6-33}$$

Substituting this into Eq. 3.6-32 gives

$$X\frac{d\overline{T}}{dt} = D\overline{T}\frac{d^2X}{dx^2}. \tag{3.6-34}$$

Separating the terms gives

$$\frac{1}{\overline{T}}\frac{d\overline{T}}{dt} = D\frac{1}{X}\frac{d^2X}{dx^2} = \alpha, \tag{3.6-35}$$

where each side of the equation is independent of the other side and can therefore be set equal to a constant α. The time dependent function is then

$$\frac{d\overline{T}}{dt} = -\frac{\overline{T}}{\tau}, \tag{3.6-36}$$

where the constant α will be written as $-1/\tau$. The solution to Eq. 3.6-36 is

$$\overline{T} = \overline{T}_o e^{-t/\tau}. \tag{3.6-37}$$

Since there is no ionization source term in Eq. 3.6-32, the plasma density decays exponentially with time from the initial state due to the diffusion.

The right-hand side of Eq. 3.6-35 has the spatial dependence of the diffusion and can be written as

$$\frac{d^2X}{dx^2} = -\frac{X}{D\tau}, \tag{3.6-38}$$

where again the constant α will be written as $-1/\tau$. This equation has a solution of the form

$$X = A\cos\frac{X}{L} + B\sin\frac{X}{L}, \tag{3.6-39}$$

where A and B are constants and L is the diffusion length given by $(D\tau)^{1/2}$. If it is assumed that X is zero at the boundaries at $\pm d/2$, then the lowest order solution is symmetric ($B = 0$) with the diffusion length equal to π. The solution to Eq. 3.6-38 is then

$$X = \cos\frac{\pi x}{d}. \tag{3.6-40}$$

The lowest-order complete solution to the diffusion equation for the plasma density is then the product of Eqs. 3.6-37 and 3.6-40:

$$n = n_o e^{-t/\tau}\cos\frac{\pi x}{d}. \tag{3.6-41}$$

Of course, higher order odd solutions are possible for given initial conditions, but the higher order modes decay faster and the lowest order mode typically dominates after a sufficient time. The plasma density decays with time from the initial value n_o, but the boundary condition (zero plasma density at the wall) maintains the plasma shape described by the cosine function in Eq. 3.6-41.

Although slabgeometry was chosen for this illustrative example due to its simplicity, situations in which slab geometries are useful in modeling electric thrusters are rare. However, solutions to the diffusion equation in other coordinates more typically found in these thrusters are obtained in a similar manner. For example, in cylindrical geometries found in many hollow cathodes and in ion thruster discharge chambers, the solution to the cylindrical differential equation follows Bessel functions radially and still decays exponentially in time if source terms are not considered.

Solutions to the diffusion equation with source or sink terms on the right-hand side are more complicated to solve. This can be seen in writing the diffusion equation as

$$\frac{\partial n}{\partial t} - D\nabla^2 n = \dot{n}, \tag{3.6-42}$$

where the source term is described by an ionization rate equation per unit volume given by

$$\dot{n} = n_a n\langle\sigma_i v_e\rangle \approx n_a n\sigma_i(T_e)\bar{v}, \tag{3.6-43}$$

and where $\bar{v}$ is the average particle speed found in Eq. 3.4-8, and $\sigma_i(T_e)$ is the impact ionization cross-section averaged over a Maxwellian distribution of electrons at a temperature T_e. Equations for the xenon and krypton ionization reaction rate coefficients averaged over a Maxwellian distribution are found in Appendix E.

A separation of variables solution can still be obtained for this case, but the time-dependent behavior is no longer purely exponential as was found in Eq. 3.6-37. In this situation, the plasma density will increase or decay to an equilibrium value depending on the magnitude of the source and sink terms.

To find the steady state solution to the cylindrical diffusion equation, the time derivative in Eq. 3.6-42 is set equal to zero. Writing the diffusion equation in cylindrical coordinates and assuming uniform radial electron temperatures and neutral densities, Eq. 3.6-42 becomes

$$\frac{\partial^2 n}{\partial r^2} + \frac{1}{r}\frac{\partial n}{\partial r} + \frac{\partial^2 n}{\partial z^2} + C^2 n = 0, \tag{3.6-44}$$

where the constant is given by

$$C^2 = \frac{n_a \sigma_i(T_e)\bar{v}}{D}. \tag{3.6-45}$$

This equation can be solved analytically by separation of variables of the form

$$n = n(0,0)f(r)g(z). \tag{3.6-46}$$

Using Eq. 3.6-46, the diffusion equation becomes

$$\frac{1}{f}\frac{\partial^2 f}{\partial r^2} + \frac{1}{\mathrm{rf}}\frac{\partial f}{\partial r} + C^2 + \alpha^2 = -\frac{1}{g}\frac{\partial^2 g}{\partial z^2} + \alpha^2 = 0. \tag{3.6-47}$$

The solution to the radial component of Eq. 3.6-47 is the sum of the zero-order Bessel functions of the first and second kind, which is written in a general form as

$$f(r) = A_1 J_o(\alpha r) + A_2 Y_o(\alpha r). \tag{3.6-48}$$

The Bessel function of the second kind Y_o becomes infinite as (αr) goes to zero, and because the density must always be finite, the constant A_2 must equal zero. Therefore, the solution for Eq. 3.6-47 is the product of the zero order Bessel function of the first kind times an exponential term in the axial direction:

$$n(r,z) = n(0,0)J_o\left(\sqrt{C^2 + \alpha^2}r\right)e^{-\alpha z}. \tag{3.6-49}$$

Assuming that the ion density goes to zero at the wall,

$$\sqrt{C^2 + \alpha^2} = \frac{\lambda_{01}}{r}, \tag{3.6-50}$$

where $\lambda_{01}$is the first zero of the zero-order Bessel function and r is now the internal radius of the cylinder being considered. Setting $\alpha = 0$, this eigenvalue results in an equation that gives a direct relationship between the electron temperature, the radius of the plasma cylinder, and the diffusion rate:

$$\left(\frac{r}{\lambda_{01}}\right)^2 n_a \sigma_i(T_e)\sqrt{\frac{8kT_e}{\pi m}} - D = 0. \tag{3.6-51}$$

The physical meaning of Eq. 3.6-51 is that particle balance in bounded plasma discharges dominated by radial diffusion determines the plasma electron temperature. This occurs because the generation rate of ions, which is determined by the electron temperature from Eq. 3.6-43, must equal the ion loss rate, which is determined by the diffusion rate to the walls, to satisfy the boundary condition. Therefore, the solution to the steady-state cylindrical diffusion equation specifies both the radial plasma profile and the maximum electron temperature once the dependence of the diffusion coefficient is specified. This result is very useful in modeling the plasma discharges in hollow cathodes and in various types of electric thrusters.

3.6.2.2 Ambipolar Diffusion Without a Magnetic Field

In many circumstances in thrusters, the flux of ions and electrons from a given region or the plasma as a whole are equal. For example, in the case of microwave ion thrusters, the ions and electrons are created in pairs during ionization by the plasma electrons heated by the microwaves, so simple charge conservation states that the net flux of both ions and electrons out of the plasma must be the same. The plasma will then establish the required electric fields in the system to slow the more mobile electrons such that the electron escape rate is the same as the slower ions loss rate. This finite electric field affects the diffusion rate for both species.

Since the expression for the flux in Eq. 3.6-29 was derived for any species of particles, a diffusion coefficient for ions (D_i) and electrons (D_e) can be designated (because D in Eq. 3.6-27 contains the particle mass) and the fluxes equated to obtain

$$\mu_i n\mathbf{E} - D_i \nabla n = -\mu_e n\mathbf{E} - D_e \nabla n, \tag{3.6-52}$$

where quasi-neutrality ($n_i \approx n_e$) in the plasma has been assumed. Solving for the electric field gives

$$\mathbf{E} = \frac{D_i - D_e}{\mu_i + \mu_e}\frac{\nabla n}{n}. \tag{3.6-53}$$

Substituting $\mathbf{E}$ into the Eq. 3.6-29 for the ion flux,

$$\Gamma = -\frac{\mu_i D_e + \mu_e D_i}{\mu_i + \mu_e}\nabla n = -D_a \nabla n, \tag{3.6-54}$$

where D_a is the ambipolar diffusion coefficient given by

$$D_a = \frac{\mu_i D_e + \mu_e D_i}{\mu_i + \mu_e}. \tag{3.6-55}$$

Equation 3.6-54 was expressed in the form of Fick's Law, but with a new diffusion coefficient reflecting the impact of ambipolar flow on the particle mobilities. Substituting Eq. 3.6-54 into the continuity equation without sources or sinks gives the diffusion equation for ambipolar flow:

$$\frac{\partial n}{\partial t} - D_a \nabla^2 n = 0. \tag{3.6-56}$$

Since the electron and ion mobilities depend on the mass:

$$\mu_e = \frac{e}{m\nu} \gg \mu_i = \frac{e}{M\nu}, \tag{3.6-57}$$

it is usually possible to neglect the ion mobility. In this case, Eq. 3.6-55 combined with Eq. 3.6-25 gives

$$D_a \approx \frac{\mathrm{kT}}{M\nu} + \frac{\mu_i}{\mu_e}\frac{\mathrm{kT}}{m\nu} = D_i + \frac{\mu_i}{\mu_e}D_e = D_i\left(1 + \frac{T_e}{T_i}\right). \tag{3.6-58}$$

Since the electron temperature in thrusters is usually significantly higher than the ion temperature ($T_e \gg T_i$), ambipolar diffusion greatly enhances the ion diffusion coefficient. Likewise, the smaller ion mobility significantly decreases the ambipolar electron flux leaving the plasma.

3.6.3 Diffusion Across Magnetic Fields

Charged particle transport across magnetic fields is described by what is called *classical diffusion* theory and non-classical or *anomalous diffusion*. Classical diffusion, which will be presented below, includes both the case of particles of one species moving across the field due to collisions with another species of particles, and the case of ambipolar diffusion across the field where the fluxes are constrained by particle balance in the plasma. Anomalous diffusion can be caused by a number of different effects. In ion and Hall thrusters, the anomalous diffusion has been described by Bohm diffusion [10], or is due to anomalous resistivity from turbulent ion acoustic oscillations [1].

3.6.3.1 Classical Diffusion of Particles across B Fields

The fluid equation of motion for isothermal electrons moving in the perpendicular direction across a magnetic field is

$$mn\frac{d\mathbf{v}_\perp}{dt} = qn(\mathbf{E} + \mathbf{v}_\perp \times \mathbf{B}) - kT_e\nabla \mathrm{n} - mn\nu\mathbf{v}_\perp \tag{3.6-59}$$

The same form of this equation can be written for ions with a mass M and temperature T_i. Consider steady-state diffusion and set the time and convective derivatives equal to zero. Separating Eq. 3.6-59 into x and y coordinates gives

$$mn\nu v_x = qnE_x + qnv_yB_o - kT_e\frac{\partial n}{\partial x}, \tag{3.6-60}$$

and

$$mn\nu v_y = qnE_y + qnv_xB_o - \mathrm{kT}_e\frac{\partial n}{\partial y}, \tag{3.6-61}$$

where $\mathbf{B} = B_o(z)$. The x and y velocity components are then

$$v_x = \pm\mu E_x + \frac{\omega_c}{\nu}v_y - \frac{D}{n}\frac{\partial n}{\partial x}, \tag{3.6-62}$$

and

$$v_y = \pm\mu E_y + \frac{\omega_c}{\nu}v_x - \frac{D}{n}\frac{\partial n}{\partial y}, \tag{3.6-63}$$

Solving Eqs. 3.6-62 and 3.6-63, the velocities in the two directions are

$$\left[1 + \omega_c^2\tau^2\right]v_x = \pm\mu E_x - \frac{D}{n}\frac{\partial n}{\partial x} + \omega_c^2\tau^2\frac{E_y}{B_o} - \omega_c^2\tau^2\frac{kT_e}{qB_o}\frac{1}{n}\frac{\partial n}{\partial y}, \tag{3.6-64}$$

and

$$\left[1 + \omega_c^2\tau^2\right]v_y = \pm\mu E_y - \frac{D}{n}\frac{\partial n}{\partial y} + \omega_c^2\tau^2\frac{E_x}{B_o} - \omega_c^2\tau^2\frac{kT_e}{qB_o}\frac{1}{n}\frac{\partial n}{\partial x}, \tag{3.6-65}$$

where $\tau = 1/\nu$ is the average collision time.

The perpendicular electron mobility is defined as

$$\mu_\perp = \frac{\mu}{1 + \omega_c^2\tau^2} = \frac{\mu}{1 + \Omega_e^2}, \tag{3.6-66}$$

where the perpendicular mobility is written in terms of the electron Hall parameter defined as $\Omega_e = eB/m\nu$. The perpendicular diffusion coefficient is defined as

$$D_{\perp} = \frac{D}{1 + \omega_c^2 \tau^2} = \frac{D}{1 + \Omega_e^2}. \tag{3.6-67}$$

The perpendicular velocity can then be written in vector form again as

$$\mathbf{v}_{\perp} = \pm \mu_{\perp} \mathbf{E} - D_{\perp} \frac{\nabla n}{n} + \frac{\mathbf{v}_E + \mathbf{v}_D}{1 + \left(\nu^2/\omega_c^2\right)}. \tag{3.6-68}$$

This is a form of Fick's law with two additional terms, the azimuthal $\mathbf{E} \times \mathbf{B}$ drift:

$$\mathbf{v}_E = \frac{\mathbf{E} \times \mathbf{B}}{B_o^2}, \tag{3.6-69}$$

and the diamagnetic drift:

$$\mathbf{v}_D = -\frac{kT}{qB_o^2} \frac{\nabla n \times \mathbf{B}}{n}, \tag{3.6-70}$$

both reduced by the fluid drag term $(1 + \nu^2/\omega_c{}^2)$. In the case of an electric thruster, the perpendicular cross-field electron flux flowing toward the wall or toward the anode is then

$$\Gamma_e = n\mathbf{v}_{\perp} = \pm \mu_{\perp} n\mathbf{E} - D_{\perp} \nabla n, \tag{3.6-71}$$

which has the form of Fick's Law but with the mobility and diffusion coefficients modified by the magnetic field.

The "classical" cross-field diffusion coefficient $D_{\perp}$, derived above and found in the literature [1, 2], is proportional to $1/B^2$. However, in measurements in many plasma devices, including in Kaufman ion thrusters and other thrusters, the perpendicular diffusion coefficient in some regions is found to be described by the Bohm diffusion coefficient:

$$D_B = \frac{1}{16} \frac{kT_e}{eB}, \tag{3.6-72}$$

which scales as $1/B$. Therefore, Bohm diffusion often progresses at orders of magnitude higher rates than classical diffusion. It has been proposed that Bohm diffusion results from collective instabilities in the plasma. Assume that the perpendicular electron flux is proportional to the $\mathbf{E} \times \mathbf{B}$ drift velocity,

$$\Gamma_e = n v_{\perp} \propto n \frac{E}{B}. \tag{3.6-73}$$

Also assume that the maximum electric field that occurs in the plasma due to Debye shielding with a potential $\phi_{\max}$ over a radius r is proportional to the electron temperature divided by the radius of the plasma:

$$E_{\max} = \frac{\phi_{\max}}{r} = \frac{kT_e}{qr}. \tag{3.6-74}$$

The electron flux to the wall is then

$$\Gamma_e \approx C \frac{n}{r} \frac{\mathrm{kT}_e}{qB} \approx -C \frac{kT_e}{qB} \nabla n = -D_B \nabla n. \tag{3.6-75}$$

where C is a constant less than 1. The Bohm diffusion coefficient has an empirically determined value of $C = 1/16$, as shown in Eq. 3.6-71, which fits most experiments with some uncertainty. As pointed out in Chen [1], this is why it is no surprise that Bohm diffusion scales as kT_e/eB.

3.6.3.2 Ambipolar Diffusion Across B Fields

Ambipolar diffusion across magnetic fields is much more complicated than the diffusion cases just covered because the mobility and diffusion coefficients are anisotropic in the presence of a magnetic field. Since both quasi-neutrality and charge balance must be satisfied, ambipolar diffusion dictates that the sum of the cross field and parallel to the field loss rates for both the ions and electrons must be the same. This means that the divergence of the ion and electron fluxes must be equal. While it is a simple matter to write equations for the divergence of these two species and equate them, the resulting equation cannot be easily solved because it depends on the behavior both in the plasma and at the boundary conditions.

A special case in which only the ambipolar diffusion toward a wall in the presence of a transverse magnetic field is now considered. In this situation, charge balance is conserved separately along and across the magnetic field lines. The transverse electron equation of motion for isothermal electrons, including electron-neutral and electron-ion collisions, can be written as

$$mn\left(\frac{\partial \mathbf{v}_e}{\partial t} + (\mathbf{v}_e \cdot \nabla)\mathbf{v}_e\right) = -en(\mathbf{E} + \mathbf{v}_e \times \mathbf{B}) - kT_e \nabla n \\ - mn\nu_{en}(\mathbf{v}_e - \mathbf{v}_o) - mn\nu_{ei}(\mathbf{v}_e - \mathbf{v}_i), \tag{3.6-76}$$

where $\mathbf{v}_o$ is the neutral particle velocity. Taking the magnetic field to be in the z-direction, and assuming the convective derivative to be negligibly small, then in steady state this equation can be separated into the two transverse electron velocity components:

$$\upsilon_x + \mu_e E_x + \frac{e}{m\nu_e}\upsilon_y B + \frac{kT_e}{mn\nu_e}\frac{\partial n}{\partial x} - \frac{\nu_{ei}}{\nu_e}\upsilon_i = 0 \tag{3.6-77}$$

$$\upsilon_y + \mu_e E_y - \frac{e}{m\nu_e}\upsilon_x B + \frac{kT_e}{mn\nu_e}\frac{\partial n}{\partial y} - \frac{\nu_{ei}}{\nu_e}\upsilon_i = 0 \tag{3.6-78}$$

where $\nu_e = \nu_{en} + \nu_{ei}$ is the total collision frequency, $\mu_e = e/m\nu_e$ is the electron mobility including both ion and neutral collisional effects, and v_o is neglected as being small compared to the electron velocity υ_e. Solving for υ_y and eliminating the $\mathbf{E} \times \mathbf{B}$ and diamagnetic drift terms in the x-direction, the transverse electron velocity is given by

$$\upsilon_e\left(1 + \mu_e^2 B^2\right) = \mu_e\left(E + \frac{kT_e}{e}\frac{\nabla n}{n}\right) + \frac{\nu_{ei}}{\nu_e}\upsilon_i, \tag{3.6-79}$$

Since ambipolar flow and quasi-neutrality are assumed everywhere in the plasma, the transverse electron and ion transverse velocities must be equal, which gives

$$\upsilon_i\left(1 + \mu_e^2 B^2 - \frac{\nu_{ei}}{\nu_e}\right) = \mu_e\left(E + \frac{kT_e}{e}\frac{\nabla n}{n}\right). \tag{3.6-80}$$

The transverse velocity of each species is then

$$\upsilon_i = \upsilon_e = \frac{\mu_e}{\left(1 + \mu_e^2 B^2 - \frac{\nu_{ei}}{\nu_e}\right)}\left(E + \frac{kT_e}{e}\frac{\nabla n}{n}\right). \tag{3.6-81}$$

In this case, the electron mobility is reduced by the magnetic field (the first term on the right-hand side of the equation), and so an electric field E is generated in the plasma to slow down the ion transverse velocity to balance the pressure term and maintain ambipolarity. This is exactly the opposite of the normal ambipolar diffusion without magnetic fields or along the magnetic field lines covered in Section 3.6.2, where the electric field slowed the electrons and accelerated the ions to maintain ambipolarity. Equation 3.6-81 can be written in terms the transverse flux as

$$\Gamma_{\perp} = \frac{\mu_e}{\left(1 + \mu_e^2 B^2 - \nu_{\text{ei}}/\nu_e\right)}(enE + kT_e \nabla n). \tag{3.6-82}$$

3.7 Sheaths at the Boundaries of Plasmas

While the motion of the various particles in the plasma is important in understanding the behavior and performance of electric thrusters, the boundaries of the plasma represent the physical interface through which energy and particles enter and leave the plasma and the thruster. Depending on the conditions, the plasma will establish potential and density variations at the boundaries to satisfy particle balance or the imposed electrical conditions at the thruster walls. This region of potential and density change is called the sheath, and understanding sheath formation and behavior is also very important in understanding and modeling electric thruster plasmas.

Consider the generic plasma in Fig. 3-2 consisting of quasi-neutral ion and electrons densities with temperatures given by T_i and T_e, respectively. The ion current density to the boundary or "wall" for singly-charged ions, to first order, is given by $n_i e v_i$, where v_i is the ion velocity. Likewise, the electron current density to the boundary wall, to first order, is given by $n_e e v_e$, where v_e is the electron velocity. The ratio of the electron current density to the ion current density going to the boundary, assuming quasi-neutrality, is

$$\frac{J_e}{J_i} = \frac{n_e e v_e}{n_i e v_i} = \frac{v_e}{v_i}. \tag{3.7-1}$$

In the absence of an electric field in the plasma volume, conservation of energy for the electrons and ions is given by

$$\frac{1}{2} m v_e^2 = \frac{kT_e}{e}.$$

$$\frac{1}{2} M v_i^2 = \frac{kT_i}{e}.$$

If it is assumed that the electrons and ions have the same temperature, the ratio of current densities to the boundary is

$$\frac{J_e}{J_i} = \frac{v_e}{v_i} = \sqrt{\frac{M}{m}}. \tag{3.7-2}$$

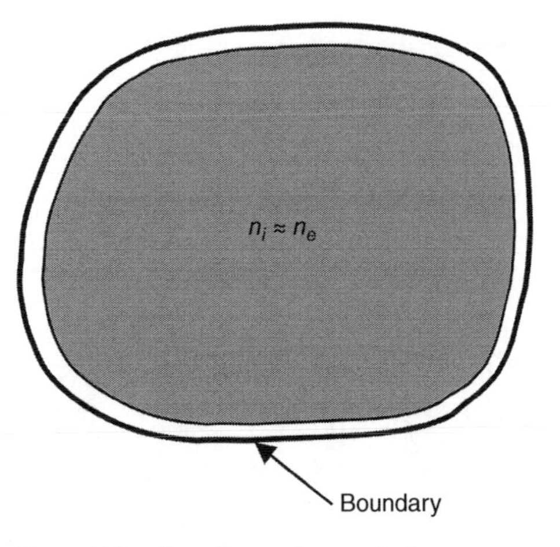

Figure 3-2 Generic quasi-neutral plasma enclosed in a boundary.

Table 3-1 shows the mass ratio M/m for several gas species. It is clear that the electron current to the boundary from of the plasma under these conditions is orders of magnitude higher than the ion current due to the much higher electron mobility. This

Table 3-1 Ion to electron mass ratios for several gas species.

Gas	Mass ratio M/m	Square root of the mass ratio M/m
Protons (H^+)	1836	42.8
Argon	73,440	270.9
Xenon	241,066.8	490.9

would make it impossible to maintain the assumption of quasi-neutrality in the plasma used in Eq. 3.7-1 because the electrons would leave the volume much faster than the ions.

If different temperatures between the ions and electrons are allowed, the ratio of the current densities to the boundary is given by

$$\frac{J_e}{J_i} = \frac{v_e}{v_i} = \sqrt{\frac{M}{m}\frac{T_e}{T_i}}. \tag{3.7-3}$$

To balance the fluxes to the wall to satisfy charge continuity (an ionization event makes one ion and one electron), the ion temperature would have to again be orders of magnitude higher than the electron temperatures. In electric thrusters, the opposite is true and the electron temperature is normally at least an order of magnitude higher than the ion temperature, which compounds the problem of maintaining quasi-neutrality in a plasma.

In reality, if the electrons left the plasma volume faster than the ions, a charge imbalance would result due to the large net ion charge left behind. This would produce a positive potential in the plasma, which creates a retarding electric field for the electrons. The electrons would then be slowed down and retained in the plasma. Potential gradients in the plasma and at the plasma boundary are a natural consequence of the different temperatures and mobilities of the ions and electrons. Potential gradients will develop at the wall or next to electrodes inserted into the plasma to maintain quasi-neutrality between the charged species. These regions with potential gradients are called sheaths.

3.7.1 Debye Sheaths

To start an analysis of sheaths, assume that the positive and negative charges in the plasma are fixed in space, but have any arbitrary distribution. It is then possible to solve for the potential distribution everywhere using Maxwell's equations. The integral form of Eq. 3.2-1 is Gauss's Law:

$$\oint_s \mathbf{E} \cdot d\mathbf{s} = \frac{1}{\varepsilon_o}\int_V \rho dV = \frac{Q}{\varepsilon_o} \tag{3.7-4}$$

where Q is the total enclosed charge in the volume V and $\mathbf{s}$ is the surface enclosing that charge. If an arbitrary sphere of radius r is drawn around the enclosed charge, the electric field found from integrating over the sphere is

$$\mathbf{E} = \frac{Q}{4\pi\varepsilon_o r^2}\hat{r}. \tag{3.7-5}$$

Since the electric field is minus the gradient of the potential, the integral form of Eq. 3.2-2 can be written

$$\phi_2 - \phi_1 = -\int_{p1}^{p2} \mathbf{E} \cdot d\mathbf{l} \tag{3.7-6}$$

where the integration proceeds along the path $d\mathbf{l}$ from point p1 to point p2. Substituting Eq. 3.7-5 into Eq. 3.2-6 and integrating gives

$$\phi = \frac{Q}{4\pi\varepsilon_o r}. \tag{3.7-7}$$

The potential decreases as *1/r* moving away from the charge.

However, if the plasma is allowed to react to a test-charge placed inside the plasma, the potential has a different behavior than predicted by Eq. 3.7-7. Utilizing Eq. 3.2-7 for the electric field in Eq. 3.2-1 gives Poisson's equation:

$$\nabla^2\phi = -\frac{\rho}{\varepsilon_o} = -\frac{e}{\varepsilon_o}(Zn_i - n_e), \tag{3.7-8}$$

where the charge density in Eq. 3.2-5 has been used. Assume that the ions are singly-charged and that the potential change around the test charge is small ($e\phi \ll kT_e$), such that the ion density is fixed and $n_i \approx n_o$. Writing Poisson's equation in spherical coordinates and using Eq. 3.5-9 to describe the Boltzmann electron density behavior gives

$$\frac{1}{r^2}\frac{\partial}{\partial r}\left(r^2\frac{\partial\phi}{\partial r}\right) = -\frac{e}{\varepsilon_o}\left[n_o - n_o e^{\left(\frac{e\phi}{kT_e}\right)}\right] = \frac{en_o}{\varepsilon_o}\left[e^{\left(\frac{e\phi}{kT_e}\right)} - 1\right]. \tag{3.7-9}$$

Since $e\phi \ll kT_e$ was assumed, the exponent can be expanded in a Taylor series:

$$\frac{1}{r^2}\frac{\partial}{\partial r}\left(r^2\frac{\partial\phi}{\partial r}\right) = \frac{en_o}{\varepsilon_o}\left[\frac{e\phi}{kT_e} + \frac{1}{2}\left(\frac{e\phi}{kT_e}\right)^2 + \ldots\right]. \tag{3.7-10}$$

Neglecting all of the higher order terms, the solution of Eq. 3.7-10 can be written as

$$\phi = \frac{e}{4\pi\varepsilon_o r} e^{\left(-r/\sqrt{\frac{\varepsilon_o kT_e}{n_o e^2}}\right)}. \tag{3.7-11}$$

By defining

$$\lambda_D = \sqrt{\frac{\varepsilon_o kT_e}{n_o e^2}} \tag{3.7-12}$$

as the characteristic Debye length, Eq. 3.7-11 can be written as

$$\phi = \frac{e}{4\pi\varepsilon_o r} e^{\left(-r/\lambda_D\right)}. \tag{3.7-13}$$

This equation shows that the potential would normally fall off away from the test charge inserted in the plasma as *1/r*, as previously found, except that the electrons in the plasma have reacted to shield the test charge and cause the potential to decrease exponentially away from it. This behavior of the potential in the plasma is, of course, true for any structure such as a grid or probe that is placed in the plasma and that has a net charge on it.

The Debye length is the characteristic distance over which the potential changes for potentials that are small compared to kT_e. It is common to assume that the sheath around an object will have a thickness of the order of a few Debye lengths for the potential to fall to a negligible value away from the object. As an example, consider a plasma with a density of 10^{17} m^{-3} and an electron

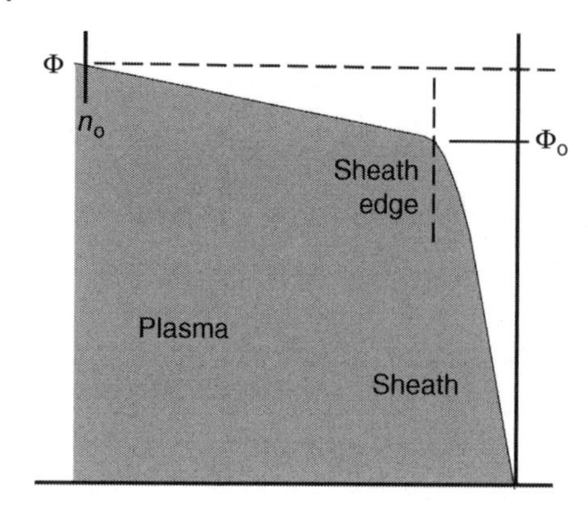

Figure 3-3 Plasma in contact with a boundary.

temperature of 1 eV. Boltzmann's constant k is 1.3807×10^{-23} J/K and charge is 1.6022×10^{-19} C, so the temperature corresponding to 1 electron volt is

$$T = 1 \cdot \left(\frac{e}{k}\right) = \frac{1.6022 \times 10^{-19}}{1.3807 \times 10^{-23}} = 11{,}604.3\,°\text{K}.$$

The Debye length, using the permittivity of free space as $8.85 \times 10^{-12} F/m$, is then

$$\lambda_D = \left[\frac{(8.85 \times 10^{-12})(1.38 \times 10^{-23})11{,}604}{10^{17}(1.6 \times 10^{-19})^2}\right]^{1/2}.$$

$$= 2.35 \times 10^{-5}\text{m} = 23.5\,\mu\text{m}$$

A simplifying step to note in this calculation is that kT_e/e in Eq. 3.7-12 has units of electron volts. A handy formula for the Debye length is

$$\lambda_D(\text{cm}) \approx 740\sqrt{T_{\text{ev}}/n_o}, \tag{3.7-14}$$

where T_{ev} is in electron volts and n_o is the plasma density in cm^{-3}.

3.7.2 Pre-sheaths

In the previous section, the sheath characteristics for the case of the potential difference between the plasma and an electrode or boundary being small compared to the electron temperature ($e\phi \ll kT_e$) was analyzed and resulted in Debye shielding sheaths. What happens for the case of potential differences on the order of the electron temperature? Consider a plasma in contact with a boundary wall, as illustrated in Fig. 3-3. Assume that the plasma is at a reference potential Φ at the center (which can be arbitrarily be set), and that cold ions fall through an arbitrary potential of ϕ_o as they move toward the boundary. Conservation of energy states that the ions arrived at the sheath edge with an energy given by

$$\frac{1}{2}Mv_o^2 = e\phi_o. \tag{3.7-15}$$

This potential drop between the center of the plasma and the sheath edge, ϕ_o, is called the pre-sheath potential. Once past the sheath edge, the ions then gain an additional energy given by

$$\frac{1}{2}Mv^2 = \frac{1}{2}Mv_o^2 - e\phi(x), \tag{3.7-16}$$

where v is the ion velocity in the sheath and $\phi(x)$ is the potential through the sheath (becoming more negative relative to center of the plasma). Using Eq. 3.7-15 in Eq. 3.7-16 and solving for the ion velocity in the sheath gives

$$v = \sqrt{\frac{2e}{M}}[\phi_o - \phi]^{1/2}. \tag{3.7-17}$$

However, from Eq. 3.7-15 $v_o = \sqrt{2e\phi_o/M}$, so Eq. 3.7-17 can be rearranged to give

$$\frac{v_o}{v} = \sqrt{\frac{\phi_o}{\phi_o - \phi}}, \tag{3.7-18}$$

which represents an acceleration of the ions toward the wall. The ion flux during this acceleration is conserved

$$n_i v = n_o v_o \\ n_i = n_o \frac{v_o}{v}. \tag{3.7-19}$$

Using Eq. 3.7-18 in Eq. 3.7-19, the ion density in the sheath is

$$n_i = n_o \sqrt{\frac{\phi_o}{\phi_o - \phi}}. \tag{3.7-20}$$

Examining the potential structure close to the sheath edge such that ϕ is small compared to the presheath potential ϕ_o, Eq. 3.7-20 can be expanded in a Taylor series to give

$$n_i = n_o \left(1 - \frac{1}{2}\frac{\phi}{\phi_o} + \dots\right), \tag{3.7-21}$$

where the higher order terms in the series will be neglected.

The electron density through the sheath is given by the Boltzmann relationship in Eq. 3.5-9. If it is also assumed that the change in potential right at the sheath edge is small compared to the electron temperature, then the exponent in Eq. 3.5-9 can be expanded in a Taylor series to give

$$n_e = n_o \exp\left(\frac{e\phi}{kT_e}\right) = n_o \left[1 + \frac{e\phi}{kT_e} + \dots\right]. \tag{3.7-22}$$

Using Eqs. 3.7-21 and 3.7-22 in Poisson's Eq. 3.7-8 for singly charged ions ($Z = 1$) in one dimension gives

$$\begin{aligned} \frac{d^2\phi}{dx^2} &= -\frac{e}{\varepsilon_o}(n_i - n_e) = -\frac{en_o}{\varepsilon_o}\left[1 - \frac{1}{2}\frac{\phi}{\phi_o} - 1 + \frac{e\phi}{kT_e}\right] \\ &= \frac{en_o\phi}{\varepsilon_o}\left[\frac{e}{kT_e} - \frac{1}{2\phi_o}\right]. \end{aligned} \tag{3.7-23}$$

In order to have a monotonically decreasing potential from the center to the wall, which avoids an inflection in the potential at the sheath edge that would slow or even reflect the ions going into the sheath, the right-hand side of Eq. 3.7-23 must always be zero or positive, which implies:

$$\frac{1}{2\phi_o} \le \frac{e}{kT_e}. \tag{3.7-24}$$

This expression can be re-written as:

$$\phi_o \ge \frac{kT_e}{2e}, \tag{3.7-25}$$

which is the Bohm Sheath Criterion [10] that states that the ions must fall through a "presheath" potential in the plasma of at least $T_{eV}/2$ before entering the sheath to produce a monotonically decreasing sheath potential. Since the ion velocity entering the sheath is $v_o = \sqrt{2e\phi_o/M}$, the Bohm Criterion in Eq. 3.7-25 can be expressed in the more familiar form as

$$v_o \ge \sqrt{\frac{kT_e}{M}}. \tag{3.7-26}$$

This is usually called the Bohm velocity for ions entering a sheath. Equation 3.7-26 states that the ions must enter the sheath with a velocity of at least $\sqrt{kT_e/M}$ (known as the acoustic velocity for cold ions) to have a stable (monotonic) sheath potential behavior. The plasma produces a potential drop of at least $T_e/2$ in the pre-sheath region before the sheath to produce this ion velocity. Although not derived here, if the ions have a finite temperature T_i, it is easy to show that the Bohm velocity will still take the form of the ion acoustic velocity given by

$$v_o = \sqrt{\frac{\gamma kT_i + kT_e}{M}}. \tag{3.7-27}$$

It is important to realize that the plasma density also decreases in the pre-sheath due to ion acceleration toward the wall. This is easily observed from the Boltzmann behavior of the plasma density, where the potential at the sheath edge has fallen to a value of $-kT_e/2e$ relative to the plasma potential at the location where the density is n_o (far from the edge of the plasma). The electron density at the sheath edge is then

$$\begin{aligned} n_e &= n_o \exp\left(\frac{e\phi_o}{\mathrm{kT}_e}\right) = n_o \exp\left[\left(\frac{e}{kT_e}\right)\left(\frac{-kT_e}{2e}\right)\right] \\ &= 0.606 n_o. \end{aligned} \tag{3.7-28}$$

Therefore, the plasma density at the sheath edge is about 61% of the plasma density in the center of the plasma.

The current density of ions entering the sheath at the edge of the plasma can be found from the density at the sheath edge in Eq. 3.7-28 and the ion velocity at the sheath edge in Eq. 3.7-26:

$$J_i = 0.61 n_o e v_i \approx \frac{1}{2} \mathrm{ne} \sqrt{\frac{kT_e}{M}}, \tag{3.7-29}$$

where n is the plasma density at the start of the pre-sheath, which is normally considered to be the center of a collisionless plasma or one collision mean free path from the sheath edge for collisional plasmas. It is common to write Eq. 3.7-29 as the Bohm current:

$$I_i = \frac{1}{2} \mathrm{ne} \sqrt{\frac{\mathrm{kT}_e}{M}} A, \tag{3.7-30}$$

where A is the ion collection area at the sheath boundary. For example, consider a xenon ion thruster with a 10^{18} m^{-3} plasma density and an electron temperature of 3 eV. The current density of ions to the boundary of the ion acceleration structure is found to be 118 A m^{-2}, and the Bohm current to an area of 10^{-2} m^2 is 1.18 A.

3.7.3 Child-Langmuir Sheath

The simplest case of a sheath in a plasma is obtained when the potential across the sheath is sufficiently large that the electrons are repelled over the majority of the sheath thickness. This will occur if the potential is very large compared to the electron temperature ($\phi \gg kT_e/e$). This means that the electron density goes to zero relatively close to the sheath edge, and the electron space charge does not significantly affect the sheath thickness. The ion velocity through the sheath is given by Eq. 3.7-17. The ion current density is then

$$J_i = n_i e v = n_i e \sqrt{\frac{2e}{M}} [\phi_o - \phi]^{1/2}. \tag{3.7-31}$$

Solving Eq. 3.7-31 for the ion density, Poisson's equation is then

$$\frac{d^2\phi}{dx^2} = -\frac{e}{\varepsilon_o}(n_i - n_e) = -\frac{J_i}{\varepsilon_o}\left(\frac{M}{2e(\phi_o - \phi)}\right)^{1/2}. \tag{3.7-32}$$

The first integral can be performed by multiplying both sides of this equation by the spatial derivative of the potential and integrating to obtain

$$\frac{1}{2}\left[\left(\frac{d\phi}{dx}\right)^2 - \left(\frac{d\phi}{dx}\right)^2_{x=o}\right] = \frac{2J_i}{\varepsilon_o}\left[\frac{M(\phi_o - \phi)}{2e}\right]^{1/2}. \tag{3.7-33}$$

Assuming that the electric field ($d\phi/dx$) is negligible at $x = 0$, Eq. 3.7-33 becomes

$$\frac{d\phi}{dx} = 2\left(\frac{J_i}{\varepsilon_o}\right)^{1/2}\left[\frac{M(\phi_o - \phi)}{2e}\right]^{1/4}. \tag{3.7-34}$$

Integrating Eq. 3.7-34, assuming $\phi \gg \phi_o$ (neglecting the initial velocity of the ions), and writing the potential across the sheath of thickness d as the voltage V gives the familiar form of the Child-Langmuir Law:

$$J_i = \frac{4\varepsilon_o}{9}\left(\frac{2e}{M}\right)^{1/2}\frac{V^{3/2}}{d^2}. \tag{3.7-35}$$

This equation was originally derived by Child [11] in 1911 and independently derived by Langmuir [12] in 1913. Equation 3.7-35 states that the current per unit area that can pass through a planar sheath is limited by space charge effects that reduce the electric field at the $x = 0$ boundary, and is proportional to the voltage to the 3/2 power divided by the sheath thickness squared. In ion thrusters, the accelerator structure can be designed to first order using the Child-Langmuir equation where d is the gap between the accelerator electrodes. The Child-Langmuir equation can be conveniently written as

$$\begin{aligned} J_e &= 2.33 \times 10^{-6}\frac{V^{3/2}}{d^2} \text{ electrons} \\ J_i &= \frac{5.45 \times 10^{-8}}{\sqrt{M_a}}\frac{V^{3/2}}{d^2} \text{ singly charged ions} \\ &= 5.95 \times 10^{-9}\frac{V^{3/2}}{d^2} \text{ krypton ions} \\ &= 4.75 \times 10^{-9}\frac{V^{3/2}}{d^2} \text{ xenon ions} \end{aligned}, \tag{3.7-36}$$

where M_a is the ion mass in atomic mass units. For example, the space charge limited xenon ion current density across a planar 1 mm grid gap with 1000 V applied is 15 mA/cm^2.

3.7.4 Generalized Sheath Solution

To find the characteristics of any sheath without the simplifying assumptions used in the above sections, the complete solution to Poisson's equation at a boundary must be obtained. The ion density through a planar sheath, from Eq. 3.7-20, can be written as

$$n_i = n_o\left(1 - \frac{\phi}{\phi_o}\right)^{-1/2}, \tag{3.7-37}$$

and the electron density is given by Eq. 3.5-9

$$n_e = n_o \exp\left(\frac{e\phi}{kT_e}\right). \tag{3.7-38}$$

Poisson's Eq. 3.7-8 for singly-charged ions then becomes

$$\frac{d^2\phi}{dx^2} = -\frac{e}{\varepsilon_o}(n_i - n_e) = -\frac{en_o}{\varepsilon_o}\left[\left(1 - \frac{\phi}{\phi_o}\right)^{-1/2} - \exp\left(\frac{e\phi}{kT_e}\right)\right]. \tag{3.7-39}$$

Defining the following dimensionless variables

$$\chi = -\frac{e\phi}{kT_e}$$

$$\chi_o = \frac{e\phi_o}{kT_e}$$

$$\xi = \frac{x}{\lambda_D}$$

Poisson's equation becomes

$$\frac{d^2\chi}{dx^2} = \left(1 + \frac{\chi}{\chi_o}\right)^{-1/2} - e^{-\chi}. \tag{3.7-40}$$

This equation can be integrated once by multiplying both sides by the first derivative of χ and integrating from $\xi_1 = 0$ to $\xi_2 = \xi$:

$$\int_0^{\xi} \frac{\partial \chi}{\partial \xi}\frac{\partial^2 \chi}{\partial \xi^2} d\xi_1 = \int_0^{\xi}\left(1 + \frac{\chi}{\chi_o}\right)^{-1/2} \partial\chi - \int_0^{\xi} e^{-\chi}\, d\chi \tag{3.7-41}$$

where ξ is a dummy variable. The solution to Eq. 3.2-7 is

$$\frac{1}{2}\left[\left(\frac{\partial \chi}{\partial \xi}\right)^2 - \left(\frac{\partial \chi}{\partial \xi}\right)^2_{\xi=0}\right] = 2\chi_o\left[\left(1 + \frac{\chi}{\chi_o}\right)^{1/2} - 1\right] + e^{-\chi} - 1. \tag{3.7-42}$$

Since the electric field ($d\phi/dx$) is zero away from the sheath where $\xi = 0$, rearrangement of Eq. 3.7-42 yields

$$\frac{\partial \chi}{\partial \xi} = \left[4\chi_o\left(1 + \frac{\chi}{\chi_o}\right)^{1/2} + 2e^{-\chi} - 2(2\chi_o - 1)\right]^{1/2}. \tag{3.7-43}$$

To obtain a solution for $\chi(\xi)$, Eq. 3.7-43 must be solved numerically. However, as was shown earlier for Eq. 3.7-23, the right-hand side must always be positive or the potential will have an inflection at or near the sheath edge. Expanding the right-hand side in a Taylor series and neglecting the higher order terms, this equation will also produce the Bohm sheath criterion and specify that the ion velocity at the sheath edge must equal or exceed the ion acoustic (or Bohm) velocity. An examination of Eq. 3.7-43 also shows that the Bohm sheath criterion forces the ion density to always be larger than the electron density through the pre-sheath and sheath, which results in the physically realistic monotonically decreasing potential behavior through the pre-sheath and sheath regions.

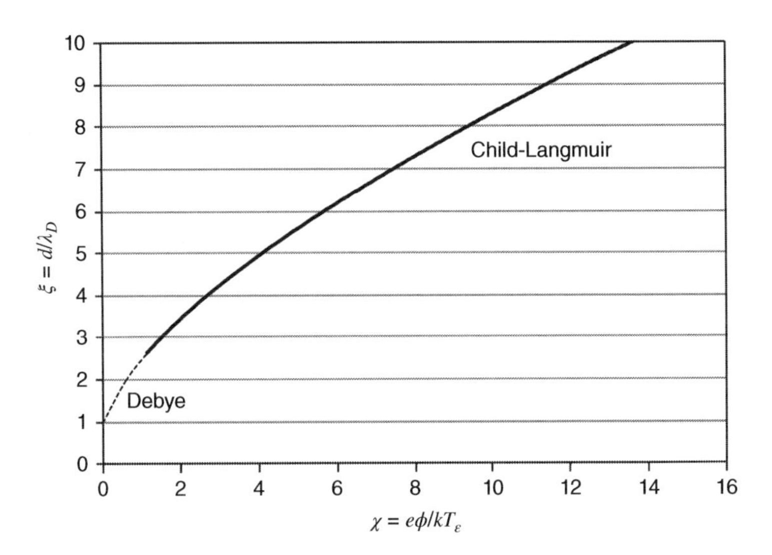

Figure 3-4 Normalized sheath thickness as a function of the normalized sheath potential showing the transition to a Child-Langmuir sheath as the potential becomes large compared to the electron temperature.

Hutchinson [13] derived an analytic expression for the sheath thickness assuming the electron density in the sheath is negligible and the incident ion current density at the sheath edge is the Bohm current given in Eq. 3.7-29. He found:

$$d \approx 1.02\,\lambda_D \left[\left(V/T_{eV}\right)^{1/2} - \frac{1}{\sqrt{2}}\right]^{1/2} \left[\left(V/T_{eV}\right)^{1/2} + \sqrt{2}\right], \tag{3.7-44}$$

where V is the total potential difference in the sheath. Figure 3-4 shows a plot of the sheath thickness d relative to the Debye length versus the total potential drop in the sheath normalized to the electron temperature. The criterion for a Debye sheath derived in Section 3.7.1 was that the potential drop is much less than the electron temperature ($e\phi \ll kT_e$), which is on the far left-hand side of the graph. The criterion for a Child-Langmuir sheath derived in Section 3.7.3 is that the sheath potential is large compared to the electron temperature ($e\phi \gg kT_e$), which occurs on the right-hand side of the graph. This graph illustrates the rule-of-thumb that the sheath thickness is several Debye lengths until the full Child-Langmuir conditions are established. Beyond this point, the sheath thickness varies as the potential to the 3/2 power for a given plasma density.

The reason for examining this general case is because sheaths with potential drops on the order of the electron temperature or higher are typically found at both the anode and insulating surfaces in ion and Hall thrusters. For example, it will be shown later that an insulating surface exposed to a xenon plasma will self-bias to a potential of about $\phi \sim 6T_e$, which is called the *floating potential*. For a plasma with an electron temperature of 4 eV and a density of 10^{18} m^{-3}, the Debye length from Eq. 3.7-12 is 1.5×10^{-5} m and the sheath thickness from Eq. 3.7-44 is about 6 Debye lengths. Once the potential is significantly greater than the electron temperature, the sheath thickness is many times this value and the sheath transitions to a Child-Langmuir sheath.

3.7.5 Double Sheaths

So far, only plasma boundaries where particles from the plasma are flowing toward a wall have been considered. At other locations in electric thrusters, such as in some cathode and accelerator structures, a situation may exist where two plasmas are in contact but at different potentials and ion and

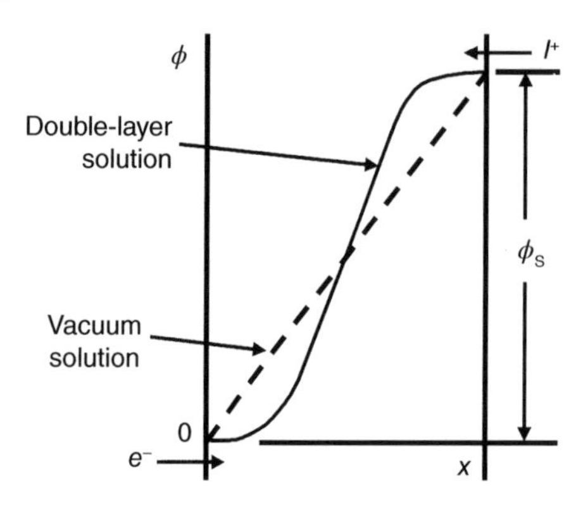

Figure 3-5 Schematic of the double-layer potential distribution.

electron currents flow between the plasmas in opposite directions. This situation is called a double sheath, or double layer, and is illustrated in Fig. 3-5. In this case, electrons flow from the zero-potential boundary on the left and ions flow from the boundary at a potential ϕ_s on the right. Since the particle velocities are relatively slow near the plasma boundaries before the sheath acceleration takes place, the local space charge effects are significant and the local electric field is reduced at both boundaries. The gradient of the potential inside the double layer is therefore much higher than in the vacuum case where the potential varies linearly in between the boundaries.

Referring again to Fig. 3-5, assume that boundary on the left is at zero potential and that the particles arrive at the sheath edge on both sides of the double layer with zero initial velocity. The potential difference between the surfaces accelerates the particles in the opposite direction across the double layer. The electron conservation of energy gives

$$\begin{aligned} \frac{1}{2} m v_e^2 &= e\phi \\ v_e &= \left(\frac{2e\phi}{m}\right)^{1/2}, \end{aligned} \tag{3.7-45}$$

and the ion energy conservation gives

$$\begin{aligned} \frac{1}{2} M v_i^2 &= e(\phi_s - \phi) \\ v_i &= \left[\frac{2e}{M}(\phi_s - \phi)\right]^{1/2} \end{aligned} \tag{3.7-46}$$

The charge density in Eq. 3.2-5 can be written

$$\begin{aligned} \rho &= \rho_i + \rho_e \\ &= \frac{J_i}{v_i} - \frac{J_e}{v_e} = \frac{J_i}{\sqrt{\phi_s - \phi}}\sqrt{\frac{M}{2e}} - \frac{J_e}{\sqrt{\phi}}\sqrt{\frac{m}{2e}}. \end{aligned} \tag{3.7-47}$$

Poisson's equation can then be written in one dimension as

$$\frac{dE}{dx} = \frac{\rho}{\varepsilon_o} = \frac{J_i}{\varepsilon_o\sqrt{\phi_s - \phi}}\sqrt{\frac{M}{2e}} - \frac{J_e}{\varepsilon_o\sqrt{\phi}}\sqrt{\frac{m}{2e}}. \tag{3.7-48}$$

Integrating once gives

$$\frac{\varepsilon_o}{2}E^2 = 2J_i\sqrt{\frac{M}{2e}}\left[\phi_s^{1/2} - (\phi_s - \phi)^{1/2}\right] - 2J_e\sqrt{\frac{m}{2e}}\phi^{1/2}. \tag{3.7-49}$$

For space-charge-limited current flow, the electric field at the right-hand boundary (the edge of the plasma) is zero and the potential is $\phi = \phi_s$. Putting that boundary condition into Eq. 3.7-49 and solving for the current density gives

$$J_e = \sqrt{\frac{M}{m}} J_i. \tag{3.7-50}$$

If the area of the two plasmas in contact with each other is the same, the electron current crossing the double layer is the square root of the mass ratio times the ion current crossing the layer. This situation is called the Langmuir Condition (1929) and describes the space-charge-limited flow of ions and electrons between two plasmas or between a plasma and an electron emitter.

For finite initial velocities, Eq. 3.7-50 was corrected by Andrews and Allen [14] to give

$$J_e = \kappa \sqrt{\frac{M}{m}} J_i, \tag{3.7-51}$$

where κ is a constant that varies from 0.2 to 0.8 for T_e/T_i changing from 2 to about 20. For typical thruster plasmas where $T_e/T_i \approx 10$, κ is about 0.5.

While the presence of free-standing double layers in the plasma volume in thrusters is often debated, the sheath at a thermionic cathode surface certainly satisfies the criteria of counter-streaming ion and electron currents and can be viewed as a double layer. In this case, Eq. 3.7-51 describes the space charge limited current density that a plasma can accept from an electron-emitting cathode surface. This is useful in that the maximum current density that can be drawn from a cathode can be evaluated if the plasma parameters at the sheath edge in contact with the cathode are known (such that J_i can be evaluated from the Bohm current), without requiring that the actual sheath thickness be known.

Finally, there are several conditions for the formation of the classic double layer described here. To achieve a potential difference between the plasmas that is large compared to the local electron temperature, charge separation must occur in the layer. This, of course, violates quasi-neutrality locally. The current flow across the layer is space-charge limited, which means that the electric field is essentially zero at both boundaries. Finally, the flow through the layer discussed here is collisionless. Collisions cause resistive voltage drops where current is flowing, which can easily be mistaken for the potential difference across a double layer.

3.7.6 Summary of Sheath Effects

It is worthwhile to summarize here some of the important equations in this section related to sheaths because these will be very useful later in describing thruster performance. These equations were derived in the sections above, and alternative derivations can be found in the referenced textbooks [1–3].

The current density of ions entering the sheath at the edge of the plasma is given by:

$$J_i = 0.6\, ne\, v_i \approx \frac{1}{2} ne \sqrt{\frac{kT_e}{M}}, \tag{3.7-52}$$

where n is the plasma density at the start of the pre-sheath far from the boundary, which was considered to be the center of the plasma by Langmuir for his collisionless plasmas. The convention of approximating the coefficient 0.6 as ½ was made by Bohm in defining what is now called the "Bohm current".

If there is no net current to the boundary, the ion and electron currents must be equal. The Bohm current of ions through the sheath is given by the current density in Eq. 3.7-52 times the wall area A:

$$I_B = \frac{1}{2} n_i e \sqrt{\frac{kT_e}{M}} A. \tag{3.7-53}$$

The electron current through the sheath is the random electron flux times the Boltzmann factor

$$I_e = \frac{1}{4}\sqrt{\frac{8kT_e}{\pi m}} n_e e A e^{\left(-\frac{e\phi}{kT_e}\right)}, \tag{3.7-54}$$

where the potential ϕ is relative to the plasma potential at the plasma center and is by convention a positive number in this formulation. Equating the total ion and electron currents ($I_i = I_e$), assuming quasi-neutrality in the plasma ($n_i = n_e$), and solving for the potential gives

$$\phi = \frac{kT_e}{e} \ln\left(\sqrt{\frac{2M}{\pi m}}\right). \tag{3.7-55}$$

This is the potential at which the plasma will self-bias to have zero net current to the walls and thereby conserve charge, and is often called the *floating potential.* Note that the floating potential is negative relative to the plasma potential.

For sheath potentials less than the electron temperature, the sheath thickness is given by the Debye length (Eq. 3.7-12):

$$\lambda_D = \sqrt{\frac{\varepsilon_o kT_e}{n_o e^2}}. \tag{3.7-56}$$

For sheath potentials greater than the electron temperature ($e\phi > kT_e$), a pre-sheath forms to accelerate the ions into the sheath to avoid any inflection in the potential at the sheath edge. The collisionless pre-sheath has a potential difference from the center of the plasma to the sheath edge of $T_{eV}/2$ and a density decrease from the center of the plasma to the sheath edge of $0.61n_o$. The $T_{eV}/2$ potential difference accelerates the ions to the Bohm velocity

$$v_{\text{Bohm}} = v_B = \sqrt{\frac{kT_e}{M}}. \tag{3.7-57}$$

The sheath thickness at the wall depends on the plasma parameters and the potential difference between the plasma and the wall, and is found from the solution of Eq. 3.7-43.

For the case of sheath potentials that are large compared to the electron temperature ($e\phi >> kT_e$), the current density through the sheath is described by the Child-Langmuir equation

$$J_i = \frac{4\varepsilon_o}{9}\left(\frac{2e}{M}\right)^{1/2} \frac{V^{3/2}}{d^2}. \tag{3.7-58}$$

Finally, for the case of double sheaths where ion and electrons are counterstreaming across the boundary between two plasmas, the relationship between the two currents is

$$J_e = \kappa\sqrt{\frac{M}{m}} J_i. \tag{3.7-59}$$

If one boundary of the double layer is the sheath edge at a thermionic cathode, Eq. 3.7-53 can be used for the Bohm current to the opposite boundary to give the maximum emission current density as

$$J_e = \frac{\kappa}{2} n_i e \sqrt{\frac{kT_e}{m}} \approx \frac{1}{4} n_e e \sqrt{\frac{kT_e}{m}}. \tag{3.7-60}$$

This is the maximum electron current density that can be accepted by a plasma due to space-charge effects at the cathode double-sheath. For example, the maximum space-charge-limited cathode emission current into a xenon plasma with a density of 10^{18} m^{-3} and an electron temperature of 5 eV is about 3.8 A cm^{-2}.

These summary equations are commonly seen in the literature on the design and analysis of ion sources, plasma processing sources, and, of course, electric thrusters.

Problems

3.1 Show that $\mathbf{E} = -\nabla\phi - \partial\mathbf{A}/\partial t$ when $\mathbf{B}$ is varying with time, where $\mathbf{A}$ is the "vector potential." How are $\mathbf{A}$ and $\mathbf{B}$ related?

3.2 Derive Eq. 3.3-21 for the force on a particle in a magnetic mirror.

3.3 Show that the magnetic moment is invariant and derive Eq. 3.3-23.

3.4 Derive the expression for ion acoustic velocity in Eq. 3.5-27.

3.5 Answer the following question that might be brought up by a student working in the lab: "In a plasma discharge set up in my vacuum chamber the other day, I measured an increase in the plasma potential with an electrostatic probe. How do I know if it is a double-layer or just a potential gradient within which the ionized gas is quasi-neutral?"

3.6 Derive Eq. 3.6-9 for the penetration distance of neutral particles in plasmas.

3.7 Derive the ambipolar diffusion coefficient with a magnetic field in terms of the electron diffusion coefficient and show the dependence of the reduced rate of electron loss under ambipolar flows.

3.8 Derive Eq. 3.6-80 for the transverse ambipolar electron velocity across magnetic field lines.

3.9 Derive Eq. 3.7-27 for the Bohm sheath criteria in the presence of finite ion temperature.

3.10 Derive an expression for the Child-Langmuir Law for the condition where the initial ion velocity entering the sheath is not neglected (ions have an initial velocity v_o at the sheath edge at $x = 0$).

3.11 A 2 mm by 2 mm square probe is immersed in a 3 eV xenon plasma,

a) If the probe collects 1 mA of ion current, what is the plasma density? (Hint: probe has two sides and is considered infinitely thin.)
b) What is the floating potential?
c) What is the probe current collected at the plasma potential?

3.12 A 2 mm diameter cylindrical probe 5 mm long in a xenon plasma with $T_e = 3$ eV collects 1 mA of ion saturation current. What is the average plasma density? How much electron current is collected if the probe is biased to the plasma potential?

3.13 An electron emitter capable of emitting up to 10 A cm^{-2} is in contact with a singly charged Xe^+ plasma with an electron temperature of 2 eV. Plot the emission current density versus plasma density over the range from 10^{10} to 10^{13} cm^{-3}.

References

1 F. F. Chen, *Introduction to Plasma Physics and Controlled Fusion*, Vol. 1, New York: Plenum Press, 1984.
2 M. Lieberman and A. Lichtenberg, *Principles of Plasma Discharges and Materials Processing*, New York: John Wiley and Sons, 1994.
3 R. J. Goldston and P. H. Rutherford, *Introduction to Plasma Physics*, London: Institute of Physics Publishing, 1995.
4 A. T. Forrester, *Large Ion Beams*, New York: John Wiley and Sons, 1988.
5 NRL Plasma Formulary. (2018). NRL/PU/6790—18-640, Naval Research Laboratory, Washington, DC, 20373-5320.
6 E. A. Mason and E. W. McDaniel, *Transport Properties of Ions in Gases*, New York: John Wiley and Sons, 1988.
7 J. S. Miller, S. H. Pullins, D. J. Levandier, Y. Chiu, and R. A. Dressler, "Xenon Charge Exchange Cross Sections for Electrostatic Thruster Models," *Journal of Applied Physics*, vol. 91, no. 3, pp. 984–991, 2002.
8 I. Katz, J. Anderson, J. Polk, and J. Brophy, "One Dimensional Hollow Cathode Model," *Journal of Propulsion and Power*, vol. 19, no. 4, pp. 595–600, 2003.
9 L. Spitzer, Jr., *Physics of Fully Ionized Gases*, New York: Interscience, pp. 127–135, 1962.
10 D. Bohm, "The Characteristics of Electric Discharges in Magnetic Fields," edited by A. Guthrie and R. Wakerling, New York: McGraw-Hill Book Company, Inc., pp. 1–76, 1949.
11 C. D. Child, "Discharge from Hot CaO," *Physical Review*, vol. 32, p. 492, 1911.
12 I. Langmuir, "The Effect of Space Charge and Residual Gases on Thermionic Currents in High Vacuum," *Physical Review*, vol. 2, p. 450, 1913.
13 I. H. Hutchinson, *Principles of Plasma Diagnostics*, 2nd edition, Cambridge: Cambridge University Press, 2002.
14 J. G. Andrews and J. E. Allen, "Theory of a Double Sheath Between Two Plasmas," *Proceedings of the Royal Society of London, Series A*, vol. 320, p. 459, 1971.

Chapter 4

Hollow Cathodes

Nearly all electric thrusters require an electron emitter for discharge current generation, propellant ionization, and ion beam charge neutralization. Most of these electron emitters are configured as hollow cathodes due to their compact size, high current capability, and long life. Hollow cathodes are straightforward to ignite with a momentary heater, rf exciter, or Paschen discharge. Once ignited they are self-heated; varying their voltage drop to provide the thermionic electron emitter heating necessary to produce the desired discharge current. They operate on a wide variety of propellants and can be scaled to produce discharge currents from fractions of an ampere to hundreds of amperes. The properties of the cathode material, the physical configuration of the hollow cathode, and structure and behavior of the cathode plasma determine, to a large extent, the performance and life of the thruster.

4.1 Introduction

Early electron-bombardment ion thrusters developed in the 1950s utilized directly heated tungsten or tantalum filaments as the cathode that emitted electrons for the plasma discharge used to produce the ions. Smaller tungsten filaments were also immersed directly into the ion beam to provide neutralizing electrons. Owing to the high work function of these refractory metals, the filaments had to be operated at temperatures of over 2400 °C to emit electron current densities in excess of 1 A/cm^2. Operation at these temperatures requires high heater power, often on the order of the discharge power, which significantly reduces the efficiency of the thruster. In addition, the life of refractory metal filament cathodes is limited by rapid evaporation of the filament material at the elevated temperatures and by ion bombardment sputtering of the surfaces exposed to the discharge plasma in the thruster or in the beam. Filament cathode life was typically limited to the order of only tens to hundreds of hours. Although the use of filament cathodes permitted early development of ion thruster discharge chambers and accelerator grids, they were inadequate for long-life space applications.

This problem was solved by the development of hollow cathodes. The first description of a discharge in a "hollow cathode" was by F. Paschen [1] in 1916. Hollow cathodes are characterized by a reentrant geometry where the internal plasma is at least partially bounded by electrode walls that are at cathode potential. Ionizing electrons produced in the discharge reflect from the cathode-potential walls, which increases their path length before loss and thereby increases the plasma

Fundamentals of Electric Propulsion, Second Edition. Dan M. Goebel, Ira Katz, and Ioannis G. Mikellides.

density and cathode emission current capability. Hollow cathodes made before the 1960s primarily used Paschen breakdowns [1] to initiate a low current glow discharge that produced higher plasma densities due to the reentrant cathode geometry.

In the early 1960s, Lidsky [2] developed what was then called a "hollow cathode arc discharge" where the reentrant cathode was a simple open-ended refractory metal tube at cathode potential with gas flowing through it. The discharge was started by a Paschen breakdown from the tip of the tube to the cylindrical anode and transitioned into a high-current thermionic discharge deeper inside the tube once the discharge produced sufficient heating of the refractory metal wall to reach thermionic emission temperatures ($\gg 2400\,°C$). These cathodes produced high discharge currents and high plasma densities, but had limited life due to the high evaporation rates of the refractory metal tube wall at the temperatures required for operation.

The first purely thermionic hollow cathode was invented by Forrester [3] in the mid-1960s for use as a neutralizer cathode for cesium surface ionization ion thrusters. This cathode featured a tungsten tube with a small orifice at the exit and an external heater to start the discharge. The cesium propellant injected into the tube produced a very low work function Cs-W surface on the inside wall, and the emitted electrons were extracted from the cesium plasma through the orifice. The plasma produced by this cathode was used as a "plasma-bridge neutralizer" that delivered electrons to the ion beam [3], eliminating the life limitations of filament electron emitters immersed in the beam that were subject to ion sputtering. Pawlik [4] modified this design for use with mercury propellants by placing a thermionic emitter "insert" inside the hollow cathode. Early plasma-bridge neutralizer cathodes were required to produce less than 1 A of current and operate for lifetimes of less than a year, which Pawlik achieved with simple oxide-coated rolled-up metal foil inserts in small diameter hollow tubes with small orifices. The first employment of such an "orificed hollow cathode" with a thermionic emitter on a spaceflight electric propulsion system was in 1970 onboard NASA's Space Electric Rocket Test (SERT 2) [5]. Rawlin [6, 7] replaced the oxide-coated insert with barium oxide dispenser cathode inserts in the 1970s to provide higher electron currents (up to 10 A) and longer life for use as the main discharge cathode in both mercury and inert gas ion thrusters. At about that same time in the 1970s, the lanthanum hexaboride hollow cathode was invented in the United States by Goebel [8] to provide even higher discharge currents from tens to hundreds of amperes. All of these cathode types have been matured since that time to produce higher discharge currents and longer life for electric propulsion applications [9, 10], but the basic geometry has remained the same.

A generic hollow cathode is shown in Fig. 4-1, where the cathode consists of a hollow refractory tube with an orifice plate on the downstream end. The tube has an insert in the shape of a cylinder

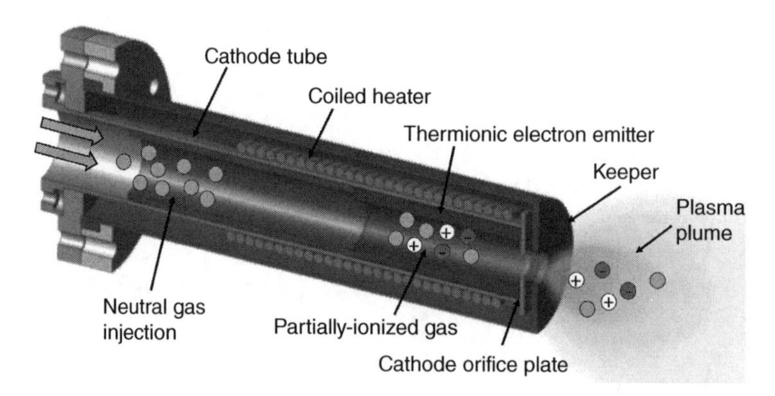

Figure 4-1 Typical hollow cathode geometry of a refractory metal tube with a thermionic insert and a heater and keeper on the outside.

that is placed inside the tube and pushed against the orifice plate. This insert is the active electron emitter and it can be made of several different materials that provide a low work function surface on the inside diameter in contact with the cathode plasma. The cathode tube is wrapped with a heater (a co-axial sheathed heater is shown in the figure) that raises the insert temperature to emissive temperatures to start the discharge. Outside the cathode tube assembly is a keeper electrode that is biased to help start the cathode and also protects the cathode from back ion bombardment sputtering. The electrons emitted from the insert ionize gas injected through the cathode tube and form a cathode plasma from which the discharge current electrons are extracted through the orifice into the thruster plasma.

A hollow cathode can be separated into three distinct plasma regions illustrated in Fig. 4-2: A dense plasma in the insert region interior to the cathode, a high current density plasma in the orifice, and a low-density plume plasma outside the cathode that completes the electrical connection to the thruster discharge plasma. The plasma ions generated throughout the device neutralize the electron space charge. As a result, hollow cathodes produce high currents at low voltages compared to vacuum cathode devices.

The structure of the hollow cathode serves three main functions. First, some fraction of the thruster propellant is injected through the hollow cathode and the discharge inside the resulting high neutral pressure region generates a cold, high-density plasma. The plasma and neutral densities are the highest of anywhere in the thruster and the electron temperature is correspondingly the lowest. This causes the plasma potential inside the hollow cathode to be very low, reducing the energy of the ions that arrive at the insert surface. This characteristic behavior is demonstrated in Fig. 4-3, which shows the measured potential and density profiles in the Nuclear-Electric Xenon Ion System (NEXIS) hollow cathode [11] discharge. Plasma densities in excess of 10^{19} m^{-3} are routinely generated inside hollow cathodes and the electron temperature is found [11, 12] to be only 1–3 eV. The low plasma potential in the insert region and high neutral scattering rates decrease the ion bombardment energy striking the insert surface to typically less than 20 eV, which essentially eliminates ion sputtering of the surface and greatly increases the life of the cathode. Second, the high-density plasma in the insert region eliminates space charge effects at the cathode surface that can limit the electron emission current density. Emission current densities of 1–10 A/cm^2 are typically employed in thruster hollow cathodes for compact size and good life, although higher current densities are achievable and sometimes used. Third, the cathode insert can be heat-shielded well in this geometry, which greatly reduces the radiation losses of the cathode at operating temperatures. This decreases the amount of power that must be deposited in the cathode to maintain the required temperature for electron emission. It also reduces the cathode heating losses to a small fraction of the discharge power, significantly reducing the discharge loss of the plasma generator.

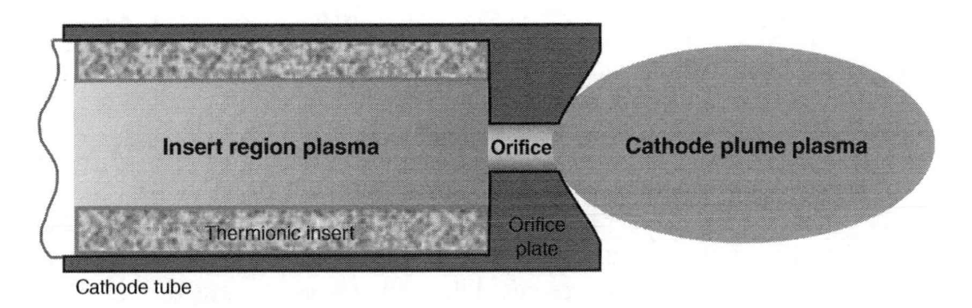

Figure 4-2 The three plasma regions in a hollow cathode discharge.

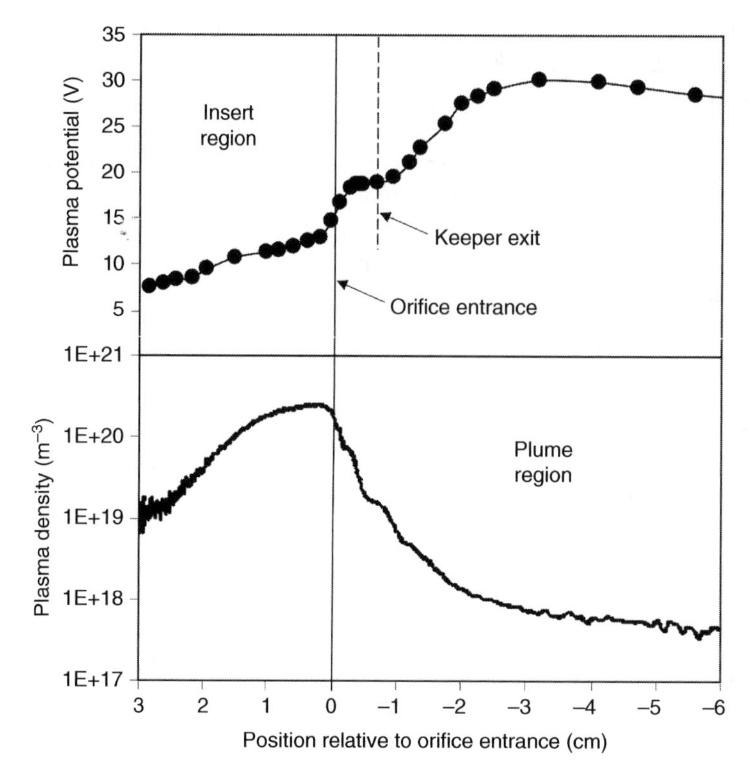

Figure 4-3 Plasma potential (top) and density (bottom) measured on axis in the NEXIS hollow cathode at 25 A of discharge current.

Since nearly the entire discharge current runs through the orifice, which is the region of the cathode with the smallest cross-sectional area, the current density there is the highest in the system and a sufficient plasma density must be generated locally to carry the current. For the 25-A discharge case shown in Fig. 4-3, the plasma density in the orifice is on the order of 10^{20} m^{-3}. The discharge current flowing through the 0.25-cm diameter orifice in this cathode is given by

$$I = nevA \tag{4.1-1}$$

where n is the plasma density, e is the electron charge, v is the electron drift velocity and A is the cross sectional area of the orifice. Solving for the drift velocity gives

$$v = \frac{I}{neA} = 7.7 \times 10^4 \text{ m/s} << v_{\text{th}}, \tag{4.1-2}$$

where the thermal electron drift velocity $v_{\text{th}} = \sqrt{kT_e/m}$ is 6×10^5 m/s for the 2 eV plasma electron temperatures measured in this location. The current is conducted through the orifice region at relatively low drift velocities, even though the electron current density exceeds 100 A/cm^2 in this case. This is typically true even at current densities exceeding 1000 A/cm^2.

In the plume region, the expanding plasma from the orifice region and ionization of the expanding neutral gas provide an ion background that neutralizes the space charge of the current carrying electrons. Hollow cathodes are normally enclosed in another electrode called a keeper, shown in Fig. 4-4. The major functions of the keeper electrode are to facilitate turning on the cathode discharge, to maintain the cathode temperature and operation in the event that the discharge or beam

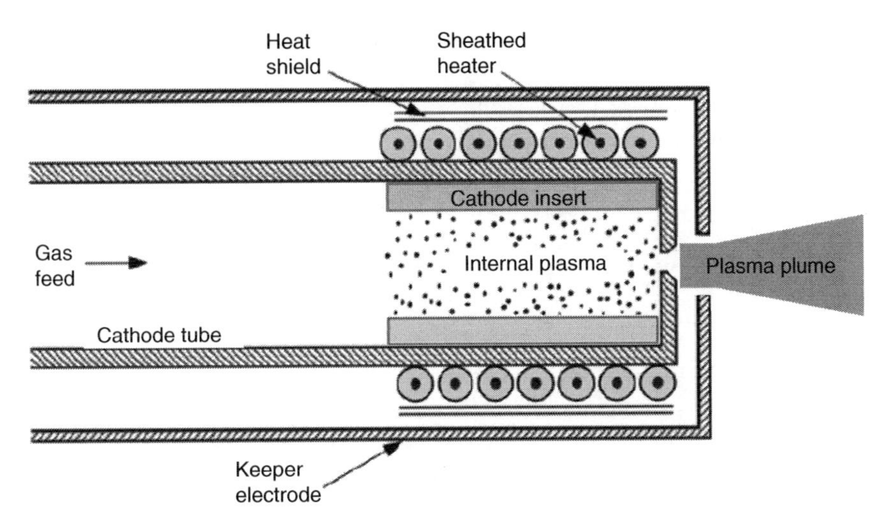

Figure 4-4 Hollow cathode schematic showing the cathode tube, insert, and heater enclosed in a keeper electrode.

current is interrupted temporarily, and to protect the cathode orifice plate and external heater from high-energy ion bombardment that might limit the cathode life. The keeper is normally biased positive relative to the cathode, which either initiates the discharge during start-up, or reduces the ion bombardment energy during normal operation. The life of the keeper electrode is very important to the life of the cathode and thruster.

Hollow cathodes operate in a "self-heating" mode where the external heater or heater-plasma discharge is turned off during operation and the cathode insert is heated by plasma bombardment from the insert-region plasma. There are three self-heating mechanisms possible in hollow cathodes: (i) orifice heating, (ii) ion heating, and (iii) electron heating. In orifice heating, the cathode is designed with a small-diameter, restrictive orifice, which produces a high internal density and pressure in both the insert and orifice regions. The plasma discharge passing through the orifice is then very resistive, causing a significant amount of power to be deposited in the orifice plasma and transferred to the orifice walls by convection. This power deposition then heats the thermionic insert by conduction and radiation. Orifice heating is used primarily in low current cathodes, like ion-beam neutralizer cathodes, where the discharge currents are very low. Ion heating is the classic mechanism for cathode heating where ions in the cathode insert-region plasma fall through the sheath potential at the insert surface and heat the surface by ion bombardment. Electron heating occurs in a regime where both the cathode internal pressure and the discharge current are relatively high, resulting in the very high plasma densities ($>10^{20}$ m^{-3}) generated in the insert region. The low electron temperatures and low sheath voltages produced in this situation result in the energetic tail of the Maxwellian electron distribution having sufficient energy to exceed the sheath potential and reach the insert surface. These electrons then deposit their energy on the insert and heat it to emission temperatures. The heating mechanism that dominates in any hollow cathode design depends on the geometry of the cathode, the internal neutral gas pressure in the insert and orifice regions, and the discharge current.

This chapter will start with a simple classification of different hollow cathode geometries to aid in the discussion of the important effects in the system, and then discuss the basics of the cathode insert that provides thermionic electron emission. The characteristics of the plasma in the insert region, the orifice, and cathode plume in the vicinity of the keeper required to extract and transmit the electrons into the thruster will then be examined. Since the neutral gas density changes all along the discharge path in hollow cathode discharges, the plasmas generated in each location (inside the insert, in the orifice, and in the cathode plume) have different properties in terms of collisionality, temperature, potential, and density. These differences determine the applicable plasma physics in each region.

4.2 Cathode Configurations

The geometry and size of the hollow cathodes depends on the amount of current they are required to emit. Discharge currents in ion thrusters are typically 5–10 times the beam current depending on the efficiency of the plasma generator, and discharge currents can range from a few amperes to over 100 A[13] depending on the size of the thruster. The hollow cathode used in a Hall thruster provides electrons for both ionization of the propellant gas and neutralization of the beam [14]. Hall thrusters also tend to run at lower specific impulse (Isp) than ion thrusters for a given thruster power level. Therefore, Hall thrusters require higher discharge currents from the cathode to achieve the same total power compared to ion thrusters, and currents of the order of 10 A to hundreds of amperes are needed. Neutralizer cathodes in ion thrusters [15] emit electrons at a current equal to the beam current. Therefore, they can be made smaller than discharge cathodes and must be designed to be self-heated and run reliably at lower currents.

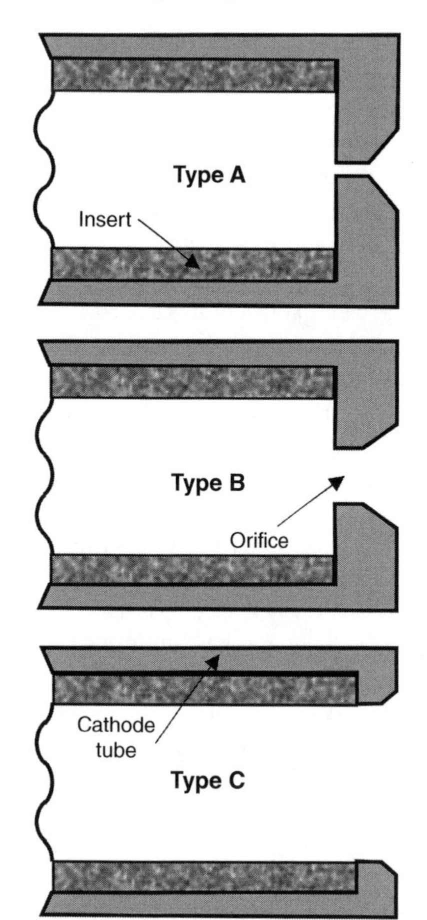

Figure 4-5 Schematics of the three types of hollow cathodes (A, B and C) depending on the orifice geometry.

Higher discharge currents require larger insert surface areas because the electron emission current density available from thermionic cathodes is limited by the cathode properties. Ultimately, this determines the diameter of the insert and the hollow cathode, which will be described in the next section. The cathode orifice diameter and length depend on many parameters. Ion thruster neutralizer cathodes have been designed with very small diameter orifices ($\leq 3\times10^{-2}$ cm), and ion thruster discharge cathodes and small Hall thruster cathodes have been designed with orifices of less than 0.1-cm diameter to over 0.3 cm in diameter. High current hollow cathodes for large ion thrusters and Hall thrusters will have even larger orifices. These cathodes are sometimes designed even without an orifice, where the insert inner diameter forms a tube exposed to the discharge plasma.

Hollow cathodes generally fall into the three categories shown schematically in Fig. 4-5, which will be useful later in describing the plasma characteristics in the three regions described above. The first type of hollow cathode is characterized by a small orifice (compared to the insert diameter) with a large length to diameter ratio, and called Type A. These cathodes typical operate at low current and relatively high internal gas pressures, and are heated primarily by cathode orifice heating. The second type of cathode features an orifice diameter typically larger than the length, shown in Fig. 4-5 as Type B, and operates at lower internal gas pressures. The heating mechanism in these cathodes is typically due to a combination of electron and ion bombardment of the insert that depends on the orifice size and operating condition. The third type of cathode, typically used in high current applications and shown in Fig. 4-5 as Type C, has a very large diameter cathode orifice or essentially no orifice at all. These cathodes typically have a reduced internal pressure overall compared to orificed cathodes. The heating mechanism for Type C cathodes is normally ion bombardment of the insert.

The geometry of the hollow cathode and conditions at which it is operated critically affect the maximum plasma density and its gradient in the cathode interior [16, 17]. Figure 4-6 shows examples of plasma density profiles measured along the cathode centerline with fast scanning probes [11] inside a 0.38 cm inside diameter (I.D.) cathode insert operating at 13 A of discharge current and a xenon flow rate of 3.7 sccm for two different orifice diameters and the case of no orifice plate at all. Small orifices, characteristic of Type A cathodes, operate at high internal pressures and produce high plasma densities but constrain the axial extent of the plasma to the order of a few millimeters. For a given emission current density, this can restrict the discharge current that is available. As the orifice is enlarged, the plasma extends farther into the insert, resulting in utilization of more insert surface area for electron emission. This is illustrated in Fig. 4-7 that shows a numerical simulation [18] of the Type A, B, and C cathodes and illustrates how the orifice selection strongly affects the internal plasma density distribution. Likewise, for a fixed orifice size and discharge current, increasing the gas flow rate increases the internal pressure and is observed to produce a steeper plasma density gradient upstream of the peak value while that peak density value moves closer to the orifice plate.

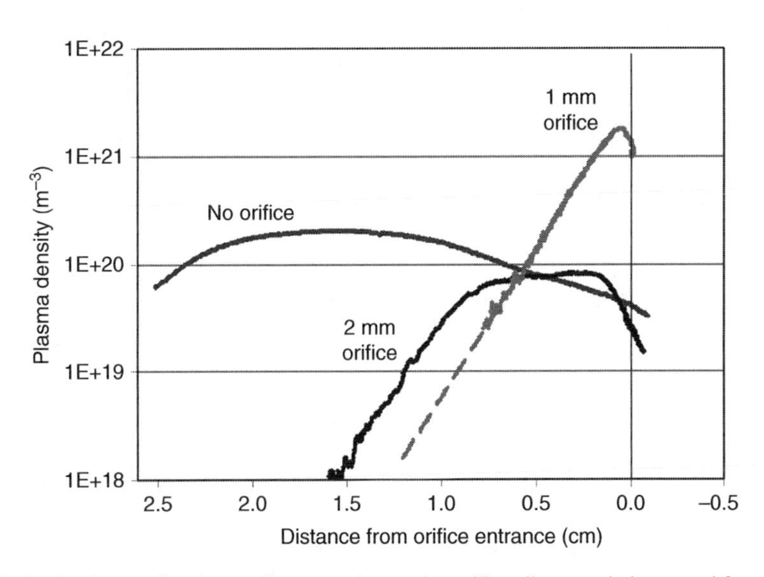

Figure 4-6 Cathode plasma density profile examples as the orifice diameter is increased for a constant discharge current of 13 A and a constant xenon flow rate of 3.7 sccm (*Source:* [18]).

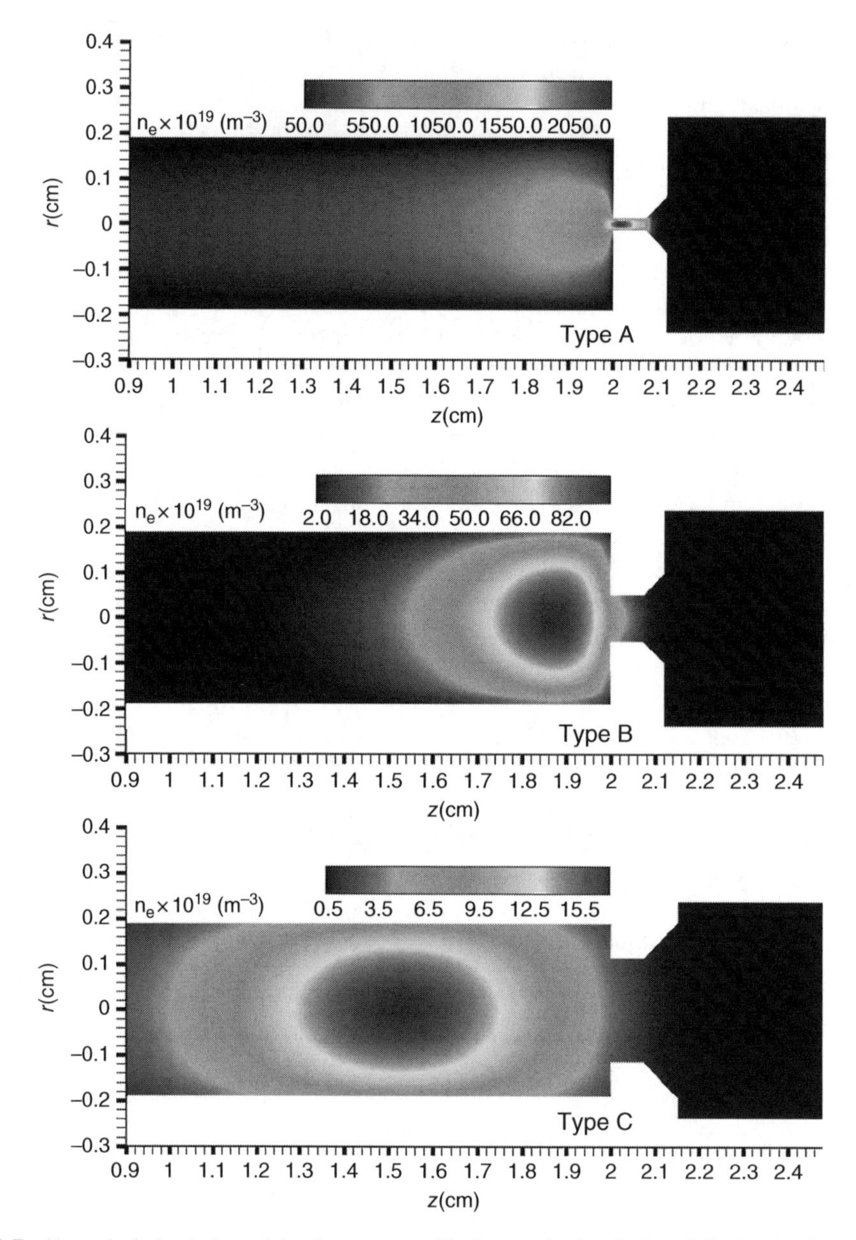

Figure 4-7 Numerical simulation of the three types of hollow cathodes (A, B and C) showing how the formation of the interior plasma depends strongly on the orifice geometry (*Source:* [18]).

The effect on the internal plasma density of increasing the discharge current in a given cathode design at a constant flow rate [19] is illustrated in Fig. 4-8. Increasing the discharge current raises the plasma density and also tends to push the plasma downstream toward the orifice plate. This motion of the plasma can become problematic because, as the plasma density upstream falls, the electron emission can become locally space charge limited [20]. This will force the cathode to emit more of the discharge current from a smaller area downstream, which will require higher insert temperatures and produce higher evaporation rates. Higher currents also typically require

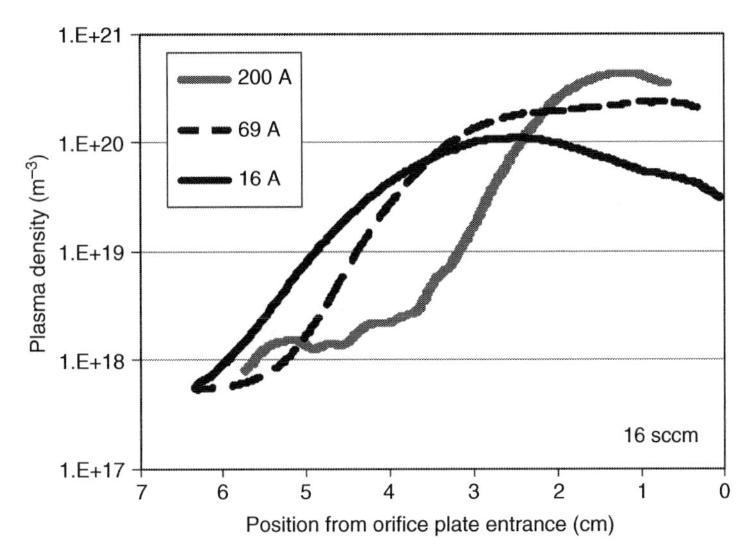

Figure 4-8 Internal cathode plasma density profiles in a larger Type B cathode showing higher peak densities and the plasma moving downstream toward the orifice as the discharge current is increased for a constant gas flow rate (*Source:* [10]).

higher gas flow rates to avoid instabilities outside the cathode in the near cathode plume (discussed in Section 4.6), which also push the plasma downstream and reduce the contact area between the plasma and the insert. For very high current hollow cathodes in excess of 100 A, this behavior drives the cathode design to larger insert diameters with more emitting surface area near the orifice plate, and to larger orifice diameters to provide as much plasma penetration and insert contact area as possible.

Determining the spatial variations of the plasma parameters inside the cathode as the operating conditions and/or geometry are changed is an inherently two-dimensional (2-D) problem that involves a number of coupled nonlinear processes, both in the plasma region and along the boundaries. Hence, a detailed description of the profiles is not possible using low-dimensional idealized models which provide only global descriptions of the plasma. A detailed discussion on this topic is included in Section 4.4.

Depending on the orifice diameter, the electron current density in the orifice can be the highest of anywhere in the system and can easily exceed 1 kA/cm^2. If the orifice is long compared with its radius, as is the case in most Type A neutralizer hollow cathodes, the physics is the same as a classical positive column plasma where an axial electric field in the collisional plasma conducts the current, and plasma resistive heating is very important. A large fraction of this ohmic power deposited in the orifice plasma goes into heating the orifice plate by ion bombardment, which contributes to the insert heating by conduction and radiation. In Type B cathodes, the cathode orifice plate is typically thin compared to the orifice diameter, and there is little local resistive heating. The plasma in the insert region is generated by ionization of the neutral gas by the discharge current flowing through the insert region into the orifice. At high cathode neutral gas flow rates (and subsequent high plasma density) in this type of cathode, the insert heating is primarily by plasma electrons. At low flow rates or with large orifices, the insert is heated predominately due to bombardment by ions accelerated by a higher sheath voltage. In Type C cathodes, there is little or no orifice and the plasma couples from a collisionally dominated region upstream inside the insert directly into the nearly collisionless cathode plume region. This creates long axial density and potential gradients and may expose some of the downstream region of the insert to higher potentials and ion bombardment. Heating in this case is predominately by ion bombardment through the higher cathode sheath potential. Without an orifice plate, the downstream end of the insert radiates more freely and runs cooler than it would if there had been an orifice plate.

Naturally, there is a continuous range of cathode operation that may demonstrate properties of all three cathode types. Indeed, a given cathode geometry can transition from low resistive heating in the orifice at low currents and low gas flow rates to substantial resistive heating and plasma generation at high currents and high gas flow rates. These three types of hollow cathodes will be discussed in more detail after the thermionic electron emitter properties are described in the next section.

4.3 Thermionic Electron Emitters

Electrons are introduced into the hollow cathode by thermionic emission from the insert surface. The thermionic current density is described by the Richardson-Dushman equation [21]:

$$J = AT^2 e^{-e\phi_{wf}/kT} \tag{4.3-1}$$

where A is, ideally, a constant with a value of 120 A/cm^2 K^2, T is the temperature in Kelvin, e in the exponent is the charge, k is Boltzmann's constant, and ϕ_{wf} is the work function of the surface. Experimental investigations of the thermionic emission of different materials reported values of A that vary considerably from the theoretical value. The cause of the deviation of A from a constant has been attributed to several different effects such as variations in the crystal structure of the surface, variations in the surface coverage (for dispenser cathodes), changes in the density of states at the surface due to thermal expansion, etc. This issue has been addressed [22] for many of the thermionic electron emitters used in hollow cathodes by introducing a temperature correction for the work function of the form

$$\phi_{\text{wf}} = \phi_o + \alpha T, \tag{4.3-2}$$

where ϕ_o is the classically reported work function and α is an experimentally measured constant. This dependence can be inserted into Eq. 4.3-1 to give

$$J = Ae^{-e\alpha/k} T^2 e^{-e\phi_{wf}/kT} = DT^2 e^{-e\phi_o/kT}, \tag{4.3-3}$$

where D is a material-specific modification to the Richardson-Dushman equation.

In the presence of strong electric fields at the surface of the cathode, the potential barrier that must be overcome by the electrons in the material's conduction band is reduced, which results effectively in a reduced work function. This effect was first analyzed by Schottky [23], and the so-called Schottky effect is included in the emission equation by the addition of a term [24] to describe the effect of the surface electric field on the emission current density:

$$J = DT^2 e^{-e\phi_o/kT} \exp\left[\left(\frac{e}{kT}\right)\sqrt{\frac{eE}{4\pi\varepsilon_o}}\right], \tag{4.3-4}$$

where E is the electric field at the emitting surface. The Schottky effect often becomes significant inside hollow cathodes where the plasma density is very high and the electric field in the sheath becomes significant.

Table 4-1 Work function and Richardson coefficients for some commonly used cathode materials.

	A	*D*	ϕ
BaO-W 411 [25]	120	—	$1.67 + 2.82\times10^{-4}\,T$
BaO-Scandate [29]	120	—	$8\times10^{-7}\,T^2 - 1.3\times10^{3}\,T + 1.96$
LaB_6 [33]	—	29	2.66
LaB_6 [34]	—	110	2.87
LaB_6 [35]	120	—	2.91
LaB_6 [22]	120	—	$2.66 + 1.23\times10^{-4}\,T$
Molybdenum [22]	—	55	4.2
Tantalum [22]		37	4.1
Tungsten [22]	—	70	4.55

The properties of the material selected for the thermionic emitter or insert determine the required operating temperature of the cathode for a given emission current. The work functions and values of *D* found in the literature for several common cathode materials are summarized in Table 4-1. The refractory metals are seen to have work functions in excess of 4 eV, and so they must operate at very high temperatures to achieve significant emission current density.

The so-called "oxide" cathodes have work functions of about 2 eV and so are capable of producing high emission current densities at temperatures under 1000 °C. Oxide layers, such as barium oxide, were first deposited on tungsten or nickel filaments to lower the work function and reduce the heater power required. However, these surface layers eventually evaporate and are easily sputtered by ion bombardment, limiting the life in vacuum applications to thousands of hours and in plasma discharges to tens of hours. This problem was mitigated by the development of dispenser cathodes, where a reservoir of the oxide material is fabricated into the tungsten substrate that continuously re-supplies the low work function surface layer. The most commonly used dispenser cathode in thrusters, the "Phillips Type-S," uses a porous tungsten matrix that is impregnated with an emissive mix of barium and calcium oxides and alumina. Different molar concentrations of the three constituents of the emissive mix are used depending on the required emission current density and life. The matrix material containing the impregnate can be directly heated by passing a current through the material, or configured as an insert placed inside a hollow cathode with a radiatively coupled heater.

In dispenser cathodes, chemical reactions in the pores of the matrix or at the surface at high temperatures reduce the emissive material and evolve a barium oxide dipole attached to an active site on the tungsten substrate. The work function of various molar concentrations of the impregnate has been measured and is summarized in Cronin [25]. The most commonly used dispenser cathode in electric thrusters has a 4 : 1 : 1 stoichiometric mix of barium calcium aluminate (4BaO: 1CaO: $1Al_2O_3$) as the impregnate, and its work function is found from Table 4-1 to be 2.06 eV at temperatures of about 1100 °C. However, the actual work function of the surface is strongly dependent on the surface coverage of the tungsten by the barium oxide dipoles. At higher temperatures, the evaporation rate of the barium on the surface will exceed the supply rate from the impregnate diffusing up through the pores, and the work function will increase. A detailed analysis by Longo [26, 27] based on extensive life modeling showed that the surface coverage θ of barium on tungsten is given by

$$\theta = \frac{E'\tau}{E'\tau + \sqrt{1+\gamma t}} = \frac{3.12 \times 10^{-8} e^{\zeta/T}}{\left(3.12 \times 10^{-8} e^{\zeta/T} + \sqrt{1+0.166t}\right)} \tag{4.3-5}$$

where $E' = E_o e^{-\zeta/T}$ is a fitting function related to the surface activation energy, τ is the surface desorption time constant, ζ is the activation energy for surface desorption, T is the surface temperature, γ is a constant with units kh^{-1}, and t is the operation time in kh. The numerical values in this equation were determined empirically from cathode life testing results. The surface coverage θ ranges from 0 to 1, where 1 corresponds to a monolayer of barium on the surface. The activation energy for barium desorption from a 4 : 1 : 1 dispenser cathode expressed as a temperature is about 26,500 K. As the cathode ages, the supply of barium at a given temperature decreases and the surface coverage decreases. This ultimately determines the life time of the cathode because as the barium impregnate is depleted and the surface coverage decreases, the work function rises. The cathode then has to be at a higher temperature to produce the same desired emission current, and the evaporation rate increases further. This causes runaway as barium is rapidly depleted from the pores and the cathode fails.

The work function of the barium-tungsten dispenser cathode as a function of the surface coverage was analyzed by Mueller [28], who found

$$\phi_{wf} = \phi_o - \frac{1.88 e\mu_o N^{3/2}}{1 + c\alpha N^{3/2}} = \phi_o - \frac{8.212\,\theta}{1 + 2.25\,\theta^{3/2}} \tag{4.3-6}$$

where ϕ_o is the work function of the pure tungsten surface ($\sim$ 4.55 eV), μ_o is the surface dipole, α is the effective polarizability of the dipole in Å^3, N is the surface coverage in 10^{15} atoms/cm^2, and c is a constant. The behavior of the barium oxide surface coverage and work function with temperature from the Longo/Mueller models for a 4 : 1 : 1 dispenser cathode is shown in Fig. 4-9. For comparison, the corresponding work function reported by Cronin [25] in Table 4-1 is also shown. The agreement at low temperatures is good, but at temperatures above 1200 °C the reduced surface

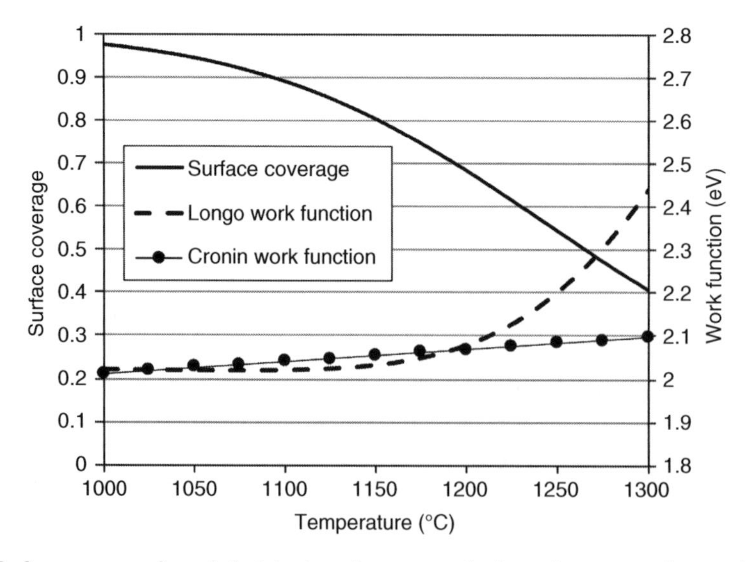

Figure 4-9 Surface coverage for a 4 : 1 : 1 barium dispenser cathode, and corresponding work function, as a function of the surface temperature.

coverage predicted by Longo increases the work function above that from the Cronin fit. This limits the maximum electron emission current density of barium oxide dispenser cathodes.

The electron emission current density, calculated using Eq. 4.3-3 for several different emitter materials given in Table 4-1, is shown in Fig. 4-10. The barium oxide on tungsten dispenser cathode provides emission current densities of 10 A/cm^2 at surface temperatures of about 1100 °C. The maximum current density for this cathode is seen to be 20 A/cm^2 at temperatures of 1200 °C, above which the barium surface coverage cannot be maintained by diffusion from the pores and the emission current density falls off.

The work function of dispenser cathodes can be further reduced by the introduction of small amounts of other refractory materials, such as iridium or osmium, in the tungsten matrix. These "mixed metal matrix" cathodes can have work functions below 1.9 eV, and they typically slow some of the chemical reactions that take place in the cathode. It was also found that the addition of scandium to the surface of the barium oxide dispenser cathode reduces the work function significantly. This is shown in Fig. 4-10 where the reported work function for a scandate cathode [29] at 1100 °C is about 1.7 eV, which is significantly less than the 2.06 eV for the BaO-W dispenser cathode. This results in much lower temperatures for a given emission current density. Although fabricating stable, long-life scandate electron emitters can be challenging, scandate-BaO-W dispenser cathodes have been developed [30] and successfully used in several different hollow cathodes in electric thrusters.

Because chemistry is involved in the formation of the low work function surface in dispenser cathodes, they are subject to poisoning that can significantly increase the work function [31, 32]. Some care must be taken in handling the inserts and in the vacuum conditions used during operation of these cathodes to avoid poisoning by impurities in the propellant that produce unreliable emission and shorten the lifetime. In addition, propellant impurities can also react with the tungsten insert at higher temperatures (above 1100 °C) causing migration and deposition of tungsten or tungstates (compounds of tungsten, barium, and oxygen) on the surface, which change the surface structure and porosity and can reduce the surface coverage of the low work function BaO layer. One of the major drawbacks of using BaO dispenser cathodes in electric propulsion applications is the extremely high propellant gas purity specified to avoid these poisoning and tungsten-material transport issues, which has resulted in a special "propulsion-grade" xenon with 99.9995% purity to be specified by most users of these cathodes for flight.

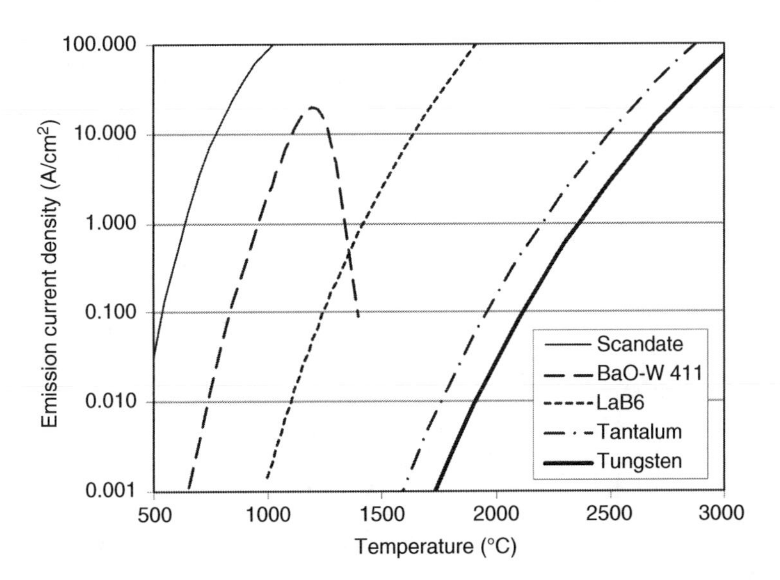

Figure 4-10 Emission current density versus emitter temperature for various cathode materials.

Another electron emitter material commonly used in electric thruster applications is lanthanum hexaboride (LaB_6) [33], which is a crystalline material first developed as an electron emitter by Lafferty in the 1950s. It is made for inserts by press sintering LaB_6 powder into rods or plates and then machining the solid material to the desired shape. Polycrystalline LaB_6 cathodes have a work function of about 2.67 eV [34, 35], depending on the surface stoichiometry [36], and will emit over 10 A/cm^2 at a temperature of 1650 °C, as shown in Fig. 4-10. Since the bulk material is emitting, there is no chemistry directly involved in establishing the low work function surface and LaB_6 cathodes are insensitive to impurities and air exposures that can destroy a BaO dispenser cathode [37]. LaB_6 cathodes can withstand gas-feed impurity levels two orders of magnitude higher than dispenser cathodes at the same emission current density. In addition, the cathode life is determined primarily by the low evaporation rate of the LaB_6 material at typical operating temperatures. The higher operating temperature of bulk LaB_6 and the need to support and make electrical contact with LaB_6 with materials that inhibit boron diffusion at the operating temperatures require some careful engineering of the cathode structure. However, LaB_6 cathodes are commonly used in Russian Hall thrusters in communications satellite applications [14, 38].

The first reported use of LaB_6 in a hollow cathode was by Goebel [8] in 1978, and the development of a high-current LaB_6 cathode for plasma sources that dealt with supporting, heating, and making electrical contact with the material was described by him in 1985 [39].The lanthanum-boron system can consist of combinations of stable LaB_4, LaB_6, and LaB_9 compounds, with the surface color determined [40] by the dominant compound. The evolution of LaB_4 to LaB_9 compounds is caused either by preferential sputtering of the boron or lanthanum atoms at the surface by energetic ion bombardment [39], or by preferential chemical reactions with the surface atoms [36]. However, a lanthanum-boride compound, when heated in excess of 1000 °C in reasonable vacuum, will evaporate its component atoms at rates that produces a stable $LaB_{6.0}$ surface [34]. Lanthanum hexaboride can be heated indirectly by internal or external radiative heaters [37, 39], or directly heated in a filament configuration [41].

Dispenser cathodes and LaB_6 cathodes offer long lifetimes in thruster applications because the evaporation rate is significantly lower than for refractory metals. Figure 4-11 shows the evaporation

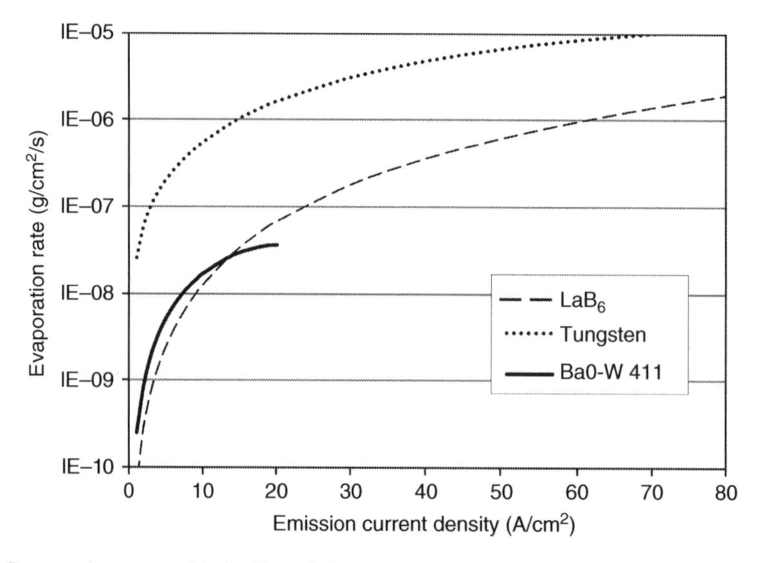

Figure 4-11 Evaporation rates of LaB_6, Type-S 4 : 1 : 1-dispenser cathodes, and tungsten.

rate as a function of the emission current density for a Type-S 4 : 1 : 1 dispenser cathode [32], LaB_6[33] and tungsten [22] (for comparison). The dispenser cathode and LaB_6 cathode evaporation rates are more than one order of magnitude lower compared to tungsten at the same emission current density. As discussed above, excessive evaporation of barium and reduced surface coverage limit the current density of dispenser cathodes to less than about 20 A/cm^2 in continuous operation. Despite operating at a significantly higher temperature than the barium cathode, the LaB_6 has a lower evaporation rate until the emission current exceeds about 15 A/cm^2 and can provide longer life. The life of these cathodes is discussed in more detail in Section 4.8.

4.4 Insert Region

The insert region of the hollow cathode, as was illustrated in Fig. 4-4, usually has a cylindrical geometry with electron emission from the interior surface of a thermionic insert material. A plasma discharge is established inside the insert region, and electrons emitted from the insert surface are accelerated through the cathode sheath that forms between the insert surface and the plasma. The insert-region plasma must have sufficient density to support the emitted electron current density from the insert and provide heating of the insert for the cathode to operate properly. The maximum electron current density into the insert-region plasma is then determined by either space charge limitations at the sheath edge or by characteristics of the surface (work function and temperature) that limit the thermionic emission. As shown in Section 3.7.5 by the double sheath analysis, ions flowing back from the plasma through the sheath to the cathode surface neutralize electron space charge and increase the extracted electron current density from the insert surface. The electrons accelerated through the sheath quickly give up their energy to the dense collisional plasma inside the insert. Electrons in the tail of the Maxwellian distribution in this plasma have sufficient energy to ionize some portion of the thruster propellant injected through the cathode, which is only a small fraction of the total propellant injected into the thruster. Plasma electrons incident on the downstream end of the cathode tube flow through the orifice and into the main discharge chamber.

The barium evaporated from dispenser cathode inserts is easily ionized in plasmas with this electron temperature because its ionization potential is only 5.2 eV. A calculation of the ionization mean free path in NSTAR-sized hollow cathodes [42] predicts about 4×10^{-5} m, which is much smaller than the interior dimensions of the cathode. The ionized barium then migrates upstream from the ionization location because the potential gradient in the hollow cathode that pulls electrons out of the cathode plasma also accelerates barium ions in the opposite direction (upstream). This means that the barium released from the insert does not leave the cathode during discharge operation, but tends to recycle in the plasma [43] and is deposited in the cooler upstream regions of the hollow cathode.

The pressure inside the hollow cathode is set primarily by the gas flow rate through the cathode and the orifice size, and must be sufficiently high to produce a collisional plasma. This condition is required to slow the ions that are backstreaming from the orifice region and from the peak plasma potential on axis (primarily by charge exchange) to avoid sputtering of the insert surface by high-energy ion bombardment. Although this condition may not necessarily be satisfied everywhere inside a Type C cathode (with no orifice), at least some fraction of the insert is protected by the collisional processes for proper cathode operation and life. The collisional plasma will also tend to have a low electron temperature, which reduces the sheath voltages and further protects the low work function insert surface from damage or modification by the plasma.

The plasma inside the insert region is inherently 2-D and has strong coupling between the neutral gas, plasma and sheath, which usually requires 2-D numerical simulations to solve for the density, potential and temperature profiles. However, important insights can be obtained by using simple particle and energy balance models and plasma diffusion models [44] to describe the plasma because the transport inside the hollow cathode is dominated by collisions and the ion motion is slow enough to neglect their inertia. In Chapter 3, the solution to the radial diffusion equation for ions in collisionally dominated plasmas in cylindrical geometry resulted in an eigenvalue equation with a unique dependence on the electron temperature:

$$\left(\frac{r}{\lambda_{01}}\right)^2 n_o \sigma_i(T_e) \sqrt{\frac{8kT_e}{\pi m}} - D = 0 \tag{4.4-1}$$

where r is the internal radius of the insert, $\lambda_{01} = 2.4048$ is the first zero of the zero-order Bessel function, n_o is the neutral density, σ_i is the ionization cross section averaged over a Maxwellian electron temperature and D is the diffusion coefficient. This means that the electron temperature is constrained to produce sufficient ions to offset the diffusion losses to the wall.

If we assume the diffusion in the radial direction in the insert region is ambipolar, and the ion mobility is limited by resonant charge exchange (CEX) with the xenon neutral atoms, the average collision frequency ν_i for the ions is then

$$\nu_i = \sigma_{\text{CEX}} n_o \upsilon_{\text{scat}} \tag{4.4-2}$$

where the effective velocity υ_{scat} for scattering of the ions in the insert region is approximated by the ion thermal speed:

$$\upsilon_{\text{scat}} = \sqrt{\frac{kT_i}{M}} \tag{4.4-3}$$

and T_i is the ion temperature in Kelvin. Since the electron mobility is much higher than the ion mobility, the ambipolar diffusion coefficient D_a from Eq. 3.6-58 for this case is then

$$D_a = D_i\left(1 + \frac{T_e}{T_i}\right) = \frac{e}{M}\frac{(T_{\text{iV}} + T_{\text{eV}})}{\sigma_{\text{CEX}}\, n_o\, \upsilon_{\text{scat}}} \tag{4.4-4}$$

where D_i is the ion diffusion coefficient derived in Section 3.6.2 and the ion and electron temperatures are given in units of eV. As an example, consider two hollow cathodes operating with xenon and having different insert inner diameters. The neutral density inside the hollow cathode is described by Eq. 2.7-1 for a given pressure P inside the cathode, which is determined by the gas flow rate and the orifice size. A simple analytical technique to estimate the neutral pressure in the insert region is given in Appendix B. Typical pressures inside discharge hollow cathodes usually range from 1 to 15 Torr, although higher pressures are often found in neutralizer cathodes. Figure 4-12 shows the electron temperature versus internal pressure found from Eq. 4.4-1 for two insert diameters assuming a charge exchange cross section of 10^{-18} m^2 [45] for low temperature xenon ions and neutrals inside the hollow cathode and a neutral gas temperature of 2500 K. The smaller NSTAR insert diameter requires a higher electron temperature to offset the higher diffusion losses to the closer wall at a given pressure. During operation at the high power TH15 throttle point at 13.1 A and 3.7 sccm, the internal pressure was measured to be about 7.5 Torr and the predicted electron temperature was then about 1.36 eV. This agrees well with probe data taken in the insert region [46] in this mode. The larger NEXIS cathode nominal discharge conditions of 25 A and 5.5 sccm

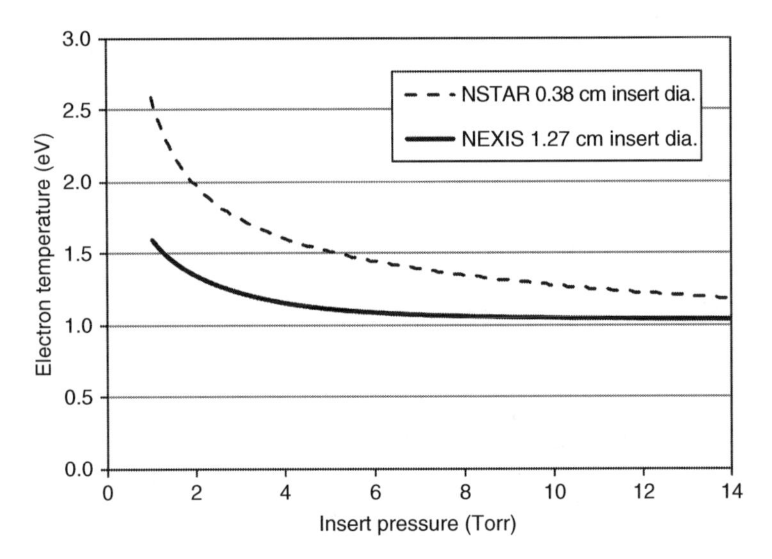

Figure 4-12 Electron temperature in the insert region of two hollow cathodes with different inner diameters as a function of internal pressure.

produce an internal pressure of 1.8 Torr, which results in a predicted electron temperature of about 1.4 eV that is also in good agreement with the measurements [46].

The radially averaged ion density in the hollow cathode is related to the ion density on the cathode centerline by

$$\overline{n} = \frac{\int_0^r n(0) J_o(\lambda_{01} r'/r) 2\pi r' dr'}{\pi r^2} = n(0) \left[\frac{2J_1(\lambda_{01})}{\lambda_{01}} \right] \tag{4.4-5}$$

where J_o and J_1 are the zero and first Bessel functions. The ion flux going radially to the wall is

$$\Gamma_i = \overline{n} v_r = -D_a \frac{\partial n}{\partial r} = n(0) D_a \frac{\lambda_{01}}{r} J_1(\lambda_{01}) = \overline{n} D_a \frac{\lambda_{01}^2}{2r}. \tag{4.4-6}$$

Using the ambipolar diffusion coefficient from Eq. 4.4-4, the effective radial drift velocity at the wall is then

$$v_r = \frac{(2.4)^2}{2\, r\, \sigma_{CEX} n_o\, v_{scat}} \frac{e}{M} (T_{iV} + T_{eV}) = \frac{2.9\, e(T_{iV} + T_{eV})}{r\, M\, \sigma_{CEX} n_o\, v_{scat}}. \tag{4.4-7}$$

In the example above, the larger diameter insert produces an electron temperature of about 1.4 eV at 1.8 Torr internal xenon pressure. The effective ion velocity found near the wall is slowed by ion-neutral CEX collisions to 3.1 m/s, which is significantly less than the 500 m/s ion thermal velocity and 1200 m/s xenon ion acoustic velocity. Since the pre-sheath potential that accelerates the ions to the Bohm velocity extends only the order of the collision mean free path into the plasma, ions diffusing to the plasma edge are accelerated to the Bohm velocity very close to the sheath due to the high collisionality in the insert-region plasma.

If the cathode has a Type B or C geometry, the density of the plasma in the insert region can be estimated to within factors of about two of the true value by a simple 0-D particle and energy balance model that assumes the plasma is uniform. In the insert region, heating of the plasma from thermionic electrons that are accelerated through the cathode sheath and the resistive heating in the plasma (terms on the left) is balanced by the power loss (terms on the right):

$$I_t \phi_s + RI_e^2 = I_i U^+ + \frac{5}{2} T_{eV} I_e + (2T_{eV} + \phi_s) I_r e^{-\phi_s/T_{eV}}, \tag{4.4-8}$$

where I_t is the thermionically emitted electron current, ϕ_s is the cathode sheath voltage, R is the insert-plasma resistance, I_e is the hollow cathode discharge current, I_i is the total ion current generated in the insert region, U^+ is the ionization potential, T_{eV} is the electron temperature (in volts), and I_r is the random electron flux at the sheath edge. The power loss from the insert-region plasma (the right-hand terms in Eq. 4.4-8) is the power it takes to make ions $I_i U^+$, the power convected out of the plasma through the orifice $\frac{5}{2} T_{eV} I_e$, and the electron power loss due to tail electrons overcoming the sheath to the wall. Ion excitation and radiation losses seen in the discharge chamber energy balance equations in Chapter 5 are ignored because the high density plasma inside the hollow cathode is optically "thick" and the radiated energy is reabsorbed by the plasma. The resistance R is the resistivity times the average conduction length l divided by the cross sectional area of the insert-region plasma:

$$R = \eta \frac{l}{\pi r^2}, \tag{4.4-9}$$

with the resistivity of the plasma given by

$$\eta = \frac{1}{\varepsilon_o \tau_e \omega_p^2}. \tag{4.4-10}$$

The collision time τ_e for electrons, accounting for both electron-ion and electron-neutral collisions, is given by

$$\tau_e = \frac{1}{\nu_{ei} + \nu_{en}}, \tag{4.4-11}$$

where ν_{ei} is the electron-ion collision frequency given in Eq. 3.6-14 and ν_{en} is the electron-neutral collision frequency given in Eq. 3.6-12.

At the insert, power balance gives:

$$H(T) + I_t \phi_{wf} = I_i \left(U^+ + \phi_s + \frac{T_{eV}}{2} - \phi_{wf} \right) + (2T_{eV} + \phi_{wf}) I_r e^{-\phi_s/T_{eV}}, \tag{4.4-12}$$

where *H(T)* is the total power lost from the insert by radiation and heat conduction, and ϕ_{wf} is again the cathode work function. Particle conservation in the discharge dictates that the discharge current is the sum of the thermionically emitted electron current, the ion current and the return electron current from the plasma:

$$I_e = I_t + I_i - I_r e^{-\phi_s/T_{eV}}. \tag{4.4-13}$$

The random electron current I_r within a collision length of the sheath edge is given by

$$I_r = \frac{1}{4} \left(\frac{8kT_e}{\pi m} \right)^{1/2} n_e \, eA, \tag{4.4-14}$$

where the plasma density n_e is evaluated at the sheath edge and A is the total cathode surface area over which the current is to be determined. The ion current is given by the Bohm current (Eq. 3.7-29), where the ion density is again evaluated within one collision length of the sheath edge.

Eqs. 4.4-12, 4.4-13 and 4.4-14 can be combined to eliminate the ion current term, which gives

$$\frac{RI_e^2 + I_e\left(\phi_s + \frac{5}{2}T_{\mathrm{eV}}\right)}{H(T) + I_e\phi_{\mathrm{wf}}} = \frac{U^+ + \phi_s + 2T_{\mathrm{eV}}\left(\frac{2M}{\pi m}\right)^{1/2} e^{-\phi_s/T_{\mathrm{eV}}}}{U^+ + \phi_s + \frac{T_{\mathrm{eV}}}{2} + 2T_{\mathrm{eV}}\left(\frac{2M}{\pi m}\right)^{1/2} e^{-\phi_s/T_{\mathrm{eV}}}}. \tag{4.4-15}$$

Since the electron temperature is given by the solution to Eq. 4.4-1 in the insert region (as shown above), Eq. 4.4-15 can be solved for the cathode sheath voltage as a function of the discharge current if the radiation and conduction heat losses are known. The insert heat losses are found from thermal models of the cathode, which will be discussed in Section 4.7. Eq. 4.4-15 can be greatly simplified by realizing that in most cases $T_{\mathrm{eV}}/2 << (U^+ + \phi_s)$, and the right-hand side is essentially equal to one. Eq. 4.4-15 then reduces to a simple power balance equation, and the cathode sheath voltage is

$$\phi_s = \frac{H(T)}{I_e} - \frac{5}{2}T_{\mathrm{eV}} + \phi_{\mathrm{wf}} - I_e R. \tag{4.4-16}$$

Figure 4-13 shows the calculated sheath voltage from Eq. 4.4-16 for the NSTAR cathode at a fixed 3.7 sccm xenon flow rate as a function of the discharge current for four values of the combined radiated and conducted power loss. The electron temperature is taken to be 1.4 eV from Fig. 4-12 for the 7.8 Torr measured at 13 A of discharge current and 3.7 sccm flow. A thermal model of this cathode [47] to be described in Section 4.7 predicts that the total self-heating power available is about 40 W at 12 A of discharge current, resulting in a sheath voltage from the figure of only about 2 V. In this case, over 20% of the random electron flux in the plasma can overcome the sheath voltage and be collected on the insert to provide heating. The balance of the power required to heat the insert in the NSTAR cathode comes from orifice plate heating [47], which will be discussed in Section 4.5.

In low-pressure Type B and C cathodes, the sheath potentials are much greater than the 3.6 V calculated for the NSTAR discharge cathode. For example, in Fig. 4-12, the NSTAR solution for the electron temperature is at the far right of the graph in excess of 7 Torr, whereas NEXIS and other large orifice cathodes are closer to the left side of the graph between 1 and 2 Torr. The sheath potential found by solving Eq. 4.4-16 for the NEXIS electron temperature is over 7 V, and so relatively few plasma electrons return to the emitter and they do little heating. Most of the insert heating in lower pressure (on the order of 1–2 Torr), lower internal plasma density cathodes is from ion bombardment of the insert surface due to the higher sheath voltage.

The insert-region plasma density can now be found from Eq. 4.4-8. The ion current term is given by

$$I_i = n_o \overline{n} e \langle \sigma_i v_e \rangle V, \tag{4.4-17}$$

where n_o is the neutral density, $<\sigma_i v_e>$ is the ionization reaction rate coefficient, V is the volume, and $\overline{n}$ is the average plasma density over the insert volume. Remembering that the plasma density in the random electron flux equation is evaluated at the plasma edge, Eq. 4.4-8 can be solved using the above equations to produce an expression for the average plasma density:

$$\overline{n} = \frac{RI_e^2 - \left(\frac{5}{2}T_{\mathrm{eV}} - \phi_s\right)I_e}{\left[f_n 2T_e\left(\frac{kT_e}{2\pi m}\right)^{1/2} eAe^{-\phi_s/T_{\mathrm{eV}}} + n_o e\langle\sigma_i v_e\rangle V(U^+ + \phi_s)\right]}, \tag{4.4-18}$$

where f_n is the edge to average plasma density ratio. Since the electrons in the insert-region plasma are Maxwellian, the value of f_n can be estimated from the potential difference between the center and the edge

$$f_n = \frac{n}{\bar{n}} \approx e^{-(\phi_{\text{axis}} - \phi_s)/T_{\text{eV}}}, \tag{4.4-19}$$

where the potential on axis ϕ_{axis} must be provided by measurements or 2-D numerical simulations. The plasma density calculated for the NSTAR discharge cathode at a constant xenon gas flow of 3.7 sccm from Eq. 4.4-18, using the electron temperature from the radial diffusion model (Fig. 4-12), the sheath potential from the power balance model (Fig. 4-13), and a measured on-axis plasma potential of about 8.5 V [16], is shown in Fig. 4-14. Good agreement with the plasma

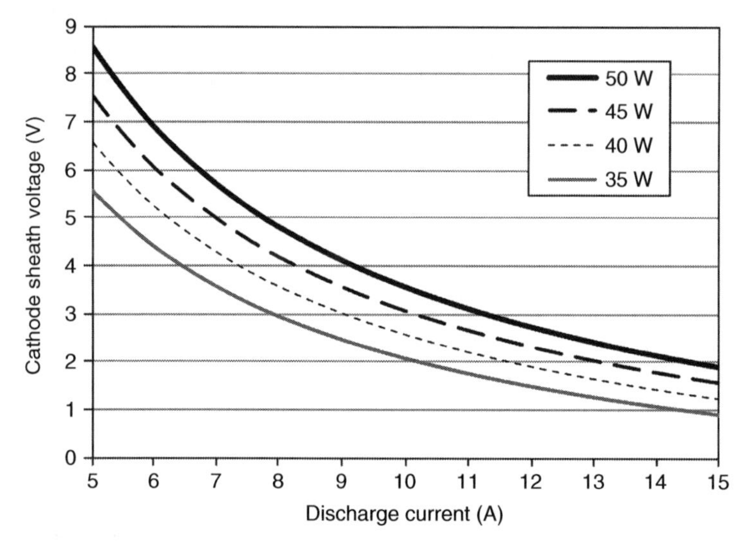

Figure 4-13 Insert sheath voltage versus discharge current for the NSTAR cathode for four values of the radiated and conducted heat loss.

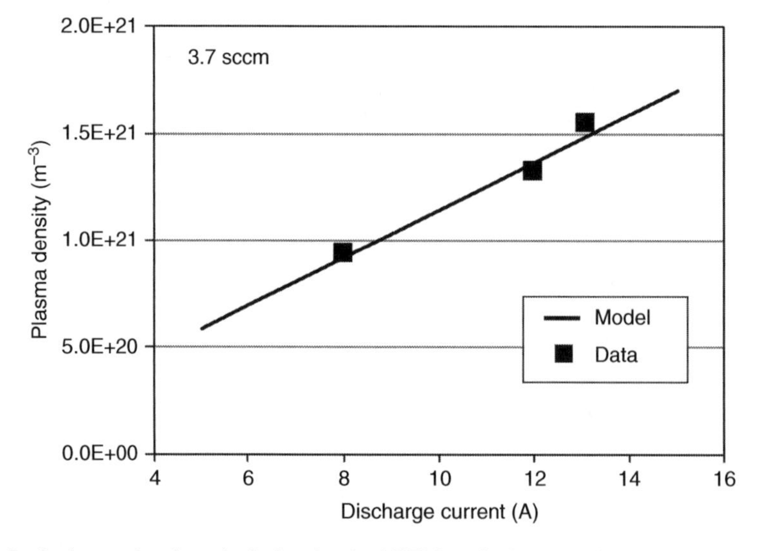

Figure 4-14 Peak plasma density calculation for the NSTAR cathode operating at constant gas flow of 3.7 sccm.

density measurements made by a miniature scanning probe in this cathode [48] is obtained, and a nearly linear dependence on discharge current is predicted by the model and shown experimentally. While the idealized 0-D cathode model described above requires insert heat loss from a cathode thermal model and on-axis potentials from probe measurements or 2-D simulations, it provides reasonable agreement with the data and illustrates the dependence of the insert-region plasma density and temperature, on the geometry and the plasma conditions inside the cathode.

The 0-D model also illuminates the heating mechanism in the hollow cathode. The ion heating power to the insert is found in Eq. 4.4-12:

$$\text{Power}_{\text{ions}} = I_i\left(U^+ + \phi_s + \frac{T_{\text{eV}}}{2} - \phi_{\text{wf}}\right), \tag{4.4-20}$$

where the ion current is given by the Bohm current at the sheath edge. Using the above parameters for the Type-B NSTAR discharge cathode shown in Fig. 4-5 at the full power TH15 operating point of 13 A and 3.7 sccm ($U^+ = 12.1$ eV, $\phi_s = 3.6$ eV, $T_e = 1.36$ eV, $\phi_{\text{wf}} = 2.06$ V, $\phi_{\text{axis}} = 8.5$ V and an ion density $n_i \approx 1.5 \times 10^{21} \text{m}^{-3}$), the ion heating power from Eq. 4.4-20 is only 4.7 W. The electron heating power to the insert is also found in Eq. 4.4-12:

$$\text{Power}_{\text{electrons}} = (2T_{\text{eV}} + \phi_{\text{wf}})I_r e^{-\phi_s/T_{\text{eV}}}, \tag{4.4-21}$$

where the random electron current is again evaluated at the sheath edge. For the same example parameters for Type B NSTAR cathode given above, the electron heating of the insert is found to be about 45 W. This particular Type B cathode is, therefore, heated predominately by electron heating of the insert, with a comparable amount coming from orifice heating (shown in Section 4.7). Similar analysis of Type B cathodes with larger orifices or lower flow rates, and also most Type C cathodes, indicates that ion heating will become the dominant heating mechanism due to the higher electron temperature and larger sheath potential drop at the insert. More extensive investigations of the power distributions and interior power characteristics of hollow cathodes can be found in the literature [17, 18, 49].

It is important to recognize that as the pressure in the hollow cathode is increased, much of the plasma heating comes from resistive heating of the current flowing through the partially ionized plasma. The higher the neutral gas background pressure, the greater the contribution of resistive heating. In cathodes with larger orifices that produce lower internal pressures, most of the heating of the insert-region plasma comes from the emitted electrons that are accelerated across the cathode sheath potential. In lower pressure cathodes, the sheath potential is higher and the plasma resistivity is lower, resulting in less Joule heating of the plasma but more ion bombardment heating of the insert surface. This is illustrated in Fig. 4-15, which shows the sheath potential and the ion and electron currents impacting the cathode as a function of the resistive Joule heating of the plasma.

The behavior shown in Fig. 4-15 can be understood by rearranging the equations in the power balance model above. Using Eqs. 3.7-29, 4.4-13 and 4.4-14 in the power balance equation (Eq. 4.4-8) and solving for the sheath potential gives

$$\phi_s = \frac{-RI_e^2 + I_i U^+ + \frac{5}{2}T_{\text{eV}}I_e + (2T_{\text{eV}} + \phi_s)I_i\sqrt{\frac{2M}{\pi m}}e^{-\phi_s/T_{\text{eV}}},}{\left(I_e - I_i\left(1 - \sqrt{\frac{2M}{\pi m}}e^{-\phi_s/T_{\text{eV}}},\right)\right)}. \tag{4.4-22}$$

The decrease in the sheath potential observed in Fig. 4-15 as the Joule heating (RI_e^2) becomes more significant follows directly from Eq. 4.4-22, because the Joule heating term enters with a negative sign. Equation 4.4-13 also shows that a decrease in the sheath potential allows for more of the electron flux to return to the emitter. Finally, if the heat loss H(*T*) is fixed, Eq. 4.4-12 shows that the increased return electron current (second term on the right-hand side) must be balanced by a

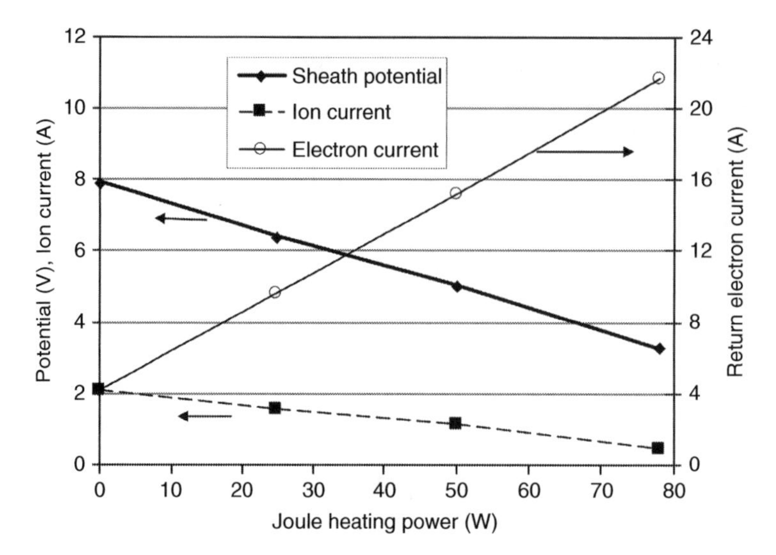

Figure 4-15 Sheath potential and currents to the insert as a function of the resistive Joule heating in the insert-region plasma.

reduced ion flux (first term on the right-hand side). This illustrates how the design and operating conditions of the hollow cathode (sizes, flow, and discharge current) determine which terms dominate in the cathode self-heating.

It is also possible to estimate the axial extent of the plasma in the insert region for Type A and some Type B cathodes with small orifices that again produce diffusion dominated plasmas. This is useful in understanding the plasma "attachment" or "contact length" with the insert, which impacts where the electron emission can take place. As was shown in Chapter 3, the solution to the 2-D diffusion equation in cylindrical geometry in Eq. 3.6-49 is the product of a zero order Bessel function radially times an exponential term in the axial direction:

$$n(r,z) = n(0,0)J_o\left(\sqrt{C^2+\alpha^2}r\right)e^{-\alpha z}, \tag{4.4-23}$$

where z is the axial distance and α is one over the e-folding distance of the plasma density from the reference location on axis at (0,0).This length can be found by considering the ion generation inside the insert. The ion current to the insert surface is the ion generation rate integrated over the volume inside the insert region:

$$I_i = 2\pi\int_0^r\int_0^L n_o\, n\, e\, \sigma_i v_e r dr dz. \tag{4.4-24}$$

Taking the axial integral in z in Eq. 4.4-24 to be approximately the e-folding distance ($L = 1/\alpha$), Eq. 4.4-24 is simply

$$I_i = \frac{\pi r^2}{\alpha} n_o \overline{n}\, e \,\langle \sigma_i v_e \rangle. \tag{4.4-25}$$

The average plasma density is found from Eq. 4.4-5:

$$\overline{n} = n(0,0)\left[\frac{2J_1(\lambda_{01})}{\lambda_{01}}\right] = n(0,0)\left[\frac{(2)(0.52)}{2.4}\right] = 0.43n(0,0). \tag{4.4-26}$$

Using Eq. 4.4-26 in Eq. 4.4-25 and solving for the value of α gives

$$\alpha = \frac{0.43\pi r^2}{I_i} n_o n(0,0) e \langle \sigma_i v_e \rangle. \tag{4.4-27}$$

For example, the axial plasma density profile from the scanning probe along the centerline inside the NSTAR hollow cathode [46] operating at 15 A and 3.7 sccm is shown in Fig. 4-16. Taking the peak plasma density from the figure of $n(0,0) = 1.6 \times 10^{21}$ m^{-3} as the reference density at position (0,0) and using the neutral density calculated inside the insert from Eq. 2.7-2 of 2.5×10^{22} m^{-3}, Eq. 4.4-27 gives a value of $\alpha = 6.0$ (corresponding to an e-folding length of $1/\alpha = 1.7$ mm) if the ion current to the insert is 0.5 A. The fit to the exponential decrease in the plasma density upstream of the orifice shown in Fig. 4-16 gives $\alpha = 6.1$. The assumed value of 0.5 A for the ion current actually results from a 2-D numerical model of the insert-region plasma that will be described later in this section [12]. This simple diffusion model shows an exponential behavior in the axial plasma density profile predicted from Eq. 4.4-23, which is consistent with the near-exponential profiles measured in the NSTAR cathode sufficiently far away from the orifice region.

A closer examination of Eq. 4.4-27 shows that the terms on the right-hand side represent the ionization rate per unit volume. If the geometry of the cathode is fixed, then the number of ions flowing to the insert (I_i in the denominator) is proportional to this ionization rate per unit volume. Therefore, the value of alpha will be constant for varying operating conditions of a given sized cathode. This behavior is illustrated in Fig. 4-17, where the density profile for an NSTAR-size cathode with two different orifice sizes operating at the same discharge current and gas flow is shown. For the larger orifice cathode, the internal pressure at the constant gas flow is lower, and the penetration of the 2-D effects associated with downstream boundary condition and the electron current funneling into the orifice extends deeper into the insert region. In this case, the plasma is still collisional enough to neglect the ion inertia terms in the momentum equation, but other properties of the plasma, like the diffusion coefficient, electron temperature, and plasma density are no longer radially uniform. Therefore, a 2-D model that includes conservation equations for the neutral gas, electron energy, Ohm's Law, and appropriate sheath boundary conditions is required to describe the

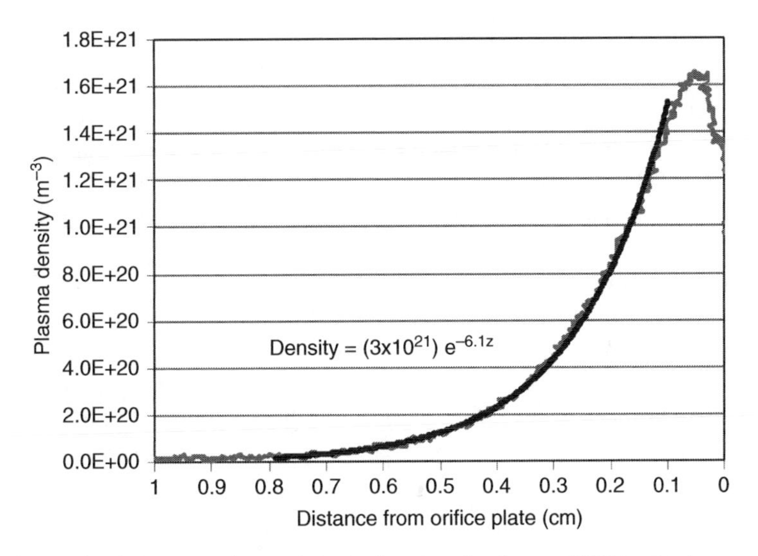

Figure 4-16 Plasma density measured on-axis in the insert region for an NSTAR cathode operating at TH15.

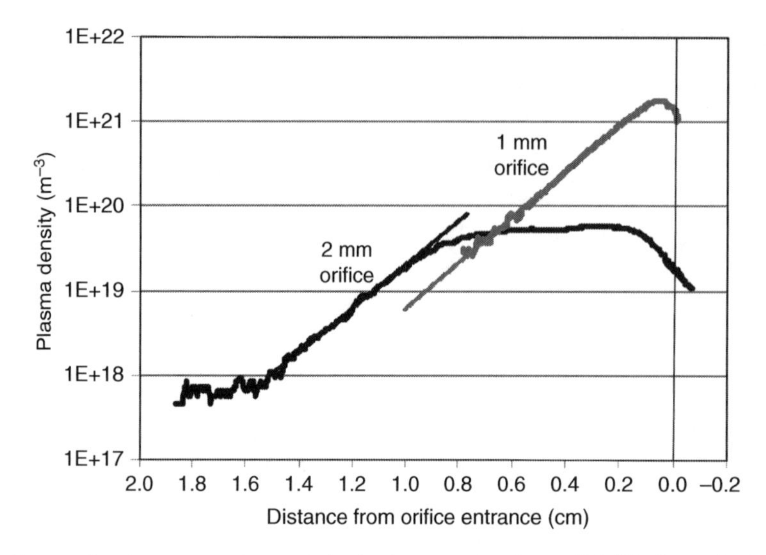

Figure 4-17 Plasma density measured on-axis in the insert region for an NSTAR-size cathode operating at TH15 with two different orifice sizes.

behavior in these cathode geometries. However, deep enough into the insert region from the orifice, where 2-D effects have diminished and diffusion-limited plasma flow to the insert dominates, then Eq. 4.4-23 is again valid and the value of alpha in the cathode is seen to be essentially constant.

It should be emphasized that the e-folding distance for the plasma density measured inside the NSTAR cathode in Fig. 4-16 is $1/\alpha = 1.7$ mm. The plasma is only in significant contact with the insert for a few e-foldings, which in this small orifice case is less than 1 cm. This rapid plasma density decrease away from the orifice is the result of the very high pressure in the NSTAR cathode [12, 15] and also occurs in most neutralizer cathodes. For high pressure cathodes like this, utilizing inserts significantly longer than 1–1.5 cm in length are not very useful because there is little plasma left beyond this distance to accept the thermionic emission from the insert.

Although the 0-D and 1-D models described above can provide some basic insight into the operation of hollow cathodes, 2-D models that solve the extensive system of governing equations are the only means to capture the spatial details of the plasma density in the insert region and near the cathode orifice. The first 2-D model that solved such an extensive set of equations with appropriate sheath boundary conditions is the 2-D Orifice Cathode (OrCa2D) code developed by Mikellides and Katz in the mid-2000s. Early versions of this code included the insert but excluded the cathode orifice [12, 50], which meant that boundary conditions at the entrance to the orifice had to be prescribed. This was no longer necessary in later versions of the code that included the orifice and keeper regions, and a sufficiently large plume region to allow simulation of different anode configurations.

The insert-region plasma energy balance in OrCa2D was derived in [12] based on the individual electron and ion energy conservation equations. These equations were also described in Section 3.5.3 of Chapter 3. Neglecting the energy terms associated with the thermal non-equilibrium between electrons and the heavy species, the steady-state electron energy equation can be written:

$$0 = -\nabla \cdot \left(-\frac{5}{2}\mathbf{J}_e \frac{kT_e}{e} - \kappa_e \frac{\nabla(kT_e)}{e} \right) + \eta J_e^2 - \mathbf{J}_e \cdot \frac{\nabla(nkT_e)}{ne} - \dot{n}eU^+ \tag{4.4-28}$$

where $\mathbf{J}_e$ is the electron current density in the plasma, κ_e is the electron thermal conductivity given by Eq. 3.5-30, η is the plasma resistivity given by Eq. 3.6-21, and U^+ is the ionization potential of the neutral gas. The steady-state ion energy equation assuming singly-charged ions only is

$$0 = -\nabla \cdot \left(-\frac{5}{2}\mathbf{J}_i\frac{kT_i}{e} - \kappa_n\frac{\nabla kT_i}{e}\right) + \mathbf{v}_i \cdot \nabla(nkT_i) + nM\nu_{\text{in}}v_i^2 + Q_T, \tag{4.4-29}$$

where $\mathbf{J}_i$ is the ion current density, κ_n is the thermal conductivity for neutrals, and it is assumed that the ions and neutrals are in thermal equilibrium ($T_n = T_i$) in the collisional insert-region plasma.

The energy balance equations are used to close the system of equations describing the plasma in the insert region. These equations also are used to describe the self-heating mechanism characteristic of hollow cathodes due to the particle flux and energy hitting the cathode walls. This effect will be discussed in Sections 4.5 and 4.7 with respect to the cathode thermal models.

Writing the steady-state momentum equations from Eq. 3.5-5 for the ions and electrons (neglecting the inertia terms):

$$0 = en\mathbf{E} - \nabla \cdot \mathbf{p}_i - Mn[\nu_{\text{ie}}(\mathbf{v}_i - \mathbf{v}_e) + \nu_{\text{in}}(\mathbf{v}_i - \mathbf{v}_n)], \tag{4.4-30}$$

$$0 = -en\mathbf{E} - \nabla \cdot \mathbf{p}_e - mn[\nu_{\text{ei}}(\mathbf{v}_e - \mathbf{v}_i) + \nu_{\text{en}}(\mathbf{v}_e - \mathbf{v}_n)]. \tag{4.4-31}$$

where $\mathbf{v}_s$ is the drift velocity of the species "s". Adding these two equations, assuming that the neutrals move slowly compared to the charged particles, and writing the result in terms of the ion and electron current densities gives

$$\mathbf{J}_i = \frac{m}{M}\frac{\nu_{\text{en}}}{\nu_{\text{in}}(1+\nu)}\mathbf{J}_e - \frac{\nabla(nkT_i + nkT_e)}{M\nu_{\text{in}}(1+\nu)}, \tag{4.4-32}$$

where $\nu = \nu_{\text{ie}}/\nu_{\text{in}}$ and ν_{in} is the collision frequency between ions and neutrals.

Subtracting the electron continuity equation from the ion continuity equation yields

$$\nabla \cdot (\mathbf{J}_e + \mathbf{J}_i) = 0 \tag{4.4-33}$$

which, when combined with the electron momentum Eq. 4.4-31 gives a particle balance equation

$$\nabla \cdot \left(\frac{\nabla\phi}{\eta}\right) = \nabla \cdot \left[\frac{\nabla(nkT_e)}{\eta ne} + \mathbf{J}_i\left(1 - \frac{\nu_{\text{ei}}}{\nu_{\text{en}} + \nu_{\text{ei}}}\right)\right], \tag{4.4-34}$$

that can be solved for the plasma potential ϕ because the electric field $\mathbf{E} = -\nabla\phi$. The total classical resistivity in the plasma in Eq. 4.4-34 is given by combining Eqs. 4.4-10 and 4.4-11:

$$\eta = \frac{m(\nu_{\text{en}} + \nu_{\text{ei}})}{ne^2}. \tag{4.4-35}$$

This system of equations was solved numerically in OrCa2D [12] to determine the plasma density, temperature, and potential in the insert region for the NSTAR discharge cathode operating conditions of 12 A and 4.25 sccm. Utilizing thermionic emission from the insert surface described by Eq. 4.3-4 with temperatures measured by Polk [51], and applying the proper boundary conditions at the wall and the entrance to the orifice, the plasma density profile along the axis of symmetry is compared with the laboratory measurement in Fig. 4-18. The 12 A net cathode current was found to result from almost 32 A electron emission by the insert countered by 20 A of plasma (thermal) electron current back to the insert and the orifice plate. The particle balance in the insert is shown in Table 4-2, where only about 0.5 A of the net cathode current is due to ionization of the xenon gas, which was used in the previous analysis to obtain the exponential density scale lengths.

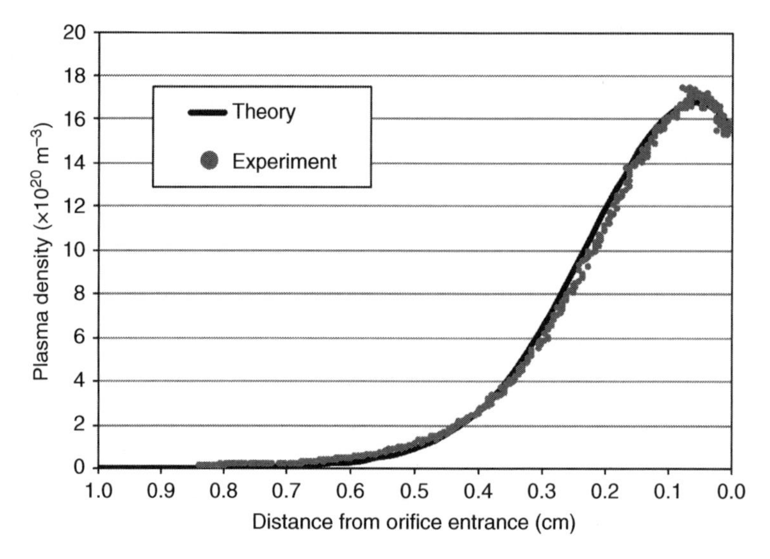

Figure 4-18 Comparison of the plasma density measured on-axis in the NSTAR-size cathode operating at 12 A and 4.25 sccm xenon flow, with predictions from OrCa2D [12] simulations.

Table 4-2 Currents from numerical simulations of the insert-region plasma in the NSTAR cathode.

Source	Current (A)
Emitted electrons	31.7
Absorbed electrons	20.2
Absorbed ions	0.5
Net current	12

OrCa2D describes accurately what is happening in the cathode insert region. For example, the numerical results in Table 4-2 capture the 2-D effects upstream of the cathode orifice as shown in Fig. 4-19, predicting a density profile that is consistent with the data [17]. The code's predictions of the electron temperature and plasma potential are also close to the measured values in the emission zone, which extends less than about 0.5 cm upstream of the orifice entrance in the NSTAR cathode. The plasma density contours plotted in Fig. 4-19a show that the density falls radially as expected toward the insert wall. The 2-D plasma potential contours for this case are also shown in Fig. 4-19b. Good agreement with the measurements has been achieved by the code for larger cathodes as well, such as with the 1.5-cm diameter NEXIS cathode [11, 12], and at different operating conditions [15, 17, 52–56] spanning all three types of cathodes in Fig. 4-5.

The 2-D numerical simulations have revealed critical insights about the main drivers behind the internal plasma density profiles. The final state of the plasma inside the cathode depends primarily on two competing characteristic penetration depths: L_n, which is associated with the penetration of the neutral gas from the cathode inlet into the ionization region and, L_j, the penetration depth of the electron current density from the orifice to the cathode interior. The former can be simply estimated as $L_n \simeq v_n / n_e \langle \sigma v_e \rangle_{iz}$ where v_n and $n_e \langle \sigma v_e \rangle_{iz}$ are the velocity and effective ionization collision

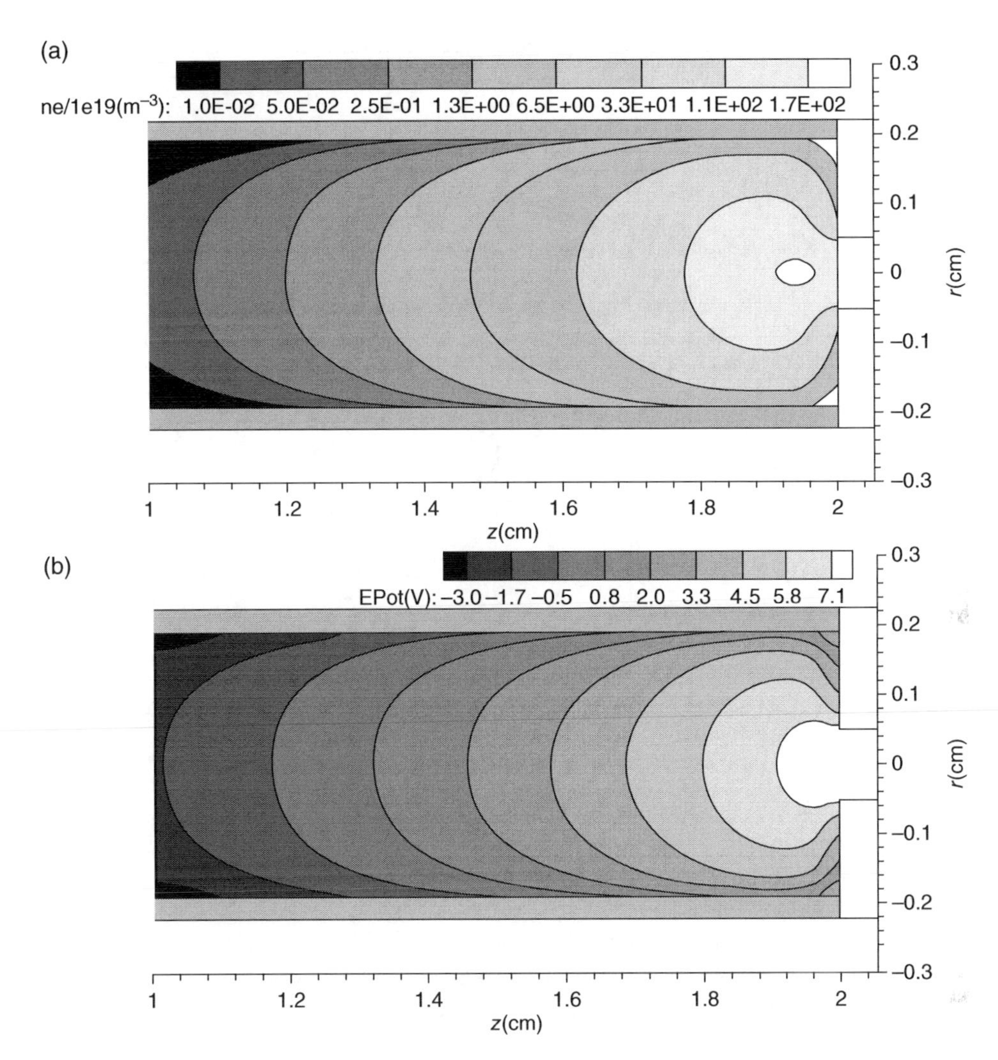

Figure 4-19 Density (a) and plasma potential (b) contours plotted for the NSTAR cathode from the OrCa2D code (*Source:* [12]).

frequency of the neutrals. Hence, any process that increases the velocity of the neutrals, such as increasing the flow rate in a given cathode geometry or decreasing the emitter I.D. at a given flow rate, will also increase the penetration of neutrals into the ionization region, thereby reducing that region's characteristic scale size. This will lead also to higher internal pressure and a steeper plasma density gradient upstream of the peak plasma density value. An example of this trend is shown in Fig. 4-20 in a cathode where only the mass flow rate was changed for two different insert diameters.

One can show from the electron and neutral continuity equations that $L_j \sim I_d/D_E J_E$ where D_E is the emitter I.D. and I_d is the discharge current. The quantity J_E, represents the *net* current density of electrons the emitter must provide to support I_d. The value of J_E is determined by the difference between the flux of electrons that are thermionically emitted and those that are generated in the plasma and have sufficient energy to overcome the sheath potential and return back to the emitter. As we discussed in Section 4.3, the flux of emitted electrons depends on the temperature of the

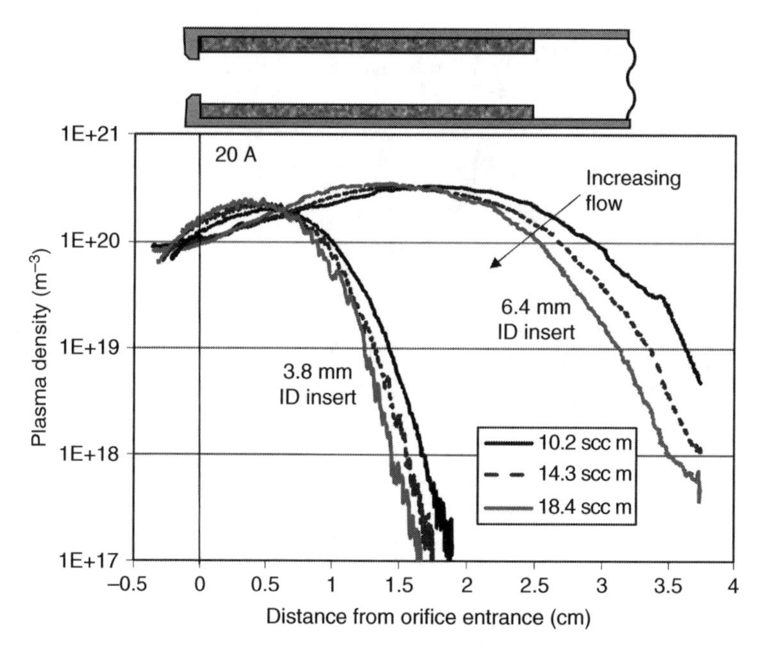

Figure 4-20 Plasma density measurements along the centerline of a LaB_6 Type B hollow cathode (top) operating at 20 A and three different xenon mass flow rates and two different insert diameters.

emitter which is driven by the plasma heat loads generated during operation and the properties of the cathode's thermal system, as described in Section 4.7. The flux of return electrons depends on the thermal flux of electrons toward the insert and the potential difference across the sheath that forms along the emitter.

It is through these two main processes that the cathode regulates the electron supply to the discharge. Therefore, though L_j appears to have a deceivingly simple dependence on I_d and J_E, it is in fact much more difficult to discern than L_n. In addition to the complexities associated with the thermal balance and plasma loads that ultimately determine the temperature at which the emitter will operate, the sheath drop at the emitter is dependent on the electric field that develops inside the cathode, which can only be determined by Ohm's law given by Eq. 4.4-36 and the appropriate boundary conditions. For simplicity Eq. 4.4-36 includes only the two dominant terms on the right-hand side, namely the resistive and the electron pressure gradient contributions to the electric field **E**:

$$\mathbf{E} = (\eta_{\text{ei}} + \eta_{\text{en}})\mathbf{J}_e - \nabla p_e / en_e. \tag{4.4-36}$$

Here, η_{ei} and η_{en} are the resistivities due to electron-ion (e-i) and electron-neutral (e-n) collisions, and p_e and n_e are electron pressure and number density, respectively.

Numerical simulations have shown that in the ionization region of most of these cathodes, the resistivity is largely because of electron-ion collisions and is therefore insensitive to changes in both the plasma and neutral gas densities. A typical comparison of the centerline electron-ion and electron-neutral resistivities determined by 2-D numerical simulations of a Type B discharge cathode operating at 25 A [55] is shown in Fig. 4-21. Also shown on the plot for reference are the neutral gas and electron number densities. This insensitivity of the electron-ion resistivity implies that the only means of regulating the current density in the ionization region is through the pressure gradient, despite the fact that $|\eta j_e| \gg |\nabla (nT_{\text{eV}})/n|$ in this region. Therefore, for a given mass flow rate, the

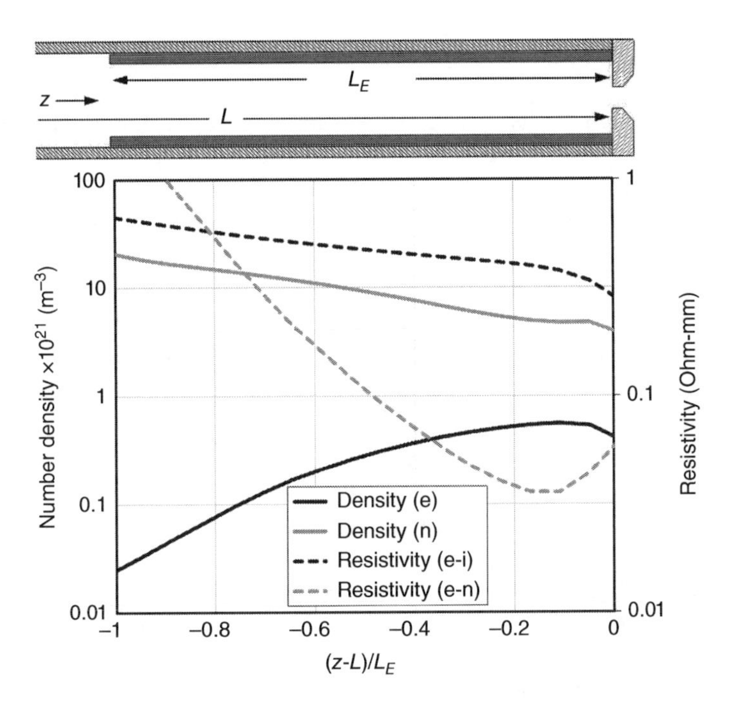

Figure 4-21 Number density of electrons (e) and neutrals (n) and resistivities for electron-ion (e-i) and electron-neutral (e-n) collisions along the centerline of a hollow cathode operating at 25 A. Computed using 2-D numerical simulations.

potential drop from the orifice to the emitter, the thermionic emission, and the features of the density profiles observed in measurements like those in Figs. 4-6, 4-8 and 4-20 are largely the response of the plasma to Ohm's law through the pressure gradient.

To further elaborate on what takes place in this region we use the following example. Consider any process that increases the thermionic emission flux of electrons at fixed discharge current and mass flow rate; say, by using a shorter emitter or an external heat source that raises the emitter temperature. Because the discharge current is fixed by the external power supply, the return electron current to the emitter must increase to produce the correct net current, which requires a lower sheath drop at the emitter and therefore a lower electric potential in the plasma region. This increases the electric field which would, in turn, also lead to an increase in the electron current density and ultimately a higher I_d. To counter that increase, the electron pressure must increase, which leads to a smaller density gradient length scale, $L_{\nabla n} \equiv (\nabla n/n)^{-1}$. This, of course, does not happen without also affecting L_n which must increase since a lower upstream plasma density increases the penetration of neutrals into the ionization region and, in turn, the interior pressure.

The density profile trends shown in Fig. 4-8, which are for a cathode in which the discharge current is changed, can be explained along these same lines, although in this case the only parameter increasing is I_d. Nevertheless, higher I_d in general still increases the emitter heating and lowers the sheath potential, which again leads to a smaller $L_{\nabla n}$. Moreover, because the emitter is warmer at higher discharge currents, ions that are neutralized at the surface and return back to the plasma as neutrals are also warmer, which increases the interior pressure. Finally, higher peak plasma densities mean increased heating rates for the heavy species through charge-exchange collisions; this also increases the interior pressure.

4.5 Orifice Region

Electrons are extracted from the insert-region plasma through the cathode orifice into the discharge chamber or ion beam. For cathodes with no orifice, a transition region exists at the end of the insert and cathode tube where the neutral gas density is sufficiently low and the flow becomes collisionless. Orificed cathodes, in most cases, also have a transition region to collisionless neutral flow, which can occur inside the orifice or slightly downstream depending on the orifice size and the gas flow rate. Inside the orifice, the electron current density is the highest in the entire system. In this region, classical electron scattering with the ions and neutral gas produces resistive heating. The hot electrons then ionize a large fraction of the xenon gas, most of which strike the orifice wall as ions and heat it. The amount of orifice-plasma resistance and orifice plate heating depends on the geometry, flow rate, and discharge current. Type A cathodes have long, narrow orifices and high pressures, which lead to high resistivity, strong ion bombardment of the orifice wall, and significant local heating. Type B cathodes tend to have smaller orifice heating unless the orifice is relatively small and the gas flow high, both because the resistance is usually lower than in Type A cathodes and because a larger fraction of the power deposited in the plasma in this region convects out into the cathode plume. For example, the 1-mm diameter orifice NSTAR cathode has significant orifice heating, but the 2.5 mm diameter orifice NEXIS cathode has lower orifice heating even at higher currents.

Under typical conditions that Type A and B cathodes are operated in electric propulsion, the ion flow in the cylindrical region of the orifice can be assumed to be diffusion limited. Their low flow speed and short mean free paths for charge exchange collisions with the neutrals mean that ions in this region do not move very far in a collision time. For example, the NSTAR discharge cathode operating in the TH15 mode has an average neutral density in the orifice region of $n_o \approx 5 \times 10^{21}\ \text{m}^{-3}$. The average ion mean-free-path for resonant charge exchange collisions with a cross section of $\sim 10^{-18}\ \text{m}^2$ is

$$\lambda = \frac{1}{\sigma_{\text{CEX}} n_o} \approx 2 \times 10^{-4} \text{m}, \tag{4.5-1}$$

which is about 5 times smaller than the orifice diameter. Thus, diffusion is a good approximation for the ion motion in these hollow cathode orifices. This is in fact true for a wide range of cathode geometries and operating conditions and was confirmed by 2-D numerical simulations with OrCa2D after the ion momentum equation was updated in the code to include the ion inertia terms $\partial(n\mathbf{v}_i)/\partial t + \nabla \cdot (n\mathbf{v}_i\mathbf{v}_i)$. For example, Fig. 4-22 shows a comparison of the electron number density with (top) and without (bottom) the ion inertia terms, for a LaB_6 cathode operating at 100 A [56]. It is clear that the differences in the two solutions are quite small in the orifice and insert regions, but become larger in the cathode plume where ions attain larger speeds and their density, as well as that of neutrals, decline.

We can take advantage of these properties in the orifice to construct a 0-D model of the cathode orifice plasma to show the dependence of the plasma density, electron temperature, and voltage drop in the orifice region. However, such a model provides only a rough estimate of these parameters because there is a large neutral pressure gradient generated going toward the orifice, whereas the model uses average parameters. It is assumed for now that the orifice is long compared to its length so that the radial ion diffusion equation applies. The solution to this equation for collisional plasmas in the orifice, described above for the insert region, results in the usual eigenvalue equation dependent on the electron temperature:

$$\left(\frac{r_o}{\lambda_{01}}\right)^2 n_o \sigma_i(T_e) \sqrt{\frac{8kT_e}{\pi m}} - D = 0, \tag{4.5-2}$$

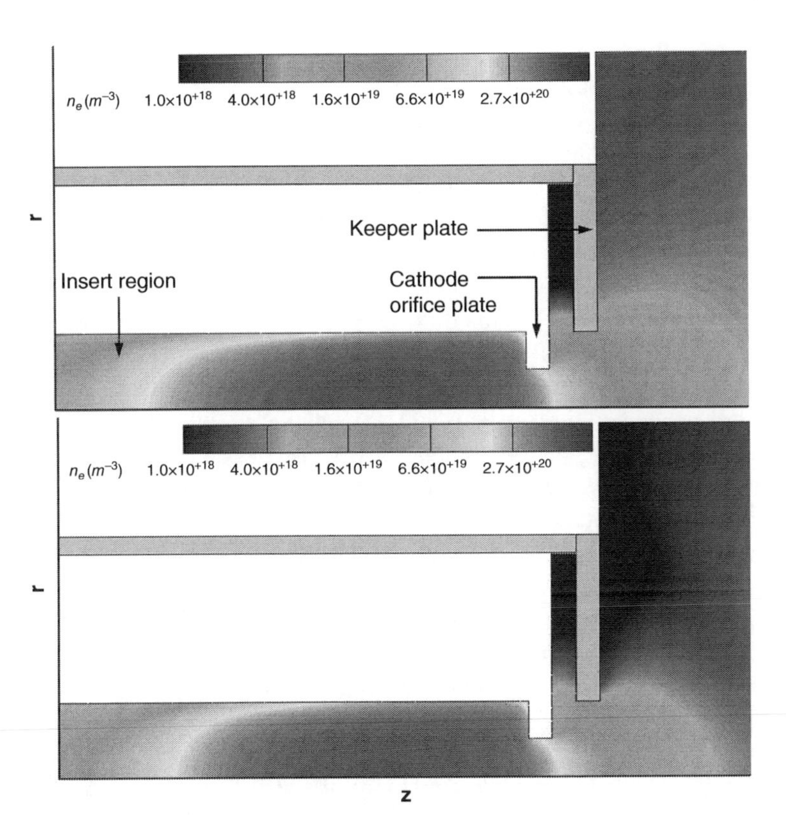

Figure 4-22 Comparison of the electron number density accounting for the ion inertia (top) and assuming diffusion-limited ion flow only (bottom) from 2-D numerical simulations with OrCa2D (*Source:* [56]).

where r_o is now the internal radius of the orifice. The electron temperature is again constrained to produce sufficient ions to offset the diffusion losses, as in the insert region analysis. Eq. 4.5-2 can be solved for the local electron temperature in the orifice using the terms evaluated in Eq. 4.4-4 through Eq. 4.4-7.

The steady-state electron energy equation (Eq. 4.4-28) is integrated over the cylindrical orifice, ignoring thermal conduction and radiation losses, to yield an equation for the average plasma density in the orifice. In this case, ohmic heating in the orifice plasma is balanced by convection of the energy deposited in the orifice plasma electrons and ionization losses:

$$I_e^2 R = \frac{5}{2} I_e \left(\frac{kT_e}{e} - \frac{kT_e^{\text{in}}}{e} \right) + n_o \overline{n}_e e \langle \sigma_i v_e \rangle U^+ \left(\pi r_o^2 l \right), \tag{4.5-3}$$

where l is the length of the orifice cylindrical section. Eq. 4.5-3 can be solved for the average plasma density in the orifice

$$\overline{n}_e = \frac{I_e^2 R - \frac{5}{2} I_e \frac{k}{e} \left(T_e - T_e^{\text{in}} \right)}{n_o e \langle \sigma_i v_e \rangle U^+ \pi r_o^2 l}. \tag{4.5-4}$$

An evaluation of the terms in Eq. 4.5-4 for the orifice region uses the same techniques previously described in Section 4.4 for the insert-region plasma. The resistance R is given by Eq. 4.4-9 where the conduction length is now simply the orifice plasma length. The input electron temperature,

T_e^{in}, is the electron temperature in the insert-region plasma that comes from another model such as the diffusion model used in Section 4.4, or from experimental measurements.

The detailed measurements of the plasma density and electron temperature in the orifice of the NSTAR discharge cathode [48] will be used as a first example to compare with the model predictions. The NSTAR discharge cathode has an orifice diameter of 0.1 cm, and the case of the full power TH15 operating point with 13 A of discharge current at a xenon gas flow rate of 3.7 sccm will be used. The pressure measured inside the insert region for this case is about 7.8 Torr [48]. Assuming a fully developed viscous (Poiseuille) flow (see Appendix B), the pressure in the orifice is estimated to fall to less than 3 Torr by the end of the 0.75 mm long cylindrical section of the orifice. Assuming a gas temperature of about 2000 K in the orifice, the solution for the electron temperature in the diffusion equation (Eq. 4.5-2) versus pressure in the orifice is shown in Fig. 4-23. The electron temperature predicted by this model varies by less than 1 eV along the orifice length, and the average in the channel is about 2.3 eV. This value is close to the experimentally measured values of 2.2–2.3 eV found in this region [46].

Using this electron temperature, the density in the orifice is calculated from Eq. 4.5-4 and plotted in Fig. 4-24 versus the discharge current for the NSTAR cathode. The agreement with the experimental data [48] taken for two discharge currents at the nominal 3.7 sccm cathode flow rate is also very good. The resistance calculated from Eq. 4.4-9 for the cylindrical orifice length is 0.31 Ω which, at 13 A, produces a voltage drop in the orifice of about 4 V. This is the same magnitude as the voltage change observed in the experimental data, which illustrates that the potential drop in the hollow cathode orifice is resistive due to the very collisional plasma that exists there in these xenon hollow cathodes. Detailed 2-D calculations described below indicate that roughly half of the power deposited in this region ($P = 4\,\text{V} \times 13\text{A} = 52\,\text{W}$) goes to the orifice wall, and the remainder is convected into the discharge chamber by the plasma.

Since the 0-D orifice model with just the average parameters has been shown to provide rough estimates of the orifice parameters, it is reasonable to use it to examine Type A cathode orifice heating. Consider the orifice of the NSTAR neutralizer cathode [48], which has an inside-diameter of 0.028 cm. The pressure measured inside the neutralizer during operation at 3.2 A, associated with

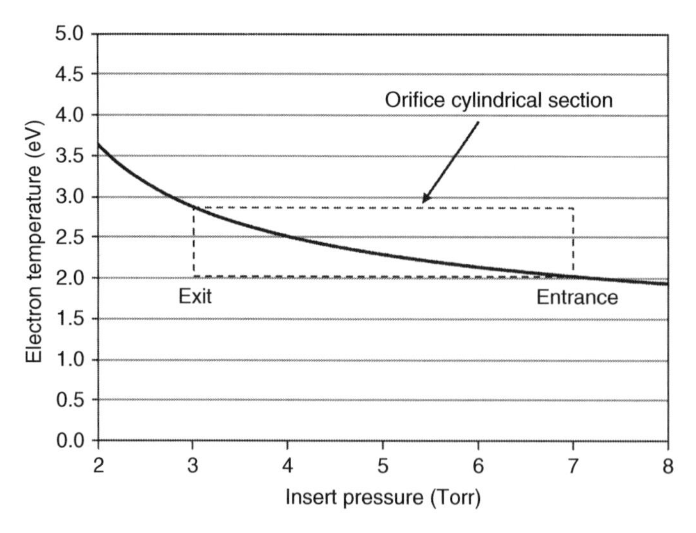

Figure 4-23 Orifice electron temperature calculated from the 0-D model through the orifice for the NSTAR discharge cathode at TH15.

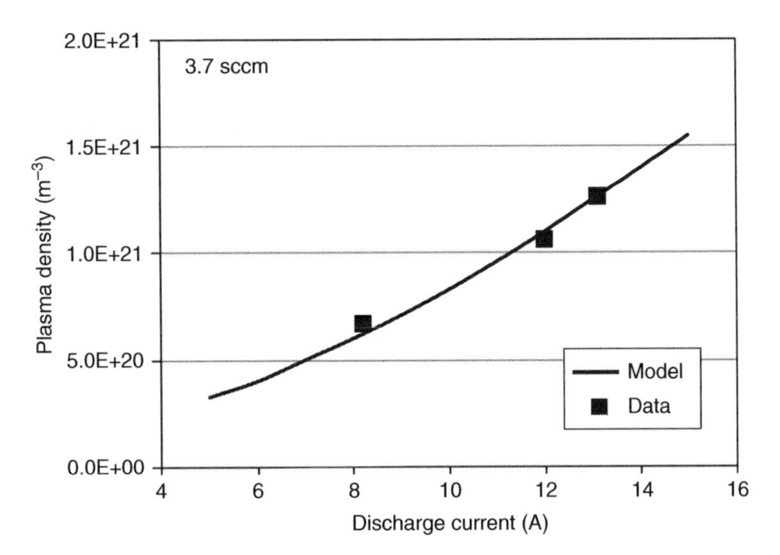

Figure 4-24 Orifice plasma density calculated from the 0-D model and measured points for the NSTAR discharge cathode at 3.7 sccm xenon gas flow.

the full power TH15 case, is 145 Torr. Assuming again simple Poiseuille flow (see Appendix B) through the 3 : 1 aspect ratio orifice channel in this cathode, the pressure is found to fall to less than 20 Torr by the end of the 0.75 mm long cylindrical section of the orifice. Assuming the same gas temperature of about 2000 K in the orifice again, the solution to the radial diffusion Eq. 4.5-2 predicts the electron temperature to vary by only 0.5 eV along the orifice, with an average value of about 1.4 eV. It is also assumed that a minimal 1 eV electron temperature exists in the insert region for the T_e^{in} in Eq. 4.5-3.

The average plasma density in the orifice is plotted in Fig. 4-25 versus the discharge current. At 3.2 A, corresponding to the neutralizer cathode producing the beam current of 1.76 A plus the keeper current of 1.5 A, the predicted plasma density is about 6×10^{22} m^{-3}. The resistance calculated from Eq. 4.4-9 for the cylindrical orifice section is 3.5 Ω, which, at 3.2 A, produces a resistive voltage drop in the orifice of about 11 V. The power deposited in the plasma ($P = 11\text{ V} \times 3.2\text{A} = 35\text{ W}$) in this case goes primarily to the orifice wall because the convection power loss is low due to the large geometrical aspect ratio of Type A orifices and the low electron temperature. This demonstrates the resistive orifice heating power characteristically found in Type A cathodes.

Although 0-D models are useful to illustrate the strong resistive effects in the orifices of all Type A and some Type B cathodes, the use of average pressures and temperatures reduces the accuracy of these models. However, it is possible to construct a 1-D model for the cathode orifice [44] to address this issue. The orifice plate is usually chamfered on the downstream side, which must be included in the analysis because the rapidly expanding gas plume in this region often transitions to the collisionless regime in which the flow is not dominated by diffusion.

In the orifice, continuity dictates that the ions that hit the orifice wall are re-emitted as neutrals and re-enter the plasma. The continuity equations for the three species (neutrals, ions, and electrons) in the cylindrically symmetric orifice region are, respectively:

$$\pi r^2 \left(-\frac{\partial n}{\partial t} + \frac{\partial v_o n_o}{\partial z} \right) + 2\pi \mathrm{r}\, v_{\text{wall}} n = 0 \tag{4.5-5}$$

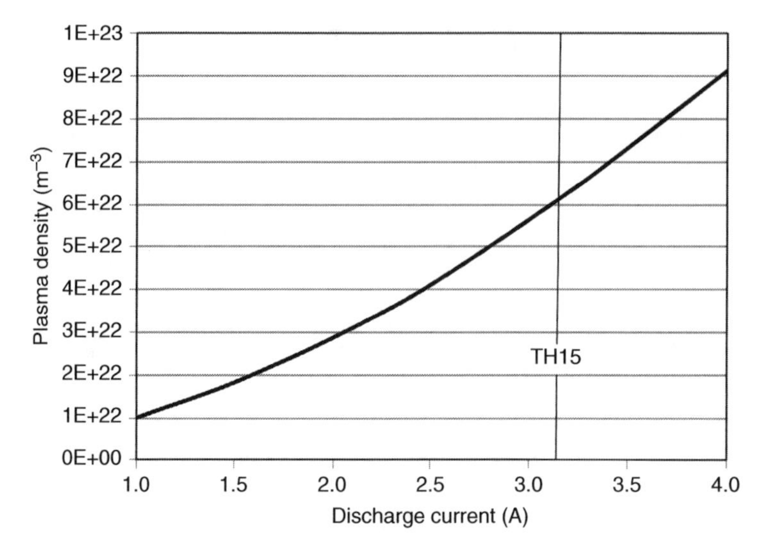

Figure 4-25 Orifice plasma density calculated from the 0-D model for the NSTAR neutralizer cathode at TH15.

$$\pi r^2 \left(\frac{\partial n}{\partial t} + \frac{\partial v_i n_o}{\partial z} \right) - 2\pi \mathrm{r}\; v_{\text{wall}} n = 0 \tag{4.5-6}$$

$$\pi r^2 \left(e \frac{\partial n}{\partial t} + \frac{\partial J_e}{\partial z} \right) = 0, \tag{4.5-7}$$

where v is the ion or neutral velocity and v_{wall} is the particle velocity at the radial boundary.

The average neutral velocity is found from Poiseuille flow for a pressure P:

$$v_o = -\frac{r^2}{8\zeta} \frac{dP}{dz} \tag{4.5-8}$$

where ζ is the temperature dependent neutral gas dynamic viscosity. For xenon, the viscosity is [57]

$$\begin{aligned} \zeta &= 2.3 \times 10^{-5} T_r^{0.945} \quad \text{for } T_r < 1 \\ &= 2.3 \times 10^{-5} T_r^{(0.71\,+\,0.29/T_r)} \quad \text{for } T_r > 1 \end{aligned}, \tag{4.5-9}$$

with units of Ns/m^2 or Pa-s and a relative temperature given by $T_r = T/289.7$ K. Since a large fraction of the ions undergo charge exchange within the orifice, the neutral gas is heated and the viscosity is increased. This is incorporated into the model [57] by assuming that the gas temperature varies as

$$T = T_{\text{wall}} + \frac{M}{k} \left[(f v_r)^2 + v_o^2 \right], \tag{4.5-10}$$

where the fraction f of the neutrals that receive the ion radial velocity via charge exchange is given by

$$f = 1 - \exp\left[-\frac{n}{n_o} \frac{\tau_{\text{wall}}}{\tau_{\text{CEX}}} \right], \tag{4.5-11}$$

where τ_{wall} is the average time between collisions with the wall for a neutral particle, and τ_{CEX} is the average ion-neutral charge exchange time. This effective heating mechanism by charge exchange has been observed in experiments and in 2-D numerical simulations (shown later) where the neutral temperatures are higher than the orifice wall temperatures.

Combining the electron and ion momentum equations (Eq. 4.4-30 and Eq. 4.4-31) to eliminate the electric field gives an expression for the particle flux in terms of the ambipolar diffusion coefficient and the ion and electron mobilities.

$$n(\upsilon_i - \upsilon_o) = -D_a \frac{\partial n}{\partial z} + \frac{\mu_i}{\mu_e} \frac{J_e}{e}. \tag{4.5-12}$$

The ambipolar diffusion coefficient for this case is given by Eq. 4.4-4. In the orifice, the radial drift velocity will often exceed the ion thermal speed due to the radial potential gradient, so the ion scattering velocity must be approximated by

$$\upsilon_{\text{scat}} = \sqrt{\upsilon_{\text{th}}^2 + (\upsilon_i - \upsilon_o)^2 + \upsilon_r^2}, \tag{4.5-13}$$

where υ_r is the radial ion velocity found from Eq. 4.4-7.

The continuity equations in the orifice (Eqs. 4.5-5 through 4.5-7) were solved using the electron energy Eq. 4.4-28 and Ohm's Law in the cylindrical orifice to produce ion density and plasma potential profiles in the orifice region. The first result from this work is that a double sheath often postulated in the orifice region [58] is not observed for xenon ion thruster cathodes. There is a potential gradient through the orifice, but this results from resistive effects associated with classical electron collisions in the orifice channel and is not due to any double layer inside the orifice channel.

As an example, Fig. 4-26 shows a plot of the neutral and plasma density along the axis of an NSTAR neutralizer cathode orifice operating at the TH15 power point, producing 3.76 A of current with a xenon gas flow rate of 3.5 sccm. The peak plasma density occurs in the cylindrical section of the orifice, and the density falls though the chamfered region because of the neutral gas density decrease. It should be noted that the peak plasma density predicted by the 1-D model in the orifice is in reasonable agreement with the 0-D model results shown above that used the average neutral density and temperature along the length of the orifice. When the assumptions remain valid and

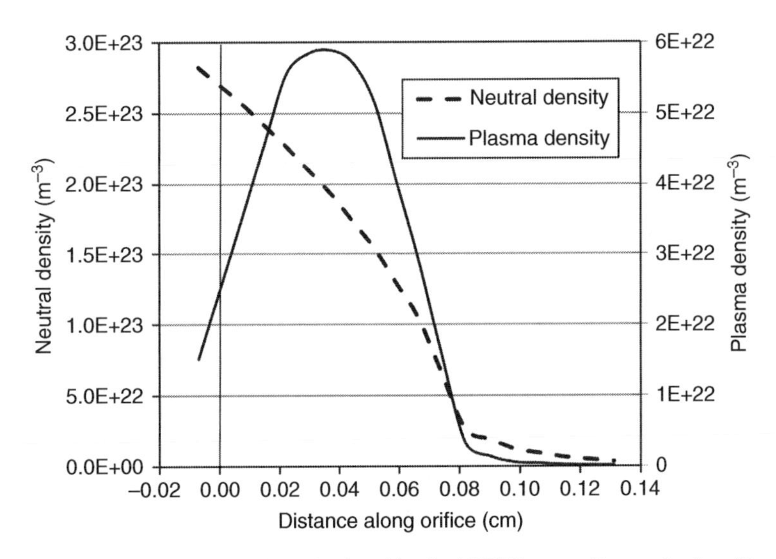

Figure 4-26 Neutral and plasma densities calculated in the NSTAR neutralizer cathode orifice at the TH15 operation point.

using inputs from other models or measurements, reasonably accurate results can be obtained using simple 0-D and 1-D models to illustrate the driving physics in this region.

An interesting result of this simplified analysis is that significant ionization occurs in the orifice, which provides electrons to the discharge. Figure 4-27 shows the electron current calculated as a function of the distance along the orifice axis. The electron current is about 50% higher exiting the orifice compared to the amount extracted from the insert-region plasma. This is because the very high neutral gas density in the neutralizer cathode orifice region causes significant ionization. Discharge cathodes have much lower electron multiplication factors in the orifice because the neutral and plasma densities are typically an order of magnitude lower.

The ion current density to the orifice wall, which naturally follows the plasma density profile in Fig. 4-26, is shown in Fig. 4-28. The ion bombardment of the orifice walls is seen to peak well before the chamfer region starts. Since the plasma potential is increasing along the axis from the insert-region plasma to the exit due to the plasma resistance, the ions in this region can have sufficient energy to sputter the wall. This effect was observed in the cross section of the NSTAR neutralizer after the 8200 Life Demonstration Test (LDT) [59] and is shown in Fig. 4-29. The orifice was observed to open up in the center cylindrical region before the chamfer, consistent with the predicted ion bombardment location in Fig. 4-28.

The time required to produce this erosion pattern is not known from the experiments because Fig. 4-29 shows a destructive analysis after the end of the test. In fact, the erosion pattern shown in the destructive analysis of the neutralizer cathode orifice after the 30,352 h Extended Life Test (ELT) [60] shown in Fig. 4-30 is nearly identical to the shorter duration LDT result. The ELT cathode experienced nearly double the operation time of the LDT cathode at the full power level, which did not further erode the orifice. The 1-D orifice model described above shows that the neutral pressure and plasma density in the orifice decrease as the diameter increases, which reduces the ion bombardment flux but cannot capture the plasma potential changes that occur along the channel. Considering the low ion energies that are generated in this region of the cathode, the sputtering

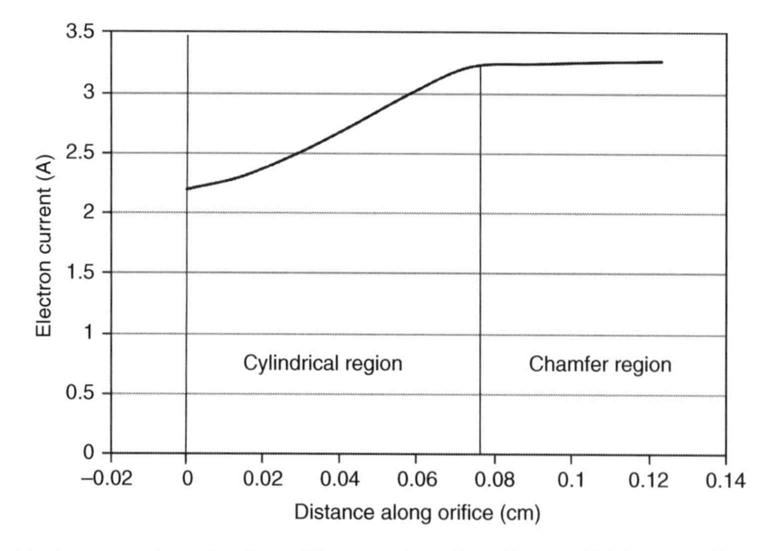

Figure 4-27 Electron current as a function of distance along the axis in the NSTAR neutralizer cathode orifice at TH15.

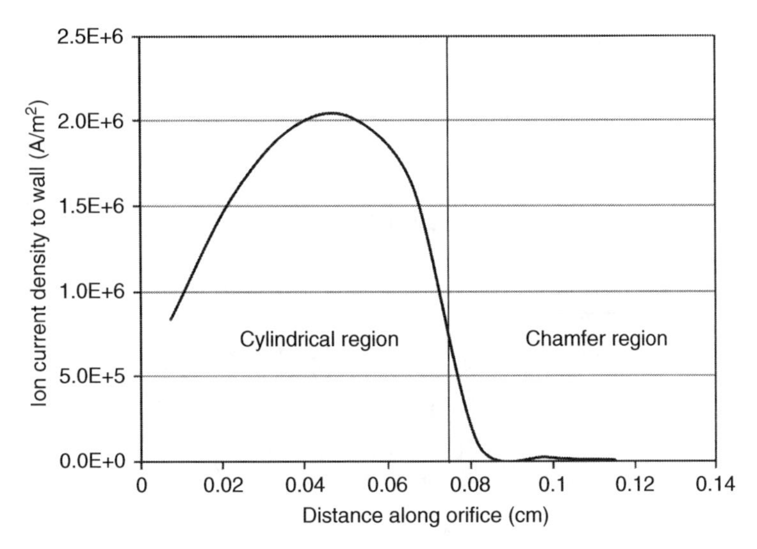

Figure 4-28 Radial ion current to the orifice wall as a function of distance along the axis in the NSTAR neutralizer cathode at TH15.

yield and, therefore, the erosion rate, can be very strongly dependent on the sheath drop so its accurate determination is of utmost importance.

Explanations for this behavior and quantitative predictions of the erosion rates were provided by detailed numerical simulations using OrCa2D [53]. The numerical simulations showed that the main mechanism responsible for the channel erosion was sputtering by singly-charged ions. These ions were accelerated by the sheath along the channel and bombarded the surface with sufficient kinetic energy to erode the walls at the beginning of the test. The density of the ions inside the neutralizer orifice was found to be as high as 2×10^{22} m^{-3}. As the orifice channel opened due to the erosion, the plasma potential, and in turn the energy of the ions bombarding the orifice walls through the sheath, diminished. The maximum energy/charge with which Xe^{+} bombarded the orifice walls at the beginning of the test was found to be about 17 V, as shown in the plasma potential contours of Fig. 4-31 (top). By the time the channel opened to a radius of 0.023 cm, the maximum ion energy had fallen below 9 V corresponding to almost a three-order-of-magnitude drop in the sputtering yield. At such low yields the erosion of the orifice material is negligible and the sputtering essentially stopped.

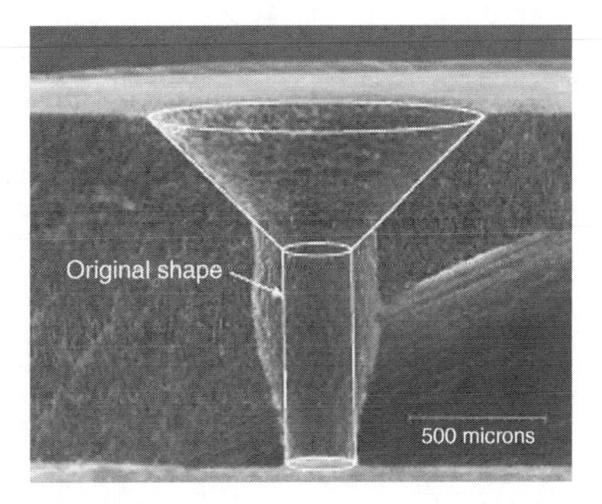

Figure 4-29 Neutralizer cathode orifice cross section showing erosion pattern after the NSTAR 8200 h test (*Source:* [59]/American Institute of Aeronautics and Astronautics, Inc.).

The potential contours from the numerical simulations of the eroded channel after 500 h of operation are shown in Fig. 4-31 (bottom). The results predicted that more than half of the erosion occurred in less than 1000 h, and that the erosion became negligible in less than a few thousand hours. A comparison of the final state of the eroded orifice from the simulations is compared with

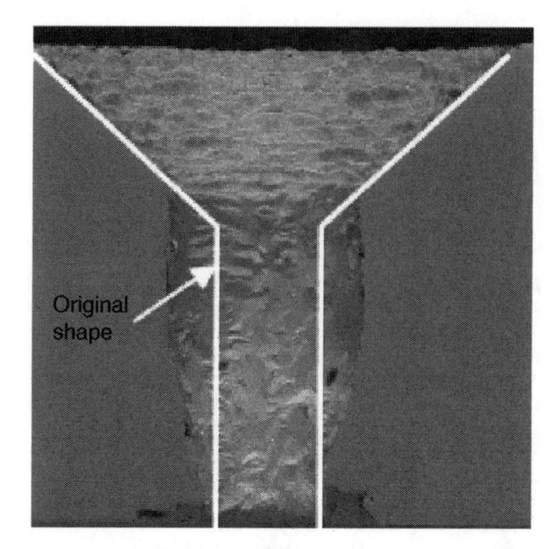

Figure 4-30 Neutralizer cathode orifice cross section showing the erosion pattern after the NSTAR 30152-h ELT (Extended Life Test) (*Source:* [60]/American Institute of Aeronautics and Astronautics).

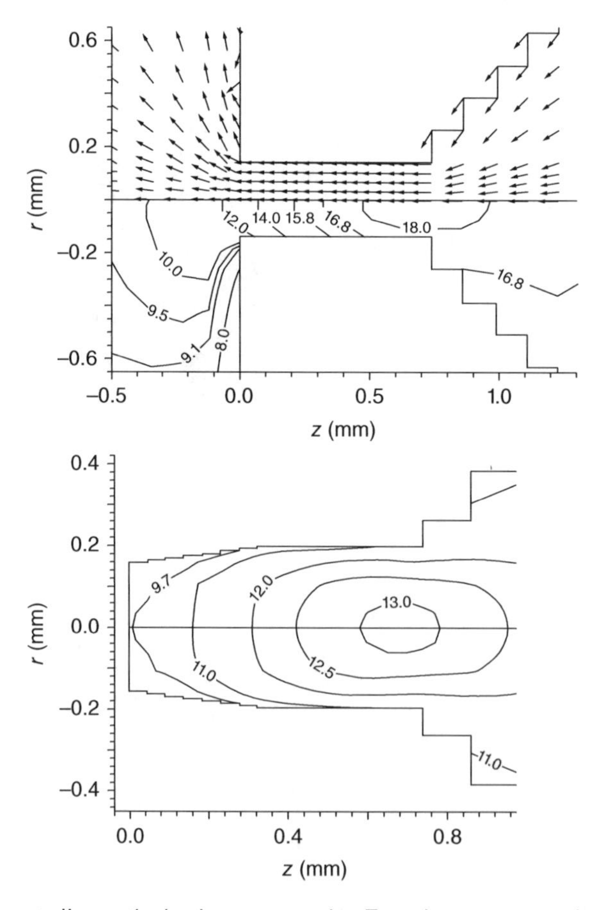

Figure 4-31 NSTAR neutralizer cathode plasma properties. Top: electron current density unit vectors (top half) and plasma potential contours in volts (bottom half). Bottom: contours of the plasma potential (in volts) in the eroded channel after 500 h of operation (*Source:* [53]).

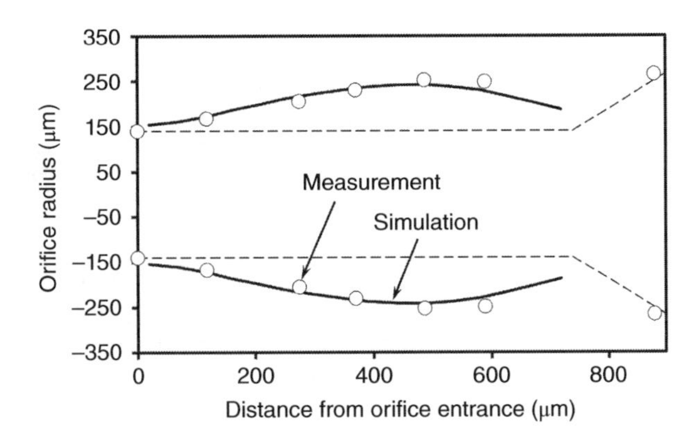

Figure 4-32 Comparison of the measured and computed erosion profiles in the NSTAR neutralizer orifice. The computed profile is from 2-D numerical simulations with OrCa2D (*Source:* [53]).

the measurement in Fig. 4-32. Modeling capabilities of this kind are critical in the design and life qualification of hollow cathodes for electric propulsion where, in the absence of long and costly wear tests, these are the only means of evaluating the life of these neutralizers since the size of their orifices prohibit access to probe diagnostics.

One might expect similar erosion behavior from discharge cathode orifices. Figure 4-33 shows the destructive analysis of the LDT discharge cathode orifice plate after 8200 h at full power where there is no discernable erosion in the cross section. The OrCa2D simulations showed quantitatively that due to the 3.5× larger diameter of the discharge cathode orifice, the resistive drop and resistive heating, which largely drive the plasma potential, electron heating and ionization inside the orifice, were both much higher in the neutralizer orifice. The plasma density inside the orifice of the discharge cathode was more than 40× lower, as shown in Fig. 4-34, and the sheath drop 7 V lower compared to the values in the neutralizer. At these conditions, singly charged xenon ions could cause no significant sputtering of the surface.

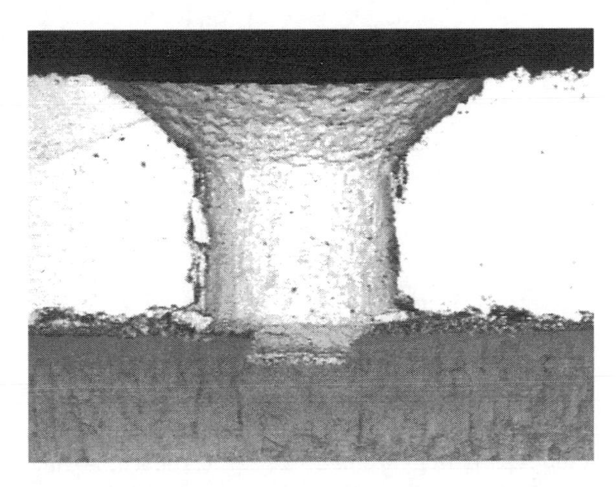

Figure 4-33 Discharge cathode orifice showing erosion pattern after the NSTAR 8200 h Long Duration Test (*Source:* [59]/American Institute of Aeronautics and Astronautics, Inc.).

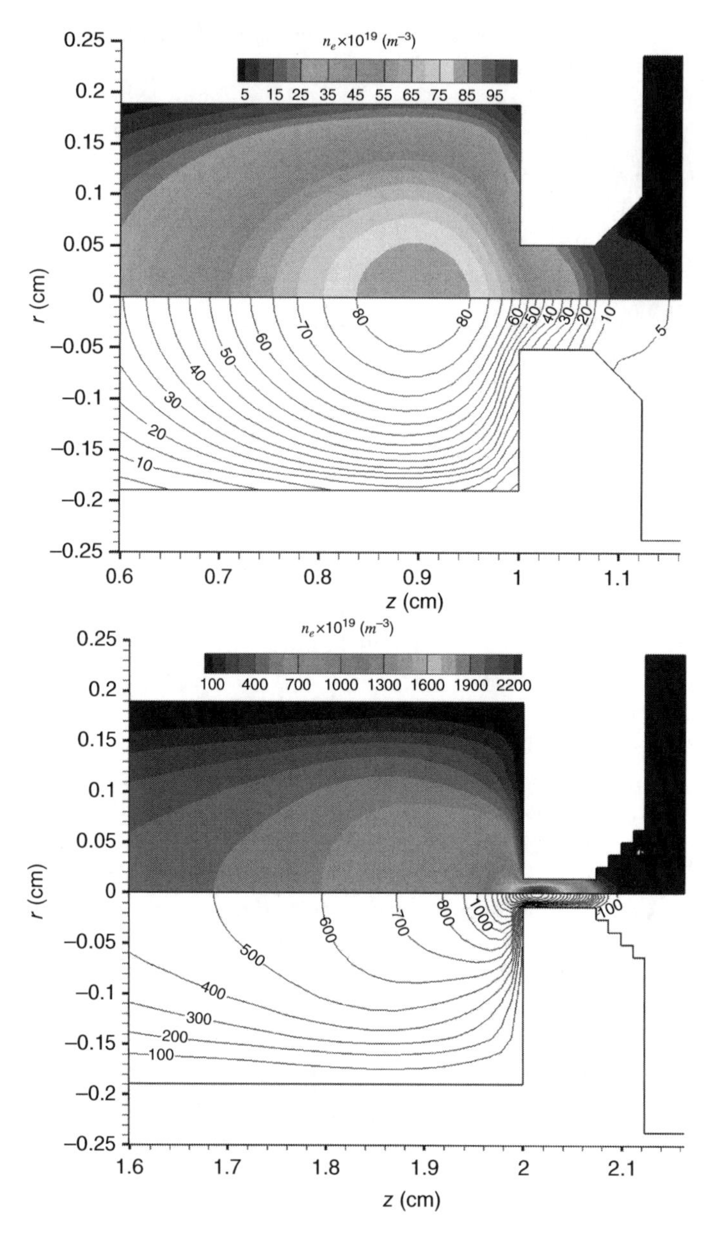

Figure 4-34 Comparison of the computed plasma density inside the NSTAR discharge (top) and neutralizer (bottom) hollow cathodes from numerical simulations with OrCa2D (*Source:* [53]).

4.6 Cathode Plume Region

The cathode insert and orifice regions were examined above with both idealized and more comprehensive 2-D models, and this information was used to provide an understanding of the plasma parameter dependence and self-heating mechanism of the cathode. The final region of the hollow cathode to cover is exterior to the cathode orifice where the cathode plume interacts with the keeper electrode and couples the cathode emission current to the thruster discharge plasma or the ion beam plasma. In this region, the neutral gas expands rapidly away from the cathode and is either collisionless from the cathode orifice region or makes the transition to collisionless near the keeper. The electrons from the cathode are accelerated by the potential difference between the cathode orifice plasma and the plasma in the near cathode plume. The total potential difference is usually near the discharge chamber anode voltage in ion thrusters or the beam plasma potential in Hall thrusters. There is usually an applied axial magnetic field on the order of 100 G in the near cathode plume region of the discharge cathode in DC ion thrusters and at the hollow cathode exit region in Hall thrusters with center-mounted cathodes. In ion thrusters this field provides a transition to the ring-cusp fields, which produces some confinement of the cathode plume electrons which generate the cathode plume plasma. In Hall thrusters with centrally mounted cathodes, this field is generated at the central pole face due to the magnetic field design to produce the radial field of the thruster near the channel exit.

The plasma stream exiting the hollow cathode is often reported as having various structures consisting of dark spaces, plasma spheres, and brightly divergent plume shapes. Two common cases are shown in Fig. 4-35, where in each photo the cathode is on the right and the anode on the left. The plasma stream consists of the electrons from the hollow cathode, neutral gas expanding from the keeper aperture in addition to more uniform background neutral gas from the thruster, and the plasma plume generated by ionization of this gas by the electrons. The on-axis potential and temperature profiles measured by scanning probes for these two cases [11] are shown in Fig. 4-36. Although the discharge current is the same in these two cases, the high gas flow reduced the discharge voltage from about 26 V at 5.5 sccm to 20 V at 10 sccm. The structure of the potential and temperature profiles is significantly different in the plume region as the gas flow and discharge voltage change. The higher gas flow case reduces the potentials and temperatures throughout the system and pushes the plasma sphere observed at the cathode exit farther downstream. The flat potential region just downstream of the cathode orifice exit is related to the keeper orifice location, and higher gas flows pushes this potential structure even further downstream into the plume.

If the processes described above were not complex enough, the electrons from the cathode are extracted at high enough velocities and low densities that electrostatic instabilities can also be generated in this region of the cathode. The topic of instabilities will be discussed further in this

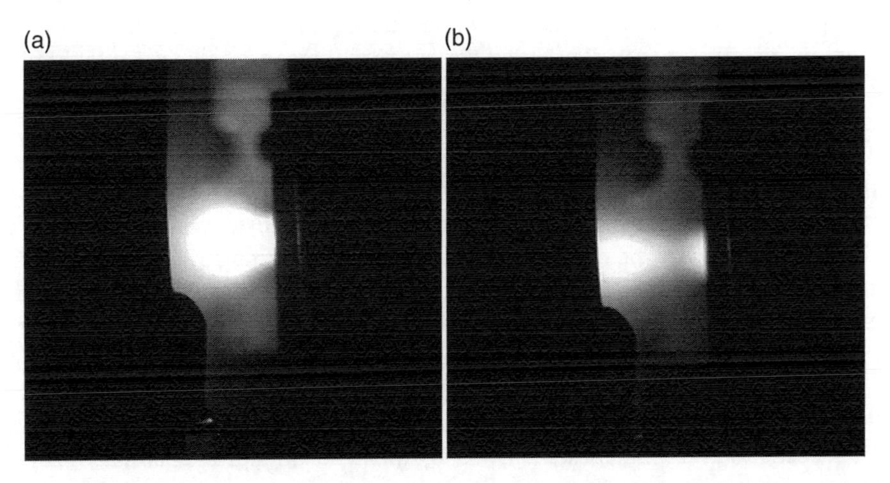

Figure 4-35 NEXIS cathode plume at 25 A with the plasma ball in (a) at 5.5 sccm and a dark space in (b) at high flow (10 sccm) (*Source:* [11]/AIP Publishing LLC).

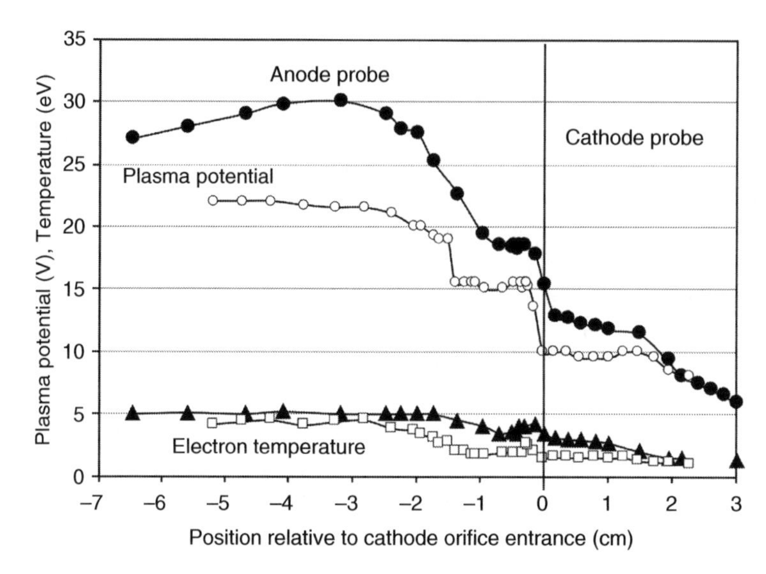

Figure 4-36 Plasma potential and electron temperature profiles at 25 A of discharge current for the two photos in Figure 4-35 where the closed symbols are the 5.5 sccm, 26.5 V case and the open symbols are the 10 sccm, 19 V case (*Source:* [11]).

section and in Section 4.10. The combination of all these processes makes this region of the cathode unamenable to idealized modeling in 0-D or 1-D. Comprehensive 2-D models and numerical simulations coupled with experimental measurements to validate the results appears to be the only means by which the complexity of these plumes can be resolved and thoroughly analyzed.

The 2-D structure of the cathode plume as it expands from the cathode orifice has been investigated experimentally by several authors [46, 48, 61]. Figure 4-37 shows plasma density contours measured [62] with a fast scanning Langmuir probe. The density is the highest on-axis and closest to the cathode orifice, which is consistent with the visual appearance of a bright plasma sphere or spot at the cathode exit that expands both radially and axially into the discharge chamber [46]. A reduction in cathode gas flow causes the sphere or spot to pull back toward the cathode orifice and the plasma to expand into what is called a plume mode [4, 10, 63]. Plume mode operation generally results in high frequency oscillations in the cathode plume that propagate into the discharge chamber and keeper region, and can couple to the power supply connection leads if of sufficient amplitude.

The plasma potential contours measured by the scanning probe for the case of Fig. 4-37 are shown in Fig. 4-38. The potential is actually a minimum on axis near the cathode, and then increases radially and axially away from the cathode exit to a value several volts in excess of the discharge voltage [46, 62]. This structure near the cathode is sometimes called a "trough" because ions generated externally to the cathode tend to funnel into the trough and toward the cathode. Large amplitude plasma potential oscillations (20–80 V) in the range of 50–1000 kHz have been observed primarily in and around the edge of the plasma sphere and in front of the keeper electrode from high speed scanning emissive probes [64, 65].

The abundance of physical processes that take place in the plume region of the hollow cathode also called for extensive 2-D numerical modeling to explain the physics and plasma behavior. Early attempts were made to develop 2-D models of the plasma discharge [66, 67], but in general these models did not account for all the pertinent physics that were later found to be important and/or

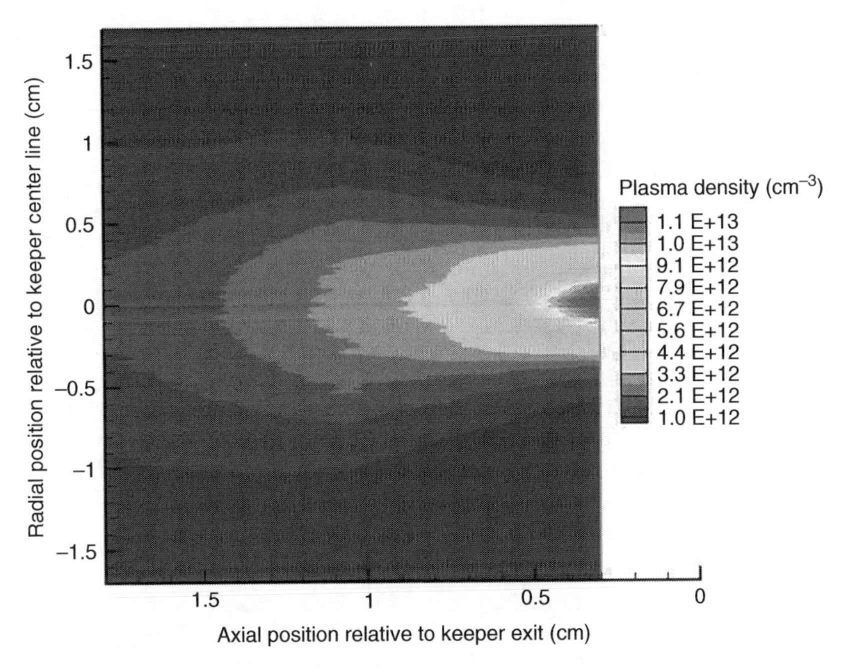

Figure 4-37 Plasma density contours measured for the NEXIS cathode at the nominal 25 A and 5.5 sccm gas flow discharge conditions (*Source:* [62]).

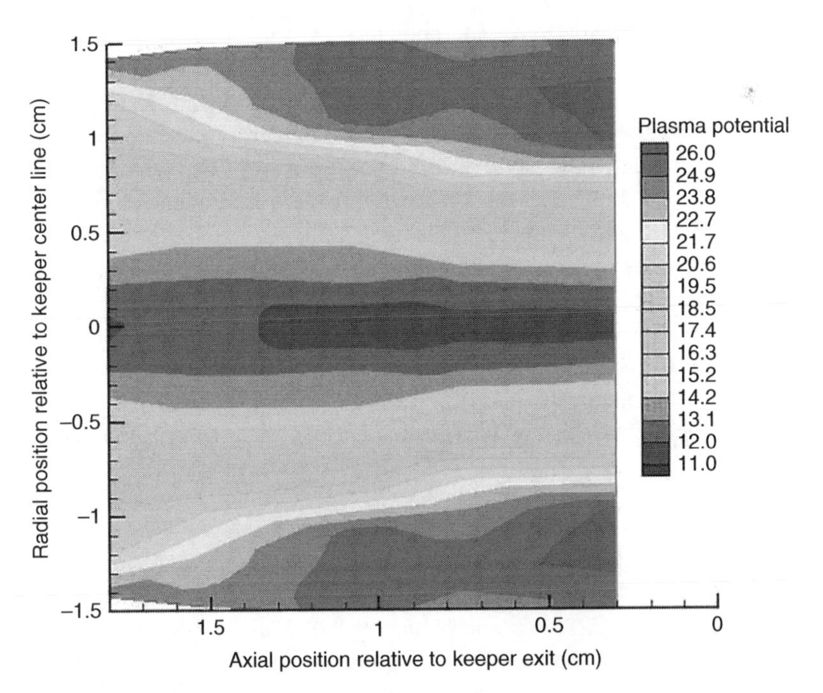

Figure 4-38 Plasma potential contours measured for the NEXIS cathode at the nominal 25 A, 5.5 sccm discharge condition (*Source:* [62]).

only encompassed portions of the cathode-plume domain [12, 68]. The first 2-D model to successfully solve an extensive system of governing laws for the cathode discharge, using numerical simulation and a computational domain that spanned all regions of the hollow cathode, namely the insert, cathode and keeper orifices, and the plume region, was the OrCa2D code [56]. The system of governing laws solved in this code is based on Braginskii's formulations for partially ionized gases [69] subject to appropriate sheath boundary conditions in the presence of thermionic emission [70]. The system also includes the Navier–Stokes equations for the neutral gas since viscosity was found to be important in the insert region [71]. The neutral gas model is smoothly transitioned from the continuum regime in the insert region to the collisionless regime in the plume. This transition typically occurs inside the cathode plate orifice.

One of the first significant findings from the global OrCa2D simulations was on plasma instabilities. The first efforts to model the plume with OrCa2D revealed that the electron resistivity due to classical collisions was orders of magnitude too low to explain the measured plasma parameter profiles. Mikellides [72] postulated that the presence of anomalous resistivity in the plume of a 25-A hollow cathode was required to explain the measured plasma potential and electron temperature profiles. It was also shown [12] that the conditions for the growth of the ion acoustic instability were met near the orifice and that this could saturate to ion acoustic turbulence (IAT) [73–75] downstream in the plume. When an idealized anomalous resistivity model for the electrons subjected to IAT was included, the agreement with the measurements near the orifice improved significantly. Ultimately, in the global version of OrCa2D the anomalous contribution to the generalized Ohm's law was incorporated through the addition of an anomalous collision frequency, ν_α, yielding a total resistivity η as follows:

$$\eta = \frac{m(\nu_e + \nu_\alpha)}{e^2 n} \tag{4.6-1}$$

where ν_e represents the classical electron-ion and electron-atom collision frequencies.

Figure 4-39 top compares measurements of the plasma potential on the centerline of the NSTAR discharge hollow cathode with the computed values produced by the model of ν_α used in the simulations, and with a version of the model that yielded one order of magnitude lower total collision frequency [54]. A steady state solution could not be obtained with classical resistivity alone ($\nu_\alpha = 0$). The comparison of the classical and anomalous collisions frequencies in those simulations in Fig. 4-39 bottom shows that the anomalous collision frequency can be approximately two orders of magnitude larger than the classical collision frequency in the plume region, and that the difference diminishes as the cathode is approached.

The idealized IAT model of the resistivity was based on the formulations of Sagdeev and Galeev [75], who assumed the IAT saturates through non-linear wave-particle interactions. This allowed them to determine the turbulent wave energy W_T and reduce the "anomalous" collision frequency ν_α to a simple algebraic function of the macroscopic plasma parameters as follows:

$$\nu_\alpha \sim \omega_{pe} \frac{W_T}{nT_e} = \beta\, \omega_{pe} \frac{T_e}{T_i} \frac{v_e}{v_{th}} \tag{4.6-2}$$

where the electron drift speed is v_e, the electron thermal speed is $v_{th} = (2\,kT_e/m)^{1/2}$, and the electron plasma frequency is ω_{pe}. Several years after the original formulations of Sagdeev and Galeev in 1969, the coefficient β was found to be approximately 0.01 [76]. In most early simulations with OrCa2D that used the Sagdeev and Galeev model [54], the heavy species were assumed to be in thermal equilibrium. Therefore, a separate ion energy equation was not solved. Consequently, the ratio T_e/T_i in Eq. 4.6-2 was assumed to be constant. This ratio was then included in a coefficient β' such that $\nu_\alpha = \beta' \omega_{pe} v_e / v_{Te}$, which was determined in the simulations by iteration based on the

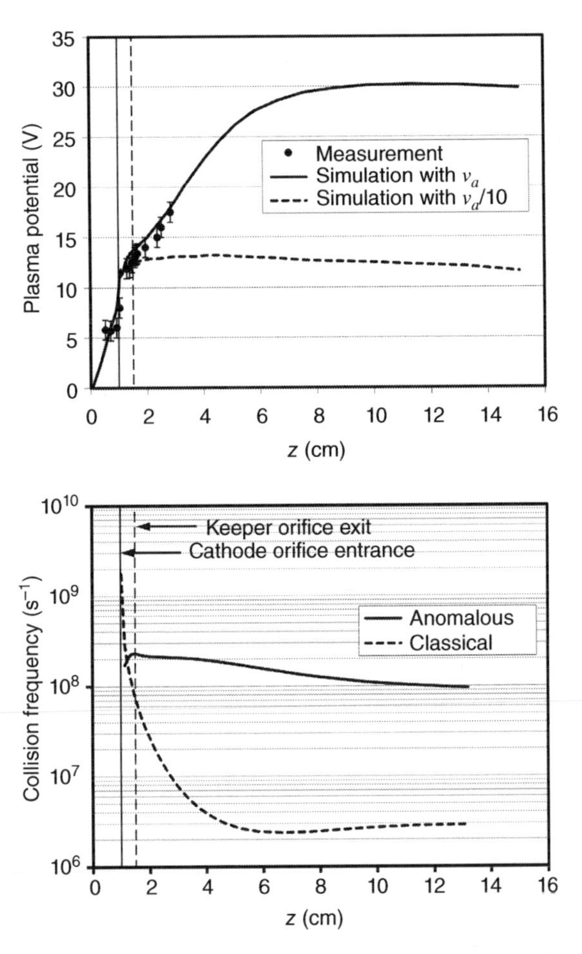

Figure 4-39 The effect of IAT-driven anomalous resistivity along the centerline of a 13-A discharge hollow cathode from 2-D numerical simulations. Top: computed and measured plasma potential. Bottom: comparison of anomalous and classical collision frequencies (*Source:* [54]).

specified discharge current. In the electron energy equation, anomalous heating of electrons is taken into account through the frictional heat exchange [55]. The IAT-based collision frequency in Eq. 4.6-2 led to results that agreed well with experimentally-measured plasma parameters along the centerlines of two discharge cathodes [54, 72] operating with discharge currents in the range of 13.3–27.5 A. Later, in an effort that combined numerical modeling and plasma measurements in a LaB_6 25-A cathode, the simulations actually allowed for the determination of the coefficient β in Eq. 4.6-2. This was made possible by the fact that more parameters about the experiment were known in these simulations than in most previous cases where β was treated as a free parameter. It was found that β was only about a factor of two lower than the estimated theoretical value of $\sim$0.01 [55].

It was argued that because anomalous heating of ions was not accounted for in the simulations, the computed ion temperatures were lower than those in the discharge, yielding a higher value of the electron-to-ion temperature and, in turn, a somewhat lower value for β. This result was nevertheless notable because it was the first time that an electric propulsion hollow cathode was used to determine the value of β through a combination of simulations and measurements. The confirmation of $\beta \sim 0.01$ also strengthened the theoretical predictions of IAT-driven anomalous resistivity in these devices when operating at tens of amperes. The application of Eq. 4.6-2 to higher discharge currents (~100 A), however, yielded only qualitative agreement with the experiments. This suggested that a more sophisticated closure relation was needed for the IAT-driven anomalous collision frequency in high-current cathodes, like that proposed by Lopez Ortega [77] who solved an additional conservation equation for the IAT wave energy density W_T in the OrCa2D global system of equations for the plasma.

The presence of IAT in the near-plume of hollow cathodes was confirmed several years later experimentally by Jorns [78] in a high-current LaB_6 hollow cathode operating over the range of 20–130 A. The authors, through direct measurements with scanning probes, confirmed that anomalous resistivity from the IAT dominates over classical values over a wide range of operating conditions, but most notably at high current to gas flow ratios. Most revealing were the measurements that allowed the authors to generate the Beall intensity plot in Figure 4-40 showing the dispersion of the waves 0.75 cm from the keeper face. The dispersion clearly exhibits a linear dependence between the frequency (ω) and the wave number (k) which strongly supports the presence of the ion acoustic instability. This conclusion was strengthened by measurements of the wave phase velocity, which the authors determined by fitting a line to the dispersion plot as shown in Fig. 4-40. They also determined directly from the measurements the anomalous collision frequency and compared it to the classical electron-ion values. The results are plotted in Fig. 4-41 and show that there are a number of operating conditions – most notably high current to flow ratios – where the anomalous collision frequency is dominant over classical values.

The theoretical predictions by Mikellides of anomalous resistivity due to IAT in the hollow cathode plume, and the experimental confirmation by Jorns of the presence there of IAT, established this as a dominant non-classical process that is fundamental to the operation of most hollow cathodes used in electric propulsion. The presence of IAT in these devices has implications not only on the electrons, but also on the heating of the ions that possibly produces erosion of the keeper. All these topics comprise an active area of research.

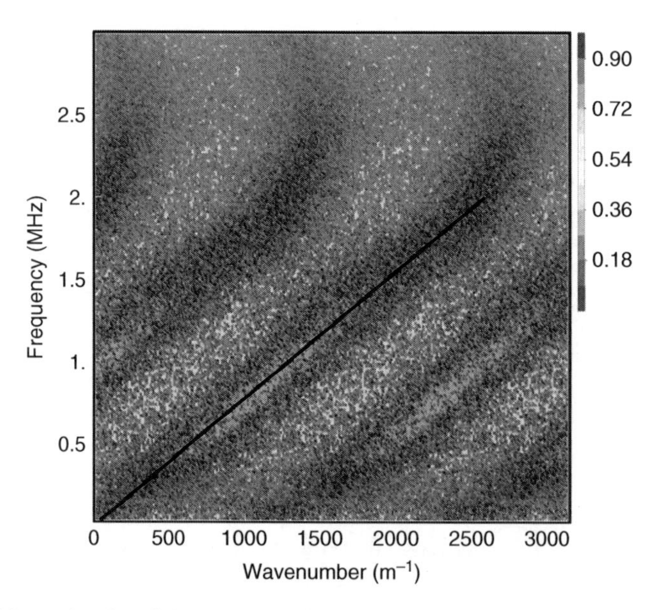

Figure 4-40 Beall intensity plot of the wave dispersion in the plume of a hollow cathode operating at a discharge current of 90 A and a flow rate of 10 sccm. The line is a best fit to the linear dispersion of the ion acoustic mode (*Source:* [78]).

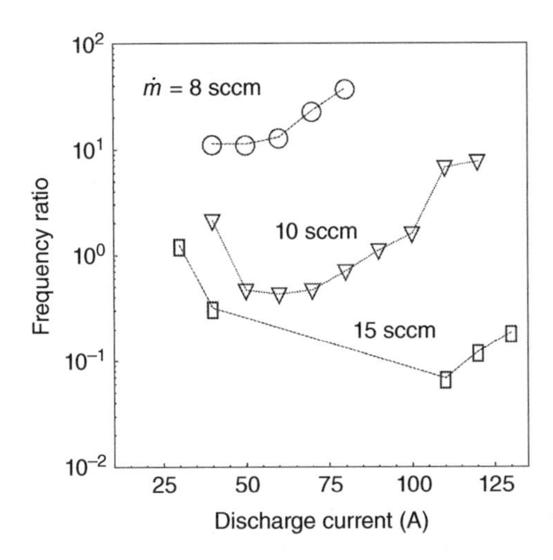

Figure 4-41 Ratio of anomalous collision frequency to classical electron-ion collision frequency versus discharge current where the anomalous frequency exceeds classical values at high current to flow ratios (*Source:* [78]).

4.7 Heating and Thermal Models

Hollow cathodes must be heated to thermionic temperatures to start emission and the plasma discharge. The heater is usually made of a refractory metal wire such as tantalum that is electrically insulated from the cathode tube by a ceramic structure. This structure can be in the form of a coaxial sheathed heater, a thin film heater on a ceramic substrate around the cathode, or a heater wire laid into a spiral or bifilar wound slot in a ceramic part placed around the cathode.

4.7.1 Hollow Cathode Heaters

Barium oxide dispenser hollow cathodes manufactured in the United States typically use a coiled coaxial tantalum sheathed heater [79, 80] that features a magnesium oxide powder insulation. This swaged heater geometry is common in industrial furnaces and can be found in the standard catalog of several companies. For space applications the tantalum heaters have been developed with various sizes and lengths to provide power levels of tens of watts to over 100 W. The MgO insulation material has a maximum operation temperature of about 1400 °C, above which chemical reactions between the oxide insulation and the heater electrode or sheath material cause a reduction in the resistance and ultimately lead to failure of the heater [80]. These heaters have demonstrated thousands of hours of on-time and over 10,000 on/off cycles in qualification tests for flight of BaO-W dispenser hollow cathodes [79].

Lanthanum hexaboride hollow cathodes typically need more heater power than a conventional BaO-W dispenser cathode and the capability to go to higher temperatures to achieve the temperatures associated with the higher work function of LaB_6. A tantalum sheathed heater that incorporates high-temperature alumina powder insulation was developed [37, 81] and has been used to heat LaB_6 hollow cathodes of many different sizes. The powdered alumina insulation has a maximum temperature of about 1800 °C, which is well in excess of the temperature required to start the LaB_6 cathode.

A 0.24 cm-dia. sheathed Al_2O_3-insulated tantalum heater routinely provides 120 W of heater power for the 0.63-cm-dia. LaB_6 cathode [81], which has accumulated over 1000 starts to date. This same size sheathed heater is used for a 1.5-cm dia. LaB_6 cathode [19, 37], which has a 50% longer

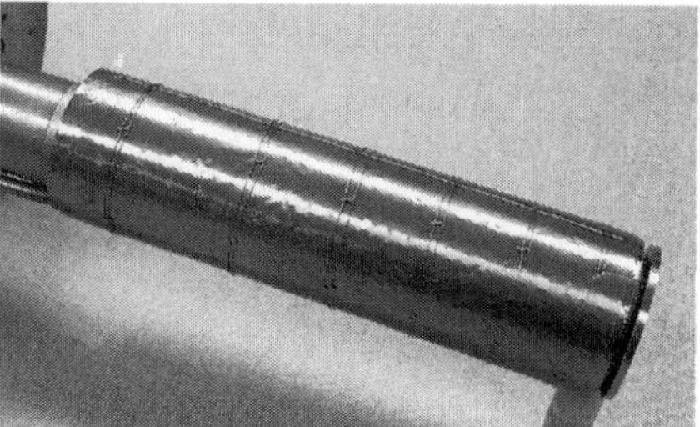

Figure 4-42 Sheathed heaters wound around the 2-cm-outside-dia. LaB_6 cathode tube (left) and the overwrapped tantalum foil heat shielding (right).

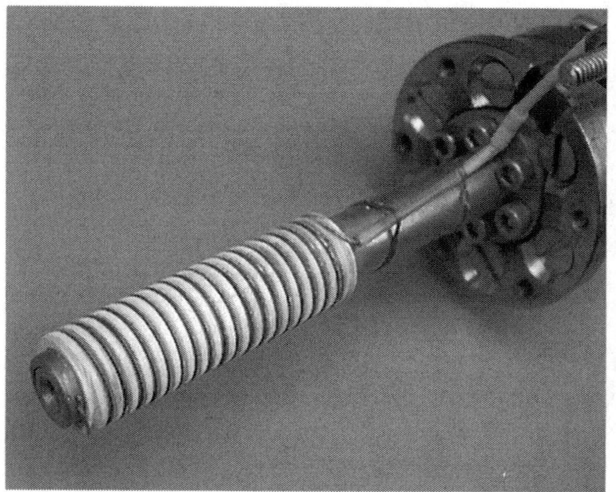
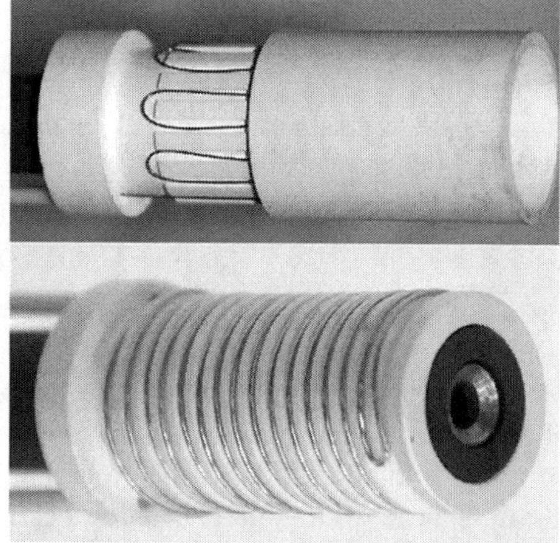

Figure 4-43 Three filament-style cathode heaters wound on ceramic mandrels (*Source:* [81]/AIP Publishing LLC).

length before winding than used in the smaller cathode to provide 280 W of heater power. A 2-cm dia. LaB_6 cathode [82, 83] typically uses 380 W of heater power from two of the sheathed heaters wound in parallel around the cathode tube, as seen in Fig. 4-42 left, to heat and ignite the larger cathode. Also seen in Fig. 4-42 right is the tantalum foil heat shield wound around the heater to reduce radiation losses from the cathode.

An alternative heater configuration consists of a filament wound in a ceramic mandrel placed around the cathode tube [84].Figure 4-43 shows several views of such a filament heater in a single coil or bifilar wound configuration. A cylindrical ceramic sheath is also normally slid over the heater filament assembly to insulate the filament from the layers of refractory metal foil heat shielding wound around the outside of the heater assembly to reduce the radiation losses out of the cathode.

4.7.2 Heaterless Hollow Cathodes

While all of the flight cathodes built to date have used an external heater, this heater is subject to failures associated with insulation breakdown, filament opening/shorting, and the electrical connection opening. To avoid these problems, heaterless ignition has been pursued where a Paschen discharge [1] is initiated between the cathode orifice plate or insert and the keeper electrode to provide the heat required to bring the emitter up to emission temperatures. The very first hollow cathodes that used straight refractory-metal tubes utilized a high voltage ignitor pulse to start a Paschen/arc discharge to heat the end of the cathode tube and ignite the cathode [2]. Called "heaterless hollow cathodes", this technology has been the subject of many development programs and a review by Lev et al. [85] summarized the progress with heaterless cathodes worldwide until 2019.

Heaterless hollow cathode ignition [86] can be divided into three phases: (i) initial electrical breakdown of the keeper discharge, (ii) heating of the cathode, and (iii) steady-state operation. Ignition is started by increasing the gas flow rate through the cathode to increase the local pressure between the cathode and the keeper. The keeper electrode is sometimes designed with a smaller diameter [85] or relatively long orifice [86] to aid in increasing the pressure just upstream. Applying high voltage between the keeper and cathode initiates a Paschen breakdown [1] that transitions into a glow discharge in the heating phase sustained by field emission and secondary electron emission. This phase heats the thermionic emitter to thermionic electron emission temperatures. The discharge then transitions to the third phase of thermionic emission and steady-state discharge operation. At this point, the gas flow through the cathode is reduced to the nominal value and the discharge current transferred to the thruster anode.

The major issue with heaterless ignition is arc formation during the high voltage (400–2000 V) Paschen breakdown phase. Arcs can cause significant damage to the cathode electrodes and limit both the number of ignitions and the cathode life. Arc initiation can be controlled by limiting the breakdown current level [87], and Robson reported [88] that a total arc energy of 0.25 J was non-destructive for a hollow cathode ignition. This typically requires limiting the peak arc current to about 0.5 A, which is below [89] the "chopping current" or "arc sustaining current" on tungsten where the arc tends to anchor on the orifice plate and causes significant surface damage. Below the self-sustaining current level, arcs are sometimes called "microarcs" because they are momentary and not anchored, which causes much less surface damage.

Care in avoiding arcing and/or limiting it to short duration microarcs are key to enabling repeated heaterless ignitions. Figure 4-44 shows a cross section photograph of a small heaterless LaB_6 hollow cathode [86] that successfully achieved 25,000 ignitions. This cathode features a long keeper orifice to raise the pressure in the cathode-keeper gap to reduce the Paschen initiation voltage to the order of only 500 V. The cathode operates in the current range of 0.5–4 A, and is designed with extremely low heat capacity and thermal conduction to the support structures in the thruster

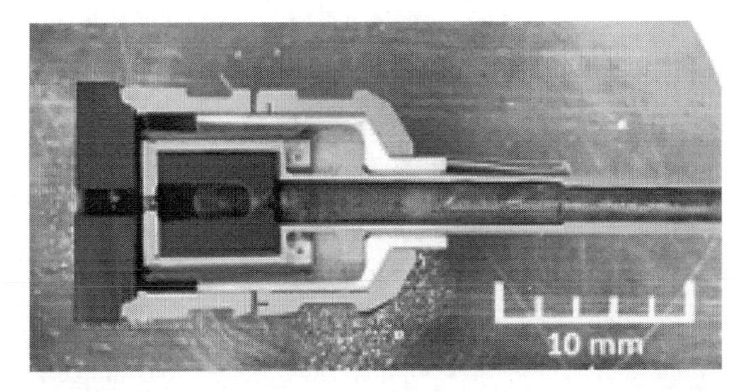

Figure 4-44 Compact LaB_6 heaterless cathode that successfully achieved 25,000 ignitions (*Source:* [86]/with permission of Elsevier).

to achieve ignition and heating to full thermionic discharge current in less than 1 s. This cathode has also operated for over 10,000 h at the full 4 A discharge current, indicating that the size and operating temperature of the LaB_6 insert is appropriate for long life.

Heaterless cathode technology has been recently extended to much larger, higher current hollow cathodes capable of discharge current over 50 A [90]. The new heaterless design exploits the tendency the Paschen discharge to strike at the longest path length between cathode potential electrodes and the keeper, and uses the capabilities of refractory metal hollow cathode tubes to act as an internal heater. The gas feed line into the upstream side of the hollow cathode is extended with a tantalum tube part way into the insert region such that when the Paschen discharge ignites between the tip of the tube and the keeper, the heat generated inside the tube (primarily in the last 1-cm where the discharge attaches) radiates directly to the insert. In this "tube-radiator" heater configuration, the Paschen discharge connects directly to the tantalum tube in a hollow cathode mode (sustained by the inner surface of the tube where the neutral gas pressure is high). The refractory metal tantalum tube is very robust and can sustain microarcing during conditioning without damage. The discharge spends very little time (on the order of seconds) in the >250 V Paschen discharge regime, reducing the significance of the sputtering and arcing erosion mechanisms of this high voltage discharge altogether. The tantalum tube reaches thermionic emission temperatures above 2300 °C within seconds, at which point the discharge transitions to an intermediate-voltage (45–65 V) tantalum thermionic discharge, which easily provides 100–300 W of heating power radiatively to the insert. Once the insert reaches thermionic emission temperatures, it takes over the majority of the electron production, the discharge voltage falls due to the lower work function emitter, and the tantalum tube temperature drops as the tube discharge falls to negligible levels. Because the insert is being radiatively heated from the inside, thermal losses to the downstream orifice plate and upstream cathode components are reduced, so a simple ignition procedure lasting only a few minutes is required to ignite large hollow cathodes [90]. This new heaterless technology has been successfully applied to larger cathodes capable of hundreds of amperes of discharge current [91], and investigations continue for electric thruster applications.

4.7.3 Hollow Cathode Thermal Models

The thermal response of the cathode to the internal plasma loads impacts how well it will perform and also how long it will last. In cases where the plasma is fairly uniform in the insert region (such as in low pressure Type-B cathodes or Type-C cathodes), it may be possible to employ reduced-order plasma models [42, 44, 57, 92] like those described in Section 4.4 to obtain general trends about the cathode's thermal response and provide some guidance for hollow cathode design. The next level of accuracy is obtained by using the plasma fluxes to the cathode surfaces from 2-D plasma simulations that are fed into a thermal model without coupling between the two models. For example, this is the approach followed by Katz [47, 49]. More advanced techniques have been developed [93, 94] that couple 2-D plasma and thermal models to provide overall cathode-system performance. Such coupled 2-D plasma and thermal models provide the best predictions of the cathode performance and behavior at the expense of added model complexity and computer run times. Key to all these approaches is having a tractable thermal model.

Figure 4-45 shows a sample input geometry for a 2-D cathode thermal model [47] in r-z coordinates that uses the ion and electron fluxes from the 2-D plasma simulations [12] as input to predict the temperature distribution in the cathode. In this figure, the positive numbers identify different materials, and negative numbers are used to identify radiative boundary conditions. The code includes thermal conduction, radiative heat losses, and radiative heat transfer within the insert

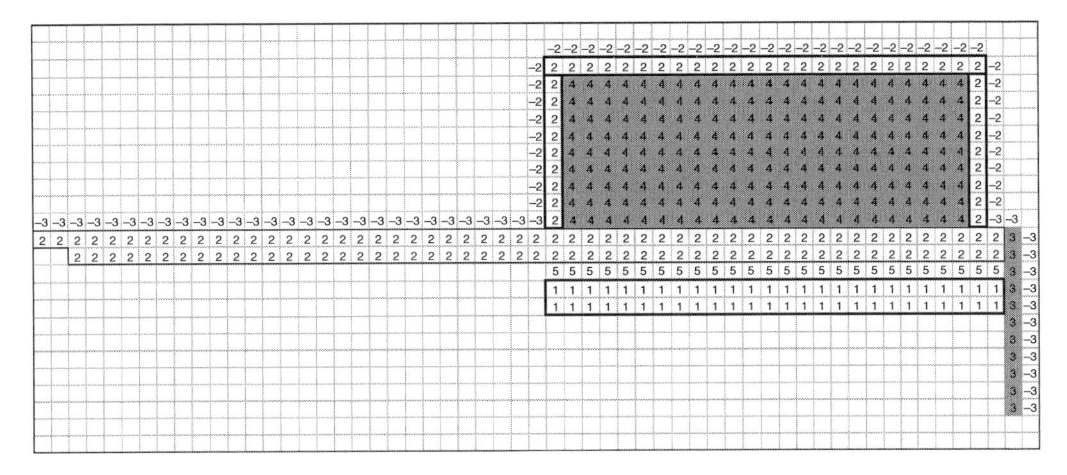

Figure 4-45 Hollow cathode thermal model input geometry with cells and boundary conditions numbered. Negative numbers indicate radiative boundaries (*Source:* [49]).

region. The plasma fluxes from the simulations also include heating of the cathode tube and insert due to power deposition in the orifice region that conducts to the tube and radiates directly to the insert.

Plasma simulations were performed of all three types of cathodes shown in Fig. 4-5 using the OrCa2D global plasma model to provide the distribution of power to the various surfaces including the cathode orifice [18]. In Type A cathodes, the plasma density peaks very close to the orifice plate, and the heating is primarily in the cathode orifice region. In Type C cathodes, the plasma peaks significantly upstream of the orifice plate, and almost all the heating occurs along the cathode insert. This upstream progression of the heating in the cathode as the orifice diameter is increased is shown graphically in Fig. 4-46. The heating reaches its maximum value very close to the orifice

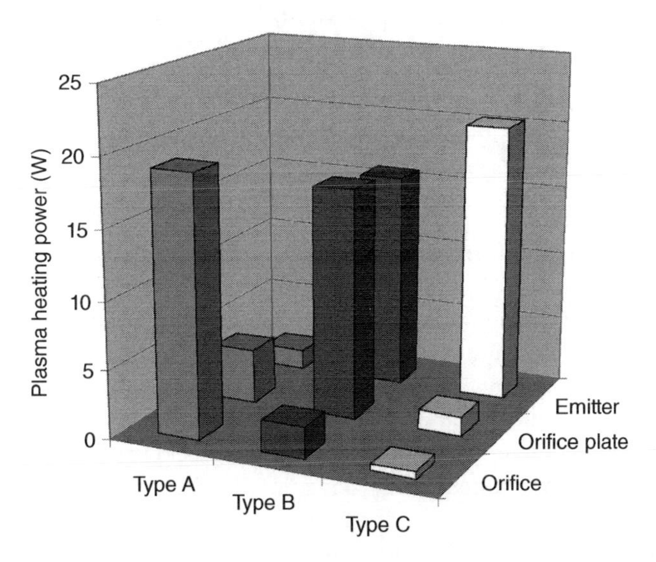

Figure 4-46 Plasma heat loads to various components of the three types of cathodes described in Figure 1 from 2-D numerical simulations (*Source:* [18]).

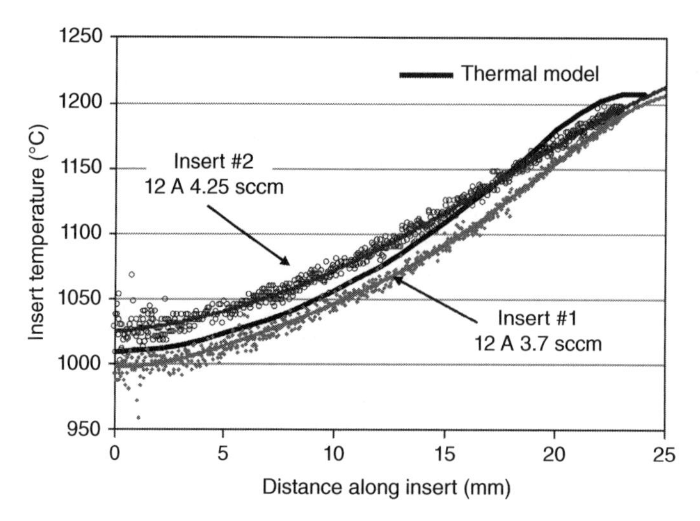

Figure 4-47 Insert temperature profile for the NSTAR cathode running at 12 A with two xenon flow rates (*Source:* [51]).

plate for the smallest orifice, Type A cathode, and moves progressively upstream as the orifice diameter increases in the Type B and Type C cathodes [18].

As an example of the results from a coupled thermal and cathode plasma model, Fig. 4-47 shows the code predictions from the thermal model [18] and measured temperatures [51] of the insert for the NSTAR cathode running at 12 A of discharge current. The 2-D code predicts an insert temperature of about 1210 °C in the first few millimeters from the orifice plate where the plasma is in good contact with the insert. The thermal model predicts a peak temperature of about 1190 °C for the heat loads from a plasma contact area of 5 mm predicted by the plasma code. The 2-D codes also show the sensitivity of the hollow cathode temperature to the emissivity of the orifice plate, the thermal contact between the emitter and the tube, and orifice heating effects (especially in neutralizer cathodes), which impact the performance and life of the cathode. These effects cannot be obtained from the simple 0-D models described above.

4.8 Hollow Cathode Life

Cathode insert life is fundamentally determined by either depletion of the barium emissive mix impregnated into the dispenser cathodes such that the surface work function is degraded [95], or by evaporation of the bulk material in refractory metal cathodes, such as tungsten and tantalum, and in crystalline cathodes such as LaB_6. The cathode mechanical structure (orifice plate, heater, cathode tube, etc.) can also be worn or degraded [53, 54] by ion-induced sputtering, which affects the cathode life. The impact on the cathode performance by these life-limiting fundamental mechanisms is important to understand in designing cathodes for electric thrusters. In addition, poisoning of inserts due to impurities in the feed gas or improper exposure to air can also increase the work function and impact cathode life.

4.8.1 Dispenser Cathode Insert-Region Plasmas

In dispenser cathodes, evaporation of the barium layer coating the cathode surface is well understood, and depletion life models can be readily constructed if this is the root cause of barium loss

[95]. However, the emitter surface is exposed to a plasma and ion bombardment of the surface by ions from the insert-region plasma that can increase the loss of barium from the surface, which will reduce the lifetime of the cathode. Although the basic concept of a hollow cathode is to reduce the erosion and modification of the low work function insert surface with a high pressure, collisional insert-region plasma, this benefit had to be validated before the cathode life could be predicted [95].

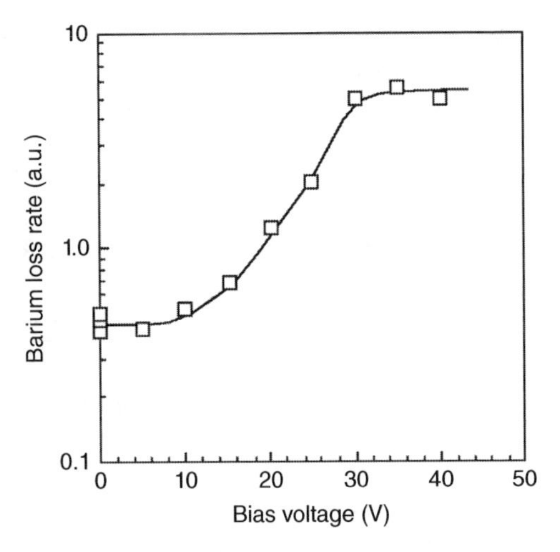

Figure 4-48 Variation of barium loss rate from the cathode surface at 725 °C with cathode bias voltage (*Source:* [96]).

An experimental and theoretical study of enhanced barium evaporation from dispenser cathode surfaces was undertaken [96] to determine the plasma conditions in the insert region under which an evaporation model could be used. The experimental arrangement measured the barium evaporation from a Type-S 4 : 1 : 1 impregnated porous tungsten cathode with an embedded heater during xenon plasma bombardment. The cathode could be biased negatively relative to the plasma to control the ion bombardment energy. The barium evaporation rate was measured by a fiber optic coupled to a visible wavelength spectrometer tuned to detect the emission intensity of the Ba-I line at 553.5 nm excited in the plasma. Since the emission intensity depends on the amount of Ba present in the plasma and the electron density and temperature, the plasma parameters were monitored with a probe and the Ba-I signal normalized to a neutral xenon line to account for any variations in plasma parameters during the measurements.

Figure 4-48 shows the barium loss rate measured at 725 °C versus the ion bombardment energy. Increasing the ion bombardment energy from 10 to 30 eV increases the barium loss rate by an order of magnitude. Figure 4-49 shows the barium loss rate as a function of temperature for two cathode

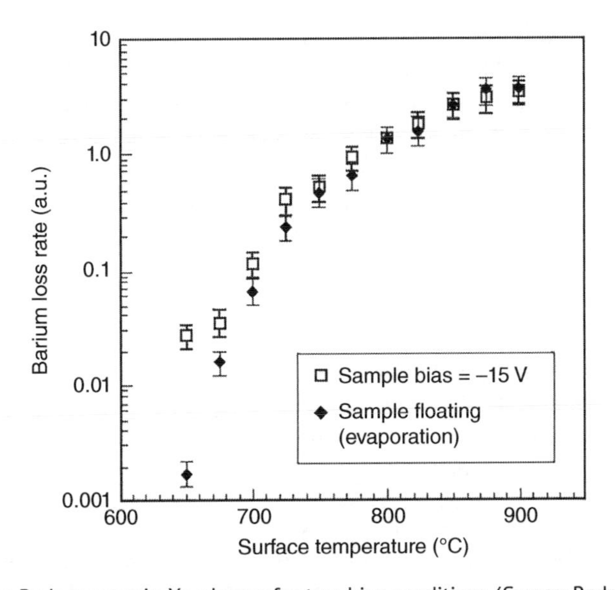

Figure 4-49 Relative Ba loss rates in Xe plasma for two bias conditions (*Source:* Redrawn from [96]).

bias energies. For the case of the cathode floating relative to the plasma, the ion bombardment energy is only a few eV and the barium loss rate is determined solely by thermal evaporation. For a bias energy of 15 eV, the barium loss rate is found to be the same as for thermal evaporation for cathode temperatures in excess of about 800 °C. Since the hollow cathodes in most thrusters operate at insert temperatures in excess of 1000 °C, this data shows that the barium loss rate is determined by thermal evaporation rates.

A model of the enhancement of barium evaporation for a surface under energetic ion bombardment was developed by Doerner [97] to explain this behavior. At elevated surface temperatures, two classes of surface particles must be considered: (i) those particles that are bound to the material lattice structure (denoted here as "lattice atoms") and (ii) atoms that have been liberated from the lattice structure, but which are still bound to the material surface with a reduced binding energy (denoted here as "adatoms"). Both species can sublimate from the material surface if an atom receives enough kinetic energy from random collisions to break free from the surface; however, because the binding energy for the two species is different, the corresponding loss rate will also be different.

The net flux of material Γ_T from the surface can be written as

$$\Gamma_T = \Gamma_i Y_{\rm ps} + K_o n_o e^{-E_o/T} + \frac{Y_{\rm ad}\Gamma_i}{\left(1 + A\exp^{(E_{\rm eff}/T)}\right)}, \tag{4.8-1}$$

where Γ_i is the plasma ion flux, n_o is the areal atom density, $K_o \approx v_{\rm th}/\lambda$, λ is the lattice scale length, $v_{\rm th}$ is the thermal velocity of the particles, $Y_{\rm ad}$ is the adatom production yield from the incident ion flux, $Y_{\rm ps}$ is the sputtered particle yield, $Y_{\rm ad}\Gamma_i$ is equal to the adatom loss rate due to both sublimation and recombination. The first term in Eq. 4.8-1 describes physical sputtering of lattice atoms (which is independent of surface temperature), the second term describes the thermal sublimation of lattice atoms, which is independent of ion flux, and the third term describes the losses due to adatom production and subsequent sublimation, which depends on both the incident ion flux and the surface temperature. For Xe ions incident on BaO at 30 eV, $Y_{\rm ps} = 0.02$, $A = 2 \times 10^{-9}$ and

$$\frac{Y_{\rm ad}}{Y_{\rm ps}} \approx 400. \tag{4.8-2}$$

These parameters can be used to model the expected net flux of Ba from a surface under bombardment with 30 eV Xe ions for various surface temperatures. The result of this model is compared with experimental measurements of Ba emissivity under these conditions in Fig. 4-50. The model compares extremely well with the experimental results. The model also qualitatively explains the key experimental observations, including the effect of ion energy on net erosion, the saturation of the adatom loss term at elevated temperatures, and the transition to losses dominated by thermal sublimation of lattice atoms at elevated temperatures. The model has been used to examine the effect of increasing the ion flux to the surface from the values in these experiments to the actual values for the hollow cathodes found from the 2-D plasma model. In this case, the model predicts that thermal evaporation dominates the barium loss rate for ion energies of less than 15 eV and cathode temperatures of over 900 °C. The model provides confidence that the barium loss rate effects in the plasma are understood and that the main result of barium loss determined by thermal evaporation rates for the plasma parameters of thruster cathodes examined here is accurate.

4.8.2 BaO Cathode Insert Temperature

Since the barium evaporation rate for the plasma conditions found in the hollow cathodes is determined by the insert surface temperature, a non-contact temperature measurement technique was developed at JPL [98] to directly measure the insert temperature of barium oxide (BaO) cathodes

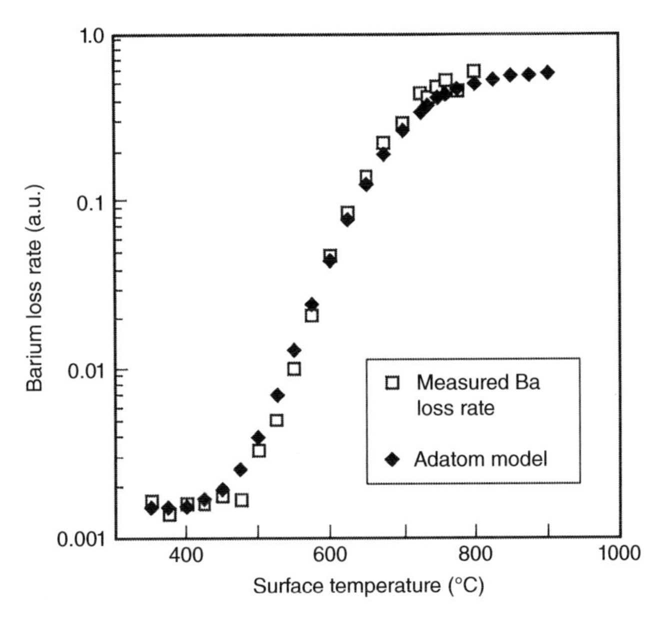

Figure 4-50 Ba concentration versus cathode surface temperature for −15 V bias. Experimental data are shown by open squares and the model predictions by the open diamonds (*Source:* Redrawn from [96]).

during cathode operation. The technique employs a stepper motor driven sapphire fiber optic probe that is scanned along the insert inside diameter and collects the light radiated by the insert surface. Ratio pyrometry is used to determine the axial temperature profile of the insert from the fiber optic probe data. Thermocouples attached on the outside of the cathode on the orifice plate provide additional temperature data during operation and are used to calibrate the pyrometer system in situ with a small oven inserted over the cathode to equilibrate the temperature.

Figure 4-51 shows temperature profiles measured for a nominal Space Station Contactor (SSC) cathode [98] operating at four different discharge currents. The peak temperature of the insert at the full 12 A current level is about 1200 °C. The insert also has approximately a 10–15% temperature

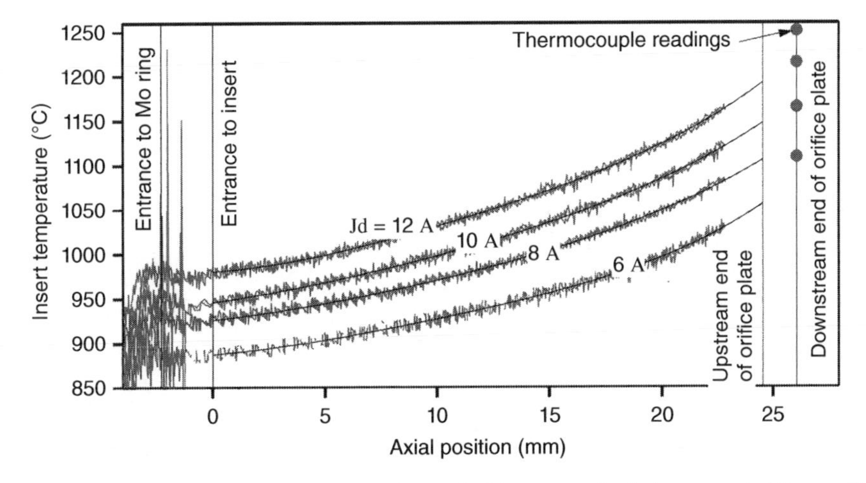

Figure 4-51 Insert temperature profile measured for a SSC hollow cathode for several different discharge currents (*Source:* [98]).

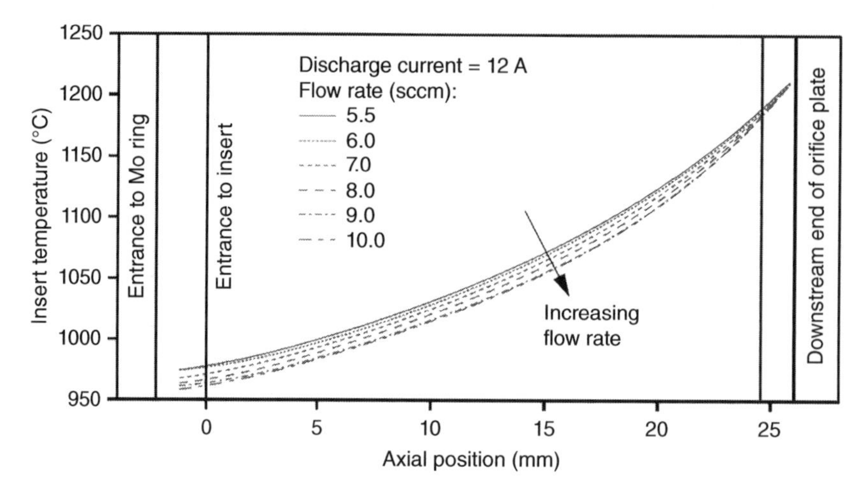

Figure 4-52 Insert temperature profile for a SSC hollow cathode operating at 12 A of current for several gas flow rates (*Source:* [98]).

gradient along its length. The change in the insert temperature with the xenon flow rate for the cathode producing 12 A of discharge current is shown in Fig. 4-52. High flow rates through the cathode is observed to reduce the insert temperature, although the effect is small.

A direct comparison of the insert temperature profile for the NSTAR discharge cathode and the Space Station Plasma Contactor cathode at identical discharge currents of 12 A and xenon flow rates of 6 sccm is shown in Fig. 4-53. The NSTAR insert temperature is higher than the SSC all along the insert. It also appears that the temperatures of the inserts tend to converge near the orifice plate. The high insert temperature for the NSTAR cathode is likely because the plasma contact area is significantly larger at the roughly 50% lower internal pressure compared to the SSC. In addition, thermocouple measurements on the orifice plate show that the smaller diameter SSC orifice plate is

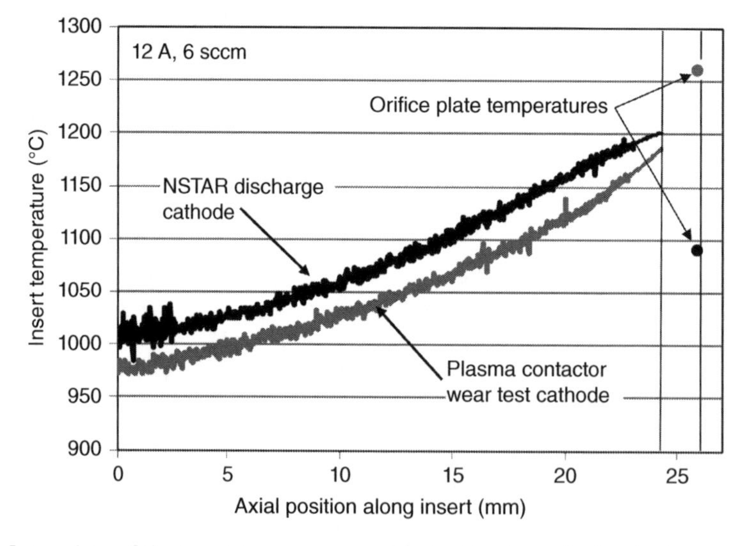

Figure 4-53 Comparison of the insert temperature profile for the Space Station Contactor cathode and the NSTAR discharge cathode at 12 A (*Source:* [51]).

significantly hotter than the NSTAR orifice plate, consistent with orifice heating effects described in Section 4.5 for smaller orifice diameters.

4.8.3 Barium Depletion Model

The previous sections showed that the barium loss rate from hollow cathode dispenser cathode inserts should be essentially the same as dispenser cathode inserts operated in vacuum if the ion bombardment energy is sufficiently low in the hollow cathode. Since plasma potentials on axis in the insert-region plasma of less than 15 V are routinely measured [11, 46] and sheath potentials of less than 10 V are found from the models discussed above, the insert life will be limited by evaporation in the same manner as in vacuum devices.

Published measurements by Palluel and Shroff [99] of the depth of barium depletion in dispenser cathodes as a function of time and temperature show that barium depletion obeys a simple diffusion law with an Arrhenius dependence on temperature. This is shown in Fig. 4-54, where the impregnate surface layer in the pore recedes with time. The activation energy in the diffusion coefficient that determines the slope of the curves in Fig. 4-54 appears to be relatively independent of the cathode type.

From data presented Fig. 4-54, the operating time to deplete impregnate from the insert material to a depth of 100 μm is

$$\ln t_{100\mu m} = \frac{eV_a}{kT} + C_1 = \frac{2.8244e}{kT} - 15.488, \tag{4.8-3}$$

where the operating time $\tau_{100\mu m}$ is in hours, e is the elementary charge, V_a is the activation energy, k is Boltzmann's constant, C_1 is a fit coefficient, and T is the insert temperature in degrees Kelvin. The activation energy was found from Fig. 4-54. Using this relationship and the fact that the depletion

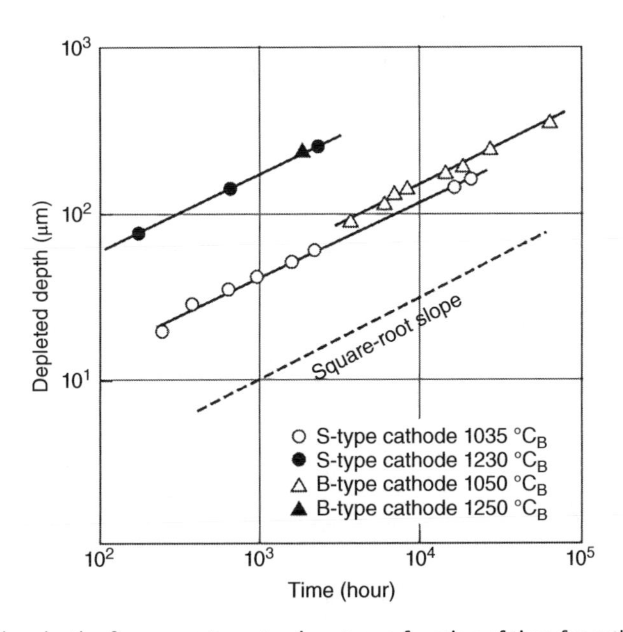

Figure 4-54 Depletion depth of a porous tungsten insert as a function of time for cathodes at different temperatures (*Source:* [99]/American Institute of Physics).

depth is proportional to the square root of the operating time [99], an equation yielding the insert lifetime due to barium depletion can be derived [95]:

$$t_{\text{life}} = t_{100\mu\text{m}} \left(\frac{y}{y_{100\mu\text{m}}} \right)^2, \tag{4.8-4}$$

where $\tau_{100\mu m}$ is the time to deplete to 100 μm in depth from Eq. 4.8-3, y is the insert thickness in μm, and $y_{100\mu m}$ is the 100 μm reference depth. Using Eq. 4.8-3 in Eq. 4.8-4, the life of a Type S dispenser cathode in hours is

$$t_{\text{life}} = 10^{-4} y^2 \exp\left(\frac{2.8244e}{kT} - 15.488 \right), \tag{4.8-5}$$

where y is the insert thickness in μm and T is the insert temperature in degrees Kelvin. Figure 4-55 shows the insert life for a 1 mm depletion depth versus the insert temperature. Insert life of over 100,000 h is readily achievable if the insert is thick enough. At around a nominal 1100 °C operating temperature, the life increases a factor of two if the temperature decreases 40 °C.

This model represents a worst-case estimate of the cathode life. In very high-density hollow cathodes, like the NSTAR cathode, the ionization mean free path for the evaporated barium is significantly less than the insert-region plasma radius. This means that a large fraction of the barium is ionized close to the insert surface. The electric field in this region is primarily radial, which means that some large fraction of the barium is recycled back to the surface. The barium surface coverage is then partially re-supplied by recycling, which can extend the life considerably.

To predict cathode life in a thruster application from an insert depletion mechanism, a relationship between the insert temperature and the discharge current at a given gas flow must be obtained. The SSC insert-temperature was measured versus discharge current by Polk [98].This data is well fit in the plasma contact region (the 3 mm closest to the orifice plate) by

$$T = 1010.6\, I_d^{0.146}\ \text{K}. \tag{4.8-6}$$

At 12 A of discharge current, this gives an insert temperature of 1453 K. Since the insert in this cathode was about 760 μm thick and we assume that the insert is depleted when the depth reaches about 2/3rds of the thickness (due to some barium diffusion out the outside diameter of the insert),

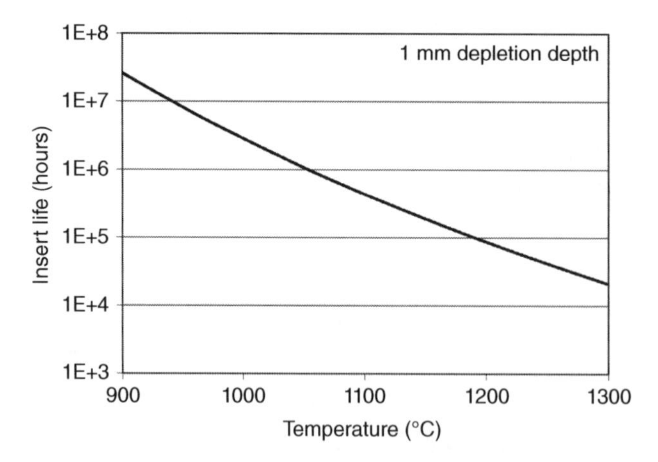

Figure 4-55 Insert life from barium evaporation calculated for a 1 mm depletion depth versus temperature.

Eq. 4.8-5 predicts a life of 30,000 h. This is in good agreement with the SSC life test data where the cathode failed to start after about 28,000 h at 12 A of discharge current [100]. In this case, barium recycling may not affect the insert life significantly because the plasma is in contact with the insert for only a couple of millimeters from the orifice plate, and the barium will tend to migrate to upstream regions that are not involved in the emission process.

For the NSTAR cathode, the insert temperature data as a function of discharge current measured by Polk [51] is well fit in the plasma contact region by

$$T = 1191.6\, I_d^{0.0988} \text{ K}. \tag{4.8-7}$$

At the full power discharge current of 13 A, and using the insert thickness of 760 μm, Eq. 4.4-5 predicts an insert life of 20,000 h. The ELT ran at full power for about 14,000 h and accumulated an additional 16 352 h at much lower discharge currents [60]. The barium depletion model indicates that the insert should have been depleted in the emission zone in less than 24,000 h. Measurements indicate partial depletion in the emission region near the orifice, but that as much as 30% of the original barium was still present. Clearly barium recycling in the plasma [43] reduced the effective evaporation rate and extended the life of the cathode significantly.

For the NEXIS hollow cathodes, the operating insert temperature profile has not yet been measured to date. Estimates of the insert temperature were made using an early version of a combined plasma and thermal model [93]. Since the discharge loss and efficiency performance of the NEXIS thruster is known, the relationships in Chapter 2 can be used to plot thruster life versus engine performance. The NEXIS ion thruster operates at 75–81% efficiency over an Isp of 6000–8000 s [101]. Figure 4-56 shows the model predicted depletion limited life of this insert versus Isp for several thruster power levels. At the nominal operating point of 7000 s Isp and 20 kW, the cathode is projected to operate for about 100,000 h. Increasing the Isp requires operation at higher beam voltages, which for a given power requires less beam current and, thereby, less discharge current. A lower discharge current reduces the insert temperature for a given cathode size, which reduces the barium evaporation rate and extends the cathode life. Likewise, lower Isp and higher power require higher discharge currents, which translate to a reduction in the cathode life. It should

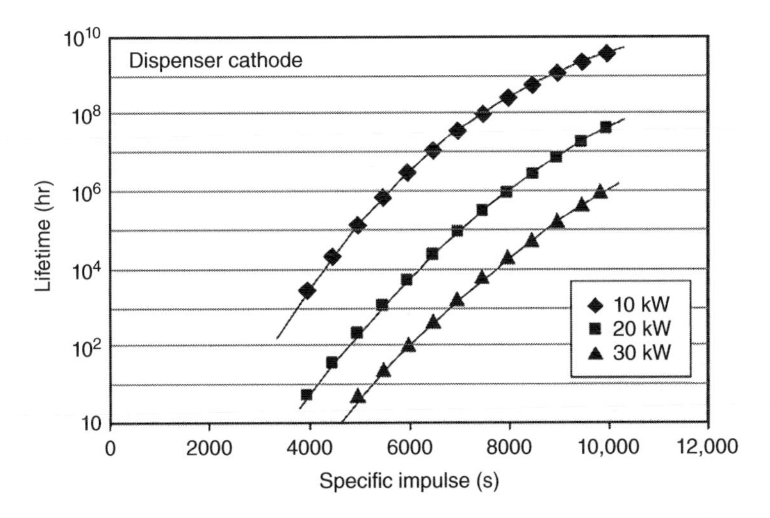

Figure 4-56 Model prediction of NEXIS dispenser cathode life versus specific impulse for several thruster powers (*Source:* [95]).

be noted that the cathode life in Eq. 4.8-5 scales as the insert thickness squared, so the life at any operating point in Fig. 4-53 can be extended simply by increasing the thickness of the insert. This may require increases in other dimensions, but proper selection of the cathode diameter and orifice size can be made to maintain the insert temperature at the desired level to provide the desired life.

4.8.4 Bulk-Material Insert Life

Cathodes that are based on bulk insert material instead of dispenser chemistry, such as LaB_6, have a lifetime that is determined by the evaporation of the insert material inside the hollow cathode [37]. In plasma discharges, sputtering of the LaB_6 surface can also impact the life [39]. However, as in dispenser hollow cathodes, the plasma potential is very low in the insert region and the bombardment energy of xenon ions hitting the surface is typically less than 20 V, which virtually eliminates sputtering of the cathode surface. It is assumed that the evaporated material leaves the cathode and does not recycle to renew the insert surface, which will provide a lower estimate of the insert life than might actually exist. Interestingly, as the insert evaporates, the inner diameter increases and the surface area enlarges. This causes the required current density and temperature to decrease at a given discharge current, which reduces the evaporation rate of the insert with time.

The life of the LaB_6 insert for three different cathode diameters versus discharge current was calculated based on the evaporation rate at the temperature required to produce the discharge current in the thermally-limited regime [33, 37]. Assuming that 90% of the insert can be evaporated, the cathode life is shown in Fig. 4-57 as a function of the discharge current. Lifetimes of tens of thousands of hours are possible, and the larger cathodes naturally tend to have longer life. Although other mechanisms, such as temperature variations along the insert or LaB_6 surface removal or material build-up due to impurities in the gas, can potentially reduce the life, re-deposition of the evaporated LaB_6 material will tend to extend the cathode life. Therefore, these life estimates for the different cathode sizes are mostly valid relative to each other, and the actual life of the cathode can be considered to be on the order of the values shown in Fig. 4-57.

To obtain an idea of the life of a LaB_6 cathode relative to a conventional dispenser cathode, the predictions from a dispenser cathode life model [95] applied to the NSTAR cathode are compared to

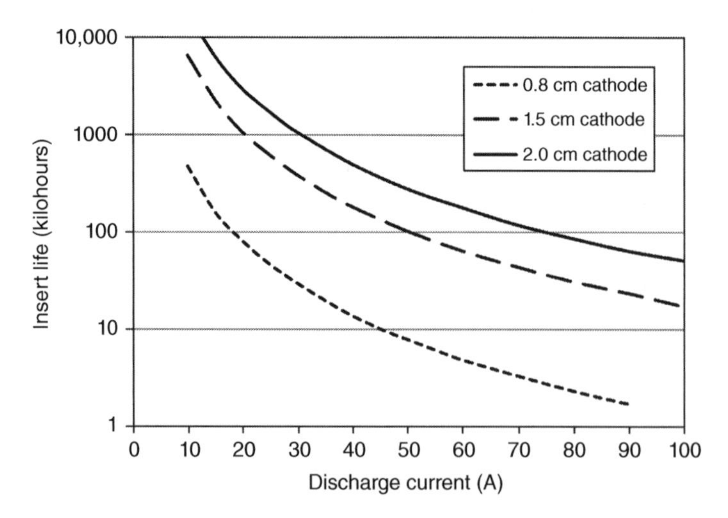

Figure 4-57 Calculated lifetime in thousands of hours versus the discharge current for three different LaB_6 cathode diameters.

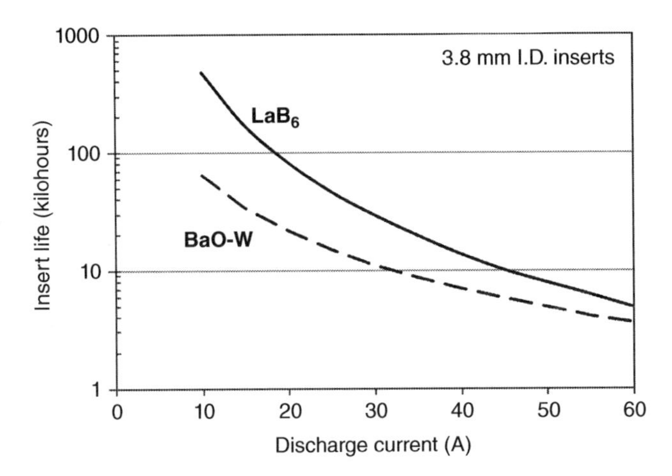

Figure 4-58 Comparison of the calculated cathode lifetime versus the discharge current for the 0.8 cm O.D. LaB6 cathode and the NSTAR dispenser cathode.

the 0.8 cm LaB_6 cathode life predictions in Fig. 4-58. These two cathodes have similar insert diameters and lengths and so a direct comparison is possible. The dispenser cathode calculation assumes that barium evaporation from the insert surface causes depletion of nearly all the barium impregnate at the end of life in the NSTAR cathode at the measured [51] insert temperature and temperature gradient. This provides an upper limit to the dispenser cathode life if other mechanisms, such as poisoning degrading the work function or impurity build-up plugging the pores, actually cause the cathode life limit. Likewise, recycling of the barium will extend the dispenser cathode life, so uncertainties in the dispenser cathode life estimates by this model have the same uncertainties due to impurities and re-deposition that are found for the LaB_6 life model (although LaB_6 is less likely to be affected by impurities). Therefore, a direct comparison of calculated life versus discharge current will be made with the understanding that the curves will likely shift together vertically due to impurity or re-deposition issues. The LaB_6 cathode life is projected to exceed the dispenser cathode life by nearly an order of magnitude at the nominal NSTAR full power currents of less than 15 A. If the NSTAR cathode is capable of producing higher discharge currents than 15 A, the LaB_6 cathode life is still projected to exceed the NSTAR life over the full current range demonstrated by the LaB_6 cathode. As was shown in Fig. 4-54, the larger LaB_6 cathodes will have even longer lifetimes, and their life significantly exceeds that projected for the NEXIS 1.5-cm dia. dispenser cathode [95] that is designed to operate up to about 35 A.

4.8.5 Cathode Poisoning

Comprehensive investigations of the poisoning of dispenser cathodes [25] and LaB_6 cathodes [102, 103] have been published in the literature. The most potent poisons for both cathodes are oxygen and water, with other gases such as CO_2 and air producing poisoning effects at higher partial pressures. Figure 4-59 shows the reduction percentage of the electron emission current density in diode tests of a Type-S 4 : 1 : 1 dispenser cathode and a LaB_6 cathode as a function of the partial pressures of oxygen and water for two different emitter temperatures. Oxygen partial pressures in the 10^{-7} Torr range can completely poison the dispenser cathode at temperatures of 1100 °C. In a similar manner, water vapor at partial pressures in the 10^{-6} Torr range will poison dispenser cathodes at temperatures below 1110 °C. For typical pressures inside hollow cathodes in excess of 1 Torr, partial pressures in

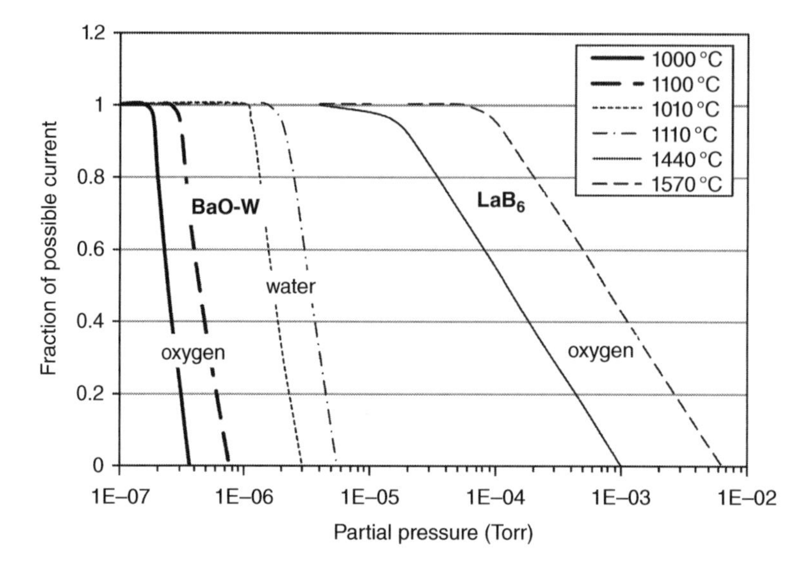

Figure 4-59 Percentage of possible thermionic emission versus partial pressure of oxygen and water showing the sensitivity of dispenser cathodes relative to LaB_6 (*Source:* [37]).

this range represent the best purity level that can be achieved by the gas suppliers, resulting in the high "propulsion-grade" purity mentioned above. This is the reason for stringent purity requirement levied on conventional dispenser hollow cathode in the United States to date.

Experiments by Polk [104] showed that oxygen poisoning observed in vacuum devices only occurs in hollow cathodes at low plasma densities (low discharge current) and high oxygen levels (>10 PPM) in the propellant gas, and that the plasma environment inside hollow cathodes tended to mitigate the poisoning of oxygen in dispenser cathodes. However, the plasma may aid in the formation of volatile tungsten oxides and tungstates from the impurity gases that contribute to tungsten migration and re-deposition on the insert surface. This modifies the dispenser cathode surface morphology, which may affect the emission capabilities. Figure 4-60 shows a photograph

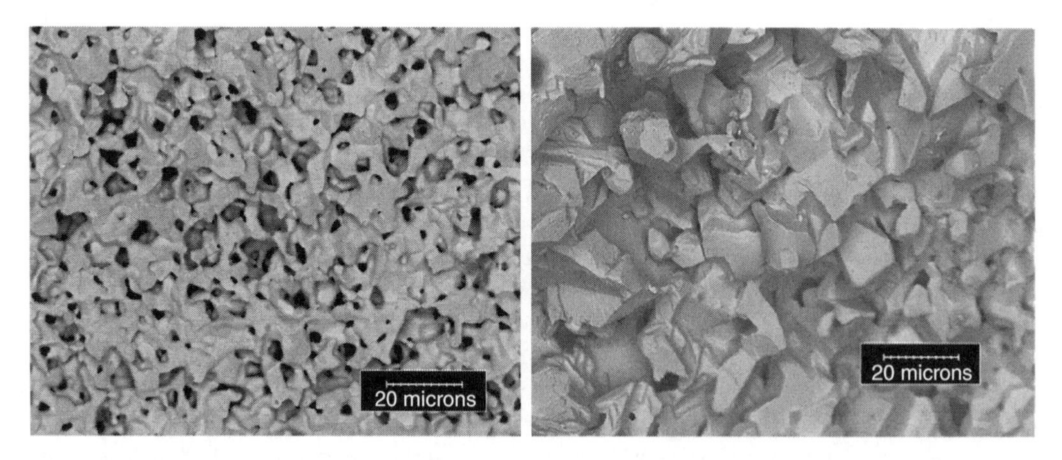

Figure 4-60 BaO-W dispenser insert surface before use in a cathode (left), and after over 30,000 h operating in the NSTAR life test thruster (right) (*Source:* [105]/Journal of Applied Physics).

of a BaO-W dispenser cathode insert surface before use in a cathode and after use for over 30,000 h in the NSTAR thruster life test [60, 105]. The surface is significantly modified and the structure on the right tended to have low barium content.

Figure 4-61 BaO-W dispenser insert surface modifications due to propellant line air impurities.

The life and emission performance of dispenser cathodes can also be degraded to some extent by propellant impurities. Figure 4-61 shows an end view of a BaO-W dispenser insert from an NSTAR-sized hollow cathode operated with a small air leak in the xenon propellant line. The porous tungsten insert material formed tungsten crystal structures on the interior surface consistent with the after-test photograph in Fig. 4-60, and the modified surface was found to be depleted in barium. This cathode insert had essentially stopped emitting electrons because of these propellant-impurity generated surface modifications, and had to be replaced to continue running the thruster.

Lanthanum hexaboride is much less sensitive to impurities that can limit the performance and life of the barium dispenser cathodes. Partial pressures of oxygen in the 10^{-5} Torr range are required to degrade the emission of LaB_6 at temperatures below 1440 °C, which was shown in Fig. 4-56. The curves for water and air poisoning of LaB_6 are at much higher partial pressures off the graph to the right. In comparison, LaB_6 at 1570 °C, where the electron emission current density is nearly the same as for the dispenser cathode at 1100 °C, can withstand oxygen partial pressures up to 10^{-4} Torr without degradation in the electron emission. This means that LaB_6 can tolerate two orders of magnitude higher impurity levels in the feed gas compared to dispenser cathodes operating at the same emission current density. For the case of xenon thrusters, LaB_6 cathodes can tolerate the crudest grade of xenon available (≈99.99% purity) without affecting the LaB_6 electron emission or life. In addition, the LaB_6 surface is not significantly modified when exposed to impurity gases compared to pure-propellant operation. The net evaporation rate simply increases as the surface temperature increases to produce the same emission current density as the work function changes slightly [106]. Wear testing of an LaB_6 hollow cathode with 10 PPM oxygen impurity in the xenon propellant line showed a 30% increase in the evaporation rate compared to "clean" xenon, and the cathode operated otherwise normally [106].

LaB_6 cathodes also do not require any significant conditioning or activation procedures that are required by dispenser cathodes. The authors have used LaB_6 cathodes emitting at currents of 5–10 A/cm^2 to produce pure oxygen plasmas in background pressures of 10^{-3} Torr of oxygen. In this case, the operating temperature of the cathode had to be increased to just over 1600 °C to avoid poisoning of the surface by the formation of lanthanum oxide, consistent with the trends in the published poisoning results shown in Fig. 4-59. The authors have also exposed hot, operating LaB_6 cathodes to atmospheric pressures of both air and water vapor. In both cases, the system was then pumped out, the heater turned back on and the cathodes started up normally. This incredible robustness makes handling and processing electric propulsion devices that use LaB_6 cathodes significantly easier than thrusters that use dispenser cathodes.

4.9 Keeper Wear and Life

In modern hollow cathodes used for electric propulsion applications, the keeper electrode typically encloses the hollow cathode and serves the functions of facilitating the starting of the cathode by bringing a high positive voltage close to the orifice and protecting the cathode from ion bombardment from the cathode plume and thruster plasmas. However, the keeper electrode is biased during normal operation at an intermediate potential between cathode and anode to collect a reduced number of electrons and, since it is below the plasma potential, it is subject to ion bombardment and wear. Cathode orifice plate and keeper electrode erosion rates measured or inferred in various experiments [107, 108] and in ion thruster life tests [59, 60, 109] have been found to be much higher than anticipated. For example, Fig. 4-62 shows the NSTAR cathode before and after the 30 352 h Extended Life Test [110]. The keeper electrode was completely eroded away by the end of the test, exposing the cathode orifice plate to the thruster discharge chamber plasma, which significantly eroded the cathode orifice plate and the sheath-heater surfaces. These results have been attributed to the high-energy ions bombarding and sputtering the cathode and keeper electrodes.

A significant effort has been expended trying to understand the mechanism for this rapid erosion. Several organizations have measured the presence of high-energy ions in ion thrusters and in the neighborhood of hollow cathodes using retarding potential analyzers (RPA) [64, 65, 111, 112] and laser induced fluorescence (LIF) [113]. For example, Fig. 4-63 shows the ion energy distribution measured [65] downstream on-axis and radially away from the plume of the NSTAR cathode. The high-energy ions are detected in both locations, with varying amounts depending on the position at which they are detected. The energy of some of the ions is greatly in excess of the 26-V discharge voltage, and if these ions were to hit the keeper or cathode orifice, they could cause significant erosion.

The source and characteristics of the high-energy ions have been the subject of much research and debate. Models of a DC potential hill [114] located inside or just downstream of the cathode orifice, or ion acoustic instabilities in a double layer postulated in the orifice of the cathode [115], have been proposed to explain the production of these ions. However, in detailed scanning probe studies to date [11, 16, 46, 62], there has been no detectable potential hill or unstable double layer at the cathode orifice or in the cathode plume that might support these mechanisms as responsible for

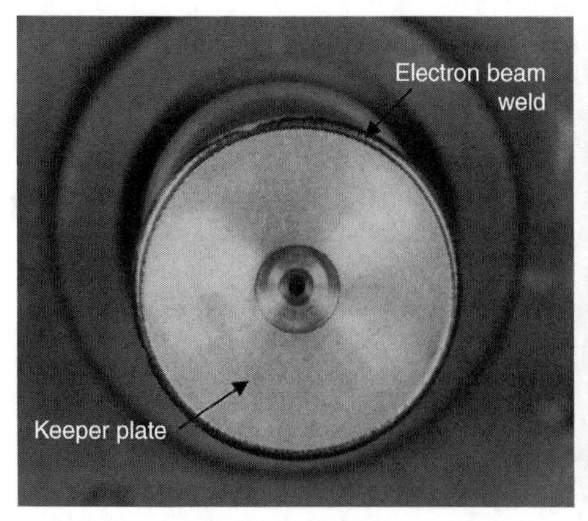

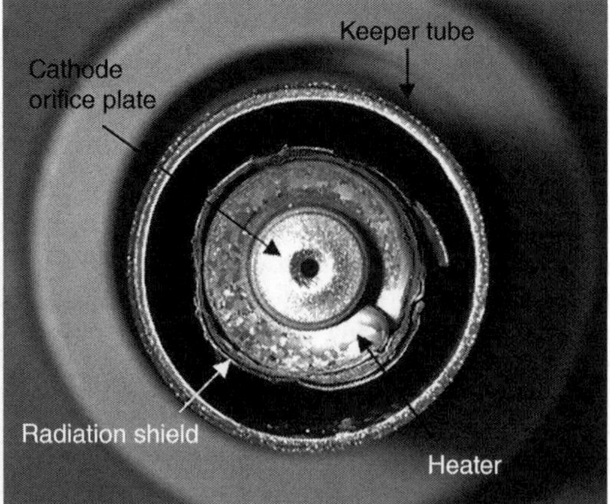

Figure 4-62 NSTAR discharge cathode before and after the 30 352 h ELT wear test showing complete sputter erosion of the keeper electrode out to the e-beam weld location on the keeper tube (*Source:* [110]/AIP Publishing LLC).

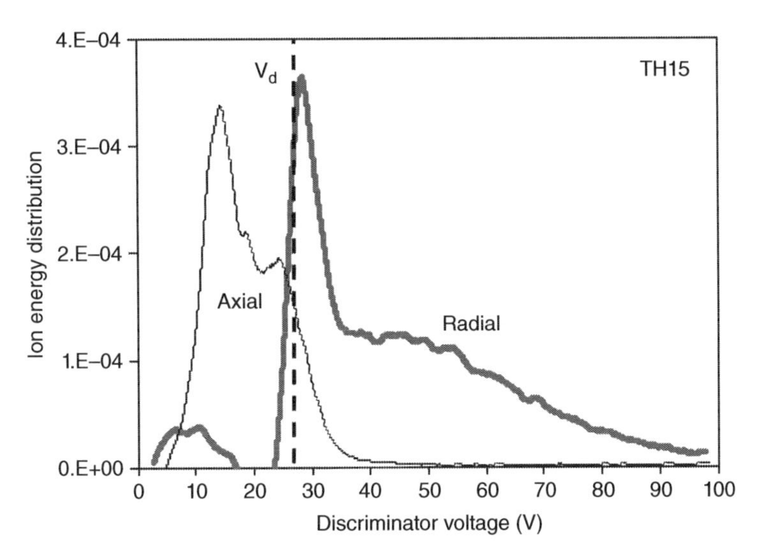

Figure 4-63 Ion energy distribution measured on-axis and radially away from an NSTAR hollow cathode at full power (13.1 A) (*Source:* [64]).

the high-energy ions or the electrode wear rates and erosion patterns. However, low frequency plasma potential oscillations in the 50–150 kHz range associated with plasma ionization instabilities have been detected in the near-cathode plume across the front of the keeper by scanning emissive probes [64, 65]. This instability is discussed more detail in the next section, and the large plasma potential fluctuations observed in the experiments have been proposed as a mechanism for accelerating ions to higher energy. In this case, ions born at the peak potential gain the full rf potential energy when striking the keeper or cathode surfaces, which can exceed 40–80 eV [65]. The fluctuations then produce sufficient ion energy to explain the keeper-face erosion reported in the literature. In some cases the ion energies measured by the RPAs approach or exceed 100 eV, which is in excess of the plasma potential fluctuation amplitudes typically measured. Mechanisms responsible for this, like that proposed by Katz [116] due to charge exchange collisions in the near-cathode plume region, can lead to ions with higher energies than the measured plasma potentials excursions, are also being investigated.

The complete mechanism for high-energy ion generation, the measured energies of these ions by various techniques, and the resulting erosion rates of the cathode and keeper electrodes are still under investigation at this time. Detailed 2-D modeling [54, 55] and additional experimental investigations are underway to understand this problem. One mitigation is to change the keeper material to graphite, which has a significantly lower sputter-yield compared to refractory metals.

Figure 4-64 Graphite keeper wear on the "TDU" cathode after testing showing texturing of the graphite keeper in the same pattern as the NSTAR cathode results but with negligible erosion (*Source:* [106]/American Institute of Aeronautics and Astronautics).

Figure 4-64 shows the texturing of the graphite keeper surface after a 2000 h wear test of the 1.5-cm-dia. LaB_6 hollow cathode [106]. This slight surface modification is in the same pattern as seen in the NSTAR cathode wear testing [59], but negligible erosion occurred and the keeper life is projected to be very long. A second mitigation is to identify the instabilities in the near-cathode plume region responsible for the energetic ion generation and develop techniques to damp them sufficiently to avoid the energetic ion production. The near-cathode plasma instabilities and damping techniques are discussed in the next section.

4.10 Discharge Behavior and Instabilities

The electron discharge from a hollow cathode can be ignited by several mechanisms. In Type A cathodes and some Type B with small orifices, the electrostatic (vacuum) potential from the keeper or anode electrode does not penetrate significantly through the relatively long, thin orifice into the insert region. In this case, electrons emitted in the insert region cannot be accelerated to cause ionization because there is no anode potential visible from inside the cathode. However, if the cathode uses a barium dispenser insert, then barium evaporated from the insert when heated can deposit on the upstream side of the cathode orifice plate and inside the orifice and diffuse onto the downstream surface of the orifice plate [117] facing the keeper electrode. The work function subsequently decreases, and the vacuum thermionic emission from the orifice plate to the keeper can be sufficient to ignite a discharge once gas is introduced. The plasma then penetrates the orifice, extending the anode potential into the insert the region and the discharge transitions directly to the insert. The orifice plate is subject to sputter erosion by the ions in the discharge, and the barium layer is removed by this ion bombardment and has to be reestablished if the cathode is turned off in order to restart [118].

For cathodes with larger orifices (typically 2-mm dia. or larger), a sufficient keeper voltage (typically 100–500 V) will cause the applied positive potential to penetrate inside the insert region at levels in excess of the ionization potential of the propellant gas. The electrons from the insert can then be accelerated locally inside the insert and cause ionization, which ignites the discharge through the orifice to the keeper or anode. This is the mechanism used in most of the LaB_6 cathodes developed by the authors to strike the discharge.

For hollow cathodes with smaller orifices or inhibited orifice plate emission (due to surface impurities, barium depletion, etc.), the arc-initiation technique described in Section 4.7.2 on heaterless hollow cathodes is typically used, even for cathodes with heaters. In this case, the keeper voltage is pulsed to a high positive value (typically $>$500 V) and the discharge starts due to either field emission of electrons from the orifice plate ionizing the injected cathode gas, or Paschen breakdown occurring in the cathode-to-keeper gap generating plasma that penetrates the orifice into the insert region. To ensure reliable thruster ignition over life, it is standard to apply both a DC keeper-voltage in the 50–150 V range and pulsed keeper voltage in the 300–600 V range. After the discharge is ignited, the heater is turned off, the keeper current is limited by the power supply and the keeper voltage falls to a value below the discharge voltage.

4.10.1 Discharge Modes

Once ignited and having transitioned to a thermionic discharge, hollow cathodes are well known to operate in distinct discharge "modes" [10]. Hollow cathodes have been historically characterized as having a quiescent "spot mode" with a broadly optimum gas flow at a given current, and a noisy

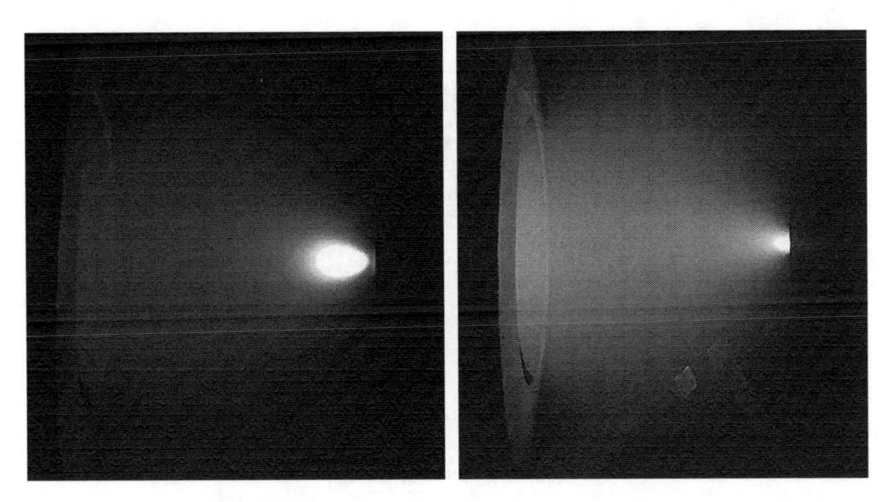

Figure 4-65 Spot mode (left) and plume mode (right) photos (*Source:* [10]/AIP Publishing LLC).

"plume mode" with the gas flow below the level at which the spot-mode is obtained [4, 10, 63]. The spot-mode, seen in Fig. 4-65 left, is visually observed as manifesting a sphere or "spot" of plasma just downstream of the cathode orifice and a slowly expanding plasma column extending from the spot to the anode. The plume mode, seen in Fig. 4-65 right, has a widely diverging plasma cone extending from the cathode, often filling the vacuum chamber with diffuse plasma with little or no spot or sphere of plasma in the cathode/keeper orifice. In neutralizer cathodes at low current there is usually a sudden transition between these modes, but in high current cathodes with larger orifices there is a continuous transition between these modes. At very high gas flows well above the optimum for the spot mode the spot moves downstream and the plume becomes more collimated and unstable because of the onset of ion acoustic turbulence [72, 77]. Very high cathode flow rates tend to suppress the discharge voltage, which adversely affects the ionization rate and discharge performance in discharge cathodes in ion thrusters. However, higher flow rates tend to also reduce the coupling voltage in both Hall and ion thruster neutralizer cathodes, which can improve the performance.

Hollow cathode discharge modes have been examined in detail in the laboratory because of the observed increases in the keeper or coupling voltages in the plume mode [4, 63, 115, 119, 120], and increases in keeper wear [50, 60] due to the production of energetic ions. At flow rates near the optimum for the spot mode, thermionic hollow cathodes can produce quiescent discharges [64, 115]. Plume mode transition [10] is usually detected by increases in the oscillation of the keeper voltage or in the magnitude of the coupling voltage. For example, plume mode onset is defined in the NSTAR neutralizer when the keeper voltage oscillation exceeds ± 5 V peak-to-peak. Transition to plume mode occurs for too low a propellant flow at a given emission current (or too high a discharge current for a given flow), and is usually detected by increases in both the keeper and discharge voltage oscillations. Transition to plume mode can also be affected by the anode location and design [10, 65, 121], with flat anodes sometimes used in cathode test facilities tending to cause plume mode onset at higher flows (for a given current) than cylindrical anodes.

In plume mode, and even in the transition to plume mode as the flow is reduced or the current raised, cathodes produce ions with energies significantly in excess of the discharge voltage that cause significant keeper and cathode orifice erosion. Figure 4-66 shows the ion distribution function measured by a compact scanning RPA [122] located at the face of the keeper on the high

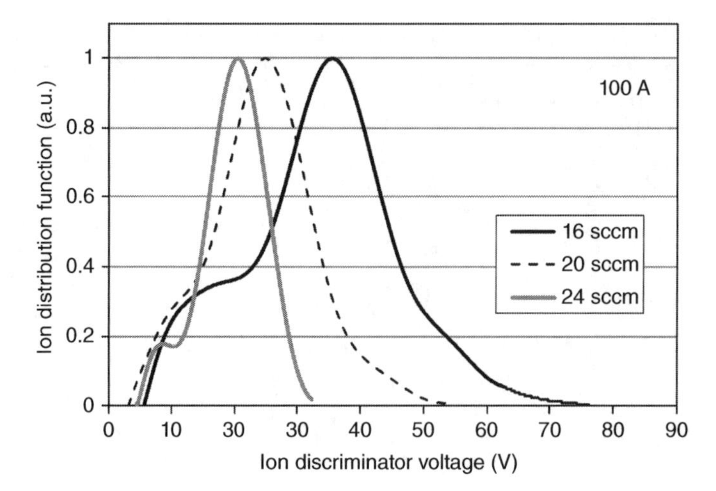

Figure 4-66 Ion energy distribution from a scanning RPA at the keeper face of the X3 cathode at JPL showing significant ion energies as the flow rate is reduced at a given current.

current X3 LaB_6 cathode [83] operating at 100 A of discharge current. Ions with energies in excess of 50 eV are seen to be incident on the keeper face, even at the nominal 16 sccm flow for this operating condition of this cathode. Higher currents are observed to produce even higher ion energies.

In terms of the plasma dynamics in the cathode exterior, plume mode oscillations have been observed in the laboratory as having three main features: (i) they are quasi-coherent and longitudinal with frequencies ranging from 10 to 150 kHz [64, 65]; (ii) they are characterized by large-scale oscillations in plasma density, plasma potential, and discharge current with amplitudes approaching 100% of the background; and (iii) the modes appear to be either dispersionless or propagating with the ion drift speed from the cathode to anode [123]. It is generally accepted that these are ionization instabilities as shown by Goebel [65]. There are additional instabilities that can be present in the plume, such as IAT and rotational modes, that will also be described below.

The dynamics of the plasma exterior to the hollow cathode in plume mode can be visually seen by high speed cameras. Figure 4-67 presents three frames from the high-speed camera footage captured when the X3 LaB_6 cathode was operated without any applied magnetic field at 100 A of discharge current and 9 sccm xenon gas flow. This condition results in the onset of plume mode for this cathode. The leftmost frame was acquired at time $t = 0$, while the following frames were taken after $t = 2.55$ μs and $t = 3.83$ μs. The images are for the camera pointing at the side with the cathode exit on the left and the anode is on the right separated by a distance of 21 mm. The plume-mode cathode

Figure 4-67 Plume mode behavior of the cathode discharge producing bursts of plasma that propagate axially at about the ion acoustic velocity (*Source:* [10]/AIP Publishing LLC).

produces plasma bursts that propagate axially to the anode. Fast camera images looking directly at the cathode through a cylindrical anode show an azimuthally uniform plume. This mode matches the behavior of ionization instabilities observed in the experiments [65] and was described as axial modes by Georgin [123].

The ionization instabilities grow at a given discharge current when the input gas flow is below a certain threshold. The ionization tends to deplete the local neutral gas, which causes the neutral density to drop and reduce the discharge current-carrying plasma density, forcing the system to oscillate on the ionization-rate time scales. Simulations and experiments show that non-classical electron heating via ion acoustic waves can couple with these instabilities, raising their frequency above the ionization rate frequency and increasing their growth [124]. The axial burst visible through the high-speed camera has some similarity with the well-known breathing mode or predator–prey ionization-instability mode characteristic of Hall thruster plumes [125].

The complexity of the physics of these modes, and how they transition from one to another, has made first-principles analyses and ab initio predictions longstanding challenges for the modeling and numerical simulation community. A breakthrough was achieved when an advanced version of the OrCa2D code was employed to simulate the transition from spot to plume modes [55] in a 25-A LaB_6 cathode. This work produced global 2-D axisymmetric simulations that yielded good agreement with the so-called plume mode margin — typically a measurement of the peak-to-peak keeper voltage fluctuations as a function of the cathode mass flow rate. The main findings from that work are outlined next.

Two separate simulations, designated as "Simulation A" and "Simulation B" in Fig. 4-68, were performed in the flow rate range of 5–20 sccm. Both simulations used Eq. 4.6-2 for the anomalous collision frequency, but in A the ratio T_e/T_i was held fixed whereas in B it was varied. The simulations captured well the measured peak-to-peak oscillations of the keeper voltage ($V_{k\text{-}pp}$) as shown in the figure. The computed spectra showed that most of the oscillation power resided in the range of 100–300 kHz. It is also worth mentioning that the values in this range are much greater than the local ionization frequency. As the flow rate was reduced the region in the plume where the neutral gas was notably depleted grew in size. This is shown in Fig. 4-69 right which depicts contours of the

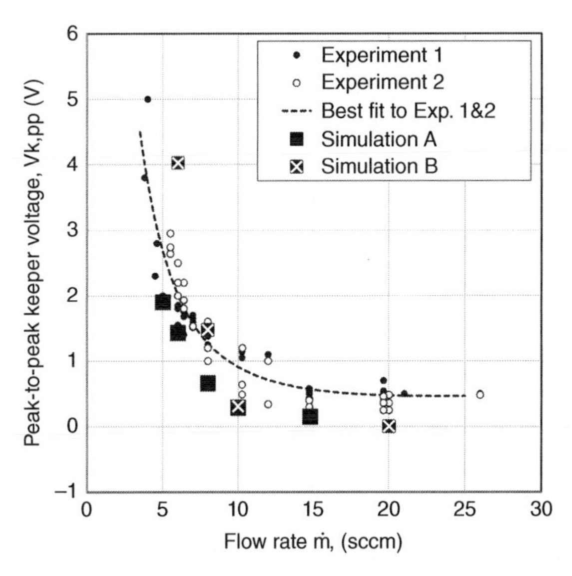

Figure 4-68 Comparisons between experiments and 2-D simulations of the peak-to-peak oscillations in the keeper voltage of a 25-A LaB_6 hollow cathode (*Source:* [55]).

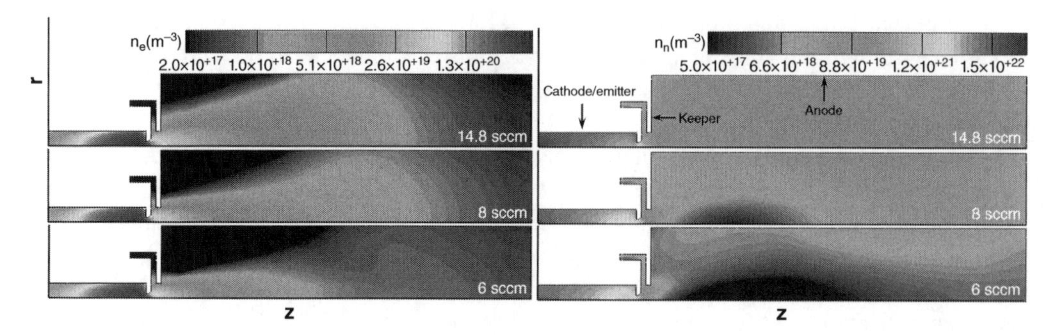

Figure 4-69 Time-averaged number densities of electrons (left) and neutrals (right) from 2-D simulations of a 25-A LaB_6 hollow cathode at three different mass flow rates (*Source:* [10]).

time-averaged neutral gas density. The spatial extent of the time-averaged plasma density column in the plume also diminished as the flow rate was decreased as shown in Fig. 4-69 left. This work revealed that there is a clear connection between the depletion of the neutral gas in the plume and the transition of the discharge from spot to plume modes. Equally important however was the finding that although the discharge dynamics are undoubtedly linked to the neutral gas density in the plume, they are also linked to the anomalous electron transport. No oscillations were found in the simulations in the absence of the IAT-driven anomalous resistivity. Moreover, in the presence of the anomalous resistivity the amplitude of the oscillations increased with decreasing density in the plume but so did the ratio of the anomalous over the classical collision frequencies.

It is important to realize that the large discharge instabilities and oscillations produced during plume mode only occur in the exterior of Type A and B hollow cathodes [65]. Figure 4-70 shows the ion saturation current measured simultaneously inside the insert region and immediately outside the keeper in the above experiments during plume mode with the ionization instability. The plasma density oscillations inside the cathode insert-region plasma are small in amplitude and

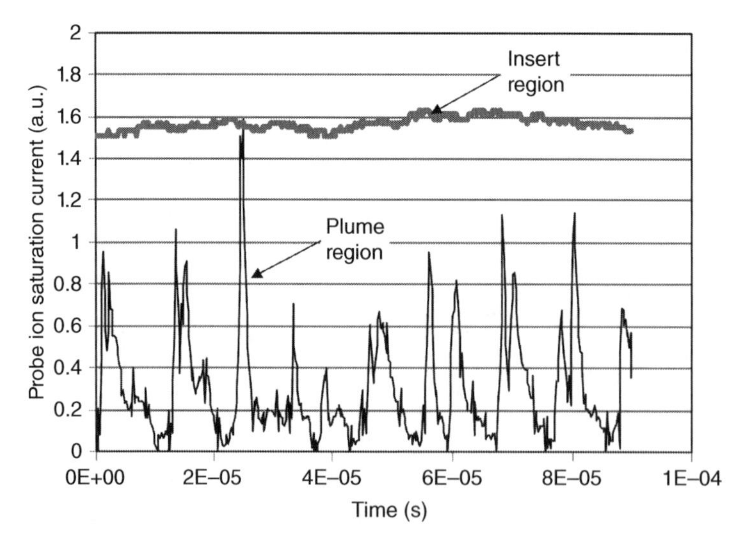

Figure 4-70 Ion saturation current oscillations inside and outside the cathode showing oscillation is exterior to the hollow cathode insert region (*Source:* [64]).

uncorrelated to the ionization and large turbulent instabilities observed outside in the near-cathode plume region. This was also the finding in the numerical simulations described above [55]. As shown in Fig. 4-69, operating at higher gas flows, which are selected to avoid plume mode, produces sufficient plasma density to carry the discharge current without plasma oscillations, and the lower electron temperatures and collisional effects in the cathode plume plasma at higher neutral densities tend to dampen or extinguish the oscillations.

The numerical simulations results shown in Figs. 4-68 and 4-69 accounted for the largely axial magnetic field that is typically applied in discharge hollow cathodes. An applied magnetic field in the cathode plume region tends to damp ionization instabilities by confining the electrons and increasing the local ionization rate. However, it may also excite other low frequency (50–100 kHz) instabilities in this region of the cathode. Rotational modes, for example, were first observed by Jorns and Hofer [126] in the plume of a hollow cathode on a Hall thruster during observations with a Fast Camera. Jorns attributed this mode to anti-drift waves [126]. Experimental work by Becatti et al. [127] showed the existence of rotating magnetohydrodynamic (MHD) kink modes in the plume of a high current hollow cathode, also using a Fast Camera. Becatti correlated the amplitude of the MHD modes with the production of energetic ions incident on the keeper face as measured by an RPA on a reciprocating probe assembly. The growth rate of these two rotational instabilities was found to be comparable [127], and it is likely that these modes compete in the cathode plume. As the discharge current increases, these rotational modes could contribute more to the production of energetic ions and increases in erosion of the keeper and thruster components.

The spot-to-plume mode simulations of the 25-A cathode also revealed oscillations in the near-plume plasma [55]. The waves were largely longitudinal, with wave vector in the axial direction and phase velocity $\omega/k > 10$ km/s, which is more than six times greater than the ion acoustic speed. Rotational modes could not be captured by the simulations since the computational model was 2-D axisymmetric. Mikellides suggested that the rotational modes observed in the experiment could in fact be coupled to the longitudinal mode they observed in their simulations, yielding a complex 3-D plasma motion. He argued that since their simulations reproduced well the measured plume mode margin, it was the longitudinal plasma motion that drove the transition from spot to plume modes at this low discharge current from the cathode where Becatti et al. demonstrated that the rotational modes are near their onset threshold [128]. At higher discharge currents, the stronger rotational modes compete with the longitudinal modes depending on the operating conditions [10], and may dominate plume mode onset.

There have been even more numerical simulations using the 2-D axisymmetric model of the Hall thruster called Hall2De [129], which is described in Chapter 7. These new results were combined with the dispersion relation of waves in the lower hybrid frequency range and showed that, radially away from the most dense region of the cathode plume, the modified two-stream instability (MTSI) and lower hybrid drift instability (LHDI) can become excited [130, 131]. In the Hall thruster-cathode arrangement, the LHDI is driven by the diamagnetic and $\mathbf{E} \times \mathbf{B}$ drifts, both of which are in a direction perpendicular to the r-z plane (rotational direction).

4.10.2 Suppression of Instabilities and Energetic Ion Production

Energetic ions have been correlated to the various instabilities (ionization, turbulent ion acoustic, and rotational) detected in the hollow cathode plume. Of critical importance is the suppression of these instabilities to minimize any impacts on cathode life. Although the configuration and operating conditions (current and gas flow rate) can be selected over a limited range to minimize the onset of these instabilities to help reduce the erosion rates [78, 132], changing the cathode operating conditions (current or flow) to throttle the thruster, or running at high discharge current for high power thrusters, causes the instabilities to grow with an increase in energetic ion production. The only technique found to be successful to date to suppress the energetic ion production, especially as the discharge current is increased, is neutral gas injection directly into the hollow cathode plume [65, 120, 133, 134]. This technique was also considered soon after early OrCa2D simulations suggested the presence of IAT-driven anomalous resistivity in the cathode plume [52].

Two techniques have been developed to inject neutral gas into the cathode plume. Gas injection external to the cathode using small electrically floating tubes positioned near the cathode exit has been the most successful. Figure 4-71 shows photos of the original testing at JPL [120] on an NSTAR-like cathode at full power (13.1 A) with a single injector above a cathode plume. The discharge on the left has a halo around the main plume column characteristic of plasma instabilities, and the injection of neutral gas into the plume in the photo on the right damps the instabilities and eliminates the halo. Figure 4-72 shows RPA data for the ion energies for these two cases. Gas injection reduces the ion energies arriving at cathode potential from over 80 eV to below about 40 eV. Experiments by Chu [133] showed that using two gas injector tubes was the most effective in reducing the ion energies with the lowest amount of gas injection. Figure 4-73 shows the high current X3 cathode with its the dual exterior gas injectors. This configuration was successfully used in the XR-100 Hall thruster testing at up to 250 A [135]. An interesting benefit of external gas injection is that the gas required to optimally run the hollow cathode insert-region plasma [20] is much less than that needed for good coupling to the Hall thruster [136]. Experiments on the XR-100 Hall thruster showed that the total cathode flow split needed to optimally run the thruster was reduced by several percent using this external injection system [133], with a corresponding increase in the total thruster efficiency.

The second method to introduce neutral gas into the plume is through the cathode to keeper gap [10, 134, 137]. In this case, the gas is injected through the cathode mounting flange on the outside the cathode tube but inside the keeper tube, and enters the discharge through the keeper orifice. The increased pressure downstream of the cathode orifice does little to change the pressure in the

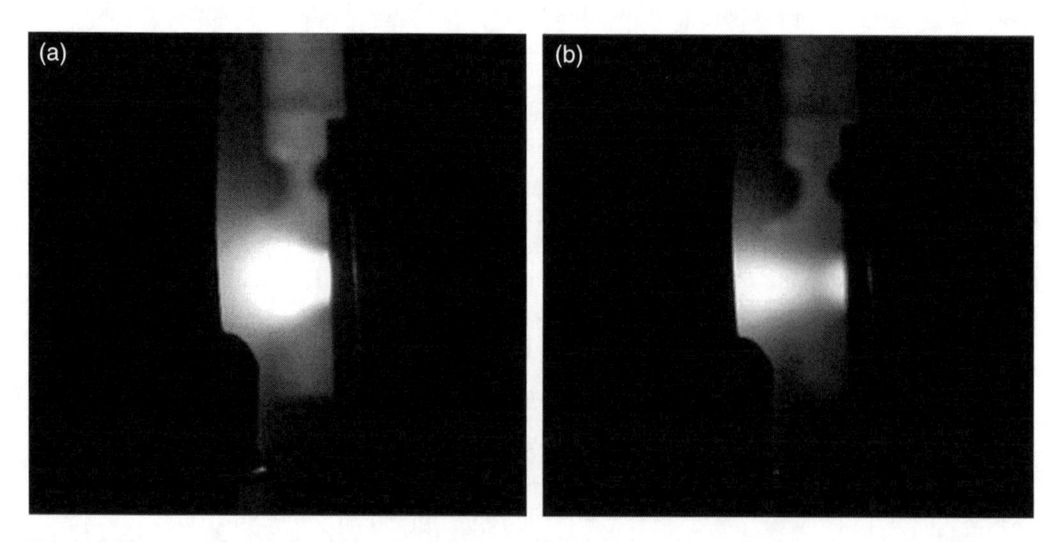

Figure 4-71 Side view of discharge with cathode on right and anode on the left just starting plume mode transition (a), and with gas injection from an external injector tube (b) (*Source:* [65]/AIP Publishing LLC).

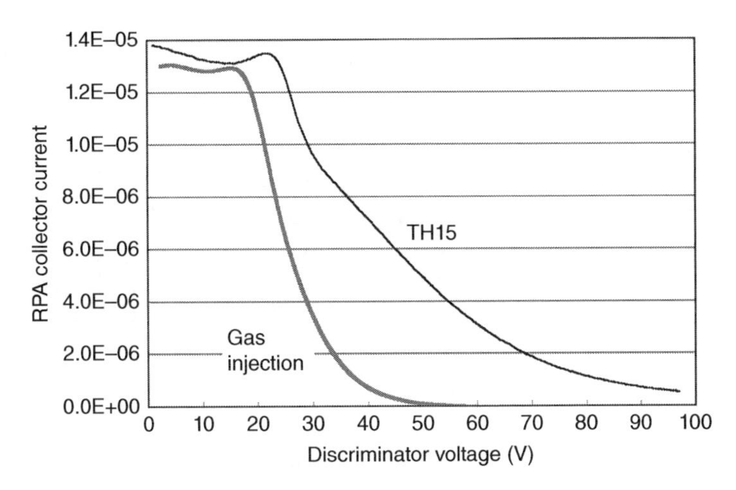

Figure 4-72 Ion current collected by the RPA for the cases with and without gas injection in the near-cathode plume shown effectiveness of external gas injection in reducing energetic ion production (*Source:* [64]).

insert region. This method avoids the mechanical complexity of the exterior gas injector tubes, and is almost as effective as that technique. Approximately 10–20% more gas in needed for this "interior" injection technique to be as effective in reducing the ion energy in the plume, likely because the gas is heated passing over the hot cathode tube and because some is ionized and depleted in the cathode plume very close to the keeper before entering the instability generation region in the near cathode plume. Nevertheless, interior injection was very effective in the XR-100 testing because the thruster requires much more gas for coupling the cathode to the thruster plasma than the cathode requires to produce the discharge current [138].

Figure 4-73 X3 cathode with external gas injectors (*Source:* [83]/American Institute of Aeronautics and Astronautics).

The technique of injecting gas between the cathode-to-keeper gap and into the cathode plume from the keeper orifice plate is effective up to very high discharge currents. The 500 A LaB_6 hollow cathode recently developed [134] used this technique, combined with a cathode orifice designed to keep the electron current density below 1 kA/cm^2, to suppress ionization instabilities and produce low energetic ion production in the cathode plume. Measurements made with the fast scanning RPA at the keeper face (collecting ions from downstream) showed that the ion energy impacting the keeper was well below 40 eV at discharge currents up to 350 A [134].

4.10.3 Hollow Cathode Discharge Characteristics

Current versus voltage characteristic curves for hollow cathode discharges always follow a flattened-*U* shape. Figure 4-74 shows an example of the discharge voltage versus current for the X3 LaB_6 hollow cathode at several different flow rates [138] of xenon gas though the cathode

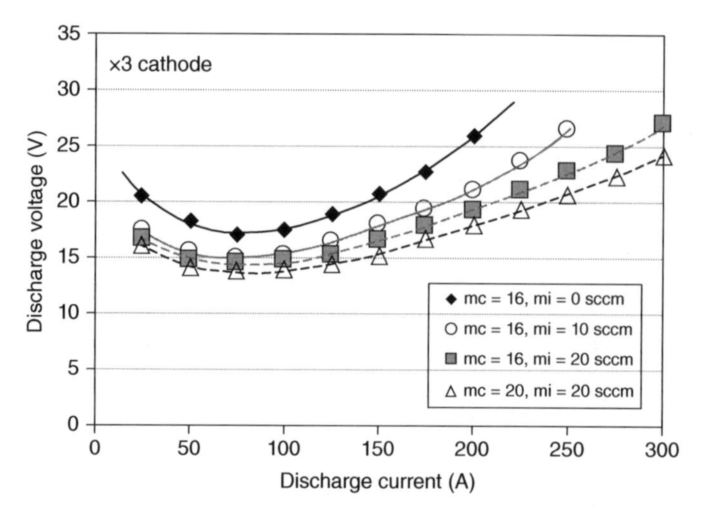

Figure 4-74 Discharge voltage versus current for the X3 cathode at different cathode flow rates. The xenon flow through the cathode is labeled "mc" and the flow through the internal injector (between cathode and keeper) is labeled "mi".

(indicated as "mc") and through the internal injector (indicated as "mi"). Increasing flow rates through both the cathode and the internal injector lower the discharge voltage at any current. The minimum of the "*U*-shaped" discharge curves is usually selected as the nominal "Spot-Mode" location with the lowest oscillation levels. On the left side of the curves, the discharge voltage increases at a lower current as the cathode increases the internal sheath voltage to provide sufficient insert heating at the lower discharge currents. On the right side of the curves, the discharge voltage increases with discharge current because of the finite resistivity of the insert, orifice and plume plasmas, and the resistance of the electrical connections to the cathode. The maximum current at a given flow was determined by the onset of plume mode oscillations up to the 300 A limit of the power supply used in this testing. The experiment used a large 10-cm-dia. cylindrical anode that could handle the high heat flux of the nearly 9 kW discharge power. The cylindrical anode also traps some of the cathode gas flow near the cathode exit to increase the plasma generation in the cathode plume and avoid plume mode and instability problems. The onset of the oscillations and the transition to plume mode is an anode-plasma effect.

There are basically four types of oscillations that occur in hollow cathode discharges [10, 65, 78, 126]: power supply oscillations; ionization-instability oscillations, turbulent ion acoustic oscillations; and rotational oscillations. The plasma discharge oscillations occur exterior to the cathode in the near-cathode plume and typically are in the frequency range of 50 to over 1000 kHz. The plasma oscillations have amplitudes that cause voltage variations on the closely coupled keeper electrode from fractions of a volt in the spot mode to tens of volts in plume mode. If sufficiently large, these oscillations can trigger regulation problems in the discharge power supply, leading to large discharge voltage oscillations occurring on the power-supply-regulation times of typically 100–1000 Hz. This behavior is shown in Fig. 4-75.

In the ionization instabilities, or so-called *predator–prey* oscillations, the plasma discharge burns out a significant fraction of the neutral gas in the near cathode plume, and the discharge collapses

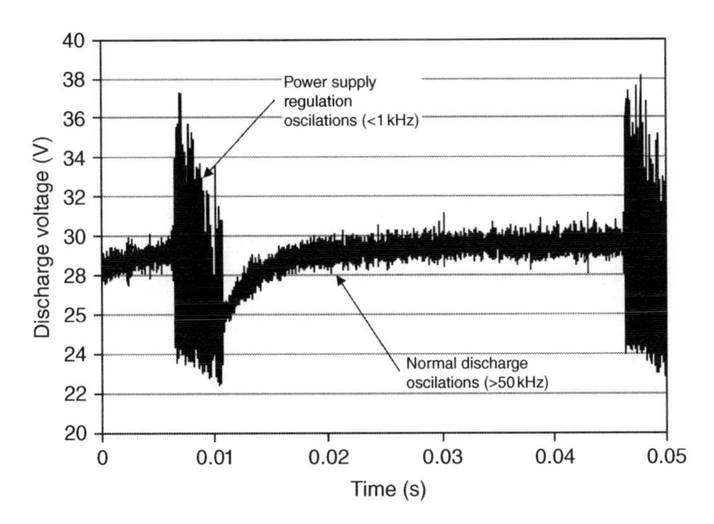

Figure 4-75 Discharge voltage oscillations showing plasma and the very low frequency power supply-induced oscillations.

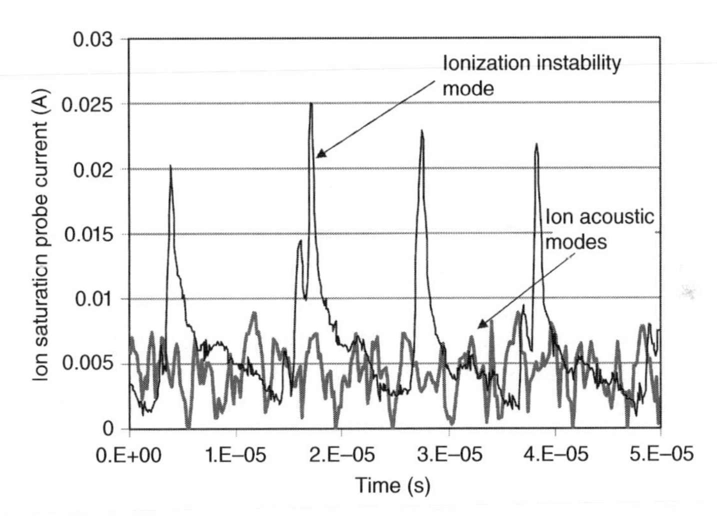

Figure 4-76 Ion saturation current from a probe in the NEXIS cathode plume showing the plume mode ionization instabilities and the broadband ion acoustic turbulent oscillations in spot mode.

on the time frame of neutral flow back into the plume region. The frequency range of these instabilities in the 50–250 kHz range for xenon depending on the physical scale lengths and size of the discharge components. Ionization instabilities are easily observed in the plasma density, which is shown by the probe's ion saturation current oscillations in Fig. 4-76 and compared to the normally present incoherent turbulent ion acoustic modes even in spot mode. Likewise, the application of an axial magnetic field introduces rotational modes (drift and lower hybrid waves as well as MHD oscillations) discussed above that compete with the ionization and ion acoustic instabilities in the near-cathode plume.

Problems

4.1 The power radiated from a hot surface to a relatively cold surface is given by $P = \sigma\varepsilon T^4 A$, where σ is the Stephan-Boltzmann constant, ε is the emissivity of the surface, T is the surface temperature in K, and A is the radiating area.

a) Design a 0.5-mm diameter tungsten filament electron emitter that emits 10 A and radiates only 200 W of power. Specifically, what is the filament length and emission current density? You can neglect axial conduction of power to the electrical connections and assume that the emissivity of tungsten is 0.3 at emission temperatures.

b) You decide to use a 0.25-mm diameter and 4 cm long filament to limit the radiated power. What is the temperature of the emitter, and how much power is actually radiated?

4.2 The insert in a BaO hollow cathode is 3 mm inside diameter and 2.5 cm long and is at a temperature of 1100 °C.

a) Using Cronin's expression for the work function, how much current is emitted by the insert?

b) If the insert has a uniform plasma with $n_e = 10^{20}\ \text{m}^{-3}$ density and $T_e = 2$ eV inside it, how much is the electron emission enhanced by the Schottky effect if the sheath is 10 V and 3 Debye lengths thick? What is the total emission current?

c) Is the emission current space charge limited?

d) Assume that the plasma density falls exponentially from the orifice entrance to $10^{18}\ \text{m}^{-3}$ in 1 cm. What is the total electron current that can be emitted into the plasma?

4.3 A 2.5 cm long BaO impregnated insert with a 2 cm inside diameter has xenon gas injected to create an internal pressure of 2 Torr at 1500 K in the hollow cathode. Assuming the insert-region plasma is infinitely long, what is the electron temperature in the insert-region plasma and the radial ion drift velocity at the wall?

4.4 For the cathode geometry in problem 4.3, what is the internal plasma density for an emission current of 30 A and a heating power of 100 W assuming a uniform plasma density with an electron temperature of 1.2 eV? (Hint: Estimate the resistivity for the sheath voltage, find plasma density, then iterate).

4.5 A lanthanum hexaboride hollow cathode 2 cm inside diameter and 2.5 cm long emits 20 A of electrons into a uniform $2 \times 10^{19}\ \text{m}^{-3}$ plasma with an electron temperature of 1.5 eV. For a heating power of 40 W and an internal xenon pressure of 1.2 Torr at 1500 K, find the ion and electron heating powers to the insert. Why is one larger than the other?

4.6 If a cylindrical discharge cathode orifice is 2 mm in diameter and has an internal pressure of 3 Torr at a temperature of 2000 K, what is the electron temperature? (neglect end losses)

4.7 A neutralizer cathode produces 3 A of electron current through a 0.6 mm diameter orifice that is 1.5 mm long. Assuming that the electron temperature in the orifice is 1.5 eV, the electron temperature in the insert region is 1.2 eV, the pressure is 50 Torr at 2000 K and the sheath voltage at the wall is 12 V, what is the plasma density, the ion heating of the orifice plate, and the axial voltage drop in the orifice plasma?

4.8 A hollow cathode has an orifice diameter of 2.5 mm and a xenon gas flow of 4 sccm with an effective temperature of 2000 K. Assume that the neutral gas density falls exponentially from the orifice exit with a characteristic length of 0.5 mm (i.e. one e-folding for every 0.5 mm of distance from the cathode). Assuming 15 V primary electrons in the cathode plume and that all of the ions generated fall back through the sheath, find the location downstream of the orifice exit where a double layer might occur (hint: where the electron current is greater than the square root of the mass ratio times the ion current).

4.9 A Hall thruster hollow cathode has a cathode orifice diameter of 3 mm and produced 20 A of 15 V primary electrons with a xenon gas flow of 10 sccm. Assume that the gas is at 2000 K and the neutral plume diverges at 45 deg. from the orifice.

a) Neglecting depletion of the electron current due to ionization, how much of the cathode gas low is ionized within 10 cm of the cathode?
b) Assume that every electron that makes an ionization collision is lost (loses most of its energy and is rapidly thermalized), and that the neutral atom is also lost (depleted). How much of the cathode gas flow is then ionized with 10 cm of the cathode?
c) If the primary electron energy is 20 eV, how much of the gas flow is ionized within 10 cm of the orifice accounting for depletion of both the primary electrons and the neutral gas due to ionization?

4.10 An ion thruster is operated at 2 A of beam current at 1500 V. The thruster has 5% double ions content, a 10° beam divergent half angle, a discharge loss of 160 eV/ion at a discharge voltage of 25 V, and uses 32 sccm of xenon gas and 20 W of power in addition to the discharge power.

a) What insert thickness is required in an NSTAR-type cathode to achieve 5 years of cathode life if barium loss is the life-limiting effect?
b) What is the thruster efficiency, Isp, and thrust?

References

1 F. Paschen, "Uber die zum Funkentibergang in Luft, Wasserstoff und Kohlensaure bei verschiedenen Drucken erforderliche Potentialdifferenz," *Annalen der Physik*, vol. 355, no. 16, 1916.
2 L. M. Lidsky, S. D. Rothleder, D. J. Rose, S. Yoshikawa, C. Michelson, and R. J. Mackin, "Highly Ionized Hollow Cathode Discharge," *Journal of Applied Physics*, vol. 33, no. 8, 1962.
3 M. P. Ernstene, A. T. Forrester, E. L. James, and R. M. Worlock, "Surface Ionization Engine Development," *Journal of Spacecraft and Rockets*, vol. 3, no. 5, pp. 744–747, 1966; doi:10.2514/3.28526.
4 V. K. Rawlin and E. V. Pawlik, "A Mercury Plasma-Bridge Neutralizer," *Journal of Spacecraft and Rockets*, vol. 5, pp. 814–820, 1968; doi:10.2514/3.29363.
5 D. C. Byers, W. R. Kerslake, and J. F. Staggs, "SERT II – Mission and Experiments," *Journal of Spacecraft and Rockets*, vol. 7, no. 1, pp. 4–6, 1970; doi:10.2514/3.29854.
6 V. K. Rawlin and W. R. Kerslake, "SERT II: Durability of the Hollow Cathode and Future Applications of Hollow Cathodes," *Journal of Spacecraft*, vol. 7, no. 1, pp. 14–20, 1970.
7 V. K. Rawlin, "Operation of the J-Series Thruster Using Inert Gas,"AIAA-82-1929, Presented at the 16th International Electric Propulsion Conference, New Orleans, November 17–19, 1982.
8 D. M. Goebel, J. T. Crow, and A. T. Forrester, "Lanthanum Hexaboride Hollow Cathode for Dense Plasma Production," *Review of Scientific Instruments*, vol. 49, pp. 469–472, 1978.
9 D. R. Lev, I. G. Mikellides, D. Pedrini, D. M. Goebe, B. A. Jorns, and M. S. McDonald, "Recent Progress in Research and Development of Hollow Cathodes for Electric Propulsion," *Reviews of Modern Plasma Physics*, vol. 3, 2019; doi:10.1063/5.0051228.

10 D. M. Goebel, G. Becatti, I. G. Mikellides, and A. L. Ortega, "Plasma Hollow Cathodes," *Journal of Applied Physics*, vol. 130, no. 050902, 2021; doi:10.1063/5.0051228.

11 D. M. Goebel, K. Jameson, I. Katz, and I. Mikellides, "Hollow Cathode Theory and Modeling: I. Plasma Characterization with Miniature Fast-Scanning Probes," *Journal of Applied Physics*, vol. 98, no. 113302, 2005; doi:10.1063/1.2135417.

12 I. G. Mikellides, I. Katz, D. M. Goebel, and K. K. Jameson, "Hollow Cathode Theory and Modeling: II. A Two-Dimensional Model of the Emitter Region," *Journal of Applied Physics*, vol. 98, no. 113303, 2005; doi:10.1063/1.2135409.

13 H. R. Kaufman, "Technology of electron-bombardment ion thrusters," in *Advances in Electronics and Electron Physics*, edited by L. Marton, New York: Academic Press, pp. 265–373, 1974.

14 B. A. Arkhopov and K. N. Kozubsky, "The Development of the Cathode Compensators for Stationary Plasma Thrusters in the USSR," IEPC-91-023, Presented at the 22nd International Electric Propulsion Conference, Viareggio, October 14–17, 1991.

15 I. G. Mikellides, D. M. Goebel, J. S. Snyder, I. Katz, and D. A. Herman, "The Discharge Plasma in Ion Engine Neutralizers: Numerical Simulations and Comparison with Laboratory Data," *Journal of Applied Physics*, vol. 108, no. 113308, 2010; doi:10.1063/1.351456.

16 D. M. Goebel, K. Jameson, R. Watkins, and I. Katz, "Cathode and Keeper Plasma Measurements Using an Ultra-Fast Miniature Scanning Probe," AIAA-2004-3430, Presented at the 40th AIAA Joint Propulsion Conference, Ft. Lauderdale, FL, July 11–14, 2004.

17 I. G. Mikellides, I. Katz, D. M. Goebel, and K. K. Jameson, "Plasma Processes Inside Dispenser Hollow Cathodes," *Physics of Plasmas*, vol. 13, no. 063504, 2006; doi:10.1063/1.2208292.

18 I. Katz, I. G. Mikellides, D. M. Goebel, and J. E. Polk, "Insert Heating and Ignition in Inert-Gas Hollow Cathodes," *IEEE Transactions on Plasma Science*, vol. 36, no. 5, pp. 2199–2206, 2008; doi:10.1109/TPS.2008.2004363.

19 E. Chu and D. M. Goebel, "High Current Lanthanum Hexaboride Hollow Cathode for 10 to 50 kW Thrusters," *IEEE Transactions on Plasma Science*, vol. 20, no. 9, pp. 2133–2144, 2014; doi:10.1109/TPS.2012.2206832.

20 D. M. Goebel, K. K. Jameson, and R. R. Hofer, "Hall Thruster Cathode flow Impacts on Cathode Coupling and Cathode Life," *Journal of Propulsion and Power*, vol. 28, no. 2, pp. 355–363, 2012; doi:10.2514/1.B34275.

21 O. W. Richardson, "Electron Theory of Matter," *Philips Magazine*, vol. 23, pp. 594–627, 1912.

22 W. H. Kohl, *Handbook of Materials and Techniques for Vacuum Devices*. New York: Reinhold, 1967.

23 W. Schottky, "Concerning the Discharge of Electrons from Hot Wires with Delayed Potential," *Annalen der Physik*vol. 44, no. 15, pp. 1011–1032, 1914.

24 A. T. Forrester, *Large Ion Beams*. New York: Wiley-Interscience, 1988.

25 J. L. Cronin, "Modern Dispenser Cathodes," *IEE Proceedings*, vol. 128, no. 1, pp. 19–32, 1981; doi:10.1049/ip-i-1.1981.0012.

26 R. T. Longo, E. A. Adler, and L. R. Falce, "Dispenser Cathode Life Model," International Electron Devices Meeting, San Francisco, December 9–12, 1984; doi:10.1109/IEDM.1984.190712.

27 R. T. Longo, "Physics of Thermionic Dispenser Cathode Aging," *Journal of Applied Physics*, vol. 94, 2003; doi:10.1063/1.1621728.

28 W. Mueller, "Computational Modeling of Dispenser Cathode Emission Properties,"International Electron Devices Meeting, Washington December 8–11, 1991: IEEE; doi:10.1109/IEDM.1991.235370.

29 J. W. Gibson, G. A. Haas, and R. E. Thomas, "Investigation of Scandate Cathodes: Emission, Fabrication and Activation Processes," *IEEE Transactions on Electron Devices*, vol. 36, pp. 209–214, 1989.

30 W. L. Ohlinger, B. Vancil, J. E. Polk, V. Schmidt, and J. Lorr, "Hollow Cathodes for Electric Propulsion Utilizing Scandate Cathodes," AIAA-2015-4009, Presented at the 51st AIAA Joint Propulsion Conference. Orlando, July 27–29, 2015.

31 R. Levi, "Improved Impregnated Cathode," *Journal of Applied Physics*, vol. 26, p. 639, 1955.

32 J. L. Cronin, (1979) Practical Aspects of Modern Dispenser Cathodes. *Microwave Journal*57–62.

33 J. M. Lafferty, "Boride Cathodes," *Journal of Applied Physics*, vol. 22, pp. 299–309, 1951.

34 D. Jacobson and E. K. Storms, "Work Function Measurement of Lanthanum-Boron Compounds," *IEEE Transactions on Plasma Science*, vol. 6, pp. 191–199, 1978.

35 J. Pelletier and C. Pomot, "Work Function of Sintered Lanthanum Hexaboride," *Applied Physics Letters*, vol. 34, pp. 249–251, 1979.

36 E. Storms and B. Mueller, "A Study of Surface Stoichiometry and Thermionic Emission Using LaB_6," *Journal of Applied Physics*, vol. 50, pp. 3691–3690, 1979.

37 D. M. Goebel, R. M. Watkins, and K. K. Jameson, "LaB6 Hollow Cathodes for Ion and Hall Thrusters," *Journal of Propulsion and Power*, vol. 23, no. 3, pp. 552–558, 2007.

38 V. Kim, G. Popov, B. Arkhipov, V. Murashko, O. Gorshkov, A. Koroteyev, V. Garkusha, A. Semenkin, and S. Tverdokhlebov. Electric Propulsion Activity in Russia," IEPC-2001-005, Presented at the 27th International Electric Propulsion Conference, Pasadena, October 14–19, 2001.

39 D. M. Goebel, Y. Hirooka, and T. Sketchley, "Large Area Lanthanum Hexaboride Electron Emitter," *Review of Scientific Instruments*, vol. 56, 1985; doi:10.1063/1.1138130.

40 E. Storms and B. Mueller, "Phase Relationship, Vaporization and Thermodynamic Properties of the Lanthanum-Boron System," *Journal of Chemical Physics*, vol. 82, pp. 51–59, 1978.

41 K. N. Leung, P. A. Pincosy, and K. W. Ehlers, "Directly Heated Lanthanum Hexaboride Filaments," *Review of Scientific Instruments*, vol. 55, pp. 1064–1068, 1984.

42 I. Katz, J. Anderson, J. Polk, and D. M. Goebel, "Model of Hollow Cathode Operation and Life Limiting Mechanisms," IEPC-2003-0243, Presented at the 28th International Electric Propulsion Conference, Toulouse March 17–21, 2003.

43 J. E. Polk, I. G. Mikellides, I. Katz, and A. Capese, "Tungsten and Barium Transport in the Internal Plasma of Hollow Cathodes," AIAA-2008-5295, Presented at the 44th AIAA Joint Propulsion Conference, Hartford, July 21–23, 2008.

44 I. Katz, J. Anderson, J. E. Polk, and J. R. Brophy, "One Dimensional Hollow Cathode Model," *Journal of Propulsion and Power*, vol. 19, no. 4, pp. 595–600, 2003.

45 J. S. Miller, S. H. Pullins, D. J. Levandier, Y. Chiu, and R. A. Dressler, "Xenon Charge Exchange Cross Sections for Electrostatic Thruster Models," *Journal of Applied Physics*, vol. 91, no. 3, pp. 984–991, 2002.

46 K. K. Jameson, D. M. Goebel, and R. M. Watkins, "Hollow Cathode and Thruster Discharge Chamber Plasma Measurements Using High-Speed Scanning Probes," IEPC-2005-269, Presented at the 29th International Electric Propulsion Conference. Princeton, November 1–4, 2005.

47 I. Katz, J. E. Polk, I. G. Mikellides, D. M. Goebel, and S. Hornbeck, "Combined Plasma and Thermal Hollow Cathode Insert Model," IEPC-2005-228, Presented at the 29th International Electric Propulsion Conference. Princeton, November 1–4, 2005.

48 K. K. Jameson, D. M. Goebel, and R. M. Watkins, "Hollow Cathode and Keeper Region Plasma Measurements," AIAA-2005-3667, Presented at the 41th AIAA Joint Propulsion Conference, Tucson, July 11–14, 2005.

49 I. Katz, I. G. Mikellides, J. E. Polk, D. M. Goebel, and S. E. Hornbeck, "Thermal Model of the Hollow Cathode Using Numerically Simulated Plasma Fluxes," *Journal of Propulsion and Power*, vol. 23, no. 3, pp. 522–527, 2007; doi:10.2514/1.21103.

50 I. G. Mikellides, I. Katz, D. M. Goebel, and J. E. Polk, Theoretical Model of a Hollow Cathode Plasma for the Assessment of Insert and Keeper Lifetimes," AIAA-2005-4234, Presented at the 41st AIAA Joint Propulsion Conference, Tucson, July 11–14, 2005.

51 J. E. Polk, A. Grubisic, N. Taheri, D. M. Goebel, R. Downey, and S. Hornbeck, "Emitter Temperature Distributions in the NSTAR Discharge Hollow Cathode," AIAA-2005-4398, Presented at the 41st AIAA Joint Propulsion Conference, Tucson, July 11–14, 2005.

52 I. G. Mikelides and I. Katz, "Numerical Simulations of a Hall Thruster Hollow Cathode Plasma," IEPC-2007-018, Presented at the 30th International Electric Propulsion Conference, Florence, September 17–21, 2007.

53 I. G. Mikellides and I. Katz, "Wear Mechanisms in Electron Sources for Ion Propulsion, 1: Neutralizer Hollow Cathode," *Journal of Propulsion and Power*, vol. 24, no. 4, pp. 855–865, 2008; doi:10.2514/1.33461.

54 I. G. Mikellides, I. Katz, D. M. Goebel, K. K. Jameson, and J. E. Polk, "Wear Mechanisms in Electron Sources for Ion Propulsion, 2: Discharge Hollow Cathode," *Journal of Propulsion and Power*, vol. 24, no. 4, pp. 866–879, 2008; doi:10.2514/1.33462.

55 I. G. Mikellides, A. L. Ortega, D. M. Goebel, and G. Becatti, "Dynamics of a Hollow Cathode Discharge in the Frequency Range of 1-500 kHz," *Plasma Sources Science and Technology*, vol. 29, no. 035003, 2020; doi:10.1088/1361-6595/ab69e4.

56 I. G. Mikellides, D. M. Goebel, B. A. Jorns, J. E. Polk, and P. Guerrero, "Numerical Simulations of the Partially Ionized Gas in a 100-A LaB6 Hollow Cathode," *IEEE Transactions on Plasma Science*, vol. 43, no. 1, pp. 173–184, 2015; doi:10.1109/TPS.2014.2320876.

57 I. Katz, J. R. Anderson, J. E. Polk, and J. R. Brophy, "A Model of Hollow Cathode Plasma Chemistry," AIAA 2002-4241, Presented at the 38th AIAA Joint Propulsion Conference, Indianapolis, July 7–10, 2022.

58 J. G. Andrews and J. E. Allen, "Theory of a Double Sheath Between Two Plasmas," *Proceedings of the Royal Society of London, Series A*, vol. 320, pp. 459–472, 1971.

59 J. E. Polk, J. R. Anderson, J. R. Brophy, V. K. Rawlin, M. J. Patterson, J. S. Sovey, and J. J. Hamley, "An Overview of the Results from an 8200 Hour Wear Test of the NSTAR Ion Thruster," AIAA Paper 99–2446, Presented at the 35th AIAA Joint Propulsion Conference, Los Angeles, June 20–24, 1999.

60 A. Sengupta, J. R. Brophy, J. R. Anderson, and C. Garner, "An Overview of the Results from the 30,000 Hour Life Test of the Deep Space 1 Flight Spare Ion Engine," AIAA-2004–3608, Presented at the 40th AIAA Joint Propulsion Conference. Ft. Lauderdale, July 11–14, 2004.

61 D. A. Herman and A. D. Gallimore, "Near Discharge Cathode Assembly Plasma Potential Measurements in a 30-cm NSTAR Type Ion Engine Amidst Beam Extraction," AIAA-2004–3958, Presented at the 40th AIAA Joint Propulsion Conference. Ft. Lauderdale, July 11–14, 2004.

62 K. K. Jameson, D. M. Goebel, I. G. Mikellides, and R. M. Watkins, "Local Neutral Density and Plasma Parameter Measurements in a Hollow Cathode Plume," AIAA-2006-4490, Presented at the 42nd AIAA Joint Propulsion Conference, Sacramento, July 10–13, 2006.

63 G. A. Csiky, "Investigation of a Hollow Cathode Discharge Plasma," AIAA-69-258, Presented at the 7th Electric Propulsion Conference. Williamsburg, March 3–5, 1969.

64 D. M. Goebel, K. K. Jameson, I. Katz, I. G. Mikellides, and J. E. Polk, "Energetic Ion Production and Keeper Erosion in Hollow Cathode Discharges," IEPC-2005-266, Presented at the 29th International Electric Propulsion Conference, Princeton, October 31–November 4, 2005.

65 D. M. Goebel, K. K. Jameson, I. Katz, and I. G. Mikellides, "Potential Fluctuations and Energetic Ion Production in Hollow Cathode Discharges," *Physics of Plasmas*, vol. 14, no. 103508, 2007; doi:10.1063/1.2784460.

66 A. Salhi and P. Turchi, "A First-Principles Model for Orificed Hollow Cathode Operation," AIAA-1992-3742, Presented at the 28th AIAA Joint Propulsion Conference, Nashville, July 6–8, 1992.
67 F. Crawford and S. Gabriel, "Microfluidic Model of a Micro Hollow Cathode for Small Ion Thrusters," AIAA-2003-3580, Presented at the 33rd AIAA Fluid Dynamics Conference, Orlando, June 23–26, 2003.
68 I. D. Boyd and M. W. Crofton, "Modeling the Plasma Plume of a Hollow Cathode," *Journal of Applied Physics*, vol. 95, no. 7, pp. 3285–3296, 2004; doi:10.1063/1.1651333.
69 S. I. Braginskii, *Transport Processes in a Plasma (Reviews of Plasma Physics)*. New York: Consultants Bureau, 1965.
70 S. Dushman, "Electron Emission from Metals as a Function of Temperature," *Physical Review*, vol. 21, no. 6, pp. 623–636, 1923.
71 I. G. Mikellides, "Effects of Viscosity in a Partially Ionized Channel Flow with Thermionic Emission," *Physics of Plasmas*, vol. 16, no. 013501, 2009; doi:10.1063/1.3056397.
72 I. G. Mikellides, I. Katz, D. M. Goebel, and K. K. Jameson, "Evidence of Nonclassical Plasma Transport in Hollow Cathodes for Electric Propulsion," *Journal of Applied Physics*, vol. 101, no. 063301, 2007; doi:10.1063/1.2710763.
73 V. Y. Bychenkov, V. P. Silin, and S. A. Uryupin, "Ion-Acoustic Turbulence and Anomalous Transport," *Physics Reports-Review Section of Physics Letters*, vol. 164, no. 3, pp. 119–215, 1988.
74 T. H. Stix, "Waves in Plasmas – Highlights from the Past and Present," *Physics of Fluids B: Plasma Physics*, vol. 2, no. 8, pp. 1729–1743, 1990.
75 R. Z. Sagdeev and A. Galeev, *Nonlinear Plasma Theory*. New York: W.A. Benjamin, 1969.
76 V. N. Tsytovich, *Lectures on Non-Linear Plasma Kinetics*. Berlin: Springer, 1995.
77 A. Lopez-Ortega, B. A. Jorns, and I. G. Mikellides, "Hollow Cathode Simulations with a First-Principles Model of Ion-Acoustic Anomalous Resistivity," *Journal of Propulsion and Power*, vol. 34, no. 4, pp. 1026–1038, 2018; doi:10.2514/1.B36782.
78 B. A. Jorns, I. G. Mikellides, and D. M. Goebel, "Ion Acoustic Turbulence in a 100-A LaB6 Hollow Cathode," *Physical Review E*, vol. 90, no. 063106, 2014; doi:10.1103/PhysRevE.90.063106.
79 G. Soulas, "Status of Hollow Cathode Heater Development for the Space Station Plasma Contactor," AIAA Paper 1994–3309, Presented at the 30th AIAA Joint Propulsion Conference. Indianapolis, July 27–29, 1994.
80 W. G. Tighe, K. Freick, and K. R. Chien, "Performance Evaluation and Life Test of the XIPS Hollow Cathode Heater," AIAA-2005-4066, Presented at the 41st AIAA Joint Propulsion Conference. Tucson, July 10–13, 2005.
81 D. M. Goebel and R. M. Watkins, "Compact Lanthanum Hexaboride Hollow Cathode," *Review of Scientific Instruments*, vol. 81, no. 083504, 2010.
82 D. M. Goebel and G. Becatti, "High Current Hollow Cathodes for High Power Electric Thrusters," IEPC-2022-101, Presented at the 37th International Electric Propulsion Conference, Cambridge, June 19–23, 2022.
83 D. M. Goebel and E. Chu, "High Current Lanthanum Hexaboride Hollow Cathode for High Power Hall Thrusters," *Journal of Propulsion and Power*, vol. 30, no. 1, pp. 35–40, 2014; doi:10.2514/1.B34870.
84 M. S. McDonald, A. D. Gallimor, and D. M. Goebel, "Improved Heater Design for High-Temperature Hollow Cathodes," *Review of Scientific Instruments*, vol. 88, no. 026104, 2017; doi:10.1063/1.4976728.
85 D. Lev, G. Alon, and L. Appel, "Low Current Heaterless Hollow Cathode Neutralizer for Plasma Propulsion-Development Overview," *Review of Scientific Instruments*, vol. 90, no. 113303, 2019; doi:10.1063/1.5097599.

86 G. Becatti, R. W. Conversano, and D. M. Goebel, "Demonstration of 25,000 Ignitions on a Compact Heaterless LaB6 Hollow Cathode," *Acta Astronauticia*, vol. 178, pp. 181–191, 2021; doi:10.1016/j.actaastro.2020.09.013.

87 V. Vekselman, Y. E. Krasik, S. Gleizer, V. T. Gurovich, A. Warshavsky, and I. Rabinovich, "Characterization of a Heaterless Hollow Cathode," *Journal of Propulsion and Power*, vol. 29, no. 2, pp. 475–486, 2013; doi:10.2514/1.B34628.

88 R. R. Robson, W. S. Williamson, and J. Santoru, "Flight Model Discharge System," Hughes Research Laboratories Report F4890. Hughes Research Laboratory, Hanscom Air Force Base, June 1988.

89 D. M. Goebel and A. C. Schneider, "High-Voltage Breakdown and Conditioning of Carbon and Molybdenum Electrodes," *IEEE Transactions on Plasma Science*, vol. 33, no. 4, pp. 1136–1148, 2005; doi:10.1109/TPS.2005.852410.

90 A. R. Payman and D. M. Goebel, "High-Current Heaterless Hollow Cathode for Electric Thrusters," *Review of Scientific Instruments*, vol. 93, no. 113543, 2022; doi:10.1063/5.0124694.

91 D. M. Goebel and A. R. Payman, "Heaterless 300-A Lanthanum Hexaboride Hollow Cathode," *Review of Scientific Instruments*, vol. 94, no. 033506, 2023; doi:10.1063/5.0135272.

92 R. Albertoni, D. Pedrini, F. Paganucci, and M. Andrenucci, "A Reduced-Order Model for Thermionic Hollow Cathodes," *IEEE Transactions on Plasma Science*, vol. 41, no. 7, pp. 1731–1745, 2013.

93 G. Sary, L. Garrigues, and J. P. Boeuf, "Hollow Cathode Modeling: I. A Coupled Plasma Thermal Two-Dimensional Model," *Plasma Sources Science & Technology*, vol. 26, no. 055007, 2017; doi:10.1088/1361-6595/aa6217.

94 P. Guerrero, I. G. Mikellides, J. E. Polk, R. C. Monreal, and D. I. Meiron, "Hollow Cathode Thermal Modelling and Self-Consistent Plasma Solution," IEPC-2019-301, Presented at the 36th International Electric Propulsion Conference, Vienna, September 15–20, 2019.

95 D. M. Goebel, I. Katz, I. G. Mikellides, and J. E. Polk, "Extending Hollow Cathode Life for Deep Space Missions," AIAA-2004-5911, Presented at the AIAA Space 2004 Conference. San Diego, September 28–30, 2004.

96 R. Doerner, G. Tynan, D. M. Goebel, and I. Katz, "Plasma Surface Interaction Studies for Next-Generation Ion Thrusters," AIAA-2004-4104, Presented at the 40th AIAA Joint Propulsion Conference. Ft. Lauderdale, July 11–14, 2004.

97 R. P. Doerner, S. I. Krasheninnikov, and K. Schmidt, "Particle-Induced Erosion of Materials at Elevated Temperature," *Journal of Applied Physics*, vol. 95, no. 8, pp. 4471–4475, 2004.

98 J. E. Polk, C. Marrese, B. Thornber, L. Dang, and L. Johnson, "Temperature Distributions in Hollow Cathode Emitters," AIAA-2004-4116, Presented at the 40th AIAA Joint Propulsion Conference, Ft. Lauderdale, July 11–14, 2004.

99 P. Palluel and A. M. Shroff, "Experimental Study of Impregnated-Cathode Behavior, Emission, and Life," *Journal of Applied Physics*, vol. 51, no. 5, pp. 2894–2902, 1980.

100 T. R. Sarver-Verhey, "28,000 Hour Xenon Hollow Cathode Life Test Results," IEPC-97-168, Presented at the 25th International Electric Propulsion Conference, Cleveland, August 24–28, 1997.

101 J. E. Polk, D. M. Goebel, J. S. Snyder, A. C. Schneider, J. R. Anderson, and A. Sengupta, "A High Power Ion Thruster for Deep Space Missions," *Review of Scientific Instruments*, vol. 83, no. 073306, 2012; doi:10.1063/1.4728415.

102 H. E. Gallagher, "Poisioning of LaB6 Cathodes," *Journal of Applied Physics*, vol. 40, no. 1, pp. 44–51, 1969; doi:10.1063/1.1657092.

103 A. A. Avdienko and M. D. Malev, "Poisoning of LaB6 Cathodes," *Vacuum*, vol. 27, no. 10/11, pp. 583–588, 1977.

104 J. E. Polk, "The Effect of Reactive Gases on Hollow Cathode Operation," AIAA-2006-5153, Presented at the 42nd AIAA Joint Propulsion Conference, Sacramento, July 9–12, 2006.

105 A. Sengupta, J. Anderson, J. R. Brophy, J. Kulleck, C. Garner, T. de Groh, T. Karniotis, B. Banks, and P. Wlaters, "The 30,000-Hour Extended-Life Test of the Deep Space 1 Flight Spare Ion Thruster," Final Report.

106 D. M. Goebel, J. E. Polk, and A. K. Ho, "Lanthanum Hexaboride Hollow Cathode Performance and Wear Testing for the Asteroid Redirect Mission Hall Thruster," AIAA-2016-4835, Presented at the 52nd AIAA Joint Propulsion Conference, Salt Lake City, July 25–27, 2016.

107 R. D. Kolasinski and J. E. Polk, " Characterization of Cathode Keeper Wear by Surface Layer Activation," AIAA-2003-5144, Presented at the 39th Joint Propulsion Conference, Huntsvilile, July 20–23, 2003.

108 M. Domonkos, J. Foster, and G. Soulas, "Wear Testing and Analysis of Ion Engine Discharge Cathode Keeper," *Journal of Propulsion and Power*, vol. 21, no. 1, pp. 102–110, 2005; doi:10.2514/1.4441.

109 M. J. Patterson, V. K. Rawlin, J. S. Sovey, M. J. Kussmaul, and J. Parkes, "2.3 kW Ion Thruster Wear Test," AIAA-95-2516, Presented at the 31st AIAA Joint Propulsion Conference, San Diego, July 10–12, 1995.

110 A. Sengupta, J. A. Anderson, C. Garner, J. R. Brophy, K. K. deGroh, B. A. Banks, and T. A. Karniotis-Thomas, "Deep Space 1 Flight Spare Ion THRUSTER 30,000 Hour Life Test," *Journal of Propulsion and Power*, vol. 25, no. 1, pp. 105–117, 2009.

111 I. Kameyama and P. J. Wilbur, "Measurement of Ions from High Current Hollow Cathodes Using Electrostatic Energy Analyzer," *Journal of Propulsion and Power*, vol. 16, no. 3, 2000; doi:10.2514/2.5601.

112 C. Farnell and J. Williams, "Characteristics of Energetic Ions from Hollow Cathodes," IEPC-2003-072, Presented at the 28th International Electric Propulsion Conference, Toulouse, March 17–21, 2003.

113 G. J. Williams, T. B. Smith, M. T. Domonkos, A. D. Gallimore, and R. P. Drake, "Laser Induced Fluorescence Characterization of Ions Emitted from Hollow Cathodes," *IEEE Transactions on Plasma Science*, vol. 28, no. 5, pp. 1664–1675, 2000.

114 I. Kameyama and P. J. Wilbur, "Potential-Hill Model of High-Energy Ion Production Near High Current Hollow Cathodes," ISTS 98-a-2-17, Presented at the 21st International Symposium on Space Technology and Science, Sonic City, May 21–24, 1998.

115 V. J. Friedly and P. J. Wilbur, "High Current Hollow Cathode Phenomena," *Journal of Propulsion and Power*, vol. 8, no. 3, pp. 635–643, 1992.

116 I. Katz, I. G. Mikellides, D. M. Goebel, K. K. Jameson, and J. E. Polk, "Production of High Energy Ions Near an Ion Thruster Discharge Hollow Cathode," AIAA-2006-4485, Presented at the 42nd AIAA Joint Propulsion Conference, Sacramento, July 10–13, 2006.

117 W. G. Tighe, K. Chien, D. M. Goebel, and R. T. Longo, "Hollow Cathode Ignition and Life Model," AIAA-2005-3666, Presented at the 41th AIAA Joint Propulsion Conference, Tucson, July 11–14, 2005.

118 W. G. Tighe, K. Chien, D. M. Goebel, and R. T. Longo, "Hollow Cathode Ignition Studies and Model Development," IEPC-2005-314, Presented at the 29th International Electric Propulsion Conference, Princeton, October 31–November 4, 2005.

119 J. R. Brophy and C. E. Garner, "Tests of High Current Hollow Cathodes for Ion Engines," AIAA-88-2913, Presented at the 24th AIAA Joint Propulsion Conference, Boston, MA, July 11–13, 1988.

120 D. M. Goebel, K. K. Jameson, I. Katz, and I. G. Mikellides, "Plasma Potential Behavior and Plume Mode Transitions in Hollow Cathode Discharges," IEPC-2007-277, Presented at the 30th International Electric Propulsion Conference, Florence, September 17–20, 2007.

121 G. C. Potrivitu, S. Mazouffre, L. Grimaud, and R. Joussot, "Anode Geometry Influence on LaB6 Cathode Discharge Characteristics," *Physics of Plasmas*, vol. 26, no. 113506, 2019.

122 D. M. Goebel and G. Becatti, “Compact Scanning Retarding Potential Analyzer,” *Review of Scientific Instruments*, vol. 92, no. 013511, 2021; doi:10.1063/5.0035964.

123 M. P. Georgin, B. A. Jorns, and A. D. Gallimore, “An Experimental and Theoretical Study of Hollow Cathode Plume Mode Oscillations,” IEPC-2017-298, Presented at the 35th Int. Electric Propulsion Conference, Atlanta, October 8–12, 2017.

124 M. P. Georgin, B. A. Jorns, and A. D. Gallimore, “Correlation of Ion Acousticturbulence with Self-Organization in a Low-Temperature Plasma,” *Physics of Plasmas*, vol. 26, no. 082308, 2019.

125 J. M. Fife, M. Martinez-Sanchez, and J. Szabo, “A Numerical Study of Low-Frequency Discharge Oscillations in Hall Thrusters,” AIAA-97-3052, Presented at the 33rd AIAA Joint Propulsion Conference, Seattle, July 6–9, 1997.

126 B. A. Jorns and R. R. Hofer, “Plasma Oscillations in a 6-kW Magnetically Shielded Hall Thruster,” *Physics of Plasmas*, vol. 21, no. 053512, 2014; doi:10.1063/1.4879819.

127 G. Becatti, D. M. Goebel, and M. Zuin, “Observation of Rotating Magnetohydrodynamic Modes in the Plume of a High-Current Hollow Cathode,” *Journal of Applied Physics*, vol. 129, no. 033304, 2021; doi:10.1063/5.0028566.

128 G. Becatti, F. Burgalassi, F. Paganucci, M. Zuin, and D. M. Goebel, “Resistive MHD Modes in Hollow Cathode External Plasma,” *Plasma Sources Science & Technology*, vol. 31, no. 015016, 2022; doi:10.1088/1361-6595/ac43c4.

129 I. G. Mikellides and I. Katz, “Numerical Simulations of Hall-Effect Plasma Accelerators on a Magnetic-Field-Aligned Mesh,” *Physical Review E*, vol. 86, no. 046703, 2012, Artn 046703.

130 I. G. Mikellides and A. Lopez-Ortega, “Growth of the Modified Two-Stream Instability in the Plume of a Magnetically Shielded Hall Thruster,” *Physics of Plasmas*, vol. 27, no. 100701, 2020; doi:10.1063/5.0020075.

131 I. G. Mikellides and A. Lopez-Ortega, “Growth of the Lower Hybrid Drift Instability in the Plume of a Magnetically Shielded Hall Thruster,” *Journal of Applied Physics*, vol. 129, no. 19, 2021, Artn 193301.

132 A. Ho, B. A. Jorns, I. G. Mikellides, D. M. Goebel, and A. L. Ortega, “Wear Test Demonstration of a Technique to Mitigate Keeper Erosion in a High-Current LaB6 Hollow Cathode,” AIAA 2016-4836, Presented at the 52nd AIAA Joint Propulsion Conference, Salt Lake City, July 25–27, 2016.

133 E. Chu and D. M. Goebel, “Reduction of Energetic Ion Production in Hollow Cathodes by External Gas Injection,” *Journal of Propulsion and Power*, vol. 29, no. 5, 2013; doi:10.2514/1.B34799.

134 G. Becatti and D. M. Goebel, “500-A LaB6 Hollow Cathode for High Power Electric Thrusters,” *Vacuum*, vol. 198, no. 110895, 2022; doi:10.1016/j.vacuum.2022.110895.

135 S. W. Shar, S. J. Hall, B. A. Jorns, R. R. Hofer, and D. M. Goebel, “High Power Demonstration of a 100 kW Nested Hall Thruster System,” AIAA-2019-3809, Presented at the AIAA Propulsion and Energy Forum, Indianapolis, August 19–22, 2019.

136 S. J. Hall, B. A. Jorns, A. D. Gallimore, and D. M. Goebel, “Operation of a High-Power Nested Hall Thruster with Reduced Cathode Flow Fraction,” *Journal of Propulsion and Power*, vol. 36, no. 6, 2020; doi:10.2514/1.B37929.

137 D. M. Goebel, G. Becatti, S. Reilly, and K. Tilley, “High Current Lanthanum Hexaboride Hollow Cathode for 20–200 kW Hall Thrusters,” IEPC-2017-303, Presented at the 35th International Electric Propulsion Conference, Atlanta, October 8–12, 2017.

138 G. Becatti and D. M. Goebel, “High Current Hollow Cathode for the X3 100-kW Class Nested Hall Thruster,” IEPC-2019-371, Presented at the 36th International Electric Propulsion Conference, Vienna, September 15–20, 2019.

Chapter 5

Ion Thruster Plasma Generators

Ion thrusters are characterized by the electrostatic acceleration of ions extracted from a plasma generator. Ion thruster geometries are best described in terms of three basic components: the ion accelerator, the plasma generator and the electron neutralizer. The ion accelerator described in Chapter 6 typically uses electrically-biased multi-aperture grids to produce the ion beam. The neutralizer cathode, which is discussed in Chapter 4, is positioned outside the thruster body to provide electrons to neutralize the ion beam and maintain the potential of the thruster and spacecraft relative to the space plasma potential. In this chapter, three types of the third component of modern flight ion thrusters, namely the plasma generator, are discussed. These plasma generators utilize DC electron discharges, rf discharges, and microwave discharges to produce the plasma. Physics-based models will be developed and used throughout the chapter to describe the performance and characteristics of these different plasma generation techniques.

5.1 Introduction

The basic geometry of an ion thruster plasma generator is illustrated well by the classic DC electron discharge plasma generator. This version of the thruster plasma generator utilizes an anode potential discharge chamber with a hollow cathode electron source to generate the plasma from which ions are extracted to form the thrust beam. A simplified schematic of a DC electron bombardment ion thruster with these components coupled to a multi-grid accelerator is shown in Fig. 5-1. Neutral propellant gas is injected into the discharge chamber and a small amount is also injected through the hollow cathode. Electrons extracted from the hollow cathode enter the discharge chamber and ionize the propellant gas. To improve the efficiency of the discharge in producing ions, some form of magnetic confinement is typically employed at the anode wall. The magnetic fields provide confinement primarily of the energetic electrons, which increases the electron path length before loss to the anode wall and improves the ionization probability of the injected electrons. Proper design of the magnetic field is critical to providing sufficient confinement for high efficiency while maintaining adequate electron loss to the anode to produce stable discharges over the operation range of the thruster.

Several power supplies are required to operate the cathode and plasma discharge. A simplified electrical schematic typically used for DC-discharge plasma generators is shown in Fig. 5-2. The cathode heater supply raises the thermionic emitter to a sufficient temperature to emit electrons,

Fundamentals of Electric Propulsion, Second Edition. Dan M. Goebel, Ira Katz, and Ioannis G. Mikellides.

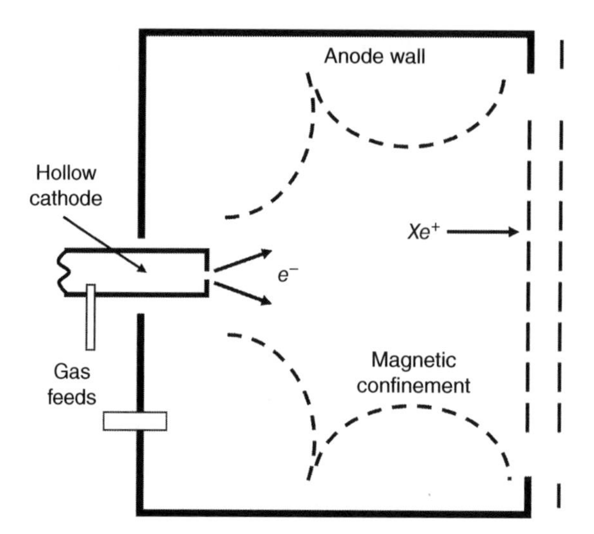

Figure 5-1 Illustration of a DC-discharge electron bombardment ion thruster.

and is turned off once the plasma discharge is ignited. The keeper electrode positioned around the hollow cathode tube is used to facilitate striking the hollow cathode discharge, and also protects the cathode from ion bombardment from the discharge chamber region. The cathode and the keeper are discussed in Chapter 4. The discharge supply is connected between the hollow cathode and the anode and is normally run in the current regulated mode to provide a stable discharge at different power levels.

RF and microwave ion thrusters utilize nearly identical ion accelerator and electron neutralizer implementations as the DC-discharge ion thruster. However, these thrusters do not employ a discharge hollow cathode or anode power supply. These components are replaced by rf or microwave antenna structures, sources of rf or microwave power and compatible discharge chambers to ionize the propellant gas and deliver the ions to the accelerator structure. These thrusters also utilize either applied or self-generated magnetic fields to improve the ionization efficiency of the system.

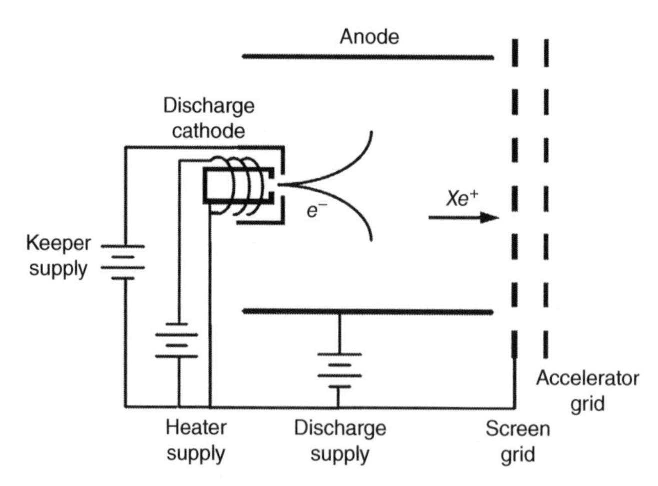

Figure 5-2 Electrical schematic of a DC-discharge ion thruster with the cathode heater, keeper and discharge power supplies.

The three thruster plasma generators to be discussed here—DC electron discharge, rf, and microwave discharge—have been successfully developed and flown in space. The principles of these different classes of plasma generators are described in the following sections after a discussion of the plasma generator efficiency that can be expected in an idealized case.

5.2 Idealized Ion Thruster Plasma Generator

It is worthwhile to examine an ion thruster in the simplest terms to provide an understanding of the dominant processes in the particle flows and energy transport required to produce the plasma. The idealized thruster model has power injected by arbitrary means into a volume filled with neutral gas to produce ionization and neutral gas excitation, with all the ions going to the accelerator grids and an equal number of plasma electrons going to the wall to conserver charge. This is illustrated schematically in Fig. 5-3. For this model, the thruster discharge chamber has a volume V that fully encloses the plasma that is produced by ionization of neutral gas by the plasma electrons. The ions from the plasma flow only to the accelerator grid structure (perfect confinement elsewhere in the discharge chamber) with a current given by the Bohm current:

$$I_i = \frac{1}{2} n_i e \, v_a A, \tag{5.2-1}$$

where n_i is the ion density in the center of the volume, v_a is the ion acoustic velocity, A is the total ion loss area which is assumed to be only the grid area, and the ions are assumed to be cold relative to the electrons. The ion beam current is then the ion current to the grids multiplied by the effective grid transparency T_g:

$$I_b = \frac{1}{2} n_i e \, v_a A T_g, \tag{5.2-2}$$

where the current lost to the accel and decel grids has been neglected as small. Ions are assumed to be produced by ionization of neutral particles by the plasma electrons in the discharge chamber at a rate given by

$$I_p = n_o n_e e \langle \sigma_i v_e \rangle V, \tag{5.2-3}$$

where n_o is the neutral gas density, n_e is the plasma electron density, σ_i is the ionization cross section, v_e is the electron velocity, and the term in the brackets is the reaction rate coefficient which is the ionization cross section averaged over the Maxwellian electron velocity distribution function. The formulation of the reaction rate coefficient was described in Chapter 3 and the values for xenon and krypton as a function of electron temperature are given in Appendix E.

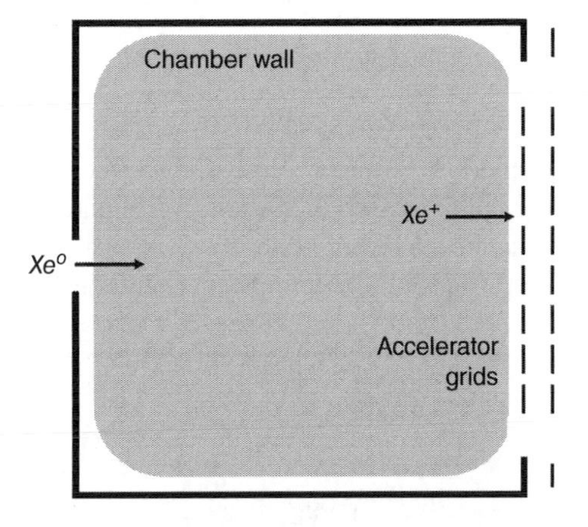

Figure 5-3 Idealized ion thruster with the ions assumed going to the grids and electrons going to the chamber wall.

Power is conserved in the system, so the power put into the plasma is equal to the power that comes out in the form of charged particles and radiation. To first order, the power injected into the plasma goes into ionization and excitation of the neutral gas, heating of the electrons, and power that is carried to the walls and the grids by the ions and electrons. The power that is put into the system is then

$$P_{\text{in}} = I_p U^+ + I^* U^* + I_i \varepsilon_i + \frac{n_e\, e\, V}{\tau} \varepsilon_e, \tag{5.2-4}$$

where U^+ is the ionization potential of the propellant gas, U^* is the excitation potential of the gas, τ is the average electron confinement time, ε_i is the ion energy carried to the walls and ε_e is the electron energy carried to the walls by the electrons leaving the plasma. The term I^* is the excited neutral production rate, given by

$$I^* = \sum_j n_o n_e e \langle \sigma_* v_e \rangle_j V, \tag{5.2-5}$$

where σ_* is the excitation cross section and the reaction rate coefficient is averaged over the electron distribution function and summed over all possible excited states j. Using Eqs. 5.2-3 and 5.2-5 in Eq. 5.2-4, the power input can then be written as

$$P_{\text{in}} = n_o\, n_e\, e \langle \sigma_i v_e \rangle V \left[U^+ + \frac{\langle \sigma_* v_e \rangle_j}{\langle \sigma_i v_e \rangle} U^* \right] + I_i \varepsilon_i + \frac{n_e\, e\, V}{\tau} \varepsilon_e. \tag{5.2-6}$$

Assuming quasi-neutrality ($n_i \approx n_e$) and that the ions and electrons leave the volume by ambipolar flow at the same rate, which is a function of the mean confinement time τ, the ion current out is given by

$$I_i = \frac{1}{2} n_i\, e\, v_a A = \frac{I_b}{T_g} = \frac{n_i\, \text{eV}}{\tau}, \tag{5.2-7}$$

where the I_b is the beam current and T_g is the grid transparency. The mean confinement time for ions and electrons is then

$$\tau = \frac{2V}{v_a A}. \tag{5.2-8}$$

The energy that an electron removes from the plasma as it goes to the wall is given by

$$\varepsilon_e = 2 \frac{kT_e}{e} + \phi, \tag{5.2-9}$$

where ϕ is the plasma potential relative to the wall. Equation 5.2-9 is derived in Appendix C. The ions fall first through the pre-sheath potential, approximated by $T_{eV}/2$ to produce the Bohm velocity, and then through the sheath potential. Since the potential in Eq. 3.7-52 is defined at the center of the plasma, each ion then removes from the plasma a total energy per ion of

$$\varepsilon_i = \frac{1}{2} \frac{kT_e}{e} + \phi_s = \phi. \tag{5.2-10}$$

The plasma potential in these two equations is found from the electron current leaving the plasma, which is given by Eq. 3.7-52:

$$I_a = \frac{1}{4} \left(\frac{8kT_e}{\pi m} \right)^{1/2} e n_e\, A_a\, e^{-e\phi / kT_e} \tag{5.2-11}$$

where A_a is the electron loss area and m is the electron mass. Since ambipolar ion and electron flow to the wall was assumed, equate Eqs. 5.2-1 and 5.2-11 and use $\sqrt{kT_e/M}$ for the ion acoustic velocity to give the plasma potential between the wall and the plasma as:

$$\phi = \phi_f = \frac{kT_e}{e} \ln\left(\frac{A_a}{A}\sqrt{\frac{2M}{\pi m}}\right). \tag{5.2-12}$$

Equation 5.2-12 is normally called the *floating potential* ϕ_f and appears in this case because there are no applied potentials in our ideal thruster to draw a net current.

The electron temperature can be found by equating the ion production and loss rates, Eqs. 5.2-1 and 5.2-3, which gives

$$\frac{\sqrt{kT_e/M}}{\langle\sigma_i v_e\rangle} = \frac{2n_o V}{A}. \tag{5.2-13}$$

The reaction rate coefficient in the denominator depends on the electron temperature, and so this equation can be solved for the electron temperature if the discharge chamber volume, neutral pressure, and ion loss area are known.

The discharge loss is defined as the power into the plasma divided by the beam current out of the thruster, which is a figure of merit for the efficiency of the plasma generation mechanism. The discharge loss for this idealized thruster, using Eq. 5.2-2 for the beam current, is then given by

$$\eta_d = \frac{P_{\text{in}}}{I_b} = \frac{2n_o\langle\sigma_i v_e\rangle V}{v_a A\, T_g}\left[U^+ + \frac{\langle\sigma_* v_e\rangle_j}{\langle\sigma_i v_e\rangle}U^*\right] + \frac{2T_{eV}}{T_g}\left[1 + \ln\left(\frac{A_a}{A}\sqrt{\frac{2M}{\pi m}}\right)\right] \tag{5.2-14}$$

As evident in Eq. 5.2-14, the grid transparency (T_g) directly affects the discharge loss, and the input power is distributed between the first term related to producing ions and excited neutrals, and the second term related to heating the electrons that are lost to the walls.

To evaluate Eq. 5.2-14, the ratio of the excitation to ionization reaction rates as a function of the Maxwellian electron temperature must be known. This is shown in Fig. 5-4 for xenon gas from data in Appendixes E. For electron temperatures below about 8 V, the excitation rate exceeds the ionization rate in xenon for Maxwellian electrons. Since the lowest excitation potential is near the ionization potential in xenon, this higher excitation rate results in more of the input power being radiated to the walls than producing ions. This effect explains at least part of the inefficiency inherent in xenon plasma generators. Excitation rates equal to or higher than the ionization rate at low electron temperatures are also generally found for other inert gas propellants.

The discharge loss from Eq. 5.2-14 for this ideal thruster example is plotted as a function of the mass utilization efficiency for a generic 20 cm diameter thruster in Fig. 5-5 with three different discharge chamber lengths, where the ionization potential of xenon of 12.13 V, an average excitation potential of 10 V, and 80% of the ions incident on the grids become beam ions ($T_g = 0.8$) are used. It was also assumed for simplicity that the plasma electrons were lost to the floating screen grid, and it will be shown later that the area of the electron loss affects the thruster discharge efficiency and stability of the discharge. The mass utilization efficiency is inversely proportional to the neutral density in the thruster, which will be derived in Section 5.3.5. In the figure, the discharge loss is shown in eV/ion, which has the equivalent units to Watts of discharge power per Ampere of beam ions (W/A). In an ideal plasma generator case with 80% of the ions that are generated and headed to

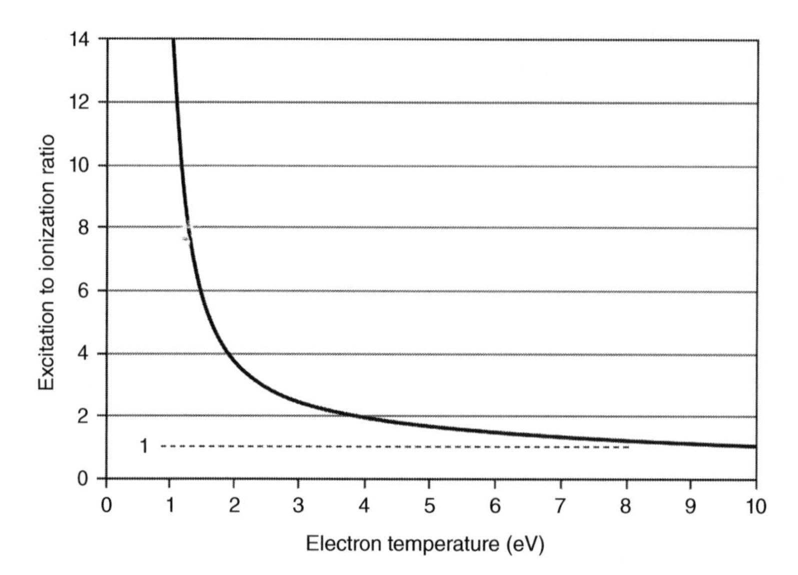

Figure 5-4 Ratio of the excitation to ionization rate coefficients for xenon as a function of the electron temperature.

the accelerator grids assumed to become beam current, the amount of power required to produce 1 A of beam current is well over 120 W. While it only takes 12.13 eV to ionize a xenon atom, even in an idealized thruster it takes about ten times this energy to produce and deliver an ion into the beam because of other losses. The dependence on discharge chamber length illustrates the surface to volume effect in the thruster where the ionization is a volume process and the loss to the walls is a surface process. The longer thruster has a smaller surface-to-volume ratio, and so has more efficient ion generation and a lower discharge loss.

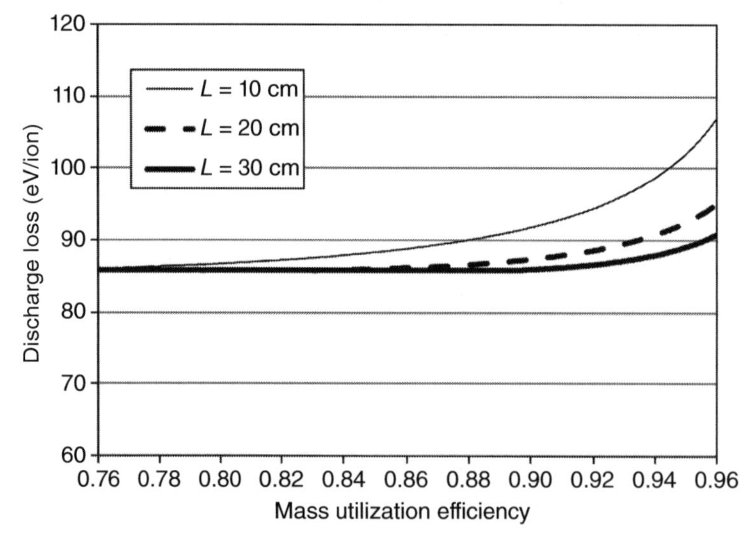

Figure 5-5 Discharge loss for an ideal ion thruster example as a function of mass utilization efficiency for three thruster lengths.

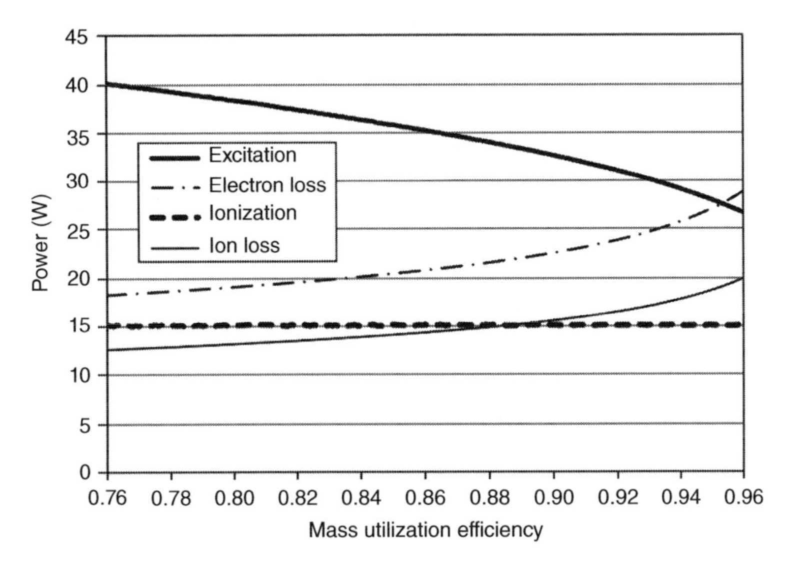

Figure 5-6 Discharge loss for each of the energy loss mechanisms for the ideal ion thruster example with a 30-cm length.

It is informative to see where the extra input power goes in the thruster. Figure 5-6 shows the power lost in each of the four energy loss mechanisms described above for an ideal thruster 30-cm long producing 1 A of beam current. The ionization power is constant in this case because this example was constrained to produce 1 A, and the power required per beam ampere is then (1/0.8)∗ 12.13 = 15.1 W. The major power loss is excitation at low mass utilization where the neutral density is high and electron temperature is low because of the higher neutral density, as shown in Fig. 5-4. However, this excitation loss decreases with mass utilization efficiency as fewer neutrals are present in the discharge chamber. The ion and electron convection losses to the wall also increase at higher mass utilization efficiencies because the neutral density is decreasing, which increases the electron temperature, raises the plasma potential and thereby increases the energy lost per electron and ion.

Many thruster design concepts use electron confinement to improve the efficiency. The impact of this can be examined in this ideal thruster model by reducing the anode area A_a. Figure 5-7 shows the four energy-loss mechanisms for the same idealized thruster example just used, but with the effective anode area collecting electrons decreased from the screen collection area to only 1 cm^2. By conservation of charge, electrons in this discharge are lost at the same rate as ions, so electron confinement does not change the number or rate of electrons lost. The reduced anode area reduces the discharge loss to below 60 eV/ion because of changes the plasma potential relative to the potential of the area that electrons are lost (the anode) to maintain charge balance, as seen from examining Eq. 5.2-11. This effect is clearly seen by comparing Figs. 5-6 and 5-7, where the energy loss rates for ionization and excitation have not changed with the better electron confinement, but the energy convected out of the plasma in the form of ion and electron power to the boundaries has decreased. This is because the plasma potential described by the last term in Eq. 5.2-14 is reduced because of the smaller anode area, which reduces the ion and electron energy loss channels. This is the fundamental mechanism for making efficiency improvements (reducing the discharge loss) in plasma generators.

The idealized thruster description illustrates that the power that must be provided to produce the plasma in a thruster is large compared to that required for ionization. In terms of the total thruster

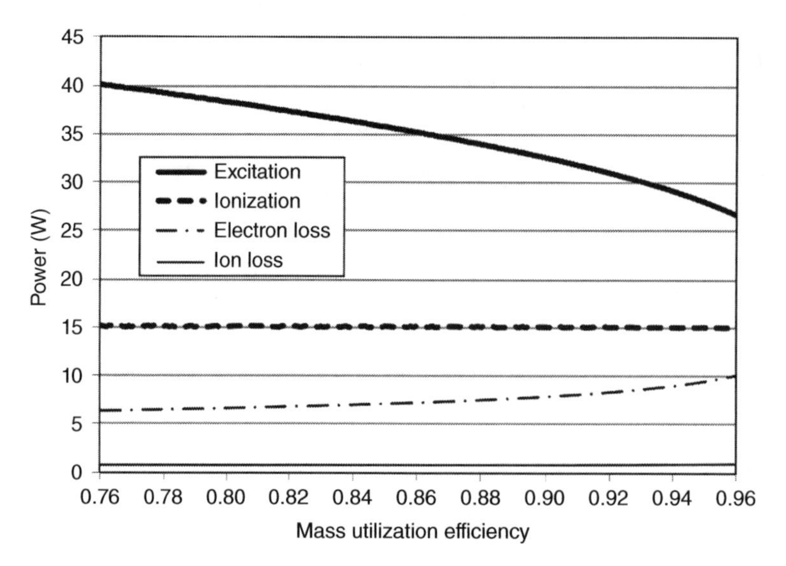

Figure 5-7 Discharge loss for each of the energy loss mechanisms for the ideal ion thruster example with reduced anode (electron loss) area.

efficiency, this is the majority of the "other" power in P_o in Eq. 2.5-1. In reality, the discharge loss is significantly higher than that found in this idealized example due to imperfect confinement of the ions and electrons in the thruster, and due to other loss mechanisms to be described below.

Finally, in most ion thrusters, such as electron bombardment thrusters and microwave-heated Electron Cyclotron Resonance (ECR) thrusters, the electron distribution function is non-Maxwellian. The higher energy electrons observed in electron bombardment thrusters are often called *primaries*, and they have been found to be either monoenergetic or have some distribution in energies depending on the plasma generator design. Primary electrons have a larger ion to excited-neutral production rate than the plasma electrons due to their higher energy, and so even small percentages of primaries in the plasma can dominate the ionization rate. The inclusion of ionization by primary electrons in particle and energy balance models such as the one just described tends to reduce the discharge loss significantly.

5.3 DC Discharge Ion Thrusters

Ion thrusters that use a DC electron-discharge plasma generator employ a hollow cathode electron source and an anode potential discharge chamber with magnetic multipole boundaries to generate the plasma and improve the ionization efficiency. Electrons extracted from the hollow cathode are injected into the discharge chamber and ionize the propellant gas introduced in the chamber. Magnetic fields applied in the discharge chamber provide confinement primarily of the energetic electrons, which increases the electron path length prior to their being lost to the anode and improves the ionization efficiency. The ions from this plasma that flow to the grids are extracted and accelerated to form the beam.

Empirical studies over the past 50 years have investigated the optimal design of the magnetic field to confine electrons and ions in thrusters. Figure 5-8 shows the evolution of the discharge chamber geometry and magnetic field shape employed in efforts primarily aimed at improving the

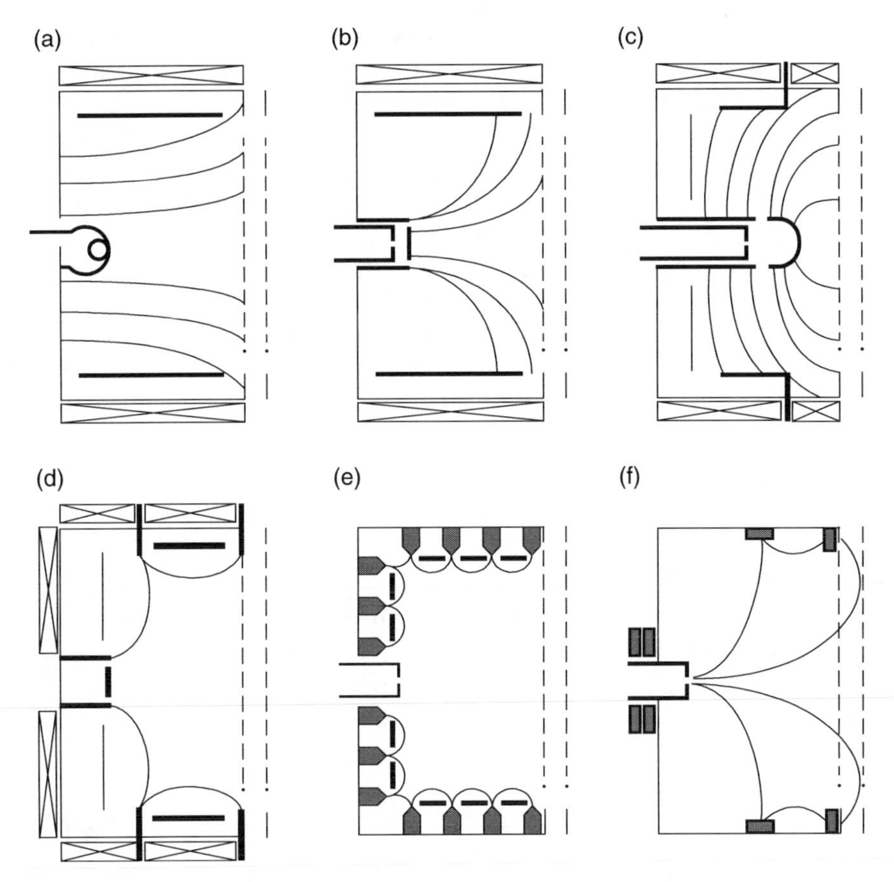

Figure 5-8 Magnetic field types of ion thrusters: (a) mildly divergent B-field, (b) strongly divergent B-field, (c) radial field, (d) cusp field, (e) magnetic multipole field, and (f) ring-cusp fields.

confinement of energetic electrons injected into the chamber from thermionic cathodes to more efficiently produce the plasma. Early thrusters pioneered by Kaufman utilized a solenoidal [1] or mildly divergent magnetic field [2] shown in (a), which requires that electrons from the on-axis thermionic filament cathode undergo collisions in order to diffuse to the anode and complete the discharge circuit. A strongly divergent magnetic field thruster [3], shown in (b), improved the primary electron uniformity in the plasma volume and resulted in a lower discharge loss and a more uniform beam profile. This thruster introduced a baffle in front of the hollow cathode electron source to further inhibit on-axis electrons. The radial magnetic field thruster [4], shown in (c), produced very uniform plasmas and good efficiencies, as did a cusp version of the divergent magnetic-field thruster shown in (d). The use of permanent magnet, multipole boundaries, first reported by Moore [5], created essentially a field free region in the center of the thruster that produced uniform plasmas. The magnets in various versions of this concept were oriented in rings or in axial lines to provide plasma confinement. Moore biased the wall and magnets at cathode potential and placed the anodes inside the cusp fields, as shown in (e), to require that electrons diffuse across the field lines by collisions or turbulent transport before being lost. The permanent magnet ring-cusp thruster of Sovey [6] is shown in (f), which has become the most widely used thruster design to date.

The divergent field Kaufman ion thruster matured in the 1970s with the development of 30-cm mercury thrusters [7, 8]. Kaufman thrusters are described in more detail in Section 5.4. Concerns with using mercury as the propellant resulted in the development of xenon ion thrusters [9, 10], which emerged at the same time as the benefits of ring-cusp confinement geometries became apparent [6, 11, 12]. The design and development of the NSTAR [13] and Xenon Ion Propulsion (XIPS) [14] flight thrusters in the 1990s was based on this early work. At this time, only two of these magnetic field geometries are still used in DC ion thrusters: the multipole magnetic field ring-cusp thrusters and the divergent solenoidal magnetic fields in Kaufman-type thrusters. Ring-cusp thrusters use alternating polarity permanent magnet rings placed around the anode-potential thruster body. Energetic electrons are injected along a weak diverging magnetic field at the cathode and demagnetize sufficiently to bounce from the surface magnetic fields until they either lose their energy by collisions or find a magnetic cusp to be lost to the anode. Kaufman thrusters inject energetic electrons along a strong diverging solenoidal magnetic field with the pole-pieces typically at cathode potential, and rely on cross-field diffusion of the electrons to an anode electrode placed near the cylindrical wall to produce ionization and create a stable discharge.

5.3.1 Generalized 0-D Ring-Cusp Ion Thruster Model

The idealized plasma generator model developed in Section 5.2 is useful in describing how the discharge produces the plasma, but neglects many of the particle flows and energy transport mechanisms found in actual thrusters. The complete particle flows in a thruster discharge chamber are shown in Fig. 5-9. The primary electron current emitted by the hollow cathode, I_e, generates ions

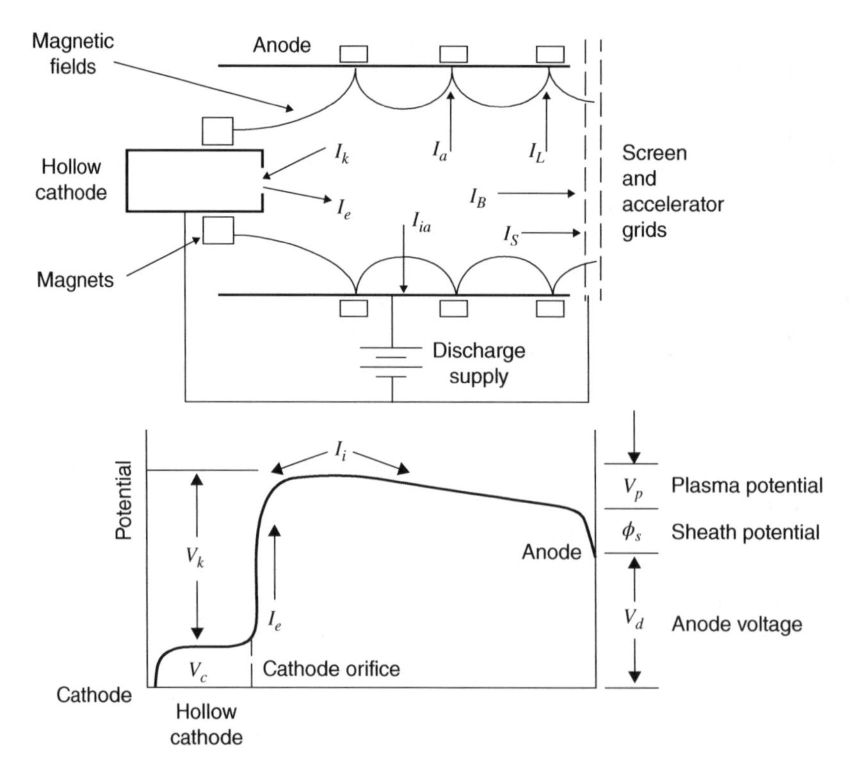

Figure 5-9 Schematic of the thruster showing particle flows and potential distribution in the discharge chamber.

and plasma electrons. The ions flow to the accelerator structure (I_s), to the anode wall (I_{ia}) and back to the cathode (I_k). Some fraction of the primary electrons is lost directly to the anode at the magnetic cusp (I_L). The plasma electrons are also predominately lost to the anode at the cusp (I_a), with only a very small fraction lost across the transverse magnetic field between the cusps corresponding to the ambipolar current flows in this region.

The particle energies are determined by the potential distribution in the thruster. Figure 5-9 also schematically shows the potential in the plasma chamber. Electrons from the plasma inside the hollow cathode at a potential V_c are extracted through the orifice and into the discharge chamber where they gain energy by passing through the potential $V_k = V_d - V_c + V_p + \phi_s = V_d - V_c + \phi$, where V_p is the potential drop in the plasma presheath region and ϕ_s is the anode sheath potential. Some of these electrons cause ionization near the hollow cathode exit, which produces a higher plasma density locally near the cathode exit that must be dispersed before reaching the grid region in order to produce the desired uniform plasma profile across the grids. The potential drop V_p in the plasma, which is assumed to be uniform and quasi-neutral, can be reasonably approximated as $kT_e/2e$ from the pre-sheath potential in the nearly collisionless plasma. Ions leaving the plasma then gain the energy $\varepsilon_i = kT_e/2e + \phi_s = \phi$, which was given in Eq. 5.2-10. Electrons in the tail of the Maxwellian distribution overcome the anode sheath and are collected by the anode at the cusps, where they remove an energy per particle of $\varepsilon_e = (2kT_e/e + \phi)$, which is given by Eq. 5.2-9 and derived in Appendix C.

Analytic models of the discharge chamber performance in ion thrusters have been described in the literature for many years [15–17]. The first comprehensive model of the discharge chamber performance using particle and energy balance equations in ring-cusp thrusters was developed by Brophy and Wilbur [18, 19] in 1984. In this model, volume averaged particle and energy balance equations including primary electrons were used to derive expressions for the discharge loss as a function of the mass utilization efficiency in the thruster. Brophy's model was extended by Goebel [20, 21] to include electrostatic ion confinement, primary confinement and thermalization, the anode sheath [22] and hollow cathode effects. This model utilizes magnetic field parameters obtained from a magnetic field solver that accurately models the magnetic boundary. Since the model assumes a uniform plasma in the volume inside the magnetic confinement in the discharge chamber, it is sometimes called a zero-dimensional or 0-D discharge chamber model.

The 0-D discharge chamber model to be described here [21] self-consistently calculates the neutral gas density, electron temperature, the primary electron density, plasma density, plasma potential, discharge current and the ion fluxes to the boundaries of the discharge chamber. Although the assumption of uniform plasma is not particularly accurate near the cathode plume, the majority of the plasma in the discharge chamber is relatively uniform and the model predictions agree well with experimental results. The 0–D model is used to solve for the discharge loss as a function of the mass utilization efficiency, which is useful in plotting performance curves that best characterize the discharge chamber performance.

The particle flows and potential distribution in the thruster used in the 0-D model are shown schematically in Fig. 5-9. Mono-energetic primary electrons with a current I_e are assumed to be emitted from the hollow cathode orifice into the discharge chamber where they ionize the background gas to produce a uniform plasma. Electrons produced in the ionization process and primary electrons that have thermalized with the plasma electrons create a Maxwellian plasma electron population that also contributes to the ionization. Owing to the relatively high magnetic field produced by the magnets near the wall, the electron Larmor radius is much smaller than the dimensions of the discharge chamber and both primary and plasma electrons are considered to be reflected from the boundary region between the magnetic cusps. The primary and plasma electrons can be lost at the magnetic cusps because the magnetic field lines are essentially perpendicular to the surface. The number of electrons lost at the cusp depends on the local sheath potential and the effective loss area at the cusp. Ions produced in the discharge chamber can flow back to the hollow cathode, to the anode wall or to the plane of the accelerator. At the accelerator, these ions are either intercepted and collected by the screen electrode with an effective transparency T_g, or are extracted from the plasma through the grids to become beam ions. The screen grid transparency depends on the optical transparency of the grid and the penetration of the high voltage fields from the accelerator region into the screen apertures. Although this transparency is an input to the discharge model, it is calculated by the ion optics codes described in Chapter 6.

In this model, the high voltage power supply that accelerates the ions, called the screen supply, is connected to the anode. This means that the ions fall from the average plasma potential in the discharge chamber to form the beam. It is also possible to connect the screen supply to the screen and cathode, which means that the ion current in the beam must pass through the discharge supply. This changes the algebra slightly in calculating the discharge performance, but it does not change the results. The components of particle and energy balance model are described in the following sections.

5.3.2 Magnetic Multipole Boundaries

Ring-cusp ion thrusters use alternating polarity permanent magnet rings oriented perpendicular to the thruster axis, with the number of rings selected and optimized for different size thrusters [20]. This configuration provides magnetic confinement of the electrons with finite loss at the magnetic cusps, and electrostatic confinement of the ions from the anode wall due to the quasi-ambipolar potentials at the boundary from the transverse magnetic fields. Line-cusp thrusters also use high field magnets, but the magnets are configured in alternating polarity axial lines that run along the chamber wall. Asymmetries at the ends of the line cusps cause plasma losses and difficulties in producing a uniform symmetric field at the cathode exit, which adversely affects the electron confinement and thruster efficiency. Ring-cusp thrusters are the most commonly used discharge chamber design at this time due to their ability to produce high efficiency and uniform plasmas at the ion accelerator surface if properly designed.

A schematic representation of a section of a ring cusp magnetic multipole boundary is shown in Fig. 5-10. In this view, a cut along the axis through a six-ring boundary at the wall is made, leaving the ends of the alternating magnets visible. The magnetic field lines terminate at the magnet face, resulting in a cusp magnetic field with field lines perpendicular to the wall at the magnet. Electrons

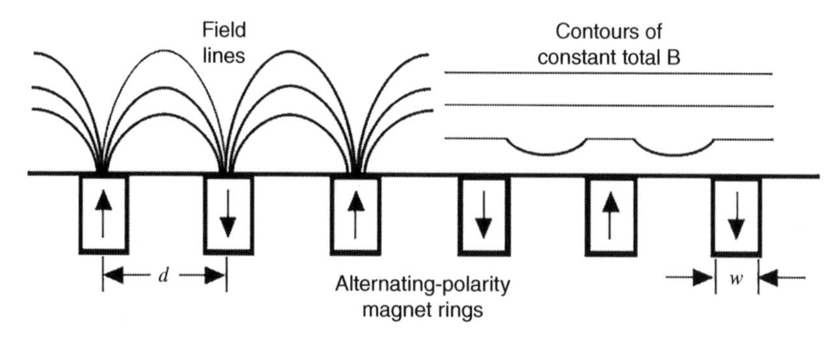

Figure 5-10 Cross section (side) view of a six-ring-cusp magnetic multipole boundary showing the magnetic field lines and examples of contours of constant magnetic field.

that are incident in this area will be either reflected by the magnetic mirror, electrostatically repelled by the sheath potential, or lost directly to the anode. Electrons that are incident between the cusps encounter a transverse magnetic field and are reflected from the boundary. The contours of constant magnetic field shown on the right in Fig. 5-10 illustrate that the total field is essentially constant across the boundary at a distance sufficiently above the magnets, although the component of the field is changing from purely perpendicular at the cusp to purely parallel between the cusps.

An analysis of the magnetic field strength for various multipole boundaries was published by Forrester [23] and discussed by Lieberman [24]. Since the divergence of the magnetic field is zero, the field satisfies Laplace's Equation and the solution for the lowest order mode at a distance from the magnets greater than the magnet separation can be expressed by a Fourier series. This gives a magnetic field strength above the magnet array described by

$$B_y(x, y) = \frac{\pi w B_o}{2d} \cos\left(\frac{\pi x}{d}\right) e^{-\pi y/d}, \tag{5.3-1}$$

where B_o is the magnetic field at the surface of the magnet, d is the distance between the magnet centers, w is the magnet width and the y-direction is perpendicular to the wall in Fig. 5-10. Owing to the localized magnet positions, the field has the periodic cosine behavior along the surface of the wall illustrated in the figure. In addition, the magnetic field decreases exponentially away from the wall all along the boundary.

At the cusp, the field actually decreases as $1/d^2$ due to the dipole nature of the permanent magnet. This rapid decrease in the field moving away from the magnet illustrates the importance of placing the magnets as close to the plasma as possible to maximize the field strength inside the discharge chamber for a given magnet size to provide sufficient field strength for primary and secondary electron confinement at the wall. Between the cusps, the dipole characteristics of the local field forces the field lines to wrap back around the magnets, which causes the magnetic field strength to have a maximum at a distance $y = 0.29*d$ from the wall, which will be derived in Section 5.3.4. The transverse maximum field strength produced between the cusps is important to provide electron and ion confinement, which improves the thruster efficiency.

Although analytic solutions to the magnetic field provide insight into the field structure, the availability of commercial computer codes to calculate the fields accurately makes it much simpler to model the entire ring-cusp field. For example, Fig. 5-11 shows the contours of constant magnetic field measured and calculated using Maxwell 3-D magnetic field solver [25] for the NEXIS ion thruster [20] with six ring cusps. The measured and calculated values are within the measurement error. This type of plot shows clearly the localized surface-field characteristic of magnetic multipole boundaries, which leaves the majority of the inner volume essentially magnetic field free. A large field free region design significantly enhances the plasma uniformity and ion current density profile [20, 26]. In this case, the 60-G magnetic field contour is closed throughout the inside surface of the thruster, which will be shown in the next section to provide good plasma confinement at the wall.

5.3.3 Electron Confinement

The primary electrons are injected into the discharge chamber from the hollow cathode. The discharge chamber can be viewed as a volume with reflecting boundaries and discrete loss areas for the electrons at the cusps where the magnetic fields lines are nearly perpendicular to the surface. The primary electrons then effectively bounce around in the chamber until they are either lost directly to the anode wall by encountering the finite loss area at the cusps, make an ionization or excitation

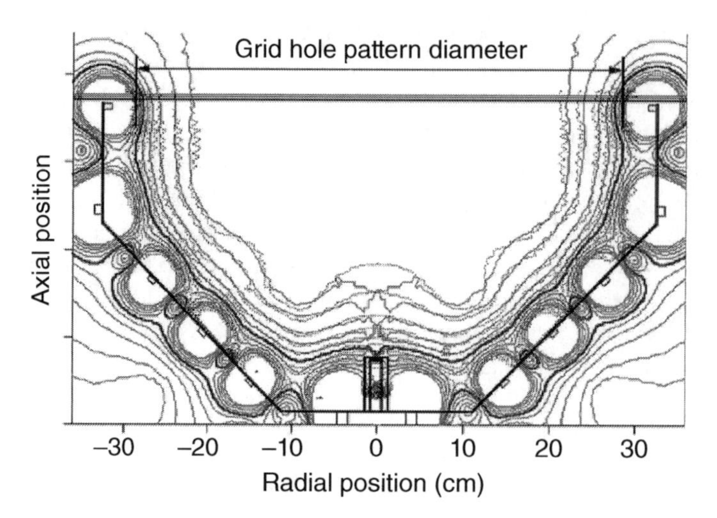

Figure 5-11 Comparison of measured (dashed) and calculated (solid) magnetic field contours in the six-ring NEXIS thruster (*Source:* [20]).

collision, or are thermalized by Coulomb interactions with the plasma electrons. The primary current lost directly to the anode cusps is given by

$$I_L = n_p e\, v_p A_p, \tag{5.3-2}$$

where n_p is the primary electron density, v_p is the primary electron velocity, and A_p is the loss area for the primaries.

The loss area for primary electrons at the cusp [27] is given by

$$A_p = 2r_p L_c = \frac{2}{B}\sqrt{\frac{2mV_p}{e}}L_c, \tag{5.3-3}$$

where r_p is the primary electron Larmor radius, B is the magnetic field strength at the cusp at the anode wall, V_p is the primary electron energy, e is the electron charge, and L_c is the total length of the magnetic cusps (sum of the length of the cusps).

Using a simple probabilistic analysis, the mean primary electron confinement time can be estimated by

$$\tau_p = \frac{V}{v_p A_p}, \tag{5.3-4}$$

where V is the volume of the discharge chamber. The mean primary electron path length prior to finding a cusp and being lost to the wall is $L = v_p{*}\tau_p$. Likewise, the ionization mean free path is $\lambda = 1/n_o\sigma$ where σ represents the total inelastic collision cross section for the primary electrons. The probability that a primary electron will make a collision and not be directly lost to the anode is then

$$P = \left[1 - e^{(-n_o\sigma L)}\right] = \left[1 - e^{(-n_o\sigma V/A_p)}\right]. \tag{5.3-5}$$

By providing strong magnetic field strengths at the cusp to minimize the primary loss area, the probability of a primary electron being lost directly to the anode can be made very small. Similarly, ion thrusters with large volumes and/or operated at higher internal gas densities will cause the

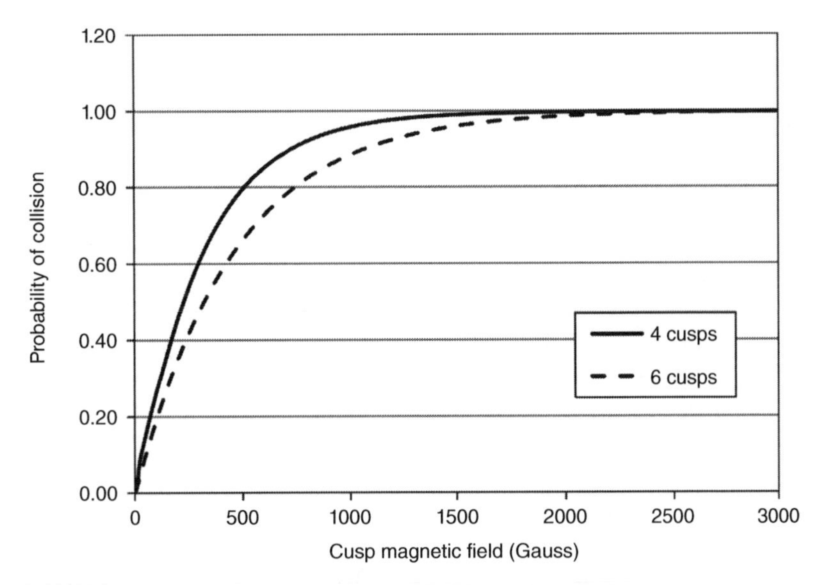

Figure 5-12 Probability of primary electron making a collision before being lost to the anode as a function of the cusp magnetic field strength for the NEXIS thruster design (*Source:* [20]).

primary electrons to undergo collisions and thermalization before being lost directly to the anode. Minimizing the energy loss associated with primaries being lost before making a collision in this way serves to maximize the efficiency of the thruster.

An example of the probability of a primary electron making a collision before finding a cusp is shown in Fig. 5-12 for the case of the NEXIS thruster designed with either four or six cusps [20]. For the design with six cusps, it is necessary to have cusp field strengths approaching 2000 G at the surface of the anode in order to minimize primary loss because the longer length of the cusp increases the loss area. Designs with a smaller number of ring cusps, corresponding to a smaller primary anode collection area from Eq. 5.3-3, require less magnetic field strength to achieve the same benefit. However, it will be shown later that the number of cusps affects efficiency and uniformity, and that maximizing the probability of a primary making a collision before being lost is only one of the trade-offs in designing an ion thruster.

Since the primary electron current lost directly to the anode is generally minimized for best efficiency, the discharge current is carried to the anode mainly by the plasma electrons. The plasma electrons are almost exclusively lost at the magnetic cusps, but their motion is affected by the presence of ions that also penetrate the cusp. Therefore, ions and plasma electrons are lost to a hybrid anode area [27] at the cusp given by

$$A_a = 4r_h L_c = 4\sqrt{r_e r_i} L_c, \tag{5.3-6}$$

where r_h is the hybrid Larmor radius, r_e is the electron Larmor radius and r_i is the ion Larmor radius. The flux of plasma electrons I_a that overcomes the sheath at the anode is

$$I_a = \frac{1}{4}\left(\frac{8kT_e}{\pi m}\right)^{1/2} en_e A_a e^{-e\phi/kT_e}, \tag{5.3-7}$$

where ϕ is the plasma potential relative to the anode wall.

The plasma in the discharge chamber obeys particle conservation in that the current injected and produced in the discharge must equal the total current that leaves the discharge:

$$\sum\left(I_{\text{injected}} + I_{\text{produced}}\right) = \sum I_{\text{out}}. \tag{5.3-8}$$

The current injected into the discharge volume is the primary electron current, and the current produced is the ion and electron pairs from each ionization collision. The current lost to the anode is the sum of the direct primary loss, the plasma electron loss, and a fraction of the ion loss. There is also ion current lost to cathode potential surfaces and the accelerator structure from the balance of the ions produced in the discharge. The plasma potential will adjust itself such that the total electron current to the anode is equal to the total ion current out of the discharge. It will be shown in the following sections that changing the anode area via the magnet strength or number of magnet rings will change the plasma potential relative to the anode (primarily the anode sheath voltage), which affects both the energy loss though the sheath and the stability of the discharge.

5.3.4 Ion Confinement at the Anode Wall

Ions are typically unmagnetized in ion thruster discharge chambers because the magnetic field is relatively low throughout the bulk of the discharge chamber, which results in a large ion Larmor radius compared to the thruster dimensions. For an unmagnetized plasma, the ion current flowing out the plasma volume in any direction is given by Eq. 3.7-30, which is the Bohm current:

$$I_i = \frac{1}{2} n_i e \sqrt{\frac{kT_e}{M}} A, \tag{5.3-9}$$

where n_i is the ion density in the center of the discharge and A is the total ion loss area. The Bohm current also describes ion flow along magnetic field lines, which will be useful later in discussing other plasma generator types.

The electrons may or may not be magnetized in the main discharge chamber volume, but they are strongly affected by the magnetic fields near the boundary in ring-cusp thrusters. The magnetized electrons then influence the ion motion near the boundaries by electrostatic effects. This causes the ion loss to the cusps to be the Bohm current to the hybrid area, given by Eq. 5.3-6, and a reduction in the Bohm current to the wall area between the cusps due to the ambipolar potentials that develop there. Since the cusp area is small compared to the rest of the anode surface area facing the plasma, the ion current to the hybrid cusp area can often be neglected. However, between the cusps the loss area is significant, and it is possible to analyze the electron and ion transport across the magnetic field to calculate the reduction in the ion velocity caused by the reduced transverse electron drift speed. This is then used to calculate the rate of ion loss to the anode compared to the unmagnetized Bohm current to the walls.

Ring-cusp thrusters are designed with various numbers of rings, distances between the rings, and magnet sizes that determine the magnetic field strength in the discharge chamber transverse to the wall. The quasi-neutral plasma flow across this magnetic field to the wall is described by the diffusion equation with an ambipolar diffusion coefficient. Ambipolar diffusion across a magnetic field was analyzed in Section 3.6.3.2. The transverse ion velocity was found to be

$$v_i = \frac{\mu_e}{\left(1 + \mu_e^2 B^2 - \dfrac{\nu_{ei}}{\nu_e}\right)} \left(E + \frac{kT_e}{e} \frac{\nabla n}{n}\right). \tag{5.3-10}$$

Setting the transverse electric field E in the plasma to zero in Eq. 5.3-10 gives the case where the ambipolar electric field exactly cancels the pre-sheath electric field that normally accelerates the ions to the

Bohm velocity. In this case, the ion velocity is just the ion thermal velocity $\left(\approx \sqrt{kT_i/M}\right)$, and the value of the magnetic field B in Eq. 5.3-10 is the minimum transverse magnetic field required to reduce the electron mobility sufficiently to produce this effect. The flux of ions passing through the transverse magnetic field is greatly reduced compared to the Bohm current due to the smaller ion velocity. The ion flux that does reach the wall is finally accelerated to the Bohm velocity close to the anode wall to satisfy the sheath criterion. Ions are conserved in this model because ions that are inhibited from flowing to the anode wall due to the transverse fields instead flow axially toward the grids where there is no confinement.

However, it is not necessary to limit this analysis to the case of $E = 0$. If the magnetic field is smaller than the critical B that causes $E = 0$, then the transverse electron mobility increases and a finite electric field exists in the magnetic diffusion length l. The ions fall through whatever potential difference is set up by this electric field, which means that the ions are accelerated to an energy given by

$$\frac{1}{2}M\, v_i^2 = eE \cdot l. \tag{5.3-11}$$

The ambipolar flow in the transverse magnetic field changes the electric field magnitude in the pre-sheath region and reduces the acceleration of the ions toward the wall.

However, in the limit of unmagnetized ions (the case for all ion thrusters), the electric field near the sheath edge must accelerate the ions to the Bohm velocity entering the sheath. This results in a ***net*** electric field in the plasma edge (the ambipolar flow and pre-sheath regions) limited to

$$E = -\frac{Mv_i^2}{el}. \tag{5.3-12}$$

Note that the electric field sign must be negative for the ion flow in the ambipolar diffusion region. Using Eq. 5.3-12 in Eq. 5.3-10, the minimum magnetic field to produce an ion velocity of v_i is

$$B = \frac{\upsilon_e m}{e}\sqrt{\frac{(kT_e - Mv_i^2)}{v_i m \upsilon_e l} - \left(\frac{\nu}{1+\nu}\right)}, \tag{5.3-13}$$

where $\nu = \nu_e/\nu_{ei}$, and $kT_e\nabla n/en$ is approximately $(kT_e/e)l$ for l representing the length the ions travel radially in the transverse magnetic field between the cusps. The value of l can be estimated from calculations of the transverse magnetic field versus the distance from the wall to the peak transverse field between the cusps (shown below), and is usually on the order of 2–3 cm.

Alternatively, the modified electric field given in Eq. 5.3-12 can be inserted into Eq. 5.3-10 to produce an expression for the ion velocity:

$$v_i^2 + \frac{el}{\mu_e M}\left(1 + \mu_e^2 B^2 - \frac{\nu_{ei}}{\nu_e}\right)v_i - \frac{kT_e}{M} = 0. \tag{5.3-14}$$

This quadratic equation can be easily solved to give

$$v_i = \frac{1}{2}\sqrt{\left[\frac{el}{M\mu_e}\left(1 + \mu_e^2 B^2 - \frac{\upsilon_{ei}}{\upsilon_e}\right)\right]^2 + \frac{4kT_e}{M}} - \left[\frac{el}{2M\mu_e}\left(1 + \mu_e^2 B^2 - \frac{\upsilon_{ei}}{\upsilon_e}\right)\right] \tag{5.3-15}$$

The collision frequencies ($\nu_e = \nu_{en} + \nu_{ei}$ and $\nu = \nu_e/\nu_{ei}$) in these equations were given in Chapter 3, where the electron-neutral collision frequency is given in Eq. 3.6-12, and the electron-ion collision frequency is given in Eq. 3.6-14. It is possible to show that in the limit that B goes to zero and the flow is essentially collisionless, Eq. 5.3-15 reverts to the Bohm velocity.

Defining an ion confinement factor

$$f_c \equiv \frac{v_i}{v_{\text{Bohm}}}, \tag{5.3-16}$$

and since the Bohm velocity is $v_{\text{Bohm}} = \sqrt{kT_e/M}$, it is a simple matter to calculate the reduction in the expected flux of ions going to the anode due to the reduction in the Bohm velocity at a given magnetic field strength B. The ion current transverse to the magnetic field between the cusps to the anode is then given by

$$I_{\text{ia}} = \frac{1}{2} n_i e \sqrt{\frac{kT_e}{M}} A_{\text{as}} f_c, \tag{5.3-17}$$

where A_{as} is the total surface area of the anode exposed to the plasma.

There are two issues with using Eq. 5.3-17 to evaluate ion loss rate reduction between the cusps. First, the magnetic field in the ring-cusp geometry is not transverse to the wall everywhere. Near the cusp shown in Fig. 5-10, the field transitions from parallel to perpendicular to the wall where the analysis above does not apply. However, the magnetic field strength in this region increases rapidly near the magnets and some fraction of the plasma electrons are reflected from the magnetic mirror. This serves to retard the ion flux electrostatically in a manner similar to the ambipolar diffusion case between the cusps described above. Ultimately, the ions are lost at the cusp at the Bohm current to the hybrid area, and it is usually found that the transition to this unimpeded ion flow to the wall occurs over an area that is small compared to the total area between the cusps.

The second issue with using Eq. 5.3-17 is that the diffusion thickness l is not known. However, this can be estimated for ring-cusp thrusters using a dipole model for the magnets. Consider the case of two rows of opposite polarity magnets, which is illustrated in part of Fig. 5-10. Each magnet has a dipole strength M per unit length and the magnets are separated in the x-direction by a distance d. The magnetic field along the line perpendicular to the midline between the magnets is

$$|B^+(y)| = \frac{q}{r} = \frac{q}{\sqrt{\frac{d^2}{4} + (y-\delta)^2}}, \tag{5.3-18}$$

where r is the length of the line from the point on the midline to the magnet, q is the number of magnetic dipoles and δ is the half height of the magnet. The magnetic field on the centerline between the magnets has only an x-component. The x-component of the field from one magnet (positive polarity) is given by

$$B_x^+(y) = |B^+(y)| \cos\theta = \frac{q\frac{d}{2}}{r^2} = \frac{q\frac{d}{2}}{\frac{d^2}{4} + (y-\delta)^2}. \tag{5.3-19}$$

The field in the x-direction from both magnets is then

$$B_x(y) = \frac{qd}{\frac{d^2}{4} + (y-\delta)^2} - \frac{qd}{\frac{d^2}{4} + (y+\delta)^2}, \tag{5.3-20}$$

and so the total field on the center line is

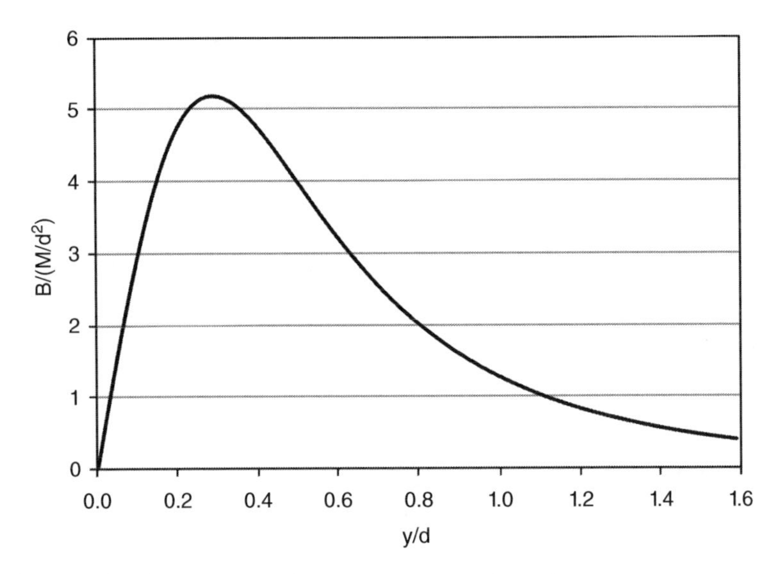

Figure 5-13 Magnetic field strength as a function of distance above the magnets.

$$B(y) = \frac{2(2\,qd)yd}{\left(\frac{d^2}{4} + y^2\right)^2} = \frac{2Myd}{\left(\frac{d^2}{4} + y^2\right)^2}, \tag{5.3-21}$$

where the magnetization M is the number of magnetic dipoles times the length of the magnet.

The maximum magnetic field strength between the magnets, found from Eq. 5.3-21, then occurs at

$$y = \frac{d}{2\sqrt{3}} = 0.29d \approx l. \tag{5.3-22}$$

It is assumed that the diffusion length l is roughly this distance. This is not an unreasonable approximation, as illustrated in Fig. 5-13. The magnetic field decreases on each side of the maximum, but is nearly the full value over the length of about 0.3 of the distance between the magnets.

The maximum transverse field strength along the centerline between the magnets, often called the "saddle-point" field, can also be calculated from this simple derivation. Using Eq. 5.3-22 in Eq. 5.3-21, the maximum magnetic field is

$$B(y_{\max}) = 5.2\frac{M}{d^2}. \tag{5.3-23}$$

The dipole strength per unit length is

$$M = \frac{B_r V_m}{4\pi w}, \tag{5.3-24}$$

where B_r is the residual magnetic field of the magnet, V_m is the volume of the magnet, and w is the width of the magnet. For example, for two rows of magnets that have a residual magnetic field of 10,000 G, a volume per width of 0.6 cm^2 and a separation of 10 cm, the maximum transverse magnetic field is 24.8 G and occurs at a distance of 2.9 cm above the boundary.

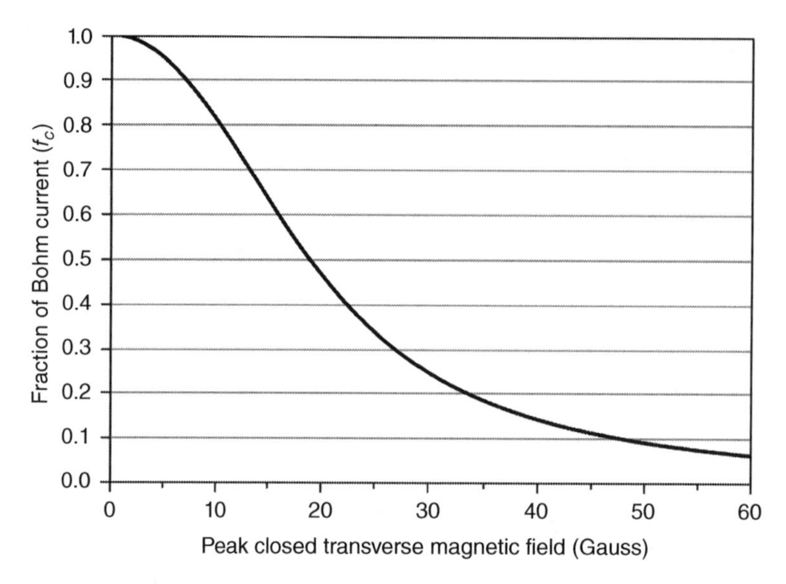

Figure 5-14 Fraction of the Bohm current density to the anode wall as a function of the transverse magnetic field strength for the NSTAR ion thruster (*Source:* [21]).

As an example of the ion loss rate to the anode, the fraction of the Bohm current to the anode ($I_{\text{ia}}/I_{\text{Bohm}}$) is plotted in Fig. 5-14 as a function of the magnetic field at the saddle point for the NSTAR ion thruster [13]. At zero transverse magnetic field, the ion flux to the anode is the Bohm current. As the transverse field increases and reduces the electron mobility, the ions are slowed and the current lost decreases. In the NSTAR design, the last closed magnetic contour is about 20 G, and so roughly half of the ions initially headed radially toward the anode are lost. For closed magnetic field contours of at least about 50 G, the ion loss to the anode is reduced by nearly a factor of ten compared to the unmagnetized Bohm current. This can make a significant difference in the efficiency of the plasma generator and the amount of discharge power required to produce the beam ions. Even though the ions are unmagnetized in these thrusters, it is clear that ambipolar effects make the ring-cusp magnetic fields effective in reducing the ion loss to the walls.

5.3.5 Neutral and Primary Densities in the Discharge Chamber

The ion and excited neutral production rates described by Eqs. 5.3-25 and 5.3-26 require knowledge of the neutral gas density in the discharge chamber. The neutral gas flow that escapes the chamber (the unionized propellant) is simply the gas injected into the discharge chamber minus the gas particles that are ionized and extracted to form the ion beam:

$$Q_{\text{out}} = Q_{\text{in}} - \frac{I_b}{e}. \tag{5.3-25}$$

The neutral gas that leaks through the grid is the neutral flux on the grids (in particles per second) times the grid optical transparency T_a and a conductance reduction term η_c known as the Clausing factor [28]:

$$Q_{\text{out}} = \frac{1}{4} n_o v_o A_g T_a \eta_c, \tag{5.3-26}$$

where v_o is the neutral gas velocity, A_g is the grid area and η_c is the Clausing factor. The Clausing factor represents the reduced conductance of the grids due to their finite thickness and results from

Clausing's original work on gas flow restriction in short tubes. For typical grid apertures with small thickness to length ratios, the Clausing factor must be calculated using Monte Carlo techniques, an example of which is given in Appendix G. In general, ion thruster grids will have Clausing factors on the order of 0.5.

The mass utilization efficiency of the thruster discharge chamber is defined as

$$\eta_{\mathrm{md}} = \frac{I_b}{Q_{\mathrm{in}} e}. \tag{5.3-27}$$

Equating 5.3-25 and 5.3-26, using Eq. 5.3-27 and solving for the neutral gas density in the discharge chamber gives

$$n_o = \frac{4Q_{\mathrm{in}}(1-\eta_{\mathrm{md}})}{v_o A_g T_a \eta_c} = \frac{4I_b}{v_o e A_g T_a \eta_c} \frac{(1-\eta_{\mathrm{md}})}{\eta_{\mathrm{md}}}. \tag{5.3-28}$$

Flow is usually given in standard cubic centimeters per minute (sccm) or milligrams per sec (mg/s), and conversions from these units to number of atoms per second useful in Eq. 5.3-28 are given in Appendix B. The neutral pressure in the discharge chamber during operation of the thruster can also be found using this expression and the conversion from density to pressure given in Eq. 2.7-2, if the neutral gas temperature is known. In general, the neutral gas atoms collide with the anode wall and grids several times before being lost, and so the neutral gas can be assumed to have the average temperature of the thruster body in contact with the plasma. This temperature typically ranges from 200 to 300 °C for operating thrusters.

5.3.6 Ion and Excited Neutral Production

Ions in the discharge chamber are produced by both the primary electrons and by the tail of the Maxwellian distribution of the plasma electrons. The total number of ions produced in the discharge, expressed as a current in Coulombs per second, is given by

$$I_p = n_o n_e e \langle \sigma_i v_e \rangle V + n_o n_p e \langle \sigma_i v_p \rangle V, \tag{5.3-29}$$

where n_o is the neutral atom density, n_e is the plasma electron density, σ_I is the ionization cross section, v_e is the plasma electron velocity, V is the plasma volume inside the discharge chamber, n_p is the primary electron density, and v_p is the primary electron velocity. The terms in the brackets are the ionization cross section averaged over the distribution of primary and Maxwellian electron energies, which is usually called the reaction rate coefficient.

An example of ionization and excitation cross sections [29, 30] used for electron impact on xenon is shown in Fig. 5-15. If it is assumed that the primary electrons are monoenergetic, then the reaction rate coefficient in Eq. 5.3-29 for primary ionization is just the cross section in Fig. 5-15 times the corresponding primary electron velocity. This data is listed for xenon and krypton in Appendix D. If the primaries have a distribution in energy, then the cross section must be averaged over that distribution. For Maxwellian electrons, this is calculated for xenon and krypton and listed in Appendix E.

Excited neutrals are also produced by both the primary electrons and the tail of the Maxwellian distribution of the plasma electrons. The total number of exited neutrals produced in the discharge in Coulombs per second is given by

$$I^* = n_o n_e e \langle \sigma_* v_e \rangle V + n_o n_p e \langle \sigma_* v_p \rangle V, \tag{5.3-30}$$

where σ_* is the excitation cross section. Again, the excitation cross section is averaged over the distribution in electron energies to produce the reaction rate coefficients in the brackets. The reaction

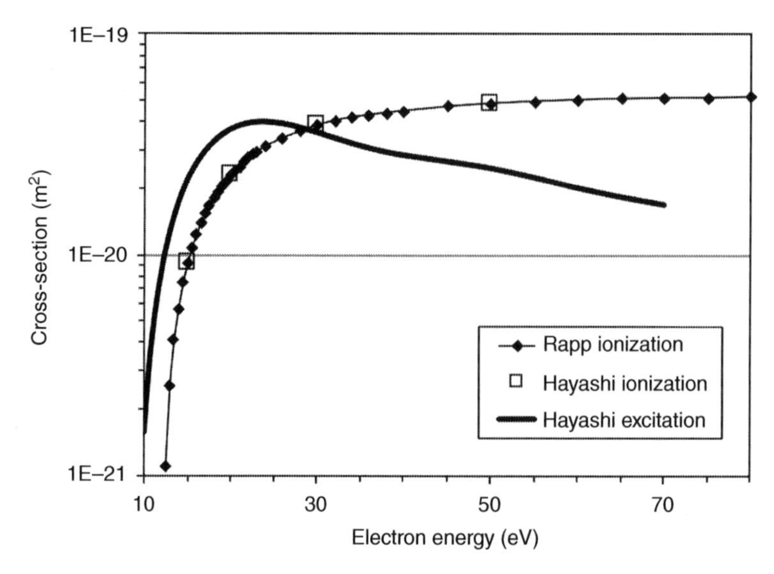

Figure 5-15 Ionization and excitation cross sections for xenon (*Source:* [29, 30]).

rate coefficients calculated by averaging the ionization and excitation cross sections over the Maxwellian energy distribution are shown in Fig. 5-16 for xenon, and are listed in Appendix E. The rate of excitation is seen to exceed that of ionization for low electron temperatures (below about 9 eV). The ratio of excitation to ionization reaction rates for xenon is shown in Fig. 5-4. As previously described, at low electron temperatures, a significant amount of the energy in the discharge goes into excitation of the neutrals at the expense of ionization. This is one of the many reasons that the cost of producing an ion in ion thrusters is usually over ten times the ionization potential.

For inert gas propellants commonly used in ion thrusters, the second ionization potential is on the order of twice the first ionization potential. For example, the first ionization potential of xenon is 12.13 eV and the second ionization potential is 21.2 eV. DC electron discharges that have electron

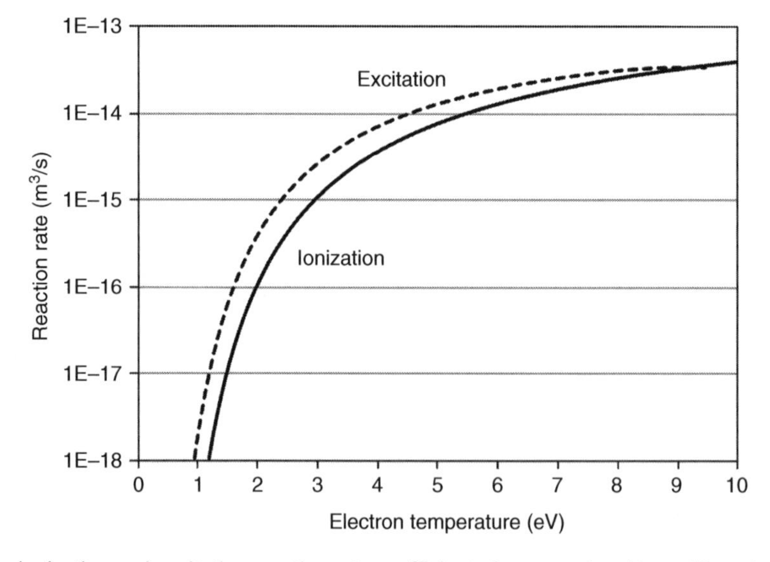

Figure 5-16 Ionization and excitation reaction rate coefficients for xenon in a Maxwellian electron temperature plasma.

energies in excess of 21.2 V can produce a significant number of double ions. In addition, the tail of the Maxwellian electron distribution will also contain electrons with an energy that exceeds the second ionization potential, and significant numbers of double ions will be produced if the electron temperature in the discharge chamber is high.

The generation rate of double ions is determined in the same manner as single ions shown above with different ionization cross sections [31]. The density of the double ions is determined by the continuity equation for that species

$$\frac{dn^{++}}{dt} + \nabla \cdot \left(n^{++} v^{+} \sqrt{2} \right) = \dot{n}^{++}, \tag{5.3-31}$$

where because of their charge the double ion velocity will be increased over the singly ionized species by a square root of two. Define γ as the ratio of the double ion beam current to the single ion beam current

$$\gamma \equiv \frac{I_b^{++}}{I_b^{+}}. \tag{5.3-32}$$

Using the Bohm current definition, the beam current of single ions from the discharge plasma boundary through the ion optics with an area A_g and a transparency T_g is

$$I_i^{+} = \frac{1}{2} n^{+} e v_a T_g A_g, \tag{5.3-33}$$

where v_a is the ion acoustic velocity at the sheath edge. The double ion current is likewise

$$I_i^{++} = \frac{1}{2} n^{++} (2e) \left(\sqrt{2} v_a \right) T_g A_g. \tag{5.3-34}$$

Combining these equations,

$$\gamma = \frac{\frac{1}{2} n^{++} (2e) \left(\sqrt{2} v_a \right) T_g A_g}{\frac{1}{2} n^{+} e v_a T_g A_g} = 2\sqrt{2} \frac{n_i^{++}}{n_i^{+}}. \tag{5.3-35}$$

The electron density is the sum of the singly and doubly ionized particle densities

$$n_e = n_i^{+} + 2n_i^{++} = n_i^{+} \left(1 + \frac{\gamma}{\sqrt{2}} \right). \tag{5.3-36}$$

If the double ion content is negligible ($\gamma << 1$), then Eq. 5.3-36 becomes the quasi-neutral situation of $n_e \approx n_i$ that neglects the primary electron density. As discussed in Chapter 2, the discharge propellant efficiency is the ratio of the propellant that becomes beam ions (of any charge) to the rate of propellant flow into the discharge chamber. Considering the effect of double ions, the propellant efficiency of the discharge chamber is then

$$\eta_{\text{md}} = \left(J_B^{+} + \frac{J_B^{++}}{2} \right) \frac{A_g}{eQ_{\text{in}}}, \tag{5.3-37}$$

where Q_{in} is the particle flow into the discharge chamber. In the event that there is a significant double ion content in the discharge plasma, the beam current and the discharge chamber mass utilization efficiency must be corrected using these equations.

5.3.7 Electron Temperature

The electron temperature in the discharge chamber can be found from the overall particle balance of the ions in the discharge chamber, where the number of ions generated must equal the total current flux of ions to the boundaries. The singly ionized propellant atoms generated are described by Eq. 5.3-29. Doubly ionized propellant atoms generated primarily by a second ionization of propellant ions by the electrons in the plasma given by

$$I^{++} = n_i^+ n_e e \langle \sigma_i^{++} v_e \rangle V + n_i^+ n_p e \langle \sigma_i^{++} v_p \rangle V, \tag{5.3-38}$$

where σ_i^{++} is the ionization cross section for this second-step process. Equating the generation rate of the ions with the total ion loss rate gives

$$\begin{aligned} n_o n_e \langle \sigma_i v_e \rangle V + n_o n_p \langle \sigma_i v_p \rangle V + n_i^+ n_e \langle \sigma_i^{++} v_e \rangle V + n_i^+ n_p \langle \sigma_i^{++} v_p \rangle V \\ = \frac{1}{2} n_i^+ v_a (A_{as} f_c + A_s) + \frac{1}{2} n_i^{++} v_a (A_{as} f_c + A_s) \\ = \frac{1}{2} n_i^+ (1 + \gamma) \sqrt{\frac{kT_e}{M}} (A_{as} f_c + A_s), \end{aligned} \tag{5.3-39}$$

where again A_{as} is the anode wall area and A_s is the screen grid area. Separating all the terms dependent on the electron temperature on the left (including the cross sections and electron velocities), we get a relationship that can be solved for the electron temperature:

$$\frac{\sqrt{\frac{kT_e}{M}}}{\langle \sigma_i v_e \rangle + \frac{n_p}{n_e} \langle \sigma_i v_p \rangle + \frac{n_e \langle \sigma_i^{++} v_e \rangle + n_p \langle \sigma_i^{++} v_p \rangle}{n_o \left(1 + \frac{\gamma}{\sqrt{2}} \right)}} = \frac{2 n_o \left(1 + \frac{\gamma}{\sqrt{2}} \right) V}{(A_{as} f_c + A_s)}. \tag{5.3-40}$$

Since modern ion thrusters typically have double ion content in the beam of only 5–10% of the single ion content, it is standard to ignore the doubles ($\gamma = 0$) and use this simpler expression to solve for the electron temperature

$$\frac{\sqrt{\frac{kT_e}{M}}}{\langle \sigma_i v_e \rangle + \frac{n_p}{n_e} \langle \sigma_i v_p \rangle} = \frac{2 n_o V}{(A_{as} f_c + A_s)} = \frac{8 V Q_{in} (1 - \eta_{md})}{v_o A_g T_a \eta_c (A_{as} f_c + A_s)}, \tag{5.3-41}$$

where Eq. 5.3-28 has been used for the neutral gas density n_o expressed in terms of the mass utilization efficiency in this equation.

If the total propellant flow into the discharge chamber and the mass utilization efficiency are specified, and the primary electron density calculated as described below, then Eq. 5.3-40 can be solved for the electron temperature. This is because the ionization reaction rate coefficient is a function of the electron temperature. Alternatively, if the beam current is specified, then the right-hand side of Eq. 5.3-28 can also be used for the neutral density in Eq. 5.3-40 to find the electron temperature. Typically, curve fits to the ionization and excitation cross section and reaction rate data shown in Figs. 5-14 and 5-15 are used to evaluate the reaction rate coefficients in a program that iteratively solves Eq. 5.3-40 for the electron temperature.

5.3.8 Primary Electron Density

The primary electron density in Eq. 5.3-40 can be evaluated from the total primary electron confinement time in the discharge chamber. The emitted current I_e from the hollow cathode is

$$I_e = \frac{n_p e V}{\tau_t}, \tag{5.3-42}$$

where τ_t is the total primary confinement time that addresses all of the primary electron thermalization and loss mechanisms. The ballistic confinement time for direct primary loss to the anode, τ_p, was given in Eq. 5.3-4. It is assumed that the primary electrons have undergone an inelastic collision with the neutral gas and have lost sufficient energy such that they are then rapidly thermalized with the plasma electrons. The mean time for a collision between the primary and a neutral gas atom to occur is given by

$$\tau_c = \frac{1}{n_o \sigma v_p}, \tag{5.3-43}$$

where σ is the total inelastic collision cross section. Using Eq. 5.3-28 for the neutral density, the mean collision time for primary electrons is

$$\tau_c = \frac{v_o e A_g g T_a \eta_c \eta_m}{4\sigma v_p I_B (1-\eta_{\text{md}})} = \frac{v_o A_g T_a \eta_c}{4\sigma v_p Q_{\text{in}} (1-\eta_{\text{md}})}. \tag{5.3-44}$$

Finally, primary electrons can also be thermalized by equilibrating with the plasma electrons. The time for primary electrons to slow into a Maxwellian electron population was derived by Spitzer [32] and is given by

$$\tau_s = \frac{w}{2A_D l_f^2 G(l_f w)}, \tag{5.3-45}$$

where $w = \sqrt{2V_{\text{pe}}/m}$, eV_{pe} is the primary energy, $l_f = \sqrt{m/2kT_e}$ is the inverse most probable speed of the Maxwellian electrons, A_D is a diffusion constant given by

$$A_D = \frac{8\pi e^4 n_e \ln\Lambda}{m^2}, \tag{5.3-46}$$

and $\ln\Lambda$ is the collisionality parameter [33] given in Eq. 3.6-15. The function $G(l_f\omega)$ is defined in Appendix F and a curve fit to Spitzer's tabulated values (in CGS units) for this function is provided.

The total primary electron confinement time can be found from

$$\frac{1}{\tau_t} = \frac{1}{\tau_p} + \frac{1}{\tau_c} + \frac{1}{\tau_s}. \tag{5.3-47}$$

Some care needs to be used in including the Spitzer slowing time because some ion thruster designs have a very non-monoenergetic primary energy distribution, which is not described well by Eq. 5.3-45. The Spitzer equilibration time must then be used in Eq. 5.3-47 if the electron distributions are Maxwellian.

The current emitted from the hollow cathode is

$$I_e = I_d - I_s - I_k, \tag{5.3-48}$$

where I_s is the screen current and I_k is the ion current back to the cathode. Using Eqs. 5.3-4 and 5.3-44, in Eq. 5.3-42, the primary electron density is given by

$$n_p = \frac{I_e \tau_t}{eV} = \frac{I_e}{eV}\left[\frac{1}{\tau_p} + \frac{1}{\tau_c} + \frac{1}{\tau_s}\right]^{-1} = \frac{I_e}{eV}\left[\frac{v_p A_p}{V} + \frac{4\sigma v_p Q_{\text{in}}(1-\eta_{\text{md}})}{v_o A_s T_a \eta_c} + \frac{1}{\tau_s}\right]^{-1}. \tag{5.3-49}$$

Assuming that the primary electron loss directly to the anode is negligible, the electron equilibration time is long, and the ion current flowing back to the cathode is small, then Eq. 5.3-49 can be written as

$$n_p = \frac{I_e v_o A_s T_a \eta_c}{4V\sigma v_p I_b} \frac{\eta_{\text{md}}}{(1-\eta_{\text{md}})} = \frac{(I_d - I_s) v_o A_s T_a \eta_c}{4V\sigma v_p I_b} \frac{\eta_{\text{md}}}{(1-\eta_{\text{md}})} \tag{5.3-50}$$

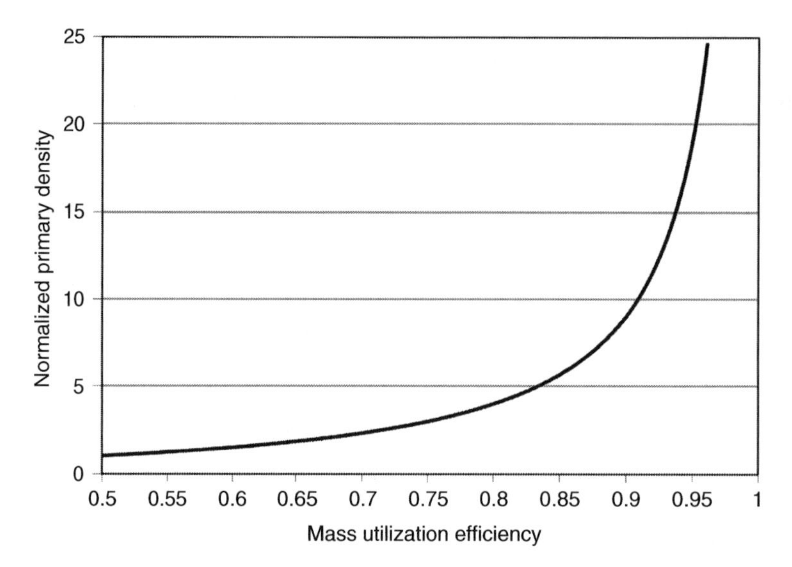

Figure 5-17 Normalized primary electron density showing the strong increase with higher mass utilization efficiency.

This equation demonstrates the characteristic behavior of the primary electron density being proportional to the mass utilization efficiency divided by one minus the mass utilization efficiency that was originally described by Brophy [18, 19]. This dependence is valid unless there are paths for the primary electrons to be lost other than just collisionally with the neutral gas, such as ballistic flow to the anode or by thermalization with the plasma electrons. The behavior of the primary electron density with changes in the mass utilization efficiency is shown in Fig. 5-17, where the primary electron density is normalized to the value at $\eta_{\text{md}} = 0$. As the neutral density decreases in the discharge chamber at higher mass utilization efficiencies, the primary electron density increases rapidly. At 90% mass utilization efficiency, the primary electron density in the discharge chamber is nine times higher than at 50% mass utilization efficiency. This strongly affects the ionization rate and the discharge loss behavior with neutral gas pressure, which will be shown later.

5.3.9 Power and Energy Balance in the Discharge Chamber

The currents and potential distributions in the ring-cusp thruster discharge were shown in Fig. 5-9. The power into the discharge chamber is the emitted current from the hollow cathode multiplied by the voltage the electrons gain in the discharge chamber (V_k in Fig. 5-9):

$$P_{\text{in}} = I_e V_k = I_e \left(V_d - V_c + V_p + \phi_s\right) = I_e \left(V_d - V_c + \phi + \frac{5}{2} T_c\right), \tag{5.3-51}$$

where V_d is the discharge voltage, V_c is the cathode voltage drop, V_p is the potential drop in the plasma, ϕ is the sum of the pre-sheath and anode sheath potentials relative to the anode wall, and the last term is the convected energy out of the insert-plasma inside the cathode from the cathode electron temperature T_c. The power into the discharge is transferred from the primary electrons from the cathode into producing ions, excited neutrals and Maxwellian electrons. The power leaving the discharge to the electrodes is from ions flowing to the anode, cathode and screen plane, and from primary and plasma electrons flowing to the anode. The power out of the discharge is then the sum of these terms given by:

$$P_{\text{out}} = I_p U^+ + I^* U^* + (I_s + I_k)(V_d + \phi) + (I_b + I_{\text{ia}})(V_p + \phi_s) + I_a \varepsilon_e + I_L(V_d - V_c), \tag{5.3-52}$$

where I_p is the total current of ions produced in the discharge, U^+ is the ionization potential of the propellant gas, I^* is the current of excited ions produced in the discharge chamber, U^* is the excitation energy, I_s is the current of ions to the screen plane, I_k is the current of ions flowing back to the cathode, I_b is the beam current, I_a is the plasma electron current to the anode, I_{ia} is the ion current to the anode, and I_L is the primary electron fraction that falls from the plasma potential to be lost to the anode. The plasma electron energy lost to the anode wall, ε_e, is equal to $2kT_e/e + \phi$, which is derived in Appendix C. The ions fall through the pre-sheath potential from the center of the plasma to the sheath edge, such that V_p can be approximated as $kT_e/2e$. The ion energy to the anode, ε_i, is then $kT_e/2e + \phi_s = \phi$, which was given in Eq. 5.2-10.

With the screen grid connected to the cathode potential, the current emitted from the hollow cathode was given in Eq. 5.3-48 in terms of the other currents in the circuit. Likewise, conservation of particles flowing to the anode gives

$$I_a = I_d + I_{\text{ia}} - I_L, \tag{5.3-53}$$

where I_d is the discharge current measured in the discharge power supply. Equating the power into the discharge to the power out, using the particle balance equations in 5.3-48 and 5.3-53, and solving for the beam current from the thruster gives:

$$I_b = \frac{I_d(V_d - V_c + 2T_{\text{eV}}) - I_p U^+ - I^* U^* - (I_s + I_k)(2V_d - V_c + 2\phi)}{\phi} - \frac{I_{\text{ia}}(2T_{\text{eV}} + 2\phi) + I_L(V_d - V_c - 2T_{\text{eV}} - \phi)}{\phi} \tag{5.3-54}$$

where T_{eV} is the electron temperature in electron volts.

The issue in evaluating Eq. 5.3-54 for the beam current produced by a given thruster design is that several of the current terms in the numerator contain the plasma density, which is not known. In addition, the beam current I_b is given by the Bohm current averaged over the screen-grid plane times the effective transparency T_s of the screen grid:

$$I_b \approx \frac{1}{2} n_i e \sqrt{\frac{kT_e}{M}} A_s T_s, \tag{5.3-55}$$

where n_i is the peak ion density at the screen grid, υ_a is the ion acoustic velocity, A_s is the screen grid area, and T_s is the effective screen transparency with high voltage applied to the accelerator grids. The screen grid current in Eq. 5.3-54 is given by

$$I_s = \frac{(1 - T_s)}{2} n_i e \, \upsilon_a A_s. \tag{5.3-56}$$

and the ion current to the anode is given by Eq. 5.3-17 and the primary electron loss to the anode is given by Eq. 5.3-2. The ion plasma density n_i in these equations is the sum of the plasma electron and primary electron densities: $n_i = n_e + n_p$. It is often assumed that the primary electron density is small and $n_e \approx n_i$ (quasi-neutrality), but the primary density can be easily included.

Equations 5.3-54–5.3.56 can be solved for the plasma density:

$$n_e = \frac{I_d(V_d - V_c + 2T_{eV}) + I_L(V_d - V_c - 2T_{eV} + \phi)}{\frac{ev_a A_s}{2}(T_s\phi + (1 + T_s)(2V_d - V_c + 2T_{eV}) + f_c(2T_{eV} + 2\phi)}, \tag{5.3-57}$$

where the ion loss back to the cathode I_k has been ignored as small. The plasma density is found to be proportional to the discharge current I_d decreased by the amount of primary electron loss (I_L) directly to the anode. This dependence illustrates why implementing sufficient cusp magnetic field strength to reduce the primary electron loss is critically important to the thruster performance. Unfortunately, this equation still contains the primary electron density in the primary electron current lost term, and so must be solved iteratively for the plasma density. Once the plasma density is known, the beam current can be calculated from 5.65. If the flatness parameter, which is defined as the average ion current density over the grids divided by the peak current density, is known, then the peak plasma density and peak beam current density can be obtained. The flatness parameter is normally found by experimental measurements of the plasma and beam profiles, or by 2-D models of the discharge that are discussed in Section 5.7.

5.3.10 Discharge Loss

The discharge loss in an ion thruster is defined as the power into the thruster divided by the beam current. This parameter then describes the power required to produce the beam current, which is a good figure of merit for the discharge chamber performance. In DC-discharge thrusters, the discharge loss for the plasma generator is given by

$$\eta_d = \frac{I_d V_d + I_{ck} V_{ck}}{I_b} \approx \frac{I_d V_d}{I_b}, \tag{5.3-58}$$

where I_b is the beam current, I_{ck} is the current to the cathode keeper electrode (if any), and V_{ck} is the keeper bias voltage. The keeper power is typically negligible in these thrusters, but it is a simple matter to include this small correction. Combining Eqs. 5.3-54 and 5.3-58, the discharge loss is

$$\eta_d = V_d \left[\frac{\frac{I_p}{I_b}U^+ + \frac{I^*}{I_b}U^* + \frac{(I_s + I_k)}{I_b}(2V_d - V_c + 2\phi) + \phi + \frac{I_{ia}}{I_b}(2T_{eV} + 2\phi) + \frac{I_L}{I_b}(V_d - V_c - 2T_{eV} - \phi)}{V_d - V_c - 2T_{eV}} \right], \tag{5.3-59}$$

where T_{eV} is in electron volts. To evaluate the first current fraction in this equation, the ions are produced by both primary electrons and the energetic tail of the Maxwellian distribution of the plasma electrons. The total number of ions produced in the discharge, I_p, is given in Eq. 5.3-29, and the total number of excited neutrals produced in the discharge, I^*, is given in Eq. 5.3-30.

Using Eqs. 5.3-29 and 5.3-30 for the particle production and excitation, Eq. 5.3-55 for the beam current, and assuming $n_i \approx n_e$, the first current fraction in Eq. 5.3-59 is

$$\begin{aligned} \frac{I_p}{I_b} &= \frac{2n_o n_e e\langle\sigma_i v_e\rangle V}{n_i e\sqrt{\frac{kT_e}{M}}A_s T_s} + \frac{2n_o n_p e\langle\sigma_i v_p\rangle V}{n_i e\sqrt{\frac{kT_e}{M}}A_s T_s} \\ &= \frac{2n_o V}{\sqrt{\frac{kT_e}{M}}A_s T_s}\left(\langle\sigma_i v_e\rangle + \frac{n_p}{n_e}\langle\sigma_i v_p\rangle\right) \end{aligned}. \tag{5.3-60}$$

The second current fraction is likewise:

$$\frac{I^*}{I_b} = \frac{2n_o V}{\sqrt{\frac{kT_e}{M}} A_s T_s}\left(\langle\sigma_* v_e\rangle + \frac{n_p}{n_e}\langle\sigma_* v_p\rangle\right). \tag{5.3-61}$$

Neglecting again the small amount of ion current backflowing to the hollow cathode I_k, the third current fraction is

$$\frac{I_s}{I_b} = \frac{1-T_s}{T_s}. \tag{5.3-62}$$

The ion current that goes to the anode wall is, again, the Bohm current reduced by the confinement factor f_c, given in Eq. 5.3-16. In this model, the value of the confinement factor must be evaluated for the particular ion thruster discharge chamber being analyzed. However, for most ion thruster designs, if the 50 G contour is closed, it is possible to assume to first order that $f_c \approx 0.1$ and the ion loss to the anode surface area is essentially one-tenth of the local Bohm current. For a given confinement factor f_c, the fourth current fraction in Eq. 5.3-59 is

$$\frac{I_{\text{ia}}}{I_b} = \frac{\frac{1}{2}n_i e\sqrt{\frac{kT_e}{M}}A_{\text{as}} f_c}{\frac{1}{2}n_i e\sqrt{\frac{kT_e}{M}}A_s T_s} = \frac{A_{\text{as}} f_c}{A_s T_s}, \tag{5.3-63}$$

where A_{as} is the surface area of the anode facing the plasma in the discharge chamber.

The primary electron current lost to the anode, I_L, is given by Eq. 5.3-2. The last current fraction in Eq. 5.3-59 is then

$$\frac{I_L}{I_b} = \frac{n_p e\, v_p A_p}{\frac{1}{2}n_i e\, v_a A_s T_s} = \frac{2n_p v_p A_p}{n_e v_a A_s T_s}. \tag{5.3-64}$$

The discharge loss can then be found

$$\eta_d = \frac{V_d\left[\frac{I_p}{I_b}U^+ + \frac{I^*}{I_b}U^* + \frac{1-T_s}{T_s}(2V_d - V_c + 2\phi) + \phi + \frac{A_{\text{as}} f_c}{A_s T_s}(2T_{\text{eV}} + 2\phi)\right]}{V_d - V_c - 2T_{\text{eV}}} + \frac{V_d\left[\frac{2n_p v_p A_p}{n_e v_a A_s T_s}(V_d - V_c - 2T_{\text{eV}} - \phi)\right]}{V_d - V_c - 2T_{\text{eV}}}. \tag{5.3-65}$$

Equation 5.3-65 illuminates some of the design features that improve the discharge efficiency. Since the discharge voltage V_d appears in both the numerator and denominator of Eq. 5.3-65, there is no strong dependence of the discharge loss on voltage shown directly in this equation. However, increases in the discharge voltage raise the primary energy strongly, which increases the ionization rate and beam current. Therefore, higher discharge voltages always result in lower discharge losses. Higher screen grid transparency T_s, smaller ion confinement factor f_c (better ion confinement), and smaller primary loss area A_p and wall surface area A_{as} all reduce the discharge loss. Lowering the plasma potential also reduces the discharge loss by reducing the energy lost to the anode by the plasma electrons, which is accomplished by reducing the anode loss area at the cusps.

The input data required to solve Eq. 5.3-65 are:

- Discharge voltage
- Discharge chamber surface area and volume
- Magnetic field strength at the cusp
- Magnetic field closed contour strength between the cusps
- Grid area
- Grid transparency
- Gas temperature
- Cathode voltage drop

It is necessary to specify either the discharge current or the beam current to calculate the plasma density in the discharge chamber. The grid transparency is obtained from the grid codes (often called "optics codes"). Several of these codes, such as the JPL CEX ion optics codes [34, 35], are described in Chapter 6. The cathode voltage drop is either measured inside the hollow cathode [36], or calculated using a separate 2-D hollow cathode plasma model [37] that was described in Chapter 4.

Discharge chamber behavior is characterized by "performance curves", which were described in Chapter 2, and are graphs of discharge loss versus mass utilization efficiency. These curves plot the electrical cost of producing beam ions as a function of the propellant utilization efficiency, and give useful information of how well the plasma generator works. Performance curves are normally taken at constant beam current and discharge voltage so that the efficiency of producing and delivering ions to the beam is not masked by changes in the discharge voltage or average plasma density at the grids.

Calculating performance curves using Eq. 5.3-65 requires iteration of the solutions for the electron temperature, discharge current and/or beam current in the above equations. To measure the discharge loss versus mass utilization in thrusters, the discharge current, total gas flow and gas flow split between the cathode and main discharge chamber are normally varied to produce a constant beam current and discharge voltage as the mass utilization efficiency changes. This means that a beam current and mass utilization operating point can be specified, which determines the neutral gas density in the discharge chamber from Eq. 5.3-28 and the average plasma density in the discharge chamber from the Bohn current in Eq. 5.3-9. If an initial discharge current is then specified, the primary electron density can be calculated from Eq. 5.3-50 and the electron temperature obtained by finding a solution to Eqs. 5.3-40 or 5.3-41. These parameters are used to solve for the discharge loss, which is evaluated from the given beam current, discharge voltage and discharge loss. A program is iterated until a discharge current is found that produces the correct discharge loss at the specified beam current.

An example of performance curves calculated using this model and compared to measured curves for the NEXIS ion thruster [38] are shown in Fig. 5-18. The discharge loss was measured for three different discharge voltages during operation at 4 A of beam current. The 180 eV/ion discharge loss data at the 26.5 V discharge voltage required that the cathode produce a discharge current of 27.5 A to generate the 4 A of ion beam current.

The discharge model also matches the discharge loss data obtained from other thrusters. Figure 5-19 shows the discharge loss measured at JPL in a laboratory-copy of the NSTAR thruster [39] operating at the full power (2.3 kW) TH15 throttle level. The model predictions agree with the thruster data if the measured 6.5 V cathode voltage drop in the NSTAR hollow cathode [40] is used for V_c. The ability of a 0-D model to match the NSTAR data is significant only in that the NSTAR plasma is not very uniform (flatness parameter ≈ 0.5) and contains over 20% double ions peaked on the axis. The 0-D model likely works in this case because the ionization is still

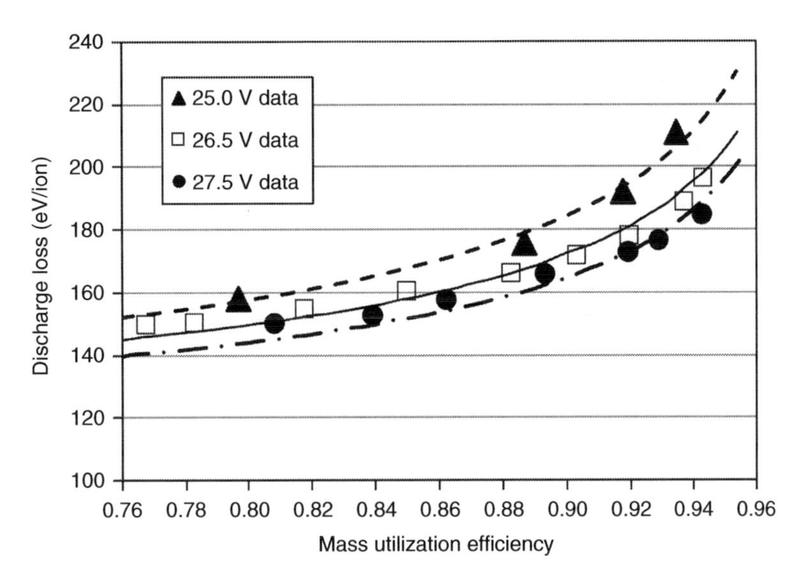

Figure 5-18 Example of the discharge loss versus mass utilization efficiency for three discharge voltages in the NEXIS ion thruster (*Source:* [38]).

dominated by the average volume effects, and the losses are still determined by the magnetic field structure at the wall, which 0-D models can capture sufficiently to give reasonably accurate results.

The shape of the performance curves is also important. As the mass utilization is increased, the neutral density in the discharge chamber decreases (see Eq. 5.3-28) and more of the primary energy goes into heating the plasma electrons and energy loss directly to the anode, as was illustrated by the simplified model for the idealized thruster case in Section 5.2. Optimal thruster designs have

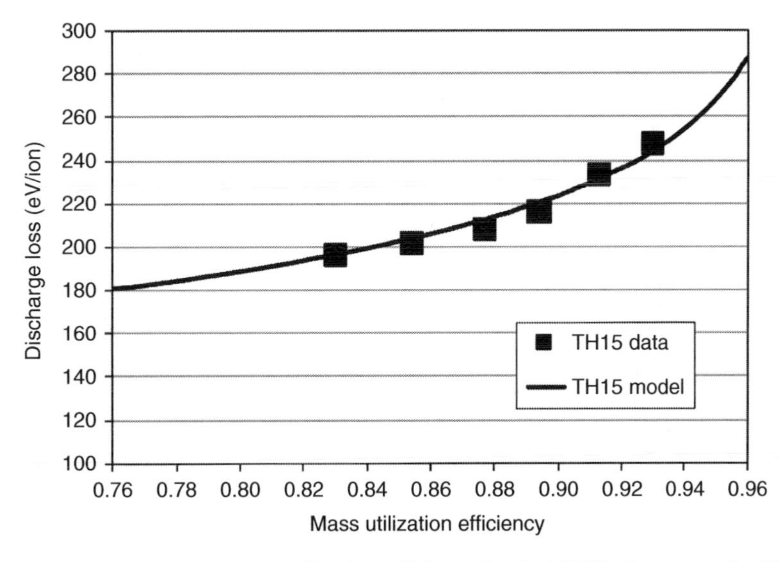

Figure 5-19 Discharge loss versus mass utilization efficiency for the NSTAR thruster at the high power TH15 throttle point.

flatter discharge performance curves that exhibit lower discharge losses as the mass utilization efficiency is increased. The model suggests that this is generally achieved in thrusters by designing for good primary and plasma electron confinement such that the convective losses are minimized at low neutral density and higher electron temperatures.

A significant challenge for most discharge models is handling the primary electrons correctly. For the case of monoenergetic primaries assumed in this model, the primary density is determined by collisional and ballistic (direct-to-anode) losses that change as a function of the neutral pressure, which is inversely proportional to the mass utilization efficiency. The primary electron density then varies strongly as the mass utilization efficiency is changed. However, if primary electrons are neglected altogether (i.e., assumed thermalized immediately in the cathode plume) so that the plasma in the discharge chamber is produced only by ionization by the high-energy tail of the Maxwellian electron population, the discharge loss is extremely high. This is shown in Fig. 5-20, where the discharge loss in the NEXIS thruster increases to over 240 eV/ion if the primary electron ionization effects are neglected. Likewise, if the primary electron density is independent of the neutral pressure, then the discharge loss curve in Fig. 5-20 has a steep slope resulting from an excessive number of primary electrons at low mass utilization (high pressure) which produces more ionization than actually occurs. Clearly, including the presence of primary electrons in the analysis is required for the model results to agree with the data, which, in turn, suggests that primary or energetic electrons and non-Maxwellian electron populations must exist in this type of thruster.

Having a representative model of the discharge permits environmental changes to the thruster to also be understood. For example, the neutral gas temperature depends on the operating time of the thruster until equilibrium is reached, which can take hours in some cases during which the discharge loss will vary [41]. The 0-D model predictions are shown in Fig. 5-21 for three different neutral gas temperatures. The discharge loss data points shown were measured for the NEXIS thruster operating at 26.5 V and 92% mass utilization efficiency at first turn on, after 1 h, and after 10 h. In this case, the thruster starts at essentially room temperature, and the model predicts that the discharge heats the thruster and neutral gas to about 470 K after about 10 h of operation. Although

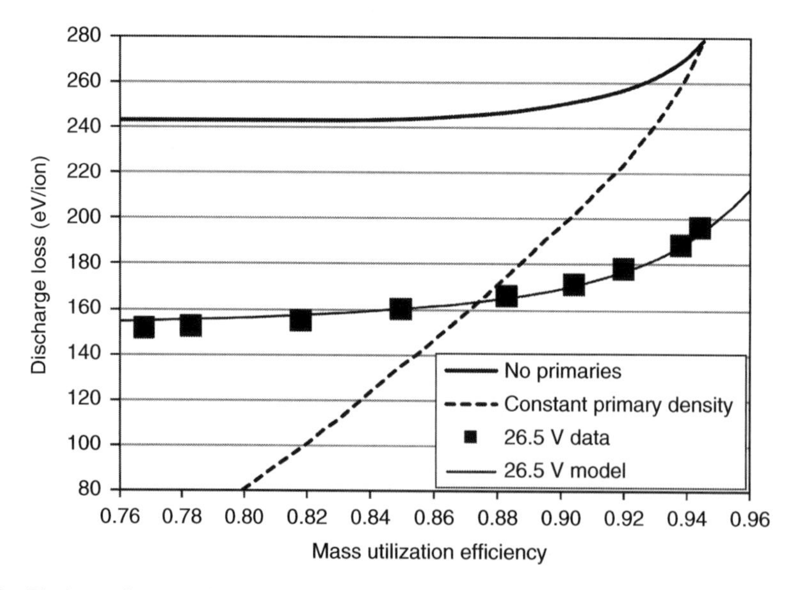

Figure 5-20 Discharge loss prediction for the cases of no primary electron density and a constant primary electron density showing these two cases produce poor agreement with the measurements.

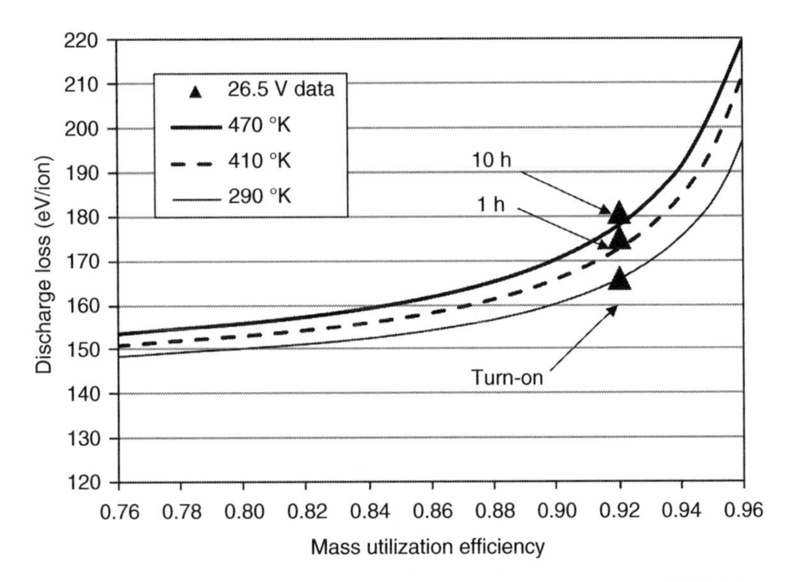

Figure 5-21 Discharge loss versus mass utilization efficiency from model for the NEXIS thruster for three neutral gas temperatures.

thruster thermal time constants are usually on the order of 1 h, this long heating time was found to be related to the facility thermal time constant. This behavior of the discharge loss with time and temperature illustrates how characterization of the thruster must always be measured in thermal equilibrium, because the performance of the discharge chamber is strongly affected by the neutral gas density, which changes with the thruster temperature for a constant input flow rate.

5.3.11 Discharge Stability

There is a strong relationship between the discharge loss and the stability of the discharge. By inspection of Eq. 5.3-65, it is clear that the efficiency increases (discharge loss decreases) if the anode collection area for primary electrons A_p is minimized. Although it is logical to assume that this is also true if the anode area for plasma electrons is minimized to reduce the energy loss from the Maxwellian-electron population, a dependence on A_a does not appear in Eq. 5.3-65. However, since the discharge current is carried to the anode primarily by the plasma electrons, the sheath potential at the anode wall in Eq. 5.3-7 is found to decrease as the anode area decreases for a given plasma electron current to the anode. This dependence on the sheath potential is seen in the discharge loss equation, which suggests that minimizing the sheath potential maximizes the efficiency. However, the anode area for plasma electrons cannot go to zero because the discharge current could not be collected by the anode, and the discharge would either interrupt or become unstable [22]. So there is some minimum anode area and plasma potential that can be tolerated for discharge stability.

The value of the plasma potential relative to the anode (essentially the anode sheath voltage drop) can be calculated using the expression for the random electron flux to the anode given in Eq. 5.3-7. From current conservation in the discharge, an expression for the discharge current can also be found from the current to the anode Eq. 5.3-53:

$$I_d = I_a + I_L - I_{ia}. \tag{5.3-66}$$

Using Eqs. 5.3-7, 5.3-2 and 5.3-17 for each of the three currents, and dividing by the beam current in Eq. 5.3-55, Eq. 5.3-66 becomes

$$\frac{I_d}{I_b} = \frac{\frac{1}{4}\left(\frac{8kT_e}{\pi m}\right)^{1/2} n_e A_a}{\frac{1}{2} n_e v_a A_s T_s} \mathrm{e}^{(-e\phi/kT_e)} + \frac{n_p v_p A_p}{\frac{1}{2} n_e v_a A_s T_s} - \frac{\frac{1}{2} n_e v_a A_{as} f_c}{\frac{1}{2} n_e v_a A_s T_s}. \tag{5.3-67}$$

Solving for the plasma potential gives

$$\phi = \frac{kT_e}{e} \ln \left[\frac{\left(\frac{2M}{\pi m}\right)^{1/2} \frac{A_a}{A_s T_s}}{\frac{I_d}{I_b} + \frac{A_{as} f_c}{A_s T_s} - \frac{2 n_p v_p A_p}{n_e v_a A_s T_s}} \right] \tag{5.3-68}$$

By inspection of Eq. 5.3-68, it is clear that as the anode area A_a decreases, the plasma potential also decreases. If the anode area is made too small, then the plasma potential will go negative relative to the anode potential. This is called a "positive going" or "electron accelerating" anode sheath, and is illustrated in Fig. 5-22. In this case, the anode area at the cusps is insufficient to collect the total discharge current by collection of the entire incident random electron flux over the anode cusp area. The plasma then biases itself to pull in electrons in the Maxwellian distribution that are not initially headed toward the anode, which delivers more current to satisfy the discharge current and charge balance requirements. The plasma electron current collected by the anode then becomes

$$I_a = \frac{1}{4}\left(\frac{8kT_e}{\pi m}\right)^{1/2} e n_e A_a e^{e\phi/kT_e} \left[1 - \mathrm{erf}\left(\frac{-e\phi}{kT_e}\right)^{1/2}\right]^{-1}, \tag{5.3-69}$$

where the potential ϕ is now a negative number. If the potential goes sufficiently negative relative to the anode, the current density can reach a factor of two higher than the one-sided random electron flux normally collected to satisfy the discharge current requirement.

However, once the potential goes sufficiently negative relative to the anode to repel the ions (about T_i), then the anode area for the plasma electron is not the hybrid area, but is just twice the plasma electron Larmor radius times the cusp length, similar to Eq. 5.3-3 for the primary loss area. This results in a significant decrease in the cusp anode area A_a in Eq. 5.3-68 for negative plasma potentials, which further lowers the plasma potential relative to the anode. Examining the potential distribution in the plasma in Fig. 5-22, the transition from the normal negative-going sheath to a negative plasma potential (positive-going anode sheath) will subtract from the primary electron energy V_{pe} at a given discharge voltage. The ionization rate then decreases, and the discharge collapses into a high impedance mode or oscillates between this mode and a positive potential typically on power supply time constants as the supply tries to reestablish the discharge by increasing the anode voltage.

The stability of the plasma discharge at a given operating point (discharge current, beam current, neutral density in the discharge chamber, etc.) is therefore determined by the magnetic field design. For example, in Fig. 5-23, plasma potential is plotted as a function of the strength of the cusp magnetic field for an arbitrary thruster design with two different numbers of ring cusps. The cusp field strength enters into the anode area A_a in Eq. 5.3-6, the primary electron loss area A_p in Eq. 5.3-3 and in the plasma potential in Eq 5.3-68. The model predicts that a four-ring design would be unstable (when the potential goes negative relative to the anode) for cusp magnetic fields greater than 2000 G. Since strong magnetic fields are desirable from a primary electron and ion confinement point of

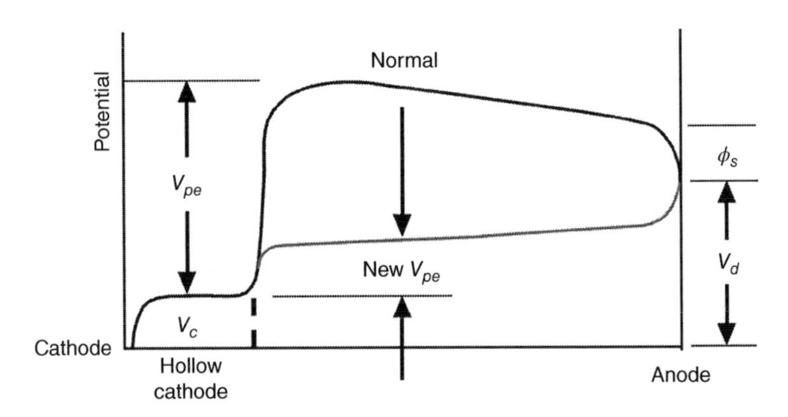

Figure 5-22 Transition of the plasma potential to negative relative to the anode due to an anode area decrease, which results in a lower primary electron energy.

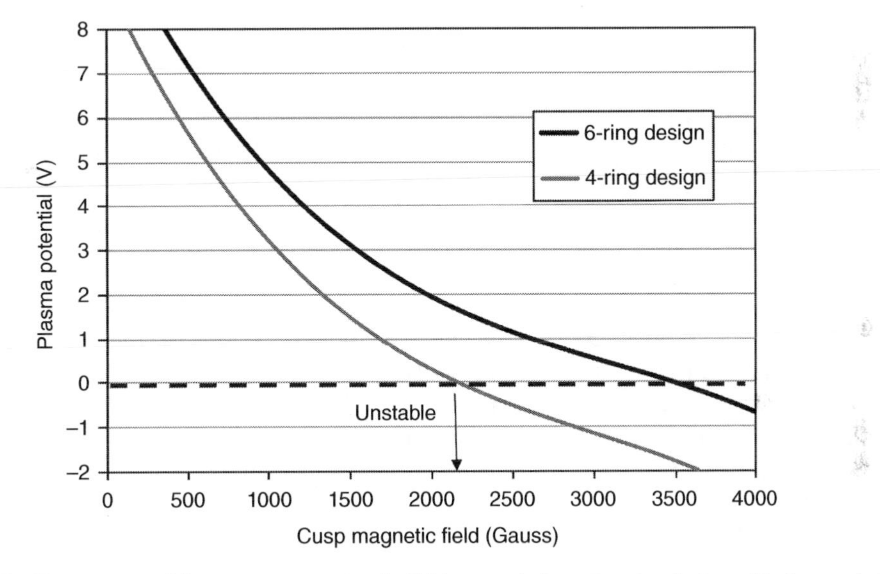

Figure 5-23 Plasma potential versus cusp magnetic field strength for a thruster design with four and six magnet-rings.

view, additional rings are required to maintain a positive plasma potential. A six-ring design increased the anode area sufficiently to raise the plasma potential at the 2000 G magnet design point. An analysis of the discharge loss from Eq. 5.3-65 indicates that the improved stability associated with the larger anode area of six-ring design comes with a loss in efficiency. The trade-off between efficiency and stability is an important aspect of ion thruster design.

5.3.12 Recycling Behavior

Ion thrusters clear momentary faults or breakdowns in the high voltage accelerator grids by momentarily turning off the high voltage, an event called recycling. To restart the thruster, the accelerator grid ("accel grid") voltage must be turned back on to avoid electron backstreaming into the thruster as the screen voltage is reapplied. If the plasma discharge is left on during this sequence, the negatively-biased accel grid collects nearly the entire ion beam current at the applied accel voltage until the screen voltage is re-established. This can lead to excessive power loading and even erosion of the accel grid if a significant number of recycles are encountered. Therefore, it is standard procedure to also either turn-off the discharge during recycling, or cut it back to a low level such that the accel grid current surge is acceptably low during reestablishing of the beam voltages. The discharge current is then raised to the desired level with the screen voltage ramp-up.

The main issue with this process is that the thruster discharge often goes into oscillation during the cutback condition or on restarting in the recycle sequence. When the high voltage is turned off in a recycle, ions that would have left the discharge chamber as beam ions now strike and neutralize on the accel grid, and some fraction flow back into the discharge chamber as neutral gas. This raises the neutral gas pressure in the discharge chamber, which has two effects. First, a higher neutral pressure collisionally thermalizes the primary electrons more rapidly, which can lead to a reduction in the plasma potential [22]. Second, lowering the discharge current while raising the neutral pressure leads to a lower impedance discharge and a lower discharge voltage. These two effects will be shown next to cause a reduction the plasma potential, and thrusters designed for low discharge loss with a minimum plasma potential at the nominal operating point can encounter negative plasma potentials and discharge instability during recycling.

The time-dependent behavior of the pressure in the discharge chamber from the high-voltage-off event can be calculated using molecular dynamics, and the subsequent time dependent plasma potential for stability evaluated using the 0-D model. The time-dependent pressure [42] in the thruster is given by

$$V\frac{dP}{dt} = Q_{\text{in}} - C\Delta P, \tag{5.3-70}$$

where V is the discharge chamber volume, P is the pressure in the thruster discharge chamber, C is the conductance of the grids and ΔP is the pressure drop across the grids. The initial pressure just before the start of the recycle, when the thruster is operating normally, is found from Eq. 5.3-28 and the conversion of neutral density to pressure in Eq. 2.7-2:

$$P_o = 4.1 \times 10^{-25} \frac{T_o Q_{\text{in}}(1-\eta_m)}{v_o e A_g T_a \eta_c}. \tag{5.3-71}$$

With the high voltage off, the ions and neutrals flow to the grid region, where a small fraction exits through the accel aperture to escape, and the majority strike the upstream side of the grids or the grid aperture barrel wall and flow back into the thruster. Since the grid conductance is defined as the flow divided by the pressure drop [42], the final pressure after steady state has been achieved is

$$P_f = (1-T_a)\frac{Q_{\text{in}}}{C}, \tag{5.3-72}$$

where C is the conductance of the grids and the downstream pressure from the grids has been neglected as small. The conductance of the grids can be estimated from the molecular conductance of a thin aperture [42] times the Clausing factor for the finite thickness grids. The conductance is then

$$C = 3.64\left(\frac{T}{M_a}\right)^{1/2} T_a A_g \eta_c \text{ [liters/ sec]}, \tag{5.3-73}$$

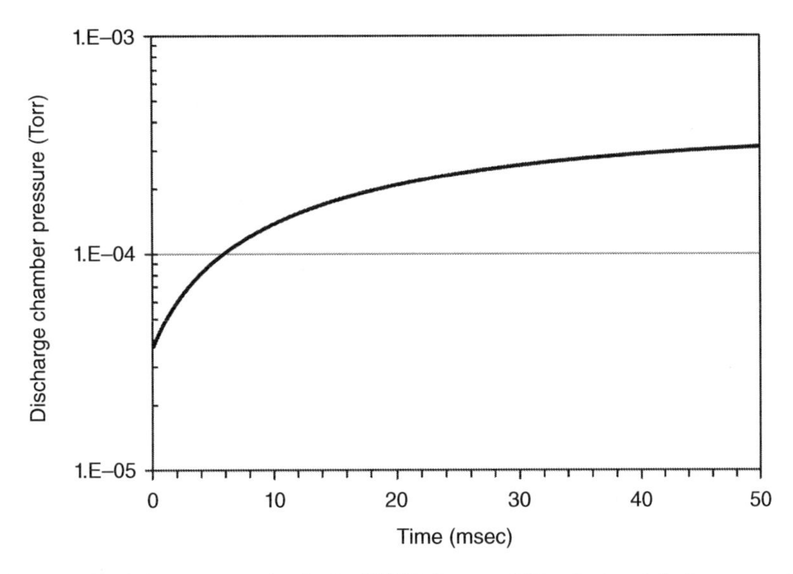

Figure 5-24 Example of the pressure rise in the NEXIS thruster [20] calculated during a recycle.

where M_a is the ion mass in AMU, and the effective open area of the grids is the optical transparency of the accel grid, T_a, times the grid area A_g. Integrating Eq. 5.3-72 from the initial pressure to the final pressure gives

$$P(t) = P_f - \left(P_f - P_o\right)e^{-t/\tau_g}, \tag{5.3-74}$$

where $\tau_g = V/C$ is the gas flow time constant for filling the thruster chamber. To use Eq. 5.3-74 to find the final pressure, the gas flow rate has to be converted from particles per second to Torr-l/sec by multiplying the neutral gas flow in Eq. 5.3-72 by 2.81×10^{-20}.

Figure 5-24 shows an example of the pressure increase with time calculated in the NEXIS thruster discharge chamber from the start of a recycle. The pressure in the discharge chamber during normal operation is in the mid-10^{-5} Torr range due to the large grid area and high mass utilization efficiency. During a recycle, the pressure in the discharge chamber reaches equilibrium in about 60 msec, with the pressure increasing almost an order of magnitude once the high voltage is turned off. This magnitude of pressure increases in the thruster once the high voltage is turned off is consistent with the ≈90% mass utilization efficiency of many thruster designs.

The plasma potential response to pressure changes in the discharge chamber calculated using the 0-D model for two different discharge voltages is shown in Fig. 5-25a for a given magnetic field design. During the recycle, the discharge current is reduced (called cutback), which reduces the discharge voltage and thereby the plasma potential. The model indicates that the plasma potential reduction and subsequent unstable operation is the result of the lower discharge voltage, and does not occur directly because of the discharge current being lower. This analysis shows that a given thruster design that produces a stable discharge under normal conditions can go unstable due to negative plasma potentials as the pressure rises and the discharge voltage decreases.

The plasma potential calculated using Eq. 5.3-68 for two magnet designs is shown in Fig. 5-25b for the 23-V NEXIS case, which illustrates the effect of the smaller anode area reducing the plasma potential at a given pressure. In this case, increasing the anode area permitted the discharge current

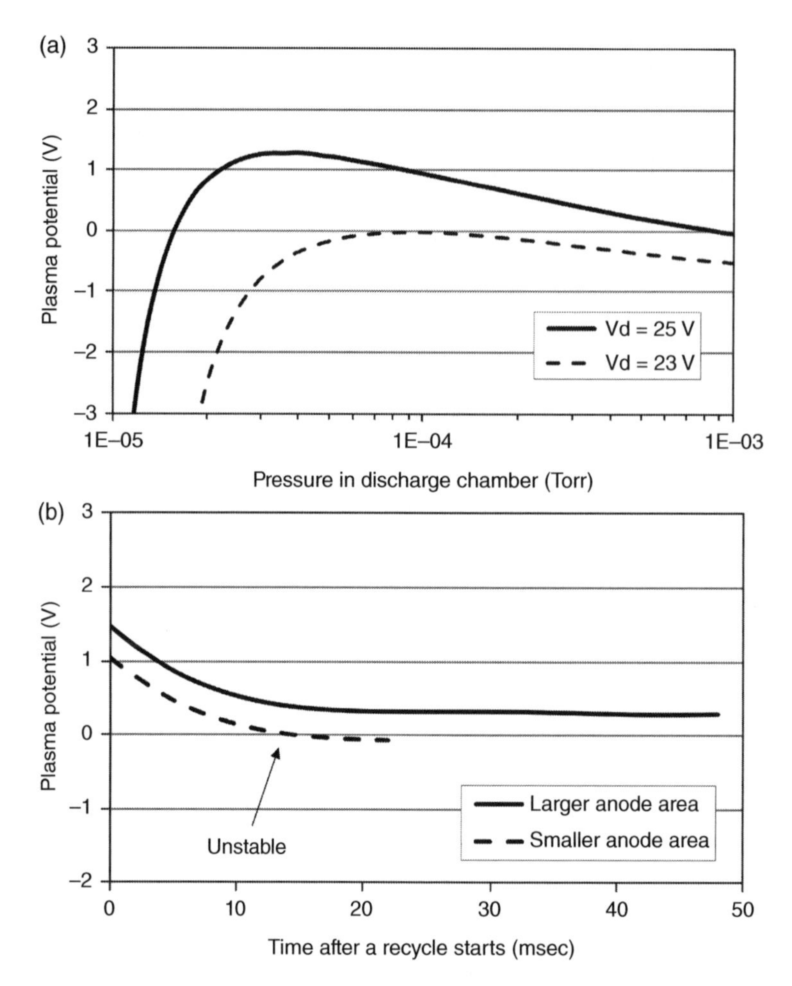

Figure 5-25 Plasma potential as a function of pressure for two discharge voltages (a), and plasma potential versus time (b), showing instability of the smaller anode area design at a given pressure.

to be cutback during the recycle to the desired level without oscillating, which facilitates re-starting the high voltage. Of course, the larger anode area increased the loss in the discharge chamber and raised the discharge loss. This tradeoff is often required to provide good performance and stable discharge operation.

5.3.13 Limitations of a 0-D Model

Although the 0-D models described in this chapter provide useful information on the design parameters of ion thrusters and give good insight into the plasma production and loss mechanisms, there are several limitations to their use. First, 0-D models assume that the electron and neutral densities are uniform and averages the ion production throughout the volume of the discharge chamber. For ion thrusters with significantly non-uniform plasmas, this leads to inaccuracies in the average plasma density and beam current calculated by the 0-D model that can only be handled by multi-dimensional discharge chamber models. Second, the source of the gas in actual discharge chambers is from the localized hollow cathode aperture and the gas manifold inside the discharge chamber. The neutral density, therefore, is never completely uniform, and variations in the neutral density can affect the transport, diffusion and ionization rates in the discharge chamber.

Third, ion thrusters with localized electron sources like hollow cathodes have strongly varying primary electron densities within the discharge chamber. As shown earlier, the primary electron density strongly affects the ionization rate, and so localized sources of primaries produce non-uniform plasmas that the 0D models cannot address. In addition, these models utilize a monoenergetic primary energy. A distribution in the primary electron energy has been measured in some ion thrusters [43, 44], which changes the ionization and primary electron thermalization rates compared to the monoenergetic calculations presented here. Although primary electron energy distributions can be incorporated in 0-D models, this has not been attempted to date.

Finally, the 0-D model assumes monoenergetic primary electrons with an energy of $e(V_d - V_c + \phi)$. For typical discharge voltages around 25 V and cathode voltage drops of 5–10 V, this means that potentially none of the primaries have sufficient energy to doubly ionize xenon which has an ionization potential of 21.2 V. Double ions can then only be produced by the tail of the plasma electron distribution. For electron temperatures of 3–5 eV, less than 1% of the electrons have sufficient energy to produce double ions. Since the double ion content in NSTAR thrusters has been reported to exceed 20%, a monoenergetic primary electron energy causes the 0-D model to not accurately address double ion production. Although including primary electrons is necessary to obtain agreement between the 0-D models and experimental results, knowledge of the correct energy distribution and even spatial variation in the primaries is required, and is better handled by 2-D models discussed in Section 5.7.

5.4 Kaufman Ion Thrusters

The formulation of particle and energy balance models just described applies to any ion thruster geometry where the electron loss can be defined by a finite anode electrode area collecting electrons at a fraction of the random electron flux depending on the sheath voltage. One class of thrusters still in use, the Kaufman ion thruster shown schematically in Fig. 5-26, features a strongly diverging axial magnetic field that shields a cylindrical anode electrode located near the wall of the discharge chamber. In this case, electron transport to the anode is determined by cross-field diffusion.

The flux of electrons due to cross-field diffusion is given by Eq. 3.6-71:

$$\Gamma_e = n\mathbf{v}_\perp = \pm \mu_\perp n\mathbf{E} - D_\perp \nabla n. \tag{5.4-1}$$

For the case of Kaufman thrusters, the perpendicular diffusion coefficient is likely to be close to the Bohm diffusion coefficient [45]:

$$D_B = \frac{1}{16}\frac{kT_e}{eB}. \tag{5.4-2}$$

The electron current collected by the anode is the flux that diffuses through the magnetic field times the Boltzmann factor at the sheath:

$$I_a = (\mu_\perp n\mathbf{E} - D_\perp \nabla n) e A_{as} e^{-e\phi/kT_e}, \tag{5.4-3}$$

where A_{as} is again the anode surface area exposed to the plasma discharge. The actual current distributions and potential distribution in a Kaufman are the same as for DC discharge thruster shown in Fig. 5-9. However, there are several terms that were analyzed for ring cusp thrusters that can be neglected in Kaufman thrusters.

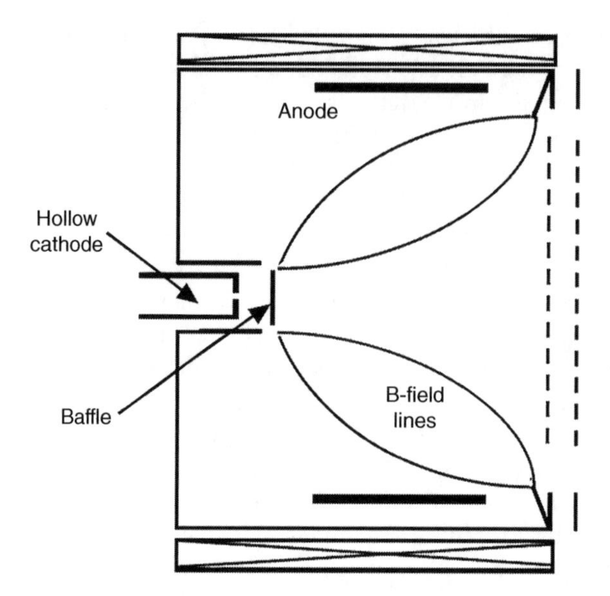

Figure 5-26 Schematic of a Kaufman ion thruster showing the hollow cathode with a baffle, and the anode protected by magnetic fields produced by an external solenoid coil.

First, if the axial magnetic field is in the discharge chamber on the order of 100 G, then, the Larmor radius for, say, 20 eV primaries is 1.5 mm. Since the magnetic field lines do not intersect the anode and primaries are too energetic to participate in the collective instabilities that drive Bohm diffusion, the primary electrons must make collisions to cross the magnetic field to be lost. That means that the fraction of the primary electron current lost directly to the anode in ring-cusp thrusters, I_L, can be neglected, which is an advantageous feature for modeling of Kaufman thrusters.

Second, the plasma flow across the magnetic field is still governed by ambipolar effects. As was shown in Section 5.3.4, if the transverse magnetic field strength is in excess of about 50 G in typical ion thruster discharge chambers, then the radial electric field in the plasma (in the magnetic field region) is near zero and the ion loss rate is on the order of one-tenth the Bohm current toward the wall. This means that the ion current to the anode term, I_{ia}, can also be neglected to first order. Since the discharge current collected through the anode leg of the discharge power supply connection was given in Eq. 5.3-66 as the plasma electron current minus the ion current plus the primary current, the discharge current is now just

$$I_d = I_a = -D_{\perp}\nabla neA_{as}e^{-e\phi/kT_e}. \tag{5.4-4}$$

Third, the ion current flowing back towards the hollow cathode was neglected in our treatment of ring-cusp thrusters because the hollow cathode exit area in contact with the plasma was so small. In Kaufman thrusters, a baffle is placed on axis in front of the cathode to force the primary electrons off axis to flatten the density profile. Since the magnetic field is strongly divergent, the axial plasma density gradient is significant and the plasma density in contact with the baffle can be high. For these reasons, the ion current to the cathode, I_k, can no longer be neglected.

The power into the plasma is given by Eq. 5.3-51. The power out of the discharge is given by

$$P_{out} = I_pU^+ + I^*U^* + I_s(V_d + \varepsilon_i) + I_k(V_d + \varepsilon_i) + I_b\varepsilon_i + I_a\varepsilon_e, \tag{5.4-5}$$

where ε_i is the ion energy leaving the plasma, which from Eq. 5.2-10 is equal to ϕ, and ε_e is the electron energy removed from the plasma, which from Eq. 5.2-9 is $2T_{eV} + \phi$. Equating the power in to the power out and solving for the discharge loss gives

$$\eta_d = V_d \left[\frac{\frac{I_p}{I_b} U^+ + \frac{I^*}{I_b} U^* + \frac{I_s}{I_b}(2V_d - V_c + 2\phi) + \phi + \frac{I_k}{I_b}(2V_d - V_c + 2\phi)}{V_d - V_c - 2T_{eV}} \right], \tag{5.4-6}$$

which is the same as Eq. 5.3-59 for DC ion thruster with the ion current to the cathode included and the primary electron and ion loss to the anode removed. The first current ratio I_p/I_b is given by Eq. 5.3-60, and the second current ratio I^*/I_b is given by Eq. 5.3-61. The current ratio I_s/I_b is given by Eq. 5.3-62, and the cathode ion current to beam current ratio is

$$\frac{I_k}{I_b} = \frac{\frac{1}{2} n_k e \sqrt{\frac{kT_e}{M}} A_k}{\frac{1}{2} n_i e \sqrt{\frac{kT_e}{M}} A_s T_s} = \frac{n_k A_{sa}}{n_e A_s T_s}, \tag{5.4-7}$$

where n_k is the plasma density at the cathode baffle. The discharge loss for Kaufman thrusters is then

$$\eta_d = \frac{V_d \left[\frac{I_p}{I_b} U^+ + \frac{I^*}{I_b} U^* + \phi + \left(\frac{1 - T_s}{T_s} + \frac{n_k A_{sa}}{n_e A_s T_s} \right)(2V_d - V_c + 2\phi) \right]}{V_d - V_c - 2T_{eV}}. \tag{5.4-8}$$

The plasma potential in Eq. 5.4-8 is found from solving Eq. 5.4-4:

$$\phi = \frac{kT_e}{e} \ln\left(\frac{-D_\perp \nabla n e A_{as}}{I_d} \right), \tag{5.4-9}$$

and the electron temperature is found from the similar iterative solution to the ion particle balance in Eq. 5.3-39 used for ring-cusp thrusters. The negative sign in Eq. 5.4-9 appears problematic in the natural log function, but the density gradient ∇n is negative going outward from the plasma. The primary electron density is calculated from Eq. 5.3-50, with the ballistic loss term neglected as described above since primaries are not lost directly to the anode. Finally, the plasma volume term in the ion and excited neutral production rates can be assumed to be the volume of a cone from the baffle to the grids because the plasma is well confined by the strong diverging magnetic field. Since the 0-D model assumes relatively uniform plasma, estimates for the radial gradient of the plasma density in the magnetic field region near the anode and the additional cathode voltage drop due to the baffle must be made for Eq. 5.4-8 to be accurate.

As an example, take a conceptual Kaufman thruster with a 20 cm diameter screen grid with 80% transparency, a 25 cm diameter anode with 25 cm between the grids and the baffle. Assuming that the average magnetic field strength in the thruster is about 50 G, the discharge loss from Eq. 5.4-8 is plotted in Fig. 5-27 for two values of the cathode voltage drop. In this case, the cathode voltage drop is higher than in a ring cusp thruster because it includes the potential drop in the baffle region. The discharge loss is strongly dependent on this value because it directly affects the primary electron energy. Discharge losses in this range at mass utilization efficiencies of about 90% have been reported in the literature for Kaufman thrusters through the years [46–48], suggesting that the 0-D model can produce reasonable predictions of the discharge loss if the cross-field diffusion is handled properly.

The need for higher discharge voltages in Kaufman thrusters, compared to ring-cusp thrusters, is illustrated in Fig. 5-28, where the discharge loss is plotted for the Kaufman thruster example above

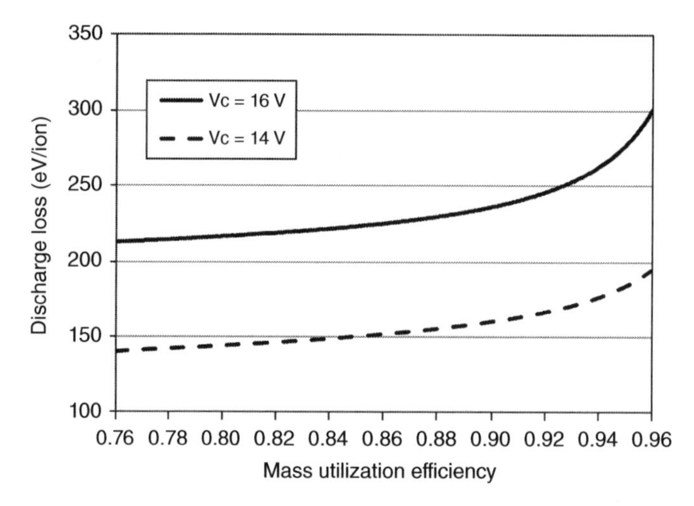

Figure 5-27 Discharge loss calculated for a Kaufman thruster example.

with two cases of the discharge voltage at a constant (total) cathode voltage drop of 16 V. Low discharge loss is achieved for the 35 V discharge voltage case, but decreasing the discharge voltage to 30 V causes the discharge loss to increase dramatically. This is because the primary electron energy in the discharge chamber is near the threshold energy for ionization at this discharge voltage, and the discharge efficiency decreases as more ionization is required from the plasma electrons. In addition, the lower discharge voltage causes the plasma potential to go significantly negative relative to the anode potential ($\approx T_e$), which will cause the discharge to become unstable.

Although Kaufman-type thrusters are considered to be the first ion thruster to achieve good discharge production performance, they now compete with ring-cusp thrusters for application in modern electric propulsion systems. This is because of several constraints in Kaufman thruster

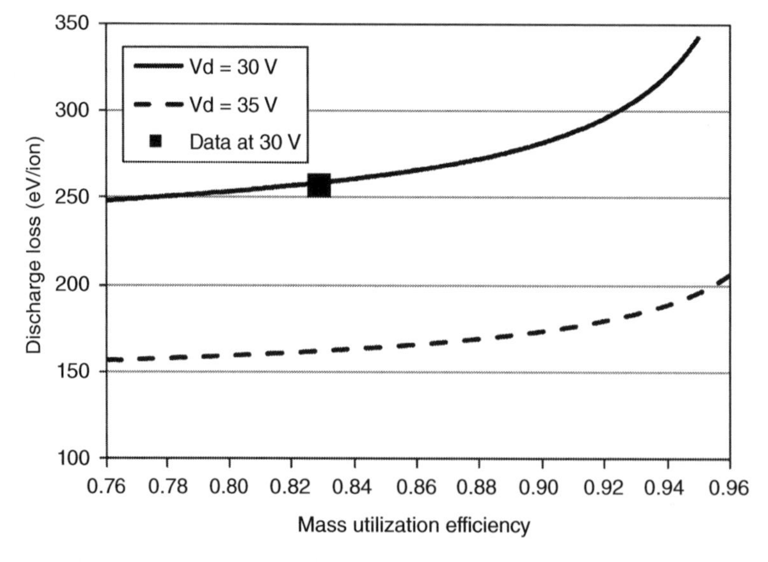

Figure 5-28 Discharge loss calculated for the Kaufman thruster example at two discharge voltages with a data point at one power level for a Kaufman thruster operated at JPL.

design. First, the strong axial magnetic field restricts electron motion to the anode to cross-field diffusion, which requires either high neutral pressures in the discharge chamber for electron-neutral collisional diffusion and, thereby, low mass utilization efficiency, or relies on collective instabilities to increase the diffusion rate to obtain sufficient electron loss to support the discharge. The instabilities are usually related to $\mathbf{E} \times \mathbf{B}$ driven instabilities and Bohm diffusion [24], which create significant noise in the discharge that can appear in the beam current. Second, the baffle required to force the primary electrons off axis to produce a more uniform plasma profile is susceptible to ion bombardment sputtering and plasma losses in the dense plasma region near the cathode. This has historically limited the life of these types of thrusters, although alternative materials can mitigate this problem. In addition, the primary electrons are injected purely off axis, which means that the plasma profile, and hence the beam profile, can be hollow or peaked depending on the cross-field diffusion and mobility throughout the discharge chamber.

Finally, the thruster size, shape and magnetic field strength is limited to regimes where the magnetic field is sufficient to confine ions by electrostatic ambipolar effects to obtain good efficiency, and yet the magnetic field is not so high that the cross-field diffusion cannot provide adequate electron current for the discharge to be stable. If the field is too strong or the anode area in contact with the plasma is too small, the plasma potential goes negative relative to the anode to pull the electrons out of the discharge. It is shown in Fig. 5-22 that when the plasma potential is negative relative to the anode, then the primary energy is decreased at a given discharge voltage, which strongly affects the discharge efficiency [22]. Since the discharge voltage cannot be arbitrarily increased due to ion sputtering of the baffle and screen electrodes, in addition to excessive double ion production, this will significantly reduce the discharge efficiency. In the case of negative plasma potentials, the electron loss to the anode has the form [22]

$$I_a = -D_\perp \nabla n e A_{as} e^{e\phi/kT_e} \left[1 - \text{erf}\left(\frac{-e\phi}{kT_e} \right)^{1/2} \right]^{-1} \tag{5.4-10}$$

where ϕ is a negative number. The negative plasma potential increases the current to the anode area A_{as} by pulling some of the electrons from the plasma population that were headed away from the anode. Although up to a factor of two more electron current can be theoretically drawn compared to the case of positive plasma potentials, in practice drawing even the random electron flux can strongly deplete or perturb the Maxwellian population and affect the plasma discharge and ionization rates. The geometry of Kaufman thrusters for good efficiency is limited to configurations where the plasma potential in the discharge chamber is not allowed to go negative relative to the anode, which constrains the design space for the electrodes and fields.

5.5 rf Ion Thrusters

The ion thrusters described in the previous sections utilize a thermionic hollow cathode and DC discharge power supply to inject hot electrons into the discharge chamber to ionize the propellant gas. To eliminate any potential life or power supply issues with the hollow cathode and DC-electron discharge, an alternative thruster design utilizes electromagnetic fields to heat the plasma electrons that, in turn, ionize the injected gas. One method to achieve this goal is to use an inductive plasma generator, which is normally called a radio-frequency or rf ion thruster. In this case, low frequency rf voltage is applied to an antenna structure around or in the plasma, and the rf energy is coupled to the electrons that perform the ionization.

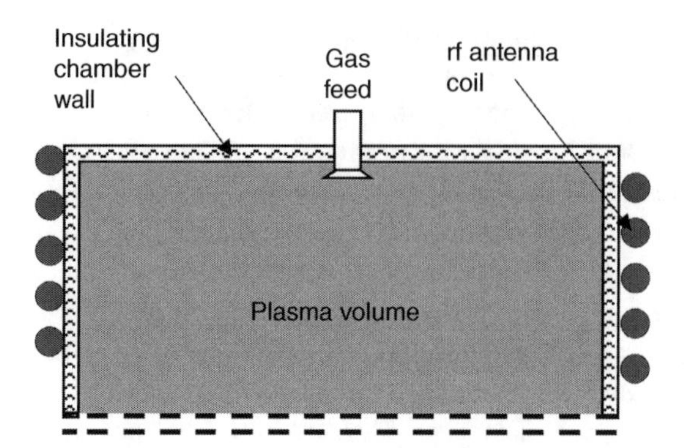

Figure 5-29 Schematic of an rf ion thruster showing the induction coil, insulating body, gas feed and two-grid accelerator structure.

The simplest configuration for an rf ion thruster is shown schematically in Fig. 5-29. The rf coil is wrapped around an insulating chamber with a gas feed. The chamber can be cylindrical, hemispherical or conical in shape and is connected to an ion accelerator structure that is the same as those used for DC electron-bombardment ion thrusters with either two or three grids. The plasma floats relative to the first grid, and the high voltage is applied between the two grids to accelerate ions that flow through the first grid and form the beam. The rf coil is connected to a rf power supply that provides the power to generate the plasma. There is usually no applied magnetic field in rf ion thrusters, although one can be applied in principle to improve the discharge performance. As in other ion thruster designs, the entire discharge chamber is enclosed in a metallic screen or structure to eliminate electron collection from the space plasma, and a neutralizer cathode is connected to provide net charge neutralization of the beam.

The coil wrapped around the insulating thruster body can be modeled as a solenoid with N turns, and the rf voltage applied to it drives rf current in the coil. Typical frequencies used in rf ion thrusters are in the range of 1 MHz. At these frequencies, the penetration of the fields from the coil at the boundary is limited by the skin depth in the plasma [24], which is on the order of, or slightly less than, the radius of most rf ion thrusters at the plasma densities required to produce xenon ion current densities in excess of 1 mA/cm^2. This produces an attenuation of the electric and magnetic fields toward the axis, and the majority of plasma interaction with the fields occurs off axis closer to the boundary.

The axial magnetic field inside the coil induced by the rf current, neglecting end effects, is

$$B_z = \frac{NI}{\mu_o} e^{iwt}, \tag{5.5-1}$$

where I is the rf current in the coil, μ_o is the permeability of vacuum, ω is the cyclic frequency ($2\pi f$) of the rf, and t is the time. From Maxwell's equation, the time varying magnetic field creates a time varying electric field:

$$\nabla \times \mathbf{E} = -\frac{\partial \mathbf{B}}{\partial t}. \tag{5.5-2}$$

The induced rf electric field in the rf thruster geometry is then in the azimuthal direction:

$$E_\theta = -\frac{iwr}{2} B_{zo} e^{iwt}, \tag{5.5-3}$$

where r is the distance from the axis and B_{zo} is the peak axial rf magnetic field from Eq. 5.5-1. A finite electric field is generated spatially off axis inside the thruster.

The induced electric field exists in one direction ($\pm\theta$ direction) for roughly half a period, which for a 1 MHz frequency is 0.5 μsec. The electrons, however, don't see the oscillating component of the electric field because they transit the interaction region close to the antenna in a time much less than this value. For example, a 5 eV electron will travel a distance of about 1 m in 1 μs, and so can traverse the electric field region many times within a half cycle. Therefore, electrons traversing the induced electric field region "see" a DC electric field and are accelerated. If they make a collision before leaving the region, they can then retain some or all of the velocity imparted by the electric field and are heated.

The criteria for the rf plasma generator to provide net heating of the electrons is that a sufficient number of electrons make a collision within the electric field interaction region. If the interaction region is, say, a few centimeters across, the mean free path should be on this order. The probability of an electron making a collision is given by

$$P = 1 - e^{-x/\lambda} = 1 - e^{-n_o \sigma x}. \tag{5.5-4}$$

Using Eq. 2.7-2 to convert from neutral density to pressure (in Torr), the minimum pressure at a temperature T in the plasma chamber of an rf thruster for breakdown to occur is

$$P_{\min}(\text{Torr}) = \frac{-1.04 \times 10^{-25}\, T}{\sigma x} \ln(1 - P). \tag{5.5-5}$$

For example, the minimum pressure for starting the rf-generated plasma is plotted in Fig. 5-30, where room temperature (290 K) xenon gas with a xenon atomic radius of 1.24 Å in a 5-cm long interaction region is assumed. If 10% of the electrons must make an electron-neutral collision within a 5 cm interaction region to provide sufficient heating for sustaining ionization and breakdown to proceed, then the minimum pressure in the thruster is about 1×10^{-3} Torr. Minimum pressures in the range of 10^{-3} to 10^{-2} Torr are commonly reported in the literature for rf plasma sources to ignite the plasma. Once the plasma source is ignited, the required electron collisions to provide the heating in the rf electric fields can be supplemented by Coulomb collisions between the plasma electrons, which reduces the operating pressure requirement and permits high mass utilization efficiency to be achieved.

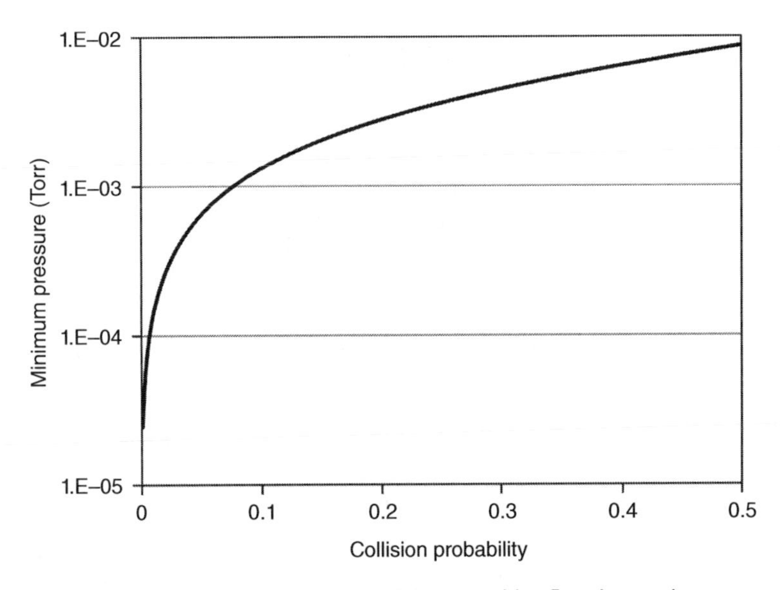

Figure 5-30 Minimum pressure for starting a xenon rf thruster with a 5 cm interaction zone as a function of the probability of an electron having a collision.

Starting an inductive plasma discharge can also be problematic because initially there are few free electrons present to interact with the rf fields and ionize the fill gas. Before the plasma ignition, there is no load on the rf circuit driving the coil; the reactive power stored in the inductive components in the rf matching network grows, which increase the voltage across the coil and the induce higher electric fields inside. If the minimum gas pressure is provided, the discharge will ignite when either the field is large enough to excite the few electrons naturally present in the chamber, or cause field emission to occur. Another method for ignition is to inject electrons from a spark generator, small cathode, or most commonly the neutralizer cathode (with the accel voltage turned off momentarily) into the discharge chamber to provide the seed electrons for interaction with the rf electric fields.

If the antenna in rf thrusters is directly exposed to the plasma, ions in the discharge can be accelerated by the rf voltage on the surfaces and sputter-erode the antenna. This can ultimately limit the life of rf thrusters. This problem is minimized by either encasing the antenna in an insulator [49], or by making the thruster body an insulating material and mounting the antenna exterior to the plasma volume [50]. In this case, the rf voltage across the coil is shielded from the plasma, and the ions are not accelerated to high energy before striking the insulator. Mounting rf antennas outside insulating-material walls such as quartz or alumina is common practice in inductive plasma generators used in the semiconductor processing industry. An example of this arrangement applied to a RIT-XT thruster [50] is shown in Fig. 5-31. In this case, the body of the thruster is constructed of a conical (or hemispherical) alumina insulator, and a high conductivity material (typically copper) antenna is coiled around the insulator. As long as the alumina body is not significantly coated by conductive layers and remains an insulator, the rf fields will couple through the wall and generate plasma.

This type of ion thruster is readily analyzed by a particle and energy balance model [51] because it does not have localized electron sources (hollow cathodes or primary electrons); the rf fields simply heat the Maxwellian electron distribution that provides the ionization, and the plasma in the

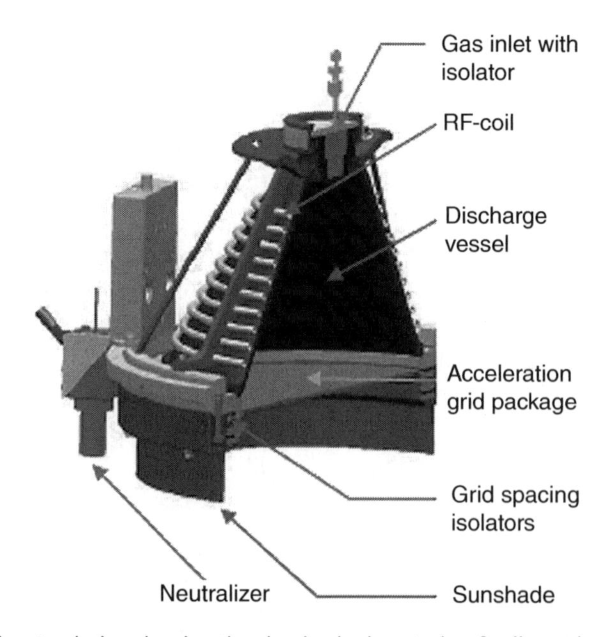

Figure 5-31 rf ion thruster design showing the alumina body, exterior rf coil, accelerator grid assembly and neutralizer cathode. The antenna system is normally enclosed in a metal "plasma shield" to eliminate electron collection from the space plasma (*Source:* [50]).

discharge chamber is very uniform. In the energy balance equation, it is assumed that the power absorbed by the plasma is simply given by the net forward rf power absorbed P_{abs}. Ions generated in the plasma volume drift to the interior surfaces in the thruster, and only electrons in the tail of the Maxwellian distribution have sufficient energy to overcome the potential difference between the plasma and the wall to be lost. The power out of the plasma equals the power absorbed, which is given by

$$P_{abs} = I^+ U^+ + I^{++} U^{++} + I^* U^* + (I_s + I_w + I_b)\phi + I_a(2T_{eV} + \phi), \tag{5.5-6}$$

where U^{++} is the second ionization potential, and the ion and electron energy loss terms are shown explicitly.

Since there are no primary electrons in rf ion thrusters, the singly ionized particle production rate I^+ is given by

$$I^+ = n_o n_e \langle \sigma_i v_e \rangle V \tag{5.5-7}$$

and the doubly ionized particle production rate I^{++} is given by

$$I^{++} = n_i^+ n_e \langle \sigma_i^{++} v_e \rangle V \tag{5.5-8}$$

where n_i^+ is the density of the singly ionized gas. Equating the ion production and loss rates gives

$$\begin{aligned} n_o n_e \langle \sigma_i v_e \rangle V &+ n_i^+ n_e \langle \sigma_i^{++} v_e \rangle V \\ &= \frac{1}{2} n_i^+ v_a (A_{as} f_c + A_s) + \frac{1}{2} n_i^{++} v_a (A_{as} f_c + A_s), \\ &= \frac{1}{2} n_i^+ (1+\gamma) \sqrt{\frac{kT_e}{M}} (A_{as} f_c + A_s) \end{aligned} \tag{5.5-9}$$

where γ is the ratio of doubles to singles ions in the beam from Eq. 5.3-32. Rearranging Eq. 5.5-9 gives an expression that can be solved for the electron temperature in the discharge chamber:

$$\frac{2n_0 V}{(A_w f_c + A_s)} \frac{\left(1 + \frac{\gamma}{\sqrt{2}}\right)}{1+\gamma} = \frac{\sqrt{\frac{kT_e}{M}}}{\langle \sigma_i^+ v_e \rangle + \frac{n_e}{n_0} \frac{\langle \sigma_i^{++} v_e \rangle}{1 + \frac{\gamma}{\sqrt{2}}}}. \tag{5.5-10}$$

Since the electron temperatures is usually low (<5 eV) in rf ion thrusters and there are no primary electrons, the double ion content is also typically very low (<5%) and usually can be ignored in rf thruster discharge models. Neglecting the double ions, Eq. 5.5-10 becomes

$$\frac{2n_0 V}{(A_w f_c + A_s)} = \frac{\sqrt{\frac{kT_e}{M}}}{\langle \sigma_i^+ v_e \rangle}. \tag{5.5-11}$$

Equating the input power to the output power and solving for the absorbed power divided by the beam current, the discharge loss is then

$$\eta_d = \frac{P_{abs}}{I_b} = \frac{I_p}{I_b} U^+ + \frac{I^*}{I_b} U^* + \left(\frac{I_s}{I_b} + \frac{I_w}{I_b} + 1\right)\phi + \frac{I_a}{I_b}(2T_{eV} + \phi) \tag{5.5-12}$$

The ionization and excitation terms are now only due to the plasma electrons, so the first current fraction in Eq. 5.5-12, using Eq. 5.3-55, again neglecting any double ions, and assuming quasi-neutrality ($n_i \approx n_e$), is

$$\frac{I_p}{I_b} = \frac{2n_o\langle\sigma_i v_e\rangle V}{\sqrt{\frac{kT_e}{M}}A_s T_s}, \tag{5.5-13}$$

and the second current fraction is likewise:

$$\frac{I^*}{I_b} = \frac{2n_o\langle\sigma_* v_e\rangle V}{\sqrt{\frac{kT_e}{M}}A_s T_s}. \tag{5.5-14}$$

The screen current to beam current ratio from Eq. 5.3-62 as $(1-T_s)/T_s$.

The ion current that goes to the wall is the Bohm current to the wall area A_w reduced by a radial confinement factor that results from any applied or induced magnetic fields. The fourth current ratio is then

$$\frac{I_w}{I_b} = \frac{\frac{1}{2}n_i v_a A_w f_c}{\frac{1}{2}n_i v_a A_s T_s} = \frac{A_w f_c}{A_s T_s}, \tag{5.5-15}$$

where f_c is again a confinement factor for the reduction in the Bohm velocity due to ambipolar effects in the ion and electron flows to the wall due to any transverse magnetic fields. Since there are no applied DC potentials in the discharge chamber and all the walls float, the electron current out is the same as the total ion current out:

$$I_a = I_s + I_w + I_b. \tag{5.5-16}$$

The plasma potential in the expression for the discharge loss Eq. 5.5-11 can be evaluated by equating the total ion and electron currents exiting the plasma:

$$\frac{n_i}{2}\sqrt{\frac{kT_e}{M}}(A_w f_c + A_s) = \frac{n_e}{4}\sqrt{\frac{8kT_e}{\pi m}}[A_w + (1-T_s)A_s]e^{-e\phi/kT_e}. \tag{5.5-17}$$

Solving for the plasma potential gives

$$\phi = \frac{kT_e}{e}\ln\left(\frac{A_w + (1-T_s)A_s}{A_w f_c + A_s}\sqrt{\frac{2M}{\pi m}}\right). \tag{5.5-18}$$

If the wall area is large compared to the screen area, or the grid transparency is small compared to 1, this turns into the normal equation for floating potential:

$$\phi = \frac{kT_e}{e}\ln\left(\sqrt{\frac{2M}{\pi m}}\right), \tag{5.5-19}$$

which for xenon is $5.97{*}T_{eV}$.

Using Eqs. 5.5-12–5.5-16, the discharge loss for rf ion thrusters can then be written

$$\eta_d = \frac{2n_o\langle\sigma_i v_e\rangle V\left(U^+ + U^*\frac{\langle\sigma_* v_e\rangle}{\langle\sigma_i v_e\rangle}\right) + \left[\frac{1-T_s}{T_s} + \frac{A_w f_c}{A_s T_s} + 1\right](2T_{eV} + 2\phi)}{\sqrt{\frac{kT_e}{M}}A_s T_s}, \tag{5.5-20}$$

where the plasma potential ϕ is given by Eq. 5.5-18 in electron-volts (eV). The electron temperature is found by solving Eq. 5.5-10.

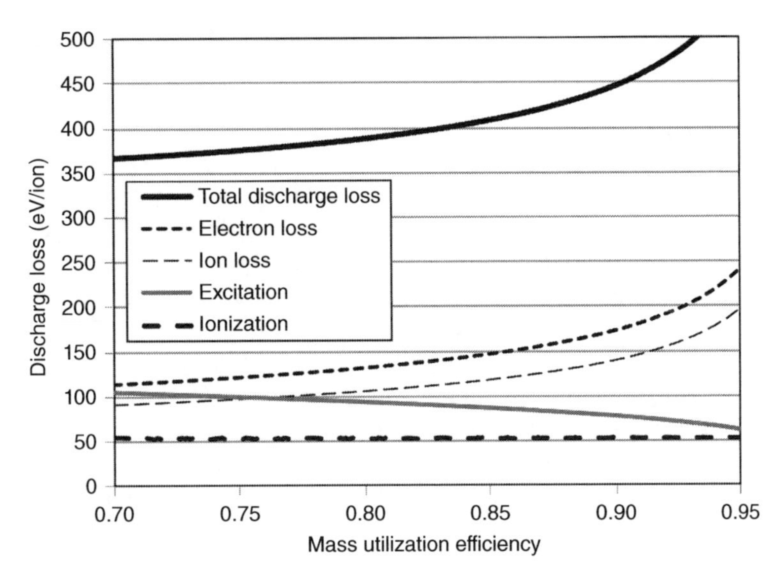

Figure 5-32 Discharge loss calculated for the example rf ion thruster and the contribution from the four energy loss channels.

As an example, assume that the rf ion thruster has a 20 cm grid diameter, an 18 cm deep conical ceramic discharge chamber, a grid transparency of 80%, and that it produces 2 A of beam current in xenon. Figure 5-32 shows the calculated discharge loss as function of the mass utilization efficiency from Eq. 5.5-20 assuming no applied or induced magnetic fields and, therefore, no plasma confinement. A discharge loss of about 450 eV/ion is predicted at 90% mass utilization efficiency. This is a very high discharge loss, and it can be seen in Fig. 5-32 that the majority of the energy loss is carried out by the ions and electrons flowing to the floating-potential walls. This is because the Maxwellian electron temperature required to produce the ions that flow to the entire interior surface area of the discharge chamber at 90% mass utilization efficiency (from the solution of Eq. 5.5-10 is 5 eV, and the plasma potential to achieve net ambipolar flow is, therefore, nearly 30 V. The high sheath potential required to self-confine the electrons for particle balance and the large plasma loss area $(A_w + A_s)$ carry significant energy to the discharge chamber wall, causing a high discharge loss.

The discharge loss performance of rf ion thrusters typically reported in the literature [51] is much lower than that found in our example. This is because even though these thrusters do not usually have an applied DC magnetic field, the rf coil forms a solenoid around the dielectric discharge chamber and the rf current flowing in the antenna coil induces an AC magnetic field in the interior of the discharge chamber with a frequency at the rf oscillator frequency. In most typical rf thrusters, this frequency is on the order of 1 MHz. The ion acoustic speed $\sqrt{kT_e/M}$ at $T_e = 5$ eV is 1.9 km/s, and so in a 1 μsec cycle, the ions can only move less than 2 mm, which implies that the ions can be considered stationary on the magnetic field cycle time. The electrons are certainly not stationary in the period, but the ion space charge will hold the electrons in place during a cycle. Therefore, the AC magnetic field from the rf coil can provide some confinement for the plasma and reduce the flux to the discharge chamber walls. The magnetic field induced by the rf coil depends on the coil size and amount of power [52]. For example, assume that the coil occupies 1 turn per centimeter (100 turns/m), and the coil impedance is 50 Ω. For an input power of 500 W, this would result in 10 A of rf current flowing in the coil. For simplicity, let's assume the rf coil forms a solenoid and the magnetic field inside a solenoid (neglecting end effects) is

$$B[\text{Gauss}] = 10^4 \mu_o N I, \tag{5.5-21}$$

where μ_0 is the permeability of free space $= 4\pi \times 10^{-7}$ Henries/m, N is the number of turns per meter and I is the coil current in Amperes. For this rf thruster example, a magnetic field of 12.6 G is produced. Although this sounds like a low field, it is an axial field induced in the majority of the interior of the thruster depending on the plasma skin depth, which is large in these low-density plasmas.

The reduction in the ion velocity flowing radially to the wall for the situation of a transverse magnetic field and ambipolar flows was analyzed in Section 5.3.4. Figure 5-33 shows the reduction in the radial Bohm current (f_c) from evaluating Eq. 5.3-16 for the condition where the diffusion length is now essentially the thruster radius. Fields on the order of 10 G throughout the thruster volume can reduce the ion and electron loss to the discharge chamber wall by over a factor of two. Although the rf magnetic field strength decreases with radius due to the finite length of the antenna coil (solenoid end effects), the field strength near the axis is still sufficient to reduce the ion loss rate [51].

The discharge loss calculated by the 0-D model for our 20 cm rf thruster example is shown in Fig. 5-34 as a function of rf magnetic field induced in the plasma. The discharge loss is reduced from the assumption of no confinement ($B = 0$) of 450 eV/ion at 90% mass utilization to a value of 230 eV/ion if 10 G is induced in the chamber. This is a significant reduction in the calculated loss and is the key to rf ion thruster discharge performance.

To produce the 2 A beam in our 20-cm thruster example at 230 eV/ion, a total input power to the antenna of 460 W is required to be absorbed by the plasma. Since the rf power supplies are typically 90% efficient in this frequency range, the input power to the thruster PPU would be about 511 W. This predicted performance is in good agreement with the data from this size rf thruster found in

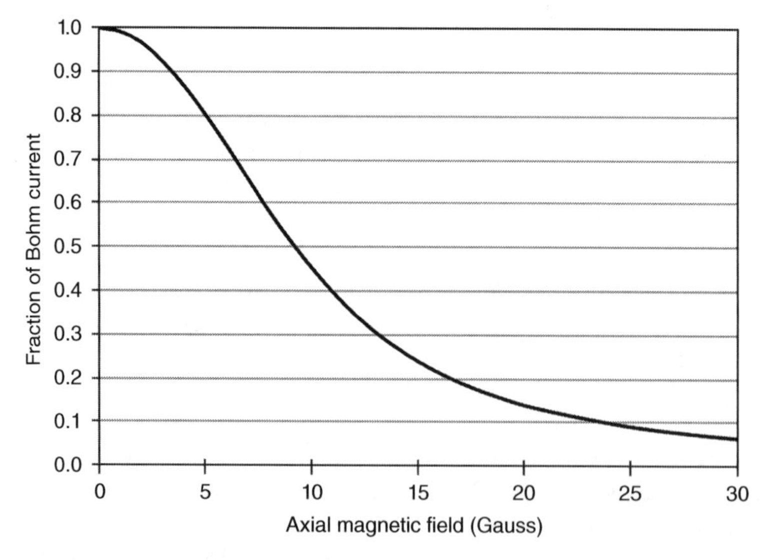

Figure 5-33 Ion confinement factor (the fraction of the Bohm current to the wall) as a function of the induced magnetic field in the discharge chamber volume.

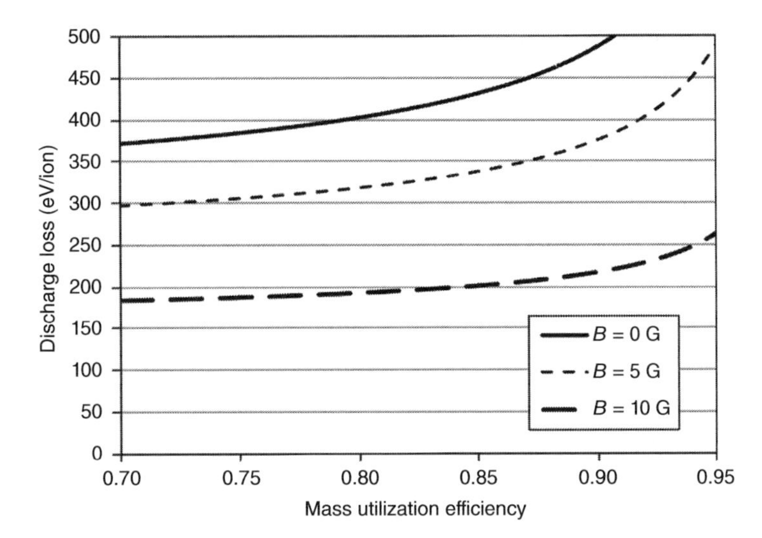

Figure 5-34 rf ion thruster discharge loss versus mass utilization efficiency for three values of the induced magnetic field in the discharge chamber.

the literature [51], suggesting that a 0-D particle and energy balance model can provide reasonably accurate performance predictions.

One advantage of rf ion thrusters is that they have only Maxwellian electrons and ambipolar ion and electron loss rates, which simplifies the discharge loss expressions and makes it easy to analyze the few geometric parameters to optimize the discharge loss. As an example of the process, first, specifying the required beam current and current density determines the grid diameter in any ion thruster. Ion optics codes then determine the grid transparency. Once the grid design is set, a Monte-Carlo gas code is used to evaluate the Clausing factor introduced in Eq. 5.3-26. Assuming a conical or cylindrical discharge chamber shape of a given length immediately specifies the loss areas and plasma volume. Then, specifying the mass utilization efficiency gives the neutral density, and the electron temperature can be found from Eq. 5.5-18 with an initial confinement factor assumption. These values are the input parameters to the discharge loss given by Eq. 5.5-20, which provides the required input rf power to the antenna assuming that the antenna efficiency and coupling (reflected power) are known. The approximate induced AC magnetic field can then be calculated from Eq. 5.5-21 and the ion confinement factor f_c found as in Section 5.3.4. A simple iteration then gives the final discharge loss and rf power.

Improvements in rf ion thruster discharge chamber models have been published in recent years. Authors have added coupled thermal models [53], heuristic plasma density profiles models [24, 54], and antenna coupling and neutral heating models [55] to 0-D models of the discharge chamber to improve the agreement with experimental data. The physics of radio-frequency discharges has also been detailed in a book [56]. Including thermal issues and antenna coupling increase the systems level predictions for this thruster type. Nevertheless, the 0-D model presented here provides good results for the discharge chamber behavior compared with the experimental data, and helps to explain the observed performance.

The scaling of rf ion thrusters to small sizes and lower power operation is problematic. As the discharge chamber length decreases, the antenna axial extent also decreases, which reduces the electric field interaction region and decreases the AC axial magnetic field strength due to end effects in the solenoid coil. The ability to breakdown the neutral gas initially and then to couple the rf energy efficiently to the electrons is compromised as the length decreases, which affects the discharge loss scaling.

Finally, a disadvantage of rf ion thrusters is that the antenna must be insulated from the plasma, and the insulator is then subject to ion bombardment and material deposition. Dielectric discharge chambers are susceptible to mechanical problems in fabrication, environmental testing and launch, and life issues from coating of the insulator surface with conducting layers. The structural issues have been addressed on some flight units by the use of a ceramic discharge chamber with an exterior mounted antenna structure to provide the rigidity required for launch survival. Although the discharge loss in rf ion thrusters is typically higher than that found for well-designed electron bombardment ion thrusters such that the total efficiency is lower, the simplified design of an rf thruster makes it easier to analyze them and predict the performance than most other ion thruster configurations. The rf thruster design concept eliminates any potential discharge cathode life issues and utilizes fewer power supplies to operate the discharge. These factors make rf ion thrusters very competitive for future space flight applications.

5.6 Microwave Ion Thrusters

An alternative to producing the plasma in the thruster with electron discharges or rf induction heating of the electron population is to generate the plasma using electromagnetic fields at microwave frequencies. This eliminates life issues associated with the discharge hollow cathode, and the lack of applied DC voltages in the discharge chamber can potentially reduce the sputter erosion of electrodes exposed to the plasma compared to DC electron discharges. However, electromagnetic waves can only propagate and be absorbed in plasmas under certain conditions. For example, if the microwave frequency is too high or the plasma density too low, the microwave radiation is reflected completely from the plasma (called "cutoff"). If the conditions are such that the microwaves do propagate in the plasma, the microwave energy is coupled to the plasma by resonant heating of the electrons in a magnetic field in the presence of collisions. The required magnetic field to achieve this resonance is significant, and the pressure required to achieve sufficient collisions to start the discharge can be relatively high. These effects impact the plasma generator design and performance.

The propagation of microwaves in a plasma can be understood by examining the dispersion relationship. The behavior of microwaves in the thruster plasma is described by Maxwell's equations:

$$\nabla \times \mathbf{E} = -\frac{\partial \mathbf{B}}{\partial t} \tag{5.6-1}$$

$$\nabla \times \mathbf{B} = \mu_o \left(\mathbf{J} + \varepsilon_0 \frac{\partial \mathbf{E}}{\partial t} \right) \tag{5.6-2}$$

The electromagnetic behavior is analyzed by linearizing these two equations using

$$\mathbf{E} = \mathbf{E}_o + \boldsymbol{E}_1 \tag{5.6-3}$$

$$\mathbf{B} = \mathbf{B}_o + \boldsymbol{B}_1 \tag{5.6-4}$$

$$\mathbf{J} = \mathbf{j}_o + \boldsymbol{j}_1, \tag{5.6-5}$$

where $\mathbf{E}_o$, $\mathbf{B}_o$ and $\mathbf{j}_o$ are the equilibrium values of the electric and magnetic fields and currents, and $\boldsymbol{E}_1$, $\boldsymbol{B}_1$ and $\boldsymbol{j}_1$ are the perturbed values in the electromagnetic fields and current. Linearizing Eqs. 5.6-1 and 5.6.2, realizing that the equilibrium values have no curl or time dependence, and that $\varepsilon_o\mu_o = 1/c^2$ in a vacuum, gives

$$\nabla \times \boldsymbol{E}_1 = -\frac{\partial \boldsymbol{B}_1}{\partial t} \tag{5.6-6}$$

$$c^2 \nabla \times \boldsymbol{B}_1 = \frac{\boldsymbol{j}_1}{\varepsilon_o} + \frac{\partial \boldsymbol{E}_1}{\partial t}. \tag{5.6-7}$$

Taking the curl of Eq. 5.6-6 gives

$$\nabla \times \nabla \times \boldsymbol{E}_1 = \nabla(\nabla \cdot \boldsymbol{E}_1) - \nabla^2 \boldsymbol{E}_1 = -\nabla \times \frac{\partial \boldsymbol{B}_1}{\partial t}, \tag{5.6-8}$$

and the time derivative of Eq. 5.6-7 gives

$$c^2 \nabla \times \frac{\partial \boldsymbol{B}_1}{\partial t} = \frac{1}{\varepsilon_o}\frac{\partial \boldsymbol{j}_1}{\partial t} + \frac{\partial^2 \boldsymbol{E}_1}{\partial t^2}. \tag{5.6-9}$$

Combining Eq. 5.6-9 with Eq. 5.6-8 results in

$$\nabla(\nabla \cdot \boldsymbol{E}_1) - \nabla^2 \boldsymbol{E}_1 = -\frac{1}{\varepsilon_o c^2}\frac{\partial \boldsymbol{j}_1}{\partial t} - \frac{1}{c^2}\frac{\partial^2 \boldsymbol{E}_1}{\partial t^2}. \tag{5.6-10}$$

Assuming that the microwaves are plane waves that vary as

$$\mathbf{E} = E_o e^{i(kx - \omega t)} \tag{5.6-11}$$

$$\mathbf{j} = j_o e^{i(kx - \omega t)}, \tag{5.6-12}$$

where $k = 2\pi/\lambda$ and ω is the cyclic frequency $2\pi f$, then Eq. 5.6-10 becomes

$$-k(k \cdot \boldsymbol{E}_1) + k^2 \boldsymbol{E}_1 = \frac{i\omega}{\varepsilon_o c^2}\boldsymbol{j}_1 + \frac{\omega^2}{c^2}\boldsymbol{E}_1. \tag{5.6-13}$$

The electromagnetic waves are transverse waves, so $k \cdot \boldsymbol{E}_1 = 0$ and Eq. 5.6-13 becomes

$$\left(\omega^2 - c^2 k^2\right)\boldsymbol{E}_1 = \frac{-i\omega}{\varepsilon_o}\boldsymbol{j}_1. \tag{5.6-14}$$

Since these waves are in the microwave frequency range, the ions are too massive to move on these fast time scales and the perturbed current $\boldsymbol{j}_1$ can only come from electron motion. The perturbed electron current density in a plasma is

$$\boldsymbol{j}_1 = -n_e e \boldsymbol{v}_{e1}, \tag{5.6-15}$$

where n_e is the plasma density and $\boldsymbol{v}_{e1}$ is the perturbed electron velocity. If the applied magnetic field is zero or the perturbed electric field is parallel to the applied magnetic field (so called "O-waves"), the equation of motion for the perturbed electron motion is

$$m\frac{\partial \boldsymbol{v}_{e1}}{\partial t} = -e\boldsymbol{E}_1. \tag{5.6-16}$$

Solving for the perturbed electron velocity, assuming plane waves, and inserting this into Eq. 5.6-15, the perturbed current is

$$\boldsymbol{j}_1 = -n_e e \frac{\varepsilon_o \boldsymbol{E}_1}{i\omega m}. \tag{5.6-17}$$

Inserting Eq. 5.6-17 into Eq. 5.6-14 and solving for the frequency gives the dispersion relation for electromagnetic waves in a plasma:

$$\omega^2 = \frac{n_e e^2}{\epsilon_o m} + c^2 k^2 = \omega_p^2 + c^2 k^2, \tag{5.6-18}$$

where the definition of the electron plasma frequency $\omega_p{}^2 = n_e e^2/\varepsilon_o m$ has been used.

This expression can be solved for the wavelength of the microwaves in the plasma

$$\lambda = \frac{2\pi c}{\sqrt{\omega_p^2 - \omega^2}} = \frac{c}{\sqrt{f_p^2 - f^2}}, \tag{5.6-19}$$

where f_p is the real plasma frequency and f is the microwave frequency. If the microwave frequency exceeds the plasma electron frequency, the wavelength becomes infinitely long and the wave becomes evanescent (it will not propagate into the plasma) and is reflected. This condition, called cutoff, determines the maximum plasma density into which a microwave source can inject power to produce the plasma. Table 5-1 shows the cutoff frequency for a range of plasma densities and the ion current density from a xenon plasma at an electron temperature of 3 eV. As an example, if the ion thruster design requires an ion current density to the grids of say 1.2 mA/cm^2, then a frequency in excess of 2.85 GHz must be used to produce the plasma or else some or all the microwave power will be reflected.

The microwave energy is coupled to the plasma by electron cyclotron resonance heating, where the microwave frequency corresponds to the cyclic frequency of the electrons in a magnetic field. The resonant frequency is the electron cyclotron frequency, which was derived in Chapter 3:

$$\omega_c = \frac{|q|B}{m}. \tag{5.6-20}$$

The cyclotron frequency is easily calculated using a convenient formula of $f_c = eB/2\pi m = 2.8$ GHz/kG. In the plasma, the microwave frequency is given in Table 5-2 for several magnetic field values. If it is assumed that the microwave energy is deposited into the volume of plasma immersed in the magnet field, the maximum plasma density (and corresponding ion current density to the grids) to avoid cutoff is shown for each of the magnetic field values. To produce current densities in excess of 1 mA/cm^2 of xenon to the accelerator grids from a 3 eV electron temperature plasma requires magnetic fields in excess of 1000 G, and values closer to 2000 G are required to avoid cutoff for slightly higher ion current densities to the grids. This is a significant magnetic field to produce over the discharge chamber volume.

Table 5-1 Cutoff frequencies for several plasma densities, and the corresponding ion current density from a xenon plasma with T_e = 3 eV.

Plasma density (m^{-3})	Cutoff frequency (GHz)	J (mA/cm^2)
10^{15}	0.285	0.0118
10^{16}	0.900	0.118
10^{17}	2.846	1.184
10^{18}	9.000	11.84
10^{19}	28.460	118.4

Table 5-2 Electron cyclotron frequencies for several magnetic field levels, the corresponding maximum plasma density before cutoff, and the maximum ion current density to the grids from a 3 eV electron temperature xenon plasma.

Magnetic field (G)	Cyclotron frequency f_c (GHz)	Maximum plasma density (m^{-3})	Maximum ion current density (mA/cm^2)
100	0.28	9.68×10^{14}	0.012
500	1.40	2.42×10^{16}	0.286
1000	2.80	9.68×10^{16}	1.146
2000	5.60	3.87×10^{17}	4.58
3000	8.40	8.71×10^{17}	10.31
4000	11.20	1.55×10^{18}	18.34

The use of microwave radiation enables direct heating of the plasma electrons, but for the wave to add energy to the electrons, collisions must occur. Otherwise, the energy received by an electron during acceleration on each half-cycle of their cyclotron motion is taken back by deceleration of the electron in the field on the next half-cycle. Therefore, there is a minimum pressure at which sufficient collisions occur to ignite the plasma and sustain the discharge. The probability of a collision occurring is

$$P = [1 - e^{-n_o \sigma x}] = \left[1 - e^{(-x/\lambda_{en})}\right], \tag{5.6-21}$$

where x is the path length of the electron in the neutral gas with a density of n_o, and λ_{en} is the electron-neutral collision mean-free-path. An electron entering the interaction region gyrates around the magnetic field lines due to its perpendicular velocity and travels along the magnetic field line due to its parallel velocity.

Although the electron cyclotron heating tends to spin-up the electron motion around the field lines, collisions tend to scatter the motion along the direction of the field lines and thermalize the electrons into a Maxwellian distribution, sometimes with a high-energy bump or tail driven by the resonance. The collisionality requirements to achieve heating can be found from examining the path length of an electron at a temperature T_e spiraling along a field line. The distance that the electron travels when gyrating around the field lines is given by the Larmor radius of Eq. 3.3-13:

$$r_L = \frac{v_\perp}{\omega_c} = \frac{m v_\perp}{qB} = \frac{1}{B}\sqrt{\frac{2mV_\perp}{e}}. \tag{5.6-22}$$

The time for an electron to leave the microwave interaction region of length L is

$$t = \frac{L}{v_\parallel}, \tag{5.6-23}$$

where $v_\parallel$ is the parallel electron velocity along the field line. The number N of gyrations that an electron makes in the interaction region is the microwave frequency f multiplied by the time in the resonant region. The path length of the perpendicular gyration of the electron is then

$$L_g = 2\pi\, r_L N = 2\pi\, r_L\, f \frac{L}{v_\parallel}. \tag{5.6-24}$$

The total path length of the helical motion of the electron is

$$L_T = \sqrt{L_g^2 + L^2} = \sqrt{\left(\frac{2\pi r_L f L}{v_{||}}\right)^2 + L^2}. \tag{5.6-25}$$

Using this value for the path length x of the electron in Eq. 5.6-21 gives the probability of a collision with the neutral gas. Figure 5-35 shows this probability calculated for xenon gas at room temperature for electrons with a temperature of 2 eV in two different interaction lengths. To achieve the order of 10% of the electrons colliding with neutral gas atoms in a 5–10-cm-long resonance region requires an internal pressure of at least 10^{-3} Torr. In reality the electrons must make multiple collisions within the interaction region because the energy gain in a single gyration is small. However, this pressure is similar to that found for rf thrusters to achieve sufficient collisions to start or sustain a discharge, for essentially the same reasons. Again, once the plasma is started, Coulomb collisions will aid in transferring the electron motion in the microwave fields into heating, which reduces the pressure required to operate the plasma generator and permits higher mass utilization efficiencies to be achieved.

As was shown in Tables 5-1 and 5-2, a high magnetic field (>1 kG) and a high microwave frequency (>2.8 GHz) are required to produce sufficient plasma density to deliver reasonable current densities (>1 mA/cm^2 in xenon) to the grids in microwave thrusters. Owing to the difficulty in producing these high magnetic fields throughout the discharge chamber volume, the resonance region is often localized to a small zone inside the thruster volume and the plasma allowed to expand to the grids along divergent magnetic field lines. Figure 5-36 shows an ECR plasma source where a stronger magnetic field region resonant with 2.45 GHz radiation (produced by commercial magnetron microwave sources) is restricted to the rear of the discharge chamber. Of course, expanding the plasma from the resonance region to the grids decreases the plasma density and current density, so even higher magnetic fields and frequencies than just mentioned are normally required in the interaction region to produce over 1 mA/cm^2 to the grids.

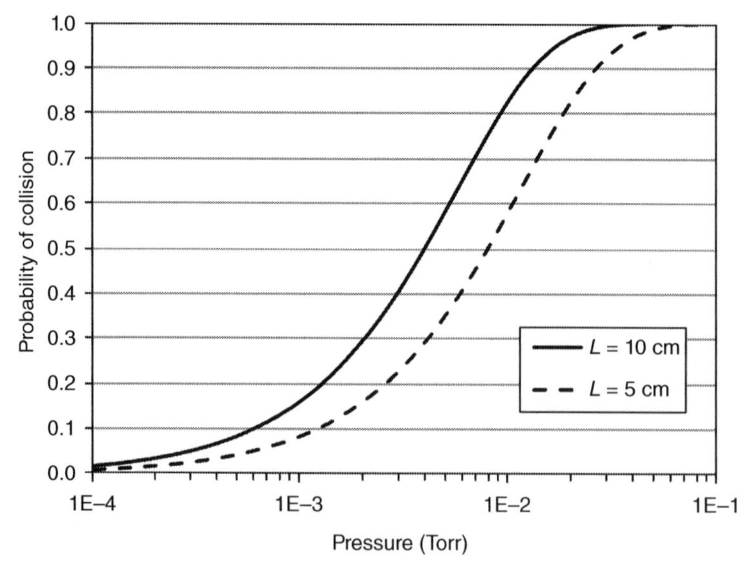

Figure 5-35 Probability of an electron-neutral collision before leaving the resonance zone length indicated as a function of the neutral pressure for 2-eV electrons.

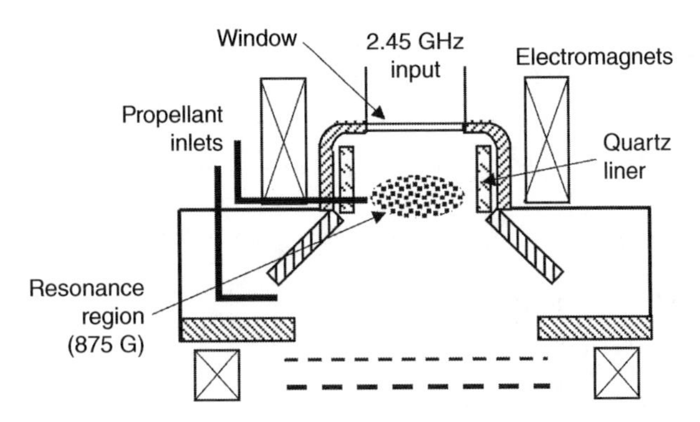

Figure 5-36 Schematic of microwave ion source with a volume-resonance zone of strong magnetic field produced by electromagnets.

The microwave radiation in this ECR plasma source is coupled into the rear of the discharge chamber through a waveguide window, and a quartz liner is used in the resonant region to ensure that the hot electrons are not lost directly to the metal walls of the chamber. The magnetic field in this geometry is produced by electromagnets, with a strong divergence in the field to spread the plasma over the grid region at the exit of the discharge chamber. This is a common geometry for industrial ion sources and plasma sources used in plasma processing, and the performance of the plasma generator is well known.

The performance of this style microwave ion thruster can be examined with a 0-D model. Assume that the magnetic field is sufficiently strong enough that radial losses can be neglected. This assumption implies that the plasma is frozen on the field lines such that the density decreases linearly with the area increase as the field expands. This simplifies the model to the case of a straight cylindrical source with no radial losses. The plasma is lost axially to both the screen area A_s and the rear wall area A_w. Since there is no DC applied electric field, the plasma floats relative to the internal surfaces, the electrons are lost to the axial rear-wall area and the collection area of the screen grid given by $(1 - T_s)A_s$. Neglecting the cost of producing the microwave radiation, the power absorbed by the plasma is equal to the power lost:

$$P_{\text{abs}} = I_p U^+ + I^* U^* + (I_s + I_w + I_b)\phi + I_a(2T_{\text{eV}} + \phi), \tag{5.6-26}$$

where I_s is the ion current collected by the screen grid, I_w is the ion current collected by the entire wall, and the ion energy loss is again just the plasma potential relative to the wall ϕ. The amount of energy lost by electrons to the wall assumes that the electrons have a Maxwellian distribution, which may underestimate the energy lost due to the high energy tail in the electron distribution generated by the resonant ECR heating. The discharge loss is the power in (or out) divided by the beam current:

$$\eta_d = \frac{P_{\text{abs}}}{I_b} = \frac{I_p}{I_b}U^+ + \frac{I^*}{I_b}U^* + \left(\frac{I_s}{I_b} + \frac{I_w}{I_b} + 1\right)\phi + \frac{I_a}{I_b}(2T_{\text{eV}} + \phi). \tag{5.6-27}$$

The first three current fractions in this equation are given by Eqs. 5.3-60, 5.3-61, 5.3-62, respectively. The fourth current fraction is given by

$$\frac{I_w}{I_b} = \frac{\frac{1}{2}n_i v_a A_w}{\frac{1}{2}n_i e\, v_a A_s T_s} = \frac{A_w}{A_s T_s}, \tag{5.6-28}$$

where the wall area A_w is the rear wall area only. The plasma potential is found again from charge conservation by equating the total ion and electron current:

$$\frac{n_i e}{2}\sqrt{\frac{kT_e}{M}}(A_w + A_s) = \frac{n_e e}{4}\sqrt{\frac{8kT_e}{\pi m}}[A_w + (1 - T_s)A_s]e^{-e\phi/kT_e}. \tag{5.6-29}$$

Assuming quasi-neutrality and solving for the plasma potential gives

$$\phi = \frac{kT_e}{e}\ln\left(\frac{A_w + (1 - T_s)A_s}{A_w + A_s}\sqrt{\frac{2M}{\pi m}}\right), \tag{5.6-30}$$

which is different than that found for rf ion thrusters because there is no ion confinement factor due to the induced magnetic fields from the antenna (the ions are assumed perfectly confined radially due to the strong magnetic field). The electrons are lost to the rear wall and the screen grid, so the final current fraction in Eq. 5.6-27 is

$$\frac{I_a}{I_b} = \frac{\frac{1}{4}\sqrt{\frac{8kT_e}{\pi m}}n_e e\,[A_w + (1 - T_s)A_s]}{\frac{1}{2}n_i e\sqrt{\frac{kT_e}{M}}A_s T_s}e^{-e\phi/kT_e}. \tag{5.6-31}$$

Using Eq. 5.6-30 for the plasma potential, this becomes

$$\frac{I_a}{I_b} = \frac{A_w + A_s}{A_s T_s}. \tag{5.6-32}$$

The discharge loss is then

$$\eta_d = \frac{2n_o\langle\sigma_i v_p\rangle V}{\sqrt{\frac{kT_e}{M}}A_s T_s}\left(U^+ + U^*\frac{\langle\sigma_* v_e\rangle}{\langle\sigma_i v_e\rangle}\right) + \left[\frac{1 - T_s}{T_s} + \frac{A_w}{A_s T_s} + 1\right]\phi + \frac{A_w + A_s}{A_s T_s}(2T_{\text{eV}} + \phi), \tag{5.6-33}$$

with the plasma potential given by Eq. 5.6-30. The electron temperature and neutral density are solved in the same manner as previously for the other types of thrusters. The discharge loss for a generic microwave ion thruster producing 1 A of xenon ions from a 20 cm diameter grid with 80% transparency is shown in Fig. 5-37 for several thruster lengths. Discharge losses on the order of 200 eV/ion are predicted. This discharge loss is comparable to that of our idealized ion thruster in Section 5.2 for short thruster lengths, but nearly twice that found for comparable long ideal thrusters. This is because both the ideal and the microwave source cases assumed ionization by Maxwellian electrons and perfect radial confinement, but the microwave source case includes plasma loss to the rear wall. Although the assumption of negligible radial loss is reasonable in microwave thrusters due to the strong magnetic fields, some additional loss is expected in this direction that will degrade the actual discharge loss somewhat. The large loss area for plasma to the beam area and rear wall tend to drive up the plasma potential to maintain net ambipolar flows and charge balance, which increases the discharge loss compared to well-designed DC discharge thrusters.

Microwave ion source designers mitigate the backwall losses by imposing a stronger magnetic field upstream of the resonance zone. This creates a magnetic mirror, which was described in Section 3.3, that confines the plasma electrons and reduces the axial losses. Because the magnetic

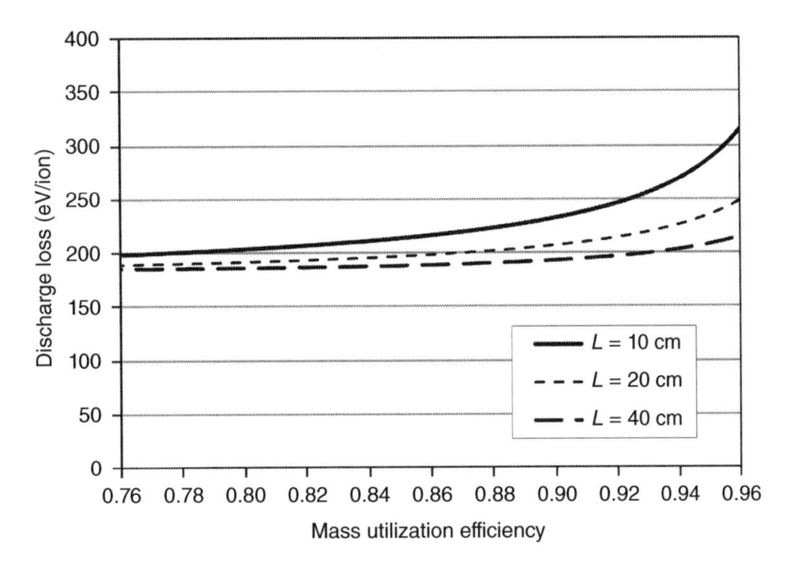

Figure 5-37 Discharge loss versus mass utilization efficiency for our microwave thruster example with perfect radial confinement.

moment (defined in Eq. 3.3-22 as $mv_{\perp}^2/2B$) is invariant along the magnetic field lines, electrons with sufficient initial perpendicular velocity are reflected from the increasing magnetic field as their parallel energy is converted into rotational energy. The electrons that are lost along the magnetic field lines have a parallel velocity of

$$v_{\|} > v_{\perp}\sqrt{R_m - 1}, \tag{5.6-34}$$

where R_m is the mirror ratio given by B_{max}/B_m. For example, if the mirror ratio is 5, only electrons with a parallel velocity twice that of their perpendicular velocity will be lost. If the electrons have a Maxwellian distribution with a temperature T_e, then the number of particles with $v_{\|} > 2v_{\perp}$ is $e^{-2} = 13.5\%$, so a large majority of the population is reflected. Since the cyclotron heating adds perpendicular energy to the electrons, mirror ratios of four to six are very efficient in confining the heated electrons that produce ionization.

The ion source shown in Fig. 5-36 utilizes electromagnets to produce the high field over a significant volume and also to create the confining mirror ratio. However, the power required to operate the electromagnets in this design increases the effective discharge loss and limits the electrical efficiency of the device in thruster applications. In addition, it is difficult to create large area plasmas with good uniformity using microwave excitation due to the strong magnetic fields that confine the plasma and influence the profile. This leads to other magnetic configurations to produce the plasma using microwave ECR techniques.

In a volume-ionization ECR sources, like that shown in Fig. 5-36, a significant fraction of the discharge chamber must be filled with a strong magnetic field to satisfy the resonance condition. If this field is produced by a solenoid the electrical power is likely a significant penalty for the thruster efficiency. Likewise, if the field is produced by permanent magnets, the weight of the magnetic material required to produce this field can represent a significant weight penalty for the thruster. This problem can be mitigated by using magnetic multipole boundaries that produce

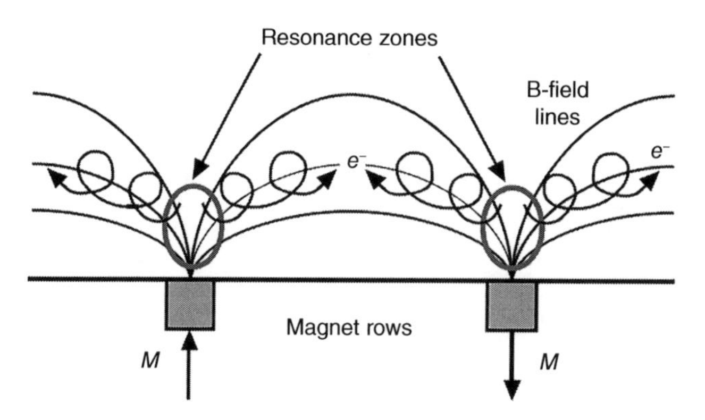

Figure 5-38 Magnetic field lines and electron cyclotron resonant zone in a ring-cusp boundary microwave ion source.

strong magnetic fields at the discharge chamber wall using ring or line-cusp magnet configurations. Figure 5-38 shows the field lines between two magnet rings and the regions of strong magnetic field close to the magnet where the resonant condition is satisfied. Injection of the microwave radiation between the cusps, either by cutoff-waveguides inserted between the rows [57], by slotted waveguides run along the rows [58], or by antenna structures placed between the rows, will couple the microwaves to the high magnetic field interaction region.

Although this geometry eliminates the solenoidal magnet coils and minimizes the size of the permanent magnets required to produce the resonant field strength, there are several issues remaining. First, the magnetic field strength in the cusp region decreases as one over the distance from the surface squared. This means that very strong magnets are required to produce the resonant field at any significant distance from the wall. Second, electrons that gain energy from the microwaves can be easily lost along the field lines to the wall due to their finite parallel velocity. This means that optimal ECR designs using permanent multipole magnets will have the resonance region as far from the wall as possible, and produce a large mirror ratio approaching the wall to reflect the electrons to avoid excessive direct loss.

Nevertheless, wall losses are a concern in this configuration because the plasma production is a surface effect that is confined to the boundary region, as is the loss. Electrons that are heated in the resonance zone sufficiently to ionize the propellant gas generate plasma on the near-surface magnetic field lines. Coupling the plasma from the resonance region or the surface magnetic layer into the volume of the thruster is problematic due to the reduced cross-field transport. In the other thruster designs discussed in this chapter, the ion production was a volume effect and convective loss a surface effect, so thruster efficiency scaled as the volume to surface ratio. This means that larger DC and rf discharge thrusters can be made more efficient than smaller ones. Microwave thrusters, on the other hand, don't scale in the same manner with size because large amounts of plasma must be produced and transported from the surface region to fill the volume of larger thrusters, which can impact the discharge loss. In addition, the plasma density is limited by both cutoff and the magnitude of the resonant field, and so high current density ion production requires very high magnetic fields and high microwave frequencies. Therefore, microwave thrusters have been limited to date to lower current densities and smaller sizes than the other thrusters discussed here. However, work continues on scaling microwave thrusters to larger sizes and higher efficiencies.

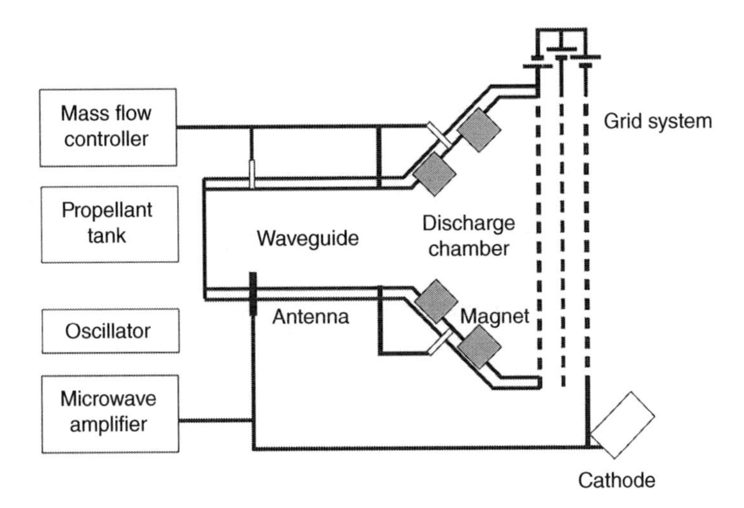

Figure 5-39 Schematic of the μ10 microwave ion source showing the strong magnets and small volume characteristic of these thrusters (*Source:* [61]).

The most successful design of a microwave thruster to date is the μ10 ECR 10-cm microwave thruster [58–60]. The μ10 ion thruster, shown schematically in Fig. 5-39 with the microwave power and flow control systems, is an improved version [61–63] where the thrust and life were increased over the original version that flew in space on the Hayabusa Mission. In this thruster, extremely strong SmCo magnets are used to close the resonance field at the operating frequency between the magnets. This produces heating away from the wall and traps the electrons on the field lines due to a mirror ratio of two to three achievable in this geometry. The thruster volume is also minimized, with the plasma production region close to the grids. This configuration produces over 1 mA/cm^2 of xenon ions over the active grid region using a 4.2 GHz microwave source with a discharge loss of about 300 eV per ion at over 85% mass utilization efficiency [58].

Finally, there are several other components intrinsic to these thrusters that contribute to the difficulty of achieving high efficiency and compact size in a microwave thruster subsystem. Sources of microwave frequencies in the gigahertz range, such as traveling wave tubes (TWT) and magnetrons, have efficiencies in the 50–70% range and the power supply to run them is usually about 90% efficient. This represents nearly a factor of two in-line loss of the electrical power delivered to the thruster that must be account for in the total discharge cost of the sub-system. The plasma is typically a difficult load to match well, and reflection of 10–30% of the microwave energy back into the recirculator (which absorbs the reflected power from the thruster in the case of mismatch or faults) is typical. The microwave source and recirculator usually represent a significant mass and volume addition to the ion thruster system. To avoid cutoff and produce ion current densities to the grids of 1–2 mA/cm^2, from an examination of Table 5-1, requires microwave sources in the 4–6 GHz range. At this time, space TWTs in this frequency range are limited in power capability to the order of a few hundred watts. For a given discharge loss, this limits the total ion current that can be produced by a microwave thruster. Although microwave thrusters hold the promise of eliminating the need for thermionic cathodes used in DC-discharge thrusters, and doing away with the requirement for dielectric discharge chambers in rf thrusters, producing high efficiency, high thrust ion propulsion systems based on this technology can be challenging. This is certainly an area for continued future research.

5.7 2-D Models of the Ion Thruster Discharge Chamber

The analytical models described above can generally explain the behavior and predict the overall discharge chamber performance of well-defined configurations, but multi-dimensional computer models are required to predict thruster performance parameters such as plasma profile and double ion content, and to examine the details of different designs. Multi-dimensional modelling of the discharge chamber requires detailed models of discharge chamber walls and magnetic fields as well as neutral propellant gas, ions, and primary and secondary plasma electrons [64–68]. Because the important physical mechanisms are different, each species (neutral gas, ions, and primary and secondary electrons) is modeled differently. For example, most neutral gas atoms travel in straight lines until they hit a wall or are ionized, so the neutral models can take advantage of simple straight-line trajectories to develop neutral density profiles. On the other hand, primary electron trajectories are dominated by rotation around magnetic field lines, and typically particle-tracking techniques are used to determine the density and spatial distributions. Ion and secondary electron behavior is obtained using fluid equations due to the relatively collisional behavior of the species. Therefore, ion thruster discharge models that require computer codes that use both fluid and particle tracking models are known as "Hybrid" codes.

Figure 5-40 shows a generic flow diagram for an ion thruster hybrid model [66]. Using the thruster inputs (geometry), a mesh is generated inside the discharge chamber. A magnetic field solver determines the field everywhere in the chamber. Depending on the type of mesh used, the mesh-generator may be iterated with the magnetic field solver to align the mesh points with the magnetic field lines. Aligning the magnetic field line simplifies the plasma diffusion

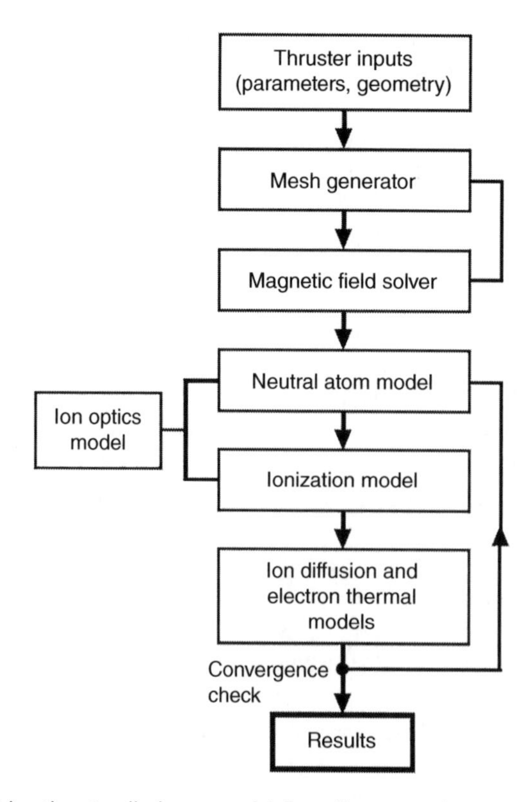

Figure 5-40 Hybrid 2-D ion thruster discharge model flow diagram and components overview.

calculations since the equations can be separated into parallel and perpendicular components, which can result in improved code accuracy for a sufficiently fine mesh. A neutral gas model, such as the "view-factor" model described below, determines the neutral density throughout the volume. The "ionization model" uses the magnetic field and electric field to compute the trajectories of primary electrons and their collisions with other plasma components (i.e. neutrals, ions, secondary electrons), which create ions and serve to dissipate the primary electron energy. The ionization model also determines the collisions due to secondary electrons. The ion optics model determines the transparency of the ion optics to neutrals and ions, as described in detail in Chapter 6. The ion diffusion model uses the magnetic field information and plasma properties to determine the motion of the plasma. The electron thermal model determines the energy balance for the electrons to find the distribution of temperatures of the secondary electron population. These processes are iterated until a convergent solution is found.

5.7.1 Neutral Atom Model

Accurate knowledge of the neutral gas is required in multi-dimensional plasma codes to predict the beam profiles, details of discharge plasma behavior, and thruster performance. For example, many thrusters utilize localized sources and sinks of the neutral gas that produce non-uniform neutral density profiles that must be considered to understand performance.

Ion thrusters operate at internal pressures on the order of 1×10^{-4} Torr or lower to achieve good mass utilization efficiency. In this pressure range, the neutral gas can be considered to be collisionless, and simple Knudsen-flow models are normally used to determine the average neutral gas density inside the thruster. Assuming surface adsorption, propellant atoms collide with the chamber walls and are re-emitted with a cosine distribution at the wall temperature. Collisions with the wall act to thermalize the gas to the wall temperature. Inside the discharge volume, the neutral atoms collide with electrons and ions. Some neutral atoms are "heated" by charge exchange that transfers the local ion energy to the neutral, but this process has little effect on the average gas temperature in the thruster discharge chamber. The spatial distribution of the neutral density is dependent on the gas injection regions (sources), gas reflux from the walls, loss of gas out the ion optic apertures, and the internal "loss" of neutral particles by ionization.

Wirz and Katz [66] developed a technique that accurately predicts the neutral gas density profiles in ion thrusters. Their model utilizes a 3-D generalization of the view factor formulation used in thermal models [69]. The view factor approach assumes that neutral particles travel in straight lines between surfaces, and, after hitting a surface, they are emitted isotopically. In this technique [70], a 3-D boundary mesh and a 2-D internal mesh in the thruster discharge chamber are created for an axisymmetric discharge. The steady state neutral fluxes are determined by balancing the injection sources, reemission from the walls, loss through the ion optics, and loss due to ionization. The local neutral density at each of the internal mesh points is calculated by integrating its view factor from the source points (all the other mesh points in the thruster), which includes the "loss" of neutrals between the source and the mesh point due to ionization by the plasma. The ionization losses affect the neutral gas analogous to absorption diminishing the intensity of a light ray. The neutral gas code and the rest of the model components, discussed below, are iterated until a stable solution for the neutral density at each mesh point is found. One advantage of this model is that the neutral gas temperature can be tracked after the gas interacts with the wall temperatures specified at the boundary mesh points. Also this technique is much faster than a Monte Carlo code since it requires a single matrix solution, allowing the coupling of the neutral and plasma codes to quickly determine both neutral and plasma density profiles.

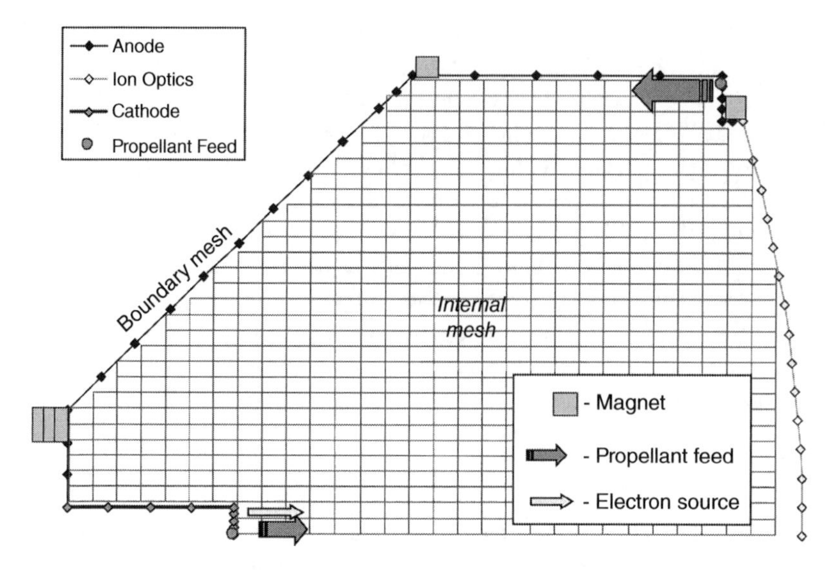

Figure 5-41 Rectangular internal mesh in an ion thruster (*Source:* [66]).

An example of the axisymmetric boundary ("wall") and internal meshes for the NSTAR ion thruster from Wirz and Katz [66] is shown in Fig. 5-41. Gas enters from the hollow cathode at the center rear and the propellant injection manifold at the front corner of the discharge chamber. The neutral gas density calculated from this code for the NSTAR thruster in its high power TH15 mode is shown in Fig. 5-42. The neutral density is highest near the injection sources at the hollow cathode and the propellant injection manifold. The neutral gas is the lowest on axis near the grids due to the NSTAR feed arrangement; however, as discussed below, the high primary electron

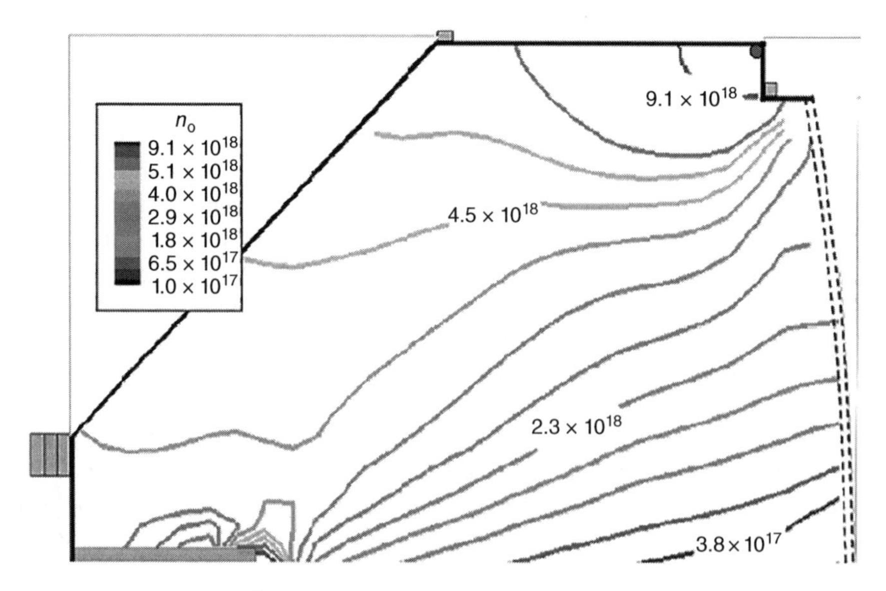

Figure 5-42 Neutral density (m^{-3}) for NSTAR throttle level TH15 (*Source:* [66]).

density found in this region of the thruster produces significant ionization, and "burns-out" the neutral gas. This result is critically important because the production of doubly-ionized atoms increases dramatically in regions where the neutral gas is burned out and most of the electron energy goes into secondary ionization of the ions in the discharge chamber.

5.7.2 Primary Electron Motion and Ionization Model

Particle simulation methods have been applied to model primary electron motion in ion thruster discharge chambers [66–68, 71, 72]. In particle simulations, the primary electrons are represented by particles, or macro-particles that represent a large number of primary electrons, that move in discrete time steps based on their initial conditions, applied boundary conditions, and internal electric and magnetic fields. Monte Carlo techniques are used to introduce the particles from the cathode exit into the computational domain at randomized velocities indicative of the cathode emission characteristics. During each time step, the local fields are recalculated based on the new particle position and velocity, and the particles move based on the local forces. Monte Carlo techniques are also typically used to handle collisions between the particles. This procedure is repeated through many time steps until the particle is lost; after which the next particle is introduced at a unique initial velocity condition.

The primary electron motion between collisions is treated as the motion of a charged particle in the presence of an electromagnetic field, which is described by the Lorentz equation

$$m\frac{\partial \mathbf{v}}{\partial t} = q(\mathbf{E} + \mathbf{v} \times \mathbf{B}). \tag{5.7-1}$$

Wirz and Katz [66] developed an improved Boris-type particle pushing algorithm [73] where the motion of the particles can be described with an implicit particle-pushing algorithm where the Lorentz forces on the particle are decomposed into electric and magnetic forces. The primary's kinetic energy is assumed to be unchanged in an elastic collision, and the particle scattering angle is estimated by a 3-D probabilistic hard sphere scattering model [66]. In an inelastic collision, some fraction of the primary energy goes into excitation or ionization of the neutrals. Additional energy loss paths exist, as previously discussed, such as Coulomb collision thermalization and anomalous processes associated with instabilities. A typical primary trajectory in the NSTAR thruster from the Wirz code is shown in Fig. 5-43, where the primaries are well confined by the strong axial magnetic field component in this thruster, and collisional effects eventually scatter the primary in to the cusp loss cone. Arakawa and Yamada's model for primary electron motion is derived from the Euler–Lagrange equations for the Lagrangian of a charge particle in a magnetic field [71]. However, this technique is computationally more intensive and does not improve the results in comparison to the improved Boris algorithm.

The primary electron density calculated by Wirz for the TH15 operating condition is shown in Fig. 5-44 and reveals that the magnetic field configuration of NSTAR tends to trap the primary electrons from the cathode on the thruster axis. This trapping of primary electrons, combined with the low neutral density on-axis, causes a relatively high rate of production of double ions along the thruster axis.

The ion and secondary electron transport may be treated by an ambipolar ion diffusion equation derived from the single ion and electron continuity and momentum equations. The steady-state continuity equation for ions is

$$\frac{\partial n}{\partial t} + \nabla \cdot (n\mathbf{v}) = \dot{n}_s, \tag{5.7-2}$$

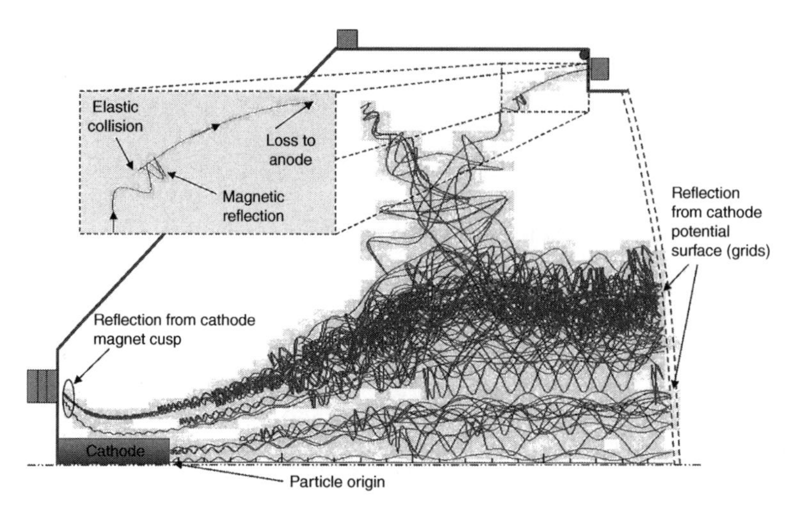

Figure 5-43 Example primary electron trajectory calculated inside the NSTAR discharge chamber (*Source:* [66]).

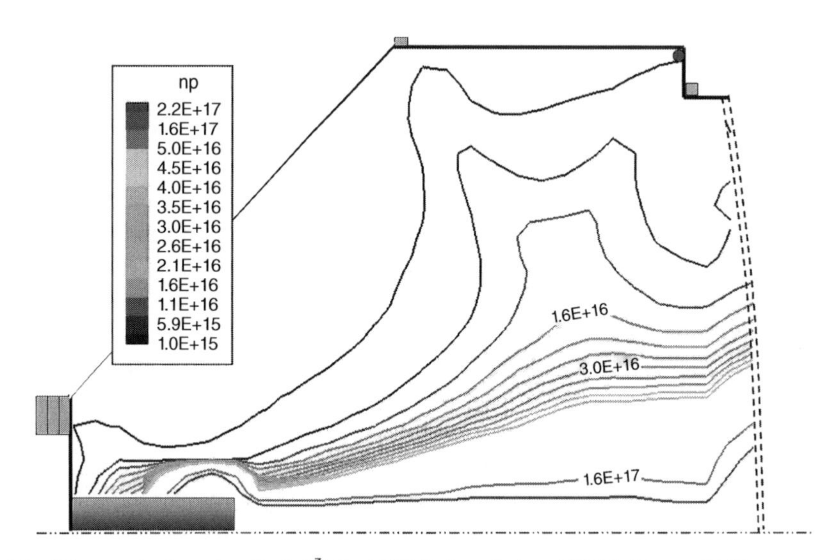

Figure 5-44 Primary electron density (m^{-3}) for NSTAR throttle level TH15 (*Source:* [66]).

where n_s is the ion source term. The momentum equation for ions and electrons is

$$m\left[\frac{\partial(n\mathbf{v})}{\partial t} + \nabla \cdot (n\mathbf{v}\mathbf{v})\right] = nq(\mathbf{E} + \mathbf{v} \times \mathbf{B}) - \nabla \cdot \mathbf{p} - nm\sum\nolimits_n \langle n_n \rangle (\mathbf{v} - \mathbf{v}_n), \tag{5.7-3}$$

where the subscript 'n' represents the other species in the plasma. Equations 5.7-2 and 5.7-3 can be combined to create a plasma diffusion equation

$$-\mathbf{D}_a \nabla^2 n = \dot{n}_s, \tag{5.7-4}$$

where $\mathbf{D_a}$ is the ambipolar diffusion coefficient. The diffusion coefficient is separated into parallel and perpendicular components, such that

$$\mathbf{D}_a = \begin{bmatrix} D_{\|a} & 0 \\ 0 & D_{\perp a} \end{bmatrix}$$

$$D_{\|a} = \frac{\mu_e D_{\|i} + \mu_i D_{\|e}}{\mu_e + \mu_i}, \tag{5.7-5}$$

$$D_{\perp a} = \frac{\mu_e D_{\perp i} + \mu_i D_{\perp e}}{\mu_e + \mu_i}$$

where the species mobilities and diffusion coefficients are determined by separately equating the parallel and cross-field fluxes of ions and electrons [74]. This simplified plasma diffusion equation assumes uniform ion and secondary electron production rates and temperatures. A derivation that does include these simplifying assumptions is given by Wirz [66].

The thermal electron energy conservation equation is derived by multiplying the Boltzmann equation by $mv^2/2$ and integrating over velocity to give

$$\frac{\partial}{\partial t}\left(\frac{nm}{2}\mathbf{v}^2 + \frac{3}{2}nkT\right) + \nabla \cdot \left[\left(\frac{nm}{2}\mathbf{v}^2 + \frac{5}{2}nkT\right)\mathbf{v} + \mathbf{q}\right], \tag{5.7-6}$$
$$= en\mathbf{E} \cdot \mathbf{v} + \mathbf{R} \cdot \mathbf{v} + Q_e + Q_c$$

where viscous effects are ignored and $\mathbf{R}$ is the mean change of momentum of electrons due to collisions with other species. This equation is combined with the electron fluxes to the boundaries and thermal conductivity to determine the total energy loss to the boundaries. Temperatures calculated from the electron energy equation are shown in Fig. 5-45 for the NSTAR thruster. The strong on-axis confinement of the primaries in NSTAR tends to locally heat the plasma electron population, generating a high on-axis plasma temperature.

5.7.3 Discharge Chamber Model Results

The 2-D discharge chamber model developed by Wirz and Katz [66] has been verified against beam profile and performance data for the 30-cm NSTAR thruster. The model results for the NSTAR

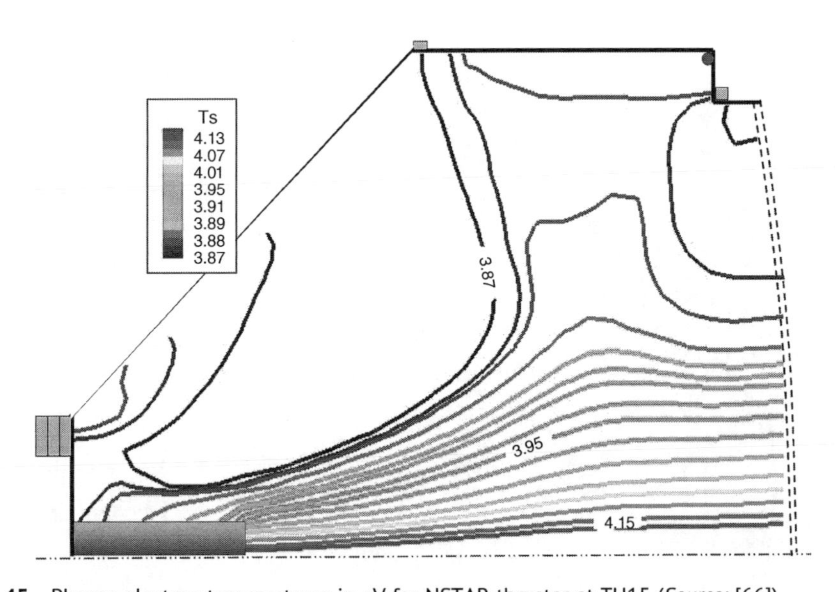

Figure 5-45 Plasma electron temperatures in eV for NSTAR thruster at TH15 (*Source:* [66]).

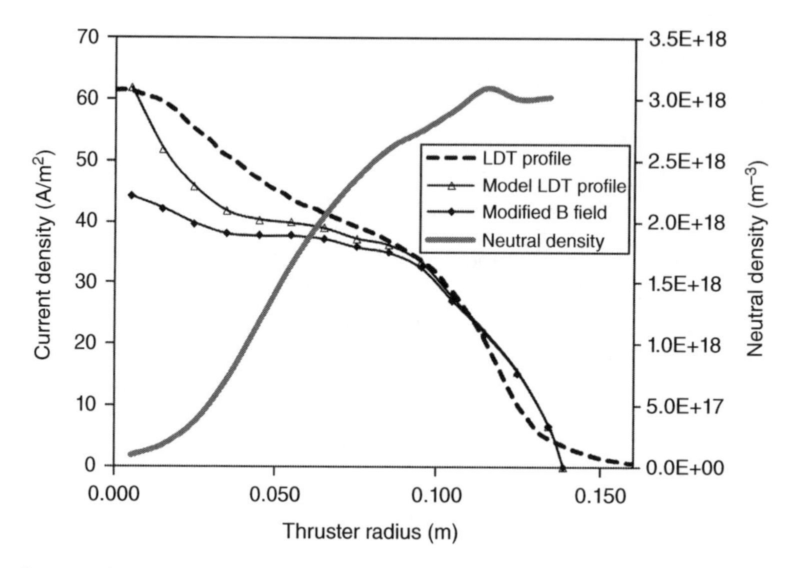

Figure 5-46 Beam and neutral density profiles at the grid for NSTAR (*Source:* [66]).

thruster at throttle condition TH15 are plotted in Fig. 5-46 where the beam current density profile calculated by the model agrees well with experimental data obtained during the 8200 hour Long Duration Test [75]. The peaked plasma profile is due to the strong confinement of the electrons from the cathode by the NSTAR magnetic configuration, which depletes the neutral gas on axis and produces a significant number of double ions. The "Modified B fields" profile in Fig. 5-46 is an example of the model prediction for the case of a modified magnetic field geometry that makes it easier for primary electrons to move away from the thruster axis. The ion density calculated by the Wirz-Katz model for the NSTAR magnetic is shown in Fig. 5-47. As suggested by the primary density and plasma electron temperatures in Figs. 5-44 and 5-45, the plasma density is strongly peaked

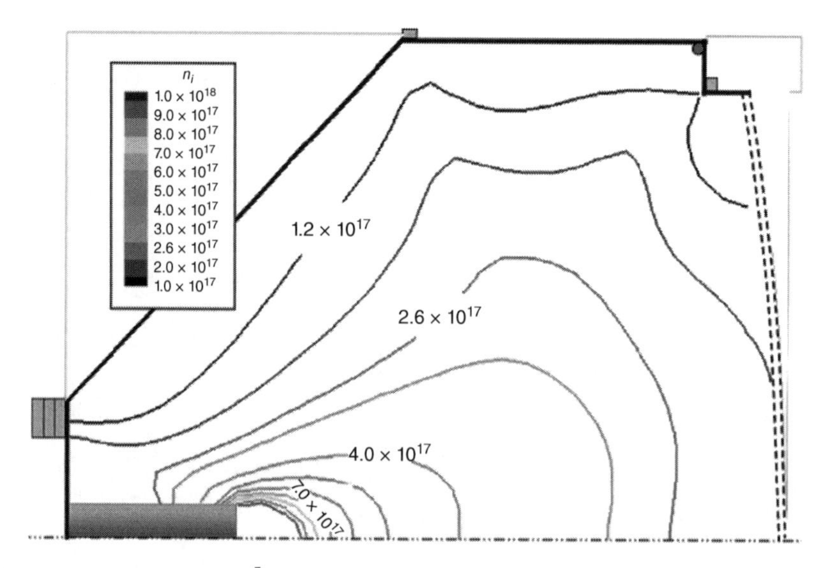

Figure 5-47 Ion plasma density (m^{-3}) for the NSTAR at throttle level TH15 (*Source:* [66]).

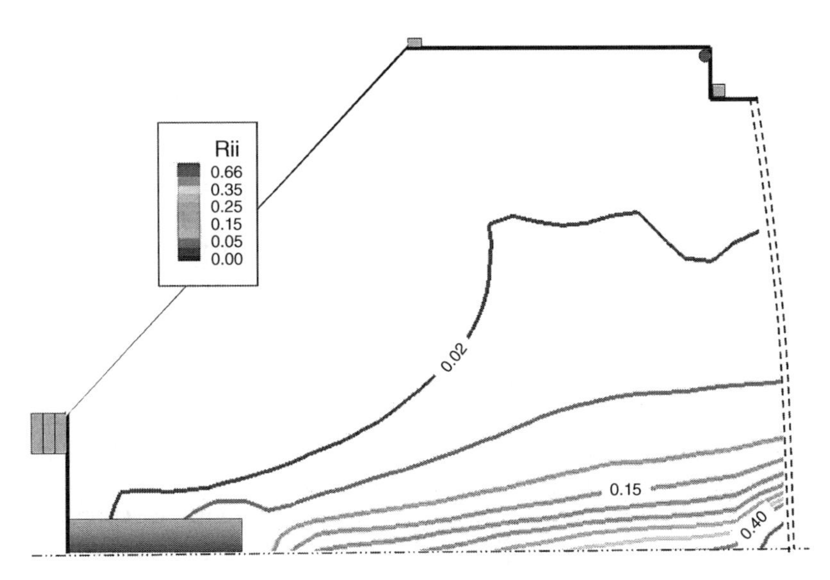

Figure 5-48 Double ion density ratio (n^{++}/n^{+}) for NSTAR operating at a power level of TH15 (*Source:* [66]).

on axis. Finally, the double-to-total ion ratio distribution throughout the discharge chamber is shown in Fig. 5-48. These results agree with experimental data that suggests the on-axis peak in the NSTAR beam profile is due to high centerline double ion content.

Analysis by the Wirz-Katz model results shows that the original NSTAR magnetic field configuration tends to trap primary electrons on-axis, which increases local electron temperature, ionization rate, and the generation of double ions in this region. This trapping of primary electrons also manifests in a neutral atom depletion on-axis, as was shown in Fig. 5-46. The "modified" configuration in this figure shows the power of a good computer model to improve ion thruster design. By allowing the primary electrons to move away from the thruster axis, the ionization is spread more uniformly throughout the discharge chamber. The flatter profile results from a decrease in primary electron density, and hence double ion content, on the thruster centerline. Wirz and Goebel [26] developed "modified" NSTAR designs that guide primary electrons away from the thruster centerline to improve the profile. These designs were validated by experiments [76], and also resulted in lower double ion content and higher neutral density along the thruster axis as predicted by the model.

Problems

5.1 Show the conditions under which the ambipolar velocity of the ions flowing to the wall in a transverse magnetic field reverts to the Bohm velocity.

5.2 An ion thruster discharge chamber has an internal pressure of 10^{-4} Torr, a plasma density of $2 \times 10^{17}\ \text{m}^{-3}$, gas and ion temperature of 500 K, electron temperature of 4 eV and a transverse magnetic field of 40 G near the wall with a diffusion length of 2 cm. What is the average transverse ion velocity and the ion confinement factor (ratio of v_i/v_{Bohm})?

5.3 In Fig. 5-16 it is shown that the reaction rate for ionization exceeds the reaction rate for excitation if the electron temperature exceeds about 9 eV. Why not run discharges with $Te \geq 9$ eV where ionization > excitation? Give a quantitative answer for an idealize thruster producing 1 A with 10 cm diameter grids on a discharge chamber 10 cm in diameter and 15 cm long with the anode being the full cylindrical and back-wall area. Assume an 80% grid transparency and neutral density of 10^{18} cm^{-3}, and plot the discharge loss as a function of electron temperature from 3 to 10 eV. Explain why (hint: examine the various loss terms).

5.4 What is the electron temperature in a xenon ion thruster that has an ion loss area of 200 cm^2, a plasma volume of 10^4 cm^3, neutral gas density of 10^{13} cm^{-3} and a 5% primary electron density at 15 eV?

5.5 A thruster plasma has a volume of 10^4 cm^3, a neutral density of 10^{12} cm^{-3}, is 10% ionized with 15 V primary electrons, 5 eV electron temperature and a primary loss area of 10 cm^2. What are the primary electron confinement time, the primary electron collision time (assume a collision cross section of 2e-16 cm^2), and the primary electron slowing down time? What is the total effective confinement time for a primary electron, and which of the three contributors to the total confinement time is the most important?

5.6 For a xenon ion thruster with a grid area of 500 cm^2 with a screen grid transparency of 70%, what is the discharge current required to produce a 2.5-A ion beam? Assume a discharge voltage of 25 V, a hollow cathode voltage drop of 10 V, a plasma potential of 5 V, a primary electron density of 5%, and an excitation energy of 10 eV. You can neglect the ion and primary electron loss to the anode, the ion current back to the cathode, and any losses to the back wall of the cylindrical discharge chamber with the same diameter as the grids.

5.7 A xenon ion thruster discharge chamber produces a 5×10^{17}m^{-3} plasma 20 cm in diameter with an electron temperature of 5.5 eV. What is the beam current and average current density if the screen grid transparency is 80%, and what flatness parameter is required to maintain the peak current density under 10 mA/cm^2?

5.8 A xenon ion thruster has a grid diameter of 20 cm with a transparency of 75%, an electron temperature of 3 eV in a 30 cm dia., 30 cm long cylindrical discharge chamber with an ion confinement factor of 0.1. What does the cusp anode area have to be to maintain the plasma potential at the sheath edge of 6 V? You can assume that the discharge current is 10 times the beam current and neglect the back wall loss area and primary electron effects. Assuming the ion temperature is 0.1 eV and that there are three magnetic rings around the cylindrical chamber, what is the magnetic field at the wall required to produce this cusp anode area?

5.9 A rf xenon ion thruster has a grid diameter of 10 cm, a grid transparency of 70% and a cylindrical discharge chamber with a diameter and length of 10 cm. Assuming an electron temperature of 4 eV, an ion confinement factor of 0.5 and a neutral density of 6×10^{18} m^{-3}, what is the plasma potential and discharge loss? If the cylindrical discharge chamber is made into a cone 10 cm long from the grid diameter, how do the plasma potential and discharge loss change?

5.10 A microwave ion thruster produces 2 A from an 80% transparent grid using a 4 GHz microwave source. If the thruster is running at 90% of cutoff with a flatness parameter of 0.6, what must the diameter of the grid be to produce this beam current?

References

1 H. R. Kaufman. An Ion Rocket With an Electron-Bombardment Ion Source. *NASA Tech. Note D-585*, January 1961.
2 P. D. Reader, "Investigation of a 10-cm Diameter Electron Bombardment Ion Rocket," NASA Technical Note D-1163, 1962.
3 R. T. Bechtel, "Discharge Chamber Optimization of the SERT-II Thruster," *Journal of Spacecraft and Rockets*, 795–800, vol. 5, no.7, 1968.
4 W. Knauer, R. L. Poeschel, and J. W. Ward, "The Radial Field Kaufman Thruster," AIAA-1969-259, 7th Electric Propulsion Conference, Williamsburg, March 3–5, 1969.
5 R. D. Moore, "Magneto-Electrostatic Contained Plasma Ion Thruster," AIAA-1969-260, 7th AIAA Electric Propulsion Conference, Williamsburg, March 3–5, 1969. https://doi.org/10.2514/6.1969-260.
6 J. S. Sovey, "Improved Ion Containment Using a Ring-Cusp Ion Thruster," *Journal of Spacecraft*, 488–495, vol. 21, no. 5, 1984.
7 J. S. Sovey and H. J. King, Status of 30-cm Mercury Ion Thruster Development. *NASA TMX-71603*, October 1974.
8 T. D. Masek, R. L. Poeschel, C. R. Collett, and D. E. Snelker, "Evolution and Status of the 30-cm Engineering Model Ion Thruster," AIAA-1976-1006, 12th International Electric Propulsion Conference, Key Biscayne, November 14–17, 1976.
9 V. K. Rawlin, "Operation of the J-Series Thruster Using Inert Gases," AIAA-1982-1929, 16th International Electric Propulsion Conference, New Orleans, November 17–19, 1982.
10 J. R. Beattie and J. N. Matossian, Inert-Gas Ion Thruster Technology. *NASA CR191093*, March 1993.
11 J. R. Beattie and R. L. Poeschel, Ring-Cusp Ion Thrusters," IEPC-84-71, 17th International Electric Propulsion Conference, Tokyo, May 28–31, 1984.
12 J. N. Matossian and J. R. Beattie, "Characteristics of Ring-Cusp Discharge Chambers," *AIAA Journal of Propulsion and Power*, 968–974, vol. 7, no. 6, 1991.
13 M. J. Patterson, T. W. Haag, V. K. Rawlin, and M. T. Kussmaul, "NASA 30-cm Ion Thruster Development Status," AIAA-1994-2849, 30th Joint Propulsion Conference, Indianapolis, June 27–29, 1994.
14 J. R. Beattie, J. N. Matossian, and R. R. Robson, "Status of Xenon Ion Propulsion Technology," *AIAA Journal of Propulsion and Power*, 145–150, vol. 7, no. 2, 1990.
15 T. D. Maske, "Plasma Properties and Performance of Mercury Ion Thrusters," *AIAA Journal*, 205–212, vol. 9, no. 2, 1971.
16 J. Ward and T. Masek, "A Discharge Computer Model for an Electron Bombardment Thruster," AIAA-1976-1009, AIAA International Electric Propulsion Conference, Key Biscayne, November 14–17, 1976.
17 J. N. Matossian and J. R. Beattie, "Model for Computing volume Averaged Plasma Properties in Electron-Bombardment Ion Thrusters," *AIAA Journal of Propulsion and Power*, 188–196, vol. 5, no. 2, 1989.
18 J. R. Brophy, Ion Thruster Performance Model, Ph.D. thesis, Colorado State University. NASA CR-174810, December 1984.
19 J. R. Brophy and P. J. Wilbur, "Simple Performance Model for Ring and Line Cusp Ion Thrusters," *AIAA Journal*, 1731–1736, vol. 23, no.11, 1985.
20 D. M. Goebel, J. E. Polk, and A. Sengupta, "Discharge Chamber Performance of the NEXIS Ion Thruster," AIAA-2004-3813, 40th AIAA Joint Propulsion Conference, Fort Lauderdale, July 11–14, 2004.
21 D. M. Goebel, R. E. Wirz, and I. Katz, "Analytical Ion Thruster Discharge Performance Model," *AIAA Journal of Propulsion and Power*, 1055–1067, vol. 23, no.5, 2007; doi:10.2514/1.26404.

22 D. M. Goebel, "Ion Source Discharge Performance and Stability," *Physics of Fluids*, 1093–1102, vol. 25, no. 6, 1982; doi:10.1063/1.863842.

23 A. T. Forrester, *Large Ion Beams*, New York: Wiley, 1988.

24 M. Lieberman and A. Lichtenberg, *Principles of Plasma Discharges and Materials Processing*, New York: Wiley, 1994.

25 Ansys Maxwell is a product of Ansys. https://www.ansys.com/products/electronics/ansys-maxwell.

26 R. E. Wirz and D. M. Goebel, "Ion Thruster Discharge Performance Per Magnetic Field Geometry," AIAA-2006-4487, 42nd Joint Propulsion Conference, Sacramento, July 10–13, 2006.

27 K. Leung, N. Hershkowitz, K. MacKenzie, "Plasma Confinement by Localized Cusps," *Physics of Fluids*, 1045–1053, vol. 19, no. 7, 1976; doi:10.1063/1.861575.

28 P. Clausing, "The Flow of Highly Rarefied Gases Through Tubes of Arbitrary Length," *Journal of Vacuum Science and Technology*, 636–646, vol. 8, no. 5, 1971; doi:10.1116/1.1315392.

29 D. Rapp and P. Englander-Golden, "Total Cross Sections for Ionization and Attachment in Gases by Electron Impact. I. Positive Ionization," *The Journal of Chemical Physics*, vol. 43, no. 5, p.1464–1479 (1965).

30 M. Hayashi, "Determination of Electron-Xenon Total Excitation Cross-Sections, From Threshold to 100-eV, From Experimental Values of Townsend's α," *Journal of Physics D: Applied Physics*, vol. 16, no. 4, p.581–589 (1983).

31 R. R. Peters, P. J. Wilbur, and R. P. Vahrenkamp, "A Doubly Charged Ion Model for Ion Thrusters," *Journal of Spacecraft and Rockets*, 461–468, vol. 14, no.8, 1977.

32 L. Spitzer, *Physics of Fully Ionized Gases*, New York: Interscience Publishers, 1956, p. 80.

33 A. S. Richardson, *NRL Plasma Formulary, U.S.* Washington: Naval Research Laboratory, 2019.

34 J. R. Brophy, I. Katz, J. E. Polk, and J. R. Anderson, "Numerical Simulations of Ion Thruster Accelerator Grid Erosion," AIAA Paper 2002-4261, 38st AIAA Joint Propulsion Conference, July 7–10, 2002.

35 J. J. Anderson, I. Katz, and D. M. Goebel, "Numerical Ssimulation of Two-Grid Ion Optics Using a 3D Code," AIAA-2004-3782, 40th AIAA Joint Propulsion Conference, Ft. Lauderdale, July 2004.

36 D. M. Goebel, K. Jameson, I. Katz, and I. Mikellades, "Hollow Cathode Theory and Modeling: I. Plasma Characterization With Miniature Fast-Scanning Probes," *Journal of Applied Physics*, 113302, vol. 98, no. 10, 2005; doi:10.1063/1.2135417.

37 I. G. Mikellades, I. Katz, D. M. Goebel, and J. E. Polk, "Hollow Cathode Theory and Modeling: II. A Two-Dimensional Model of the Emitter Region," *Journal of Applied Physics*, 113303, vol. 98, no. 10, 2005; doi:10.1063/1.2135409.

38 J. E. Polk, D.M. Goebel, J. S. Snyder, A. C. Schneider, J. R. Anderson, and A. Sengupta, "A High Power Ion Thruster for Deep Space Missions," *The Review of Scientific Instruments*, 073306, vol. 83, 2012; doi:10.1063/1.4728415.

39 J. R. Brophy NASA's Deep Space 1 Ion Engine, *Review of Scientific Instruments*, 1071–1078, vol. 73, no. 2, 2002; doi:10.1063/1.1432470.

40 K. K. Jameson, D. M. Goebel, and R. M. Watkins, "Hollow Cathode and Keeper-Region Plasma Measurements," AIAA-2005-3667, 41th AIAA Joint Propulsion Conference, Tucson, July 10–13, 2005.

41 P. J. Wilbur and J. R. Brophy. The Effect of Discharge Chamber Wall Temperature on Ion Thruster Performance," *AIAA Journal*, 278–283, vol. 24, no.2, 1986.

42 J. M. Lafferty, *Foundations of Vacuum Science and Technology*, New York: Wiley, 1998.

43 Y. Hayakawa, K. Miyazaki, S. Kitamura, "Measurements of Electron Energy Distributions in an Ion Thruster," *AIAA Journal of Propulsion and Power*, 118–126, vol. 8, no. 1, 1992.

44 D. Herman, "Discharge Cathode Electron Energy Distribution Functions in a 40-cm NEXT-Type Ion Engine," AIAA-2005-4252, 41th AIAA Joint Propulsion Conference, Tucson, July 10–13, 2005.

45 D. Bohm, *The Characteristics of Electric Discharges in Magnetic Fields*, edited by A. Guthrie and R. Wakerling, New York: McGraw-Hill, pp. 1–76, 1949.

46 N. L. Milder, "A Survey and Evaluation of Research on the Discharge Chamber Plasma of a Kaufman Thruster," AIAA-69-494, 5th Propulsion Joint Specialist Conference, U.S. Air Force Academy, CO, June 9–13, 1969.

47 H. R. Kaufman, "Performance of Large Inert-Gas Thrusters," AIAA-81-0720, 15th International Electric Propulsion Conference, Las Vegas, April 21–23, 1981.

48 T. Ozaki, E. Nishida, Y. Gotoh, A. Tsujihata, and K. Kajiwara, "Development Status of a 20 mN Xenon Ion Thruster," AIAA-2000-3277, 36th Joint Propulsion Conference, Huntsville, July 16–19, 2000.

49 W. F. Divergilio, H. Goede, and V. V. Fosnight, High Frequency Plasma Generators for Ion Thrusters. *NASA #CR-167957*, 1981.

50 H. J. Leiter, R. Killinger, H. Bassner, J. Miller, R. Kulies, and T. Fröhlich, "Evaluation of the Performance of the Advanced 200 mN Radio Frequency Ion Thruster RIT_XT," AIAA-2002-3836, 38th Joint Propulsion Conference, Indianapolis, July 7–10, 2002.

51 D. M. Goebel, "Analytical Discharge Performance Model for rf Ion Thrusters," *IEEE Transactions on Plasma Science*, 2111–2121, vol. 36, no. 5, 2008; doi:10.1109/TPS.2008.2004232.

52 M. Tuszewski, "Inductive Electron Heating Revisited," *Physics of Plasmas*, 1922–1928, vol. 4, no. 5, 1997; doi:10.1063/1.872335.

53 M. Dobkevicius and D. Feili, "Multiphysics Model for Radio-Frequency Gridded Ion Thruster Performance," *AIAA Journal of Propulsion and Power*, 939, vol. 33, no. 4, 2017; doi:10.2514/1.B36182.

54 V. H. Chaplin and P. M. Bellan, "Battery-Powered Pulsed High Density Inductively Coupled Plasma Source for Preionization in Laboratory Astrophysics Experiments," *Review of Scientific Instruments*, 073506, vol. 86, 2015; doi:10.1063/1.4926544

55 P. Chabert, J. A. Monreal, J. Bredin, L. Popelier, and A. Aanesland, "Global Model of a Gridded Ion Thruster Powered by a Radiofrequency Inductive Coil," *Physics of Plasmas*, 073512, vol. 19, 2012.

56 P. Chabert and N. Braithwaite, *Physics of Radio-Frequency Plasma*, Cambridge: Cambridge University, 2011.

57 H. Goede, "30-cm Electron Cyclotron Plasma Generator," *Journal of Spacecraft*, 437–443, vol. 24, no. 5, 1987; doi:10.2514/3.25936.

58 H. Kuninaka and S. Satori, "Development and Demonstration of a Cathode-Less Electron Cyclotron Resonance Ion Thruster," *AIAA Journal of Propulsion and Power*, 1022–1026, vol. 14, no. 6, 1998; doi:10.2514/2.5369.

59 H. Kuninaka, I. Funaki, and K. Toki, "Life Test of Microwave Discharge Ion Thrusters for MUSES-C in Engineering Model Phase," AIAA-99-2439, 35th Joint Propulsion Conference, Los Angeles, June 20–24, 1999.

60 H. Kuninaka, I. Funaki, K. Nishiyama, Y. Shimizu, and K. Toki, "Results of 18,000 Hour Endurance Test of Microwave Discharge Ion Thruster Engineering Model," AIAA-2000-3276, 36th Joint Propulsion Conference, Huntsville, July 16–19, 2000.

61 R. Tsukizaki, T. Ise, H. Koizumi, H. Togo, K. Nishiyama, and H. Kuninaka, "Thrust Enhancement of a Microwave Ion Thruster," *AIAA Journal of Propulsion and Power*, 1383–1389, vol. 30, no. 5, 2014; doi:10.2514/1.B35118.

62 G. Coral, R. Tsukizaki, K. Nishiyama, and H. Kuninaka, "Microwave Power Absorption to High Energy Electrons in the ECR Ion Thruster," *Plasma Sources Science and Technology*, 095015, vol. 27, 2018; doi:10.1088/1361-6595/aadf04

63 R. Shirakawa, Y. Yamashita, D. Koda, R. Tsukizaki, S. Shimizu, M. Tagawa, and K. Nishiyama, "Investigation and Experimental Simulation of Performance Deterioration of Microwave Discharge Ion Thruster μ10 During Space Operation" *Acta Astronautica*, vol. 174, 2020; doi:10.1016/j.actaastro.2020.05.004.

64 Y. Arakawa and K. Ishihara, “A Numerical Code for Cusped Ion Thrusters,” IEPC-91-118, 22nd International Electric Propulsion Conference, Viareggio, October 14–17, 1991.

65 Y. Arakawa and P. J. Wilbur, “Finite Element Analysis of Plasma Flows in Cusped Discharge Chambers,” *AIAA Journal of Propulsion and Power*, 125–128, vol. 7, no.1, 1991.

66 R. Wirz and I. Katz, “Plasma Processes in DC Ion Thruster Discharge Chambers,” AIAA-2005-3690, 41th AIAA Joint Propulsion Conference, Tucson, July 10–13, 2005.

67 S. Mahalingam and J. A. Menart, “Particle-Based Plasma Simulations for an Ion Engine Discharge Chamber,” *AIAA Journal of Propulsion and Power*, 673–688, vol. 26, no. 4, 2010; doi:10.2514/1.45954.

68 C. E. Huerta and R. Wirz, “DC-Ion Model Validation and Convergence Studies,” AIAA-2018-4813, AIAA Joint Propulsion Conference, Cincinnati, July 9–11, 2018.

69 R. Siegel, and J. R. Howell, *Thermal Radiation Heat Transfer*, fourth edition, New York: Taylor & Francis, 2002, 226–227.

70 R. Wirz, Discharge Plasma Processes of Ring-Cusp Ion Thrusters, Ph.D. dissertation, Aeronautics, Caltech, 2005.

71 Y. Arakawa and T. Yamada, “Monte Carlo Simulation of Primary Electron Motions in Cusped Discharge Chambers,” AIAA-1990-2654, 21st International Electric Propulsion Conference, Orlando, July 18–20, 1990.

72 S. Mahalingam and J. A. Menart, “Primary Electron Modeling in the Discharge Chamber of an Ion Engine,” AIAA-2002-4262, 38th Joint Propulsion Conference, Indianapolis, July 7–10, 2002.

73 J. P. Boris, “Relativistic plasma simulations-optimization of a hybrid code,” Proc. 4th Conf on Numerical Simulation of Plasmas, NRL Washington, Washington, November 2, 1970.

74 S. I. Braginskii, “Transport Processes in Plasmas,” *Reviews of Plasma Physics*, 205–311, vol. 1, 1965.

75 J. Polk, J. R. Anderson, J. R. Brophy, V. K. Rawlin, M. J. Patterson, J. Sovey, and J. Hamley, “An Overview of the Results From an 8200 Hour Wear Test of the NSTAR Ion Thruster,” AIAA Paper 99–2446, 35th Joint Propulsion Conference, Los Angeles, June 20–24, 1999.

76 A. Sengupta, “Experimental Investigation of Discharge Plasma Magnetic Confinement in an NSTAR Ion Thruster,” AIAA-2005-4069, 41st Joint Propulsion Conference, Tucson, July 10–13, 2005.

Chapter 6

Ion Thruster Accelerators

Ion thrusters are characterized by the electrostatic acceleration of ions extracted from the plasma generator [1]. An illustration of a DC electron bombardment ion thruster showing the ion accelerator, the plasma generator, and the neutralizer cathode was shown in Fig. 1-1. The ion accelerator in "gridded ion thrusters" consists of electrically biased multi-aperture grids, and this assembly is often called the *ion optics*. The design of the grids is critical to the ion thruster operation and is a trade between performance, life, and size. Since ion thrusters need to operate for years in most applications, life is often the major design driver. However, performance and size are always important to satisfy the mission requirements for thrust and Isp and to provide a thruster size and shape that fits onto the spacecraft.

There are many factors that determine the grid design in ion thrusters. The grids must extract the ions from the discharge plasma through the "screen" grid and focus them through the downstream acceleration ("accel") grid and the deceleration ("decel") grid (if one is used). This focusing has to be accomplished over the range of ion densities produced by the discharge chamber plasma profile that is in contact with the screen grid, and also over the throttle range of different power levels that the thruster must provide for the mission. Since the screen grid transparency was shown in Chapter 5 to directly impact the discharge loss, the grids must minimize ion impingement on the screen grid and extract the maximum number of the ions that are delivered by the plasma discharge to the screen grid surface. In addition, the grids must minimize neutral atom loss out of the discharge chamber to maximize the mass utilization efficiency of the thruster. High ion transparency and low neutral transparency drives the grid design toward larger screen grid holes and smaller accel grid holes, which impacts the optical focusing of the ions and the beam divergence. The beam divergence should also be minimized to reduce thrust loss and plume impact on the spacecraft or solar arrays, although some amount of beam divergence can usually be accommodated. Finally, grid life is of critical importance and often drives thruster designers to compromises in performance or alternative grid materials. In this chapter, the factors that determine grid design and the principles of the ion accelerators used in ion thrusters will be described.

6.1 Grid Configurations

To accelerate ions, a potential difference must be established between the plasma produced inside the thruster plasma generator and the ambient space plasma. As shown in Chapter 3, simply biasing the anode of a DC plasma generator or the electrodes of a rf plasma generator relative to a spacecraft or plasma in contact with the space potential does not result in ion beam generation because

Fundamentals of Electric Propulsion, Second Edition. Dan M. Goebel, Ira Katz, and Ioannis G. Mikellides.

the voltage will just appear in the sheath at the plasma boundary with the walls. If the potential is small compared to the electron temperature T_e, then a Debye sheath is established, and if the potential is very large compared to T_e, then a Child-Langmuir sheath exists. Therefore, to accelerate ions to high energy, it is necessary to reduce the dimension of an aperture at the plasma boundary to the order of the Child-Langmuir distance to establish a sheath that will accelerate the ions with reasonable directionality (good focusing) and reflect the electrons from the plasma. Figure 6-1 shows the Child-Langmuir length calculated from Eq. 3.7-35 for two singly charged ion current densities at an acceleration voltage of 1500 V. For xenon, the characteristic aperture dimension at this voltage is on the order of 2–5 mm and will decrease if the applied voltage is reduced or the current in the aperture is increased.

The ion current obtainable from each grid aperture is then limited by space charge. For a 0.25 cm diameter aperture extracting the space charge limited xenon current density of about 5 mA/cm^2 at 1500 V (according to the Child-Langmuir Eq. 3.7-35), the total ion current per aperture is only 0.25 mA. Assuming this produces a well-focused beamlet, the thrust produced by this current and voltage from Eq. 2.3-9 is only about 16 micro-Newtons. Therefore, multiple apertures must be used to obtain higher beam currents from the ion engine to increase the thrust. For example, to extract a total of 1 A of xenon ion current for this case would require over 4,000 apertures, which would produce over 60 mN of thrust. In reality, for reliable high voltage operation and due to non-uniformities in the plasma generator producing varying ion current densities to the boundary, the current density is usually chosen to be less than the Child-Langmuir space charge maximum and an even larger number of apertures are required. This ultimately determines the size of the ion thruster.

Figure 6-2 shows a simplified 1-D view of one of these biased apertures facing the thruster plasma. The Child-Langmuir sheath is established by the bias potential between the thruster plasma and the accelerator grid (called the "accel grid") and is affected by the current density of the xenon ions arriving at the sheath edge from the Bohm current. Ions that arrive on axis with the aperture are accelerated through to form the beam. However, ions that miss the aperture are accelerated into the accel grid and can erode it rapidly. For this reason, a "screen" grid with apertures aligned with the accel grid is placed upstream of the accel grid to block these ions. This is the classic two-grid accelerator system [1, 2]. The screen grid is normally allowed to either float electrically or is

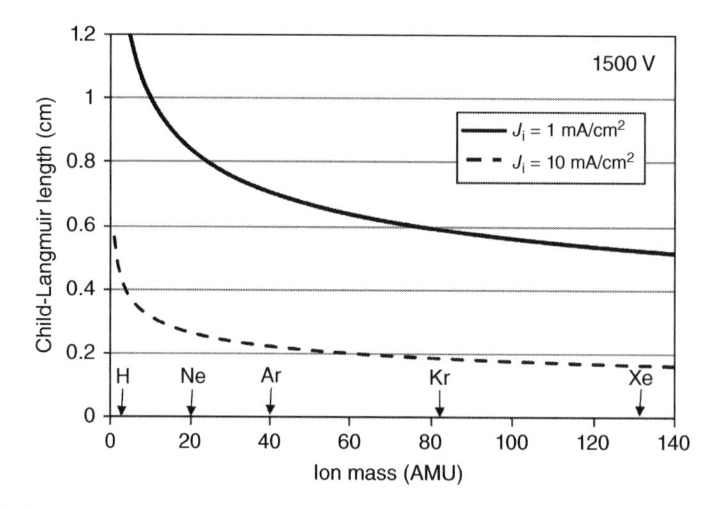

Figure 6-1 Child-Langmuir sheath length versus ion mass for two ion current densities at 1500 V acceleration voltage.

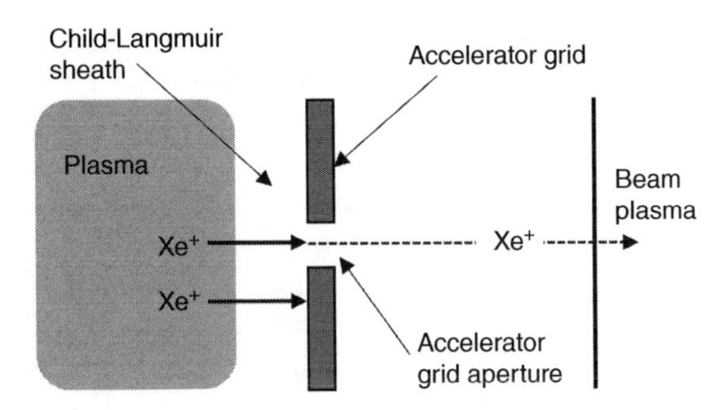

Figure 6-2 Simplified 1-D view of an accelerator aperture in contact with a plasma.

biased to the cathode potential of the plasma generator to provide some confinement of the electrons in the plasma, and so ions that strike it have a relatively low energy that can still cause some ion sputtering. In practice, the grids are made of refractory metals or carbon-based materials, and the apertures are close-packed in a hexagonal structure to produce a high transparency to the ions from the plasma generator. These grids are also normally dished to provide structural rigidity to survive launch loads and to ensure that they expand uniformly together during thermal loading [1, 3].

The electrical configuration of an ion thruster accelerator is shown schematically in Fig. 6-3. The high voltage bias supply (called the screen supply) is normally connected between the anode and

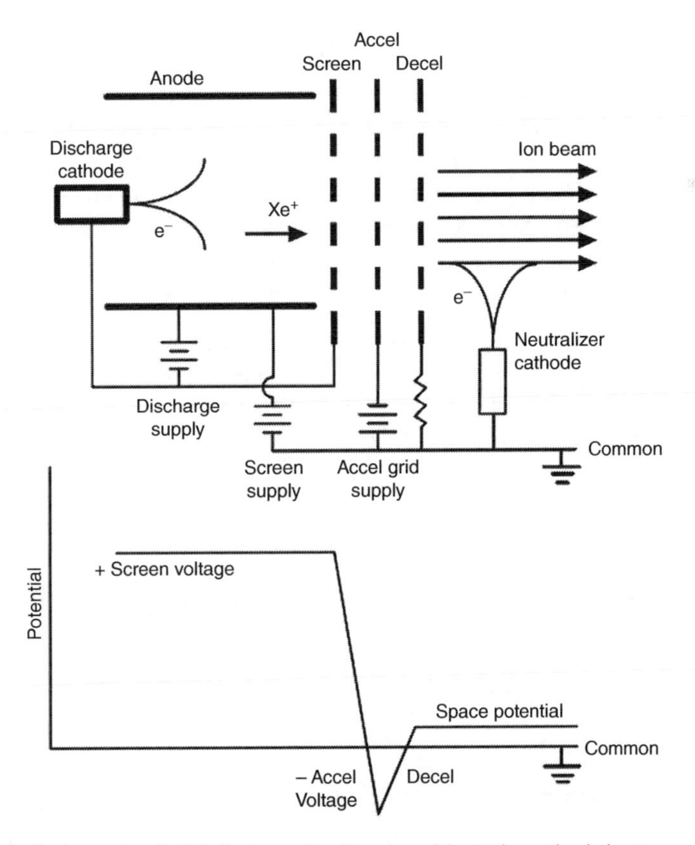

Figure 6-3 Electrical schematic of a DC discharge ion thruster without the cathode heater and keeper supplies

the common of the system, which is usually connected to the neutralizer cathode (called "neutralizer common") that provides electrons to neutralize the beam. Positive ions born in the discharge chamber at high positive voltage are then accelerated out of the thruster. The accel grid is biased negative relative to the neutralizer common to prevent the very mobile electrons in the beam plasma from back-streaming into the thruster, which would produce localized heating in the discharge chamber by energetic electron bombardment, and would ultimately overloads the screen supply if the backstreaming current becomes large. The ion beam is current neutralized and quasi-neutral (nearly equal ion and electron densities) by the electrons extracted from the neutralizer cathode. Fortunately, the thruster self-biases the neutralizer common potential sufficiently negative relative to the beam potential to produce the required number of electrons to current neutralize the beam.

Figure 6-3 has a generic thruster that includes a three-grid accelerator system where a final grid called the "decel grid" is placed downstream of the accel grid. This grid shields the accel grid from ion bombardment by charge-exchanged ions produced in the beam back flowing toward the thruster, and eliminates the downstream "pits and groves erosion" that will be discussed in Section 6.6. Three grid systems then potentially have longer accel grid life than two grid systems and generate less sputtered material into the plume that can deposit on the spacecraft. These benefits are offset by the increased complexity of including the third grid.

In actual design, the diameter of each accel grid aperture is minimized to retain unionized neutral gas in the plasma generator, and the screen grid transparency is maximized so that the grids extract the maximum number of ions from the plasma as possible. The electrode diameters and spacing are then optimized to eliminate direct interception of the beam ions on the accel grid, which would cause rapid erosion due to the high ion energy. A schematic example of a three-grid system showing the ion trajectories calculated by a 2D ion–optics code [4] is given in Fig. 6-4, where the figure shows half the beamlet mirrored on the axis. The ions are focused sufficiently by this electrode design to pass through the accel grid without direct interception. On the downstream side of the accel grid, the negative accel-grid bias applied to avoid electron backstreaming results in a relatively small deceleration of the ions before they enter the quasi-neutral beam potential region. This high transparency, strong "accel-decel" geometry typical of ion thrusters results in some beamlet divergence, as suggested by the figure. However, this small beamlet angular divergence of typically a few degrees causes negligible thrust loss because the loss scales as $\cos\theta$, and because most of the beam

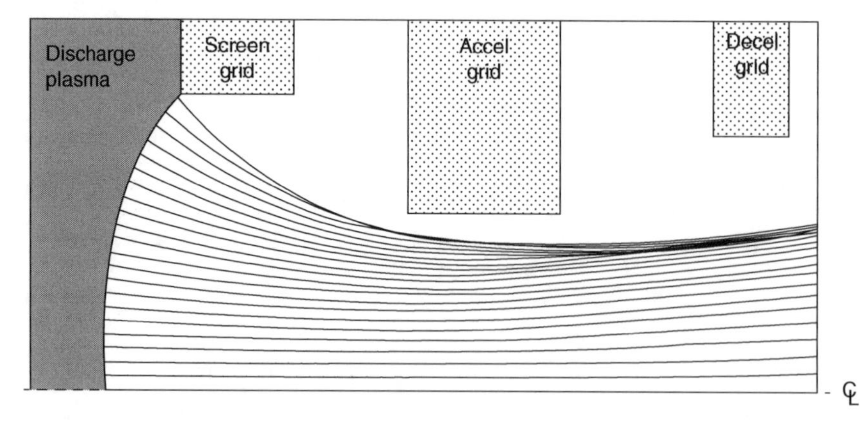

Figure 6-4 Ion trajectories from a plasma sheath (on left) in a half-beamlet inside an example three-grid accelerator.

divergence discussed in Chapter 2 related to the thrust correction factor is due to the dishing of the grids.

The amount of current that an ion accelerator can extract and focus into a beam for a given applied voltage is related to the space charge effects characterized by the Child-Langmuir equation, and is called the perveance:

$$P \equiv \frac{I_b}{V^{3/2}}. \tag{6.1-1}$$

The maximum perveance that can be achieved by an accelerator is given by the coefficient in the Child-Langmuir equation:

$$P_{\max} = \frac{4\varepsilon_o}{9}\sqrt{\frac{2q}{M}}. \tag{6.1-2}$$

For an electron accelerator, this coefficient is the familiar value of 2.33×10^{-6} A/V$^{3/2}$, and for singly charged xenon ions is 4.8×10^{-9} A/V$^{3/2}$. For round apertures, the Child Langmuir equation can be written

$$J = \frac{I_b}{\left(\pi d_b^2/4\right)} = \frac{4\varepsilon_o}{9}\sqrt{\frac{2q}{M}}\frac{V^{3/2}}{d^2}\ \left[\mathrm{A/m^2}\right], \tag{6.1-3}$$

where d is the effective grid gap and d_b is the beamlet diameter. Inserting Eq. 6.1-3 into Eq. 6.1-1, the maximum perveance for round apertures is

$$P_{\max} = \frac{\pi\varepsilon_o}{9}\sqrt{\frac{2q}{M}}\frac{d_b^2}{d^2}\left[\frac{A}{V^{3/2}}\right]. \tag{6.1-4}$$

Therefore, to maximize the perveance of the accelerator, it is desirable to make the grid gap smaller than the aperture diameters, as illustrated in the example configuration shown in Fig. 6-4.

The ion trajectories plotted in Fig. 6-4 that do not intercept either of the grids, and the minimal beamlet divergence, result from operating at or near the optimal ion current density and voltage for the grid geometry shown. Operating at significantly less than the optimal perveance, called “under-perveance” and corresponding to higher voltages or lower beamlet currents than the optimal combination, increases the CL length and pushes the sheath to the left farther into the plasma. In the extreme case, this situation can launch ions at a very large angle from the edge region near the screen aperture and cause “cross-over” trajectories, which can then produce excessive erosion of the accel grid by direct ion impingement. Likewise, operating at higher than the optimal perveance, corresponding to higher beamlet currents or lower voltages than optimal, reduces the Child-Langmuir sheath thickness and the plasma boundary pushes toward the screen aperture. This “over-perveance” condition flattens the sheath edge and accelerates ions directly into the accel grid, again causing excessive erosion. The optical performance and life of any grid design, therefore, is acceptable only over a limited range in voltage and current density, which will be discussed in Section 6.3. For this reason, the uniformity of the plasma over the grid area is important to avoid either cross-over or direct interception in different regions of the ion optics that strongly degrade the life of the grids.

In the two or three grid configurations, the geometry of the grid apertures and gaps is intended to eliminate or at least minimize direct impingement by beam ions on the most negative potential electrode in the system; namely, the accel grid. This is required to minimize sputtering of the grid by the high-energy beam ions. The screen grid does receive ion bombardment from the discharge plasma because of its finite transparency, but the ions arrive with only energy of the order of the discharge voltage in DC discharge thrusters or the floating potential in rf or microwave thrusters. Sputter erosion of the screen grid then only becomes an issue at higher discharge voltages or because of the production of high-energy ions in the hollow cathode region [5, 6] that can bombard the screen grid. Likewise, the decel grid is biased near the beam plasma potential and back flowing ions produced in the beam by charge exchange impact with very low energy, which causes little or no sputtering. For two grid systems, the back flowing ions bombard the accel grid with essentially the grid bias voltage. This can cause significant sputtering of the downstream face of the accel grid and may determine the grid life.

The decelerating field produced downstream of the accelerator grid by the accel grid bias acts as a weak defocusing lens for the ions, but keeps electrons emitted by the neutralizer from entering the high field region and backstreaming at high energy into the discharge chamber. This decelerating field is set up either by applying a potential between the accelerator grid and the decel grid, or by applying the bias between the accelerator grid and the hollow cathode neutralizer and allowing the low energy plasma downstream of the accelerator grid to act as a virtual anode. Unfortunately, ions generated between the grids by either charge exchange with unionized neutral gas escaping the plasma generator or by ionization from the most energetic backstreaming electrons do strike the accel grid and erode it. Charge exchange ion erosion of the accel grid ultimately limits the grid life, which will be discussed in Section 6.6.

6.2 Ion Accelerator Basics

The thruster ion optics assembly serves three main purposes:

1) Extract ions from the discharge chamber
2) Accelerate ions to generate thrust
3) Prevent electron backstreaming

The ideal grid assembly would extract and accelerate all the ions that approached the grids from the plasma while blocking the neutral gas outflow, accelerate beams with long life and with high current densities, and produce ion trajectories that are parallel to the thruster axis with no divergence under various thermal conditions associated with changing power levels in the thruster. In reality, grids are non-ideal in each of these areas. Grids have finite transparency; thus, some of the discharge chamber ions hit the upstream "screen grid" and are not available to become part of the beam. The screen grid transparency, T_s, is the ratio of the beam current I_b to the total ion current I_i from the discharge chamber that approaches the screen grid:

$$T_s = \frac{I_b}{I_i}. \tag{6.2-1}$$

This ratio is determined by comparing the ion beam current with the screen grid current. The transparency depends on the plasma parameters in the discharge chamber because the hemispherical sheath edge is normally pushed slightly into the plasma by the applied voltage if the screen grid is relatively thin. The pre-sheath fields in the plasma edge then tend to steer some ions that would have gone to the screen grid into the beam. For this reason, the effective transparency of the screen grid typically exceeds the optical transparency for relatively large apertures and thin grid thicknesses. In addition, the screen grid current must be measured with the screen grid biased negative relative to cathode potential to reflect energetic electrons in the tail of the Maxwellian distribution in the plasma. The goal for screen grid design is to maximize the grid transparency to ions by minimizing the screen thickness and the webbing between screen grid holes to that required for structural rigidity.

The maximum beam current density is limited by the ion space charge in the gap between the screen and accelerator grids [2], which was discussed above with respect to the perveance that was specified by the Child-Langmuir equation in which the sheath was considered essentially planar. The problem is that the sheath shape in the screen aperture is not planar, as seen in Fig. 6-4, and the exact shape and subsequent ion trajectories have to be solved by 2-D axial-symmetric codes. However, a modified sheath thickness can be used in the Child-Langmuir equation to approximately account for this effect, in which the current density is written as

$$J_{\max} = \frac{4\varepsilon_o}{9}\sqrt{\frac{2e}{M}}\frac{V_T^{3/2}}{l_e^2}, \tag{6.2-2}$$

where V_T is the total voltage across the sheath between the two grids and the sheath thickness l_e is given by

$$l_e = \sqrt{\left(l_g + t_s\right)^2 + \frac{d_s^2}{4}}. \tag{6.2-3}$$

The grid dimensions in Eq. 6.2-3 are defined in Fig. 6-5. As illustrated in the figure, the sheath is allowed to expand essentially spherically through the screen grid aperture. The sheath thickness l_e accounts for this non-planar condition and has been found to be useful in predicting the space charge limited current in ion thruster grid configurations [1, 7]. Note that the value of l_g is the "hot grid gap" that occurs once the grids have expanded into their final shape during operation at a given beam current and voltage. For xenon ions, the maximum current density is

$$J_{\max} = 4.75 \times 10^{-9}\frac{V_T^{3/2}}{l_e^2}. \tag{6.2-4}$$

The units of the current density in the Child-Langmuir equation are Amperes divided by the dimension used for the sheath thickness l_e squared.

The maximum thrust per unit area possible from an ion thruster can also be found. Thrust was defined in Chapter 2 for electric thrusters as

$$T = \frac{d(mv)}{dt} = \gamma \dot{m}_i v_i. \tag{6.2-5}$$

Assuming the ions start at rest, the ion velocity leaving the accelerator is

$$v_i = \sqrt{\frac{2eV_b}{M}}, \tag{6.2-6}$$

where eV_b is the net beam energy. Using Eq. 2.3-7 for the time rate of change of the mass, the thrust per unit area of the grids becomes

$$\frac{T}{A_g} = \frac{\gamma I_{\max} T_s M v_i}{A_g\, e} = \frac{J_{\max}\,\gamma\, T_s M v_i}{e}, \tag{6.2-7}$$

where A_g is the active grid area (with extraction apertures) and T_s accounts for the grid transparency defined in Eq. 6.2-1. The effective electric field in the acceleration gap is

$$E = \frac{V_T}{l_e}, \tag{6.2-8}$$

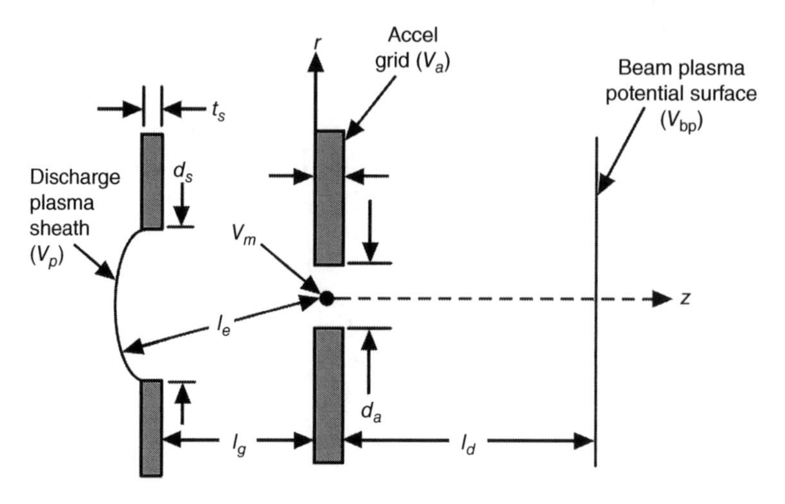

Figure 6-5 Non-planar sheath model approximation for a two-grid system.

where V_T is the total voltage across the accelerator gap (the sum of the screen and accel voltages):

$$V_T = V_s + |V_a| = \frac{V_b}{R}, \tag{6.2-9}$$

and R is the ratio of the net beam voltage to the total voltage. Using Eq. 6.2-2 for the space charge limited current density and the electric field from Eq. 6.2-8, the maximum achievable thrust density is

$$\frac{T_{\max}}{A_g} = \frac{4}{9}\frac{\varepsilon_o \gamma T_s}{e}\sqrt{\frac{2e}{M}}\frac{V_T^{3/2}}{l_e^2} M\sqrt{\frac{2eV_b}{M}} = \frac{8}{9}\epsilon_o \gamma T_s \sqrt{R} E^2. \tag{6.2-10}$$

The maximum thrust density from an ion thruster increases with the screen grid transparency and the square of the electric field [8]. Ion thrusters with thin, high transparency grids operating near the perveance limit and at the maximum possible electric field in the acceleration gap will produce the most thrust for a given grid area. A key feature of ion thrusters illustrated by Eq. 6.2-10 is that the thrust density is independent of propellant mass.

The net-to-total voltage ratio from Eq. 6.2-9 is given by

$$R = \frac{V_b}{V_T} = \frac{V_b}{V_s + |V_a|}. \tag{6.2-11}$$

This equation describes the relative magnitude of the accel grid bias relative to the screen potential. Operating with small values of R increases the total voltage between the screen and accel grids which, from Eq. 6.2-2, results in a higher current density of ions accelerated from the thruster. While it appears desirable to operate with very small values of R (large accel grid negative bias) to increase the current capability of a grid set, this results in higher energy ion bombardment of the accel grid and shortens grid life. Operating with small values of R will also change the beam divergence, but this is a relatively small effect in ion thrusters for most grid designs. For applications where thruster life is important, the magnitude of accel grid bias voltage is usually minimized to the value required to just avoid electron backstreaming, and the value of R typically ranges from 0.8 to 0.9. Finally, Eq. 6.2-10 suggests that the thrust density depends on the square root of R and would increase slowly with higher beam to total voltage ratios. This is misleading because the total voltage also appears in the electric field term ($E = V_T/l_e$), and so higher thrust densities actually occur with more negative accel grid bias because of the higher voltage applied across the screen to accel gap for a given net (beam) voltage.

Aside from mechanical tolerances, the minimum "hot-gap" grid separation, l_g, is limited is by the vacuum breakdown field of the grid material:

$$E = \frac{V}{l_g} < E_{\text{breakdown}}. \tag{6.2-12}$$

In practice, grid breakdowns initiated by arcing or small micro-discharges between the grids cause "recycles" in which the voltages are temporarily removed to extinguish the arc and then reapplied. It is common to also decrease the discharge plasma density during a recycle so that the reapplication of the acceleration voltages corresponds with ramping up the discharge current such that the accelerator approximately tracks the right perveance during start up. This minimizes ion bombardment of the accel grid during a recycle. To obtain reliable operation and avoid frequent recycles, the maximum field strength in the ion thruster is typically set to less than half the vacuum breakdown field. For example, if the grid spacing were a millimeter and the acceleration potential between the grids a thousand volts, the theoretical maximum xenon ion beam current density would be 15 mA/cm^2. A 25 cm diameter, uniform-profile beam with a 75% transparent grid system would then produce about 5.5 A of beam current. In practice, because of high voltage breakdown considerations, the maximum beam current obtainable from grid sets is typically about half the theoretical maximum.

The ion thruster size is determined by the perveance limit on the beam current density and practical considerations on the grids such as maximum grid transparency and electric field [1]. For this reason, ion thruster beam current densities are typically on the order of a tenth that found in Hall thrusters, resulting in a larger thruster footprint on the spacecraft. Alternatively, the maximum Isp that is achievable is limited by the voltage that can be applied to the grids to extract a given current density before electrical breakdown or electron backstreaming occurs [9]. Very high Isp thrusters (>10,000 s), with a size that depends on the thrust requirement, have been built and successfully tested.

6.3 Ion Optics

While the simple formulas above provide estimates of the ion accelerator optics performance, a number of computer simulation codes have been developed [4, 10–17] to more accurately evaluate the ion trajectories produced by thruster grids. Ion optics codes solve in two or three dimensions the combined ion charge density and Poisson's equations for the given grid geometry and beamlet parameters [18]. These codes have been used for the design and analysis of two- and three-grid systems, and were extended to four-grid systems [19] to examine "two-stage" ion optics performance [20] for very high voltage, high Isp applications.

6.3.1 Ion Trajectories

There are a number of codes that calculate ion trajectories and grid performance in ion thrusters, and an extensive analysis of ion optics behavior in thrusters was recently completed by Farnell [21]. An example of a multi-dimensional code is CEX, which is an ion optics code developed at JPL that calculates ion trajectories and charge exchange reactions between beam ions and un-ionized propellant gas in two [4] and three [17] dimensions. The CEX-2D code solves Poisson's equation, given by Eq. 3.7-8 in Chapter 3, on a regular mesh in cylindrical geometry. The code models a single set of screen and accel grid holes, and assumes cylindrical symmetry. The computational space is divided into a grid of rectangular cells with up to 400 increments radially and 600 axially. The radial grid spacing is uniform; the axial spacing is allowed to increase in the downstream direction. The computational region is typically a few millimeters radially and up to 10 cm along the axis downstream of the final grid. With a few exceptions, the code uses a combination of algorithms used in earlier optics codes for ion thrusters [11–15].

Upstream of the accelerator grid, the electron density is obtained analytically from Eq. 3.5-10 assuming a Maxwellian electron distribution:

$$n_e(V) = n_e(0)e^{\left(\frac{\phi-\phi_o}{T_e}\right)}. \tag{6.3-1}$$

The upstream reference electron density, $n_e(0)$, is set equal to the input discharge chamber ion density. Downstream of the accelerator grid, the electron population is also assumed to be a Maxwellian with a different reference potential:

$$n_e(V) = n_e(\infty)e^{\left(\frac{\phi-\phi_\infty}{T_e}\right)} \tag{6.3-2}$$

where the downstream reference electron density, $n_e(\infty)$,is set equal to the calculated average downstream ion beam density. As a result, downstream potentials are determined self-consistently; there is no need to assume a neutralization plane. These codes include focusing effects and the fact that the aperture dimensions are usually significantly larger than the gap size such that the electric fields are reduced from the ideal maximum.

The potential distributions are calculated using an optimized pre-conditioned least square conjugate gradient sparse matrix solver. Results for a given upstream plasma number density, n, are found by starting from zero density and iterating. At each iteration, i, a fraction, α, of the desired discharge chamber ion density is blended into the code:

$$\begin{aligned} n^0 &= 0 \\ n^{i+1} &= (1-\alpha)n^i + \alpha n \end{aligned} \tag{6.3-3}$$

The density that the code uses asymptotically approaches the final density:

$$n - n^i = n(1-\alpha)^i. \tag{6.3-4}$$

If α is sufficiently small, approximate results for all upstream densities less then n can be obtained in a single run:

$$n^i = \left[1-(1-\alpha)^i\right]n. \tag{6.3-5}$$

By saving the intermediate results, only a single run is needed to estimate the performance of an optics design over a wide range of discharge chamber densities. However, since the calculation is fully converged only at the final density, separate runs with different final densities may be necessary to obtain accurate results over the full range of discharge chamber ion densities. A typical CEX-2D calculation takes a few minutes on a personal computer. Ion optic assemblies designed using the CEX-2D code have met the predicted performance very closely [4], illustrating that grid design techniques are very mature.

The ion density in the beamlet is obtained in the codes by tracking representative ion trajectories and accounting for charge exchange collisions that alter the ion energy. Ions enter the computational region from the upstream boundary at the Bohm velocity, and their charge density is found by following their trajectories in a stationary electric field. This is in contrast to the time dependent Particle-In-Cell (PIC) technique generally used in plasma physics simulations.

An example of ion trajectories calculated by CEX-2D is shown in Fig. 6-6, which shows the computational space with the dimensions given in meters used for three values of beam perveance for half a beamlet in a three-grid configuration. In this figure, ions from the discharge chamber enter from the left and are accelerated by the electric field between the screen and accel grids. The horizontal boundaries represent lines of symmetry such that an ion crossing at these boundaries has another ion coming in from outside the domain. The top figure (a) shows an *over- perveance* condition representing too high a beamlet current for the applied voltage, or too low a voltage for the plasma density and ion current provided. In this case, ions directly impinge on the upstream face of the accel grid. This situation is considered to be the *perveance limit*, where excessive ion current

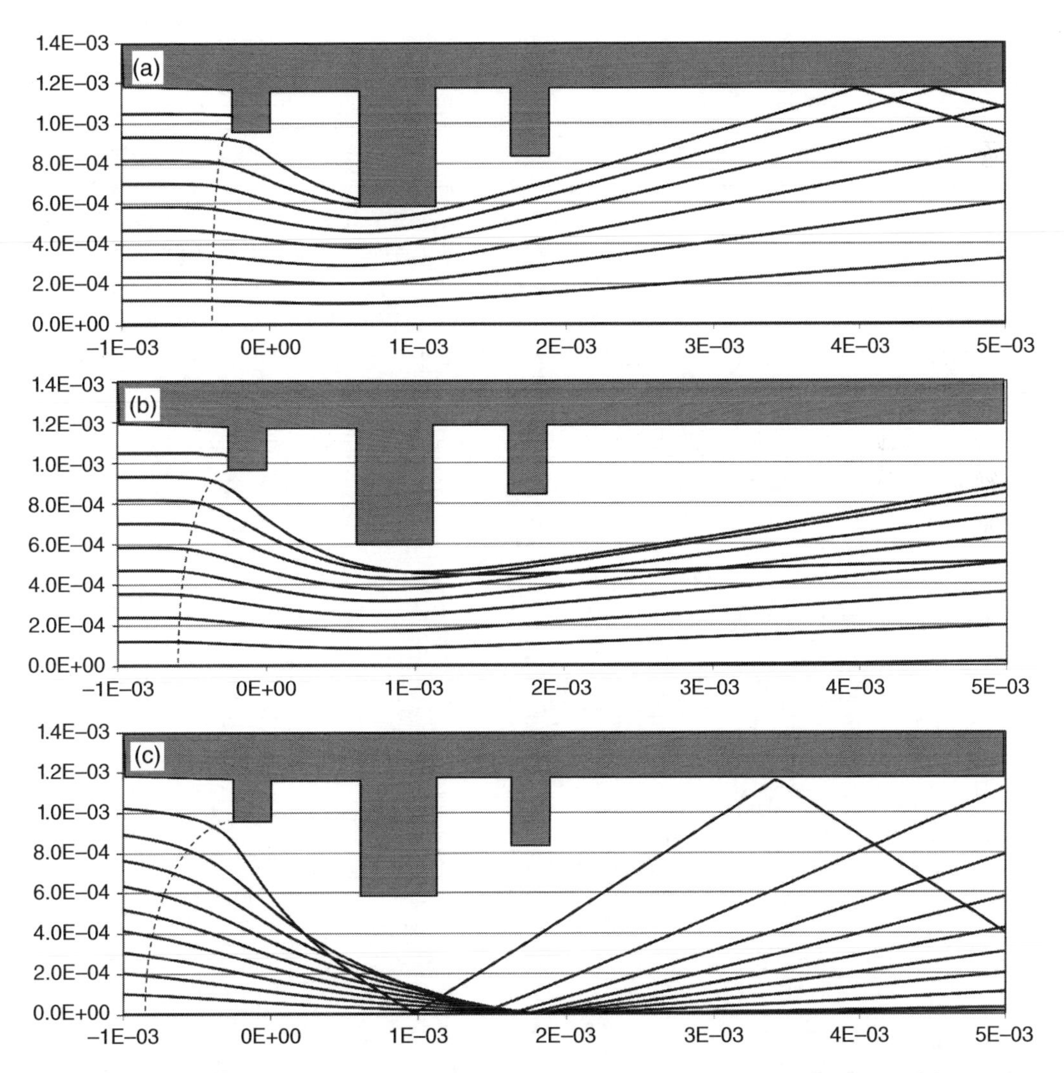

Figure 6-6 Representative ion trajectories from a CEX2D calculation for three perveance conditions: (a) over-perveance with direct accel grid interception, (b) optimal perveance, and (c) under-perveance that can produce cross-over interception.

strikes the accel grid. The middle figure (b) shows a near-optimum perveance condition where the ions are well focuses through the accel and decel grid apertures, and do not directly intercept any downstream grid. Finally, the bottom figure (c) shows an *under-perveance* condition where the ions are over focused and cross over in the accel gap. In this case, ions can directly intercept the accel grid and, eventually, the decel grid as the apertures wear open. Note that the length of the computational region shown must be long compared to its radius and is usually chosen so that neighboring beamlets will overlap.

A fraction of the ions from the plasma at the largest radii run directly into the screen grid, as seen in Fig. 6-6, and do not enter into the thrust beam. These ions represent the effect of the finite screen grid transparency that was so important in the discharge loss calculations in Chapter 5. For the near-optimal and under-perveance conditions, the screen grid transparency is greater than its geometric open area fraction, as mentioned above, because the self-consistent electric fields actually extract some of the ions at large radii that would have hit the screen grid instead of going into the screen aperture.

6.3.2 Perveance Limits

Figure 6-6 demonstrated that electrostatic accelerators produce focused ion trajectories when operated near a given design perveance and avoid grid interception or large beam divergence angles over a limited range of voltages and currents that are related by space charge considerations in the grid gap. In ion thrusters, operating sufficiently away from the perveance design of the grids results in beam interception on the downstream accel and (eventually) decel grids. Figure 6-7 shows an illustration of the accel grid current as a function of the current in a beamlet (a single aperture) for three different beam voltages. In this case, the optics were designed to run at about 2 kV and 0.8 mA of beamlet current, and the design demonstrates low grid interception over about ±50% of this current. As the beamlet current is increased, by raising the plasma density in the discharge chamber, the sheath thickness in the acceleration gap decreases, which flattens the sheath and causes the accel grid interception to increase. Eventually, the system becomes under-focused at the *perveance limit* where a large fraction of the beamlet is intercepted, as shown in Fig. 6-6a. The accel grid

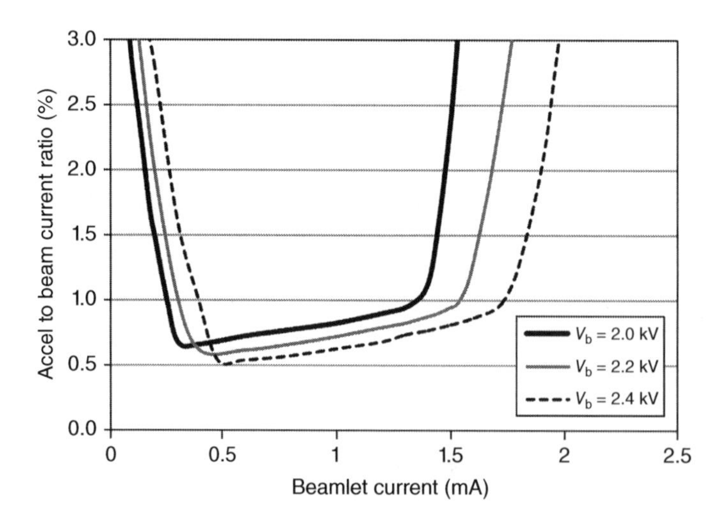

Figure 6-7 Accel grid current to beam current ratio as a function of the beamlet current for three values of the beam voltage.

current then increases rapidly with beamlet current due to the system running at too high a perveance. At low discharge chamber plasma densities, which produce low beamlet currents, the beam is over-focused and interception of the ions on the accel grid due to cross-over trajectories increases the accel grid current. The ion trajectories for this case are shown in Fig. 6-6c.

At the nominal beam voltage of 2 kV, this system can be run from about 0.4–1.2 mA of beamlet current between the cross-over and perveance limits without producing excessive accel grid current. If the ion thruster has a current profile of greater than about 3:1 peak to edge over the grid diameter (due to a poor plasma density uniformity), then grid interception will occur either in the center or at the edge of the beam. Since the grids are normally designed to deal with the high perveance condition at the peak current density near the axis, poor plasma profiles usually result in significant erosion of the edge holes due to cross-over interception. This will impact the life of the thruster and must be compensated by either changing the grid gap or beamlet aperture sizes as function of the radius or modifying the plasma generator to produce more uniform profiles.

Increasing the beam voltage shifts the curves in Fig. 6-7 to higher beamlet currents. This is clear from the dependence in the Child-Langmuir equation (Eq. 6.3-2) where the current scales as $V^{3/2}$ if the sheath thickness and grid dimensions are held constant. In Fig. 6-7, the perveance-limited beamlet current, where direct grid interception occurs, increases as $V^{3/2}$ as the beam voltage is raised. Figure 6-7 also illustrates that in situations where the thruster power must decrease, which is typical of deep space solar electric propulsion missions where the power available decreases as the spacecraft moves away from the sun, the beam voltage and Isp of the thruster must eventually decrease as the current is reduced to avoid grid interception.

The voltage range available from a given accelerator design at a fixed (or nearly constant) beam current has similar limitations as the current dependence just discussed. However, the minimum voltage at a given current is of special interest in an ion thruster because this is related to the minimum Isp of the engine for a given thrust. The *perveance-limit* of a thruster is usually defined relative to the rate at which the accel current increases due to interception as the beam voltage is decreased:

$$\text{Perveance limit} \equiv -0.02\frac{I_A}{V_{\text{screen}}}\left[{}^{mA}/_{V}\right]. \tag{6.3-6}$$

This is related to the optics situation illustrated in Fig. 6-6a where the current at a given voltage is too high for the designed gap and aperture size and the under-focused beamlet starts to directly intercept accel grid. Figure 6-8 shows the behavior of the accel grid current for the NSTAR engine operating at the full power parameters of TH15 but with the screen voltage decreasing. In this case, the perveance limit is found to be at 688.8 V, compared to the nominal 1100 V of the screen voltage at this throttle level. The perveance limit can also be defined by a given percentage increase in the accel current. However, the screen grid transparency usually decreases as the screen power supply voltage is decreased, which reduces the beam current and accel current during this measurement. The magnitude of the percentage increase in the accel current due to direct ion impingement then needs to be defined for the ion optics assembly.

6.3.3 Grid Expansion and Alignment

A significant issue in ion thrusters that utilize refractory metal grids is thermal expansion of the grids during thruster operation changing the acceleration gap dimension between the screen and accel grids. This will directly affect the ion trajectories and the perveance of the ion optics. Since the screen grid is heated by direct contact with the discharge plasma, and is usually dished outwards

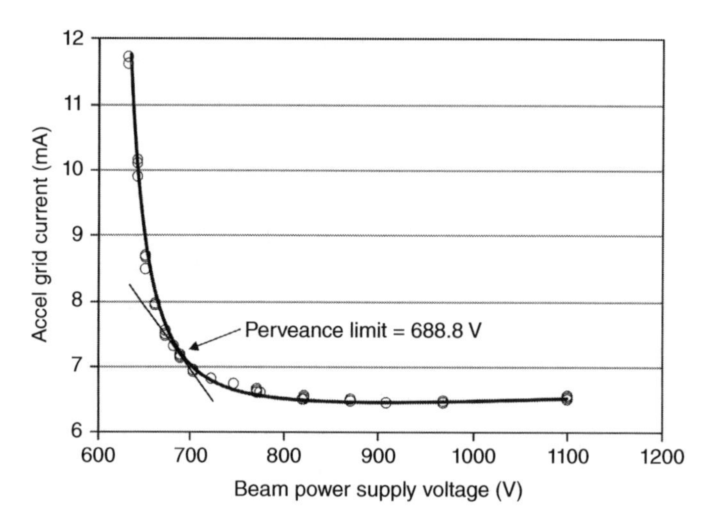

Figure 6-8 Accel grid current versus the screen supply voltage for the NSTAR thruster at TH15 showing the perveance limit.

and designed with a minimum thickness to increase the effective transparency, the screen grid expansion is usually larger than the accel grid and the gap tends to decrease as the thruster heats up. This shift from the *cold gap* to the *hot gap* causes the perveance of the optics to increase for convex grid curvature (grids domed outward from the thruster body) and changes the beamlet trajectories at the given operating point. In addition, for grids designed to hold the applied voltage across the cold gap, the hot gap may be so small that field emission and high voltage breakdown become problems. For ion thrusters with refractory metal grids designed with concave grid curvature (grids domed into the thruster body), the screen grid expands away from the accel grid and the perveance decreases as the gap gets larger. In addition, concave grids have a smaller discharge chamber volume for a given thruster size, which adversely affects the discharge loss.

Ideally, the ion optics design would have sufficient margin to operate at full power over the range that the grid gap changes. This is possible for smaller thrusters and/or lower power levels where the grid deflection is a small fraction of cold gap. For thrusters with grid diameters greater than 15–20 cm operating at power levels in excess of 1 kW, it is often necessary to design the optics for the highest power case with the small hot gap, and start the thruster in the diode-mode (discharge only) or at lower beam powers to pre-heat the grids to avoid breakdown during thermal motion and establish the grid gap dimension within the range the optics can tolerate for high power operation with minimal grid interception. It should be noted that grids fabricated from the various forms of carbon (graphite, carbon-carbon composite, or pyrolytic) have smaller or negligible thermal expansion than refractory metal grids and will have smaller grid gap changes. Ion optics sets that utilize grids made of two different materials have to deal with this issue of different thermal expansion coefficients and potentially larger grid gap changes.

Another significant grid issue is alignment of the grid apertures. The ion trajectories shown in Fig. 6-6 assumed perfect alignment of the screen and accel grid apertures, and the resultant trajectories are then axisymmetric along the aperture centerline. Displacement of the accel grid aperture relative to the screen grid centerline causes an off-axis deflection of the ion trajectories, commonly called *beam steering*. The effect of aperture displacement on the beamlet steering has been investigated for many years in both ion sources and ion thrusters [22–25]. The beamlet is steered in the direction opposite to the aperture displacement due to the higher focusing electric field induced at the accel grid aperture edge. Studies of this effect in ion thruster grid geometries [24] show that small aperture displacements ($\approx$10% of the screen aperture diameter) cause a deflection in the beamlet angle of up to about 5°. This phenomenon can be used to compensate for the curvature of the grids to reduce the overall beam divergence, which is called *compensation*. However, the perveance of the aperture is reduced in this case and interception of edge ions on the accel grid due to the non-uniform electric fields can be an issue. Mechanical misalignment of the grids due to manufacturing tolerances or thermal deformation can also produce aperture displacement and unintended beamlet steering. This problem has been identified as the cause of thrust vector variations observed as thrusters heat up [24], and for the production of azimuthal ion velocities that produce "swirl torque" on the spacecraft. For this reason, precise alignment of the grid apertures and grid support mechanisms that minimize non-uniform thermal deformation are generally required to provide stable ion optics performance with minimal beam divergence.

6.4 Electron Backstreaming

Downstream of the accelerator grid, the ion beam is charge and current neutralized by electrons from the neutralizer hollow cathode. Since electrons are much more mobile than ions, a potential barrier is needed to stop neutralizer electrons from flowing back into the positively biased discharge chamber. In the absence of a potential barrier, the electron current would be several hundred times the ion current, wasting essentially all of the electrical power. The potential barrier is produced by the negatively biased accel grid. The minimum potential established by the accel grid prevents all but the highest energy electrons from traveling backwards from the beam plasma into the discharge chamber. The so-called "backstreaming" electron current is not only a parasitic power loss since these electrons do not add thrust, but they can damage the thruster by overheating the internal component of the discharge chamber such as the cathode.

The accel grid bias voltage required to limit the electron backstreaming current to a small value (typically <1% of the beam current) can be determined by evaluating Poisson's equation in the grid aperture in the presence of the beamlet ion current with 2-D computer codes. An example of such a calculation is shown in Fig. 6-9 where the potential between the electrodes and on the axis of the half-beamlet is shown. Note that the potential minimum in the center of the beamlet is only a small fraction of the applied accel grid voltage in this example, which is due to the beam's space charge. The actual value of this minimum potential determines the margin to backstreaming, which should be set well above the value at which excessive backstreaming occurs.

Examining electron backstreaming in more detail shows that the minimum potential in the accel grid is determined by three factors: The electrostatic potential from the bias voltages applied to the different grids, the beamlet space charge in the accel grid aperture, and the required potential difference between the beam plasma and minimum voltage to reduce the backstreaming electron current to insignificant levels. Each of these factors can be evaluated analytically using simplifying approximations to help in understanding backstreaming physics.

As stated above, the backstreaming electron current results from the tail of the beam Maxwellian electron distribution overcoming the potential barrier established in the accel grid aperture. The current of electrons backstreaming into the thruster plasma is just the beam plasma random electron flux times the Boltzmann factor for the potential difference between the beam plasma and the minimum potential in the accel grid region [26]:

$$I_{\text{eb}} = \frac{1}{4} ne A_a \left(\frac{8\, kT_e}{\pi m} \right)^{1/2} e^{\frac{-\left(V_{\text{bp}} - V_m\right)}{T_{eV}}}, \tag{6.4-1}$$

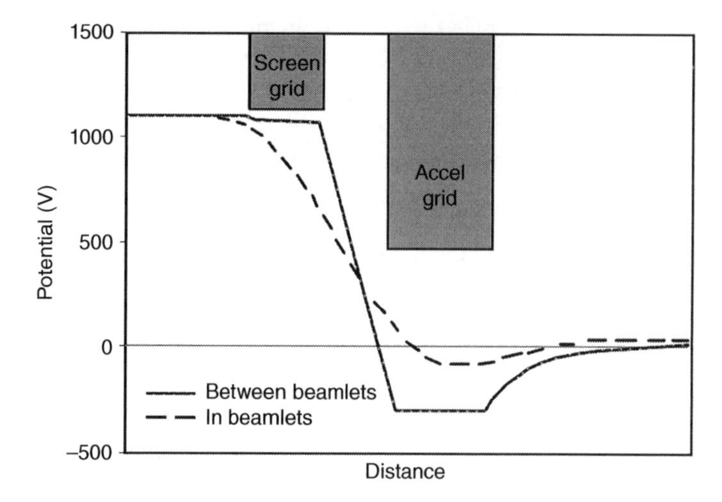

Figure 6-9 Potentials on-axis in an individual beamlet and between the beamlets intersecting the grids.

where I_{eb} is the electron backstreaming current, V_{bp} is the beam plasma potential, V_m is the minimum potential in the grid aperture and A_a is the beamlet area in the grid aperture. The current of ions in the beamlet flowing through the grid aperture is

$$I_i = n_i e v_i A_a, \tag{6.4-2}$$

and the ion velocity through the system is

$$v_i = \sqrt{\frac{2e\left(V_p - V_{\text{bp}}\right)}{M}}, \tag{6.4-3}$$

where V_p is the plasma generator plasma potential at the sheath edge. Combining Eqs. 6.4-1–6.4-3, the minimum potential is

$$V_m = V_{\text{bp}} + T_{eV} \ln\left(\frac{2I_{\text{eb}}}{I_i}\sqrt{\pi\frac{m}{M}\left(\frac{V_p - V_{\text{bp}}}{T_{eV}}\right)}\right). \tag{6.4-4}$$

This equation describes the required potential difference between the beam potential and the minimum potential in the beamlet to produce a specified amount of electron backstreaming current relative to the beam current. Note that this equation is independent of the grid geometry because it deals solely with the potential difference between a given value of V_m (independent of how it is produced) and the beam plasma potential. The required potential difference ($V_{\text{bp}} - V_m$) between the beam plasma and the minimum voltage in the grids to produce a given ratio of backstreaming current to beam current is plotted from Eq. 6.4-4 in Fig. 6-10 for several values of the beam plasma electron temperature in a thruster plume with a net accelerating voltage of $V_p - V_{\text{bp}} = 1500$ V. For an electron temperature of 2 eV in the beam, which is consistent with values found in NSTAR thruster plumes [27], a potential difference between the minimum potential in the beamlet and the beam plasma of only 12.5 V is required to reduce the backstreaming current to 1% of the beam current.

The actual minimum potential in the beamlet is determined by the grid geometry, the applied grid potentials, and the beam's space charge. The minimum potential in the two-grid

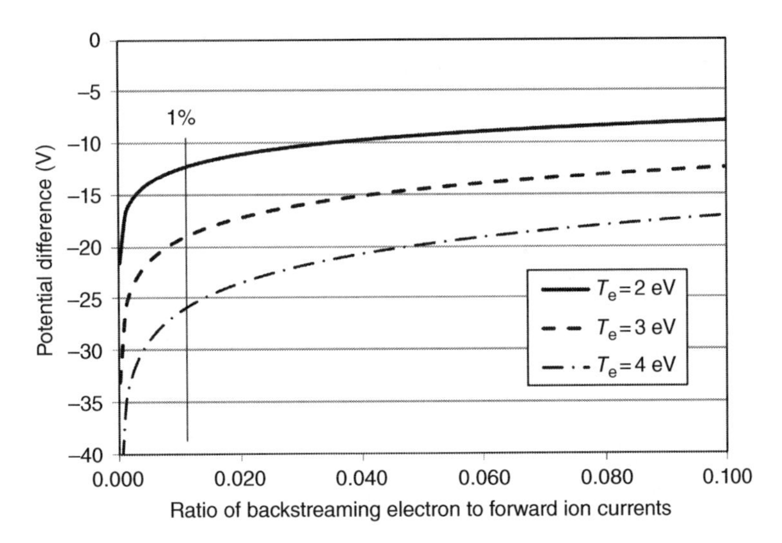

Figure 6-10 Potential difference between the beam plasma and the beamlet potential minimum required to achieve a given ratio of the electron backstreaming current to forward ion current for several beam electron temperatures.

arrangement shown in Fig. 6-5 was first found without considering space charge effects by an analytic solution to Laplace's Equation by Spangenberg [28] for thin grids in vacuum tubes. Spangenberg's expression was simplified by Williams [26] and Kaufman [1] for most ion thruster grid configurations to

$$V_m^* = V_a + \frac{d_a(V_P - V_a)}{2\pi l_e}\left[1 - \frac{2t_a}{d_a}\tan^{-1}\left(\frac{d_a}{2t_a}\right)\right]e^{-t_a/d_a}, \tag{6.4-5}$$

where V_m^* indicates the minimum potential with the ion space charge neglected, V_a is the applied accel grid potential, the grid dimensional terms are defined in Fig. 6-5, t_a is the accel grid thickness, and l_e is given by Eq. 6.3-3. Eq. 6.4-5 provides the dependence on the geometry of the grids, but is only useful if the beam space charge is negligible (very low current density beamlets).

The reduction in the magnitude of the minimum beam potential due to the presence of the ion space charge in the beamlet can be calculated [26] using the integral form of Gauss's Law:

$$\oint_S \mathbf{E} \cdot d\mathbf{A} = \frac{1}{\varepsilon_o}\int_V \rho dV \tag{6.4-6}$$

where **E** is the electric field, d**A** is the differential surface area element, ε_o is the permittivity of free space, and ρ is the ion charge density within the Gaussian surface which has a surface area S and encloses volume V. This equation is solved first in the beamlet, then in the charge-free space between the beamlet and the accel aperture inside diameter, and then adding the two potentials together gives the total potential between the grid and the beamlet centerline.

Assume that the beamlet has a radius $d_b/2$ inside the accel grid aperture with a radius of $d_a/2$. Integration of the left-hand-side of Eq. 6.4-6 over a cylindrical "Gaussian pillbox" aligned with the beamlet axis yields

$$\oint_S \mathbf{E} \cdot d\mathbf{A} = \int_0^{2\pi}\int_0^{r_a} E_r r d\theta dz = E_r\, 2\pi rz \tag{6.4-7}$$

where it has been assumed that E_r is constant in the axial direction over a distance z. If it is also assumed that the ion charge density is uniform in the volume of the pillbox, the right-hand side of Eq. 6.4-6 can also be integrated to obtain

$$\frac{1}{\varepsilon_o}\int_V \rho\, dV = \frac{1}{\varepsilon_o}\int_V \rho\, r\, dr\, d\theta\, dz = \frac{\rho}{\varepsilon_o}\pi r^2 z \tag{6.4-8}$$

Equating Eqs. 6.4-7 and 6.4-8, an expression for the radial electric field in the beamlet (E_{r1}) from the accel hole centerline to the outer edge of the beamlet is obtained:

$$E_{r1} = \frac{\rho r}{2\varepsilon_o}, \left(0 < r < \frac{d_b}{2}\right) \tag{6.4-9}$$

From the edge of the beam to the wall, Gauss's law is again used, but in this case the entire beam charge is enclosed in the Gaussian surface. The radial electric field in this "vacuum region" outside the beamlet (E_{r2}) is then found in a similar manner to be

$$E_{r2} = \frac{\pi d_a^2}{8\varepsilon_o r}\left(\frac{d_b}{2} < r < \frac{d_a}{2}\right) \tag{6.4-10}$$

The voltage difference ΔV from the centerline to the accel grid barrel due to the ion space charge is obtained by integrating the electric field between these limits. Hence:

$$\Delta V = -\int_0^{d_b/2} E_{r1} dr - \int_{d_b/2}^{d_a/2} E_{r2} dr = -\int_0^{d_b/2} \frac{\rho\, r}{2\,\varepsilon_o} dr - \int_{d_b/2}^{d_a/2} \frac{\rho\, d_b^2}{8\,\varepsilon_o r} dr \tag{6.4-11}$$

The total potential from the accel wall to the center of the beamlet due to ion space charge is then

$$\Delta V = \frac{\rho d_b^2}{8\varepsilon_o}\left[\ln\left(\frac{d_a}{d_b}\right) + \frac{1}{2}\right]. \tag{6.4-12}$$

The beam current density in the accel aperture is the charge density times the beam velocity, so the ion charge density ρ is

$$\rho = \frac{4\, I_i}{\pi d_b^2\, v_i}, \tag{6.4-13}$$

where v_i is the ion velocity evaluated at the minimum potential point:

$$v_i = \sqrt{\frac{2e(V_p - V_m)}{M}}. \tag{6.4-14}$$

Substituting Eqs. 6.4-13 and 6.4-14 into Eq. 6.4-12 gives

$$\Delta V = \frac{I_i}{2\pi\varepsilon_o v_i}\left[ln\left(\frac{d_a}{d_b}\right) + \frac{1}{2}\right]. \tag{6.4-15}$$

Since scalar potentials can be added, the sum of Eq. 6.4-15 and Eq. 6.4-5 gives the total of the potential minimum in the accel grid aperture.

$$V_m = V_a + \Delta V + \frac{d_a(V_{bp} - V_a)}{2\pi l_e}\left[1 - \frac{2t_a}{d_a}\tan^{-1}\left(\frac{d_a}{2t_a}\right)\right]e^{-t_a/d_a}. \tag{6.4-16}$$

To calculate the backstreaming current as a function of grid voltage, Eq. 6.4-16 must be equated to Eq. 6.4-4 and solved for the current:

$$\frac{I_{\text{be}}}{I_i} = \frac{e^{\left(V_a + \Delta V + \left(V_{\text{bp}} - V_a\right)C - V_{\text{bp}}\right)/T_e}}{2\sqrt{\frac{\pi\, m}{M}\frac{\left(V_p - V_{\text{bp}}\right)}{T_e}}} \tag{6.4-17}$$

where the geometric term C is given by

$$C = \frac{d_a}{2\pi l_e}\left[1 - \frac{2t_a}{d_a}\tan^{-1}\left(\frac{d_a}{2t_a}\right)\right]e^{-t_a/d_a}. \tag{6.4-18}$$

In practice, the onset of backstreaming is determined by two techniques. One method is to monitor the increase in the screen power supply current as the magnitude of the accel grid voltage is decreased. Increases in the measured current are due to backstreaming electrons, and a 1% increase is defined as the minimum accel grid voltage to avoid backstreaming: the so-called *backstreaming limit*. For example, the power supply current from Eq. 6.4-17, normalized to the initial beam current, is plotted in Fig. 6-11 as a function of the accel grid voltage for the NSTAR ion optics [29] for the maximum power throttle point TH15 at the beginning of life (BOL). In this figure, the beam potential and electron temperature were assumed to be 12 V and 2 eV respectively, consistent with measurements made on this thruster. The onset of backstreaming occurs at about −150 V on the accel grid, which is consistent with the data from tests of this engine [30, 31].

A second method to determine the backstreaming limit is to monitor the ion production cost, which is the discharge power required to produce the ion beam current divided by the beam current. This is an effective method for use in thrusters operating in the beam current regulated mode where the discharge power supply is controlled to fix the beam current. Backstreaming then appears as a decrease in the ion production cost. This method is shown in Fig. 6-12 for the experimental data taken from the NSTAR thruster at TH15. As the magnitude of the accel voltage is decreased, a 1% decrease in the ion production cost represents the defined onset of backstreaming.

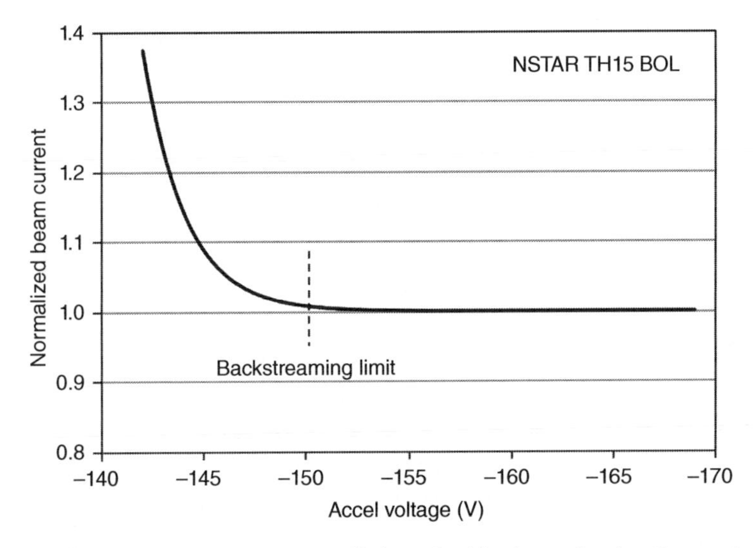

Figure 6-11 Normalized beam current versus applied accel grid voltage showing the onset of electron backstreaming as the voltage is decreased.

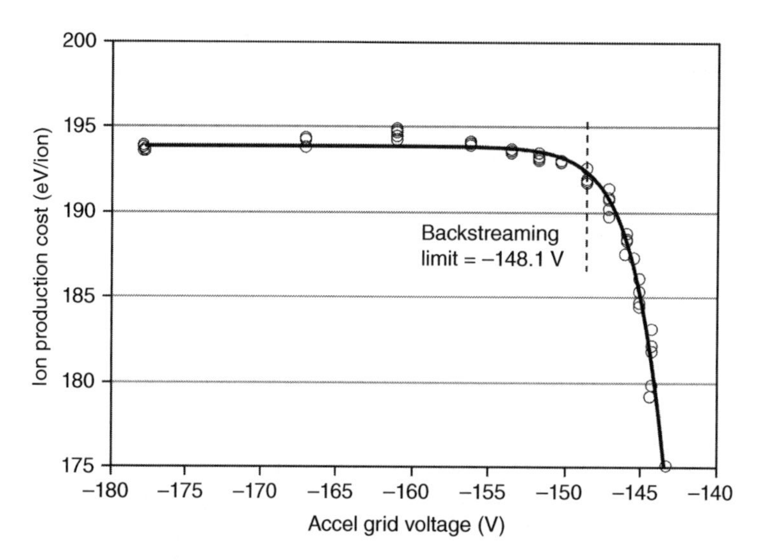

Figure 6-12 Ion production cost for NSTAR TH15 versus applied accel grid voltage showing the onset of electron backstreaming as the grid bias voltage is decreased.

In this case, the backstreaming limit was determined to be about -148 V, consistent with the above analytical model.

Equations 6.4-17 and 6.4-18 show that the electron backstreaming is a function of the accel grid hole diameter. Increases in the accel hole diameter will reduce the penetration of the applied grid bias voltage to the center of the aperture and reduce the minimum potential on axis. The larger accel hole diameter increases either the backstreaming current at a given voltage, or the backstreaming limit at a given current. The effect of accel grid hole enlargement due to grid-wear is illustrated in Fig. 6-13, where the grid voltage at which backstreaming started is plotted versus accel

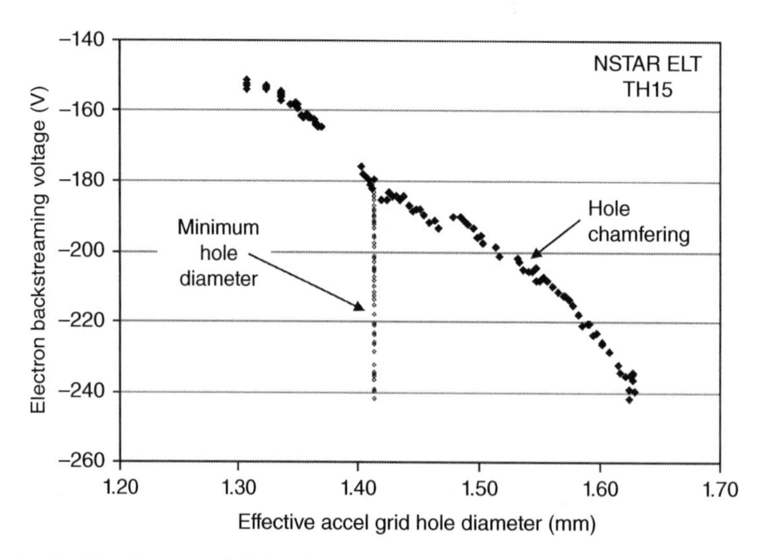

Figure 6-13 Accel grid voltage at which backstreaming occurs in the NSTAR thruster at TH15 power level versus the effective accel grid aperture diameter.

grid hole diameter for the NSTAR TH15 case measured during the ELT life test [31]. Larger grid-hole diameters required more negative biasing of the accel grid to avoid the onset of backstreaming.

Figure 6-13 also shows an interesting effect in that the shape of the grid hole is important. Early in life, the grid aperture diameter eroded due to sputtering, and the barrel diameter was adequately described by the minimum hole diameter observed optically during running of the test. However, as the test progressed, the erosion of the upstream aperture edge essentially stopped and the aperture was observed to be chamfered on the downstream portion. An effective grid diameter had to be calculated to take into account the non-uniform hole erosion in determining the backstreaming onset, shown on the right-hand-side of Fig. 6-13. While the analytical model above accounts for grid diameter and thickness, additional terms would have to be added to account for this conical erosion shape. This situation is best handled by 2-D models that determine both the time dependent shape of the grid hole and calculate the potential on axis appropriately.

It should be noted that while the analytical model shown above illustrates the mechanisms involved in electron backstreaming and provides reasonable agreement with the experimental data shown, the results are very sensitive to the dimensions and beam parameters assumed in the calculation. This is largely because the potential minimum is the difference between two large numbers representing the contributions of the electrostatic fields and the space charge fields. Therefore, this backstreaming model actually provides only an estimate of the backstreaming voltage and current levels, which can easily be off 10–20%. The 2-D grid codes described above that solve Poisson's equation exactly provide more accurate calculations of the backstreaming limit.

Finally, electron backstreaming occurs first in the region of the highest beamlet current where the ion space charge is the highest in the ion optics assembly. Thrusters with non-uniform beam profiles, such as NSTAR with a flatness parameter (defined as average to peak current density) of about 0.5 and therefore a 2:1 peak to average current density profile [30], will tend to backstream primarily from the center beamlets. This localized backstreaming accelerates electrons on-axis and can overheat components such as the cathode at the center-back of the thruster. Thrusters designed to have flat profiles, such as NEXIS with better than a 0.9 flatness parameter [32], will tend to not backstream easily because of a lower peak ion current density for a given total beam current, and also, if backstreaming starts, it will be over a larger area that minimizes the localized heating issue in the discharge chamber.

6.5 High Voltage Considerations

As shown in Section 6.3, the maximum thrust that can be produced by an ion thruster is a function of the electric field that can be sustained between the screen and accelerator grids:

$$T_{\max} = \frac{8}{9} \varepsilon_o \gamma T_s A_g \sqrt{R} E^2. \tag{6.5-1}$$

From Eq. 6.5-1, the maximum space charge limited (sometimes called *perveance-limited*) thrust of the accelerator system is directly proportional to the intra-grid electric field squared. To produce compact ion thrusters with the highest possible thrust, it is necessary to maximize the electric field between the grids. The maximum thrust in ion engines is then limited primarily by the voltage hold-off capability of the grids.

The ability of the accelerator grids to hold off high voltage reliably and to withstand occasional breakdowns without significant damage or loss of voltage standoff capability is therefore of critical importance for ion thrusters. The high voltage behavior of vacuum-compatible materials has been summarized in recent books on high-voltage engineering [33, 34]. In plasma devices [35], electric fields of up to 40 kV/cm were found useful for refractory metal electrodes and the order of 25 kV/cm for carbon materials. Degradation of the voltage hold-off due to surface damage incurred during breakdowns has been investigated for molybdenum and carbon electrodes [35] commonly used in ion thruster applications. The surfaces of these materials can be carefully prepared to withstand high electric fields required to produce the highest thrust density. However, sputter erosion over time and electrical breakdowns between grids causes some fraction of the stored energy in the power supply to be deposited on the grid surface. The formation of an arc at the cathode electrode (the accel grid) and the deposition of a significant amount of electron power from discharge into the anode electrode (the screen grid) can cause both the screen and accel grid surfaces to be modified and/or damaged. The breakdown events usually impact the subsequent voltage hold-off capability of the grid surfaces, which affects the long-term performance of the thruster.

6.5.1 Electrode Breakdown

The grids in ion thrusters have high voltages applied across small grid gaps, which can lead to high voltage breakdown and unreliable thruster operation. High voltage breakdown is usually described in terms of the electric field applied to the surface that causes an arc or discharge to start. Arc initiation is well correlated to the onset of field emission [36, 37]. If sufficient field emission occurs due to excessive voltage or a modification to the surface that enhances field emission, the gap breaks down. Physical damage to arced surfaces during the breakdown is attributed to localized energy deposition on the electrode that causes melting or evaporation of the material. On the cathode surface (the accel grid), the energy is deposited primarily by ion bombardment from the arc-plasma. On the anode surface (the screen grid), the energy is deposited from the plasma or electron stream that crosses the gap and results in localized surface heating and vaporization. The energy provided to the arc from the power supply is distributed between any series resistance in the electrical circuit, the voltage drop at the cathode surface, and the voltage drop in the plasma discharge and anode sheath. These voltage drops can be modeled using discrete series resistances in the energy balance of the system. Engineers often rate the possibility of a power supply damaging the electrodes by the amount of stored energy in the power supply. However, the amount of material removed from the surfaces and the lifetime of high voltage electrodes is usually characterized [35] by the amount of current that passes through the arc. This "Coulomb-transfer rating" is related to the energy deposition in the electrodes in a simple manner. The power running in the arc is $P = I\,V_{arc}$, where I is the discharge current and V_{arc} is the voltage drop in the arc. Assuming that most of the voltage drop is in the cathode sheath, the energy E deposited by the arc on the cathode surface is

$$E = \int P\,dt = \int IV_{arc}dt. \tag{6.5-2}$$

The voltage drop of refractory metal and graphite arcs is nearly independent of the amount of current running in the arc up to several hundred Amperes[38, 39]. Therefore the arc voltage can be considered to be essentially a constant, and the energy deposited by the arc on the cathode is

$$E = V_{arc}\int Idt = V_{arc}\,Q, \tag{6.5-3}$$

where Q is the total charge transferred in the arc. The arc energy deposited on the cathode surface for a given electrode material is characterized by the total charge transferred by the thruster power supplies during the arc time, and not just the stored energy in the power supply. Assuming that the arc remains lit during the entire time required to discharge the output capacitor in the power supply, the total charge transferred through the arc is $Q = CV$, where C is the capacitance and V is the capacitor charging voltage. If the arc current falls below the minimum value to sustain the arc, called the "chopping current", and is prematurely extinguished, then the total charge transferred is reduced.

It should be emphasized that the amount of energy delivered to the cathode surface by the arc and the amount of damage to the surface incurred by material removal are independent of any series resistance in the circuit as long as the current is stable for the duration of the event (i.e. the current is above the chopping current). This means that simply adding a series resistor to one leg of the high voltage power supply circuit or the accel grid circuit will not reduce the surface damage due to an arc unless the arc current drops to less than the chopping current. The only mechanism that reduces surface damage if the current is large compared to the chopping current is to limit the total charge transfer. This requires either reducing the power supply capacitance at a given voltage (which reduces the total stored energy), or actively shunting or opening the circuit to reduce the arc duration.

6.5.2 Molybdenum Electrodes

Molybdenum is a standard electrode material used in ion thrusters due to its low sputter erosion rate, ability to be chemically etched to form the aperture array, and good thermal and structural properties. The surface of the molybdenum grid is often slightly texturized to retain sputtered material to avoid flaking of the sputter-deposited material [40]. The threshold voltage for the onset of field emission versus the gap spacing measured for molybdenum electrodes using a standard "plate-and-ball" test arrangement in a high vacuum facility [41] is shown Fig. 6-14. The data shows a classic power-law dependence of the threshold voltage with gap spacing for small gaps, which is sometimes called the "total voltage effect"[42]. Although there are numerous possible mechanisms for the total-voltage effect, the increased gap reduces the surface electric field and the field emission current but increases the probability of an atom or particulate being ionized while traversing the gap. The ionized atom or particle is then accelerated into the cathode potential electrode and produces secondary electrons. If sufficient ionizations and secondary electrons are produced, the

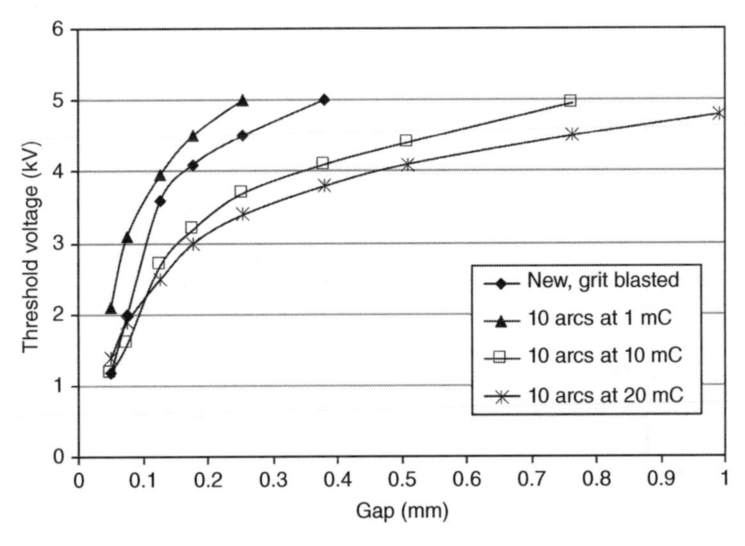

Figure 6-14 Threshold voltage versus gap for breakdown of molybdenum after 10 arcs of varying charge transfer (*Source:* [35]).

process cascades and the gap breaks-down. Therefore, the voltage that can be held across a gap does not increase linearly with the gap dimension. This is equivalent to the Paschen breakdown [34] mechanism in gas filled devices and is caused by the release of gases or particulates from the surfaces in vacuum gaps. After 10 arcs of 1 mC in charge transfer, the threshold voltage was measured again and the threshold voltage is observed to increase for every gap tested, indicating that the surface is being conditioned. Improving voltage standoff of electrodes with a series of low Coulomb-transfer arcs is common practice in the high voltage industry and often historically called "spot-knocking". This process removes small field emitters and tends to clean oxides and impurities off the surface without damaging the surface, which reduces the onset of field emission. Higher Coulomb transfer arcs on molybdenum (10 and 20 mC) improve the voltage hold-off by cleaning larger areas of the surface and removing field emission sites. This effect will continue until the surface is well conditioned, or the arc anchors in one spot and causes damage to the surface.

As the gap between the electrodes increases, the threshold voltage curves become more linear and the surface asymptotes to a constant threshold electric field. Figure 6-15 shows the threshold electric field for large gaps for a flat molybdenum surface texturized by grit blasting and actual texturized grid material with apertures chemically etched into the material. In this case, high Coulomb transfer arcs tend to damage and degrade the voltage standoff of the grids. Scanning electron

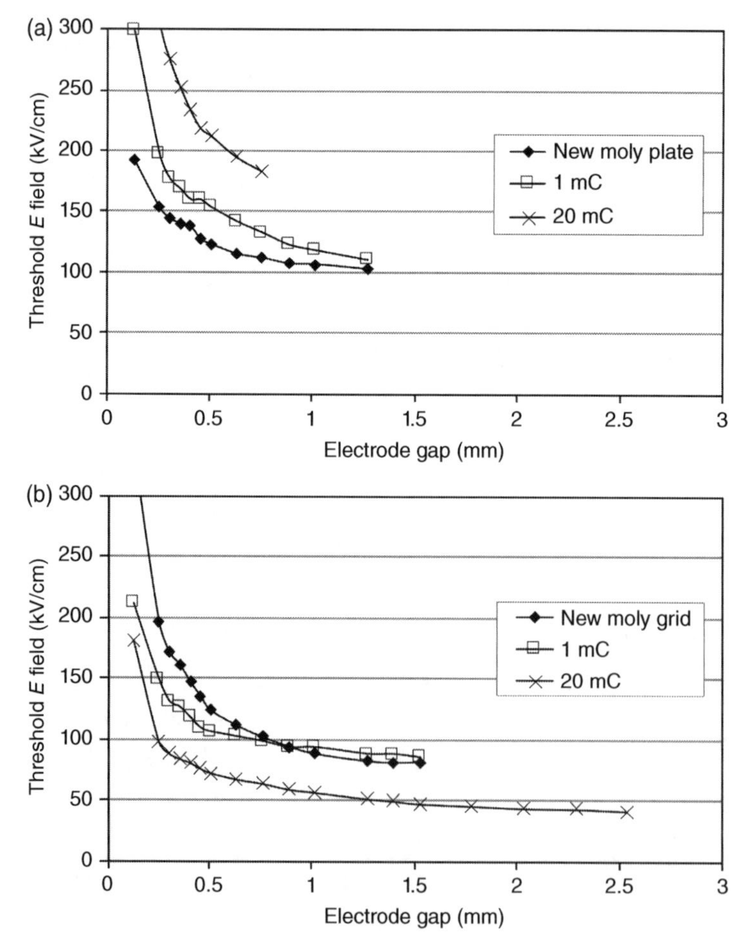

Figure 6-15 Threshold electric field versus gap for breakdown of textured molybdenum plate (a), and textured molybdenum grid material (b) (*Source:* [35]).

microscope photographs show localized damage to the edge of the beam apertures, resulting in more field emission sites. The molybdenum surfaces are initially capable of holding electric fields of well over 200 kV/cm, but the surface roughening to retain flakes and the aperture edges associated with real grids cause the voltage hold off to decrease. For molybdenum material with apertures, the resulting surface is susceptible to breakdown at electric fields of 40–50 kV/cm, which should be considered the maximum electric field for designing molybdenum grids.

6.5.3 Carbon-Carbon Composite Materials

Carbon is a desirable material for ion thruster grid electrodes because of its low sputtering yield under xenon ion bombardment [43] compared to most refractory grid materials. However, the structural properties of graphite are usually insufficient for thin graphite grids of any reasonable size (greater than about 20 cm dia.) to survive launch vibrations. Relatively thick accel grids made of high density graphite are used in the 22-cm-dia. T6 ion thruster [44, 45], but screen grids of molybdenum had to be used in that thruster because thin graphite grids were found to be too fragile to survive launch loads. Producing thin graphite screen grids and grid sets with large diameters are made possible by using carbon material such as carbon-carbon composites and pyrolytic graphite with better structural properties. Carbon-carbon (CC) composite grids have been fabricated with over 50-cm diameter [46], demonstrated low erosion in life tests and flown successfully [47]. However, the more complex structure of these materials leads to lower thresholds for field emission and less voltage standoff for grids made of these materials.

Carbon-carbon composite material used for both screen and accel grid electrodes [48] is based on carbon fibers woven into a matrix with the fibers oriented in one or two dimensions. This material has enhanced strength and flexural modulus compared to pure graphite due to the carbon fiber properties. The carbon-fiber weave is impregnated with a resin and built up to the desired shape by progressive laminate layers on a mold. The resulting material is usually densified and graphitized at high temperature, and may be further impregnated or over-coated with a thin CVD carbon layer after this process to fill any voids or smooth the final surface. High voltage breakdown tests were conducted with and without this final surface graphite coating.

The threshold voltage of the carbon-carbon composite samples is shown in Fig. 6-16, where the threshold for field emission is plotted as a function of the electrode gap for various levels of Coulomb-transfer arcing. New material (without arcing) with a fresh CVD layer has a high threshold for field emission, and therefore holds voltage well. High Coulomb-transfer arcs (>1 mC) tend to damage that surface and return it to the state of the material without the CVD over-layer. Higher Coulomb-transfer arcs also tended to damage the surface. In fact, in this example, the 10 mC arcs resulted in damage to the opposite anode electrode, which evaporated and re-deposited material back on the cathode potential surface, improving its voltage hold off capability. For this reason, the Coulomb-transfer limit for CC grids should be set to about 1 mC such that conditioning and no damage to either screen or accel grid occurs during any breakdowns.

The threshold electric field for CC material with grid apertures is shown in Fig. 6-17 for new material and after a series of arcs. After the initial characterization with 10 arcs of 1 mC each, 10 arcs of 10 mC were delivered to the surface, which degraded the voltage standoff. However, the application of 4 sets of 10 arcs of only 1 mC re-conditioned the surface. The threshold electric field was found to asymptote to just below the same 40 kV/cm field at larger gap sizes observed for low Coulomb-transfer arcs of flat material, suggesting that the aperture edges function in a similar manner as material roughness. These results suggest that carbon-carbon composite grids can be designed for reliable high voltage standoff utilizing a field emission threshold of about 35 kV/cm, even for large gaps and voltages in excess of 10 kV, provided that the Coulomb transfer is limited

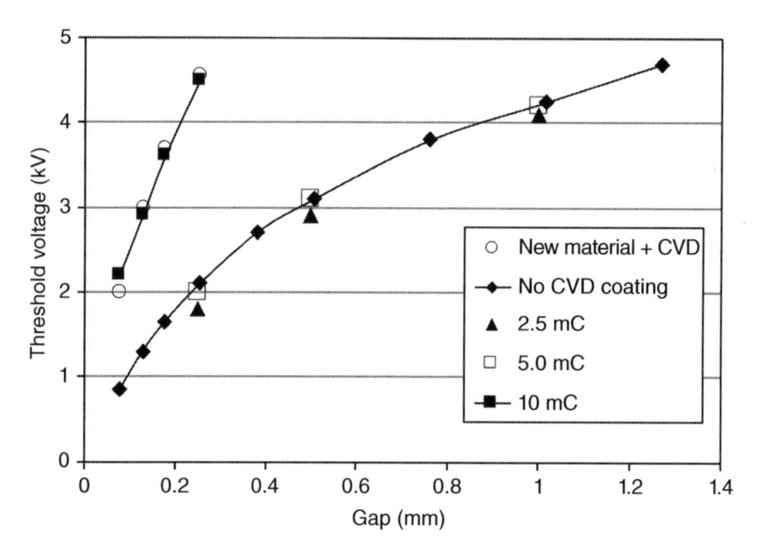

Figure 6-16 Threshold voltage for carbon-carbon composite material breakdown after 10 arcs at various Coulomb-transfers (*Source:* [35]).

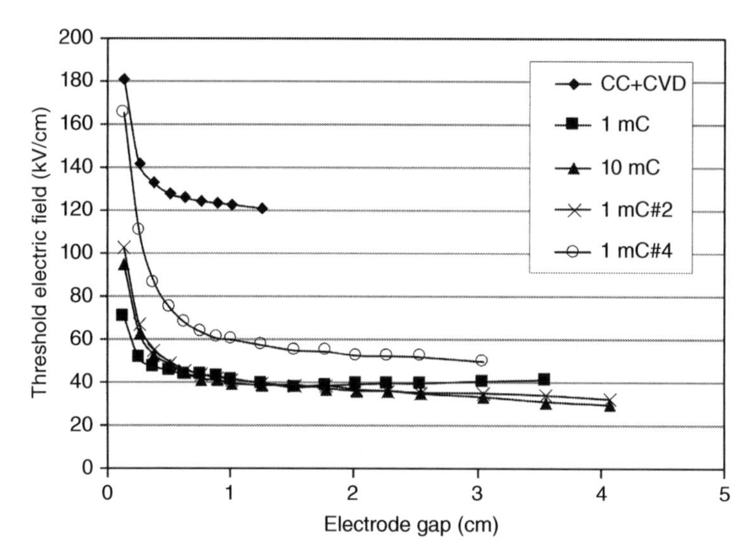

Figure 6-17 Threshold electric field for breakdown versus electrode gap for CC grid material with apertures (*Source:* [35]).

by the power supply to less than about 1 mC. This 35-kV/cm field-limit is the highest voltage stress that should be allowed, and conservative design practices suggest that a 50% margin (to ≈23 kV/cm) should be considered in designing these types of grids.

6.5.4 Pyrolytic Graphite

Pyrolytic graphite (PG) is also a candidate for accelerator grid electrodes in ion thrusters [49]. This material is configured with the carbon crystal planes parallel to the surface. Pyrolytic graphite is grown a layer at a time to near the desired shape on a mandrel and then finished machined to

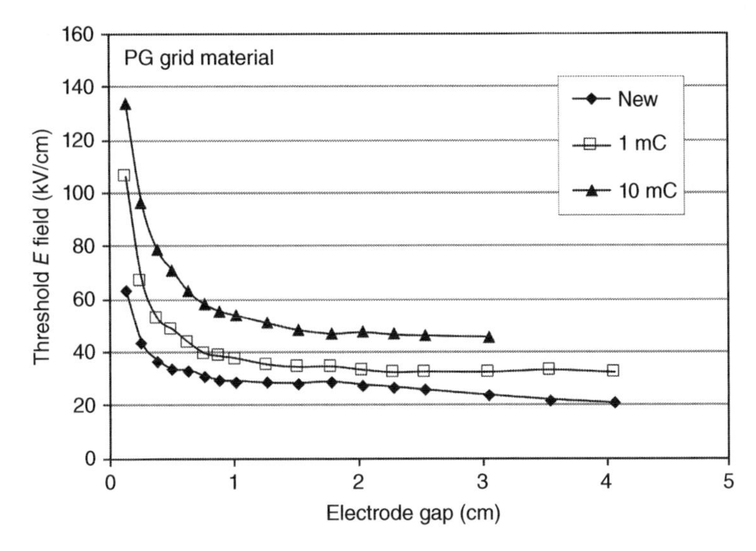

Figure 6-18 Threshold electric field for pyrolytic graphite breakdown with grid apertures (*Source:* [35]).

the final configuration. Flat test coupons were fabricated in this manner, but featured small surface bumps and depressions that were residual from the growth process. Figure 6-18 shows the behavior of a PG grid sample that had apertures laser-machined into it and then the surface lightly grit-blasted. The as-new PG material demonstrated threshold electric fields of 20–30 kV/cm for gaps of 1 mm or larger, which is lower than that found for the CC grid material. However, a series of ten 1 mC arcs tend to smooth and condition the surface and raise the threshold electric field to the order of 30 kV/cm. Higher Coulomb arcs (up to about 10 mC) also improve the voltage stand off to about 40 kV/cm. The pyrolytic graphite is more susceptible to field emission and breakdown that the carbon-carbon material, but appears to tolerate higher Coulomb-transfer arcs.

6.5.5 Voltage Hold-off and Conditioning in Ion Accelerators

Tests have shown that the arc initiation voltage is directly related to the threshold voltage and electric field for field emission in the above figures [35]. Arc initiation voltages tend to be less than 10% higher than the threshold values for field emission shown here. This is consistent with experimental observations that low levels of field emission and/or corona can be tolerated before full arc breakdown occurs, but arcing and recycling tend to increase once significant field emission starts. Molybdenum has been found to have a good tolerance for high Coulomb-transfer arcs, and grids can be designed to reliably hold electric fields well in excess of 40 kV/cm. Carbon based materials have more structure than the refractory metals and tend to form field emitters if excessive charge transfers are allowed. Nevertheless, grids utilizing carbon-based materials can be designed with electric fields in excess of 20 kV/cm if the Coulomb transfer during breakdowns is limited to about 1 mC or less. Detailed investigations of the voltage hold-off and conditioning of carbon-carbon thruster grids were performed by Martinez [8], who documented the effect for larger area grid sets. Figure 6-19 shows their reduction in field emission from carbon-carbon grids plotted on a Fowler-Nordheim plot [42] for increasing numbers of 1-mC arcs. This work shows that even if the surface of carbon-carbon grids evolve field emitters over time due to erosion from ion bombardment, proper design of the power supply to limit the Coulomb-transfer rate will result in reconditioning of the grid surfaces with every recycle event.

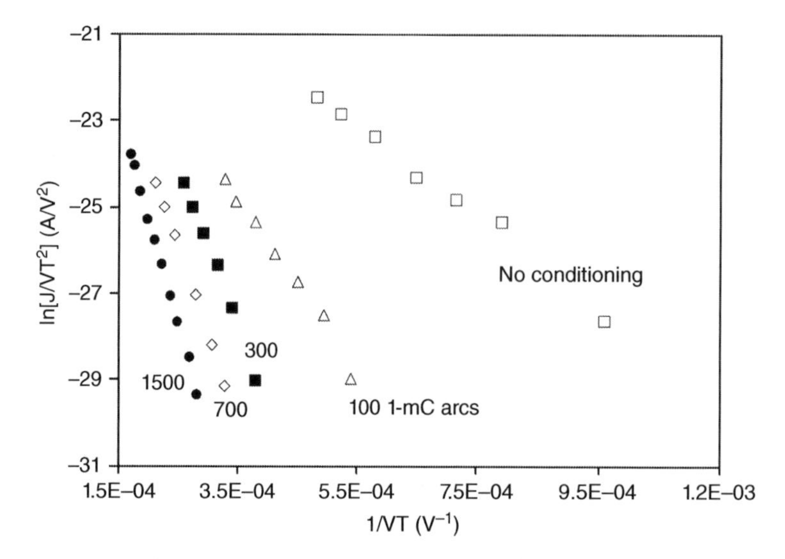

Figure 6-19 Fowler-Nordheim plots of field emission showing conditioning of carbon-carbon grids by increasing numbers of 1mC arcs from none to 1500 arcs (*Source:* [8]).

6.6 Ion Accelerator Grid Life

The most import wear mechanism in modern ion thrusters is accelerator grid erosion. Even though properly designed optics attempt to make all of the ions extracted from the discharge chamber focus through the accelerator grid apertures, a current of secondary ions generated downstream of the discharge chamber impacts the accelerator grid. These secondary ions are generated by resonant charge exchange (CEX) between beam ions and neutral propellant gas escaping from the discharge chamber. The cross section for resonant charge exchange, that is the transfer of an electron from a propellant atom to a beamlet ion, is very large; on the order of a hundred square Angstroms [50] for xenon. This process results in a fast neutral atom in the beam and a slow thermal ion. These slow ions are attracted to the negatively charged accelerator grid, and most hit with sufficient energy to sputter material from the grid. Eventually the accelerator grid apertures become too large to prevent electron backstreaming, or enough material is sputtered away that the grids fail structurally.

The erosion geometry is naturally divided into two regions. The first region, barrel erosion, is caused by ions generated between the screen grid aperture sheath and the downstream surface of the accelerator grid as shown in Fig. 6-20. Charge exchange ions generated in this region impact

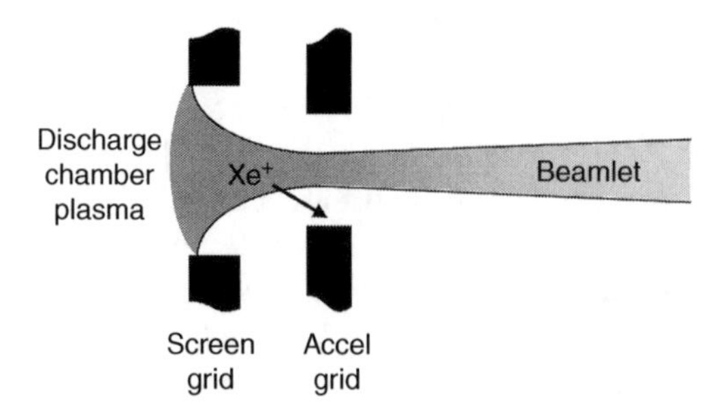

Figure 6-20 Ions that cause barrel erosion are generated by charge exchange upstream and within accelerator grid aperture.

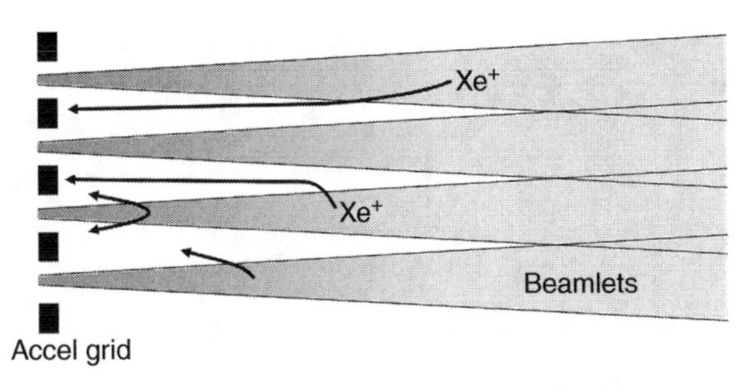

Figure 6-21 Ions that cause pits and grooves erosion are generated between the downstream surface of the accel grid and where the beamlets overlap.

the inside surface of accelerator grid aperture and results in enlargement of the aperture barrel. As the barrel diameter increases, the grid must be biased more and more negatively to establish the minimum potential required in the aperture to prevent neutralizer electrons from backstreaming into the discharge chamber. Thruster failure occurs when, at its maximum voltage, the accelerator grid power supply is unable to stop electron backstreaming.

The second region of grid erosion is caused by charge exchange ions generated downstream of the accelerator. Since the beamlets are long and thin, inside each beamlet the radial electric forces dominate and expel the slow, charge-exchange ions into the gaps between the beamlets. Charge exchange ions generated in the region before the beamlets merge to form a continuous ion density are then attracted back to the accelerator grid by its large negative potential. This is illustrated in Fig. 6-21. On impact, these ions sputter away material from the downstream surface of the accelerator grid. Sputter erosion by these backstreaming ions results in a hexagonal "pits and groves" erosion pattern on the downstream grid surface, which can lead to structural failure of the grids if the erosion penetrates all the way through the grid. Erosion of the accel grid aperture edge by backstreaming ions can also effectively enlarge the accel grid aperture diameter, leading to the onset of electron backstreaming.

Erosion of the accelerator grid by charge exchange ion sputtering was the major life limiting mechanism observed during the Extended Life Test (ELT) of the NSTAR flight spare thruster [51] for operation at the highest power TH15 level. Photographs of center holes in the grid at the beginning and the end of the 30,000-h test are shown in Fig. 6-22 where barrel-erosion enlargement of the aperture diameters is evident. Note that the triangle patterns where the webbing intersects in the end-of-test picture are locations where the erosion has completely penetrated the grid. The SEM photograph shown in Fig. 6-23 illustrates the deep erosion of the pit-and-groves pattern and that full penetration of the grid had occurred when the test was stopped. Continued operation would have eventually resulted in structural failure of the grid, but this was not considered imminent at the end of the test.

6.6.1 Grid Models

As discussed above, the primary erosion mechanism of the accelerator grid is caused by sputtering from charge exchange ions. At the simplest level, all that is needed to predict erosion rates is to calculate the number of ions generated in the beamlets, find where they hit the grids, and then to determine the amount of material that they sputter. The total calculated charge exchange ion current accounts for nearly all the measured accelerator grid current in a properly designed ion thruster (i.e. no direct interception of the beam current). The measured accelerator grid current in flight on NASA's NSTAR thruster during the Deep Space 1 Mission [52, 53] ranged from 0.2

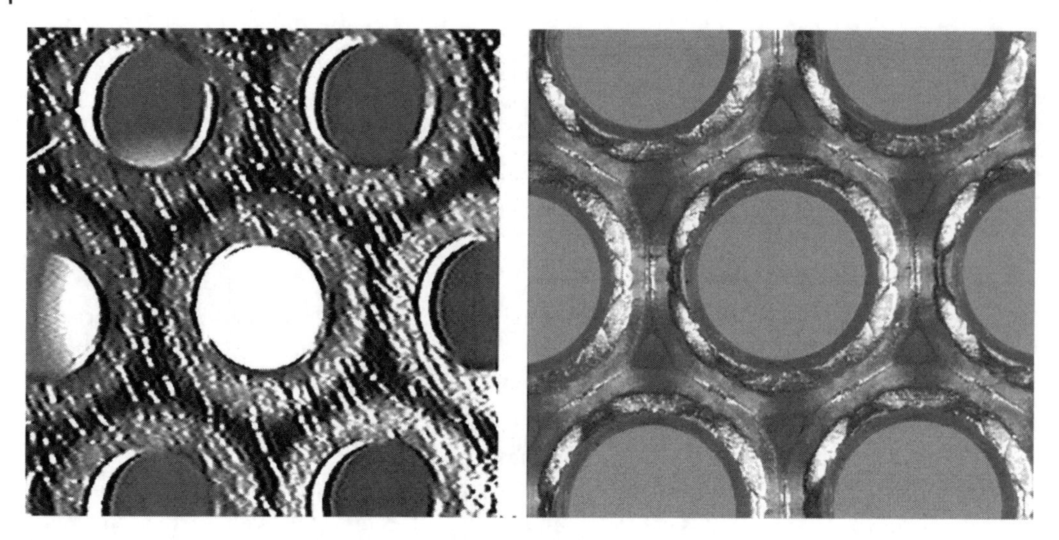

Figure 6-22 NSTAR thruster accelerator grid at 125 h (left) and 30,352 h (right) (*Source:* [51]/American Institute of Aeronautics and Astronautics, Inc.).

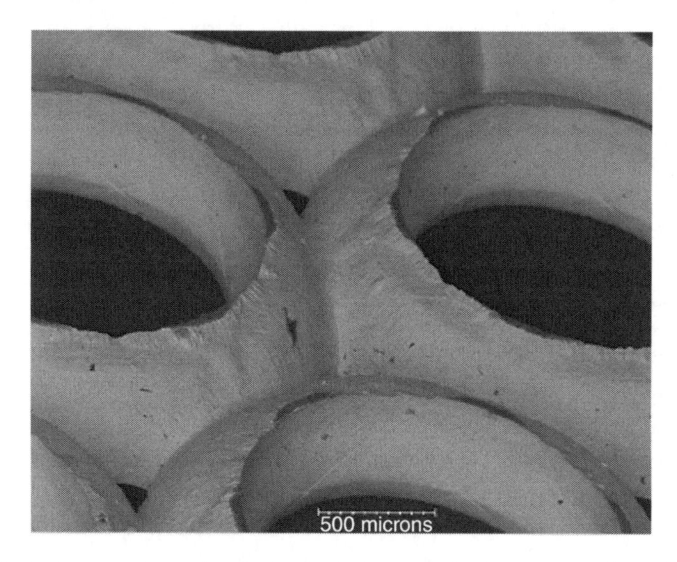

Figure 6-23 SEM photograph shows that sputtering in the webbing between the holes had almost destroyed the structural integrity of the NSTAR grids.

to 0.3% of the total beam current, which is shown in Fig. 6-24. Accel grid currents on the order of 1% or less of the beam current are standard in most ion thrusters.

Calculating the ion generation rate in the grid region due to charge exchange is relatively straightforward. The charge exchange currents generated by a single aperture's beamlet is given by

$$I_{\mathrm{CEX}} = I_{\mathrm{Beamlet}} n_o \sigma_{\mathrm{CEX}} l_d, \tag{6.6-1}$$

where l_d is the effective collection length downstream of the accel grid from which ions flow back to the grid and n_o is the average neutral density along this length. The charge exchange cross section, σ_{CEX}, is well known and varies slowly with beam energy [50]. The average neutral density along the

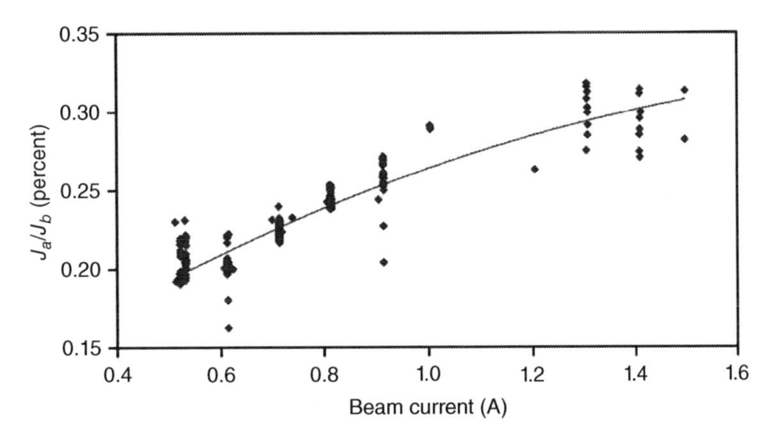

Figure 6-24 Ratio of the accel grid current to the beam current as a function of the beam current in NSTAR showing that the accel current is typically less than 1% of the beam current (*Source:* [52]).

path length l_d is estimated from the thruster propellant flow rate utilization fraction, which is the difference between the neutral atom flow rate and the beam ion current over the open area fraction of the accel grid. The neutral density is usually assumed to remain constant in the accel grid hole and decreases as the gas expands downstream of the grid surface. The neutral gas density is normally highest in holes near the edge of the grid and lower at the center where nearly all the gas has been "burned up" through ionization in the discharge chamber. The effective path length, l_d, is a basic result of the ion optics calculations and is essentially the distance downstream at which the beamlets have completely merged to form a beam-plasma with a uniform potential across the beam diameter. An estimate of the effective path length is needed when setting up a grid erosion calculation to make certain that the computational region is long enough to include all the charge exchange ions that can return to the grid.

Using Eq. 6.6-1 and the current ratio from Fig. 6-24, an estimate can be made of the effective path length (l_d) for the NSTAR thruster. If the measured accel grid current is all due to charge exchange (i.e. no direct interception), then Eq. 6.6-1 can be rewritten as

$$l_d = \frac{I_{\text{accel}}}{I_{\text{Beam}} \sigma_{\text{CEX}} n_o}. \tag{6.6-2}$$

Assuming the effective charge exchange path length is much longer than the gap between the screen and accelerator grids, the average neutral gas density can be estimated from the grid diameter, the flow of neutral gas out of the thruster, and the thruster beam current. The neutral gas density downstream of the grids close to the thruster is then

$$n_o = \frac{\Gamma_o}{v_o \pi r_{\text{grid}}^2}, \tag{6.6-3}$$

where v_o is the neutral velocity, and Γ_o is the flux of unutilized propellant escaping from the discharge chamber. Using the parameters for the NSTAR at TH15 from [29], the total neutral flow into the thruster is 28 sccm. The thruster discharge chamber has a mass utilization efficiency of about 88%, so the neutral gas flow escaping the thruster is about 3.4 sccm, which corresponds to 1.5×10^{18} particles per second. Assuming the gas exits the thruster at about the thruster operating temperature of 250°C, the neutral velocity $\bar{c}/2$ is about 110 m/s. The average neutral density from Eq. 6.6-3 is then about 2.3×10^{17} m^{-3} and neutral density varies over the grid by more than factor of two. Using the data in Fig. 6-24 extrapolated to the beam current of 1.76 A in TH15, and a charge exchange cross section of 5×10^{-19} m^2, the average effective path length from Eq. 6.6-2 becomes

$$l_d = \frac{(0.003)}{(5 \times 10^{-19})(2.3 \times 10^{17})} = 0.03[m]. \tag{6.6-4}$$

The path length is more than an order of magnitude larger than the grid gap, consistent with our assumption. The very long path length compared with grid hole spacing means that the computational space in ion optics codes is very long (several centimeters), and so the computer codes must allow for the axial zone sizes to increase downstream of the grids.

6.6.2 Barrel Erosion

As was illustrated in Fig. 6-20, charge exchange ions generated between the screen grid and the upstream surface of the accel grid can impact the interior surface of the accel grid holes. These ions sputter away grid material, increasing the barrel radius. While computer codes, such as CEX-2D [4], are normally used to calculate the erosion rate, it is instructive to derive an analytical estimate. The following calculation is based on published performance and erosion data for NASA's NSTAR thruster operating at its highest power TH15 level [29, 52].

Assume that any ions generated downstream of the discharge chamber are not focused through the hole in the accelerator grid. For barrel erosion, the path length is taken as the sum of the grid gap and the accelerator grid thickness, which for NSTAR is about 1 mm. The upstream gas density is estimated by dividing the downstream density by the grid open area fraction, f_a, and the Clausing [54] factor η_c, which reduces the gas transmission due to the finite thickness of the accel grid. The Clausing factor depends only on the aperture length to radius ratio. The neutral gas density is then

$$n_o = \frac{\Gamma_o}{v_o \pi r_{\text{grid}}^2} \frac{1}{f_a \eta_c}. \tag{6.6-5}$$

The neutral gas density in the accelerator grid apertures is higher than the gas density downstream of the accelerator grid, which was calculated using Eq. 6.6-2, because of the effects of the open area fraction and the Clausing factor. For an open area fraction of 0.24 and a Clausing factor of 0.6, the neutral density in the grid gap is about 9×10^{18} m^{-3}.

The number of grid apertures is approximately the grid open area divided by the area per aperture:

$$N \approx \frac{f_a \pi r_{\text{grid}}^2}{\pi r_{\text{aperture}}^2}. \tag{6.6-6}$$

The average aperture current is the total beam current divided by the number of apertures,

$$\bar{I}_{\text{aperture}} = \frac{I_b}{N_{\text{aperture}}}. \tag{6.6-7}$$

The maximum aperture current is obtained using the definition of beam flatness, which is given as

$$f_b \equiv \frac{\text{Average current density}}{\text{Peak current density}} = \frac{\bar{I}_{\text{aperture}}}{I_{\text{aperture}}^{\text{max}}}. \tag{6.6-8}$$

The published value of NSTAR beam flatness from Polk [30] is 0.47. Using Eqs. 6.6-6, 6.6-7 and 6.6-8, the maximum current per aperture is 2.5×10^{-4} A. Charge exchange ions that can hit the accel grid are generated in between the screen grid exit and the accel grid exit. The distance d between the screen grid exit and the accel grid exit is about 1.12 mm [4]. The charge exchange ion current to the central aperture barrel is then

$$I_{\text{CEX}} = I^{\max}_{\text{aperture}} n_o \sigma_{\text{CEX}} d = 1.4x10^{-6}\, A. \tag{6.6-9}$$

The CEX-2D computer code simulations [4] show that charge exchange ions hit the accelerator grid with about three-tenths of the beam potential. For NSTAR, the beam potential is 1100 V; thus, the average charge exchange ion energy is about 330 V. Using the curve fit in the same reference for sputtering yield Y, the aperture atom sputter rate is obtained:

$$\dot{n}_{\text{sputter}} = \frac{I_{\text{CEX}}}{e} Y \approx 3.5 \times 10^{12} [\#/\text{s}]. \tag{6.6-10}$$

This atom sputtering rate can be used to find an initial wall erosion rate by first calculating the volumetric erosion rate

$$\dot{V}_{\text{aperture}} = \frac{\dot{n}_{\text{sputter}}}{\left(\dfrac{\rho_{\text{Mo}}}{M_{\text{Mo}}}\right)}, \tag{6.6-11}$$

where the density of molybdenum is $\rho_{\text{Mo}} = 1.03 \times 10^4$ and the mass of molybdenum is $M_{\text{Mo}} = 95.94$ AMU $= 1.6 \times 10^{-25}$ kg. The volumetric erosion rate from Eq. 6.6-11 is then

$$\dot{V}_{\text{aperture}} = \frac{\dot{n}_{\text{sputter}}}{\left(\dfrac{\rho_{\text{Mo}}}{m_{\text{Mo}}}\right)} = \frac{3.5 \times 10^{12}}{\left(\dfrac{1.03 \times 10^4}{1.6 \times 10^{-25}}\right)} \approx 5.5 \times 10^{-17} \left[\text{m}^3/\text{s}\right]. \tag{6.6-12}$$

Assuming the erosion rate is uniform throughout the barrel, the rate of increase of the aperture radius is just the volumetric erosion rate divided by the barrel area

$$\dot{r}_{\text{aperture}} = \frac{\dot{V}_{\text{aperture}}}{2\pi\, r_a w_{\text{accel}}} \approx 3 \times 10^{-11} [\text{m/s}], \tag{6.6-13}$$

where the accel grid aperture radius r_a is 0.582 mm and the accel grid thickness w_{accel} is 0.5 mm. For the 8200 h NSTAR wear test results described by Polk [30], this corresponds to an increase in diameter of about 0.2 mm; roughly what was observed.

More accurate predictions of the accel grid barrel erosion rate are found using the 2D computer simulations. The CEX-2D code [4] from JPL has been benchmarked against NSTAR grid erosion observed in several wear tests [51–53], and used to predict the grid life for flight missions such as BepiColombo [55]. Multidimension simulation codes such as this use the same basic technique as shown here analytically to determine the amount of material removed by the charge exchange sputtering. The better predictions from the codes result from more accurate calculations of the neutral density and ion current densities across the grid surfaces and through the grid apertures.

6.6.3 Pits and Groves Erosion

Using three-dimensional ion optics codes, it is possible to reproduce the details of the "pits and groove" geometry of accelerator grid downstream surface erosion. The JPL CEX-3D code was developed [17] to solve for potentials and ion trajectories in a two-grid ion optics system, and was later modified to include a third grid [56]. The computational domain, illustrated in Fig. 6-25, is a triangular wedge extending from the axis of a hole-pair to the midpoint between two aperture pairs. The wedge angle of 30° is chosen to give the smallest area that can be used to model the ion optics to

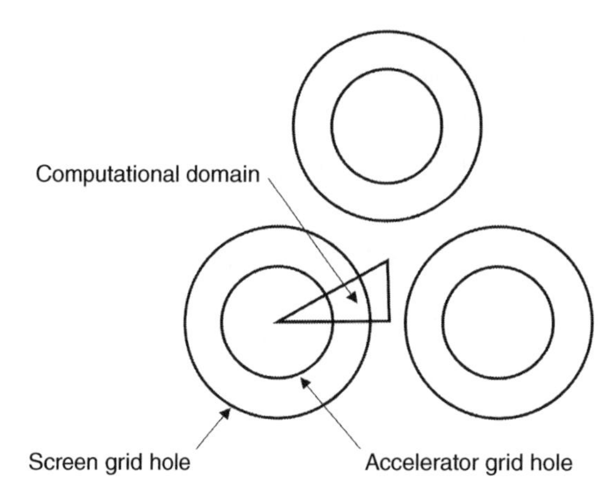

Figure 6-25 Computational domain of the CEX-3D code (*Source:* [17]).

minimize computational time. Similar triangles will cover each aperture pair by a combination of reflections and rotations. The computational domain extends from a few millimeters into the discharge chamber through the grids to few centimeters downstream of the final grid.

In addition to tracking the beam-ion trajectories, the code calculates charge exchange ion production rates and charge exchange trajectories in three dimensions. Erosion of the accel grid barrel and downstream face is caused by these charge exchange ions. The location, kinetic energy, incidence angle, and current of each particle is recorded and used to compute the rate at which the grid material is removed. As shown above, charge exchange ions that strike the downstream surface of the accelerator grid can come from several centimeters downstream of the grid. Therefore, the computations domain is usually extended to 5–10 cm downstream of the final grid.

An example of the accel-grid downstream face erosion pattern on the NSTAR thruster predicted by CEX3D is shown in Fig. 6-26. The triangular patches (the "pits"), where the grid webbing intersects, were shown in the photograph of the NSTAR ELT grid at the end of the test [51] and are

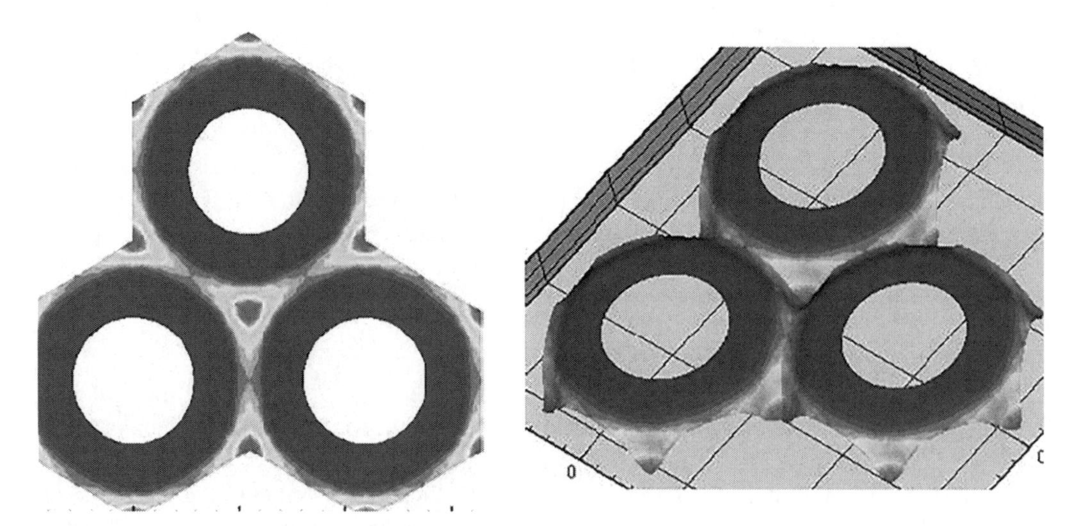

Figure 6-26 CEX3D calculation of the "pits and grooves" erosion wear patterns that match the experimental patterns shown in Figure 6-22.

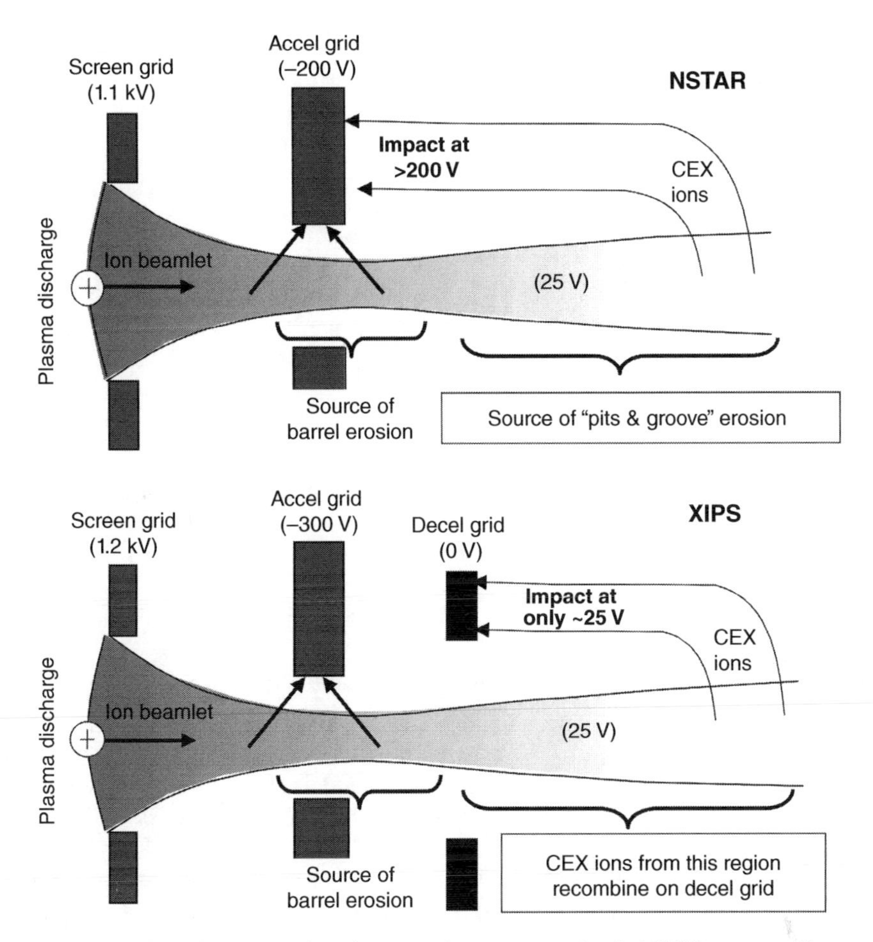

Figure 6-27 Grid cross section comparing charge exchange generation in NSTAR, a two-grid system, with XIPS, a three-grid system.

predicted by the code in the plot on the left. In addition, the depth of the ring of erosion around the aperture ("the groves") is also seen in the plot on the right from the code predictions. The distribution of wear between the pits regions and the groves depends on the operating parameters of the ion accelerator, and CEX3D clearly shows the dependence on different ion thrusters such as NEXT and T6 [55].

Accelerator grid pits and groove erosion can be almost eliminated by the use of a third decelerator grid [43]. The XIPS© thruster [57] is an example of an ion thruster that uses a three-grid ion optics system. As shown in Fig. 6-27, the third grid reduces from centimeters to millimeters the length of the region where charge exchange ions are generated that can hit the accelerator grid. This causes a dramatic reduction in the pits and grooves erosion between the two thrusters, shown in Fig. 6-28 as calculated using CEX-3D.

Although the three-dimensional code CEX-3D is used to predict erosion of the accelerator grid downstream surface, the simpler, two-dimensional, CEX-2D code is typically used for accelerator grid aperture barrel erosion calculations because the apertures are cylindrical and the CEX2D code can produce these results more quickly. CEX-2D and CEX-3D use the same algorithms for the discharge chamber plasma and for beam ion trajectories. The codes have been benchmarked with each other, and for round beamlets that can be handled by CEX-2D their results are within a few percent.

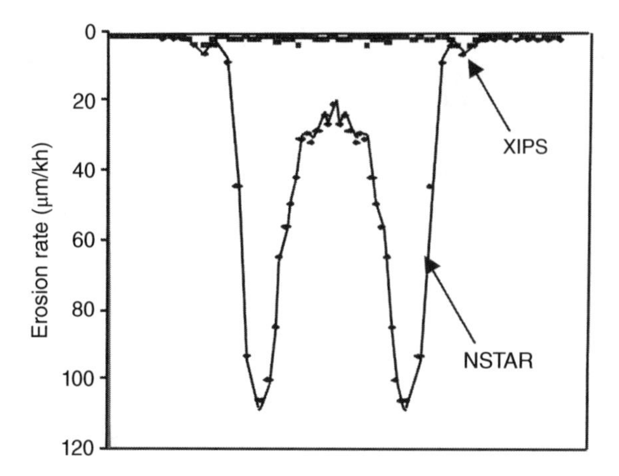

Figure 6-28 CEX-3D results showing the XIPS third grid almost eliminates pits and grooves erosion evident in the NSTAR thruster (*Source:* [56]).

Problems

6.1 Derive the dependence of the minimum Isp on the beam voltage for a given accelerator design and thrust level.

6.2 A 1 kV ion accelerator has a grid spacing of 1 mm and a screen aperture diameter of 1 mm.
a) What is the space charge limited beamlet current density for Xe^+ assuming a very thin screen grid and a planar sheath?
b) If the screen grid is 0.25 mm thick, what is the maximum beamlet current density for a non-planar sheath? How does this compare to the classic planar Child Langmuir result?

6.3 An ion thruster with a grid diameter of 20 cm has a beam current density that varies with radius as kr^2, where k is a constant.
a) If the peak current density on axis is J_p and the current density at the edge of the grid is $J_p/10$, find the expression for $J(r)$.
b) If the peak current density is 5 mA/cm^2, what is the total beam current?
c) What is the flatness parameter?
d) What is the percent reduction in the beam current compared to case of a uniform beam current density of the peak value (the flatness is 1)?

6.4 An ion thruster has a beam plasma potential of 20 V and an electron temperature in the beam of 5 eV.
a) For a plasma potential at the screen grid sheath edge of 1000 V, what potential must be established in the accel grid aperture to keep the electron backstreaming current to 1% of the beam current?
b) Neglecting space charge in the beamlet, what voltage must be applied to the accel grid to achieve the minimum potential in (a) for the case of a 3 mm screen grid diameter, 0.25 mm screen grid thickness, 2 mm accel grid diameter and 0.5 mm accel grid thickness with a 1 mm grid gap.
c) If the beamlet current is 0.2 mA and the beamlet has a diameter in the accel grid aperture of 1 mm, what must the accel grid voltage be to maintain the 1% backstreaming current specification?

6.5 One of the first ion thrusters to fly in space was a cesium surface ionization thruster where cesium ions are pulled from a hot surface by the acceleration electric field. Model the thruster as a diode with cesium ions at 7.5 mA/cm^2 coming from one surface and the other electrode is the accel grid with 80% transparency and a grid gap "d" from the ion source.

a) Assuming 100% mass utilization efficiency, neglecting the angular divergence of the beam, and using a 200 V negative bias on the accel grid, what is the voltage, current, thruster diameter and gap size required to produce 5 mN of thrust at an Isp of 3000 s?
b) If the thruster has 95% mass utilization efficiency and a total angular divergence of the beam of 10°, how does that change the results of part (a)?
c) If it takes 100 W of heater power to heating the cesium ion emitting surface to the required surface temperature of about 1350 °K, what is the total efficiency of the thruster. (use the parameters from part b).

References

1 H. R. Kaufman, "Technology of Electron-Bombardment Ion Thrusters," in *Advances in Electronics and Electron Physics*, vol. 36, edited by L. Marton, New York: Academic Press, 1974.

2 A. T. Forrester, *Large Ion Beams*, New York: John Wiley and Sons, 1988.

3 G. R. Brewer, *Ion Propulsion Technology and Applications*, New York: Gordon and Breach, 1970.

4 J. R. Brophy, I. Katz, J. E. Polk, and J. R. Anderson, "Numerical Simulations of Ion THRUSTER Accelerator Grid Erosion", AIAA-2002-4261, 38th AIAA Joint Propulsion Conference, Indianapolis, July 7–10, 2002).

5 V. J. Friedly and P. J. Wilbur, "High Current Hollow Cathode Phenomena," *AIAA Journal of Propulsion and Power*, vol. 8, no. 3, pp. 635–643, 1992; doi:10.2514/3.23526.

6 I. Kameyama and P. J. Wilbur, "Measurement of Ions from High Current Hollow Cathodes Using Electrostatic Energy Analyzer," *Journal of Propulsion and Power*, vol. 16, no. 3, pp. 529–535, 2000; doi:10.2514/2.5601.

7 P. J. Wilbur, J. R. Beattie, and J. Hyman, Jr., "An Approach to the Parametric Design of Ion Thrusters," *AIAA Journal of Propulsion and Power*, vol. 6, no. 5, pp. 575–583, 1990; doi:10.2514/3.23258.

8 R. A. Martinez, J. D. Williams, and D. M. Goebel, "Electric Field Breakdown Properties of Materials Used in Ion Optics Systems," AIAA-2006-5004, 42nd Joint Propulsion Conference, Sacramento, July 10–13, 2006.

9 P. Wilbur, "Limits on High Specific Impulse Ion Thruster Operation," AIAA-2004-4106, 40th Joint Propulsion Conference, Fort Lauderdale, July 11–14, 2004.

10 J. H. Whealton, R. W. McGaffey, and P. S. Meszaros, "A Finite Difference 3-D Poisson-Vlasov Algorithm for Ions Extracted from a Plasma," *Journal of Computational Physics*, vol. 63, no. 1, pp. 20–32, 1986; doi:10.1016/0021-9991(86)90082-3.

11 Y. Arakawa and M. Nakano, "An Efficient Three Dimensional Optics Code for Ion Thruster Research," AIAA-96-3198, 32nd Joint Propulsion Conference, Lake Buena Vista, July 1–3, 1996.

12 M. Nakano and Y. Arakawa, "Ion Thruster Lifetime Estimation and Modeling Using Computer Simulation," IEPC-99-145, 27th International Electric Propulsion Conference,Pasadena, October 15–19, 2001.

13 Y. Nakayama and P. Wilbur, "Numerical Simulation of High Specific Impulse Ion Thruster Optics," IEPC-01-099, 27th International Electric Propulsion Conference, Pasadena, October 15–19, 2001.

14 I. Boyd and M. Crofton, "Grid EROSION ANAlysis of the T5 Ion Thruster," AIAA-2001-3781, 37th AIAA Joint Propulsion Conference, Salt Lake City, July 8–11, 2001.

15 Y. Okawa, H. Takegahara, and T. Tachibana, "Numerical Analysis of Ion Beam Extraction Phenomena in an Ion Thruster," IEPC-01-097, 27th International Electric Propulsion Conference,Pasadena, October 15–19, 2001.

16 J. Wang, J. E. Polk, J. R. Brophy, and I. Katz, "Three-Dimensional Particle Simulations of NSTAR Ion Optics," IEPC-01-085, 27th International Electric Propulsion Conference, Pasadena, October 15–19, 2001.

17 J. J. Anderson, I. Katz, and D. Goebel, "Numerical Simulation of Two-Grid Ion Optics Using a 3D Code," AIAA-2004-3782, 40th AIAA Joint Propulsion Conference, Ft. Lauderdale, July 11–14, 2004.

18 I. Brown, *The Physics and Technology of Ion Sources*, New York: John Wiley& Sons, 1989.

19 Y. Nakayama and P. J. Wilbur, "Numerical Simulations of Ion Beam Optics for Many-Grid Systems," *Journal of Propulsion and Power*, vol. 19, no. 4, pp. 607–613, 2003; doi:10.2514/2.6148.

20 J. Kim, J. H. Wealton, and G. Shilling, "A Study of Two-Stage Ion Beam Optics," *Journal of Applied Physics*, vol. 49, no. 2, pp. 517–524, 1978; doi:10.1063/1.324676.

21 C. Farnell, "Performance and Life Simulation of Ion Thruster Optics," PhD dissertation, Colorado State University, 2007.

22 L. D. Stewart, J. Kim, and S. Matsuda, "Beam Focusing by Aperture Displacement in Multiampere Ion Sources," *Review of Scientific Instruments*, vol. 46, no. 9, pp. 1193–1196, 1975; doi:10.1063/1.1134443.

23 J. H. Whealton, "Linear Optics Theory of Ion Beamlet Steering", *Review of Scientific Instruments*, vol. 48, no.11, pp. 1428–1429, 1977; doi:10.1063/1.1134911.

24 M. Tarz, E. Hartmann, R. Deltschew, and H. Neumann, "Effects of Aperture Displacement in Broad-Beam Ion Extraction Systems," *Review of Scientific Instruments*, vol. 73, no. 2, pp. 928–930, 2002; doi:10.1063/1.1428785.

25 Y. Okawa, Y. Hayakawa, and S. Kitamura, "Experiments on Ion Beam Deflection Using Ion Optics with Slit Apertures," *Japanese Journal of Applied Physics*, vol. 43, no. 3, pp. 1136–1143, 2004.

26 J. D. Williams, D. M. Goebel, and P. J. Wilbur, "Analytical Model of Electron Backstreaming for Ion Thrusters," AIAA-2003-4560, 39th Joint Propulsion Conference, Huntsville, July 20–23, 2003.

27 V. A. Davis, I. Katz, M. J. Mandell, D. E. Brinza, M. D. Henry, and D. T. Young, "Ion Engine Generated Charge Exchange Environment: Comparison Between NSTAR Flight Data and Numerical Simulations," AIAA-2001-0970, 39th Aerospace Sciences Meeting, Reno, January 8–11, 2001.

28 K. R. Spangenberg, *Vacuum Tubes*, New York: McGraw-Hill, p. 348, 1948.

29 J. R. Brophy, "NASA's Deep Space 1 Ion Engine," *Review of Scientific Instruments*, vol. 73, no. 2, pp. 1071–1078, 2002; doi:10.1063/1.1432470.

30 J. Polk, J. R. Anderson, J. R. Brophy, V. K. Rawlin, M. J. Patterson, J. Sovey, and J. Hamley, "An Overview of the Results from an 8200 Hour Wear Test of the NSTAR Ion Thruster," AIAA-99-2446, 35th AIAA Joint Propulsion Conference, Los Angeles, June 20–24, 1999.

31 A. Sengupta, J. R. Brophy, and K. D. Goodfellow, "Status of the Extended Life Test of the Deep Space 1 Flight Spare Ion Engine After 30,352 Hours of Operation," AIAA-2003-4558, 39th AIAA Joint Propulsion Conference, Huntsville, July 20–23, 2003.

32 J. E. Polk, D. M. Goebel, I. Katz, J. S. Snyder, A. C. Schneider, L. Johnson, and A. Sengupta, "Performance and Wear Test Results for a 20-kW Class Ion Engine with Carbon-Carbon Grids," AIAA-2005-4393, 41st AIAA Joint Propulsion Conference, Tucson, AZ, July 10–13, 2005.

33 E. Kuffel and W. S. Zaengl, *High Voltage Engineering Fundamentals*, Oxford: Pergamon Press, 1984.

34 W. H. Kohl, *Handbook of Materials and Techniques for Vacuum Devices*, New York: Reinhold Pub, 1967.

35 D. M. Goebel and A. C. Schneider, "High Voltage Breakdown and Conditioning of Carbon and Molybdenum Electrodes," *IEEE Transactions on Plasma Science*, vol. 33, no. 4, pp. 1136–1148, 2005; doi:10.1109/TPS.2005.852410.

36 P. A. Chatterton, "Theoretical Study of the Vacuum Breakdown Initiated by Field Emission," *Proceedings of the Physics Society*, vol. 88, no. 1, pp. 231–243, 1966.

37 F. R. Schwirzke, "Vacuum Breakdown on Metal Surfaces," *IEEE Transactions on Plasma Science*, vol. 19, no.5, pp. 690–696, 1991; doi:10.1109/27.108400.

38 W. D. Davis and H. C. Miller, "Analysis of the Electrode Products Emitted by DC Arcs in a Vacuum Ambient," *Journal of Applied Physics*, vol. 40, no.5, pp. 2212–2221, 1969; doi:10.1063/1.1657960.

39 A. Anders, B. Yotsombat, and R. Binder, "Correlation Between Cathode Properties, Burning Voltage, and Plasma Parameters of Vacuum Arcs," *Journal of Applied Physics*, vol. 89, no. 12, pp. 7764–7771, 2001; doi:10.1063/1.1371276.

40 J. S. Sovey, J. A. Dever, and J. L. Power, "Retention of Sputtered Molybdenum on Ion Engine Discharge Chamber Surfaces," IEPC-01-86, 27th International Electric Propulsion Conference, Pasadena, October 15–19, 2001.

41 D. M. Goebel, "High Voltage Breakdown Limits of Molybdenum and Carbon-Based Grids for Ion Thrusters," AIAA-2005-4257, 41st AIAA Joint Propulsion Conference, Tucson, July 10–13, 2005.

42 M. J. Druyvesteyn and F. M. Penning, "The Mechanism of Electrical Discharge in Gases," *Reviews of Modern Physics*, vol. 12, no. 2, pp. 87–174, 1940.

43 R. Doerner, D. White, and D. M. Goebel, "Sputtering Yield Measurements During Low Energy Xenon Plasma Bombardment," *Journal of Applied Physics*, vol. 93 no. 9, pp. 5816–5823, 2003; doi:10.1063/1.1566474.

44 N. C. Wallace, D. G. Fearn, and R. E. Copleston, "The Design and Performance of the T6 Ion Thruster," AIAA-1998-3342, 36th AIAA Joint Propulsion Conference, Cleveland, July 13–15, 1998.

45 J. S. Snyder, D. M. Goebel, R. R. Hofer, J. E. Polk, N. C. Wallace, and H. Simpson, "Performance Evaluation of the T6 Ion Engine," *AIAA Journal of Propulsion and Power*, vol. 28, no. 2, pp. 371–379, 2012; doi:10.2514/1.B34173

46 J. E. Polk, D. M. Goebel, J. S. Snyder, A. C. Schneider, J. R. Anderson, and A. Sengupta," A High Power Ion Thruster for Deep Space Missions," *The Review of Scientific Instruments*, vol. 83, 073306, 2012; doi:10.1063/1.4728415.

47 H. Kuninaka, I. Funaki, K. Nishiyama, Y. Shimizu, and K. Toki, "Results of 18,000 Hour Endurance Test of Microwave Discharge Ion Thruster Engineering Model," AIAA-2000-3276. 36th AIAA Joint Propulsion Conference, Huntsville, July 16–19, 2000.

48 J. S. Snyder and J. R. Brophy, "Performance Characterization and Vibration Testing of 30-cm Carbon-Carbon Ion Optics," AIAA-2004-3959, 40th AIAA Joint Propulsion Conference, Ft. Lauderdale, July 11–14, 2004.

49 M. De Pano, S. Hart, A. Hanna, and A. Schneider, "Fabrication and Vibration Results of 30-cm Pyrolytic Graphite Ion Optics," AIAA-2004-3615, 40th AIAA Joint Propulsion Conference, Fort Lauderdale, July 11–14, 2004.

50 J. S. Miller, S. H. Pullins, D. J. Levandier, Y. Chiu, and R. A. Dressler, "Xenon Charge Exchange Cross Sections for Electrostatic Thruster Models," *Journal of Applied Physics*, vol. 91, no. 3, pp. 984–991, 2002; doi:10.1063/1.1426246.

51 A. Sengupta, J. R. Brophy, J. R. Anderson, C. Garner, K. de Groh, T. Karniotis, and B. Banks, "An Overview of the Results from the 30,000 Hour Life Test of the Deep Space 1 Flight Spare Ion Engine," AIAA-2004-3608, 40th AIAA Joint Propulsion Conference, Fort LauderdealeJuly 11–14, 2004.

52 J. E. Polk, R. Y. Kakuda, J. R. Anderson, J. R. Brophy, V. K. Rawlin, M. J. Patterson, J. Sovey, and J. Hamley, "Performance of the NSTAR Ion Propulsion System on the Deep Space One Mission," AIAA-2001-965, 39th Aerospace Sciences Meeting and Exhibit, Reno, January 8–11, 2001.

53 J. E. Polk, D. Brinza, R. Y. Kakuda, J. R. Brophy, I. Katz, J. R. Anderson, V. K. Rawlin, M. J. Patterson, J. Sovey, and J. Hamley, "Demonstration of the NSTAR Ion Propulsion System on the Deep Space One Mission," IEPC-01-075, 27th International Electric Propulsion Conference, Pasadena, October 15–19, 2001.

54 P. Clausing, "The Flow of Highly Rarefied Gases Through Tubes of Arbitrary Length," *Journal of Vacuum Science and Technology*, vol. 8, no. 5, pp. 636–646, 1971; doi:10.1116/1.1315392.

55 V. Chaplin, D. M. Goebel, R. A. Lewis, F. Lockwood Estrin, and P. N. Randall, "Accelerator Grid Life Modeling of the T6 Ion Thruster for BepiColombo," *AIAA Journal of Propulsion and Power*, vol. 37, no. 3, 2021; doi:10.2514/1.B37938

56 R. Wirz and I. Katz, "XIPS ion thruster grid erosion predictions for Deep Space missions," IEPC-2007-265, 30th International Electric Propulsion Conference, Florence, September 17–20, 2007.

57 J. R. Beattie, J. N. Matossian, and R. R. Robson, "Status of Xenon Ion Propulsion Technology," *AIAA Journal of Propulsion and Power*, vol. 6, no. 2, pp. 145–150, 1990; doi:10.2514/3.23236.

Chapter 7

Conventional Hall Thrusters

7.1 Introduction

Hall thrusters are relatively simple-appearing devices consisting of a cylindrical channel with an interior embedded anode, a magnetic circuit that generates primarily a radial magnetic field across the channel, and a cathode that injects electrons into the near-thruster region external to the channel. However, Hall thrusters rely on much more complicated physics compared to ion thrusters because of the $\mathbf{E} \times \mathbf{B}$ plasma discharge that accelerates ions and produces thrust. The details of the channel structure and magnetic field shape determine the performance, efficiency and life of the thruster [1–5]. The efficiency and specific impulse of flight-model Hall thrusters are approaching that achievable in ion thrusters [6, 7], but the thrust to power ratio and the beam current density are higher, and the total impulse can be comparable. Hall thrusters are simpler to build than ion thrusters because of the lack of precision-aligned close-space grids, and they require fewer power supplies and propellant flow controllers to operate.

Hall thrusters were originally developed in Russia in the 1960s [8] and first flown in the early 1970s on Russian communications satellites [9]. Two varieties of Hall thrusters emerged in Russia during this period [4]: the Stationary Plasma Thruster (SPT) and the Thruster with Anode Layer (TAL). The basic geometry of both these "conventional" Hall thrusters has not changed in the past 50 years. The performance and power level have been improved incrementally over the years, but the life of both the SPT and TAL varieties in terms of hours of operation has remained less than about 10,000 h (much shorter than ion thrusters) because of channel wall erosion from energetic ion bombardment in the acceleration region. This relative short life has limited the use of Hall thrusters in deep space science mission prime propulsion applications, but is sufficient for orbit raising and station keeping applications in earth-orbiting satellites. This situation resulted in the development of Hall thruster technology in the United States over the past 25 years primarily for use in modern communications satellites. In this chapter, the physics and technology of these "conventional" Hall thrusters will be described.

The design and performance of Hall thrusters was revolutionized in 2010 by a new technique called "Magnetic Shielding" that protects the channel wall from ion bombardment, thereby eliminating channel erosion as the primary failure mode in these thrusters. This improvement was further advanced by the observation that plasma-wall interactions that dominate the "conventional" Hall thrusters was reduced sufficiently to allow replacement of the dielectric walls with metallic conductors such as graphite. These Magnetically Shielded Hall thrusters will be described in Chapter 8.

Fundamentals of Electric Propulsion, Second Edition. Dan M. Goebel, Ira Katz, and Ioannis G. Mikellides.

7.1.1 Discharge Channel with Dielectric Walls (SPT)

The terms Hall thruster, Hall-Effect Thruster (HET), Stationary Plasma Thruster (SPT), Dielectric Wall Hall Thruster, and Magnetic-Layer Thruster are all names for essentially the same device that is characterized by the use of a dielectric insulating wall in a relatively long plasma channel, as illustrated in Fig. 7-1. The hollow cathode can be installed outside the discharge channel (as shown in the figure), or on the centerline of the thruster. The discharge channel wall is typically manufactured from dielectric materials such as boron nitride (BN) or borosil ($BNSiO_2$) in flight thrusters, and also sometimes alumina (AL_2O_3) in laboratory thrusters. These dielectric materials have a low sputtering yield and relatively low secondary electron emission coefficients under xenon ion bombardment. In this thruster geometry, the electrically biased metallic anode is positioned at the base of the channel where the majority of the propellant gas is injected into the thruster. The remainder of the propellant gas used by the thruster (typically <10% of the total) is injected through the hollow cathode. In the second version of this type of thruster, called Thrusters with Anode Layer (TAL), the dielectric channel wall is replaced by a metallic conducting wall, as illustrated in Fig. 7-2. This geometry considerably shortens the electric field region in the channel where the ion acceleration occurs near the anode; hence the name "Thruster with Anode Layer" from the Russian literature [1]. However, this configuration does not change the basic ion generation or acceleration method. The channel wall, which is usually also part of the magnetic circuit, is biased negatively (usually cathode potential) to repel electrons in the ionization region and reduce electron-power losses. The defining differences between these two types of Hall thrusters have been described in the literature [4].

In the conventional Hall thruster with dielectric walls illustrated in Fig. 7-1, an axial electric field is established between the anode at the base of an annular channel and the hollow-cathode plasma produced outside the thruster channel. A transverse (radial) magnetic field prevents electrons from this cathode plasma from streaming directly to the anode. Instead, the electrons spiral along the magnetic field lines (as illustrated) and in the $\mathbf{E} \times \mathbf{B}$ azimuthal direction (into the page) around the channel, and they diffuse by collisional processes and electrostatic fluctuations to the anode and channel walls. The plasma discharge generated by the electrons in the crossed electric and

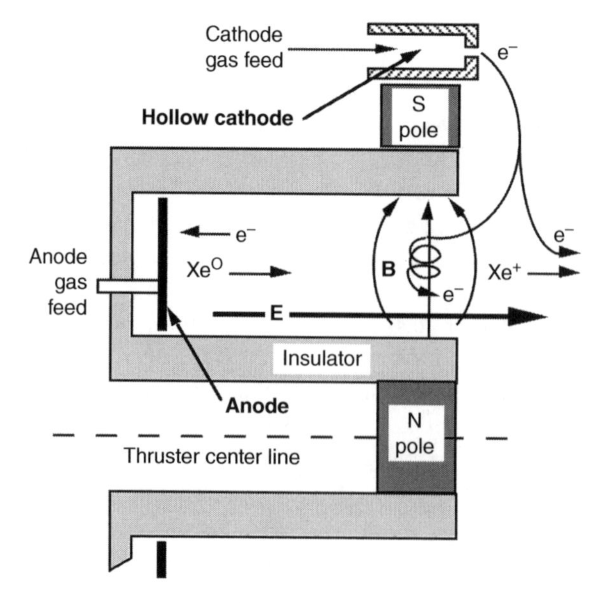

Figure 7-1 Hall thruster cross section schematic with external cathode showing the crossed electric and magnetic fields, and the ion and electron paths.

magnetic fields efficiently ionizes the propellant injected into the channel from the anode region. Ions from this plasma bombard and, near the channel exit, sputter-erode the dielectric walls, which ultimately determines the life of the thruster. Electrons from this plasma also bombard the dielectric wall, depositing a significant amount of power in this region. The reduced axial electron mobility produced by the transverse magnetic field permits the applied discharge voltage to be distributed along the channel axis in the quasi-neutral plasma, resulting in an axial electric field in the channel that accelerates the ions to form the thrust beam. Therefore, Hall thrusters are typically regarded as electrostatic devices [1] because the ions are accelerated by the applied electric field, even though a magnetic field is critical to the process (see Section 2.2.2). However, since the acceleration occurs in the plasma region near the channel exit, space charge is not an issue and the ion current density and the thrust density can be considerably higher than that achievable in gridded ion thrusters. The external hollow cathode plasma is not only the source of the electrons for the discharge, but it also provides the electrons to neutralize the ion beam. The single hollow cathode in Hall thrusters serves the same function as the two cathodes in DC-electron discharge ion thrusters that produce the plasma and neutralize the beam.

7.1.2 Discharge Channel with Metallic Walls (TAL)

The TAL thruster with metallic walls, illustrated in Fig. 7-2, has the same functional features of the dielectric-wall Hall thruster; namely the axial electric field is established between the anode in the annular channel and the plasma potential outside the thruster channel. This field accelerates ions from the ionization region near the anode out of the channel. The transverse (radial) magnetic field again prevents electrons from streaming directly to the anode, and the electron motion is the same as in the dielectric-wall Hall thruster. However, the channel walls at the exit plane have metallic guard rings biased at cathode potential to reduce the electron loss along the field lines. These rings

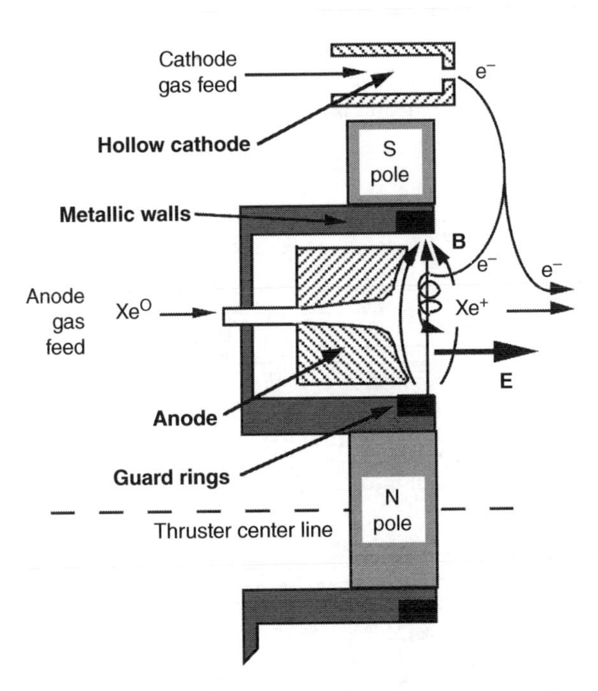

Figure 7-2 TAL thruster cross section schematic showing the crossed electric and magnetic fields, and the ion and electron paths.

represent the major erosion source in the thruster because of ion bombardment from the plasma, and the guard ring material and design often determine the thruster life. The anode typically extends close to the thruster exit and is often funnel-shaped and curved to constrain the neutral gas and plasma to the center of the channel (away from the guard rings) and to not intercept the magnetic field lines which would cause large electron losses. However, the anode is in close proximity to the high electron-temperature region of the plasma, and electrons collected by the anode can deposit a significant amount of power. The channel width in TAL thrusters is typically twice the channel depth (including the anode shaping). The external hollow cathode plasma provides the electrons for the discharge and for neutralization of the ion beam, the same as for dielectric-wall Hall thrusters.

The azimuthal drift of the electrons around the channel in the crossed electric and magnetic fields in the cylindrical thruster geometry is reminiscent of the *Hall Current* in magnetron type devices, which has caused many authors to call this generically a "closed-drift" thruster [1, 2, 4]. However, King [10] correctly points out that the orientation of the fields in magnetrons (axial magnetic and radial electric) provides a restoring force to the centrifugal force felt by the electrons as they rotate about the axis, which produces the closed-drift electron motion in magnetrons. There is no corresponding restoring force associated with the different orientation of the crossed fields (radial magnetic and axial electric required to produce axial thrust) in Hall thrusters. The closed-drift behavior of the electron motion in Hall thrusters only occurs because of wall sheath electric fields and the force associated with the magnetic gradient in the radial direction in the channel. In this case, the electrons in the channel encounter an increasing magnetic field strength as they move toward the wall, which acts as a magnetic mirror to counteract the radial centrifugal force.

The radial magnetic field gradient in the channel also forms an "ion lens" which tends to deflect the ions away from the channel walls and focus the ions out of the channel into the beam. Fig. 7-3 shows an example of the magnetic field lines in the NASA-173Mv Hall thruster [11] developed at NASA-GRC. The curvature of the field lines in the channel approaching the exit is found to significantly improve the efficiency, especially for higher voltage (high Isp) Hall thrusters [11, 12]. The strength of the radial magnetic field in the center along the channel [13] is shown in Fig. 7-4. The radial field peaks near the channel exit, and is designed to be essentially zero at or near the anode surface.

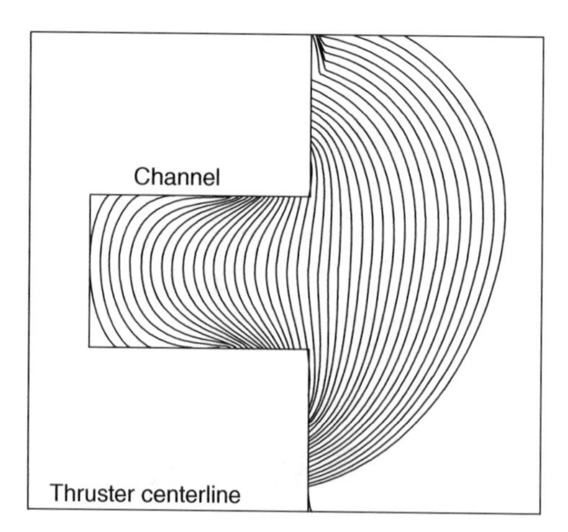

Figure 7-3 Magnetic field lines in the channel region of the NASA-173Mv Hall thruster (*Source:* [11]).

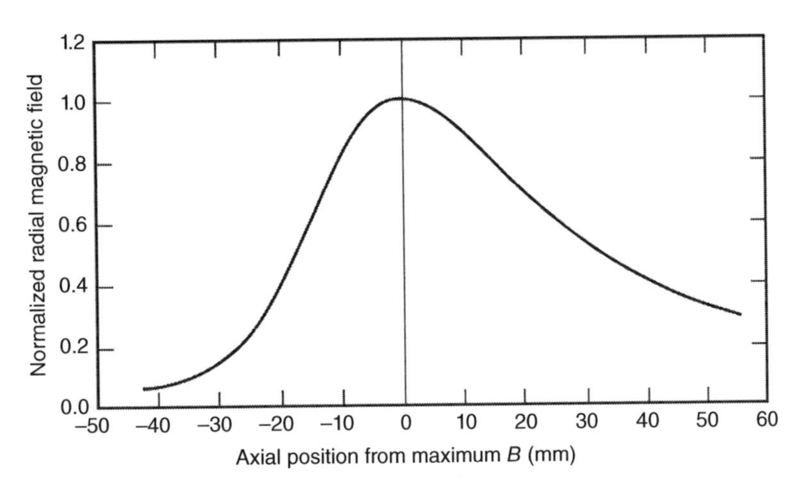

Figure 7-4 Axial variation centerline radial magnetic field normalized to the peak radial field in the NASA-173Mv Hall thruster (*Source:* Adapted from Hofer [13]).

7.2 Operating Principles and Scaling

The operating principles of both types of conventional Hall thrusters and some scaling rules for the geometries can be obtained from a simplified picture of the thruster discharge. Consider a generic Hall thruster channel shown schematically in cross section in Fig. 7-5. The propellant gas is injected from the left through the anode region and is incident on the plasma generated in the channel. An axial scale length L is defined over which the crossed-field discharge is magnetized and produces a significant plasma density of width w, which is essentially the channel width. Ions exiting this plasma over the cylindrically symmetric area A_e form the beam. The applied magnetic field is primarily vertical in the plasma region in this depiction.

7.2.1 Crossed-field Structure and the Hall Current

The electrons entering the Hall thruster channel from the exterior cathode spiral around the radial magnetic field lines with a Larmor radius derived in Chapter 3 and defined by Eq. 3.3-13. The electron Larmor radius must be less than the characteristic scale length L so that the electrons are magnetized and their mobility to the anode is reduced. If the electron velocity is characterized by their thermal velocity, then the electron Larmor radius is

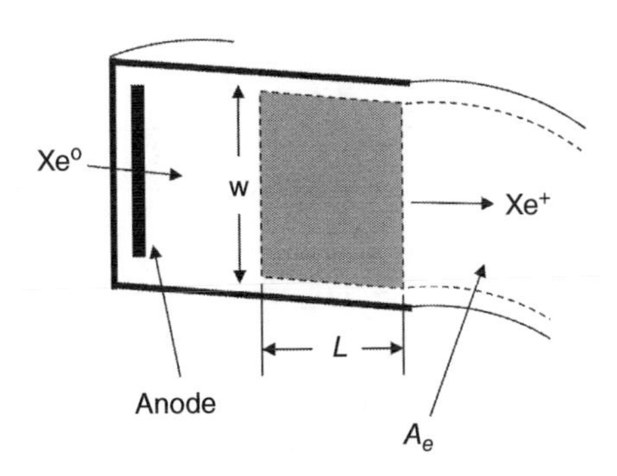

Figure 7-5 Schematic cross section of the plasma in the Hall thruster channel.

$$r_e = \frac{v_{\text{th}}}{\omega_c} = \frac{m}{eB}\sqrt{\frac{8kT_e}{\pi m}} = \frac{1}{B}\sqrt{\frac{8}{\pi}\frac{m}{e}T_{\text{eV}}} << L, \tag{7.2-1}$$

where T_{eV} is the electron temperature in eV (electron volts) and L is the magnetized plasma depth (the ionization region) in the channel. For example, the electron Larmor radius at a temperature of 25 eV and a typical radial magnetic field strength of 150 G is 0.13 cm, which is much smaller than typical channel width and plasma length in Hall thrusters. The electrons must also be considered *magnetized*, meaning that they make many orbits around a field line before a collision with a neutral or ion occurs that results in cross-field diffusion. This is normally described by stating that the square of electron Hall parameter must be large compared to unity:

$$\Omega_e^2 = \frac{\omega_c^2}{\nu^2} >> 1, \tag{7.2-2}$$

where ν is the total collision frequency. The effect of this criterion is clear in the expression for the transverse electron mobility in Eq. 3.6-66 where a large value for the Hall parameter significantly reduces the cross-field electron mobility.

In a similar manner, the ion Larmor radius must be much greater than the characteristic channel length so that the ions can be accelerated out of the channel by the applied electric field:

$$r_i = \frac{v_i}{\omega_{ci}} = \frac{M}{eB}\sqrt{\frac{2eV_b}{M}} = \frac{1}{B}\sqrt{\frac{2M}{e}V_b} >> L, \tag{7.2-3}$$

where the ion energy is approximated as the beam energy. The ion Larmor radius, for example, in the 150 G radial field and at 300 eV of energy is about 180 cm, which is much larger than the channel or plasma dimensions. These equations provide a general range for the transverse magnetic field in the thruster channel. Even if the radial magnetic field strength doubles or ion energy is half of the example given, the criteria in Eqs. 7.2-1 and 7.2-3 are still easily satisfied.

As mentioned above, the magnetic and electric field profiles are important in the thruster performance and life, and the magnetic field configuration strongly impact the plasma flows in the Hall thruster [14]. The radial magnetic field is typically a maximum near the thruster exit plane, as shown in Fig. 7-4, and it is designed to fall near zero at the anode in dielectric-wall Hall thrusters [15]. Electrons from the cathode experience joule heating in the region of maximum transverse magnetic field, providing a higher localized electron temperature and ionization rate. The reduced

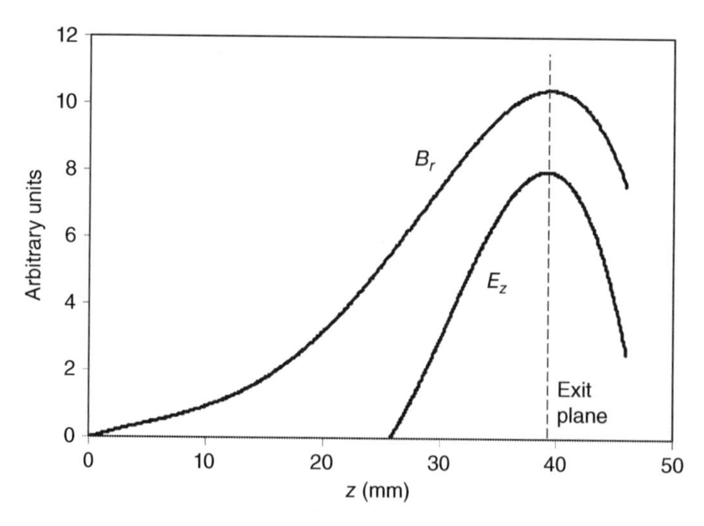

Figure 7-6 Typical Hall thruster radial magnetic field and axial electric field along the channel length.

electron mobility and high electron temperature in the strong magnetic field region causes the axial electric field to also be maximized near the exit plane, as illustrated in Fig. 7-6. Since the neutral gas is injected from the anode region and the mass utilization is very high (nearly every neutral is ionized before reaching the channel exit), it is common to describe an "ionization region" that is located upstream of the electric field peak. Of course, the ions are accelerated directly by the electric field that peaks near the exit plane, which is sometimes called the "acceleration region". The characteristic scaling length L then spans these regions and is a significant fraction of the total channel depth. The ionization and acceleration regions overlap, which leads to dispersion in the ion velocity and some angular divergence in the resultant beam. This is in contrast to ion thrusters, which have a distinct ionization region in the plasma chamber and a finite acceleration region in the grids that produces nearly monoenergetic beams with low angular divergence determined by the optics and curvature of the grids.

In the crossed electric and magnetic field region of the channel, the electrons move in the azimuthal direction because of the $\mathbf{E} \times \mathbf{B}$ force with a velocity given by Eq. 3.3-16. The magnitude of the azimuthal electron velocity was found in Chapter 3 to be

$$v_E = \frac{|\mathbf{E} \times \mathbf{B}|}{B^2} \approx \frac{E_r}{B_z} \ (m/s). \tag{7.2-4}$$

The current in the azimuthal direction, called the Hall current, is then the integral of the electron plasma density and this velocity over the characteristic thickness L [4, 5]:

$$I_H = n_e e \left(\int_0^L v_E \, dz \right) w = n_e e \left(\int_0^L \frac{E}{B} dz \right) w \tag{7.2-5}$$

where w is the plasma width (shown in Fig. 7-5) that essentially fills the channel. The axial electric field in the plasma channel is, approximately, the discharge voltage divided by the plasma thickness, so the Hall current is

$$I_H \approx n_e e \, w \, \frac{V_d}{B}. \tag{7.2-6}$$

Equation 7.2-6 shows that the Hall current increases with the applied discharge voltage and with the channel width provided that the magnetic field is unchanged. Hofer [12] showed that in Hall thrusters optimized for high efficiency, the optimal magnetic field was proportional to the discharge voltage. This implies that the Hall current is approximately constant for a given plasma density or beam current in high efficiency Hall thrusters.

The ion current leaving the plasma to form the beam through the area A_e is approximately

$$I_i = n_i e \, v_i A_e \approx n_i e \sqrt{\frac{2eV_d}{M}} \, 2\pi R w, \tag{7.2-7}$$

where R is the average radius of the plasma channel. Since the plasma is quasi-neutral ($n_i \approx n_e$), even in the magnetized region, the Hall current can be expressed using Eq. 7.2-7 as

$$I_H \approx \frac{I_i}{2\pi R B} \sqrt{\frac{M V_d}{2e}}. \tag{7.2-8}$$

Increasing the beam current in a fixed thruster design will increase the circulating Hall current for a given magnetic field and discharge voltage. From Chapter 2, the total thrust produced by a Hall thruster is

$$T = \int (\mathbf{J_H} \times \mathbf{B}) dA = I_H B \approx I_i \sqrt{\frac{M V_d}{2e}}. \tag{7.2-9}$$

This expression for the thrust has the same form as Eq. 2.3-8 derived in Chapter 2, but the force is coupled magnetically to the Hall thruster body instead of electrostatically to the ion thruster grids.

7.2.2 Ionization Length and Scaling

It is clear from the description of the Hall thruster operation above that the electrons must be magnetized to reduce their axial mobility to the anode, but the ions cannot be significantly magnetized so that the axial electric field can efficiently accelerate them to form the thruster beam. In addition, a large majority of the ions must be generated in the channel to permit acceleration by the field in that region and to produce high mass utilization efficiency [16]. This provides some simple scaling rules to be established.

The neutral gas injected from the anode region will by ionized by entering the plasma discharge in the crossed-field "ionization" region. Consider a neutral gas atom at a velocity υ_n incident on plasma of a density n_e, electron temperature T_e, and thickness L. The density of the neutral gas, n_n, will decrease with time because of ionization:

$$\frac{dn_n}{dt} = -n_n n_e \langle \sigma_i \upsilon_e \rangle, \tag{7.2-10}$$

where $<\sigma_i \upsilon_e>$ is the ionization reaction rate coefficient for Maxwellian electrons found in Appendix E. The flux of neutrals incident on the plasma is

$$\Gamma_n = n_n \upsilon_n, \tag{7.2-11}$$

and the neutral velocity is $\upsilon_n = dz/dt$, where z is the axial length. Equation 7.2-10 then becomes

$$\frac{d\Gamma_n}{\Gamma_n} = -\frac{n_e \langle \sigma_i \upsilon_e \rangle}{\upsilon_n} dz. \tag{7.2-12}$$

This equation has a solution of

$$\Gamma_n(z) = \Gamma(0) e^{-z/\lambda_i}, \tag{7.2-13}$$

where $\Gamma(0)$ is the incident flux on the ionization region and the ionization mean free path λ_i is given by

$$\lambda_i = \frac{\upsilon_n}{n_e \langle \sigma_i \upsilon_e \rangle}. \tag{7.2-14}$$

This expression for the ionization mean free path is different from the one given in Eq. 3.6-6 that applies for the case of fast particles incident on essentially stationary particles. This is because the neutral gas atoms are moving slowly as they traverse the plasma thickness, and the fast electrons can move laterally to produce an ionization collision before the neutral leaves the region. Therefore, the ionization mean free path depends on the velocity of the neutral, which determines the time it spends in the plasma thickness before a collision. The mean free path also varies inversely with the electron density because a higher number of electrons in the slab will increase the probability of one of them encountering the neutral.

The percentage of the neutral flux exiting the plasma of length L that are ionized is

$$\frac{\Gamma_{\text{exit}}}{\Gamma_{\text{incident}}} = 1 - e^{-L/\lambda_i}. \tag{7.2-15}$$

For example, to have 95% of the incident neutral flux on the plasma be ionized before it leaves the plasma, Eq. 7.2-15 gives

$$L = -\lambda_i \ln(1-0.95) \approx \frac{3v_n}{n_e\langle\sigma_i v_e\rangle}, \tag{7.2-16}$$

or the plasma thickness must be at least three times the ionization mean free path. Since some of the ions generated in the plasma hit the channel side walls and re-enter the plasma as neutrals instead of exiting as beam ions, the plasma thickness should significantly exceed the ionization mean free path to obtain high mass utilization efficiency. This leads to one of the Hall thruster scaling rules of

$$\frac{\lambda_i}{L} = \text{constant} << 1. \tag{7.2-17}$$

In this example, this ratio should be less than 0.33.

The channel's actual physical depth in dielectric-wall Hall thrusters is given by the sum of the magnetized plasma thickness (L) and the geometric length required to demagnetize the plasma at the anode. This is illustrated schematically in Fig. 7-6 where the channel depth is nearly twice the magnetized plasma length. The axial magnetic field gradient has been found to be critical for thruster performance [15]. A decreasing radial magnetic field strength going toward the anode, as shown in Fig. 7-6, results in higher thruster efficiency [5, 15]. At the anode, the plasma is largely unmagnetized and an anode sheath forms to maintain particle balance, similar to the DC plasma generator case discussed in Chapter 5. The anode sheath polarity and magnitude depend on the local magnetic field strength and direction, which affects the axial electron mobility, and on the presence of any insulating layers on the anode that affects the particle balance [17–19]. Maintaining the local plasma near the anode close to the anode potential is important in applying the maximum amount of the discharge voltage across the plasma for the acceleration of ions. In addition, the magnetic field profile near the thruster exit strongly affects both the ability to achieve closed electron drifts in the azimuthal direction [10] and focusing of the ions in the axial direction as they are accelerated by the electric field [11]. Optimal magnetic field design in the exit region reduces the ion bombardment of the walls and improves the ion trajectories leaving the thruster [20].

Additional information on the thruster operation can be obtained by examining the ionization criteria. Properly designed Hall thrusters tend to ionize essentially all the propellant gas incident on the plasma from the anode, so that

$$n_n n_e\langle\sigma_i v_e\rangle A_e L \approx n_n v_n A_e. \tag{7.2-18}$$

Using Eq. 7.2-6 for the Hall current, Eq. 7.2-18 becomes

$$L = \frac{v_n e V_d w}{I_H\langle\sigma_i v_e\rangle B}. \tag{7.2-19}$$

The length of the ionization region naturally must increase with neutral velocity and can decrease with the ionization reaction rate coefficient, as seen in Eq. 7.2-16. This is important to achieve high mass utilization when propellants with a lower mass than xenon, such as krypton, are used to increase the specific impulse (Isp) of the thruster [21, 22].

Studies of optimized Hall thrusters of different sizes [23–29] have resulted in some scaling laws. Detailed comparisons of the scaling laws in the literature with experimental results have been performed by Daren et al. [23]. Assuming that the thruster channel inner to outer diameter ratio and the ionization mean free path to plasma length ratio are constants, they found

$$\begin{aligned} &Power \propto Thrust \propto R^2 \\ &I_d \propto R^2 \\ &\dot{m} \propto R^2 \\ &w \approx R(1 - constant) \\ &A_e = \pi\left(R^2 - r^2\right) \end{aligned} \tag{7.2-20}$$

where R is the outside radius of the channel and r is the inside radius of the channel. These scaling rules indicate that the optimum current density is essentially constant as the thruster size changes. The current density in conventional Hall thrusters is typically in the range of 0.1–0.2 A/cm^2. Thus, at a given discharge voltage, the power density in a Hall thruster is also constant. Higher power densities are achieved by increasing the voltage, which has implications on the life of the thruster in these conventional unshielded designs.

7.2.3 Plasma Potential and Current Distributions

The electrical schematic for a Hall thruster is shown in Fig. 7-7. The power supplies are normally all connected to the same reference called the *cathode common*. The hollow cathode requires the same

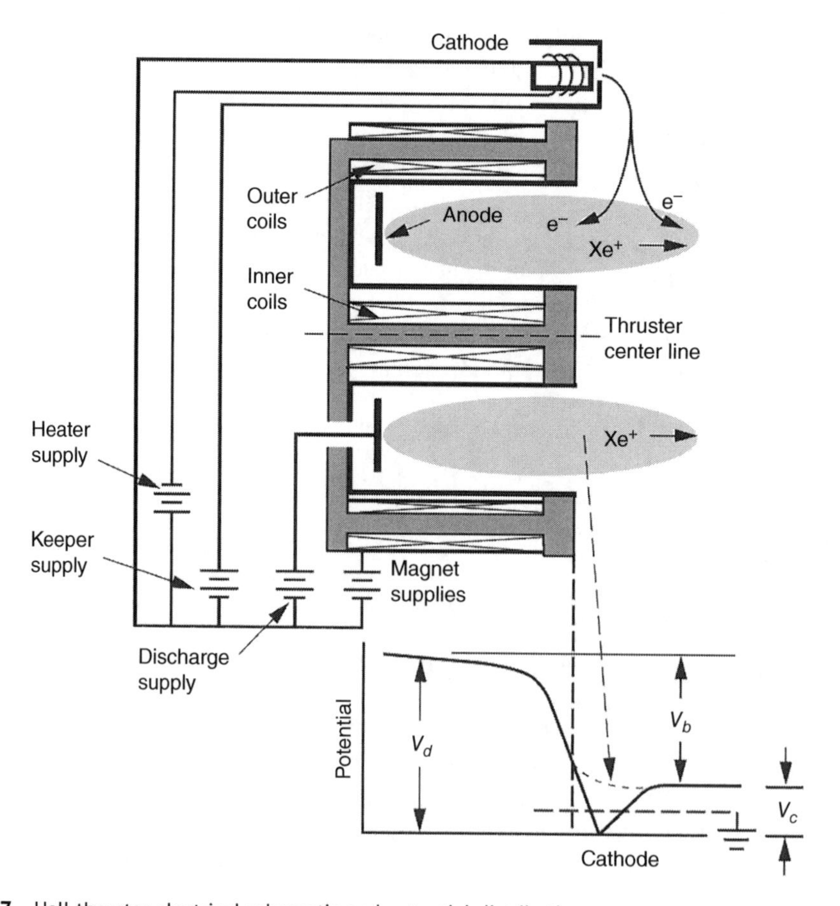

Figure 7-7 Hall thruster electrical schematic and potential distribution.

power supplies as an ion thruster, namely a heater supply to raise the emitter to thermionic emission temperatures, and a keeper supply for ignition and to ensure stable cathode operation at very low currents. The discharge supply is connected between the cathode-common (typically also connected to the thruster body or magnetic circuit) and the anode located in the bottom of the channel. As in ion thrusters, the cathode heater is turned off once the discharge supply is turned on, and the cathode runs in a self-heating mode. The keeper is also normally used only during start-up and is turned off once the thruster is ignited. Also shown are the inner and outer magnetic field coils and their associated power supplies. Hall thrusters have been built with the cathode positioned on-axis (not shown), but this does not change the electrical schematic.

The potential distribution in a Hall thruster is also illustrated in Fig. 7-7 where zero volts is cathode potential, which typically floats negative relative to the spacecraft or facility "common" potential. In the upstream region of the channel (on the left) where the transverse magnetic field is low, the plasma is weakly magnetized and the electron mobility is high. The plasma potential is then close to the anode potential. The plasma potential decreases toward the cathode potential near the thruster exit plane as the magnetic field increases (shown in Fig. 7-6) and limits the electron mobility. The difference between the cathode potential and the beam potential (on the right side of the plot) is the coupling voltage V_c, which is the voltage required to extract current from the hollow cathode. The beam voltage is then

$$V_b = V_d - V_c. \tag{7.2-21}$$

It is common in laboratory experiments to sometimes ignore the difference in potential between the beam and ground/common potential as small (typically 10–20 V) and write the beam voltage as

$$V_b \propto V_d - V_{cg}, \tag{7.2-22}$$

where V_{cg} is the cathode to ground voltage.

The on-axis potential, shown schematically by the dashed line in Fig. 7-7, decreases from the ionization and acceleration regions to the thrust-beam plasma potential. Ions are generated all along this potential gradient, which causes a spread in the ion energy in the beam. Since the majority of the ions are generated upstream of the exit plane (in the "ionization region"), the average velocity of the ion beam can then be expressed as

$$\langle v_b \rangle = \sqrt{\frac{2e\overline{V}_b}{M}}, \tag{7.2-23}$$

where $\overline{V}_b$ represents, in this case, the average potential across which the singly charged ions are accelerated. The actual spread in the beam energy can be significant [30, 31], and must be measured by plasma diagnostics.

The beam from the Hall thruster is charge neutral (equal ion and electron currents). As in ion thrusters, the thruster floats with respect to either spacecraft common in space, or vacuum chamber common on the ground. The common potential normally floats between the cathode and the beam potentials, and can be controlled on spacecraft by a resistor between the spacecraft common and the cathode common. The actual beam energy cannot be measured directly across the power supplies because the potential difference between the beam and ground or spacecraft common is unknown and must be measured by probes or energy analyzers. The coupling voltage is typically the order of twenty volts to operate the cathode discharge properly, which usually ranges from 5 to 10% of the discharge voltage for Hall thrusters with moderate Isp.

In a Hall thruster, the measured discharge current is the net current flowing through the discharge supply. The current flowing in the connection between the anode and the power supply in Fig. 7-7 is the electron and ion current arriving to the anode:

$$I_d = I_{\text{ea}} - I_{\text{ia}}. \tag{7.2-24}$$

The ion current is typically small because of its higher mass, and so the discharge current is essentially the electron current collected by the anode. Likewise, the current flowing in the cathode leg (neglecting any keeper current), is

$$I_d = I_e + I_{\text{ic}}, \tag{7.2-25}$$

where I_e is the emitted current and I_{ic} is the ion current flowing back to the cathode. As with the anode, the ion current to the cathode is typically small, so the discharge current is essentially just the cathode electron emission current. Therefore, the discharge current is approximately

$$I_d \approx I_e \approx I_{\text{ea}}. \tag{7.2-26}$$

Figure 7-8 shows a simplified picture of the currents flowing through the plasma, where the ion currents to the anode and cathode are neglected as small and the ion and electron currents to the dielectric walls are equal and are not shown. Ions are produced in the plasma by ionization events. The secondary electrons from the ionization events, I_{ei}, go to the anode, along with the primary electrons from the cathode, I_{ec}. Primary electrons either ionize neutrals or contribute energy to the plasma electrons so that the energetic electron distribution can produce the ionization. Since it is assumed that the discharge current is essentially the total electron current collected by the anode (the ion current is small), the discharge current can be written as

$$I_d = I_{\text{ei}} + I_{\text{ec}}, \tag{7.2-27}$$

The discharge current is also, essentially, the electron current emitted by the cathode:

$$I_d = I_e = I_{\text{ec}} + I_{\text{eb}}. \tag{7.2-28}$$

Using the fact that one electron and one ion are made in each ionization event such that $I_{ei} = I_{ib}$, Eq. 7.2-27 becomes

$$I_d = I_{\text{ib}} + I_{\text{ec}}. \tag{7.2-29}$$

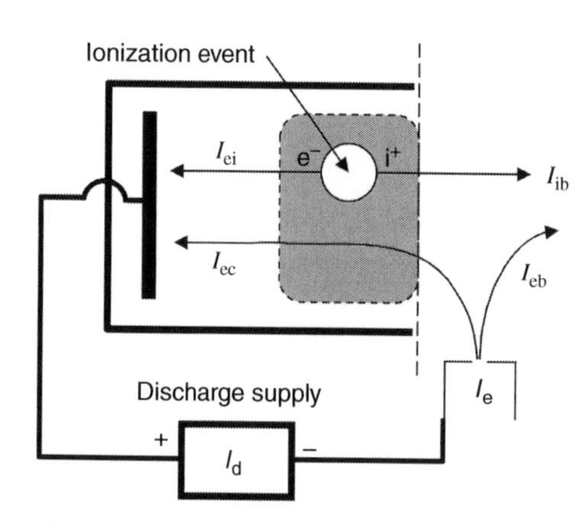

Figure 7-8 Electrical schematic for the currents flowing through the discharge plasma, neutralizer cathode and power supply.

This relationship describes the net current crossing the exit plane, and so it is commonly stated in the literature that the discharge current is the ion beam current plus the backstreaming electron current crossing the exit plane [11].

Depending on the plasma conditions, it is possible for some fraction of the secondary electrons produced near the channel exit to diffuse into the beam. Equation 7.2-28 is still valid in this case-because for every secondary electron that diffuses into the beam, another electron from the cathode plasma must cross the exit plane in the opposite direction to maintain the net discharge current. The discharge current is still the net ion beam current plus the backstreaming electron current across the exit plane. Finally, the ion beam current is equal to the current of electrons entering the beam:

$$I_{ib} = I_{eb}. \tag{7.2-30}$$

Since there is no current return path for the beam ions and electrons because the thruster floats relative to the spacecraft or the grounded vacuum system, the particles in Eq. 7.2-30 do not directly contribute to the discharge current measured in the discharge power supply.

7.3 Performance Models

The efficiency of a generic electric thruster was derived in Chapter 2. Since the beam current and ion energy in Hall thrusters are not directly measured as in ion thrusters, it is useful to develop an alternative expression for the efficiency that incorporates characteristics of Hall thruster discharges. Total efficiency is always defined as the jet power, which is one-half the thrust T times the exhaust velocity v_{ex}, divided by the total input power P_{in}:

$$\eta_T = \frac{T\, v_{ex}}{2P_{in}}. \tag{7.3-1}$$

For any electric thruster, the exhaust velocity is given by Eq. 2.3-6, the Isp by Eq. 2.4-1, and the thrust by Eq. 2.3-1, which can be combined to give

$$v_{ex} = \text{Isp} \cdot g = \frac{g}{2} \frac{v_{ex}}{g} \frac{\dot{m}_i}{\dot{m}_p} = \frac{1}{2} \frac{T}{\dot{m}_p}. \tag{7.3-2}$$

The total efficiency is then

$$\eta_T = \frac{T^2}{2\dot{m}_p P_{in}}. \tag{7.3-3}$$

7.3.1 Thruster Efficiency Definitions

The efficiency of a Hall thruster can be decomposed into a product of terms accounting for the conversion of the electrical power and propellant flow into thrust [32]. The propellant flow in Hall thrusters is split between the anode inside the discharge channel and the hollow cathode:

$$\dot{m}_p = \dot{m}_a + \dot{m}_c, \tag{7.3-4}$$

where $\dot{m}_a$ is the anode flow rate and $\dot{m}_c$ is the cathode flow rate. Since the cathode gas flow is injected exterior to the discharge channel ionization region and is, thereby, largely lost, the "cathode efficiency" is defined as

$$\eta_c = \frac{\dot{m}_a}{\dot{m}_p} = \frac{\dot{m}_a}{\dot{m}_a + \dot{m}_c}. \tag{7.3-5}$$

The total power into the thruster is

$$P_{in} = P_d + P_k + P_{mag}, \tag{7.3-6}$$

where P_d is the discharge power, P_k is the cathode keeper power (normally equal to zero during operation), and P_{mag} is the power used to generate the magnetic field. The electrical utilization efficiency for the other power used in the Hall thruster is defined as

$$\eta_o = \frac{P_d}{P_{in}} = \frac{P_d}{P_d + P_k + P_{mag}}. \tag{7.3-7}$$

Using Eqs. 7.3-5 and 7.3-7 in Eq. 7.3-3 gives a useful expression for the total efficiency of a Hall thruster:

$$\eta_T = \frac{1}{2}\frac{T^2}{\dot{m}_a P_d}\eta_c \eta_o. \tag{7.3-8}$$

By placing the Hall thruster on a thrust stand to directly measure the thrust, knowing the flow rates and flow-split between anode and cathode, and knowing the total power into the discharge, keeper and magnet, it is then possible to accurately calculate the total efficiency.

While Eq. 7.3-8 provides a useful expression for evaluating the efficiency, it is worthwhile to further expand this equation to examine other terms that affect the efficiency. Thrust is given from Eq. 2.3-18:

$$T = \gamma\sqrt{\frac{2M}{e}} I_b \sqrt{\overline{V}_b}, \tag{7.3-9}$$

where the average or effective beam voltage is used because of the spread in ion energies produced in the Hall thruster acceleration region. The fraction of the discharge current that produces beam current is

$$\eta_b = \frac{I_b}{I_d}. \tag{7.3-10}$$

Likewise, the fraction of the discharge voltage that becomes beam voltage is

$$\eta_v = \frac{\overline{V}_b}{V_d}. \tag{7.3-11}$$

Inserting Eqs. 7.3-9–7.3-11 into Eq. 7.3-8 gives

$$\eta_T = \gamma^2 \frac{M}{e}\frac{I_d}{\dot{m}_a}\eta_b^2 \eta_v \eta_c \eta_o. \tag{7.3-12}$$

Equation 7.3-12 shows that the Hall thruster efficiency is proportional to the ion mass and the discharge current, because these terms dominate the thrust production, and inversely proportional to the anode mass flow which dominates the mass utilization efficiency. This equation can be further simplified by realizing that

$$\frac{M}{e} I_d \eta_b = \dot{m}_i, \tag{7.3-13}$$

and that the total mass utilization efficiency can be expressed as

$$\eta_m = \frac{\dot{m}_i}{\dot{m}_p} = \frac{\dot{m}_i}{\dot{m}_a + \dot{m}_c}. \tag{7.3-14}$$

The total efficiency then becomes

$$\eta_T = \gamma^2 \eta_b \eta_v \eta_m \eta_o. \tag{7.3-15}$$

This expression contains the usual gamma-squared term associated with beam divergence and multiply charged ion content, and also the mass utilization and electrical utilization efficiencies. However, this expression also includes the efficiencies associated with generating beam ions and imparting the discharge voltage to the beam voltage. This shows directly that Hall thruster designs that maximize beam current production and beam energy, and minimize the cathode flow, produce the maximum efficiency provided that the beam divergence and double ion content are not adversely affected. Expressions like Eq. 7-3-15 appear in the Hall thruster literature [5, 7], because they are useful in illustrating how the efficiency depends on the degree that the thruster converts power supply inputs (such as discharge current and voltage) into the beam current and beam voltage that impart thrust. Understanding each efficiency term is critical to fully optimize the Hall thruster performance.

The efficiency of a Hall thruster is sometimes expressed in terms of the anode efficiency:

$$\eta_a = \frac{1}{2}\frac{T^2}{\dot{m}_a P_d} = \frac{\eta_T}{\eta_o \eta_c}, \tag{7.3-16}$$

which describes the basic thruster performance without considering the effects of the cathode flow or power used to generate the magnetic field. This is usually done to separate out the cathode and magnet losses so that trends in the plasma production and acceleration mechanisms can be discerned. The anode efficiency should not be confused with the total efficiency of the thruster given by Eq. 7.3-3.

It is useful to show an example of the relative magnitude of the efficiency terms derived above. Figure 7-9 shows the anode efficiency that was defined in Eq. 7.3-16 and the other efficiency terms discussed above for the laboratory-model NASA-173Mv2 Hall thruster operating at 10 mg/s versus the discharge voltage. In this figure, the charge utilization efficiency is the net efficiency decrease because of multiply charged ions [12], the voltage utilization efficiency (η_V) is the conversion of voltage into axially directed ion velocity, the current utilization efficiency (η_b) is the fraction of ion current contained in the discharge current, and mass utilization efficiency (η_m) is the conversion of neutral mass flux into ion mass flux. The anode efficiency increases with discharge voltage, largely because the voltage efficiency and current efficiency increase with voltage. The current

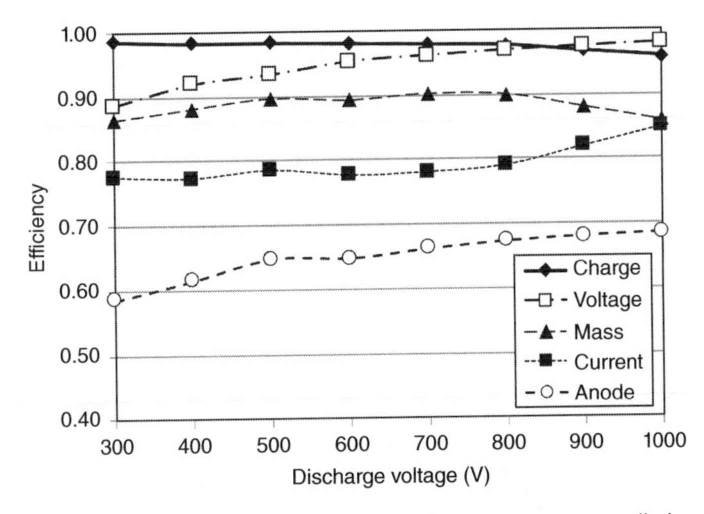

Figure 7-9 Optimized anode efficiency and the individual efficiency terms versus discharge voltage for the NASA-173Mv2 Hall thruster operating at 10 mg/s (*Source:* Adapted from Hofer et al. [11]).

utilization is always lower than the other efficiency terms, suggesting that the ultimate efficiency of Hall thrusters is dominated by the electron dynamics involved in producing the plasma and neutralizing the beam. This emphasizes the importance [11, 12] of optimizing the magnetic field design to maximize the thruster efficiency.

The value of γ in Eq. 7.3-15 that is typically found for Hall thrusters can be evaluated using Eq. 2.3-11 and the data in the literature. For example, a 10% double ion content gives a thruster correction factor in Eq. 2.3-17 of $\alpha = 0.973$. The thrust loss due to the beam angular divergence of Hall thrusters is given by Eq. 2.3-12 ($F_T = \cos\theta$). For both SPT-100 Hall thrusters [6] and TAL thrusters [33], a half angle divergence θ equal to about 20° is observed, producing $F_T = 0.94$. The total correction factor is then $\gamma = \alpha F_T = 0.915$ for typical Hall thruster conditions. Values for γ of about 0.9 have been reported.

The equivalent discharge loss for a Hall thruster can also be calculated [5, 6] to provide information on how the thruster design impacts the cost of producing the beam ions. The average energy cost for producing a beam ion is the discharge power divided by the number of beam ions minus the beam power per beam ion:

$$\varepsilon_b = \frac{I_d V_d}{I_b} - \frac{I_b V_b}{I_b} = \frac{I_d V_d}{I_b} - V_b = \frac{P_d(1 - \eta_b \eta_v)}{I_b}, \tag{7.3-17}$$

where Eqs. 7.3-10 and 7.3-11 were used. Equation 7.3-17 has the usual units for discharge loss of watts per beam-amp or electron-volts per ion. As expected, maximizing the current and voltage efficiencies minimizes the discharge loss. As an example of discharge loss in a Hall thruster, consider the SPT-100 thruster operating at the nominal 1.35 kW discharge power and 300 V. The discharge current is then 1350/300 = 4.5 A. The thruster is reported [3, 5, 6] to have values of $\eta_b \approx 0.7$ and $\eta_v = 0.95$. The cost of producing beam ions is then

$$\varepsilon_b = \frac{P_d(1 - \eta_b \eta_v)}{I_b} = 144 \text{ eV/ion}.$$

This is on the same order as the discharge loss for DC-discharge ion thrusters.

7.3.2 Multiply Charged Ion Correction

In Hall thrusters operating at higher power levels (high mass flow rate and high discharge voltages >300 V), a significant number of multiply charged ions can be generated, and their effect on the performance may be noticeable. Following the analysis by Hofer [13], the performance model from the previous section can be modified to address the case of partially-ionized thruster plasmas with an arbitrary number of ion species.

The total ion beam current is the sum of all ion species i

$$I_b = \sum_{i=1}^{N} I_i \tag{7.3-18}$$

and the current fraction of the ith species is

$$f_i = \frac{I_i}{I_b}. \tag{7.3-19}$$

Likewise, the total plasma density in the beam is the sum of the individual species densities

$$n_b = \sum_{i=1}^{N} n_i \tag{7.3-20}$$

and the density fraction of the ith species is

$$\zeta_i = \frac{n_i}{n_b}. \tag{7.3-21}$$

The total beam current is then

$$I_b = \sum_i n_i q_i \langle v_i \rangle A_e = \sum_i n_b e A_e \sqrt{\frac{2e\overline{V}_b}{M}} \zeta_i Z_i^{3/2} \tag{7.3-22}$$

where Z_i is the charge state of each species. The mass flow rate of all the beam ion species is

$$\dot{m}_b = \frac{I_b M}{e} \sum_i \frac{f_i}{Z_i}. \tag{7.3-23}$$

Using the current utilization efficiency defined in Eq. 7.3-10, the mass utilization efficiency in Eq. 7.3-14 then becomes

$$\eta_m = \frac{\dot{m}_b}{\dot{m}_p} = \frac{\eta_b I_d M}{\dot{m}_p e} \sum_i \frac{f_i}{Z_i}. \tag{7.3-24}$$

If the current utilization efficiency is the same for each species, then the mass utilization efficiency for arbitrary species can be written as

$$\eta_m = \eta_m^+ \sum_i \frac{f_i}{\zeta_i} \tag{7.3-25}$$

where η^+_m is the usual mass utilization for a singly-charged species. This is an easily implemented correction in most models if the species fractions are known. Likewise, the thrust obtained for multiple species can be generalized from Eq. 2.3-16 for Hall thrusters to

$$T_m = \sum_i T_i = \eta_b I_d \sqrt{\frac{2M\eta_v V_d}{e}} \sum_i \frac{f_i}{Z_i} \cos\theta. \tag{7.3-26}$$

7.3.3 Dominant Power Loss Mechanisms

In preparation for examining the terms that drive the efficiency of Hall thrusters, it is helpful to examine the dominant power loss mechanisms in the thruster. Globally, the power into the thruster comes from the discharge power supply. The power out of the thruster, which is equal to the input power, is given to first order by

$$P_d = P_b + P_w + P_a + P_R + P_{\text{ion}}, \tag{7.3-27}$$

where P_b is the beam power given $I_b V_b$, P_w is the power to the channel walls due of ion and electron loss, P_a is the power to the anode due of electron collection, P_R is the radiative power loss from the plasma, and P_{ion} is the power to produce the ions that hit the walls and become the beam. The additional loss terms, such as the power that electrons take into the beam, the ion power to the anode, etc., are relatively small and can usually be neglected.

In Hall thrusters with dielectric walls, the power loss because of ion and electron currents flowing along the radial magnetic field through the sheath to the channel walls (P_w) represents the most significant power loss. The current deposition and power lost to the walls can be estimated from the sheath potentials and electric fields in the plasma edge. Since the wall is insulating, the net ion and electron currents to the surface must be equal. However, ion and electron bombardment of common insulator materials, such as boron nitride, at the energies characteristic of Hall thrusters produces a significant number of secondary electrons, which reduces the sheath potential at the wall and increases the power loading.

The requirement of local net current equal to zero and particle balance for the three species gives

$$I_{iw} = I_{ew} - \gamma I_{ew} = I_{ew}(1-\gamma), \tag{7.3-28}$$

where γ is the secondary electron yield from electron bombardment. Using Eq. 3.7-53 for the Bohm current of ions to the wall, Eq. 3.7-54 for the electron current to the wall, and neglecting the secondary electron velocity, Eq. 7.3-28 can be solved for the sheath potential ϕ_s including the effect of secondary electron emission by the wall up to a secondary yield of one:

$$\phi_s = -\frac{kT_e}{e}\ln\left((1-\gamma)\sqrt{\frac{2M}{\pi m}}\right). \tag{7.3-29}$$

This expression is slightly different than that found in the literature [34, 35] because we have approximated $e^{-1/2} = 0.61 \approx 0.5$ for the coefficient in the expression for the Bohm current. Nevertheless, as the secondary electron yield increases from zero, the sheath potential decreases from the classic floating potential described in Chapter 3 toward the plasma potential. If the secondary electron yield exceeds one, this simple model breaks down and the potentials will change to obtain floating walls and zero net current.

Secondary electron yields reported in the literature [34, 36, 37] for several materials used for the walls of Hall thrusters are shown in Fig. 7-10. In this figure, the measurements were made using a monoenergetic electron gun. Generalizing this data for incident Maxwellian electron temperatures is accomplished by integrating the yield over the Maxwellian electron energy distribution function, which results in multiplying the secondary emission scaling by the Gamma

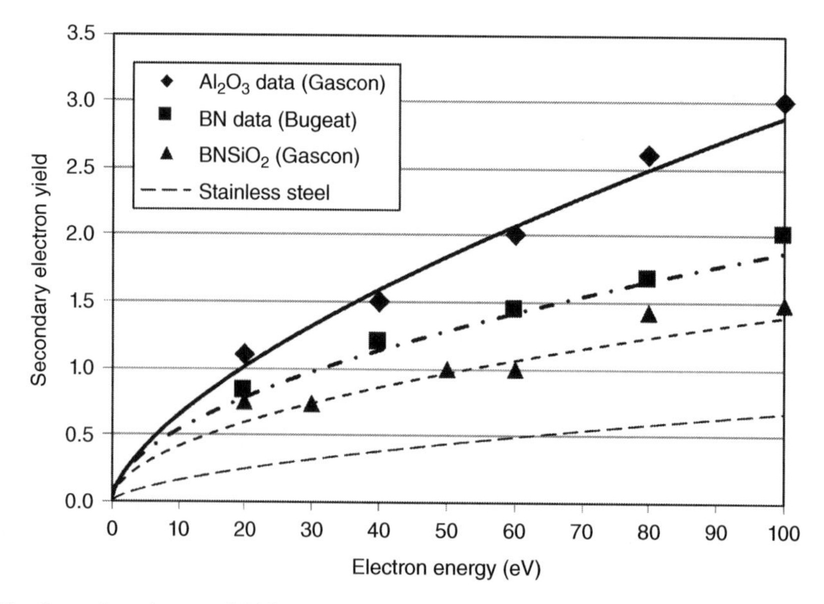

Figure 7-10 Secondary electron yield for several wall materials used in Hall thrusters measured with a monoenergetic electron beam.

Function [34]. An expression for the secondary electron yield from electron bombardment of materials is then

$$\gamma = \Gamma(2 + b)\, a\, T_{eV}^{b}, \tag{7.3-30}$$

where the electron temperature is in electron volts and $\Gamma(x)$ is the Gamma Function and the coefficients a and b are found from fits to the data in Fig. 7-10. Values of the coefficients in Eq. 7.3-30 can be found in Table 7-1 for these materials, and the actual secondary electron yield for the Hall thruster walls is plotted versus plasma electron temperature in Fig. 7-11. It should be noted that because of reflection at the wall, the effective secondary electron yield does not go to zero for zero electron energy. This effect is accommodated by linear fits to the data that result in finite yield at low electron energy. Figure 7-12 shows the data for BN and $BNSiO_2$ with the two different fitting choices. In the evaluation of the sheath potential in the presence of the secondary electron emission below, the choice of whether to use a linear or power fit does not make a significant difference in the ionization and acceleration regions for electron temperatures above about 10 eV.

Measurements of the electron temperature in the channel of Hall thrusters by a number of authors [38–40] show electron temperatures in the channel well in excess of 20 eV. Equation 7.3-29 predicts that the sheath potential will go to zero and reverse from negative going (electron repelling) to

Table 7-1 Fitting parameters for secondary electron yield data.

	a	*b*	$\Gamma(2 + b)$
Alumina (Al_2O_3)	0.145	0.650	1.49
Boron nitride (BN)	0.150	0.549	1.38
Borosil ($BNSiO_2$)	0.123	0.528	1.36
Stainless steel	0.040	0.610	1.44

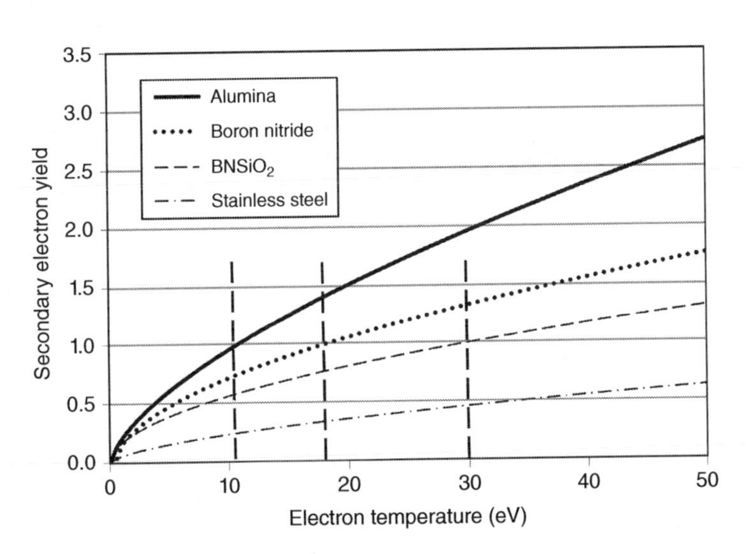

Figure 7-11 Secondary electron yield from the power-curve fits versus electron temperature showing the cross-over value at which the yield equals one.

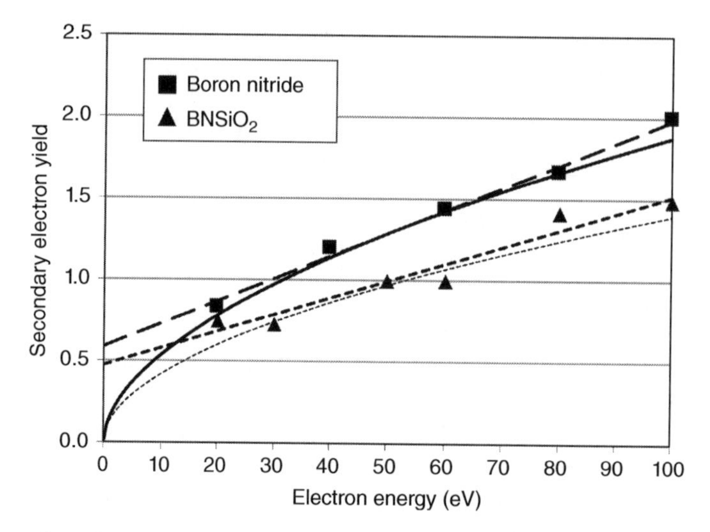

Figure 7-12 Secondary electron yield versus electron energy showing linear curve fits to the data producing finite yield at low incident energy.

positive-going (electron attracting) as the secondary electron yield approaches unity for some of the materials. The value at which this occurs for each of the materials shown in Table 7-1 is indicated in Fig. 7-11. For boron nitride and alumina walls this occurs at electron temperatures below 20 eV, and for BN-SiO_2 walls it occurs at electron temperatures on the order of 30 eV. In addition, some of the secondary electrons can pass completely through the plasma to strike the opposite wall of the channel, depending on the collision mean free path. The possibility of the sheath potential reversing to electron-attracting was used to predict very high electron power losses to the walls in some early analysis of Hall thrusters at high electron temperatures [34, 35] because the incident electron flux can then equal or exceed the random electron flux along the magnetic field lines in the plasma.

In reality, the sheath potential for a floating boundary in conventional Hall thrusters can never go more positive than the local plasma potential [41, 42] for two reasons. First, the secondary electrons are ejected from the wall with very low energy (typically 1–2 eV). Any positive going sheath (where the plasma is negative by one or two volts relative to the wall) will repel the secondary electrons and return them to the wall. This clamps the sheath potential to within a few volts positive with respect to the plasma. Second, the secondary electron emission is space charge limited in the sheath. This effect was analyzed by Hobbs and Wesson [43], who showed that space charge limits the secondary electron current from the wall independent of the secondary electron yield. The local electron space charge in the sheath clamps the sheath voltage to a maximum value that is always negative relative to the plasma.

The effects of space charge on the sheath potential at the wall can be analyzed [43] by solving Poisson's Equation for the potential in the sheath:

$$\frac{\partial^2 \phi}{\partial x^2} = \frac{1}{\varepsilon_o}(n_e + n_s - n_i), \tag{7.3-31}$$

where n_s is the secondary electron density. Using a Maxwellian distribution for the electrons, the plasma density in the channel is

$$n_e = (n_o - n_{so})e^{\left(e\phi/kT_e\right)}, \tag{7.3-32}$$

where n_o is the ion density at the sheath edge, n_{so} is the secondary electron density at sheath edge and ϕ is the potential relative to the potential ϕ_o at the wall. The ions are assumed to be cold and to have fallen through the pre-sheath to arrive at the sheath edge with an energy of

$$\mathcal{E} = \frac{1}{2} m v_o^2, \tag{7.3-33}$$

where v_o is the Bohm velocity modified for the presence of the secondary electrons. The ion density through the sheath is then

$$n_i = n_o \left(\frac{\mathcal{E}}{\mathcal{E} - e\phi} \right)^{1/2}. \tag{7.3-34}$$

The secondary electrons are assumed to be emitted with an energy that is small compared to the plasma electron temperature and are accelerated through the sheath. The equation of continuity for current at the sheath edge gives

$$n_s v_s = \frac{\gamma}{1-\gamma} n_o v_o. \tag{7.3-35}$$

where v_s is the secondary electron velocity. The secondary electron density through the sheath is then

$$n_s = n_o \frac{\gamma}{1-\gamma} \left(\frac{m}{M} \frac{\mathcal{E}}{\phi - \phi_o} \right). \tag{7.3-36}$$

Equations 7.3-32, 7.3-34 and 7.3-36 are inserted into Poisson's Eq. 7.3-31 and evaluated by the usual method of multiplying through by $d\phi/dx$ and integrating to produce

$$\begin{aligned} \frac{1}{2\varepsilon_o n_o k T_e} \left(\frac{d\phi}{dx} \right)^2 = {} & \frac{2\mathcal{E}}{kT_e} \left[\left(1 - \frac{e\phi}{\mathcal{E}} \right)^{1/2} - 1 \right] \\ & + \frac{2\gamma}{1-\gamma} \left(- \frac{m}{M} \frac{\mathcal{E}}{kT_e} \frac{e\phi_o}{kT_e} \right)^{1/2} \left[\left(1 - \frac{\phi}{\phi_o} \right)^{1/2} - 1 \right]. \\ & + \left[1 - \frac{\gamma}{1-\gamma} \left(- \frac{m}{M} \frac{\mathcal{E}}{e\phi_o} \right)^{1/2} \right] \left[\exp\left(\frac{e\phi}{kT_e} \right) - 1 \right] \end{aligned} \tag{7.3-37}$$

A monotonic sheath potential is found [43] for

$$\mathcal{E} = \frac{kT_e}{2} + \frac{\gamma}{1-\gamma} \left(\frac{m}{M} \right)^{1/2} \left(\frac{-\mathcal{E}}{e\phi_o} \right)^{3/2} \left(\frac{kT_e}{2} - e\phi_o \right). \tag{7.3-38}$$

For the case of no secondary electron emission (γ going to zero), the Bohm criteria solution of $\varepsilon \geq kT_e/2e$ is recovered. Owing to the small electron to ion mass ratio for xenon, the right-hand term is always small and the ion velocity at the sheath edge for the case of finite secondary electron emission will be near the Bohm velocity. Hobbs and Wesson evaluated this minimum ion energy at the sheath edge for the case of space charge limited emission of electrons at the wall ($d\phi_o/dx|_{x=0} = 0$ in Eq. 7.3-37, and they found

$$\mathcal{E}_o = 0.58 \frac{kT_e}{e}. \tag{7.3-39}$$

Equation 7.3-39 indicates that the Bohm sheath criterion will still approximately apply (within about 16%) in the presence of secondary electron emission.

The value of the sheath potential for the space charge limited case can be found by setting the electric field at the wall equal to zero in Eq. 7.3-37 and evaluating the potential using Eq. 7.3-38 and the current continuity equation:

$$\frac{1}{4}\left[1-\frac{\gamma}{1-\gamma}\left(-\frac{m}{M}\frac{\mathcal{E}}{e\phi_o}\right)^{1/2}\right]e^{\frac{e\phi_o}{kT_e}}\left(\frac{8kT_e}{\pi m}\right)^{1/2}=\frac{1}{1-\gamma}\left(\frac{2\mathcal{E}}{M}\right)^{1/2}. \tag{7.3-40}$$

The space charge limited sheath potential for xenon is found to be

$$\phi_o=-1.02\frac{kT_e}{e}. \tag{7.3-41}$$

The secondary electron yield at which the sheath becomes space charge limited [43] is approximately

$$\gamma_o=1-8.3\left(\frac{m}{M}\right)^{1/2}, \tag{7.3-42}$$

which for xenon is 0.983.

This analysis shows that the sheath potential for a xenon plasma decreases from $-5.97\mathrm{T_e}$ for walls where the secondary electron yield can be neglected, to $-1.02\mathrm{T_e}$ for the case of space charge limited secondary electron emission that will occur at high plasma electron temperatures. The value of the sheath potential below the space charge limit can be found exactly by evaluating the three equations Eqs. 7.3-37, 7.3-38 and 7.3-40 for the three unknowns (ϕ, γ and ε).

However, the value of the sheath potential relative to the plasma edge in the presence of the secondary electron emission can be estimated by evaluating Eq. 7.3-28 while accounting for each of three species [42]. Quasi-neutrality for the three species in the plasma edge dictates that $n_i=n_e+n_s$, where n_s is the density of the secondary electrons, and the flux of secondary electrons is the secondary electron yield times the flux of plasma electrons. Equating the ion flux to the net electron flux to the wall gives

$$I_{\mathrm{iw}}=n_i e v_i A=I_{\mathrm{ew}}(1-\gamma)=\frac{1}{4}n_e(1-\gamma)e\left(\frac{8kT_e}{\pi m}\right)^{1/2}Ae^{\left(\frac{e\phi_s}{kT_e}\right)}, \tag{7.3-43}$$

where the ion and electron densities are evaluated at the sheath edge. The sheath potential ϕ_s relative to the plasma potential is then

$$\phi_s=-\frac{kT_e}{e}\ln\left(\sqrt{\frac{M}{2\pi m}}\frac{n_e}{n_e+n_s}\frac{v_B}{v_i}(1-\gamma)\right), \tag{7.3-44}$$

where v_i is the modified ion velocity at the sheath edge due of the presence of the secondary electrons and the ion density is the sum of the plasma and secondary electrons. This equation is useful up to the space charge limited potential of $\phi_o=-1.02T_{\mathrm{eV}}$ and provides good agreement with the results for xenon described above for $n_e v_B/n_i v_i\approx 0.5$. The sheath potential predicted by Eq. 7.3-44 is plotted in Fig. 7-13 for two wall materials. In the limit of no secondary electron emission ($\gamma=0$), the classic value for the sheath floating potential is obtained from Eq. 3.7-53. Once the electron temperature is sufficiently high to produce a yield approaching and even exceeding one, then the space charge limited case of $\phi_o=-1.02T_{\mathrm{eV}}$ is obtained. In between, the sheath potential depends on the

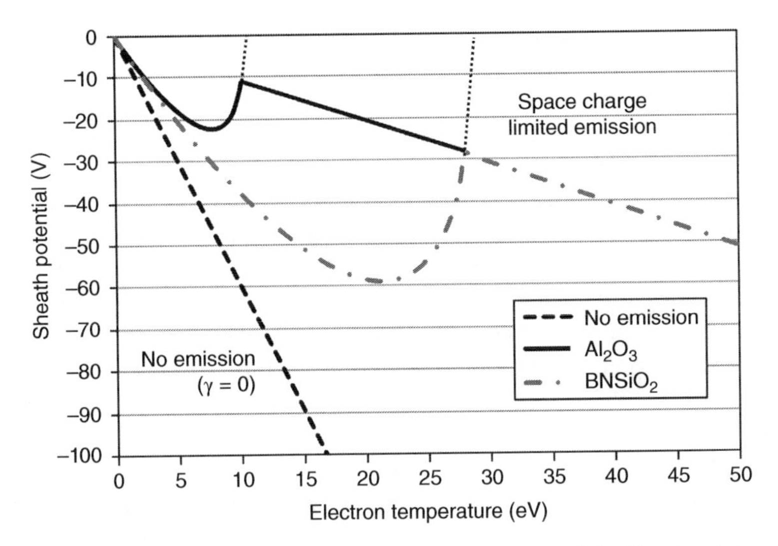

Figure 7-13 Sheath potential versus electron temperature for two materials. The sheath transitions to space charge limited where the dashed lines intersect the potential curves.

electron temperature and material of the wall. Without the space charge limited sheath regime predicted by Hobbs and Wesson, the potential would have continued along the thin dashed lines for the two cases and incorrectly resulted in very low sheath potentials and high-power loadings at the wall.

The total power to the wall of the Hall thruster is

$$P_w = \frac{1}{4}\left(\frac{8kT_e}{\pi m}\right)^{1/2} en_o A\, e^{\frac{e\phi_s}{kT_e}}\left(2\frac{kT_e}{e}\right) + n_o e\, v_i A(\mathcal{E} - \phi_s), \tag{7.3-45}$$

where the first term is because of electrons overcoming the repelling sheath potential and depositing $2T_e$ on the wall, and the second term is because of ions that have fallen through the pre-sheath potential and then the full sheath potential. Note that n_o in this equation is the plasma density at the sheath edge, and is roughly half the average plasma density in the center of the channel due to the radial pre-sheath. The cooling of the wall by the secondary electron emission has been neglected. Equation 7.3-45 can be rewritten in terms of the total ion current to the wall as

$$P_w = I_{iw}\left[\left(\frac{M}{2\pi m}\right)^{1/2} e^{\frac{e\phi_s}{kT_e}}\left(2\frac{kT_e}{e}\right) + (\mathcal{E} - \phi_s)\right]. \tag{7.3-46}$$

For the case of space charge limited secondary electron emission, the sheath potential $\phi_s = \phi_o = -1.02T_{eV}$, and the ion energy is $\varepsilon = 0.58T\text{eV}$ to satisfy the Bohm condition. Equation 7.3-45 predicts the maximum heat loading to the wall in the presence of a Maxwellian electron distribution and secondary electron emission from the wall, which is the dominant power loss mechanism in dielectric-wall Hall thrusters. If the electron distribution function is non-Maxwellian, the heat load to the wall can differ from that predicted by Eq. 7.3-45.

In the case of TAL thrusters, the channel wall is metallic and biased to cathode potential. This eliminates the zero-net current condition found on the insulating walls of dielectric-channel Hall thrusters and used to determine the local heat flux in Eq. 7.3-45. The electron flux to the cathode-biased TAL channel wall is negligible, and the secondary yield for metals is much lower than for insulators, so the secondary electron emission by the wall in TAL thrusters has little effect on the thruster operation. In addition, the plasma tends to be localized near the channel center by the anode design and gas feed geometry. The plasma then tends to be in poor contact with the guard rings at the wall that also have a small exposed area to the plasma, resulting in low radial ion currents to the wall. This is evidenced by the erosion pattern typically observed on TAL guard rings [33], which tends to be on the downstream face from particles outside the thruster instead of on the inside diameter from the channel plasma. While the ion and electron currents and power deposition to the inside diameter of the metallic guard ring are likely smaller than in the dielectric-wall thruster case (where the power loss due to the electrons is dominant), the erosion on the face of the guard ring indicates energetic ion bombardment is occurring. This effect is significant in determining the life of the TAL.

However, TAL thrusters are characterized by having the anode in close contact with the magnetized plasma near the channel exit, in contrast to the dielectric-wall Hall thrusters. The magnetized plasma has a high electron temperature, which causes a significant amount of power to be deposited from the discharge current on the anode. It is possible to evaluate this power loss mechanism based on the current and sheath potential at the anode.

As described above, the discharge current is essentially equal to the electron current collected at the anode. For the TAL thruster to transfer a large fraction of the discharge voltage to the ions, the potential of the plasma near the anode must be close to the anode potential. Assuming the local plasma potential is then equal to or slightly positive relative to the anode, the electron current to the anode, I_a, deposits $2T_{\text{eV}}$ in energy from the plasma (see Appendix C). The power deposited on the anode, P_a, is then given by

$$P_a = 2T_{\text{eV}} I_a \approx 2T_{\text{eV}} I_d, \tag{7.3-47}$$

where the electron temperature is at the anode and Eq. 7.2-26 has been used. If the plasma potential is negative relative to the anode, the thruster efficiency will suffer because of the loss of discharge voltage available to the ions, and the anode heating will increase because of the positive going sheath potential accelerating electrons into the anode. Equation 7.3-47 then represents a reasonable, but not worse case, heat flux to the anode.

This power loss to the anode can be related to the beam current using the fraction of the discharge current that produces beam current, which is defined as

$$\eta_b = \frac{I_b}{I_d}. \tag{7.3-48}$$

Therefore, the power to the anode is

$$P_a = 2T_{\text{eV}} \frac{I_b}{\eta_b}. \tag{7.3-49}$$

In well-designed Hall thrusters, η_b ranges typically from 0.6 to 0.8. Therefore, the power loss to the anode is 3–4 times the product of the electron temperature in the near-anode region and the beam current. This is the most significant power loss mechanism in TAL thrusters.

7.3.4 Electron Temperature

The electron temperature in the channel must be known to evaluate the power loss mechanisms described above. The peak electron temperature in the plasma channel can be found using the power balance described by Eq. 7.3-27. This method provides reasonable estimates because the power loss in the thruster will be shown to be a strong function of the electron temperature. Even though the plasma density and electron temperature peak in different locations along the channel associated with the different ionization and acceleration regions, the strong axial electron temperature profile in Hall thrusters causes the majority of the power loss to occur in the region of the highest electron temperature. This occurs near the channel exit where the magnetic field across the channel is the strongest. Evaluating the plasma parameters and loss terms in this region, which is bounded by the channel width and magnetic axial field extent in the channel, establishes the electron temperature that is required to satisfy the power balance in the plasma for a given thruster current and voltage.

The individual terms in Eq. 7.3-27 will now be evaluated. The input power to the thruster is the discharge current times the discharge voltage ($P_d = I_d V_d$). The power in the beam, using Eq. 7.3-48, is

$$P_b = \eta_b \eta_v I_d V_d = \eta_v I_b V_d, \tag{7.3-50}$$

where the current utilization and voltage utilization efficiencies have to be known or evaluated by some means. The difference between the beam power and the discharge power is the power remaining in the plasma channel to produce the plasma and offset the losses:

$$P_p = (1 - \eta_b \eta_v) I_d V_d = I_{ec} V_d, \tag{7.3-51}$$

where P_p is the power into the plasma. The plasma is produced and heated essentially by the collisional transport of the electrons flowing from the cathode plasma in the near-plume region to the anode inside the thruster. The power into channel walls, from Eq. 7.3-45, can be written as

$$P_w = n_e \mathrm{e} \mathrm{A} \left[\left(\frac{\mathrm{kT}_e}{e} \right)^{3/2} \left(\frac{e}{2\pi m} \right)^{1/2} e^{\frac{e\phi_s}{kT_e}} + \frac{1}{2} \sqrt{\frac{\mathrm{kT}_e}{M}} (\mathcal{E} - \phi_s) \right], \tag{7.3-52}$$

where A is the total area of the inner and outer channel walls in contact with the high temperature plasma region, n_e is the plasma density in the channel center, and the sheath potential ϕ_s is given by Eq. 7.3-44. Equation 7.3-52 shows the wall power varies linearly with the density, but with the electron temperature to the 3/2 power. This is why the dominant power loss to the wall occurs in the region of the highest electron temperature.

The power into the anode, from Eq. 7.3-47, can be written as

$$P_a = 2 I_d T_{eV}(\text{anode}). \tag{7.3-53}$$

where the electron temperature in this case is evaluated near the anode. The power radiated is

$$P_R = n_o n_e \langle \sigma_* v_e \rangle V, \tag{7.3-54}$$

where the excitation reaction rate coefficient $\langle \sigma_* v_e \rangle$ is given in Appendix E as a function of the electron temperature, and V is the volume of the high temperature plasma region in the channel which can be taken to be the channel cross-sectional area times the axial thickness L. Equations 7.3-52 and 7.3-54 require knowledge of the plasma density in the high temperature region in the channel. This can be found to first order from the beam current

$$n_e = \frac{I_b}{e v_b A_c} \approx \frac{\eta_b I_d}{e A_c \sqrt{\frac{2\eta_b e V_d}{M}}}, \tag{7.3-55}$$

where A_c is the area of the channel exit. Finally, the power to produce the ions in the thruster is the sum of the beam current and the ion current to the walls times the ionization potential:

$$P_{\text{ion}} = (I_b + I_{\text{iw}})\,U^+ = [\eta_b I_d + I_{\text{ew}}(1-\gamma)]\,U^+, \tag{7.3-56}$$

where I_{iw} is given by Eq. 7.3-28 and I_{ew} is given by the left-hand side of Eq. 7.3-52 divided by $2T_e$ (because the electron energy hitting the wall is already included in this equation).

The peak electron temperature is found by equating the input power to the plasma in Eq. 7.3-51 with the sum of the various loss terms described above, and then iterating to find a solution. For example, the SPT-100 Hall thruster has a channel outside diameter of 10 cm, a channel inside diameter of 7 cm, and runs nominally at a discharge of 300 V at 4.5 A with a current utilization efficiency of 0.7 and a voltage utilization efficiency of 0.95 [6]. From Eq. 7.3-55, the plasma density at the thruster exit is about $1.6 \times 10^{17}\ \text{m}^{-3}$. The power into the plasma, from Eq. 7.3-51 is about 433 W. Taking the electron temperature at the anode to be 5 eV and the hot-plasma thickness L to be about 1 cm, the power balance equation is satisfied if the electron temperature in the channel plasma is about 25 eV.

It is a common rule-of-thumb in conventional Hall thrusters to find that the electron temperature is about one-tenth the beam voltage [39]. The result in the example above of $T_{\text{eV}} \approx 0.08V_{\text{d}}$ is consistent with that observation. It is also important to note that nearly 70% of the power deposited into the plasma goes to the dielectric-channel walls in the form of electron heating, and that the radiation losses predicted by Eq. 7.3-54 are negligible for this case because the electron temperature is so high. Finally, the ion current to the wall for this example from the solution to Eq. 7.3-28 is 0.52 A, which is about 12% of the discharge current and 8% of the beam current in this thruster. This amount agrees well with the 10% of the ion current going to the wall calculated by Baranov [44] in analyzing Hall thruster channel wear.

7.3.5 Efficiency of Hall Thrusters with Dielectric Walls

The efficiency of a conventional Hall thruster with a dielectric-wall can be estimated by evaluating the terms in the thruster efficiency given by Eq. 2.5-7, which requires evaluating the total power-loss terms in Eq. 7.3-27 to obtain a value for the effective electrical efficiency. This also illustrates the dominant loss mechanisms in the thruster.

The first term in Eq. 7.3-27, the beam power due to the accelerated ions, P_b, is just $I_b V_b$, where the effective beam voltage will be used. The power loss to the dielectric-wall will be estimated for the SPT-100 Hall thruster [3, 5, 6] using the analysis of Hobbs and Wesson [43] described in Section 7.3.3. The power to the wall was given by Eq. 7.3-46:

$$P_w = I_{\text{iw}}\left[\left(\frac{2M}{\pi m}\right)^{1/2} e^{\frac{e\phi_s}{kT_e}}\left(\frac{kT_e}{e}\right) + (\mathcal{E}-\phi_s)\right], \tag{7.3-57}$$

where I_{iw} is the ion flux to the wall. Following Hobbs and Wesson, the modification to the Bohm criterion is small and $\varepsilon \approx T_{\text{eV}}/2$ from the Bohm criterion. From Eq. 7.3-44, the sheath potential for xenon and BNSiO2 walls in the SPT-100 thruster, assuming an average electron temperature along the channel wall of 25 eV, is about −54 V. Inserting these values into Eq. 7.3-57 gives

$$P_w = 45.8 I_{\text{iw}} T_{\text{eV}} + 2.65 I_{\text{iw}} T_{\text{eV}} = 48.5 I_{\text{iw}} T_{\text{eV}}. \tag{7.3-58}$$

The first term on the right-hand side is again the electron power loss to the wall (written in terms of the ion current to the dielectric surface), and the second term is the ion power loss. The power loss to the channel wall because of the electron loss term is an order of magnitude larger than the power loss due to ions.

It is convenient in evaluating the efficiency of the thruster to relate the ion current to the wall in Eq. 7.3-58 to the beam current. In the plasma, there is a electric field toward the wall because of the presheath of approximately $T_{eV}/2r = T_{eV}/w$. There is also the axial electric field of V_b/L producing the beam energy. It is common in Hall thrusters to find that the electron temperature is about one-tenth the beam voltage [39], and the channel width is usually approximately L [5, 23]. Therefore, the axial electric field is on the order of 10 times the radial electric field. On average, then, the ion current to the channel walls will be about 10% of the beam current. This very simple argument agrees with the SPT-100 results given in the previous section and the results of Baranov [44].

Using Eq. 7.3-58 with the above estimates for the ion current and electron temperature, the power loss to the insulator walls is

$$P_w = 48.5 I_{iw} T_{eV} = 48.5(0.1 I_b)(0.1 V_b) = 0.49 I_b V_b. \tag{7.3-59}$$

The power loss to the anode is due to the plasma electrons overcoming the sheath potential at the anode surface. From Eq. 7.2-24, the anode electron current is

$$I_{ea} = I_d + I_{ia}. \tag{7.3-60}$$

Neglecting the ion current to the anode as small (because of the mass ratio), and realizing that each electron deposits $2kT_e/e$ to the anode for positive plasma potentials (from Appendix C), the power to the anode is

$$P_a = 2T_{eV} I_d. \tag{7.3-61}$$

The electron temperature near the anode is very low, typically less than 5 eV [38–40]. Using the thruster current utilization efficiency and assuming $\eta_b = 0.7$ and $T_{eV} = 0.01V_b$ near the anode, this can be written as

$$P_a = 2\eta_b I_b (0.01 V_b) = 0.014 I_b V_b. \tag{7.3-62}$$

The power required to produce the ions is given by Eq. 7.3-56. This can be written as

$$P_{ion} = (I_b + I_{iw})U^+ = (1 + \eta_b) I_d U^+. \tag{7.3-63}$$

Taking the beam utilization efficiency as 0.7 and estimating that the ionization potential is roughly 5% of the beam voltage, the power required to produce the ions is approximately $P_{ion} = 0.09 I_d V_b$. The radiation power and other power loss mechanisms are small and will be neglected in this simple example.

The total discharge power into the thruster is then

$$P_d = I_b V_b + 0.49\, I_b V_b + 0.014\, I_b V_b + 0.09\, I_b V_b = 1.59\, I_b V_b. \tag{7.3-64}$$

The electrical efficiency of the dielectric-wall thruster is then

$$\eta_e = I_b V_b / (1.59\, I_b V_b) = 0.63. \tag{7.3-65}$$

The total thruster efficiency, assuming the same beam divergence and double ion content as evaluated above and a mass utilization efficiency of 95% reported for SPT thrusters [5], is

$$\eta_T = (0.915)^2\,(0.63)\,(0.95) = 0.5. \tag{7.3-66}$$

The SPT-100 thruster is reported to run at about 50% efficiency [5, 6]. Since the power loss is dominated by the electron wall losses, this analysis illustrates how critical the wall material selection is to minimize the secondary electron yield and maintain a sufficient wall sheath potential for good efficiency. For example, if the wall had been made of alumina and the electron temperature is about 20 V, the sheath potential would be $-1.02T_{eV}$ in the space charge limited regime. The wall power from Eq. 7.3-57 would then be about three times higher than in the $BNSiO_2$ case:

$$P_w = 142 I_{iw} T_{eV} = 1.4 I_b V_b. \tag{7.3-67}$$

The electrical efficiency of the thruster, assuming the same anode loading and energy loss to ionization, would be $\eta_e \approx 0.40$ and the total efficiency

$$\eta_T = (0.915)^2 (0.40)(0.95) \approx 0.32. \tag{7.3-68}$$

Recent parametric experiments in which different wall materials were used in the SPT-100 [36] showed that changing from $BNSiO_2$ to alumina reduced the efficiency to the order of 30%, consistent with the increased secondary electron yield of the different wall material.

The agreement of this simple analysis with the experimentally measured efficiencies is somewhat fortuitous because the predictions are very sensitive to the secondary electron yield of the wall material and the actual sheath potential. Small errors in the yield data, changes in the wall material properties during thruster operation, and inaccuracies in the empirical values for the electron temperature and ion flux with respect to the beam parameters, will significantly affect the calculated results. Other effects may also be significant in determining the thruster efficiency. The analysis of the sheath potential assumed a Maxwellian electron distribution function. It was recognized several years ago [41, 45, 46] that the electron distribution may not be Maxwellian. Detailed kinetic modeling of the Hall thruster channel plasma indicates [47] that the electron velocity distribution is depleted of the high-energy tail electrons that rapidly leave the plasma along the magnetic field lines and impact the wall. This is especially true near the space charge limit where the sheath voltage is small and a large fraction of the electron tail can be lost. The collision frequencies and thermalization rates in the plasma may be insufficient to re-populate the Maxwellian tail. This will result in effectively a lower electron temperature in the direction parallel to the magnetic field toward the walls [48], which can increase the magnitude of the sheath potential and reduce the electron heat loss to the wall. In addition, re-collection of the secondary electrons at the opposite wall [49, 50], because of incomplete thermalization of the emitted secondary electrons in the plasma, modifies the space charge limits and sheath potential, which also can change the electron heat flux to the wall.

These effects are difficult to model accurately because of the presence of several different electron populations, several collision/thermalization processes, the effect of magnetization on the electrons, and the presence of plasma instabilities. Understanding what determines the electron temperature and velocity distribution as a function of the discharge voltage and current, and uncovering the effects that determine the wall power flux and finding techniques to minimize them, are continuing areas of research at this time.

7.3.6 Efficiency of TAL Thrusters with Metallic Walls

As with the 1.35 kW SPT-100 Hall thruster example above, an estimate will be made of the power loss terms in Eq. 7.3-27 to obtain an electrical efficiency for the 1.4 kW D-55 TAL thruster [33]. Equation 2.5-7 will then be used to obtain an estimate for the thruster efficiency. The beam power P_b is, again, just $I_b V_b$. As stated in the previous section, the wall losses (P_w) are essentially negligible in TAL thrusters and the power to the anode is given by Eq. 7.3-49

$$P_a = 2T_{\text{eV}}\frac{I_b}{\eta_b} = 0.29 I_b V_b. \tag{7.3-69}$$

In Eq. 7.3-69, it is again assumed $\eta_b = 0.7$ and $T_{\text{eV}} = 0.1V_b$, although these values may be somewhat different in TAL thrusters. The power required to produce the ions is again approximately $0.09I_bV_b$.

The total discharge power, Eq. 7.3-27, then becomes

$$P_d = I_bV_b + 0.29\, I_bV_b + 0.09\, I_bV_b = 1.4\, I_bV_b. \tag{7.3-70}$$

Neglecting the power in the cathode keeper (if any) and the magnet as small compared to the beam power, the electrical utilization efficiency from Eq. 2.5-1 is then

$$\eta_e = \frac{P_d}{1.4P_d} = 0.72. \tag{7.3-71}$$

The total thruster efficiency, assuming a 10% double ion content, a 20° angular divergence [33, 51], and a 90% mass utilization efficiency reported for TAL thrusters [33, 51, 52], is then from Eq. 2.4-7:

$$\eta_{\text{T}} = (0.915)^2\,(0.72)\,(0.9) = 0.54. \tag{7.3-72}$$

This result on the same order as that reported in the literature [33, 51, 52] for this power level TAL and is essentially the same as the SPT-100 efficiency in this simple example if the wall losses had been included. However, the power loss to the anode is seen as the dominant energy loss mechanism in the TAL efficiency. This is the main issue with TAL thrusters in that the anode heating is usually significant.

7.3.7 Comparison of Conventional Hall Thrusters with Dielectric and Metallic Walls

It is interesting to make a few direct comparisons of dielectric-wall Hall thrusters with metallic-wall TAL thrusters. Similar discussions have appeared in the literature [1, 4, 35], often with conflicting opinions. The basic plasma physics in the channel described above applies to both the dielectric-wall Hall thruster and the TAL. The maximum electron temperature occurs in both thrusters near the channel exit in the region of strongest magnetic field where the Hall current is a maximum. The different interaction of the thruster walls with this plasma determines many of the characteristics of the thruster, including life. Conventional dielectric-wall thrusters have a significant amount of their input power deposited as loss on the dielectric channel walls because of electron bombardment. In the example efficiency calculation above, approximately 25% of the power going into the thruster was deposited on the channel walls. The metallic walls in TAL thrusters collect a smaller electron current because they are biased to cathode potential, and also tend to have a small exposed area in poor contact with the plasma which limits the amount of ion and power lost to these surfaces. However, the anode is positioned very close to the high electron temperature region, and receives a significant amount of power deposition in collecting the discharge current. In the example TAL efficiency calculation above, over 20% of the power going into the thruster was deposited on the anode.

The deep channel in dielectric-wall Hall thrusters, with a low magnetic field strength and low electron temperature near the anode, tends to minimize the power deposition on the anode. In the simple example above, only 1% of the thruster input power was deposited on the anode. Nevertheless, the anode is normally electrically isolated from the thruster body (and therefore thermally isolated), and so anode overheating is sometimes an issue, especially at high power density. The anode in TAL thrusters can also have heating issues because the loading is much higher, even though the view-factor for the anode to radiate its power out of the thruster is better than the deep channel in the insulating-wall configuration. In addition, with the anode positioned physically close to the thruster exit in TALs, impurity deposition and material buildup problems can occur. This has been an issue in ground testing of some TAL thrusters [33] where carbon deposition on the anode from backsputtering from the beam dump became significant over time. TAL thrusters with deeper channels can be designed and operated [4]. The performance of the thruster is likely different in this configuration, and ion bombardment and sputtering of the metallic channel walls can become significant and affect the thruster life.

Dielectric-wall Hall thrusters are often described in terms of an ionization zone upstream of the exit plane and an acceleration zone in the region of the exit plane. TAL thrusters have a similar ionization region near the magnetic field maximum, which is now closer to the anode because the magnetic field gradient is greater. The TAL acceleration zone is described as being a layer close to the anode [1, 4] and can extend outside of the thruster [51]. The higher electron temperatures associated with TAL thrusters support higher electric fields in the quasi-neutral plasma, which compresses these zones relative to dielectric-wall thrusters. In addition, the metallic walls and higher electric fields are conducive to multiple acceleration stages, which can improve thruster performance and produce higher Isp than a conventional single stage TAL thruster [53, 54]. Multiple-stage dielectric-wall Hall thrusters that operate at high Isp have also been investigated (see [20] and the references cited therein).

Finally, the difference between dielectric-wall Hall thrusters and TAL thrusters is sometimes attributed to the secondary electron coefficients of the different wall materials. The discussion above shows that this is not the dominant difference. Instead, the proximity of the TAL anode electrode to the high temperature plasma region and the thruster exit plane is what changes the electric field profile, power deposition and sputtering characteristics compared to the dielectric-wall Hall thruster.

7.4 Discharge Dynamics and Oscillations

Depending on their size and operating characteristics, Hall thrusters can produce a wide range of plasma oscillations, with frequencies from 1 kHz to 10's of MHz. A survey of these discharge dynamics, covering waves and instabilities that can be excited in a SPT was compiled by Choueiri [55]. Estimates of several characteristic frequencies of relevance in the SPT discharge chamber is provided in Fig. 7-14. The values were determined using as initial input probe measurements taken inside the channel by Bishaev and Kim [56, 57]. The SPT in that study had a diameter of 10 cm and operated with xenon propellant at a discharge voltage and current of 200 V and 3 - to - 3.2 A, respectively. The estimates encompass several natural frequencies associated with electrons and ions, specifically: the cyclotron, plasma and lower-hybrid frequencies as well as several collision frequencies. Both elastic and inelastic collisions were evaluated, between not only the two charged species but also between those species and neutrals.

Owing to their strong connection to thruster performance and stability, the most widely studied oscillations have been those associated with ionization instabilities in the annular discharge channel. They are observed to occur at frequencies in the tens of kHz. An example of a measurement made in an SPT-100 exhibiting a ~17 kHz discharge current oscillation is shown in Fig. 7-15 from [59]. These types of oscillations are typically seen in the discharge current when the thruster is

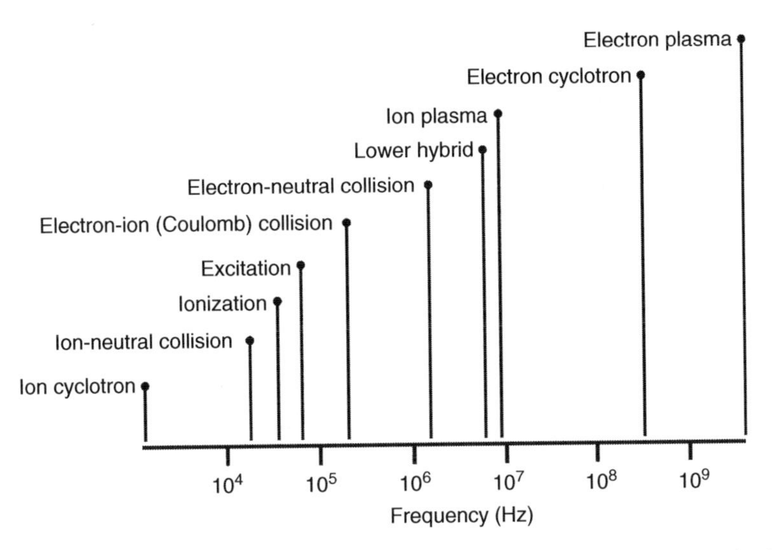

Figure 7-14 Relevant frequencies in an SPT estimated from probe measurements inside the discharge chamber (*Source:* [55]).

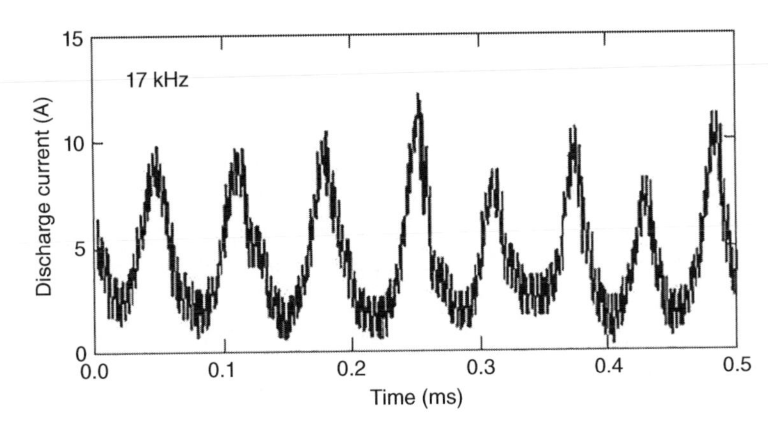

Figure 7-15 Measured evolution of the discharge current for the SPT-100 (*Source:* Adapted from Mikellides [59]).

operated in a voltage regulated mode and have been called "breathing modes" [58] and "predator-prey modes" [34].

The time dependence of the low-frequency oscillations can be seen from a simple analytical model [34] that describes the ion and neutral behavior. The ion conservation equation is written as

$$\frac{\partial n_i}{\partial t} = n_i n_n \langle \sigma_i v_e \rangle - \frac{n_i v_i}{L}, \tag{7.4-1}$$

and the neutral particle conservation equation is

$$\frac{\partial n_n}{\partial t} = -n_i n_n \langle \sigma_i v_e \rangle + \frac{n_n v_o}{L}, \tag{7.4-2}$$

where v_o is the neutral velocity and L is the axial length of the ionization zone. The perturbed behavior of the ion and neutral densities with time is linearized such that:

$$\begin{aligned} n_i &= n_{i,o} + \delta n_i' \\ n_n &= n_{n,o} + \delta n_n' \end{aligned}, \tag{7.4-3}$$

where the "o" subscript denotes the unperturbed state and primed quantities are the density perturbations. Combining Eqs. 7.4-1–7.4-3 to first order in δ gives

$$\frac{\partial^2 n_i'}{\partial t^2} = n_{i,o} n_{n,o} n_i' \langle \sigma_i v_e \rangle^2. \tag{7.4-4}$$

This equation represents an undamped harmonic oscillator with a frequency given by

$$f_i = \frac{1}{2\pi}\sqrt{n_{i,o} n_{n,o} \langle \sigma_i v_e \rangle^2} \approx \frac{\sqrt{v_i v_o}}{2\pi L}, \tag{7.4-5}$$

The low frequency oscillatory behavior of Hall thrusters is proportional to the velocities of the ions and neutrals relative to the scale length of the ionization zone. The periodic depletion of the neutral gas in the ionization region causes the ion density to fall, and the frequency is then dependent on the neutral velocity driven time needed for the neutral flow from the periphery to replenish the region and restart the ionization. The ion density then oscillates, which impacts the electron conductivity through the transverse magnetic field and thereby the discharge current. The ionization region location will also oscillate axially in the channel on the time scale of neutral replenishment time. The specific frequency these breathing modes exhibit depends on several factors including the operating conditions, applied magnetic field and channel geometry.

Using an electrostatic High-speed Dual Langmuir Probe (HDLP) system, Lobbia and Gallimore [60] surveyed the behavior of these modes in a 600-W xenon Hall thruster under various operating conditions. The measurements revealed the rich dynamic behavior of the discharge in this frequency range. Figure 7-16 shows the power spectral density (PSD) of the discharge current

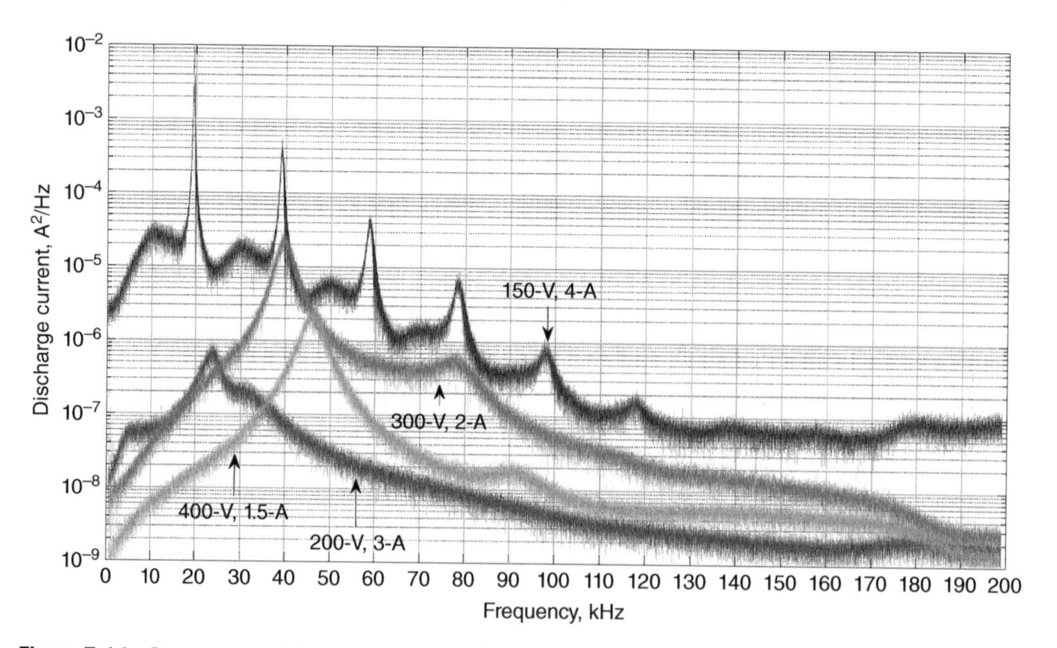

Figure 7-16 Power spectral density measurements in a 600-W Hall thruster operating with xenon (*Source:* Adapted from Lobbia and Gallimore [60]).

oscillations observed at different discharge voltage and current settings. The dominant lowest-frequency peak of the breathing mode can be seen in all traces. As the thruster discharge voltage or the mean discharge current increased, the breathing mode frequency also increased. At the lower discharge voltages (150 V and 200 V), at least five harmonics of the mode's fundamental frequency also were observed, spanning frequencies up through 100 kHz.

Due to the significance of the breathing mode on thruster performance and stability, the study of these oscillations in the laboratory has been supported by extensive physics-based analyses using models of various levels of complexity. The modeling of these and other oscillations in the Hall thruster, as well as its time-averaged behavior comprise the topic of the next section.

7.5 Channel Physics and Numerical Modeling

Significant progress has been made in the last couple decades in understanding the physics of $\mathbf{E} \times \mathbf{B}$ discharges on which Hall thruster performance depends [61]. Reviews on how these physics are modeled in Hall thrusters and the state of that understanding have been recently reported in [62–64]. However, several aspects of how Hall thrusters work remain poorly understood. For example, the true electron distribution function in the vicinity of the acceleration region, the mechanisms responsible for electron transport across the applied magnetic field, and the role of oscillations on the particle transport and discharge stability all need further investigation.

In this section we provide a basic set of the conservation equations used in the modeling of these thrusters. Due to their characteristic operational principle, which forces electrons to be highly magnetized and ions to be unmagnetized, we focus more on the electron equations. Of course, to properly account for all the physical processes in a model of a Hall thruster all the appropriate equations governing conservation of mass, momentum and energy for the heavy species (ions and neutrals) must be solved. Moreover, all these equations must be subject to appropriate boundary conditions, which can be quite complex and markedly consequential in Hall thrusters [36, 65–68].

The subsection on the basic model equations is followed by a few examples of numerical models and simulations. We are well aware that an extensive amount of work has been performed since the early numerical modeling efforts of the 1990s. Our goal is not to provide an extensive review of this quite vibrant area of Hall thruster research, but rather to touch on some of the more salient aspects of physics-based modeling in these devices.

7.5.1 Basic Model Equations

Many Hall thruster models utilize a steady state fluid electron momentum equation and a time dependent electron energy equation to solve for the electron temperature and plasma potential in the channel and plume. The electron momentum equation after neglecting electron inertia and the thermoelectric effect can be expressed as

$$\mathbf{E} + \mathbf{u}_e \times \mathbf{B} = \frac{\mathbf{R}_e}{en_e} - \frac{\nabla p_e}{en_e} \tag{7.5-1}$$

where the electron pressure tensor has been simplified as $\boldsymbol{p}_e = p_e \mathrm{I}$ (with I being the delta tensor). The drag force density $\mathbf{R}_e$ represents changes in the electron momentum as a result of collisions with ions and neutrals and leads to the resistive contribution to the electric field $\mathbf{E}$ as follows:

$$\frac{\mathbf{R}_e}{en_e} = (\eta_{\mathrm{ei}} + \eta_{\mathrm{en}})\mathbf{J}_e + \eta_{\mathrm{ei}}\mathbf{J}_i \tag{7.5-2}$$

where η_{ei} and η_{en} are the resistivities due to electron-ion and electron-neutral collisions and $\mathbf{J}_e$ and $\mathbf{J}_i$ are the electron and ion current densities respectively. In general, for momentum transfer collisions between electrons and species "s" that occur at a mean frequency of $<\nu_{es}>$ the resistivity is given by

$$\eta_{es} = \frac{m\,\nu_{es}}{e^2 n_e}. \tag{7.5-3}$$

Here we have dropped the angle brackets $< >$ for convenience. Let us now define an effective electric field

$$\mathbf{E}' \equiv \mathbf{E} + \frac{\nabla p_e}{e n_e} - \eta_{ei}\mathbf{J}_i, \tag{7.5-4}$$

which allows us to write Eq. 7.5-1 in the form of Ohm's law:

$$\mathbf{J}_e = \frac{\mathbf{E}'}{\eta} - \Omega_e \mathbf{J}_e \times \hat{\boldsymbol{\beta}} \tag{7.5-5}$$

where we have combined the two collisional contributions to the resistivity into a single value $\eta = \eta_{ei} + \eta_{en}$. Also, $\mathbf{B} = B\hat{\boldsymbol{\beta}}$ with $\hat{\boldsymbol{\beta}}$ being the unit vector in the direction of the magnetic flux density $\mathbf{B}$. With the electron cyclotron frequency denoted by ω_c, the Hall parameter can be written as

$$\Omega_e = \frac{B}{e n_e \eta} = \frac{\omega_c}{\nu_e}. \tag{7.5-6}$$

As with the resistivity η, in the presence of only ions and neutrals it is implied that we have combined the momentum transfer collision frequencies for electrons into a single frequency

$$\nu_e = \nu_{ei} + \nu_{en}. \tag{7.5-7}$$

We can now express the electron current density in the parallel, $\mathbf{J}_{e\parallel} \equiv \left(\mathbf{J}_e \cdot \hat{\boldsymbol{\beta}}\right)\hat{\boldsymbol{\beta}}$ and perpendicular directions, $\mathbf{J}_{e\perp} \equiv \hat{\boldsymbol{\beta}} \times \left(\mathbf{J}_e \times \hat{\boldsymbol{\beta}}\right)$, relative to $\mathbf{B}$ as follows:

$$\begin{aligned}
\mathbf{J}_{e\parallel} &= \frac{\mathbf{E}'_{\parallel}}{\eta} \\
\mathbf{J}_{e\perp} &= \frac{\mathbf{E}'_{\perp}}{\eta} - \Omega_e \mathbf{J}_e \times \hat{\boldsymbol{\beta}} \\
\mathbf{J}_e \times \hat{\boldsymbol{\beta}} &= \frac{\mathbf{E}' \times \hat{\boldsymbol{\beta}}}{\eta} + \Omega_e \mathbf{J}_{e\perp}.
\end{aligned} \tag{7.5-8}$$

The transverse components are sometimes expressed using the subscript "$\wedge$", that is, $\mathbf{a}_{\wedge} \equiv \mathbf{a} \times \hat{\boldsymbol{\beta}}$.

7.5.1.1 Electron Motion Perpendicular to the Magnetic Field

In the absence of an applied electric field in the transverse direction or any induced components in that direction that permit a non-zero average value (such as instability driven fluctuations), then $\mathbf{E}' \times \hat{\boldsymbol{\beta}} = 0$ and we can combine the last two expressions in Eqs 7.5-8 to write the effective electric field in the perpendicular direction as

$$\mathbf{E}'_{\perp} = \eta\left(1 + \Omega_e^2\right)\mathbf{J}_{e\perp}. \tag{7.5-9}$$

Equation 7.5-9 illustrates the fundamental distinction between Hall thrusters and gridded ion engines regarding the generation of the accelerating force on ions. In the latter the electric field on the ions is generated through direct application of a voltage difference in the grid system, whereas in the former, $\mathbf{J}_e \times \hat{\boldsymbol{\beta}}$ is responsible for the force (see Eq. 7.5-8) in compliance with Ohm's law. It is this operational principle that in some cases puts Hall thrusters in the category of electromagnetic (versus electrostatic) thrusters in the literature. Most physics models we will be discussing in the ensuing sections attempt to solve numerically or analytically Eq. 7.5-9 or some close variant for the electric field after assuming the discharge is quasi-neutral.

It is also common to use the electron mobility instead of the resistivity. In the perpendicular direction, with $\eta_\perp = \eta(1 + \Omega_e^2)$ the two are related as follows:

$$\mu_\perp = \frac{\mu}{1 + \Omega_e^2} = \frac{1}{en_e\eta_\perp}. \tag{7.5-10}$$

This expression for the perpendicular electron mobility accounts for both electron-ion and electron-neutral collisions. The perpendicular electron current density may then be written as

$$\mathbf{J}_{e\perp} = \mu_\perp(en_e\mathbf{E}_\perp + \nabla_\perp p_e) - \frac{\mu_\perp}{\mu_{ei}}\mathbf{J}_{i\perp} \tag{7.5-11}$$

and the electron mobility due only to electron-ion collisions is given by $\mu_{\text{ei}} = e/m_e\nu_{\text{ei}}$. In many cases it is possible to neglect the last term on the right-hand side which simplifies Eq. 7.5-11 further.

Since the electrons are strongly magnetized near the exit of the channel where the magnetic field strength is the highest, the electron Hall parameter (Eq. 7.5-6) is much greater than unity. Hence the perpendicular resistivity, which follows the proportionality $\eta_\perp \propto B^2/\nu_e n_e$ under these conditions, can be quite high. In fact, calculations of the electron collision frequency ν_e based on classical collisions only (Eq. 7.5-7) show it is too high. This leads to insufficient cross-field transport of electrons to support the discharge current passing through the thruster (see for example [61] and references therein). Therefore, in what is likely an over-idealized approximation, an effective or "anomalous" collision frequency ν_α is simply added to the total to reduce the resistance on the electrons imposed by the magnetic field, as follows:

$$\nu_e = \nu_{\text{ei}} + \nu_{\text{en}} + \nu_\alpha. \tag{7.5-12}$$

Two mechanisms were originally proposed to capture the "anomalous" cross-field electron transport. Morozov [3] postulated that electron-wall interactions in the channel region will scatter electron momentum and introduce secondary electrons, which can increase the effective cross-field transport. This effect is introduced into the effective collision frequency by a wall scattering frequency ν_w, that is, $\nu_\alpha = \nu_w$. With ζ being an adjustable parameter used to match the experimental data, the wall scattering frequency is either given by $\zeta \times 10^7\ \text{s}^{-1}$ [58] or the wall collision frequency of electrons is determined directly in a numerical simulation [69]. While this effect does increase the electron transport in the channel, it has in many cases been found to provide insufficient enhancement of the electron transport. In addition, in the plume of the thruster there are no walls and the neutral density is very low, which precludes the use of ν_w to increase the cross-field transport sufficiently to explain the experimental data. Yet, as we will show later, much of the anomalous collision frequency is needed in the vicinity of the acceleration channel exit and near-plume of the thruster.

Additional cross-field transport was then proposed by invoking Bohm diffusion both inside and outside the thruster channel. As discussed in Chapter 3, Bohm diffusion likely arises from $\mathbf{E} \times \mathbf{B}$

driven drift instabilities, postulated to occur naturally in these thrusters because of the Hall current. Using the Bohm diffusion coefficient from Eq. 3.6-72 and the Einstein relationship of Eq. 3.6-28, a Bohm mobility can be defined as

$$\mu_B = \frac{1}{\beta B} = \frac{e}{\beta m \omega_c} \tag{7.5-13}$$

where β is an adjustable coefficient. If the discharge is subject to full Bohm diffusion, then $\beta = 16$. The effective Bohm collision frequency is then $\nu_B = \beta\omega_c$ and the total "anomalous" collision frequency is

$$\nu_\alpha = \nu_w + \nu_B, \tag{7.5-14}$$

with ν_w neglected in the plume. The inclusion of the "Bohm" anomalous collision frequency gained significant popularity in the early numerical simulations of Hall thrusters largely because it allowed for the prediction of general discharge properties and thruster performance to within acceptable levels of accuracy. As both numerical simulations and plasma diagnostics advanced however, it became more evident that the diffusion of electrons in the acceleration region and into the channel was far from "Bohm-like." This is discussed further in Section 7.5.2.2. Before ending this section, perhaps of historical interest is the fact that the earlier hydrodynamic models proposed by Morozov et al. did not incorporate Bohm diffusion at all. In fact, in his 2001 monograph [3], referring to Bohm's formula for the anomalous electrical conductivity $\sigma = 1/\eta$, Morozov stated: *The appearance of Bohm's formula (in the 1940')] convinced practically all gas-discharge physicists that $\sigma_\perp \sim 1/H$ (with H being the magnetic field strength)]. However, experiments carried out at Kurchatov AEI (Atomic Energy Institute) in 1963-64 in the Shchepkin-Morozov laboratory showed that this is not the case. The value of $\sigma_\perp$ proved to be at least an order of magnitude smaller than that given by Bohm's formula. Several years later the same result was obtained in tokamaks.*"

7.5.1.2 Electron Motion Parallel to the Magnetic Field

When the Hall parameter $\Omega_e^2 >> 1$, the ratio of electron heat ($\mathbf{q}_e$) and charge fluxes in the parallel and perpendicular directions to the magnetic field are proportional to

$$\frac{\mathbf{q}_{e\parallel}}{\mathbf{q}_{e\perp}} \propto \Omega_e^2 \frac{\nabla_\parallel T_e}{\nabla_\perp T_e} \qquad \frac{\mathbf{J}_{e\parallel}}{\mathbf{J}_{e\perp}} \propto \Omega_e^2 \frac{\mathbf{E}'_\parallel}{\mathbf{E}'_\perp} \tag{7.5-15}$$

with $\mathbf{E}'$ given by Eq. 7.5-4 after neglecting the contribution from the ion current. The fluxes in the parallel direction are finite. In fact, the left-hand side ratios in Eq. 7.5-15 above may attain, in principle, any value along a given magnetic field line. For example, consider such a line that is perpendicular to the insulator in the acceleration region of the channel. Under typical conditions in this region, $n_e \sim 10^{17}\ \text{m}^{-3}$, $T_e \sim 20$ eV and sheath drop $\Delta\phi_s \sim T_e$, the parallel (mean) current density of electrons to the sheath, $\sim \frac{1}{4} e n_e \langle c_e \rangle \exp(-\Delta\phi_s/T_e)$ is ~0.5 A/cm^{-2}, whereas the perpendicular current density is $(\sigma_0/\Omega_e^2)E'_\perp \sim 0.05$ A/cm^{-2}. Thus, $J_{e//}/J_{e\perp} \sim 10$ even though Ω_e^2 may exceed 10^4 in this region. This implies that in regions where $\Omega_e^2 >> 1$ the gradient forces in the parallel direction are much smaller than those in the perpendicular direction. The approximation then is to assume that these forces vanish along magnetic fields lines

$$\Omega_e^2 \gg 1: \quad \nabla_\parallel T_e = 0 \quad \mathbf{E}'_\parallel \approx \frac{\nabla_\parallel p_e}{e n_e} - \nabla_\parallel \phi = 0 \tag{7.5-16}$$

leading to the following idealized algebraic equations:

$$T_e = T_{eR} \qquad \phi = \phi_R + T_e \ln\left(\frac{n}{n_R}\right), \tag{7.5-17}$$

where subscript "R" denotes reference values along the line. The equation for the potential was derived in Section 3.5.1 and represents the Boltzmann relation for plasmas with Maxwellian electron distribution function. The right-hand side is often called the *thermalized potential* [3] in Hall thruster literature and is sometimes denoted by ϕ^*.

In thrusters with a largely radial magnetic field inside the acceleration channel it is usually the case that the density gradient along the magnetic field is relatively small, so the potential change along a line is essentially zero since $T_e \ln(n/n_R) \approx 0$. Therefore, such magnetic field lines also represent equipotential lines in the plasma. The thermalized potential has therefore been used for many years [3, 4] in the design of Hall thrusters to relate the magnetic field shape to the electric field in the plasma. In thrusters with more curved magnetic field lines, like those employing magnetic shielding topologies that will be discussed in Chapter 8 or in the near-thruster and cathode plumes, where the Hall parameter can be significantly smaller, such simplifying approximations can often introduce significant error. This is discussed further in ensuing sections.

7.5.1.3 Electron Continuity and Energy Conversation

Mass continuity for the charged species in the plasma can be expressed in terms of the total current density $\mathbf{J} = \mathbf{J}_e + \mathbf{J}_i$ as follows

$$\nabla \cdot \mathbf{J} = 0. \tag{7.5-18}$$

In addition, charge balance at the insulating wall dictates that

$$J_i = J_e - J_{se}, \tag{7.5-19}$$

where J_{se} is the secondary electron current density, which is equal to the secondary electron yield times the incident electron flux. The boundary conditions related to the sheath that forms along these boundaries are very important in the modeling of Hall thrusters and have therefore been discussed quite extensively in the literature. This topic is beyond the scope of this chapter however so we refer the reader to the non-exhaustive collection of references in [36, 43, 67–69].

The electric field in the perpendicular direction can be solved for using Eq. 7.5-11. The solution then allows us to determine the voltage drop V along a path l of length Λ from the anode to the cathode applied across the plasma

$$V = -\int_0^{\Lambda} \mathbf{E}_\perp \cdot dl. \tag{7.5-20}$$

The electron energy equation expressed in terms of the electron temperature is given by

$$\frac{\partial}{\partial t}\left(\frac{3}{2} e n_e T_{eV}\right) + \nabla \cdot \left(\frac{5}{2} T_{eV} \mathbf{J}_e - \mathbf{q}_e\right) = \mathbf{E} \cdot \mathbf{J}_e - R - S - P_w \tag{7.5-21}$$

where $\mathbf{E} \cdot \mathbf{J}_e$, R, S and P_w are the volumetric power terms representing the work done on the electrons by the electric field (which include ohmic heating), radiative and ionization losses and losses to the walls, respectively. The volumetric radiative power loss is

$$R = U^* n_e n_o \langle \sigma_* v_e \rangle \tag{7.5-22}$$

and the ionization energy loss is given by

$$S = U^+ n_e n_o \langle \sigma_i v_e \rangle. \tag{7.5-23}$$

The radiative (excitation) and ionization reaction rate coefficients in Eqs. 7.5-22 and 7.5-23 are given in Appendix E. Finally, the thermal conduction heat flux tensor $\mathbf{q}_e$ can be expressed in terms of its parallel and perpendicular components: $\mathbf{q}_{e\parallel} = -\kappa_{e\parallel}\nabla_{\parallel}T_e$ and $\mathbf{q}_{e\perp} = -\kappa_{e\perp}\nabla_{\perp}T_e$ with κ_e being the electron thermal conductivity. Based on the arguments made in Section 7.5.1.2, $\mathbf{q}_{e\parallel}$ is very small and can be neglected thereby preserving the isothermalization of the magnetic field line in regions where the Hall parameter $\Omega_e^2 \gg 1$.

7.5.1.4 Heavy Species: Ion and Neutrals

The treatment of ions has been wide-ranging in the modeling of Hall thrusters. This is illustrated by a few representative estimates of the relevant characteristic length scales using plasma conditions from a 6-kW laboratory thruster and an SPT, identified in Fig. 7-17 as "1" and "2", respectively.

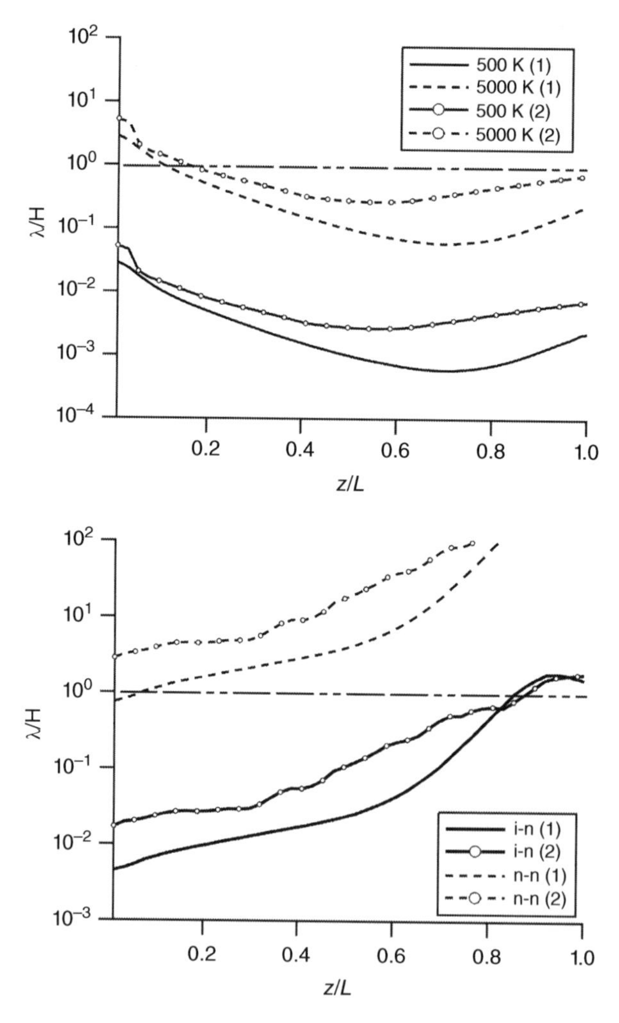

Figure 7-17 Ratio of mfp (λ) for collisions over the channel height H, along the channel centerline in two different Hall thrusters. Top: Ion-ion Coulomb collisions. Bottom: Ion-neutral charge exchange and neutral-neutral collisions (*Source:* [70]/American Physical Society).

Given an acceleration channel of length L, the ion transit time $\tau_u = L/u_i$ for a ion drift velocity u_i can range approximately from (0.03 m)/(2×10^4 m/s^{-1}) = 1.5 μs, for those ions that are accelerated downstream of the acceleration channel, to (0.01 m)/(5×10^2 m/s^{-1}) = 10 μs for those generated near the anode region and lost to the walls. For comparison, the thermal equilibration time between electrons and ions, $\tau_{ei} \approx \tau_e\, M/2m$ (with τ_e being the electron relaxation time and M the ion mass) ranges 0.03–0.5 s inside the channel, which implies that the ions remain "cold" relative to the electrons. The (thermal) mean free path (mfp) for ion-ion collisions $\lambda_{ii} = \upsilon_i\tau_i$ with $\upsilon_i = (2kT_i/M)^{½}$ being the ion thermal speed, is plotted in Fig. 7-17 (top) along the middle of the acceleration channel in the two different Hall thrusters for various values of the ion temperature [70]. The axial direction and channel length are z and L, respectively. It is noted that although a case of 5000 K ions is plotted here, this is an extreme value since the channel walls typically do not exceed 1000 K. It is also the case that the ion density can in fact be substantially higher in the anode region compared to the values used here. This would suggest even smaller collision mfps for ions in this region than those plotted in Fig. 7-17. Also, Laser-Induced Fluorescence (LIF) measurements of Xe^+ inside the 6-kW Hall thruster have shown that ions follow very closely the equilibrium distribution function [71], which further strengthens the continuum assumption for the ions in this region.

Depicted in Fig. 7-17 (bottom) is the charge-exchange collision mfp for ions colliding with atoms of number density n_n as estimated by $\lambda_{in} = (\sigma_{in} n_n)^{-1}$. The mfp is plotted for two values of the ion-neutral charge-exchange cross section σ_{in}, 50 Å^2 and 100 Å^2. Based on the measurements of Miller, *et al.* [72], the two values cover the range of typical ion energies attained in the acceleration channel, <1–300 eV, with the highest value of the cross section representing the lowest energy ions. For comparison, the characteristic mfp for self-collisions between neutrals $\lambda_{nn} = (\pi n_n D^2 \sqrt{2})^{-1}$ is also plotted in Fig. 7-17 (bottom) using a mean atomic diameter for Xe of D = 2.6 Å. It is worthwhile noting that the addition of charge-exchange collisions can become increasingly important in the anode region since the electric force can be negligibly small there [73].

It is apparent then from the estimates in Fig. 7-17 that in many cases, depending on the specific thruster conditions and geometry, ions can be treated using continuum approximations inside the channel without significantly degrading the accuracy of the solution. In such an approach the momentum equation is solved, expressed below in conservative form as

$$\frac{\partial}{\partial t}(nm\mathbf{u})_i + \nabla \cdot (nm\mathbf{uu})_i = en_i\mathbf{E} - \nabla p_i + \mathbf{R}_i. \tag{7.5-24}$$

By comparison to the same conservation equation for the electrons Eq. 7.5-1, the absence of the $\mathbf{u}_i \times \mathbf{B}$ term is noted since the ions are considered unmagnetized. Also, in Eq. 7.5-24 the viscous terms have been neglected and the ion pressure tensor has again been simplified as $\mathbf{p}_i = p_i\mathrm{I}$ (with I being the delta tensor). The drag force density $\mathbf{R}_i$ represents changes in the ion momentum as a result of collisions between ions and other species and, in general, may be composed of both elastic and inelastic components [70].

Even though the continuum approximation for ions can be valid in certain regions of Hall thrusters, the range of ion mean free paths and energies in these devices can vary enough that, many times, the use of discrete-particle methods is more accurate or simpler to use. For example, though ion collisions can be frequent enough inside the channel, they can become quite rare as ions exit the channel. Under such conditions, multifluid [70, 74] or direct-kinetic methods [75] have also been used, separately or in some combination with discrete-particle methods. For example, in [58] an idealized form of Boltzmann's equation for the ion distribution function has been used to solve for the ion generation and motion. This has primarily been applied for investigating low frequency oscillations on the order of the ion characteristic time scales. In one dimension, the idealized equation can be written as

$$\frac{\partial f}{\partial t} + v_x \frac{\partial f}{\partial t} + \frac{e}{M} E \frac{\partial f}{\partial v_x} = n_e n_o \langle \sigma_i v_e \rangle \delta(v_x - v_o) \tag{7.5-25}$$

where f is the ion distribution function and $\delta(v_x - v_o)$ is the Dirac delta function evaluated for the ion velocity v_x relative to the neutral velocity v_0. The ion density is then found from

$$n_i = \int f(x, v_x, t) dv_x. \tag{7.5-26}$$

For neutrals, the modeling method is more straightforward since the collision mean free paths between them are consistently much larger than the characteristic size of the channel in Hall thrusters. It is therefore most accurate to use particle [76–78] or view factor methods [70, 79] for these species but in a manner that properly takes in to account loss of neutrals by ionization collisions with electrons and interactions with the walls. Both these processes can affect markedly the spatiotemporal characteristics of the discharge.

To properly capture the evolution of the heavy species the above equations should be accompanied by conservation laws for mass continuity and boundary conditions that appropriately capture their interactions with the thruster surfaces.

7.5.2 Numerical Modeling and Simulations

The numerical simulation of Hall thrusters now spans more than three decades. Although the first theoretical models of the partially-ionized gas in an SPT were, in fact, reported in the 1970s by Morozov et al. [80–82], it was not until the 1990s that a considerable effort began. These early efforts focused on the development of fluid, kinetic, particle-in-cell (PIC) and/or hybrid models for predicting performance and for explaining some of the observed behavior in these thrusters. Naturally, it was easier to begin with solutions in one dimension before moving to multiple dimensions, an effort that proved to be quite insightful for many aspects of Hall thruster physics.

7.5.2.1 Modeling in One Dimension

It was recognized very early that modeling the flow field inside the channel would be extremely challenging, owing largely to the complexity of the electron physics and the inherently wide range of temporal and spatial scales of the light (electrons) and heavy species (neutrals and ions). Thus, the natural evolution from 0-D scalings and pure empiricism involved the numerical solutions of idealized models in one dimension, most often along the axis of symmetry. It is interesting to note, despite the popularity of more modern numerical methodologies that incorporate discrete-particle methods for the heavy species, the earliest approach for these species followed purely hydrodynamic formalisms (e.g. see [82] reported in the early 1980s). It was not until the mid-1990s that the so-called "hybrid" approach, which ultimately became the most commonly-used approach in the 2-D numerical modeling of Hall thrusters, was attempted in 1-D simulations (for example, see Morozov and Savelyev in [83]). The word "hybrid" here is used in reference to the different methods followed to obtain solutions for the charged species: the ion motion is determined by solving a kinetic equation for the ion distribution function or by using discrete-particle methods whereas the electrons are treated using the continuum equations.

Despite the complexity of the physics in these thrusters, 1-D numerical models showed they could provide useful insight into both steady-state and transient trends. For example, Fig. 7-18 plots

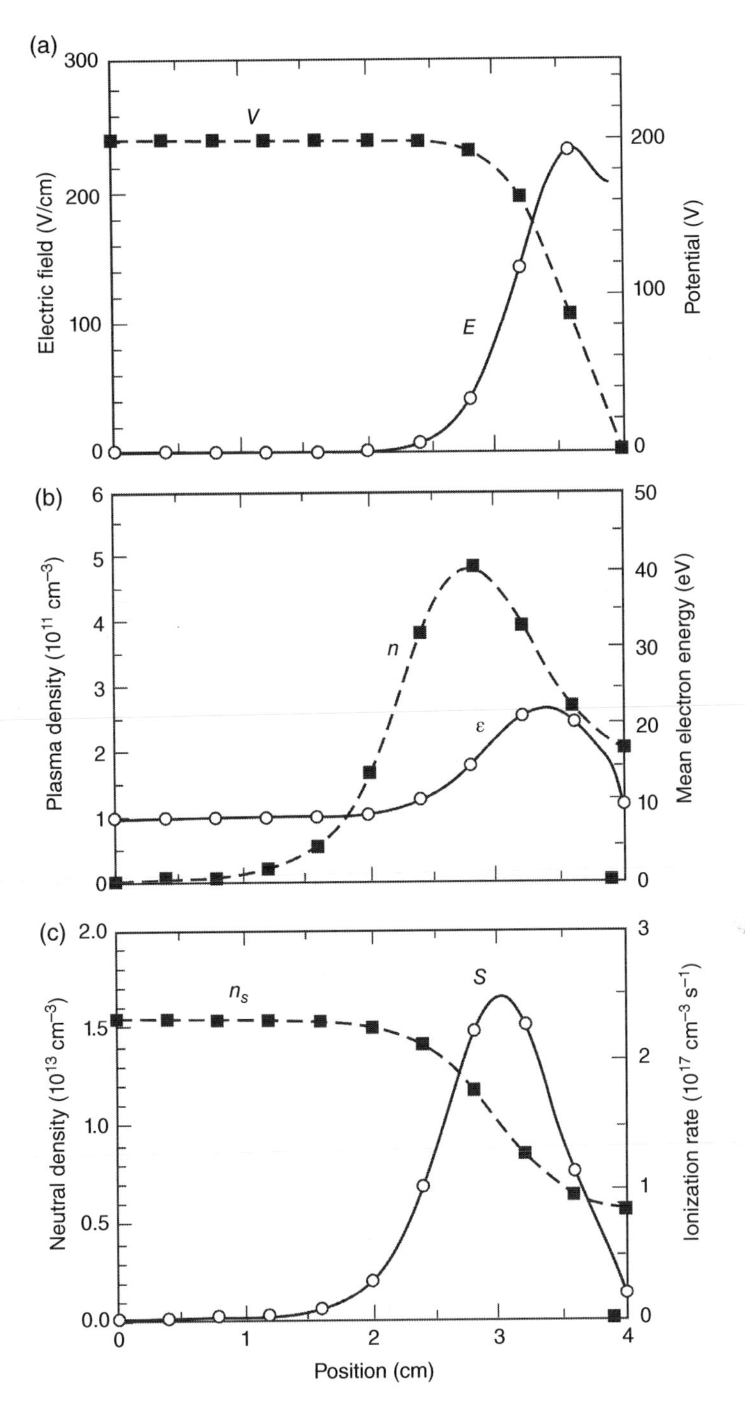

Figure 7-18 Potential and electric field (a), plasma density and electron energy (b), and neutral density and ionization rate (c) from a 1-D Hall thruster code for the SPT-100 (*Source:* Adapted from Boeuf et al. [58]).

along the channel axis the time-average profiles of the potential, electric field, plasma density, mean electron energy, neutral density and ionization rate for the SPT-100 predicted by one of the eariiest 1-D numerical models, developed by Boeuf and Garrigues [58] in the late 1990s. The model assumed that the discharge is quasi-neutral and used a transient hybrid treatment for the electron and ion transport in the device; electrons were treated as a fluid and ions were described by a collision-less kinetic equation. In Fig. 7-18 the position of the channel exit is at 4 cm. The average plasma density peaks upstream of the exit for the thruster channel. There is a characteristic peak in the plasma density upstream of the channel exit in the ionization region, and a decreasing plasma density is seen moving out of the channel as the ions are accelerated in the electric field of the acceleration region. The profiles in Fig. 7-18 suggest that three overlapping but distinct regions exist in the plasma channel of a Hall thruster. Near the anode, the potential drop is small because of the low magnetic field in this region, resulting in good plasma conduction to the anode but small ionization. The ionization zone occurs upstream of the channel exit where the neutral gas density is still high and the electrons are well confined and have significant temperature. The acceleration zone exists near the channel exit where the electric field is a maximum, which occurs at this location because the magnetic field is a maximum and the transverse electron mobility is significantly reduced as described above. Outside the channel, the electric field, plasma density and electron temperature drop as the magnetic field strength decays and the Hall current decreases.

Interestingly, the study also showed that this simple model could not predict experimental observations if the electron mobility in the perpendicular direction was based only on classical electron-neutral collisions. In fact, including Bohm conductivity did not improve the comparison with experiments either. Qualitatively better results were obtained only after the contribution of electron-wall interactions was included. As it will be discussed in the next section, this is in contrast to the results from some of the first 2-D simulations performed around the same time. It turns out that now, decades later, although good progress has been made on the likely anomalous processes and how to model them [84–88], a first-principles universal model of the anomalous transport across the magnetic field remains elusive. Nevertheless, such reduced-dimension models served as useful and fast modeling tools, not only in providing insight on steady-state trends but also on some of the well-observed transient behavior of these thrusters. For example, the 1-D numerical model predictions in [58] for the total current, electron and ion current at the thruster exit are shown in Fig. 7-19 for the SPT-100. In this case, a breathing frequency of 16 kHz is predicted, in good agreement with the aforementioned measurements in the same thruster.

Owing to the relative simplicity and need for fast engineering guidance in the development of these thrusters, low-dimensionality models like the one by Boeuf and Garrigues have continued to be pursued by the community (e.g. see [41, 59, 89–93]. Their limited scope has also become more evident however since the solution must depend on a wide range of model inputs that, themselves, can vary significanly. One major challenge, for example, is a model's assumption of the anomalous collision frequency. To illustrate why this is a challenge, the results from a sensitivity study performed by Hara and Mikellides [94] using a 1-D fluid model of a 6-kW Hall thruster are shown in Fig. 7-20. The top figure plots the prescribed anomalous collision frequency as a function of axial distance z from the anode, with L being the channel length. The bottom figure depicts the dynamic response of the solution for the discharge current. The legend lists different values of the coefficient used to prescribe the anomalous collision frequency in the channel interior. The profiles are similar to those reported in [99, 104] and were determined based on a combination of 2-D axisymmetric simulations and plasma measurements. It was found that the details of the anomalous collision frequency in the acceleration channel, which as we mentioned before remain elusive today, can affect appreciably the dynamic behavior of the discharge in the simulation. The dynamics can also

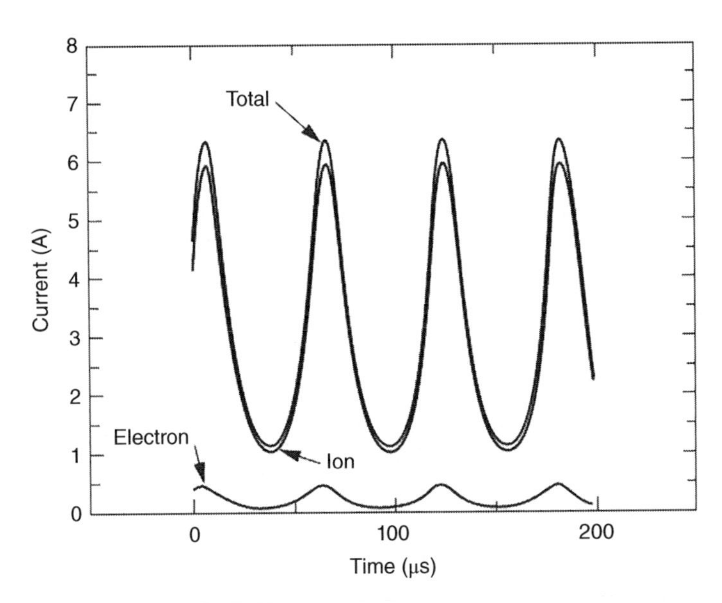

Figure 7-19 Oscillating current predictions from the 1-D code for the SPT-100 (*Source:* Adapted from Boeuf et al. [58]).

be sensitive to the numerical method(s) employed to discretize and solve the equations [94]. Other "free" or uncertain parameters, such as those used to model the interactions of both the heavy and light species with the solid boundaries, can affect the solution as well. This strong sensitivity of 1-D models is one of the many reasons that higher-dimensionality models are frequently pursued despite their increased complexity and computational cost.

7.5.2.2 Modeling in Multiple Dimensions

Some of the earliest simulations in two dimensions (2-D) were in fact performed using the PIC method for all species [95]. It soon became evident however that modeling both electrons and the heavy species using such methods at the time was prohibitively costly in computational time and could only offer limited insight. In response, a hybrid computational approach was adopted allowing simulations to capture the bulk plasma phenomena and ion kinetics in the thruster, in 2-D, and within reasonable computational times. For that reason, hybrid approaches became quite common in the simulation of Hall thrusters and their popularity has continued to this day. Since the mid-1990s an extensive body of work has been published in this area. Attempting to cover these efforts here, in a manner that covers all the important advancements while also properly acknowledging the many researchers who have contributed to this area, would be a futile effort. Instead, we have selected to discuss in some detail only two examples of computational codes developed specifically for the modeling of Hall thruster physics because they encompass many of the most common approaches employed in the simulation of these devices today. Wherever suitable however we provide additional references to other important methods and findings.

Hybrid-PIC Hall (HPHall) One of the first 2-D models to follow the hybrid approach was developed at the Massachusetts Institute of Technology by Fife and Martínez-Sánchez in the late 1990s [34, 38]. The code called "Hybrid-PIC Hall" (HPHall) had a seminal impact on the numerical simulation of

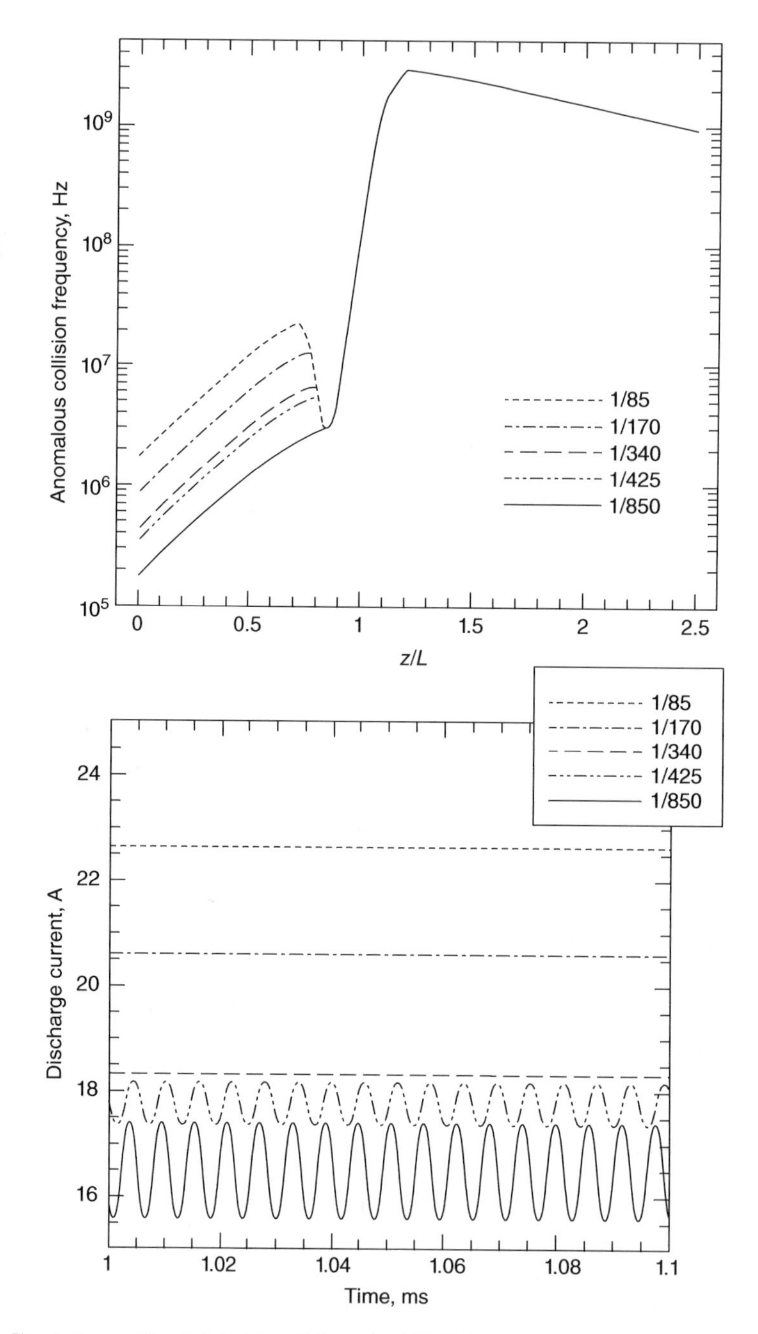

Figure 7-20 Simulations with a 1-D fluid model of a 6-kW Hall thruster showing the effects of the anomalous collision frequency in the channel interior on the discharge current dynamics (*Source:* Adapted from Hara and Mikellides [94]).

Hall thrusters, in part because it was the first to reproduce in a radial-axial (r-z) computational domain the so-called breathing mode oscillations in Hall thrusters. The low-frequency oscillation behavior of Hall thrusters from the 2D HPHall simulations are shown in Fig. 7-21, where the anode current and beam current are plotted as a function of time. The computed frequency in this

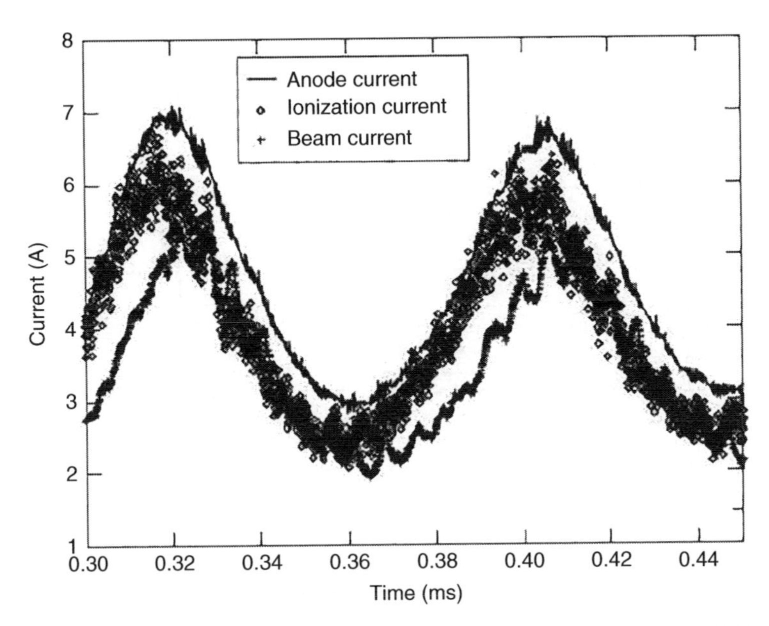

Figure 7-21 Anode current, ionization and beam current calculated by HPHall for the SPT-70 Hall thruster (*Source:* [34]/American Institute of Aeronautics and Astronautics).

simulation is 11 kHz. The characteristic ionization processes driving the oscillations were found to be the same as those obtained by the 1-D simulations described in the previous section [58]. These low frequency oscillations can reach 100% of the discharge current depending on the voltage and mass flow (current) for a given thruster design. However, more modern designs typically have much lower oscillation amplitudes. In the HPHall simulations Bohm diffusion was invoked for the electron fluid across the magnetic field (Eq. 7.5-13), which yielded 2-D plasma profiles that were similar to those observed in some (but not all) experiments [96]. Typical 2-D time-averaged contours of the plasma density from an SPT-100 simulation are shown in Fig. 7-22. As in the 1-D results of Fig. 7-15, the average plasma density peaks upstream of the exit of the thruster channel. In both cases, there is a characteristic maximum upstream of the channel exit in the ionization region, and a decreasing plasma density is seen moving out of the channel as the ions are accelerated by the electric field in the acceleration region. Similar hybrid models ensued by others [76, 78]. The early success of HPHall motivated additional development of the code later that led to improved sheath and channel erosion models [13, 50, 97, 98].

It is worthwhile noting here that despite the fundamentally different anomalous mobility models used in the two simulations, the peak plasma density in the 1-D code ($\sim 5 \times 10^{17}$ m^{-3}) is only slightly lower than the 2-D HPHall result ($\sim 8 \times 10^{17}$ m^{-3}). This exposes a major hurdle in Hall thruster modeling that continues to be prevalent today: the lack of a first-principles model for the anomalous resistivity challenges our ability to validate unambiguously these models and impairs our capacity to produce truly predictive simulations. The reader may recall the example in Section 7.5.2.1 related to the sensitivity of breathing mode 1-D models on the details of the anomalous resistivity profile. Similar challenges are encountered in the validation of the 2-D time-averaged results also, as discussed in more detail by Mikellides and Lopez Ortega [99]. The subject is revisited later in this section.

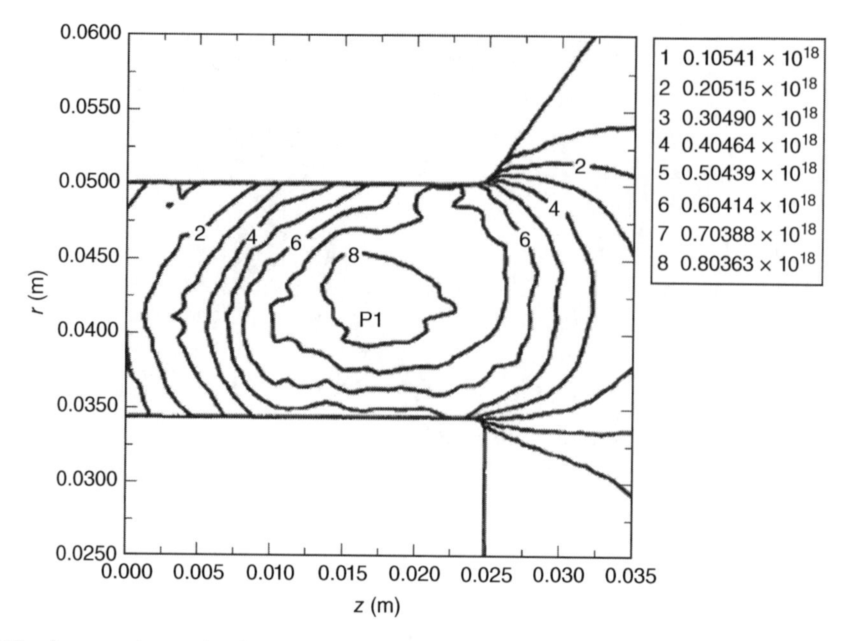

Figure 7-22 Average plasma density computed by HPHall for the SPT-100, which the peak plasma density at P1 = 8×10^{17} m^{-3} Fife et al. (*Source:* [34]/American Institute of Aeronautics and Astronautics).

One of the novel approaches employed in HPHall was based on the fundamental principle in Hall thrusters that the acceleration of ions is achieved by operating the discharge at high electron Hall parameter ($\Omega_e > 100$). Under these conditions, the resistance to the transport of mass and heat in the electron flow in the direction perpendicular to the magnetic field is much greater (by $\sim\Omega_e^2$) than that in the parallel direction for most of the acceleration channel (Section 7.5.1.2). This allows for the solution of the electron transport equations only in the direction that is perpendicular to the magnetic field. Numerically, this so-called "quasi-1-D assumption" allowed for the discretization of the electron equations in a quadrilateral computational element that is bounded by two adjacent lines of force, rather than an element with arbitrary dimensions as shown in Fig. 7-23. The simplification reduced both the complexity and computational cost of numerical simulations in 2-D.

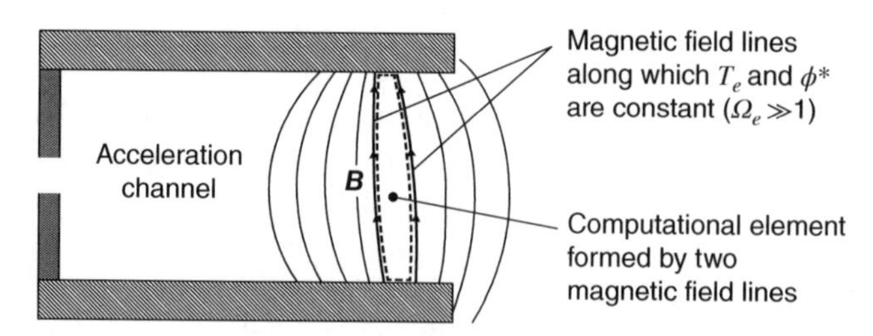

Figure 7-23 The quasi-1D assumption allows two magnetic field lines and the wall segments they intersect to form a computational element on which the electron temperature T_e and thermalized potential ϕ^* can be determined in the numerical simulation of regions where $\Omega_e >> 1$.

Hall Thruster with 2-D Electrons (Hall2De) In conventional Hall thrusters that employ a largely radial magnetic field (like SPTs), the quasi-1D assumption does not significantly degrade the accuracy of the solution inside the acceleration channel and in the very near-plume regions of the domain where the Hall parameter is quite large. The assumption is challenged however in regions with more complex magnetic fields. Examples are topologies near eroded walls and in the new generation of these devices called magnetically shielded thrusters that will be discussed in Chapter 8. The assumption also fails in regions with lower Hall parameter such as those near the anode and in the cathode plume.

To properly model these more complex cases, a solution to at least the 2-D form of Ohm's cannot be avoided. However, because of the inherently large disparity of the electron transport in the two directions relative to the applied magnetic field, this can be numerically very challenging. Employment of magnetic field-aligned meshes (MFAM) is a long-standing computational approach for simulating highly anisotropic plasmas, and is widely used nowadays especially by the sustained fusion energy community [100–103]. The first code to pursue this path in Hall thruster simulations was named "Hall with 2D electrons" (Hall2De) and was developed by Mikellides and Katz [70, 79] in the late 2000s. Hall2De solved the axisymmetric vector form of Ohm's law in the r-z domain and avoided the employment of discrete-particle or kinetic methods in the treatment of the heavy species in the first versions of the code. Using predefined energy bins, the ions were modeled as multiple fluids [70, 74] and the neutrals were treated using line-of-sight formulations that account for ionization collisions and collisions with walls [79]. More recently, PIC/DSMC modules were incorporated to capture the details of the ion velocity distributions functions making Hall2De a hybrid (multifluid/PIC-ion fluid-electron) code for the charged species [104, 105].

The magnetic field-aligned mesh was necessary to eliminate excessive numerical diffusion that would otherwise be caused by the anisotropy of the electron transport coefficients when solving the equations on a regular mesh. Referring to Fig. 7-24 (middle), in this approach the plasma potential ϕ_i is solved for on each computational element "i" after combining Ohm's law for the electrons (Eq. 7.5-5) with current conservation Eq. 7.5-18. At each new time step t+Δt and at each face "k" of area ΔA_k of element "i", the dot products of the current density needed in a finite volume discretization of the equations, $\sum_{k=1}^{4}\left[\left(\mathbf{J}_{\parallel}+\mathbf{J}_{\perp}\right)\cdot\hat{\mathbf{n}}_k\Delta A\right]_k$, can be determined by

$$\mathbf{J}_k^{t+\Delta t}\cdot\hat{\mathbf{n}}_k=\left(\frac{-\nabla\phi_k^{t+\Delta t}+\boldsymbol{\varepsilon}_k^t}{\eta_k^t}\right)\cdot\overline{\mathbf{n}}_k \tag{7.5-27}$$

where $\overline{\mathbf{n}}\equiv\overline{n}_r\hat{\mathbf{r}}+\overline{n}_z\hat{\mathbf{z}}$, η is the resistivity and ε represents all terms in Ohm's law other than the electric field $\mathbf{E}=-\nabla\phi$. Here, the usual expressions for the parallel and perpendicular components of the current density have been implied: $\mathbf{J}_{\parallel}=\left(\mathbf{J}\cdot\hat{\boldsymbol{\beta}}\right)\hat{\boldsymbol{\beta}}$ and $\mathbf{J}_{\perp}=-\hat{\boldsymbol{\beta}}\times\left(\hat{\boldsymbol{\beta}}\times\mathbf{J}\right)$, with the unit vector in the direction of the magnetic field given by $\hat{\boldsymbol{\beta}}=\beta_r\hat{\mathbf{r}}+\beta_z\hat{\mathbf{z}}$. The novel part of this approach is that $\overline{n}_r$ and $\overline{n}_z$ can be simplified considerably on a MFAM because the element faces are either parallel or perpendicular to the magnetic field lines. Numerical diffusion because of the disparity between the terms that depend on Ω_e and those that do not is practically eliminated in this way. The accuracy of the solution then depends on the extent of the spatial deviations of the mesh from the true lines of constant potential and stream functions χ and ψ. Here χ and ψ are the set of conjugate harmonic functions satisfying the Cauchy-Riemann conditions for the radial and axial components of the magnetic field. A set of such lines in the acceleration channel of a typical Hall thruster are shown in Fig. 7-24 (left). The corresponding MFAM is shown in Fig. 7-24 (right).

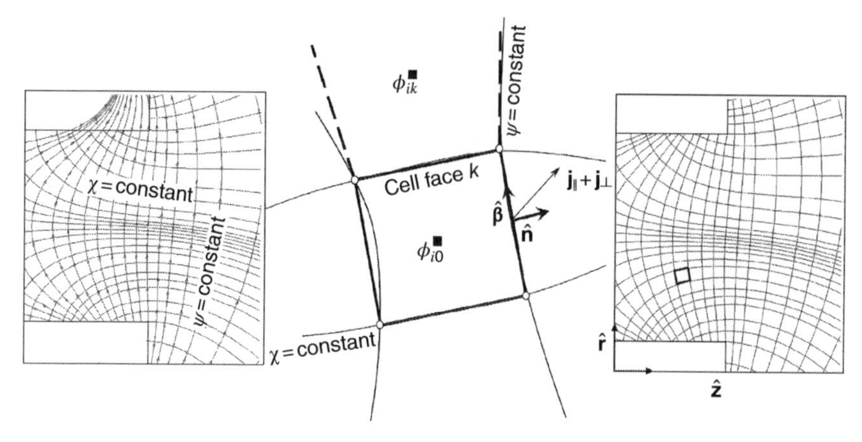

Figure 7-24 A set of lines of constant stream function ψ that define the magnetic vector field overlaid by lines of constant potential function χ, in the vicinity of the acceleration channel of a typical Hall thruster. Middle: Each face of a computational cell in Hall2De is closely aligned with either a χ-line or a ψ-line. Right: corresponding MFAM computational mesh (*Source:* Adapted from Mikellides and Katz [70]).

Hall2De was the first code to use a MFAM in the simulation of Hall thrusters. A similar MFAM code was developed later by Domínguez-Vázquez, et al. [106]. The employment of an MFAM in Hall2De, with other advancements made in the code based on the many lessons learned from HPHall and other similar codes, led to several new insights. Most well-known is the code's critical role in the derivation of the fundamental principles of magnetic shielding [69, 107] to be described in Chapter 8. Considering the impact of magnetic shielding in Hall thrusters, a separate chapter is dedicated solely to this topic so we will not pursue this here any further. There have been a few other notable insights gained from the Hall2De simulations over the years, however, that are worth summarizing here.

The generalization to a full 2-D solution of Ohm's law allowed for a much larger computation area than previous codes (like HPHall). An example of a typical simulation domain, its corresponding MFAM and results for the ion density are shown in Fig. 7-25. Among others, this larger MFAM domain allowed simulations to encompass the cathode boundary where the lines of force can become non-isothermal. In Fig. 7-26 the computed electron temperature contours are shown in the entire computational region (left) and in the vicinity of the acceleration channel (right) of a typical Hall thruster operating with a center-mounted cathode. The contours are overlaid by selected arrowed lines tangent to the magnetic vector field (called hereinafter "streamlines" and loosely associated with the appropriate definition of the term in fluid velocity vector fields) to illustrate that in the majority of the computational domain the isothermalization of the lines of force is indeed preserved. Deviations do occur however in the cathode region because of the high collisionality of the plasma there. This allowed simulations to employ cathode boundary conditions unambiguously thereby capturing the interactions of the cathode-thruster plasma flows more accurately [108]. Therefore, unlike HPHall and other codes that rely on the quasi-1D assumption, Hall2De does not assume that the electron temperature remains fixed along magnetic field lines.

Perhaps equally important have been the insights gained from Hall2De simulations on the question of the anomalous electron transport. As it was noted earlier, the first numerical models of Hall thrusters assumed some combination of wall collisions and Bohm diffusion (Eq. 7.5-14). As more plasma measurements were performed however, it became clear that neither the two processes applied separately nor their combination could reproduce accurately these measurements in the

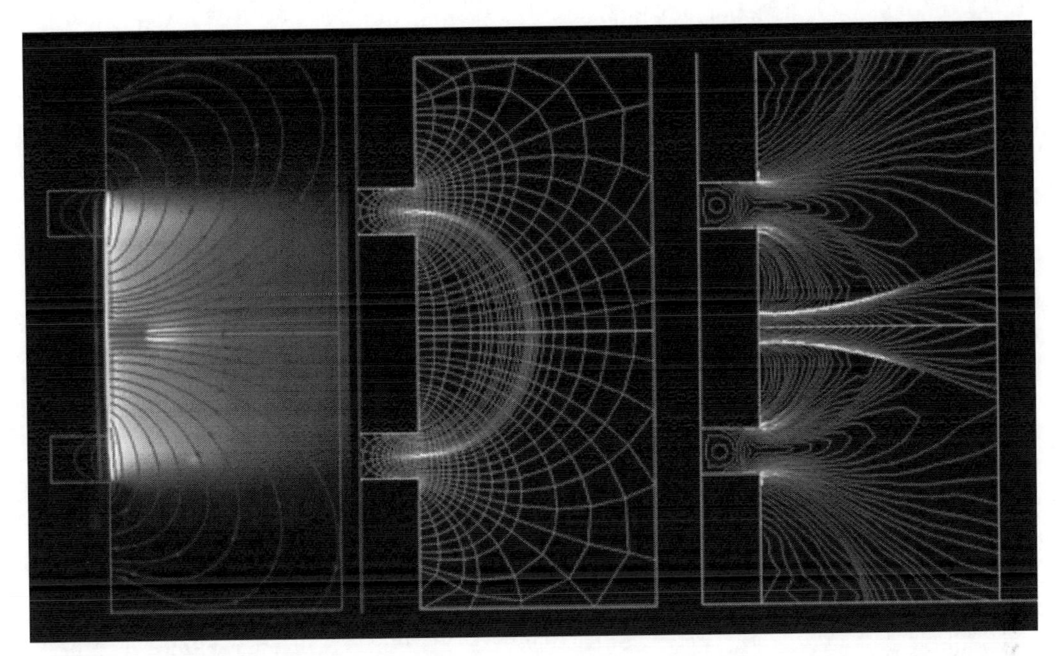

Figure 7-25 The Hall2De domain used in the simulation of a 6-kW Hall thruster with center-mounted cathode, showing magnetic field lines overlaid on a photograph of the thruster operating in a vacuum facility (left), the magnetic field aligned computational mesh (middle) and contours of the ion density (right).

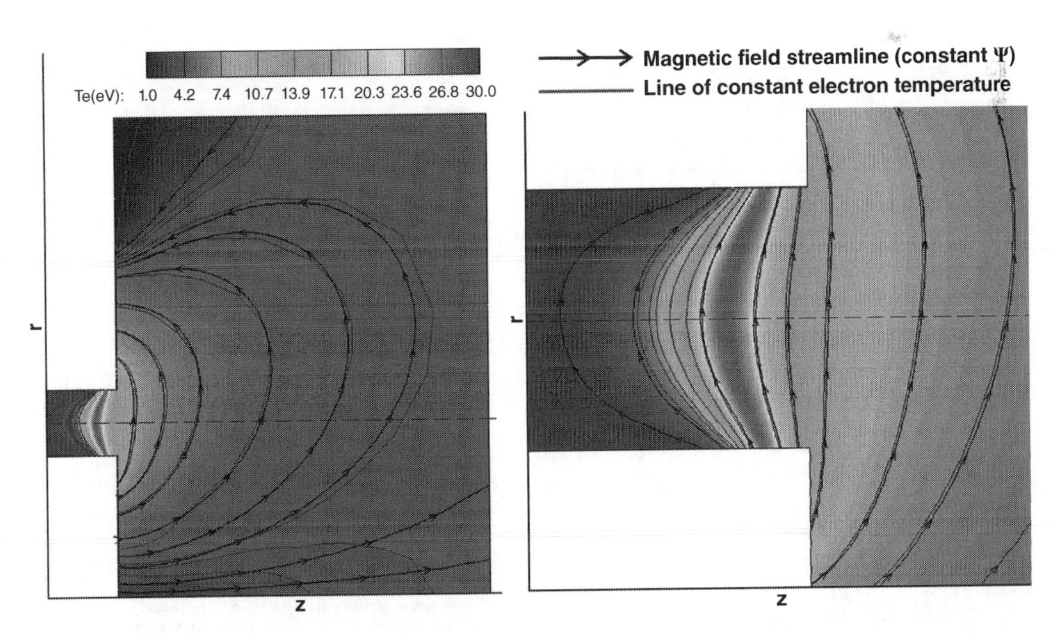

Figure 7-26 Electron temperature contours from a Hall2De simulation overlaid by selected magnetic field "streamlines" (*Source:* Adapted from Mikellides and Katz [70]).

vicinity of the acceleration region. This prompted some researchers to divide the computational domain in 2–3 different regions and apply a different coefficient β (see Eq. 7.5-13) in each region (e.g. [109]). In retrospect, this was simply the beginning of the realization that Bohm diffusion physics were simply not present in the Hall thruster. Ultimately, the strongest argument against Bohm-driven transport was established after several years of Hall2De comparisons with plasma and wall erosion measurements in different thrusters. To be more specific, Mikellides and Lopez Ortega [99] incorporated in Hall2De a multi-variable function for ν_α and combined it with LIF measurements of the ion velocity to determine the function's coefficients. This allowed them to obtain a semi-empirically derived profile of the anomalous collision frequency in the acceleration channel and near-plume. A similar effort to determine the anomalous collision frequency from a combination of probe measurements and 2-D simulations (using a code similar to HPHall) was pursued several years earlier at Stanford University [110, 111]. In the more recent effort with Hall2De however a much more resolved and accurate set of measurements was attained from LIF diagnostics, that also spanned the acceleration region and the near-plume. This allowed the simulations to capture the spatial variation of the collision frequency much more accurately in these regions. An example of comparisons between Hall2De simulations and LIF measurements for the ion velocity field is provided [112] in Fig. 7-27. The mathematical formulation of the piecewise function for ν_α is described in [99].

Typical Hall2De solutions for ν_α along the channel centerline of a magnetically shielded Hall thruster and comparisons with ion velocity measurements for two different magnetic flux densities (B_1 and B_2) are illustrated in Fig. 7-28. The results are plotted as a function of the axial direction z,

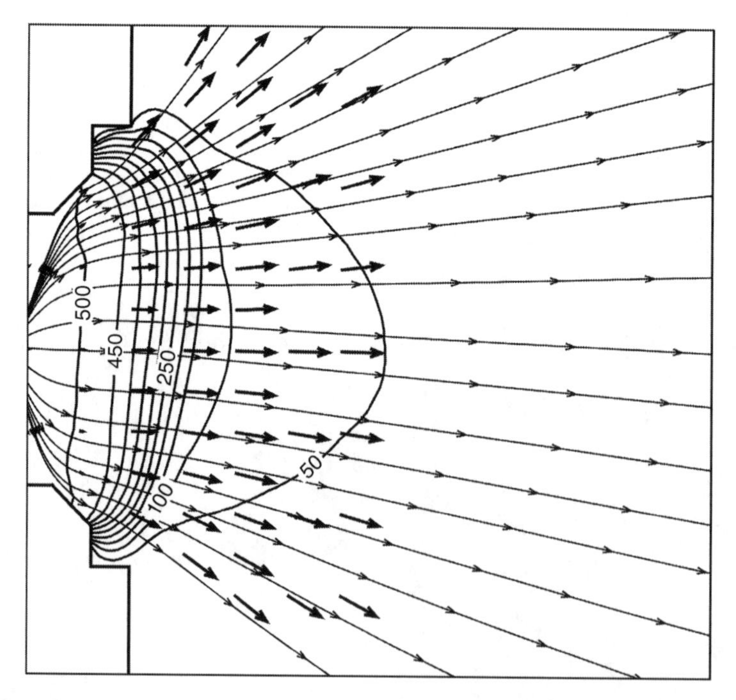

Figure 7-27 Comparison of the ion velocity fields from simulations (continuous arrowed traces) and LIF measurements (discrete vectors) in the near-plume of the MaSMi Hall thruster operating at 1 kW and 500 V (*Source:* Adapted from Lopez Ortega et al. [112]). Also shown are contours of the computed plasma potential with values given in volts.

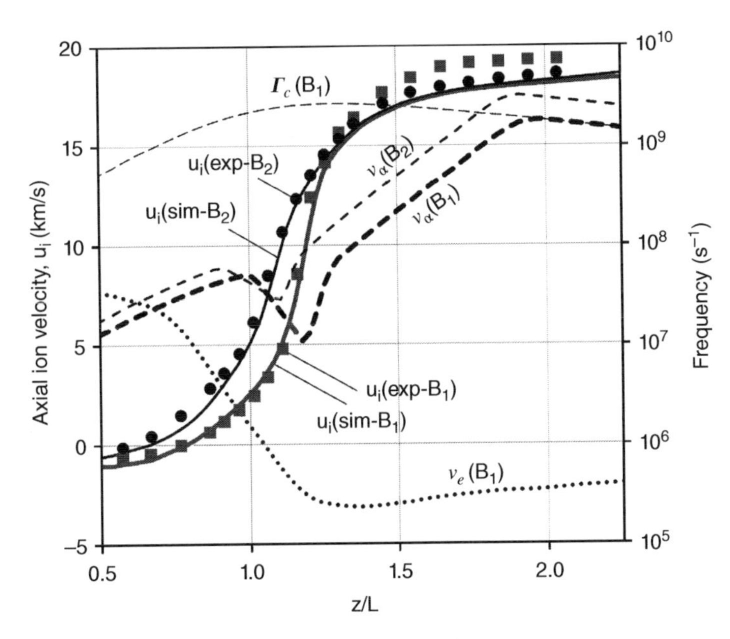

Figure 7-28 Results from Hall2De simulations ("sim") of a magnetically shielded Hall thruster for the anomalous (ν_α) and classical ($\nu_e = \nu_{ei} + \nu_{en}$) collision frequencies, and the ion velocities (u_i), at two different magnetic flux densities ($B_1 < B_2$). Overlaid for comparison are ion velocity measurements ("exp") obtained by LIF diagnostics and the electron cyclotron frequency ω_c for B_1.

along the centerline of the acceleration channel with length L. The channel exit is at $z/L = 1$. Also shown for direct comparison with ν_α is the electron cyclotron frequency ω_c. The comparisons underscore the following critical findings: (i) the marked differences between Bohm-diffusion scaling $\nu_B \propto \omega_c$ and the empirical-numerical result for ν_α and, (ii) the dominance of the anomalous contribution to the total collision frequency in a large portion of the channel interior over that from only (classical) electron-ion (e-i) and electron-neutral (e-n) collisions. Although these trends have been found in more than one Hall thrusters (e.g. [104, 113, 114]), in the absence of a first-principles model of ν_α the possibility of a different behavior cannot be dismissed. Nevertheless, profiles like the ones shown in Fig. 7-28 serve as some of the most accurate estimates of the spatial variation of the anomalous collision frequency we have to-date for these thrusters (at least along the channel centerline). The significance of this is that numerical simulations may be employed to test the validity of theories associated with the anomalous transport in these devices, before they are implemented as first-principles models in r-z electron-fluid codes.

The approach of combining detailed non-intrusive measurements and 2-D simulations to determine the anomalous collision frequency does not yet allow for a fully-predictive modeling capability. However, in addition to providing a testbed for electron transport models and insight into to the physics of a variety of other processes that occur in these devices, such capability *can* (and does) also guide thruster design. Perhaps even more important is the support such modeling capabilities can provide in thruster qualification for deep-space missions. A unique challenge with these missions is that because the Hall thruster is usually required to operate for many thousands of hours in space, many times it is simply impossible (due to time and cost constraints) to demonstrate its life in the laboratory for the entire throttle range and under true space vacuum conditions. The path to space qualification therefore, inevitably, must cross both laboratory testing and physics-based modeling, and it is the latter to which codes like Hall2De offer the greatest contribution. Specifically, once validated and informed by plasma diagnostics, these codes can provide predictions of thruster wear and throughput, for the entire mission profile and in true space vacuum.

Simulations of wave propagation and instabilities in multiple dimensions. The elusiveness of the anomalous transport coefficients in Hall thrusters has led to a wide range of computational approaches over the last three decades, all aimed at capturing the driving spatiotemporal scales in these devices. Here, we mentioned only a few representative examples and cite some notable results from the wide range of dynamic behaviors that has been captured by numerical simulations. An in-depth synopsis of ongoing activities in this vibrant area of Hall thruster modeling is provided by Kaganovich et al. [61] and the references therein.

A major challenge in the modeling of anomalous transport in Hall thrusters is that the discharge is, in principle, three-dimensional (3-D) and, as already inferred in previous sections, the driving physics span multiple scales. Moreover, it is now also widely accepted that deviations from the equilibrium distribution function likely occur in the charged species, at least in some regions of the thruster. This allows for the possibility of kinetic instabilities in the discharge. The solution for the growth and saturation of such instabilities requires simulations that can follow the evolution of the relevant velocity distributions functions. This can be very complex and computationally expensive. Although computing processing power has been steadily increasing, kinetic simulations with realistic parameters that span the entire Hall thruster domain remain quite challenging. Although particle methods in 2-D (z-θ [84, 85, 115, 116] and r-θ [86, 117, 118]) and even in 3-D [119, 120] have indeed been possible, they are limited to restricted spatial and/or temporal domains and, therefore, are unable to account for all pertinent processes. To be more specific, simulations in the z-θ plane can provide detailed insight into instabilities with wave vectors in the $\mathbf{E} \times \mathbf{B}$ and axial directions but they provide no information about plasma-wall interactions. On the other hand, r-θ simulations cannot account for the variation of the plasma properties in the axial direction and therefore cannot capture how the transport physics evolve across the magnetic field in the different regions of the channel.

Despite these limitations there has been considerable progress in our understanding of many aspects of the Hall thruster discharge related to instabilities, especially in the vicinity of the acceleration region. For example, Adam and Héron [84, 86] reported strong evidence that the turbulence generated by the electron cyclotron drift instability (ECDI) [121] is the cause of the anomalous electron transport in a Hall thruster. Their conclusions were based on the results of 2-D particle simulations in the z-θ plane and on measurements [122–124]. Coche and Garrigues [85] also captured ECDI-induced fluctuations in the azimuthal electric field, in the frequency range of MHz and wave number of 3000 rad s^{-1} (seen in Fig. 7-29), from simulations with a different 2-D, z-θ particle code. They reached conclusions similar to those by Adam, et al. Their simulations were also able to reproduce the breathing mode without the introduction of any additional anomalous processes. Further insight was provided by the PIC simulations of Janhunen and Smolyakov [125] who found that for finite values of the wave number ($\boldsymbol{k}$) in the direction of **B**, the ECDI may transition to the Modified Two-Stream Instability (MTSI) [126]. Simulations by Katz [115] in the z-θ plane have suggested that ionization can also play a significant role, especially immediately downstream of the acceleration region where both the plasma potential and electric field diminish. Growth of the MTSI and another instability in the lower hybrid frequency range, the lower hybrid drift instability (LHDI) [127, 128], also have been postulated to exist but in the near-plume region, between the main ion beam and the thruster centerline [129, 130] with possible implications on the wear of the front pole surfaces facing the plume [105].

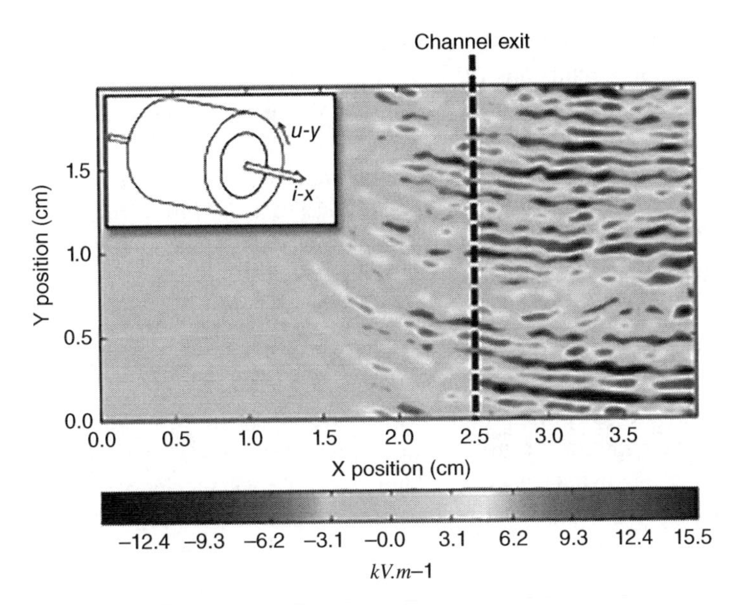

Figure 7-29 Computed azimuthal electric field from PIC simulations in the z-θ domain, denoted as x-y in the plot (*Source:* Adapted from Coche and Garrigues [85]).

7.6 Operational Life of Conventional Hall Thrusters

The operating time and total impulse of a conventional Hall thruster is determined primarily by erosion of the discharge channel wall and the life of the cathode. Hollow cathode wear-out has not represented a life limitation to date because thruster lifetimes based on the discharge chamber wall erosion of less than 10,000 h are typical, and robust LaB_6 hollow cathodes have been used in all the Russian Hall thrusters. Other issues such as deposited material build-up on the electrodes, conductive-flake production, electrical shorting, etc. are also of concern in evaluating the life of a Hall thruster. However, the erosion of the channel wall by ion bombardment sputtering is a very visible process that changes the channel dimensions and ultimately exposes the magnetic circuit, which, when eroded, can degrade the thruster performance. In addition, life tests of flight thrusters such as the SPT-100 and the PPS-1350 show that it can take hundreds to thousands of hours for magnetic circuit erosion to significantly alter the thruster performance. Of greater concern, in this case, is the sputtering of iron from the magnetic circuit, which would have a significantly higher impact if deposited on most spacecraft components. Therefore, understanding the wall erosion rate and its dependence on thruster materials and operating parameters is of importance in predicting the thruster life and performance over time and its potential impact on the spacecraft.

The erosion rate, given by the rate of change of the wall thickness "w", is

$$\dot{\mathcal{R}} = \frac{\partial w}{\partial t} = \frac{J_i W}{\rho q A_v} Y(\varepsilon_i), \tag{7.6-1}$$

where j_i is the ion current density, W is the atomic weight, ρ is the material density, q is the ion charge, A_v is Avogadro's Number, and $Y(\varepsilon_i)$ is the sputtering yield of the material which is dependent on the ion type and energy ε_i. Since the material properties are known, the issue becomes one of knowing the ion flux, ion energy and sputtering yield of the wall.

Several models of the Hall thruster have been developed and applied to this problem [41, 98, 131, 132]. The most accurate predictions have been achieved using 2-D codes such as HPHall and Hall2de to obtain the ion fluxes and energies. The sputtering yield of boron nitride compounds used in dielectric-wall Hall thrusters has been measured by Garnier [133] versus incidence angle and ion energy, and is used in several of these models. However, the Garnier data is at only a few energies and in excess of 300 V. Gamero [97] extrapolated this data to lower energies using the semi-empirical sputtering law scaling of Yamamura and Tawara [134], obtaining the following expression for the sputtering yield in units of mm^3/Coulomb:

$$Y = \left(0.0099 + \alpha^2 6.04 \times 10^{-6} - \alpha^3 4.75 \times 10^{-8}\right)\sqrt{\varepsilon_i}\left(1 - \sqrt{\frac{58.6}{\varepsilon_i}}\right)^{2.5}, \tag{7.6-2}$$

where α is the incident angle of the ion on the surface. In Eq. 7.6-2, the value 58.6 represents the estimated threshold energy for sputtering required by Yamamura's model. Figure 7-30 shows an example of the yield predicted by Eq. 7.6-2 for two different incidence angles. Equation 7.6-2 was shown to accurately fit the data of Garnier [133] and provides projections of the sputtering yield down to low ion energies predicted by HPHall deeper in the channel.

Figure 7-31 shows the predicted [97] and experimentally measured erosion profiles [135] for the SPT-100 thruster inner and outer channel walls. Reasonable agreement with the observed channel erosion is seen near the thruster exit, and the profiles have the correct functional shape. It is likely that inaccuracies in the extrapolated sputtering yield at low energies and plasma predictions by HPHall caused the disagreement with the data deep in the channel.

Significant improvements to the erosion predictions by 2-D codes are illustrated by the recent use of Hall2De's modeling of the channel erosion to provide mission critical predictions of the thruster throughput and life prediction for the SPT-140 Hall thruster. These thrusters will provide main propulsion for NASA's mission to the asteroid Psyche for the heliocentric cruise from Earth [136] and, once at the asteroid, they will also be used for orbit transfers and maintenance. The mission to Psyche will be the first ever to use Hall thrusters beyond cis-lunar space. In addition to validating Hall2De simulations of the SPT-140 with data from plasma diagnostics, channel erosion measurements from a wear test in a ground facility were also used to confirm the code's simulations under the same facility background pressure [113]. The comparisons with the measurements for

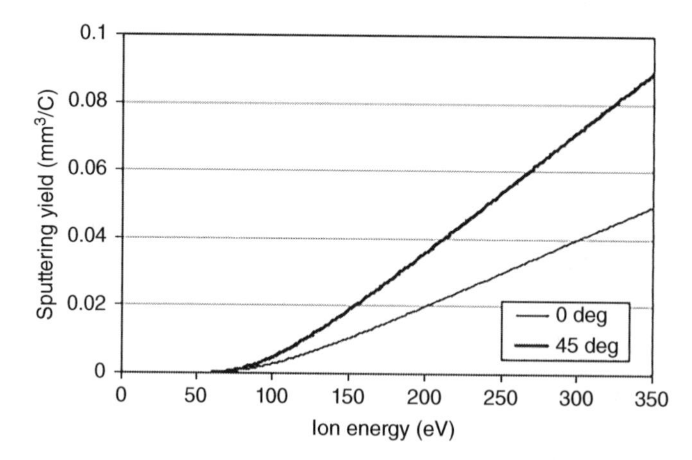

Figure 7-30 Sputtering yield calculated for singly ionized xenon on $BNSiO_2$ versus ion energy for two incidence angles.

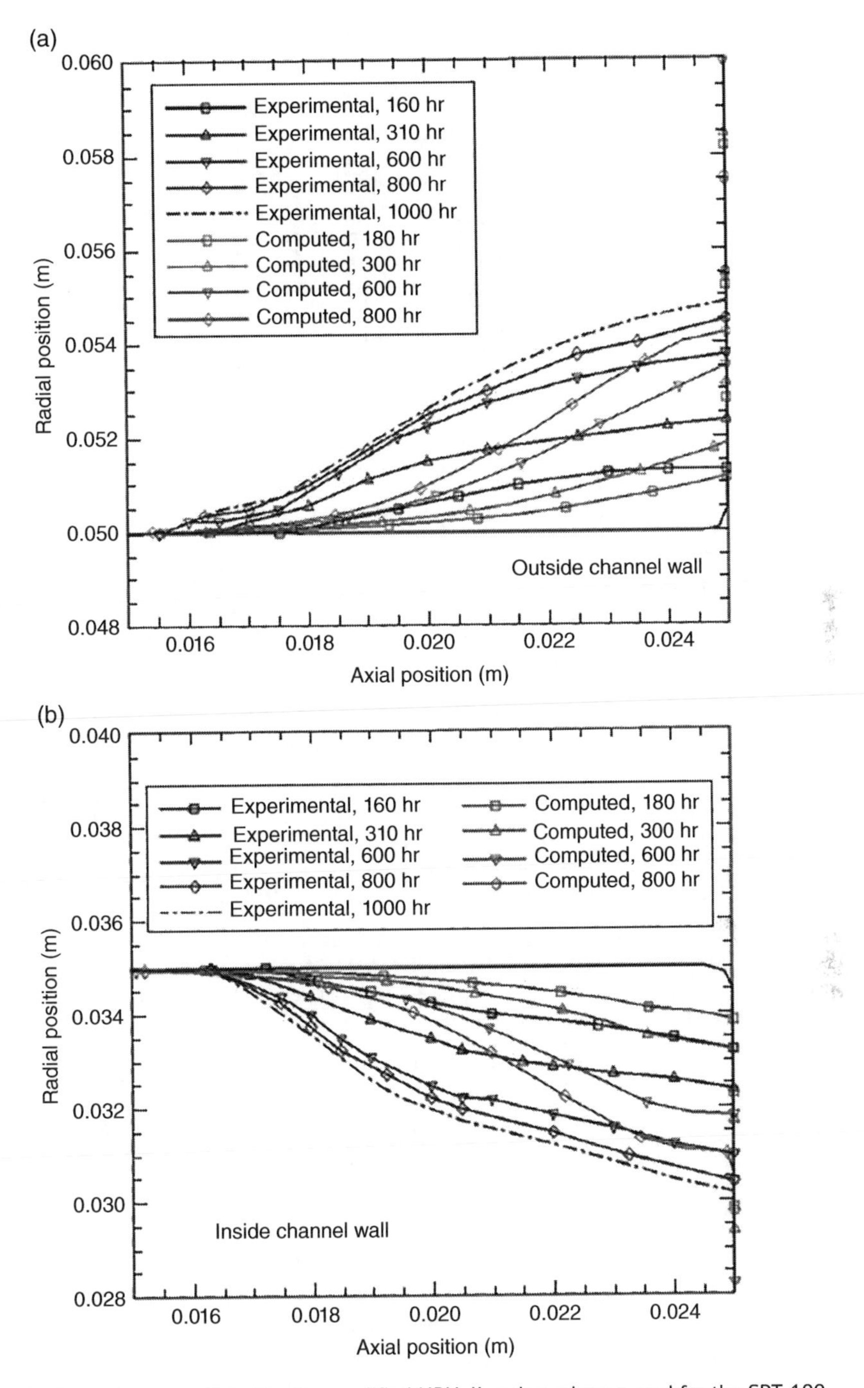

Figure 7-31 Erosion pattern predicted by the modified HPHall code and measured for the SPT-100 thruster (*Source:* [97]/IEPC).

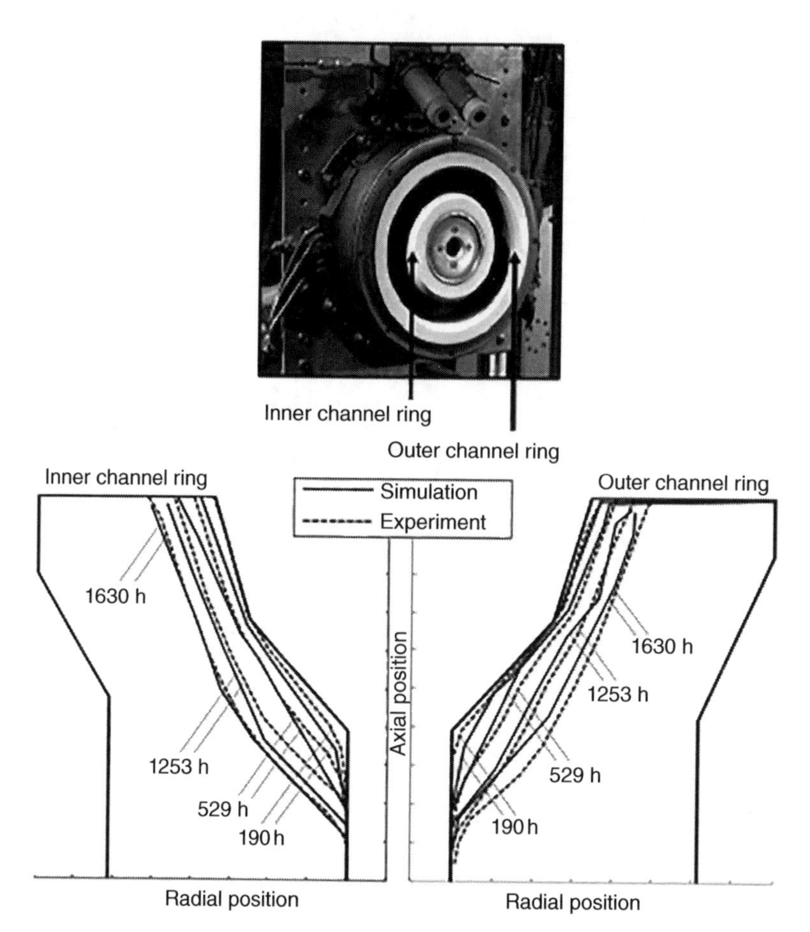

Figure 7-32 Top: Photograph of a demonstration model thruster designated SPT-140 DM4 showing the inner and outer channel rings (*Source:* [137]/California Institute of Technology). Bottom: Comparison of wear profiles between numerical simulations with Hall2De and measurements made at different times during a wear test in a vacuum facility (*Source:* Adapted from Lopez Ortega et al. [113]).

the SPT-140 are shown in Fig. 7-32 (bottom). Figure 7-32 (top) shows a photograph of a demonstration model of the SPT-140 showing the inner and outer channel rings [137]. The combination of wear tests and simulations validated the thruster life for the larger power range and the mission duration [138]. The simulations were also used in support of thruster performance modeling and the determination of thruster swirl torque [139]. The latter was particularly important because it drove the system and operation designs in ways that Earth-orbiting missions do not.

An area of continuing investigation in conventional Hall thruster erosion is the appearance of scallops or striations on the eroded wall azimuthally around the axis of the thruster. Figure 7-33 shows the wall erosion observed around the channel in STP-100 [140] and PPS-1350 [141] wear tests. Similar patterns were observed after the >10,400 h wear test of the BPT-4000 Hall thruster [142]. Similar patterns with different scalloping period around the channel have also been observed in other life tests. The cause of this structure is still unexplained.

It is possible to develop some simple scaling rules for conventional Hall thruster discharge wall erosion in the magnetized plasma region near the exit plane. It was estimated in Section 7.3.4 above

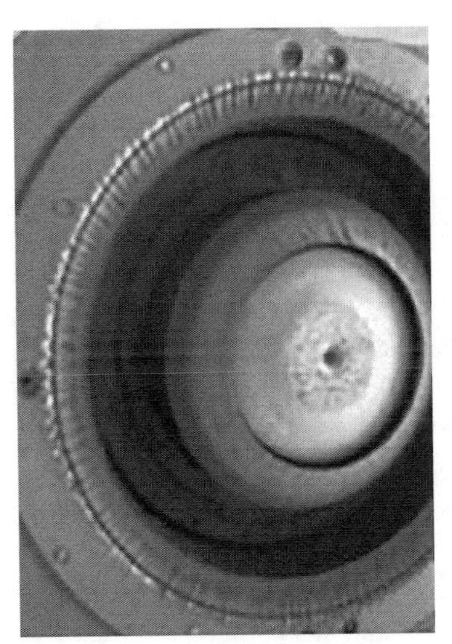

Figure 7-33 Photograph of the SPT-100 (left) (*Source:* [140]/ALCATEL SPACE) and the PPS-1350 (right) (*Source:* [141]/with courtesy of Safran), after their respective wear testing showing discharge channel edge scallop erosion.

that the ion flux to the wall in dielectric-wall Hall thrusters was about 10% of the beam current. It can be assumed that the energy of the ion flux to the wall is related to the beam energy, which is proportional to the discharge voltage. An examination of Fig. 7-30 shows that the sputtering yield above threshold is essentially a linear function of the ion energy. The erosion rate in Eq. 7.6-1 then becomes

$$\dot{\mathcal{R}} \propto K \frac{I_b}{A_w} V_d = K \frac{I_d V_d}{\eta_b A_w}, \tag{7.6-3}$$

where K is a constant, A_w is the wall area and Eq. 7.3-10 has been used for the beam current efficiency. Equation 7.6-3 shows that the erosion rate of the thruster wall is proportional to the power density in the accelerator channel [5]. This indicates that larger Hall thrusters are required to increase the power for a given operation time as determined by the allowable erosion of the insulator wall thickness. A good rule-of-thumb for the relationship of operation-time over a reasonable throttle range of a given Hall thruster design is

$$\text{Power} \times (\text{Operation Time}) \approx \text{constant}.$$

Over a limited range, the thrust from a Hall thruster is proportional to the discharge power, and so

$$\text{Thrust} \times (\text{Operation Time}) \approx \text{constant}.$$

This suggests that the total impulse is essentially a constant for a given thruster design. Therefore, operating at lower thrust in throttled mission profiles will result in longer thruster operation time. However, if the throttling is too deep, the thruster performance will degrade (requiring higher input power to produce a given thrust) and the relationship above is no longer valid. Hall thruster throttle ranges of over 10:1 have been demonstrated with good performance, depending on the thruster design.

Finally, the life of TAL thrusters has not been as extensively investigated as the Russian SPT thrusters. The erosion of the channel guard rings has been attributed to be the primary life limiting mechanism [35], and alternative materials were suggested to extend the thruster life by reducing the sputtering yield. Since the wall/guard-ring is biased at cathode potential, the incident ion energy along the wall depends on the potential profile in the thruster channel and past the exit plane. This certainly influenced the selection of the TAL anode placement and the design of the anode/channel region to minimize the ion energy (and flux) to the walls. The dielectric-wall Hall thrusters limited the ion energy to the floating potential ($\approx 6T_e$ for xenon) for wall materials with very low secondary electron yield, and to lower energies with materials that have secondary electron yields approaching or exceeding one at the electron temperatures of typical operation. The sheath potential at the wall is likely on the order of $3T_e \approx 0.3V_d$ because of space charge and non-Maxwellian electron distribution function effects. However, the lower sheath potential at the wall increases the electron flux, which results in increased power loading at the wall.

The wall material selection, therefore, is a tradeoff between efficiency and life. Dielectric walls reduce the bombarding ion energy of the wall at the expense of higher electron fluxes and higher power loading. Metallic-wall Hall thrusters have higher ion energies to the wall and therefore sputter-erosion life issues, and so they have to compensate with geometry changes to obtain the desired life. This results in higher heat fluxes to the anode, which dominates the TAL efficiency. An increase in the power of both types of thrusters also requires increases in the thruster size to obtain the same or longer lifetimes. Therefore, Hall thruster design, like ion thruster design, is a tradeoff between performance and life.

Problems

7.1 You want to design an experimental Hall thruster to operate from 100 to 800 V and from 100 Gauss to 300 Gauss. Assuming that the electron temperature is always about 10% of the discharge voltage, what are the minimum and maximum lengths of the magnetized region in the channel to have a factor of five margin against electron and ion orbit limits. Neglect collisions.

7.2 Derive Equation 7.3-42.

7.3 A Hall thruster has a plasma channel with a 15 cm outer diameter and a 10 cm inner diameter. Measurements made on the thruster indicate that the xenon plasma density in the channel is 5×10^{17} ions per m^3, the electron temperature T_e is 20 eV, and the radial magnetic field B_r is 200 Gauss (0.02 Tesla). If the thruster is operated at a discharge voltage of 300 V:

a) What is the beam power?
b) What is the electron Larmor radius r_L?
c) What is the electron Hall parameter Ω_e?
d) If the thrust correction factor $\gamma = 0.9$ and the mass utilization efficiency $\eta_m = 0.8$, what is the thrust and Isp?
e) What is the Hall current?

7.4 A xenon Hall thruster has boron nitride walls with a linearly varying secondary electron yield with a value of 0.5 at zero electron energy and 2 for an electron energy of 100 eV.

a) What is the equation for the secondary electron yield in terms of the electron energy?
b) Find the equation for the secondary electron yield for a Maxwellian distribution of electron energies (hint, use Eq. C-5) in terms of the electron temperature T_e.
c) What is the electron temperature at which the electron flow to the wall is space charge limited?
d) Assuming $n_e v_B / n_i v_i = 0.5$, what is the maximum sheath potential for non-space charge limited flow (T_e less than the value found in part b)?

7.5 Assume that all the ions in a Hall thruster are produced by the Hall current ionizing the neutral gas in the channel.
a) Neglecting the ion current to the wall as small so that all the ions produced become beam ions, what is the ratio of the Hall current to the beam current if the average electron temperature is 25 eV? (hint: write the ion production rate in terms of the Hall current and use Appendix E for ionization and excitation collision cross sections).
b) For a xenon ion thruster with a mean radius of 9 cm, a radial magnetic field of 150 G and a discharge voltage of 300 V, what is the ratio of the Hall current to the beam current?

7.6 A xenon Hall thruster has a channel outside diameter of 10 cm and a channel width of 3.5 cm with $BNSiO_2$ walls. Assume a plasma density of 2×10^{17} m^{-3} and an electron temperature of 20 eV in the channel with the majority of the plasma in contact with 1 cm of the wall axially.
a) What is the electron current to the wall?
b) What is the net electron current to the wall?
c) What is the power deposited on the wall associated with this electron current?
d) What is the power deposited on the wall associated with ion current?

7.7 Assume that the thruster in Problem 7.6 has alumina walls and produces 3.5 A of beam current at 400 V with an electron temperature in the channel of 15 eV. The thruster also has a beam current utilization efficiency $\eta_b = 0.5$.
a) What is the power into the discharge?
b) What is the total power into the alumina walls for a contact length of L-2 cm?
c) Assuming that the electron temperature at the anode is 5 eV, the mass utilization efficiency is 90% and the thrust correction factor $\gamma = 0.9$, and neglecting all other power loss channels, what is the thruster efficiency?
d) For a beam voltage utilization efficiency of 0.9, how much thrust and Isp is produced?

7.8 The electron current to the anode in a Hall thruster can be estimated from the perpendicular electron flux diffusing through the plasma channel.
a) Neglecting the pressure gradient terms, derive an expression for the current to the anode in terms of the collision frequency in the channel plasma.
b) For the thruster in Problem 7.7 with a transverse magnetic field of 150 G and an axial electric field of 3×0^4 V/m, what is the transverse electron current if only classic electron-ion collisions are considered?
c) The effective wall collision frequency can be estimated as the electron current to the wall times the secondary electron yield and divided by the total number of particles in the plasma ($\nu_w = \gamma I_{ew}/N$, where N is approximately the plasma density times the channel cross sectional area times the plasma length L). Derive an expression for the anode current because of the electron-wall collisions in terms of the electron current to the wall.

d) What is the total transverse electron current for this thruster example using $L = 2$ cm for the bulk of the plasma density?
e) If the walls are made of $BNSiO_2$, what is the anode current? Why does it depend so strongly on the wall material?

7.9 Calculate the power lost to the wall in a xenon TAL thruster with stainless steel walls that has a plasma density at the sheath edge of 2×10^{17} m^{-3} and an electron temperature of 20 eV. The channel is 12 cm outside diameter, 8 cm inside diameter and 0.5 cm long. Which power loss channel (ions or electrons) is larger?

7.10 The life of a TAL thruster is limited primarily by the ion sputtering of the metallic guard rings next to the thruster exit. Assume a TAL has a plasma density near the wall of 10^{17}m^{-3} and an electron temperature of 25 eV.
a) For stainless steel walls, what is the ion current density to the walls (the guard rings) and the sheath potential?
b) Assuming that the stainless-steel sputtering yield is about 0.1 atoms per incident ion at the sheath voltage found in (a), what is the life in hours of the TAL if 2 mm thickness of the stainless-steel guard ring material can be eroded away?
c) Assume that the wall material has been changed to graphite with a secondary electron yield of about 0.5. What is the sheath potential at the wall?
d) Assuming that the graphite sputtering yield is about 5×10^{-3} atoms per incident ion at the sheath voltage found in (c), what is the life in hours of the TAL if 1 mm thickness of the graphite guard ring material can be eroded away?

References

1 S. D. Grishin and L. V. Leskov, *Electrical Rocket Engines of Space Vehicles*, Moscow: Mashinostroyeniye Publishing House, 1989. (in Russian).

2 H. R. Kaufman, "Technology of Closed-Drift Thrusters," *AIAA Journal*, 78–87, vol. 23, no. 1, 1985.

3 A. I. Morozov and V. V. Savelyev, "Fundamentals of Stationary Plasma Thruster Theory," in *Reviews of Plasma Physics*, vol. 21, edited by B. B. Kadomtsev and V. D. Shafranov, Boston: Springer, pp. 203–391, 2000.

4 V. V. Zhurin, H. R. Kaufman, and R. S. Robinson, "Physics of Closed Drift Thrusters," *Plasma Sources Science T*, R1–R20, vol. 8, no. 1, 1999; doi:10.1088/0963-0252/8/1/021.

5 V. Kim, "Main Physical Features and Processes Determining the Performance of Stationary Plasma Thrusters," *AIAA Journal of Propulsion and Power*, R1–R20, vol. 8, no. 1, 1999.

6 J. R. Brophy, J. W. Barnett, J. M. Sankovic, and D. A. Barnhart, "Performance of the Stationary Plasma Thruster: SPT-100," AIAA-92-3155, Presented at the 28th AIAA Joint Propulsion Conference, Nashville, July 6–8 1992.

7 B. A. Arhipov, L. Z. Krochak, S. S. Kudriavcev, V. M. Murashko, and T. Randolph, "Investigation of the Stationary Plasma Thruster (SPT-100) Characteristics and Thermal Maps at the Raised Discharge Power," AIAA-98-3791, Presented at the 34th AIAA Joint Propulsion Conference, Cleveland, July 13–15, 1998.

8 A. S. Bober, V. P. Kim, A. S. Koroteev, L. Latyshev, A. I. Morozov, G. A. Popov, Y. P. Rylov, and V. V. Zhurin, "Status of Works on Electrical Thrusters in the USSR," IEPC-91-003, Presented at the 22nd International Electric Propulsion Conference, Viareggio, October 14–17, 1991.

9 V. Kim, G. Popov, B. Arkhipov, V. Murashko, O. Gorshkov, A. Koroteyev, V. Garkush, A. Semenkin, and S. Tverdokhlebov, "Electric Propulsion Activity in Russia," IEPC-2001-005, Presented at the 27th International Electric Propulsion Conference, Pasadena, October 15–19, 2001.

10 L. B. King, "A Re-examination of Electron Motion in Hall Thruster Fields," IEPC-2005-258, Presented at the 29th International Electric Propulsion Conference, Princeton, October 31–November 4, 2005.

11 R. R. Hofer, R. S. Jankovsky, and A. D. Gallimore, "High-Specific Impulse Hall Thrusters, Part 1: Influence of Current Density and Magnetic Field," *AIAA Journal of Propulsion and Power*, vol. 22, no. 4, pp. 721–731, 2006.

12 R. R. Hofer and A. D. Gallimore, "High-Specific Impulse Hall Thrusters, Part 2: Efficiency Analysis," *AIAA Journal of Propulsion and Power*, vol. 22, no. 4, pp. 732–740, 2006.

13 R. R. Hofer, "Development and Characterization of High-Efficiency, High-Specific Impulse Xenon Hall Thrusters," PhD. Aerospace Engineering, University of Michigan, 2004.

14 T. Andreussi, V. Giannetti, A. Leporini, M. M. Saravia, and M. Andrenucci, "Influence of the Magnetic Field Configuration on the Plasma Flow in Hall Thrusters," *Plasma Physics and Controlled Fusion* vol. 60, no. 014015, 2018; doi:10.1088/1361-6587/aa8c4d.

15 A. I. Morozov, Y. V. Esipchuk, A. M. Kapulkin, V. A. Nevrovskii, and V. A. Smirnov, "Effect of the Magnetic Field on a Closed-Electron-Drift Accelerator," Soviet Physics *Technical Physics*, vol. 17, no. 3, pp. 482–487, 1972.

16 Y. Raitses, J. Ashkenazy, and M. Guelman, "Propellant Utilization in Hall Thrusters," *AIAA Journal of Propulsion and Power*, vol. 14, no. 2, pp. 247–253, 1998.

17 L. A. Dorf, Y. F. Raitses, A. N. Smirnov, and N. J. Fisch, "Anode Fall Formation in a Hall Thruster," AIAA-2004-3779, Presented at the 40th AIAA Joint Propulsion Conference, Ft. Lauderdale, July 11–14, 2004.

18 L. Dorf, Y. Raitses, and N. J. Fisch, "Experimental Studies of Anode Sheath Phenomena in a Hall Thruster Discharge," *Journal of Applied Physics*, vol. 97, no. 103309, 2005.

19 L. Dorf, Y. Raitses, and N. J. Fisch, "Effect of Magnetic Field Profile on the Anode Fall in a Hall-Effect Thruster Discharge," *Physics of Plasmas*, vol. 13, no. 057104, 2006.

20 R. R. Hofer, P. Y. Peterson, A. D. Gallimore, and R. S. Jankovsky, "A High Specific Impulse Two-Stage Hall Thruster with Plasma Lens Focusing," IEPC-2001-036, Presented at the 27th International Electric Propulsion Conference, Pasadena, October 15–19, 2001.

21 J. A. Linnell and A. D. Gallimore. Krypton Performance Optimization in High-Voltage Hall Thrusters," *AIAA Journal of Propulsion and Power*, vol. 22, no. 4, pp. 921–925, 2006.

22 J. A. Linnell and A. D. Gallimore, "Efficiency Analysis of a Hall Thruster Operating with Krypton and Xenon Thrusters," *AIAA Journal of Propulsion and Power*, vol. 22, no. 6, pp. 1402–1412, 2006.

23 Y. Daren, D. Yongjie, and Z. Shi, "Improvement on the Scaling Theory of the Stationary Plasma Thruster," *AIAA Journal of Propulsion and Power*, vol. 21, no. 1, pp. 139–143, 2005.

24 M. Andrenucci, F. Battista, and P. Piliero, "Hall Thruster Scaling Methodology," IEPC-2005-187, Presented at the 29th International Electric Propulsion Conference, Princeton, October 31–November 4, 2005.

25 K. Dannenmayer and S. Mazouffre, "Elementary Scaling. Relations for Hall Effect Thrusters," *AIAA Journal of Propulsion and Power*, vol. 27, no. 1, pp. 236–245, 2011; doi:10.2514/1.48382.

26 A. I. Bugrov, N. Maslennikov, and A. I. Morozov, "Similarity Laws for the Global Properties of a Hall Accelerator," *Journal of Technical Physics*, vol. 61, no. 6, pp. 45–51, 1991.

27 V. Khayms and M. Martinez-Sanchez, "Design of a Miniaturized Hall Thruster for Microsatellites," AIAA-1996-3291, Presented at the 32nd AIAA Joint Propulsion Conference, Lake Buena Vista, July 1–3, 1996.

28 O. Gorshkov, "Russian Electric Propulsion Thruster Today," *Russian Space News*, vol. 61, no. 7, pp. 5–11, 1999.

29 A. Smirnov, Y. Raitses, and N. J. Fisch, "Parametric Investigations of Miniaturized Cylindrical and Annular Hall Thrusters," *Journal of Applied Physics*, vol. 92, no. 10, pp. 5673–5679, 2002.

30 L. B. King and A. D. Gallimore, "Ion Energy Diagnostics in the Plasma Exhaust Plume of a Hall Thruster," *AIAA Journal of Propulsion and Power*, vol. 85, no. 13, pp. 2481–2483, 2004.

31 J. E. Pollard, K. D. Diamant, V. Khayms, L. Werthman, D. Q. King, and K. H. de Grys, "Ion Flux, Energy and Charge State Measurements for the BPT-4000 Hall Thruster," AIAA-2001-3351, Presented at the 37th AIAA Joint Propulsion Conference, Salt Lake City, July 8–11, 2001.

32 D. L. Brown, C. W. Larson, and B. E. Beal, "Methodology and Historical Perspective of a Hall Thruster Efficiency Analysis," *AIAA Journal of Propulsion and Power*, vol. 25, no. 6, pp. 1163–1177, 2012, doi:10.2514/1.38092.

33 C. E. Garner, J. R. Brophy, J. E. Polk, S. Semenkin, V. Garkusha, S. Tverdokhlebov, and C. Marrese, "Experimental Evaluation of Russian Anode Layer Thrusters," AIAA-1994-3010, Presented at the 30th AIAA Joint Propulsion Conference, Indianapolis, June 27–29, 1994.

34 J. M. Fife, M. Martinez-Sanchez, and J. Szabo, "A Numerical Study of Low-Frequency Discharge Oscillations in Hall Thrusters," AIAA-1997-3052, Presented at the 33rd AIAA Joint Propulsion Conference, Seattle, July 6–9, 1997.

35 E. Y. Choueiri, "Fundamental Difference Between Two Variants of Hall Thrusters: SPT and TAL," *Physics of Plasmas*, vol. 8, no. 11, pp. 5025–5033, 2001.

36 N. Gascon, M. Dudeck, and S. Barral, "Wall Material Effects in Stationary Plasma Thrusters. I. Parametric Studies of an SPT-100," *Physics of Plasmas*, vol. 10, no. 10, pp. 4123–4136, 2003; doi:10.1063/1.1611880.

37 J. P. Bugeat and C. Koppel, "Development of a Second Generation of SPT," IEPC-95-035, Presented at the 24th International Electric Propulsion Conference, Moscow, September 19–23, 1995.

38 J. M. Fife and M. Martinez-Sanchez, "Comparison of Results from a Two-Dimensional Numerical SPT Model with Experiment," AIAA-96-3197, Presented at the 32nd AIAA Joint Propulsion Conference, Lake Buena Vista, July 1–3, 1996.

39 J. M. Haas, "Low-Perturbation Interrogation of the Internal and Near-Field Plasma Structure of a Hall Thruster Using a High-Speed Probe Positioning System," PhD. University of Michigan, 2001.

40 J. A. Linnell and A. D. Gallimore, "Internal Langmuir Probe Mapping of a Hall Thruster with Xenon and Krypton Propellant," AIAA-2006-4470, Presented at the 42nd AIAA Joint Propulsion Conference, Sacramento, July 9–12, 2006.

41 E. Ahedo, P. Martinez-Cerezo, and M. Martinez-Sanchez, "One-Dimensional Model of the Plasma Flow in a Hall Thruster," *Physics of Plasmas*, vol. 8, no. 6, pp. 3058–3068, 2001; doi:10.1063/1.1371519.

42 E. Ahedo, "Presheath/Sheath Model with Secondary Electron Emission from Two Parallel Walls," *Physics of Plasmas* vol. 9, no. 10, pp. 4340–4347, 2002; doi:10.1063/1.1503798.

43 G. D. Hobbs and J. A. Wesson, "Heat Flow Through a Langmuir Sheath in Presence of Electron Emission," *Plasma Physics*, vol. 9, no. 1, pp. 85–87, 1967; doi:10.1088/0032-1028/9/1/410.

44 V. Baranov, Y. Nazarenko, and V. Petrosov, "The Wear of the Channel Walls in Hall Thrusters," IEPC-2001-005, Presented at the 27th International Electric Propulsion Conference, Pasadena, October 15–19, 2001.

45 E. Ahedo, J. M. Gallardo, and M. Martinez-Sanchez, "Effects of the Radial Plasma-Wall Interaction on the Hall Thruster Discharge," *Physics of Plasmas*, vol. 10, no. 8, pp. 3397–3409, 2003; doi:10.1063/1.1584432.

46 J. M. Fife and S. Locke, “Influence of Channel Insulator Material on Hall Thruster Discharges: A Numerical Study,” AIAA-2001-1137, Presented at the 39th Aerospace Sciences Meeting, Reno, January 8–11, 2001.

47 I. D. Kaganovich, Y. Raitses, D. Sydorenko, and A. Smolyakov, “Kinetic Effects in Hall Thruster Discharge,” AIAA-2006-3323, Presented at the 42th AIAA Joint Propulsion Conference, Sacramento, July 9–12, 2006.

48 N. B. Meezan and M. A. Capelli, “Kinetic Study of Wall Collisions in a Coaxial Hall Discharge,” *Physical Review E*, vol. 66, no. 036401, 2002.

49 E. Ahedo and F. I. Parra, “Partial Trapping of Secondary Electron Emission in a Hall Thruster Plasma,” *Physics of Plasmas*, vol. 12, no. 073503, 2005; doi:10.1063/1.1943327.

50 E. Ahedo and V. DePablo, “Effects of Electron Secondary Emission and Partial Thermalization on a Hall Thruster,” AIAA-2006-4328, Presented at the 42th AIAA Joint Propulsion Conference, Sacramento (9–12 July 2006).

51 M. T. Domonkos, A. D. Gallimore, C. M. Marrese, and J. M. Haas, “Very-Near-Field Plume Investigation of the Anode Layer Thruster,” *AIAA Journal of Propulsion and Power*, vol. 16, pp. 91–98, 2000.

52 A. V. Semenkin, S. O. Tverdokhlebov, V. I. Garkusha, A. V. Kochergin, G. O. Chislov, B. V. Shumkin, A. V. Solodukhin, and L. E. Zakharenkov, “Operating Envelopes of Thrusters with Anode Layer,” IEPC-2001-013, Presented at the 27th International Electric Propulsion Conference, Pasadena, October 15–19, 2001.

53 D. T. Jacobson, R. S. Jankovsky, and R. K. Rawlin, “High Voltage TAL Performance,” AIAA-2001-3777, Presented at the 37th AIAA Joint Propulsion Conference, Salt Lake City, July 8–11, 2001.

54 A. E. Solodukhin, A. V. Semenkin, S. O. Tverdohlebov, and A. V. Kochergin, “Parameters of D-80 Anode Layer Thruster in One and Two-Stage Operation Modes,” IEPC-2001-032, Presented at the 27th International Electric Propulsion Conference, Pasadena, October 15–19, 2001.

55 E. Y. Choueiri, “Plasma Oscillations in Hall Thrusters,” *Physics of Plasmas*, vol. 8, no. 4, pp. 1411–1426, 2001; doi:10.1063/1.1354644.

56 A. Bishaev and V. Kim, “Local Plasma Properties in a Hall-Current Accelerator with an Extended Acceleration Zone,” *Zhurnal Tekhnicheskoi Fiziki*, vol. 48, pp. 1411–1426, 2001.

57 A. M. Bishaev, V. M. Gavryushin, A. I. Burgova, V. Kim, and V. K. Kharchvniko, “The Experimental Investigations of Physical Processes and Characteristics of Stationary Plasmathrusters with Closed Drift of Electrons,” RGC-EP-92-06, Presented at the 1st Russian-German Conference on Electric Propulsion Engines and their Technical Applications, Giessen, 1992.

58 J. P. Boeuf and L. Garrigues, “Low Frequency Oscillations in a Stationary Plasma Thruster,” *Journal of Applied Physics*, vol. 84, no. 7, pp. 3541–3554, 1998.

59 I. G. Mikellides, I. Katz, M. J. Mandell, and J. Snyder, “A 1-D Model of the Hall-Effect Thruster with an Exhaust Region,” AIAA-2001-3505, Presented at the 37th AIAA Joint Propulsion Conference, Salt Lake City, July 8–11, 2001.

60 R. Lobbia and A. D. Gallimore, “Two-Dimensional Time-Resolved Breathing Mode Plasma Fluctuation Variation with Hall Thruster Discharge Settings,” IEPC-2009-106, Presented at the 31st International Electric Propulsion Conference, Ann Arbor, September 20–24, 2009.

61 I. Kaganovich, A. Smolyakov, Y. Raitses, E. Ahedo, I. Mikellides, B. A. Jorns, F. Taccogna, R. Gueroult, S. Tsikata, A. Bourdon, J. P. Boeuf, M. Keidar, A. Tasman Powis, M. Merino, M. Cappelli, K. Hara, J. Carlsson, N. Fisch, P. Chabert, I. Schweigert, T. Lafleur, K. Matyash, A. Khrabrov, R. Boswell, and A. Fruchtman, “Physics of ExB Discharges Relevant to Plasma Propulsion and Similar Technologies,” *Physics of Plasmas*, vol. 27, no. 120601, 2020; doi:10.1063/5.0010135.

62 J. P. Boeuf, "Tutorial: Physics and Modeling of Hall Thrusters," *Journal of Applied Physics*, vol. 121, no. 011101, 2017; doi:10.1063/1.4972269.

63 K. Hara, "An Overview of Discharge Plasma Modeling for Hall Effect Thrusters," *Plasma Sources Sci T*, vol. 28, no. 4, 2019; doi:10.1088/1361-6595/ab0f70.

64 F. Taccogna and L. Garrigues, "Latest Progress in Hall Thrusters Plasma Modelling," *Reviews of Modern Plasma Physics*, vol. 3, no. 1, p. 12, 2019; doi:10.1007/s41614-019-0033-1.

65 E. Ahedo and D. Escobar, "Two-Region Model for Positive and Negative Plasma Sheaths and Its Application to Hall Thruster Metallic Anodes," *Physics of Plasmas*, vol. 15, no. 3, 2008; doi:10.1063/1.2888523.

66 A. Dominguez-Vazquez, F. Taccogna, P. Fajardo, and E. Ahedo, "Parametric Study of the Radial Plasma-Wall Interaction in a Hall Thruster," *Journal of Physics D: Applied Physics*, vol. 52, no. 47, 2019; doi:10.1088/1361-6463/ab3c7b.

67 Y. Raitses, I. D. Kaganovich, A. Khrabrov, D. Sydorenko, N. J. Fisch, and A. Smolyakov, "Effect of Secondary Electron Emission on Electron Cross-Field Current in E x B Discharges," *Ieee T Plasma Sci*, vol. 39, no. 4, pp. 995–1006, 2011; doi:10.1109/Tps.2011.2109403.

68 Y. Raitses, A. Smirnov, D. Staack, and N. J. Fisch, "Measurements of Secondary Electron Emission Effects in the Hall Thruster Discharge," *Physics of Plasmas*, vol. 13, no. 1, p. 014502, 2006; doi:10.1063/1.2162809.

69 I. G. Mikellides, I. Katz, R. R. Hofer, D. M. Goebel, K. de Grys, and A. Mathers, "Magnetic Shielding of the Channel Walls in a Hall Plasma Accelerator," *Physics of Plasmas*, vol. 18, no. 3, p. 033501, 2011; doi:10.1063/1.3551583.

70 I. G. Mikellides and I. Katz, "Numerical Simulations of Hall-Effect Plasma Accelerators on a Magnetic-Field-Aligned Mesh," *Physical Review E*, vol. 86, no. 4, p. 046703, 2012; doi:10.1103/Physreve.86.046703.

71 W. Huang, B. Drenkow, and A. D. Gallimore, "Laser-Induced Fluorescence of Singly-Charged Xenon Inside a 6-kW Hall Thruster," AIAA-2009-5355, Presented at the 45th AIAA Joint Propulsion Conference, Denver, August 2–5, 2009.

72 J. S. Miller, S. H. Pullins, D. J. Levandier, Y. Chiu, and R. A. Dressler, "Xenon Charge Exchange Cross Sections for Electrostaticthruster Models," *Journal of Applied Physics*, vol. 91, no. 3, pp. 984–991, 2002.

73 B. M. Reid and A. D. Gallimore, "Langmuir Probe Measurements in the Discharge Channel of a 6-kW Hall Thruster," AIAA-2008-4920, Presented at the 44th AIAA Joint Propulsion Conference, Hartford, July 21–23, 2008.

74 A. Lopez Ortega and I. G. Mikellides, "A new cell-centered implicit numerical scheme for ions in the 2-D axisymmetric code Hall2De," AIAA-2014-3621, Presented at the 50th AIAA Joint Propulsion Conference, Cleveland, July 28–30, 2014.

75 K. Hara, I. D. Boyd, and V. I. Kolobov, "One-Dimensional Hybrid-Direct Kinetic Simulation of the Discharge Plasma in a Hall Thruster," *Physics of Plasmas*, vol. 19, no. 11, 2012; doi:10.1063/1.4768430.

76 E. Fernandez, M. Cappelli, and K. Mahesh, "2D Simulations of Hall Thrusters," *Center for Turbulence Research (CTR) Annual Research Briefs*. [Online]. Available: http://www.stanford.edu/group/ctr/ResBriefs/ARB98.html, 1998

77 J. M. Fife, "Hybrid-PIC Modeling and Electrostatic Probe Survey of Hall Thrusters," PhD. Massachusetts Institute of Technology, 1998.

78 G. J. M. Hagelaar, J. Bareilles, L. Garrigues, and J. P. Boeuf, "Two-Dimensional Model of a Stationary Plasma Thruster," *Journal of Applied Physics*, vol. 91, no. 9, pp. 5592–5598, 2002; doi:10.1063/1.1465125.

79 I. Katz and I. G. Mikellides, "Neutral Gas Free Molecular Flow Algorithm Including Ionization and Walls for Use in Plasma Simulations," *Journal of Computational Physics*, vol. 230, no. 4, pp. 1454–1464, 2011; doi:10.1016/j.jcp.2010.11.013.

80 A. Morozov, "Plasma Accelerators," Plasma Accelerators, Translated into English from the Monograph Plazmennye Uskoriteli, *USSR*, 1973, pp. 5–15.

81 A. I. Morozov and I. V. Melikov, "Similarity of Processes in Plasma Accelerator with a Closed Electron Drift in Presence of Ionization," *Zhurnal Tekhnicheskoi Fiziki*, vol. 44, no. 3, pp. 544–548, 1974.

82 B. I. Volkov, A. I. Morozov, A. G. Sveshnikov, and S. A. Iakunin, "Numerical Simulation of Ion Dynamics in Closed-Drift Systems," *Fizika Plazmy*, vol. 7, pp. 245–254, 1981, March-April.

83 A. I. Morozov and V. V. Savelyev, "Numerical Simulation of Plasma Flow Dynamics in SPT," IEPS 95-161, Presented at the 24th International Electric Propulsion Conference, Moscow, September 19–25, 1995.

84 J. C. Adam, A. Heron, and G. Laval, "Study of Stationary Plasma Thrusters Using Two-Dimensional Fully Kinetic Simulations," *Physics of Plasmas*, vol. 11, no. 1, pp. 295–305, January 2004; doi:10.1063/1.1632904.

85 P. Coche and L. Garrigues, "A Two-Dimensional (Azimuthal-Axial) Particle-in-Cell Model of a Hall Thruster," *Physics of Plasmas*, vol. 21, no. 2, 2014; doi:10.1063/1.4864625.

86 A. Héron and J. C. Adam, "Anomalous Conductivity in Hall Thrusters: Effects of the Non-Linear Coupling of the Electron-Cyclotron Drift Instability with Secondary Electron Emission of the Walls," *Physics of Plasmas*, vol. 20, no. 8, p. 082313, 2013; doi:10.1063/1.4818796.

87 T. Lafleur, S. D. Baalrud, and P. Chabert, "Theory for the Anomalous Electron Transport in Hall Effect Thrusters. II. Kinetic Model," *Physics of Plasmas*, vol. 23, no. 5, 2016; doi:10.1063/1.4948496.

88 T. Lafleur, S. D. Baalrud, and P. Chabert, "Theory for the Anomalous Electron Transport in Hall Effect Thrusters. I. Insights from Particle-in-Cell Simulations," *Physics of Plasmas*, vol. 23, no. 5, 2016; doi:10.1063/1.4948495.

89 S. Barral and E. Ahedo, "Low-Frequency Model of Breathing Oscillations in Hall Discharges," *Physical Review E*, vol. 79, no. 4, 2009; doi:10.1103/Physreve.79.046401.

90 O. Chapurin, A. I. Smolyakov, G. Hagelaar, and Y. Raitses, "On the Mechanism of Ionization Oscillations in Hall Thrusters," *Journal of Applied Physics*, vol. 129, no. 23, 2021; doi:10.1063/5.0049105.

91 V. Giannetti, M. M. Saravia, L. Leporini, S. Camarri, and T. Andreussi, "Numerical and Experimental Investigation of Longitudinal Oscillations in Hall Thrusters," *Aerospace-Basel*, vol. 8, no. 6, 2021; doi:10.3390/aerospace8060148.

92 K. Hara, "Non-Oscillatory Quasineutral Fluid Model of Cross-Field Discharge Plasmas," *Physics of Plasmas*, vol. 25, no. 12, 2018; doi:10.1063/1.5055750.

93 T. Lafleur, P. Chabert, and A. Bourdon, "The Origin of the Breathing Mode in Hall Thrusters and Its Stabilization," *Journal of Applied Physics*, vol. 130, no. 5, 2021; doi:10.1063/5.0057095.

94 K. Hara and I. G. Mikellides, "Characterization of Low Frequency Ionization Oscillations in Hall Thrusters Using a One-Dimensional Fluid Model," AIAA-2018-4904, Presented at the AIAA Joint Propulsion Conference 2018, Cincinnati (9–11 July 2018.

95 M. Hirakawa and Y. Arakawa, "Particle Simulation of Plasma Phenomena in Hall Thrusters," IEPC-1995-164, Presented at the 24th International Electric Propulsion Conference, Moscow, September 19–23, 1995.

96 J. M. Fife and M. Martinez-Sanchez, "Two-Dimensional Hybrid Particle-in-Cell (PIC) Modeling of Hall Thrusters," IEPC-1995-240, Presented at the 24th International Electric Propulsion Conference, Moscow, September 19–23, 1995.

97 M. Gamero-Castaño and I. Katz, "Estimation of Hall Thruster Erosion Using HPHall," IEPC-2005-303, Presented at the 29th International Electric Propulsion Conference, Princeton, October 31–November 4, 2005.

98 F. I. Parra, E. Ahedo, J. M. Fife, and M. Martinez-Sanchez, "A Two-Dimensional Hybrid Model of the Hall Thruster Discharge," *Journal of Applied Physics*, vol. 100, no. 2, 2006; doi:10.1063/1.2219165.

99 I. G. Mikellides and A. L. Ortega, "Challenges in the Development and Verification of First-Principles Models in Hall-Effect Thruster Simulations that Are Based on Anomalous Resistivity and Generalized Ohm's Law," *Plasma Sources Sci T*, vol. 28, no. 1, 2019; doi:10.1088/1361-6595/aae63b.

100 R. Marchand and M. Dumberry, "CARRE: A Quasi-Orthogonal Mesh Generator for 2D Edge Plasma Modelling," *Computer Physics Communications*, vol. 96, no. 2–3, pp. 232–246, 1996.

101 Z. Li, T. S. Hahm, W. W. Lee, W. M. Tang, and R. B. White, "Turbulent Transport Reduction by Zonal Flows: Massively Parallel Simulations," *Science*, vol. 281, no. 5384, pp. 1835–1837, 1998.

102 A. M. Dimits, "Fluid Simulations of Tokamak Turbulence in Quasiballooning Coordinates," *Physical Review E*, vol. 48, no. 5, pp. 4070–4079, 1993.

103 M. J. Lebrun, T. Tajima, M. G. Gray, G. Furnish, and W. Horton, "Toroidal Effects on Drift Wave Turbulence," *Physics of Fluids B: Plasma Physics*, vol. 5, no. 3, pp. 752–773, 1993.

104 A. Lopez Ortega, I. G. Mikellides, and V. Chaplin, "Numerical Simulations for the Assessment of Erosion in the 12.5-kW Hall Effect Rocket with Magnetic Shielding (HERMeS)," IEPC 2017-154, Presented at the 35th International Electric Propulsion Conference, Atlanta, October 2017.

105 A. Lopez Ortega, I. G. Mikellides, V. Chaplin, H. W. Huang, and J. D. Frieman, "Anomalous Ion Heating and Pole Erosion in the 12.5-kW Hall Effect Rocket with Magnetic Shielding (HERMeS)," AIAA-2020-3620, Presented at the 2020 AIAA Propulsion and Energy Forum, New Orleans, August 2020.

106 J. Pereles-Diaz, A. Dominguez-Vazquez, P. Fajardo, E. Ahedo, F. Faraji, M. Reza, and T. Andreussi, "Hybrid Plasma Simulations of a Magnetically Shielded Hall Thruster," *Journal of Applied Physics*, vol. 131, no. 10, 2022; doi:10.1063/5.0065220.

107 I. G. Mikellides, I. Katz, R. R. Hofer, and D. M. Goebel, "Magnetic Shielding of Walls from the Unmagnetized Ion Beam in a Hall Thruster," *Applied Physics Letters*, vol. 102, no. 2, 2013; doi:10.1063/1.4776192.

108 A. Lopez Ortega and I. G. Mikellides, "The Importance of the Cathode Plume and Its Interactions with the Ion Beam in Numerical Simulations of Hall Thrusters," *Physics of Plasmas*, vol. 23, no. 4, 2016; doi:10.1063/1.4947554.

109 R. R. Hofer, I. Katz, I. G. Mikellides, D. M. Goebel, K. K. Jameson, R. M. Sullivan, and L. K. Johnson, "Efficacy of Electron Mobility Models in Hybrid-PIC Hall Thruster Simulations," AIAA-2008-4924, Presented at the 44th AIAA Joint Propulsion Conference, Hartford, July 21–23, 2008.

110 M. K. Scharfe, N. Gascon, M. A. Cappelli, and E. Fernandez, "Comparison of Hybrid Hall Thruster Model to Experimental Measurements. *Physics of Plasmas*, vol. 13, no. 8, 2006; doi:10.1063/1.2336186.

111 E. Sommier, M. K. Scharfe, N. Gascon, M. A. Cappelli, and E. Fernandez, "Simulating Plasma-Induced Hall Thruster Wall Erosion with a Two-Dimensional Hybrid Model," *Plasma Science, IEEE Transactions*, vol. 35, no. 5, pp. 1379–1387, 2007.

112 A. Lopez Ortega, I. G. Mikellides, R. W. Conversano, R. Lobbia, and V. H. Chaplin, "Plasma simulations for the assessment of pole erosion in the magnetically shielded miniature Hall thruster (MaSMi)," IEPC-2019-281, Presented at the 36th International Electric Propulsion Conference, Vienna, September 15–19, 2019.

113 A. Lopez Ortega, I. G. Mikellides, V. H. Chaplin, J. S. Snyder, and G. Lenguito, "Facility Pressure Effects on a Hall Thruster with an External Cathode, I: Numerical Simulations," *Plasma Sources Science and Technology*, vol. 29, no. 035011, 2020; doi:10.1088/1361-6595/ab6c7e.

114 A. Lopez Ortega, I. G. Mikellides, M. J. Sekerak, and B. A. Jorns, "Plasma Simulations in 2-D (r-z) Geometry for the Assessment of Pole Erosion in a Magnetically Shielded Hall Thruster," *Journal of Applied Physics*, vol. 125, no. 033302, 2019; doi:10.1063/1.5077097.

115 I. Katz, V. H. Chaplin, and A. L. Ortega, "Particle-in-Cell Simulations of Hall Thruster Acceleration and Near Plume Regions," *Physics of Plasmas*, vol. 25, no. 123504, 2018; doi:10.1063/1.5054009.

116 T. Lafleur, S. D. Baalrud, and P. Chabert, "Characteristics and Transport Effects of the Electron Drift Instability in Hall-Effect Thrusters," *Plasma Sources Sci T*, vol. 26, no. 2, 2017; doi:10.1088/1361-6595/aa56e2.

117 F. Taccogna, R. Schneider, S. Longo, and M. Capitelli, "Kinetic Simulations of a Plasma Thruster," *Plasma Sources Science and Technology*, vol. 17, no. 2, p. 024003, 2008.

118 C. Vivien, L. Trevor, B. Zdeněk, B. Anne, and C. Pascal, "2D Particle-in-Cell Simulations of the Electron Driftinstability and Associated Anomalous Electron Transport in Hall-Effect Thrusters," *Plasma Sources Science and Technology*, vol. 26, no. 3, p. 034001, 2017.

119 K. Matyash, R. Schneider, S. Mazouffre, S. Tsikata, and L. Grimaud, "Rotating Spoke Instabilities in a Wall-Less Hall Thruster: Simulations," *Plasma Sources Sci T*, vol. 28, no. 4, 2019; doi:10.1088/1361-6595/ab1236.

120 F. Taccogna and P. Minelli, "Three-Dimensional Particle-in-Cell Model of Hall Thruster: The Discharge Channel," *Physics of Plasmas*, vol. 25, no. 061208, 2018.

121 D. W. Forslund, R. L. Morse, and C. W. Nielson, "Electron Cyclotron Drift Instability," *Physical Review Letters*, vol. 25, no. 18, 1970; doi:10.1103/PhysRevLett.25.1266.

122 J. Cavalier, N. Lemoine, G. Bonhomme, S. Tsikata, C. Honore, and D. Gresillon, "Hall Thruster Plasma Fluctuations Identified as the ExB Electron Drift Instability: Modeling and Fitting on Experimental Data," *Physics of Plasmas*, vol. 20, no. 8, 2013; doi:10.1063/1.4817743.

123 A. Ducrocq, J. C. Adam, A. Heron, and G. Laval, "High-Frequency Electron Drift Instability in the Cross-Field Configuration of Hall Thrusters," *Physics of Plasmas*, vol. 13, no. 10, 2006; doi:10.1063/1.2359718.

124 S. Tsikata, N. Lemoine, V. Pisarev, and D. M. Gresillon, "Dispersion Relations of Electron Density Fluctuations in a Hall Thruster Plasma, Observed by Collective Light Scattering," *Physics of Plasmas*, vol. 16, no. 033506, 2009; doi:10.1063/1.3093261.

125 S. Janhunen, A. Smolyakov, D. Sydorenko, M. Jimenez, I. Kaganovich, and Y. Raitses, "Evolution of the Electron Cyclotron Drift Instability in Two-Dimensions," *Physics of Plasmas*, vol. 25, no. 8, 2018; doi:10.1063/1.5033896.

126 J. B. Mcbride, E. Ott, J. P. Boris, and J. H. Orens, "Theory and Simulation of Turbulent Heating by Modified 2-Stream Instability," *Physics of Fluids*, vol. 15, no. 12, pp. 2367–2383, 1972; doi:10.1063/1.1693881.

127 R. C. Davidson and N. T. Gladd, "Anomalous Transport Properties Associated with Lower-Hybrid-Drift Instability," *Physics of Fluids*, vol. 18, no. 10, pp. 1327–1335, 1975; doi:10.1063/1.861021.

128 N. A. Krall and P. C. Liewer, "Low-Frequency Instabilities in Magnetic Pulses," *Physical Review A*, vol. 4, no. 5, pp. 2094–2103, 1971; doi:10.1103/PhysRevA.4.2094.

129 I. G. Mikellides and A. Lopez Ortega, "Growth of the Modified Two-Stream Instability in the Plume of a Magnetically Shielded Hall Thruster," *Physics of Plasmas*, vol. 27, no. 10, p. 100701, 2020; doi:10.1063/5.0020075.

130 I. G. Mikellides and A. Lopez Ortega, "Growth of the Lower Hybrid Drift Instability in the Plume of a Magnetically Shielded Hall Thruster," *Journal of Applied Physics*, vol. 129, no. 19, 2021; doi:10.1063/5.0048706.

131 D. Manzella, J. Yim, and I. Boyd, "Predicting hall thruster operational lifetime," AIAA-2004-3953, Presented at the 40th AIAA Joint Propulsion Conference, Ft. Lauderdale, July 11–14, 2004.

132 V. Abgaryan, H. Kaufman, V. Kim, D. Ovsyanko, I. Shkarban, A. Semenov, A. Sorokin, and V. Zhurin, "Calculation Analysis of the Erosion of the Discharge Chamber Walls and Their Contamination During Prolonged SPT Operation," AIAA-94-2859, Presented at the 30th AIAA Joint Propulsion Conference, Indianapolis, June 27–29, 1994.

133 Y. Garnier, V. Viel, J. F. Roussel, and J. Bernard, "Low Energy Xenon Ion Sputtering of Ceramics Investigated for Stationary Plasma Thrusters," *Journal of Vacuum Science and Technology A*, vol. 17, pp. 3246–3254, 1999; doi:10.1116/1.582050.

134 Y. Yamamura and H. Tawara, "Energy Dependence of Ion-Induced Sputtering from Monatomic Solids at Normal Incidence," *Atomic Data and Nuclear Data Tables*, vol. 62, no. 2, pp. 149–253, 1996; doi:10.1006/adnd.1996.0005.

135 S. K. Absalamov, V. Andrew, T. Colberi, M. Day, V. V. Egorov, R. U. Gnizdor, H. Kaufman, V. Kim, A. I. Korakin, K. N. Kozubsky, S. S. Kudravzev, U. V. Lebedev, G. A. Popov, and V. V. Zhurin, "Measurement of Plasma Parameters in the Stationary Plasma Thruster (SPT-100) Plume and Its Effect on Spacecraft Components," AIAA-1992-3156, Presented at the 28th AIAA Joint Propulsion Conference, Nashville, July 6–8, 1992.

136 D. Y. Oh, S. Collins, D. M. Goebel, W. Hart, T. Imken, K. Larson, J. Maxwell, M. Maxwell, C. A. Polanskey, J. S. Snyder, T. Weise, L. Elkins-Tanton, and P. Lord, "Preparation for Launch of the Psyche Spacecraft for NASA's Discovery Program," IEPC-2022-452, Presented at the 37th International Electric Propulsion Conference, Cambridge, June 19–23, 2022.

137 C. E. Garner, B. Jorns, R. R. Hofer, R. Liang, and J. Delgado, "Low-Power Operation and Plasma Characterization of a Qualification Model SPT-140 Hall Thruster," AIAA-2005-3720, Presented at the 51st AIAA Joint Propulsion Conference, Orlando, July 27–29, 2015.

138 J. S. Snyder, D. M. Goebel, V. Chaplin, H. A. Lopez Ortega, I. G. Mikellides, F. Aghazadeh, I. Johnson, T. Kerl, and G. Lenguito, "Electric propulsion for the psyche mission," IEPC-2019-244, Presented at the 36th International Electric Propulsion Conference, Vienna, September 15–19, 2019.

139 J. S. Snyder, V. H. Chaplin, D. M. Goebel, R. R. Hofer, A. L. Ortega, I. G. Mikellides, T. Kerl, G. Lenguito, F. Aghazadeh, and I. Johnson, "Electric Propulsion for the Psyche Mission: Development Activities and Status," AIAA-2020-3607, Presented at the AIAA Propulsion and Energy 2020 Forum, Virtual Event, August 24–28, 2020.

140 P. Garnero and O. Dulau, "Alcatel Space Plasma Propulsion Subsystem Qualification Status," IEPC-2003-131, Presented at the 28th International Electric Propulsion Conference, Toulouse, March 17–21, 2003.

141 B. Laurent, A. Rossi, M. Öberg, S. Zurbach, G. Largeau, P. Lasgorceix, D. Estublier, and C. Boniface,"High Throughput 1.5 kW Hall Thruster for Satcoms," IEPC-2019-274, Presented at the 36th International Electric Propulsion Conference, Vienna, September 15–20, 2019.

142 K. H. DeGrys, A. Mathers, B. Welander, and V. Khayms, "Demonstration of >10,400 Hours of Operation on 4.5kW Qualification Model Hall Thruster," AIAA-2010-6698, Presented at the 46th AIAA Joint Propulsion Conference, Nashville, July 25–28, 2010.

Chapter 8

Magnetically Shielded Hall Thrusters

8.1 Introduction

Erosion of the discharge chamber in Hall thrusters can expose its magnetic circuit to bombardment by energetic beam ions, which may lead to the failure of the engine. This was recognized as a potential limitation of Hall thrusters early in their history. Although propulsive performance dominated their development at that time, techniques to reduce or eliminate channel erosion were considered as early as the 1960s. In an extensive review of stationary plasma thrusters (SPT), Morozov and Savelyev stated [1]: ... *at the beginning of the 1960s magnetic force line equipotentialization became known, and the chosen geometry of force lines (convex toward the anode) provided repulsion of ions from the walls by the electric field, thus reducing the channel erosion.* Many advancements in the applied magnetic field design followed those early efforts, eventually leading to improvements in both thruster performance and wear (e.g. see [2–5]). However, channel erosion was never understood well enough to eliminate it or reduce it adequately to achieve the required thruster lifetime and retire the risk for deep-space science missions. It has been largely due to this risk that, as of the writing of this book, Hall thrusters have not yet flown beyond lunar orbit despite their enabling propulsive capabilities and demonstrated performance in Earth-orbiting station keeping and orbit raising applications.

In contrast to gridded ion thrusters in which the ion beam can be well controlled with the proper arrangement of the electrode apertures, the focusing of unmagnetized ions away from the discharge chamber walls in gridless devices like the Hall thruster is significantly more challenging. This is mainly because the induced electric field must conform to the generalized Ohm's Law. In conventional Hall thrusters, this allows for deviations from "magnetic force line equipotentialization" due to the finite electron pressure. Such deviations increase with higher electron temperature and can occur not only inside the sheath that forms near the channel walls, but also in the pre-sheath and in the plasma. Therefore, a component of the electric field along the applied magnetic field can be, and usually is, established that accelerates some high energy beam ions toward the walls causing sputtering.

As a result of this breakdown in the orthogonality of the electric and magnetic fields, a topology of magnetic field lines with convex curvature toward the anode cannot effectively control the electric field near surfaces if the near-wall lines are not also truly equipotential. This is true even in the more advanced "plasma lens" configurations that evolved from the older SPT-like designs. Morozov had proposed that, alternatively, such equipotentialization may be achieved either by virtue of the internal conductivity of the channel or through the use of special anular electrode holders [6]. Although

Fundamentals of Electric Propulsion, Second Edition. Dan M. Goebel, Ira Katz, and Ioannis G. Mikellides.

unconventional electrode arrangements can indeed alter the relevant plasma properties favorably (e.g. see [7]), no thruster design with or without atypical electrode configurations had demonstrated zero or near-zero channel erosion throughout the entirety of long wear or life tests. Since the inception of Hall thrusters over 60 years ago and through the early 2000s, efforts to eliminate channel erosion remained unsuccessful [8].

Around 2010, a new technique to protect the channel from ion bombardment was discovered. In doing so, wall erosion was eliminated as the primary failure mode of Hall thrusters, which extended their applicability to space missions far beyond lunar orbit. The events that led to the solution of this longstanding problem in Hall thrusters was a fortuitous combination of an unforeseen laboratory test result and the theoretical investigations that ensued to explain it [9] – a combination that is not at all uncommon in science and engineering breakthroughs. Ultimately, the basic principles of the technique were the outcome of theoretical arguments and physics-based numerical simulations [9, 10]. However, the significance of the laboratory tests that preceded [11] and followed [12] such derivations cannot be understated, for they instigated and ultimately validated the theory on which the technique is founded.

Termed *magnetic shielding* [10], the fundamental impetus behind this wall-erosion elimination technique is the achievement of near-ideal equipotentialization of the lines of force near the walls without the use of additional electrodes. By considering the isothermal properties of the magnetic field lines in these thrusters, it was recognized that this can be accomplished by a magnetic field topology that extends field lines adjacent to the walls deep into the anode region where plasma electrons are cold (~1–3 eV) [9, 10]. In doing so, not only is the contribution of the electron pressure to the parallel electric field marginalized, but the energy that ions can gain through the sheath is limited significantly. This ideal equipotentialization then allows for a near-total control of the electric field near the walls by the applied magnetic field because it preserves closely the orthogonality between them.

Magnetic shielding can increase the propellant throughput capability of Hall thrusters by at least tenfold for a wide range of power levels and thruster sizes. It has, therefore, revolutionized the way these devices are designed worldwide. Since 2010, there have been programs developing and investigating magnetically shielded Hall thrusters in several countries including France [13, 14], Italy [15, 16], China [17, 18], India [19], and Japan [20]. In the United States, there have been several R&D programs involving thrusters across a wide range of discharge powers, from 20 kW [21] down to the sub-kW level [22, 23]. One of the longest programs has been the development of the 600-V, 12.5-kW Hall Effect Rocket with Magnetic Shielding (HERMeS) [24], which began at NASA in 2012 and ultimately led to partnerships with the private sector where the thruster technology HERMeS was incorporated into the Advanced Electric Propulsion System (AEPS) [25]. The HERMeS program has produced an extensive body of experimental [26–28] and theoretical work [29–33] (and references therein) on magnetic shielding. The first flight of this technology is planned for 2025 on the Power and Propulsion Element of the lunar outpost Gateway, a part of NASA's Artemis program [34].

8.2 First Principles of Magnetic Shielding

We begin this section by revisiting a few elementary physics of Hall thrusters from Chapter 7 that will prove critical in magnetic shielding. A schematic illustrating the basic features of the device is shown in Fig. 8-1a. The beam is produced by the formation of an azimuthal electron current that

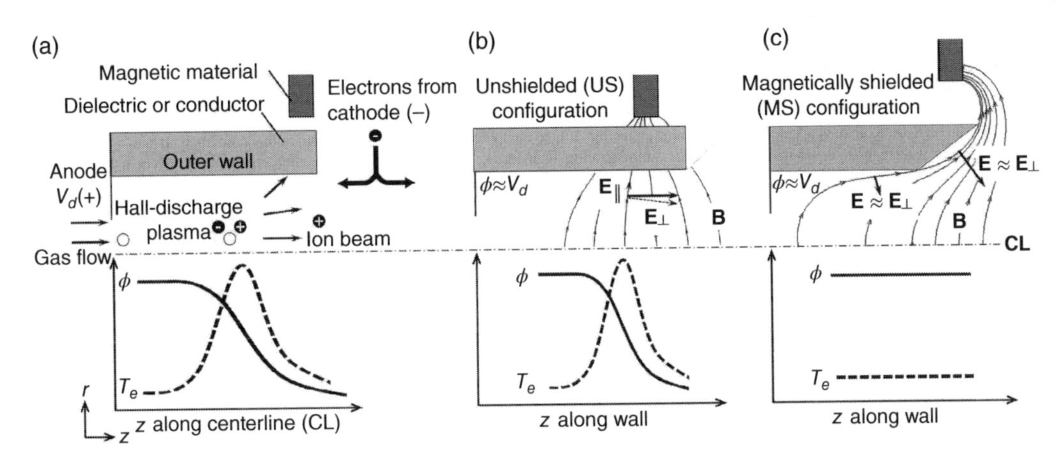

Figure 8-1 Schematics of the downstream portion of an annular channel containing a Hall discharge (top) and typical profiles of ϕ and T_e (bottom) established during ion acceleration. (a) Basic features of the accelerator and typical profiles along the channel centerline. (b) B-lines and profiles along the wall in an unshielded configuration. (c) B-lines and profiles along the wall in a magnetically shielded configuration (*Source:* [10]/AIP Publishing).

interacts with an applied radial magnetic flux density **B** to produce a largely axial force on the ions. The electron number density (n_e) is low enough that classical collisions in the azimuthal direction seldom impede their $\mathbf{E} \times \mathbf{B}$ drift, inducing a Hall current. Operation under these conditions implies a high Hall parameter for the electrons, $\Omega_e \equiv \omega_c/\nu_m >> 1$, where ω_c and ν_m are the electron cyclotron and total momentum-transfer collision frequencies, respectively. As the Hall current crosses **B**, the induced electric field **E** is in the direction perpendicular ($\perp$) to **B** and proportional to ~ $\eta \Omega_e^2 \mathbf{j}_{e\perp}$ where the electron current density and resistivity have been denoted by $\mathbf{j}_e$ and η, respectively. $q_i \mathbf{E}_\perp$ serves as the main force on the ions where q_i is the ion charge. Also, $\partial \mathbf{B}/\partial t = 0$ thus $\mathbf{E} = -\nabla\phi$, with ϕ being the plasma potential. The increased resistive heating of electrons in the region of high **E** also leads to an increase in the electron temperature, T_e. Typical ϕ and T_e profiles along the channel centerline are sketched in Fig. 8-1a.

Under these discharge conditions the resistance to the electron transport of heat and mass parallel (||) to **B** is much smaller (by ~Ω_e^2) than that in the $\perp$ direction. In the limit $\Omega_e \rightarrow \infty$, T_e is constant along the magnetic field lines

$$\nabla_\parallel T_e = 0 \tag{8.2-1}$$

and the electron momentum equation simplifies to

$$\mathbf{E}_\parallel = -T_e \nabla_\parallel \ln(n_e). \tag{8.2-2}$$

Equations 8.2-1 and 8.2-2 yield two well-known conditions along lines of force: $T_e = T_{eR}$ and $\phi = \phi_R + T_e \ln(n_e/n_{eR})$ where T_{eR}, ϕ_R and n_{eR} are constants of integration. Hence, although each line is isothermal it is not also equipotential, that is, $\mathbf{E}_\parallel \neq 0$.

Erosion of the channel walls occurs when ions bombard them with sufficient energy to sputter off material. The erosion rate is proportional to the product of the incident ion flux and the sputtering yield of the material. The latter is a strong function of the total ion energy, which consists of energy gains made in the plasma and those made inside the sheath. Henceforth, we shall use the terms "kinetic" and "sheath" to distinguish between the two energy contributions. For dielectric materials, the electric potential drop in the sheath is dependent on the electron temperature

[35]. In these sheaths, higher T_e typically implies higher sheath drop. A few representative **B**-lines in an SPT-like thruster channel are illustrated in Fig. 8-1b. In this configuration, the variation of ϕ and T_e near the walls is similar to that along the centerline because the lines are nearly radial. Consequently, the higher $\mathbf{E}_{\parallel}$ and T_e there can produce a flux of high-energy ions toward the material. For reasons that will become apparent shortly we designate this arrangement of channel geometry and magnetic field as an "unshielded (US)" configuration.

The above arguments then imply that if a magnetic field can by constructed such that $T_e \rightarrow 0$ and $\phi \rightarrow V_d$ near the channel walls, with V_d being the discharge voltage, the kinetic and sheath energies of an incident ion can be marginalized. Moreover, if the channel geometry relative to **B** is designed properly, **E** can be controlled to be both nearly perpendicular to the surface and large in magnitude, thereby accelerating ions away from walls and without loss of propulsive performance. This also reduces the wall-incident ion flux. A schematic of the arrangement that accomplishes all the above – hereinafter termed a magnetically shielded configuration – is depicted in Fig. 8-1c. The key principle behind magnetic shielding lies in the recognition that the electron pressure (right hand side of Eq. 8.2-2) forces **E** and **B** to no longer form an orthogonal set (Fig. 8-1b). Thus, a geometry of **B**-lines with convex curvature toward the anode (first proposed by Morozov [1] and references therein) cannot effectively control $\boldsymbol{E}$ near surfaces (and, in turn, the erosion) if the near-wall lines are not also truly equipotential. By contrast, the magnetically shielded configuration in Fig. 8-1c eliminates the contribution of the electron pressure by exploiting those **B**-lines that extend to the anode. Because these lines are associated with high ϕ_R and low T_{eR} the contribution of $T_e \ln(n_e)$ is marginalized.

8.3 The Protective Capabilities of Magnetic Shielding

The theoretical underpinnings of magnetic shielding were first derived by Mikellides et al. [9, 10] based on the insight revealed by 2-D numerical simulations of a commercial Hall thruster [9]. These were followed by additional simulations [36] and dedicated experiments [12] that validated the theory and demonstrated in the laboratory the predicted reductions of the erosion rates along the channel walls. We outline in this section some of the more salient parts of that work for the purpose of expanding further on the first principles of the technique, and for illustrating more quantitatively the protection it provides against ion sputtering.

8.3.1 Numerical Simulations

The first numerical simulations of magnetic shielding were performed with the 2-D axisymmetric code Hall2De developed by Mikellides and Katz at the Jet Propulsion Laboratory [37, 38]. The simulations were undertaken to explain never-before-seen erosion trends observed in a qualification life test (QLT) of a commercial Hall thruster [9, 11]. Specifically, the channel walls eroded in the beginning of the QLT at rates that are typical of other conventional (or "unshielded") thrusters but then, after about 5600 h, the walls reached a "zero-erosion" state. The test result was unprecedented and potentially very significant, but it also lacked a physics-based explanation. Therefore, the possibility that facility effects or other inconsequential processes were behind the observed wear trend could not be excluded. It was later shown by Hall2De simulations [36] and validation experiments [12], that the physics that led to this outcome were complex enough that they could not have been predicted by previous models or basic thruster design tools.

Hall2De was the only code during that time that solved an extensive system of equations governing the physics of Hall thrusters with complex magnetic field topologies near solid boundaries. It was, therefore, uniquely suited to elucidate the QLT result. The simulations revealed that as portions of the magnetic field slowly became exposed to the discharge by the recession of the channel material, several changes were triggered in the local plasma that led to the formation of an effective "shielding" of the channel walls from ion bombardment. Because all these changes were found to be driven by the magnitude and topology of the magnetic field there, the authors termed this wear-reducing mechanism "magnetic shielding" [9].

The significance of these findings prompted a focused proof-of-principle experimental investigation shortly thereafter, which established and validated the protective capabilities of magnetic shielding. The effort involved the modification of the channel geometry and magnetic field of an existing thruster, a 6-kW laboratory Hall thruster called the H6, to yield a magnetically shielded variant dubbed the H6MS. A main objective of the campaign was to show that much lower erosion rates could be achieved by the H6MS compared to the original, unshielded, version of the thruster, which was called the H6US. But it was also recognized that lower erosion rates alone would not be sufficient to demonstrate unambiguously that magnetic shielding was achieved in the laboratory. Proof of the magnetic shielding physics that actually led to the wear reductions would have to be established as well. Therefore, thruster testing was accompanied by a wide range of plasma and erosion diagnostics to allow for one-on-one comparisons with the plasma simulations by Hall2De.

Representative results from Hall2De simulations in the vicinity of the acceleration channel are compared in the H6US and H6MS configurations in Fig. 8-2. Comparisons that are pertinent to wall erosion also are plotted in Fig. 8-3. The effect of magnetic shielding on the plasma potential is evident in Fig. 8-2-top. Only a 4–15 V reduction was computed along the magnetically shielded diverging walls compared to a drop that was as high as 230 V along the unshielded inner wall near the channel exit. The kinetic energy of the ions was therefore reduced significantly in the magnetic shielding configuration as shown in Fig. 8-3a. If **B**-lines are isothermal then the lines that *graze* the corner formed by the cylindrical and diverging sections of the magnetically shielded channel must also be associated with low T_e because they extend deep into the acceleration channel where the electrons are cold (Fig. 8-1-right). Because of their critical role in magnetic shielding, these special lines of force in the overall magnetic field topology of the thruster are termed the "grazing lines" [9, 10]. Indeed, the comparison of the two configurations in Fig. 8-2-middle shows a significant reduction of T_e in these highly shielded regions of the walls. Because T_e is reduced, a decrease of the sheath fall is also achieved. In the downstream ~20% of the channel a reduction in the sheath energy of about 4–10 times that in the unshielded configuration (Fig. 8-3b) was computed. Finally, because $\mathbf{E}_{||}$ is practically eliminated near the walls of the magnetically shielded thruster, ion acceleration occurs mostly away from the walls. This leads to lower wall incident ion flux. Referring to Fig. 8-3c, the perpendicular component of the Xe^+ ion current density ($\mathbf{J}_{i\perp}$) near the channel exit was found to be 7.7 and 49.1 times lower at the magnetically shielded outer and inner walls, respectively. In regions where the magnetically shielded fluxes begin to become comparable to the unshielded fluxes, the ion energy was found to be below the energy threshold for sputtering. In fact, the total ion energy at the H6MS outer wall is below the threshold (25 V) for sputtering along the entire outer wall, and thus zero erosion was found.

8.3.2 Laboratory Experiments and Model Validation

The two H6 thruster configurations (with and without magnetic shielding) were operated with xenon at discharge voltage and power of 300 V and 6 kW, respectively, to assess the effect of

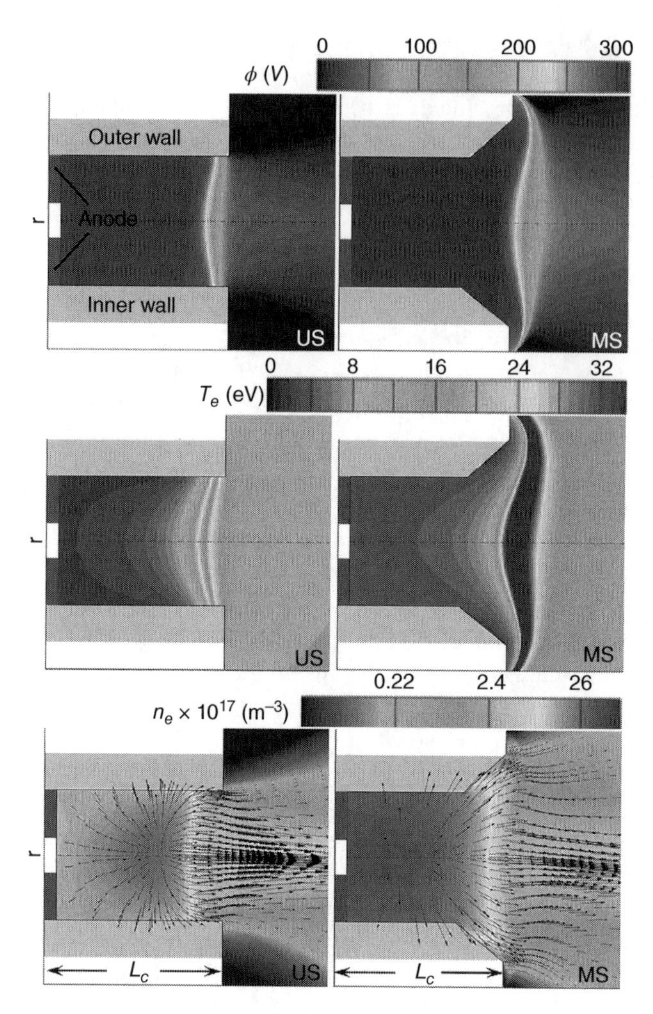

Figure 8-2 2-D numerical simulations of the Hall thruster plasma in the unshielded (left) and magnetically shielded (right) accelerator configurations. Top: Plasma potential (ϕ). Middle: Electron temperature (T_e). Bottom: Electron number density (n_e) overlaid by vectors of the current density of singly-charged xenon ions (Xe^+) (*Source:* [36]).

magnetic shielding on the channel wall erosion. The channel walls in both these thruster configurations were made of boron nitride. Sixteen different diagnostics were employed in the first proof-of-principle magnetic shielding experiments with the H6 Hall thruster [12]. These diagnostics are listed in Table 8-1 to illustrate all of the parameters that needed to be measured or determined to validate the physics of magnetic shielding described by the simulations. The most revealing results were likely those from flush-mounted probes imbedded in the channel walls. The objective with these probes was to measure the plasma properties near the magnetically shielded walls and compare them with those from the unshielded version of the thruster. Profilometry measurements of the channel erosion were also critical to demonstrate the erosion reductions. We summarize in this section some of these measurements and discuss their significance in the validation of the magnetic shielding theory.

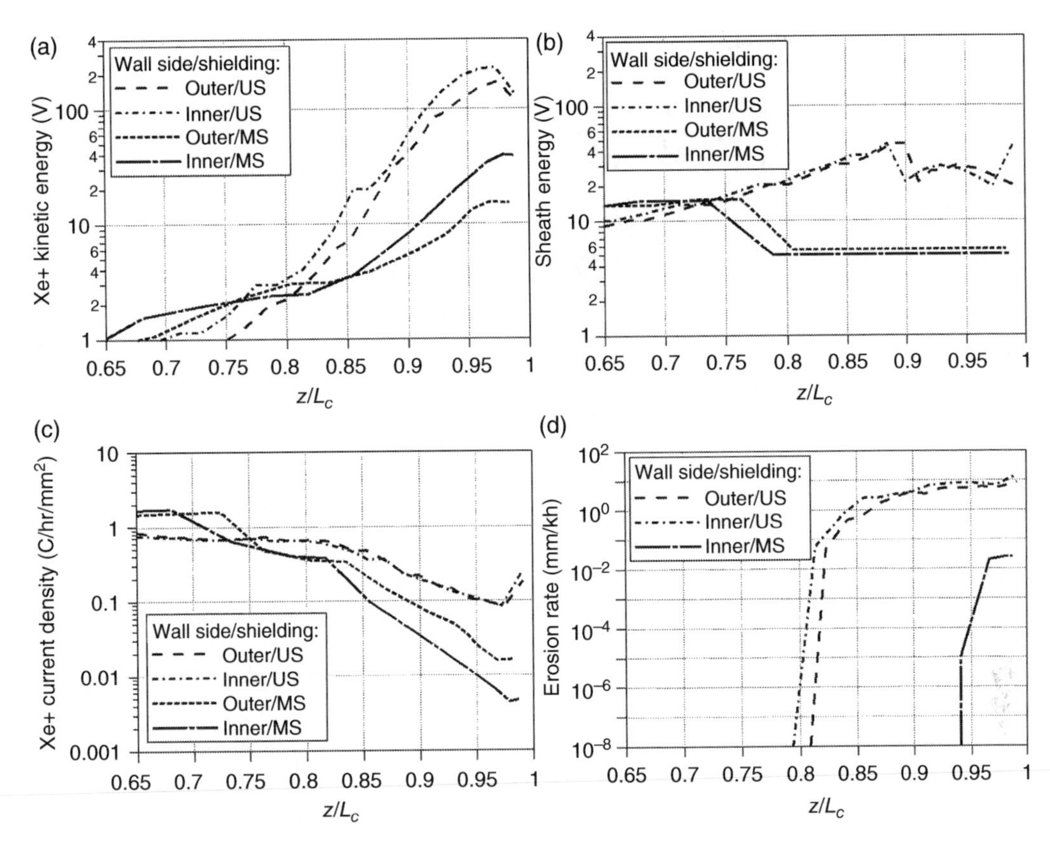

Figure 8-3 Computed plasma properties along the channel walls from 2-D numerical simulations of the H6US and H6MS. (a) impact kinetic energy of Xe^+. (b) Sheath energy of Xe^+. (c) Current density of incident Xe^+. (d) Total erosion rate accounting for all three ion charges states (*Source:* [36]).

Table 8-1 Diagnostics used to validate magnetic shielding.

	Diagnostic	Purpose
1	Thrust stand	Direct thrust measurement
2	Calibrated flow meters	Flow rate determining the Isp
3	Faraday probes (near and far-field in plume)	Beam ion flux and profile
4	Emissive probes (near and far-field in plume)	Plume plasma potential
5	RPA in the plume (near and far-field)	Beam ion energy
6	Mass analyzer (near and far-field in plume)	Beam species fractions
7	Axial reciprocating Langmuir probes	Thruster ϕ and T_e profiles
8	Flush-mounted probes embedded in BN walls	Thruster ϕ and T_e profiles
9	Discharge current probes	Anode current oscillations
10	Coordinate measuring machine (CMM)	Discharge chamber erosion rate
11	Thermocouples embedded in the chamber walls	Thruster body temperature
12	Thermal camera	Surface temperature profiles
13	Digital camera	Plasma location in channel
14	Residual gas analyzer (RGA)	Facility background gases
15	Quartz crystal microbalance (QCM)	Carbon back-sputter rate
16	3-axis gauss meter	Magnetic field profiles

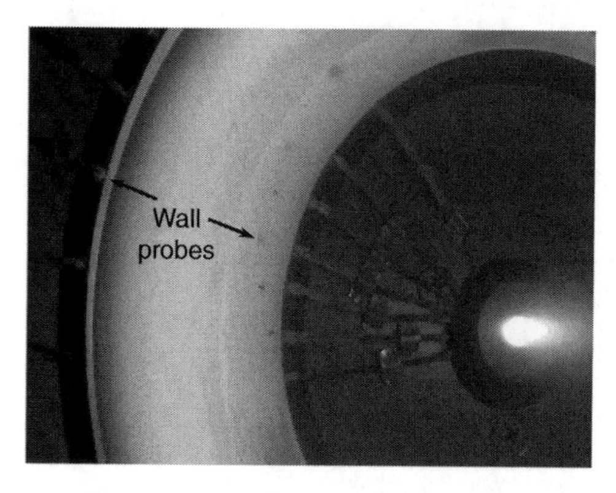

Figure 8-4 Photograph taken during operation of the magnetically shielded 6-kW laboratory Hall thruster H6MS showing flush-mounted probes installed at the inner and outer channel walls (*Source:* [12]/AIP Publishing LLC).

To measure the plasma parameters at the discharge chamber wall in the acceleration region, an array of flat tungsten probes was embedded in the inner and outer rings of the H6MS, as shown in Fig. 8-4. The H6US was also outfitted with these probes to allow for direct comparisons with the H6MS and the simulations. A representative set of comparisons between computed and simulated values of ϕ at the inner wall (top) and T_e at the outer wall (bottom) is shown in Fig. 8-5. The full set of probe results at both walls is reported in [36]. These comparisons confirmed the magnetic shielding theory outlined in the previous sections. Specifically, the plasma potential ϕ was found to be nearly constant along the magnetically shielded walls and near the anode potential, in contrast to a decrease in the H6US configuration by more than 150 V. The electron temperature in the magnetically shielded thruster remained at near-anode values along the wall in the acceleration region. Of particular importance is the agreement at the z/L_c location where the plasma potential begins to decrease appreciably from the discharge voltage since in the unshielded configurations it is beyond this point that ions near the walls begin to acquire significant kinetic energy. This is also the region where ions can gain additional energy from the sheath, as suggested by the rise of T_e in this location (Fig. 8-5-bottom). Hence, it was expected that this region of the walls would exhibit the highest erosion rates. Indeed, visual inspection of the channel rings after the thruster was tested in a vacuum facility confirmed this expectation [12]. A photograph of the H6US inner ring after 28 h of operation is provided in Fig. 8-7-left. The photograph shows clearly a texturing of the wall material, which was caused by the impingement of high-energy ions on the surface. Along this "erosion band" the wear rate was found to exceed the carbon deposition rate from the vacuum facility. It is worthwhile noting that the extent of the band was measured to be $z/L_c \approx 0.1$. This length scale is in-agreement with the profiles of Fig. 8-3 where the ion energy gains were found to exceed the sputtering yield energy threshold. This threshold was estimated to be in the range of 25–50 V [36] for the H6 channel material. The visual observation of the erosion zone was also found to be consistent with profilometry measurements of the erosion, as will be shown later in this section.

The observed trends in T_e were found to be in agreement with the theoretical predictions of a low electron temperature along the discharge chamber wall as well (Fig. 8-5 bottom). The electron temperature was measured to be about ~2.5–3 times lower along the H6MS walls compared to those in the H6US. Another prediction from the simulations confirmed by these experiments was that the location of the acceleration region should move slightly downstream in the magnetically shielded configuration. This was first validated by utilizing fast scanning probe measurements of the axial potential and electron temperature profiles on the discharge channel centerline [12] plotted in Fig. 8-6. The plasma potential data was obtained using an emissive probe, and the electron temperature data was from a Langmuir probe installed on the scanning probe tip. The electron temperatures are observed to be higher in the magnetically shielded configuration. Both the plasma

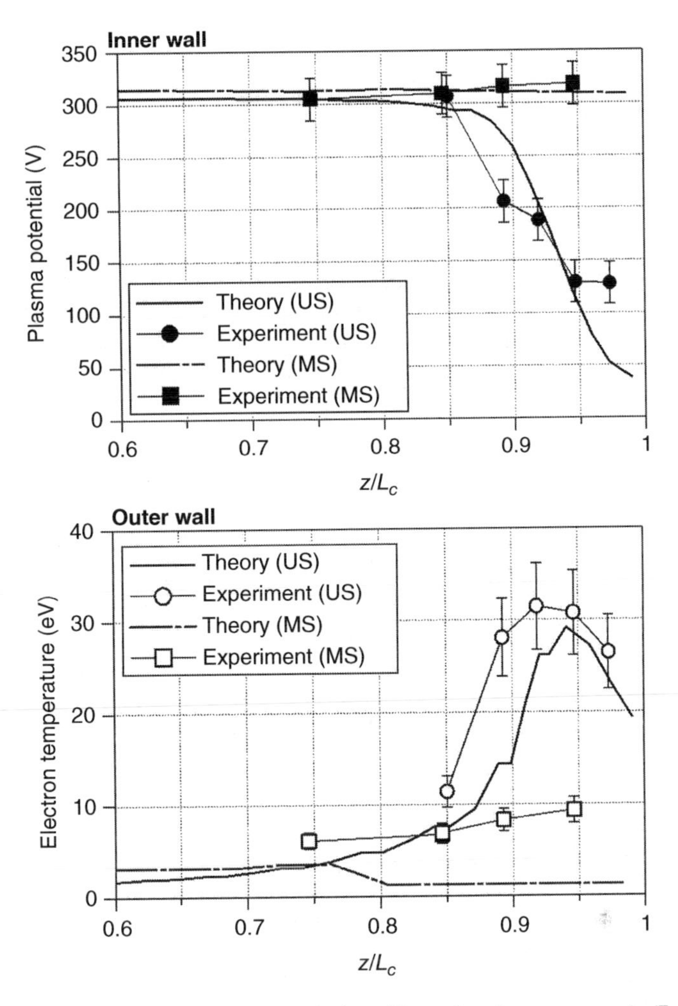

Figure 8-5 Comparisons between numerical simulations (Theory) and measurements (Experiment) from flush-mounted wall probes. The comparisons validate for magnetic shielding: (i) plasma potential values are near the discharge voltage and, (ii) electron temperatures at near-anode values (*Source:* [10]/AIP Publishing).

potential profile and the location of the peak electron temperature show that the acceleration region in the magnetic shielding case shifted downstream by some fraction of a centimeter. Subsequent non-intrusive measurements of the potential profile by laser-induced fluorescence (LIF) confirmed this plasma movement trend [39], and showed that the acceleration region was located 2- 3 mm downstream of the location suggested by the probe data due to perturbation of the plasma position by the probe [40].

Inspection of the H6MS rings after operation of the thruster for tens of hours supported the conclusion that the low temperatures and constant potential along the discharge chamber wall results in low erosion rates. A photograph of the inner magnetically shielded wall after 19 h of operation is provided in Fig. 8-7-right and shows clearly that the region of net carbon deposition spanned the entire wall, including the chamfer. To demonstrate the significance of the magnetic field topology

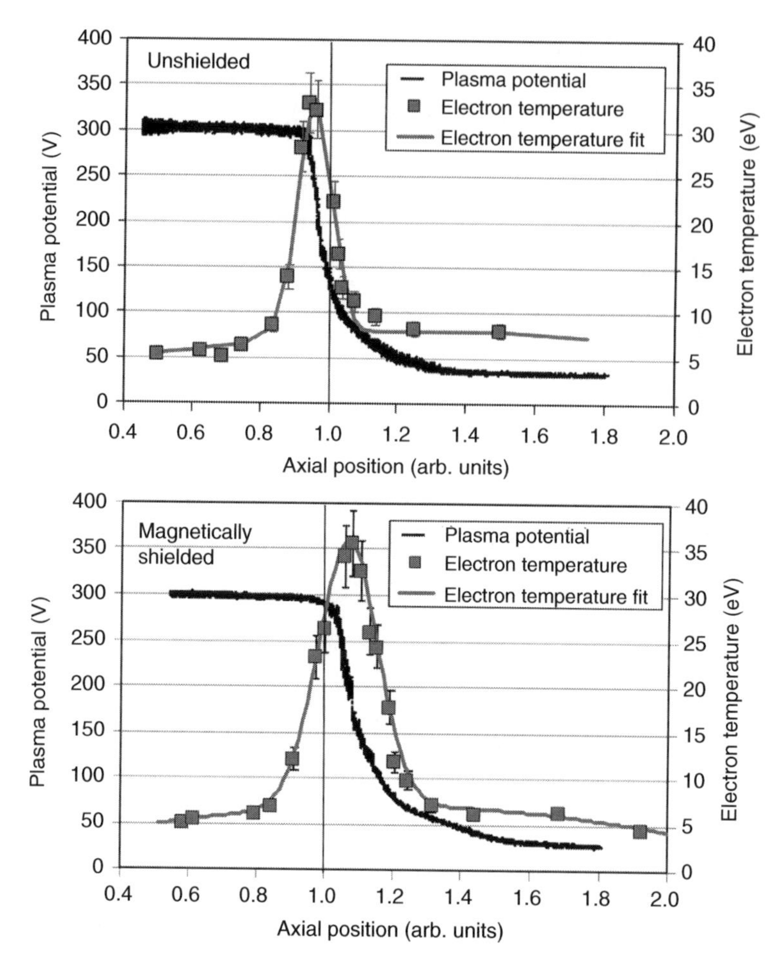

Figure 8-6 Plasma potential and electron temperature profiles for the unshielded H6 thruster (top) and magnetically shielded H6 thruster (bottom) showing shift in location of acceleration region (*Source:* [12]/AIP Publishing).

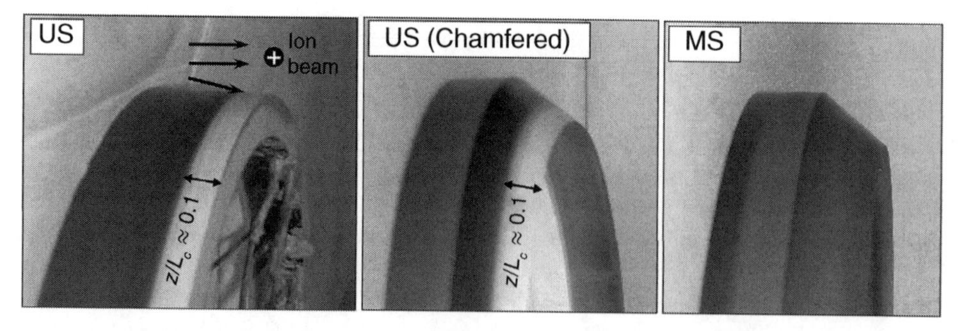

Figure 8-7 Photographs of the inner wall in the unshielded and magnetically shielded configurations of the H6 Hall thruster after operation in a vacuum facility. Dark-colored surfaces depict carbon-coated regions. Left: After 28 h of operation a white-colored band was visible in the H6US that was well correlated with profilometry measurements of the erosion zone. Middle: The H6MS channel after 10 h of operation using the H6US magnetic field to demonstrate that chamfering of the channel wall without a properly design magnetic field does not reduce erosion. Right: The H6MS ring after a 14-h test was fully coated with carbon (*Source:* [12]/AIP Publishing LLC).

in magnetic shielding, a different experiment was conducted in which the H6MS rings were used but with the magnetic field topology of the H6US. A photograph of the inner wall after completion of a 10-h test is shown in Fig. 8-7-middle. Clearly visible is the same "erosion band" that was observed in the H6US in Fig. 8-7-left with the same spatial extent of ~$0.1L_c$.

All theoretical and experimental evidence discussed thus far indicates that channel erosion should have been found to be significantly reduced in the H6MS configuration after the wear test. Indeed, profilometry ultimately confirmed this expectation [12]. The erosion measurements were conducted using a coordinate measuring machine (CMM). The deposition rate of back-sputtered carbon was also monitored during hot-fire testing using a quartz crystal microbalance (QCM) that was mounted adjacent to the thruster. Erosion rates from the simulations are compared in Fig. 8-8 with net rates measured by the CMM. The latter were determined by averaging CMM measurements taken at four circumferential locations around the rings [12]. Two simulation results are plotted that correspond to different sputtering yield models: one with threshold ion energy $E_K = 25$ V and the other with $E_K = 50$ V. For the H6MS only the first model ($E_K = 25$ V) yields a non-zero erosion rate at the inner wall; at the outer wall, the computed rate is zero because the impact energy of ions is found to be less than E_K. The chamfered walls in the H6MS span $0.72 \lesssim z/L_c \leq 1$. The measurements in Fig. 8-8 were obtained after the two thruster configurations were operated for

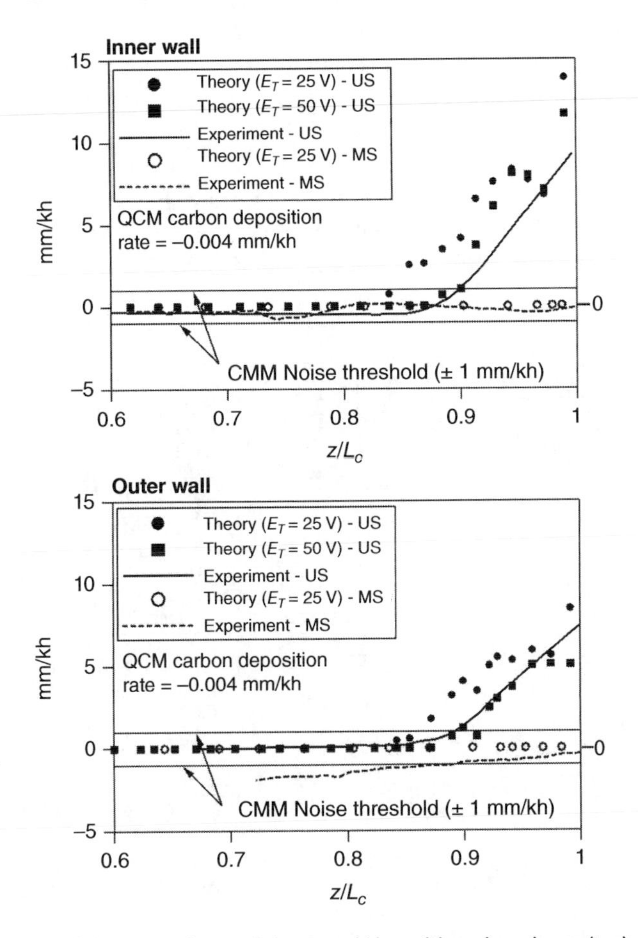

Figure 8-8 Comparisons of the rates of material removal/deposition along inner (top) and outer (bottom) channel walls in the unshielded and magnetically shielded configurations of a 6-kW laboratory Hall thruster (*Source:* [36]).

greater than 19 h. Also depicted on these plots are the CMM noise threshold limits (±1 mm/kh) and the carbon deposition rate (−0.004 ± 0.001 mm/kh) measured by the QCM. Negative and positive values indicate net deposition and net erosion, respectively.

Based only on the wall probe plasma data and an intermediate sputtering yield model with $E_K = 30.5$ V (that is, between the two used in the simulations), the erosion rate was found to be at least ~1000 times (±60%) lower at the H6MS inner wall compared to the highest value in the H6US. At the outer wall the ion energy was below E_K. This is consistent with the simulations which predicted a reduction of ~600 times at the magnetic shielding inner wall and ion energies below E_K at the outer wall (Fig. 8-3d). Collectively, the results from the numerical simulations and the experiments agreed that the erosion was reduced in the magnetic shielding configuration by at least 2–3 orders of magnitude. The photographs of the magnetically shielded thruster taken before and after the test show that the walls are fully coated with carbon after operation, as shown in Fig. 8-9-right, which has now become one of the most widely recognized symptoms of magnetically shielded Hall thrusters after testing in laboratory facilities. However, coated walls are not in and of themselves sufficient evidence of magnetic shielding because facilities with very high carbon backsputter rates will coat even unshielded thruster walls. Additional diagnostics tests as described above are required to fully validate that magnetic shielding has been achieved.

The ability of magnetic shielding to protect the discharge chamber from ion sputtering would, of course, not be very meaningful if the technique also degraded thruster performance in any consequential way. One of the major contributions of these first proof-of-principle experiments was that they demonstrated both near erosion-free operation and high thruster performance. The total efficiency decreased by only 1.7% whereas specific impulse increased by 2.9% compared to the unshielded thruster performance. The thrust decrease was found to be due to plume divergence angle increases in the magnetically shielded configuration due to the field shape and the movement of the plasma downstream. The specific impulse increase was found to be due to a larger amount of higher ionized ions in the plume, which increase the ion velocity and therefore the specific impulse.

Figure 8-9 The magnetically shielded version of the 6-kW H6 Hall thruster installed in the vacuum chamber before (left) and after (right) testing. The boron nitride rings are coated during thruster operation with a carbon backsputtered film from the vacuum chamber walls because magnetic shielding reduced the erosion rates below the carbon deposition rates (*Source:* [12]/AIP Publishing LLC).

The thermal camera diagnostic showed that the temperatures of the insulator rings were reduced by 12–16%, indicating the potential for higher power operation without overheating the thruster components. Discharge current oscillations, while increasing 25%, did not affect the global stability of the discharge [12]. Magnetically shielded thrusters developed and tested later also showed high performance for a wide range of operating conditions and thruster sizes [41–43].

8.4 Magnetically Shielded Hall Thrusters with Electrically Conducting Walls

The experiments described above showed that magnetic shielding significantly reduces the plasma interactions with the discharge chamber walls and that, when properly designed, these thrusters tend to build up carbon layers on the boron nitride insulators during operation in ground test facilities. The experiments demonstrated that the thruster performance was not significantly affected by the presence of a somewhat resistive surface on the discharge chamber walls in contact with the plasma due to the thin carbon deposition layer. This suggests that the use of ceramic walls is not needed in magnetically shielded Hall thrusters [44].

The ceramic walls in conventional unshielded Hall thrusters are there to provide three main features:

1) Insulating surfaces to avoid shorting out the axial ion-accelerating electric field in the thruster acceleration region,
2) Low sputtering yield to minimize the erosion and extend thruster life,
3) Low secondary electron yield to minimize the electron power loss to the wall

Conventional Hall thrusters that used wall materials other than boron nitride, such as higher secondary electron yield ceramics (Al_2O_3, SiC, Macor, quartz, etc.), or conductors (graphite, carbon velvet, stainless steel, etc.) [45–48] are significantly less efficient and are predicted to have shorter life times and lower throughput capabilities. The poorer performance of these thrusters using alternative materials has been largely attributed to the higher secondary yield [45, 49–51] or higher conductivity of these materials compared to the now standard boron nitride or borosilicate walls.

Magnetic shielding thus opens the design space on the discharge chamber in a Hall thruster. Making the discharge chamber from electrically conducting materials simplifies the construction of the thruster that is required to withstand launch vibrations by eliminating large, fragile ceramics and their mechanical supports. Reducing support structure and utilizing lightweight materials such as graphite will also lead to significant reductions in the thruster mass and cost. Conducting wall designs could also lead to factors of two to three increase in the thruster power density due to the lower power loading on the discharge chamber walls and the use of higher emissivity surfaces to radiate that power away. Fig. 8-10 shows a photograph of the H6 thruster modified to have graphite walls in the discharge chamber in contact with the plasma in the acceleration region.

The first experiments with conducting walls in a magnetically shielded Hall thruster were performed by Goebel et al. [44] who showed that the performance was not significantly affected by the use of graphite walls, in stark contrast to that found in unshielded Hall thrusters. A comparison is shown in Fig. 8-11, which plots the thrust and efficiency measured on the H6 thruster at 300 V and 20 A of current (6 kW of power) with the three different wall configurations (unshielded, magnetically shielded with boron nitride walls, magnetically shielded with graphite walls as shown

Figure 8-10 Photograph of the H6MS with graphite walls. The white wire seen across the pole faces was for electrical connections to measure the wall potential (*Source:* [44]/American Institute of Aeronautics and Astronautics).

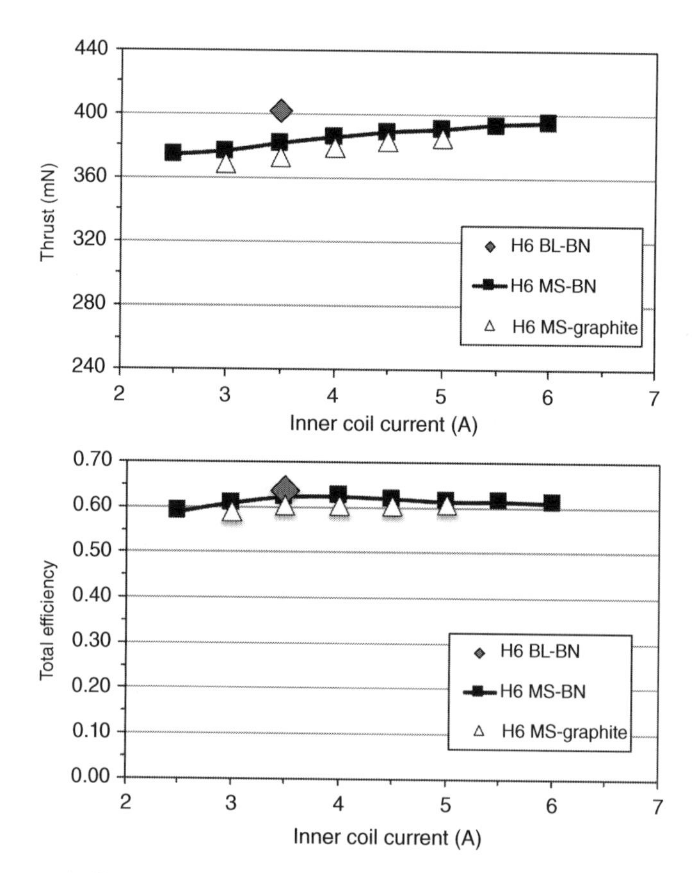

Figure 8-11 Thrust and efficiency versus inner coil current for the H6 thruster unshielded, magnetic shielding with BN walls, and magnetic shielding with graphite walls (*Source:* [44]).

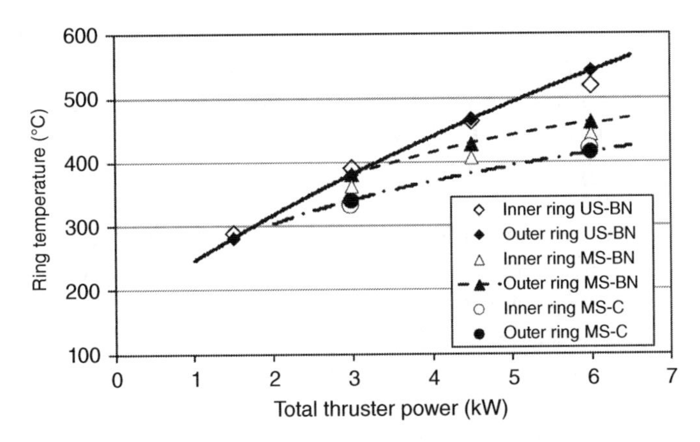

Figure 8-12 H6 wall temperatures versus thruster power for the three cases tested of unshielded, magnetically shielded with BN rings, and with the magnetically shielded with carbon (graphite) walls showing lower temperature operation (*Source:* [44]).

in Fig. 8-7). The data is plotted as a function of the inner magnetic field coil current, which illustrates the dependence of the performance on the magnetic field strength near the exit plane (a design parameter in Hall thrusters). The plot shows clearly that the magnetically shielded thruster has nearly the same efficiency as the baseline unshielded thruster (within 2%), but the thrust is reduced slightly.

The experiments also showed that graphite wall Hall thrusters can operate at a higher power than one with BN walls. Figure 8-12 shows the temperatures of the inner and outer rings (the discharge chamber wall on the downstream end in the H6) for the three configurations (unshielded with BN wall, magnetically shielded with BN walls, and magnetically shielded with graphite walls) described above. The circular data points show that the graphite surfaces are 100-to-200°C lower than the BN walls at the same power level. The lower wall temperatures are roughly half because of lower heat fluxes in the magnetically shielded configuration, and half because of the higher emissivity graphite wall material. Higher power thrusters, or more compact Hall thrusters, can then be made using this configuration. Since this original work, conducting wall Hall thruster operation has been extended by Hofer [52] to 800 V at over 12 kW in the H9MS at JPL, demonstrating reliable operation at 3000 s specific impulse. Conducting wall Hall thrusters have also been investigated in France [13, 53] and China [54], and work continues on this advancement in Hall thruster design elsewhere in the world.

8.5 Magnetic Shielding in Low Power Hall Thrusters

Conventional unshielded low-power (sub-kW) Hall thrusters typically have lower efficiencies (less than 50%) and shorter lifetimes (on the order of a few thousand hours) compared to the state-of-the-art Hall thrusters designed for operation in excess of about 1.5 kW. This is largely because of the inherent problem of scaling Hall thrusters to smaller sizes where the surface-to-volume ratio decreases considerably. Since favorable processes like ionization and acceleration occur in the volume of the plasma, and unfavorable processes like particle losses and power deposition occur at the surface of the plasma, smaller Hall thrusters with lower surface-to-volume ratios tend to show lower performance and shorter life. In addition, small Hall thrusters tend to have externally mounted hollow cathodes because of a lack of space on axis to insert a centrally-mounted cathode. Externally mounted cathodes have been shown to reduce the efficiency of the thruster [55, 56] and create significant sensitivity to positioning (relative to the thruster's discharge channel) and background pressure.

Magnetic shielding has been found to mitigate both the performance and life issues in low power Hall thrusters. As expected, the erosion of the discharge chamber wall in low power Hall thrusters with magnetic shielding was found to be reduced by orders of magnitude compared to conventional unshielded thrusters. In addition to this longer life, the MaSMi (Magnetically Shielded Miniature) Hall thruster was originally developed by Conversano [57] to also demonstrate the higher performance and efficiency afforded by magnetic shielding in a sub-kilowatt Hall thruster.

A photograph of the 12-cm-O.D. MaSMi Hall thruster is shown in Fig. 8-13. The thruster uses an additively manufactured (3D printed) Hiperco magnetic circuit to produce the field shape and intensity needed for magnetic shielding. To simplify the PPU to a single magnet power supply, the inner and outer magnet coils are connected in series internally. The magnetic circuit was designed to minimize shifts in the magnetic field topology across the accessible range of magnetic field strengths. Testing across the full range of magnetic field settings demonstrated that the field profile remained centered in the channel and any small shifts did not significantly alter the thruster behavior. The discharge chamber is fabricated from a high strength blended ceramic material that features exceptional mechanical and thermal shock resistance and essentially no water or air uptake behavior when the thruster is in atmosphere. Therefore, the thruster performance is stable immediately after turn on following extended periods of exposure to air (no outgassing required). Further, the performance and behavior of the MaSMi thruster were demonstrated to be invariant over life [58].

Figure 8-14 shows the thrust, Isp, and total efficiency of the MaSMi thruster as a function of power [43]. At 1.35 kW and 300 V, equivalent to the nominal operating point of the SPT-100 Hall thruster, the MaSMi thruster produced 94.5 mN of thrust at a specific impulse of 1800 s and a total efficiency of 60.3%. Each of these values is nearly 20% higher than those produced by the SPT-100 at this operating point [59], demonstrating that properly designed low-power Hall thrusters utilizing magnetically shielded can have higher performance than conventional unshielded thrusters. The MaSMi thruster recently completed a >7200 h wear test and exceeded 100 kg of xenon propellant throughput and more than 1.55 MN-s of total delivered impulse [58] when the test was voluntarily terminated. The thruster's ultimate lifetime is projected to be in excess of 10 kh, with corresponding increases in throughput and total impulse capability. This life capability greatly exceeds that achieved in unshielded low-power Hall thrusters.

Figure 8-13 Photograph of the 12-cm-O.D. MaSMi Hall thruster operating at 1 kW of discharge power (*Source:* NASA/JPL, Public Domain).

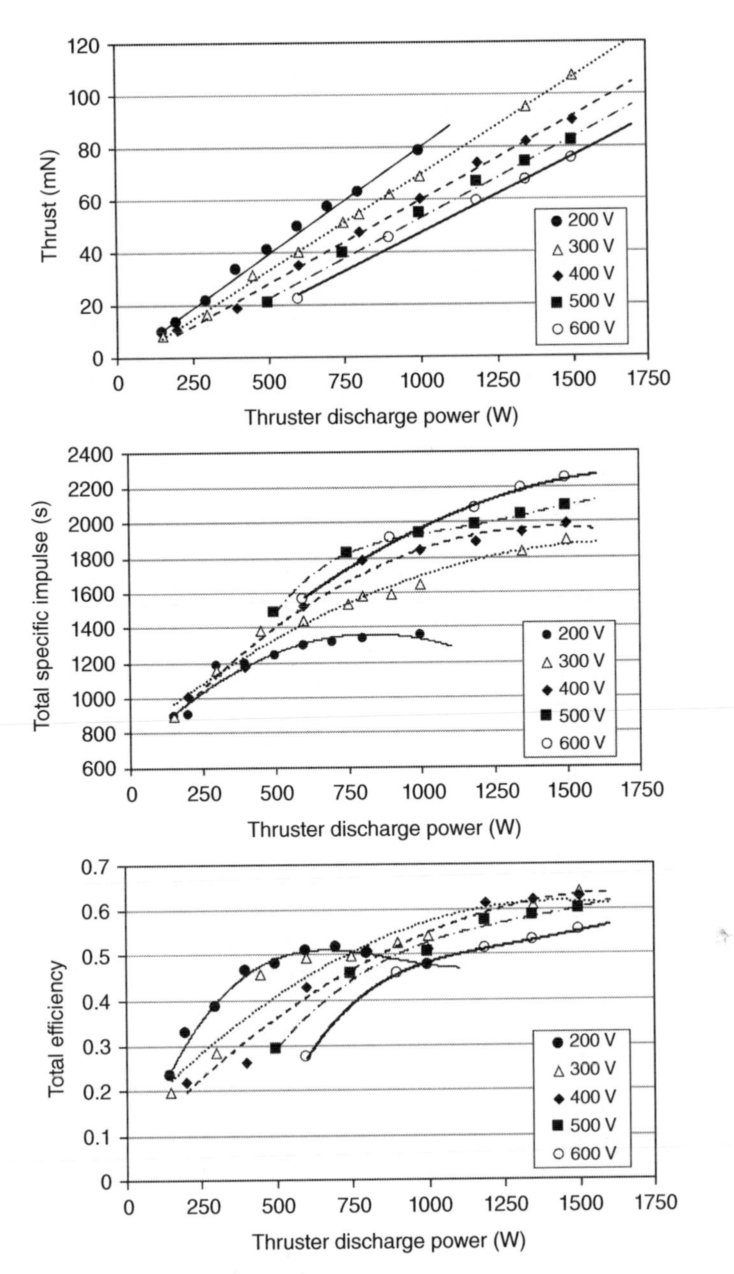

Figure 8-14 Thrust, Isp and total efficiency as a function of discharge power up to 1.5 kW for the MaSMi-DM Hall thruster.

8.6 Final Remarks on Magnetic Shielding in Hall Thrusters

During the early development of the principles of magnetic shielding, misconceptions emerged, which is not uncommon when a solution to a very longstanding problem is found. Now, more than a decade after the first papers on magnetic shielding were published, most of these misconceptions have been elucidated in the peer reviewed literature, but it is useful to discuss a few of them here in this final section of the chapter.

Due to its electrode configuration, a magnetically shielded Hall thruster is distinctly different from the so-called "Thruster with Anode Layer" (TAL). (Chapter. 7.1.2, [60]). The discharge chamber walls in a TAL are metallic and are typically biased to the cathode potential in order to minimize electron energy losses. In a magnetically shielded Hall thruster with a conventional channel, the insulating walls tend to float near the anode potential because the ion and electron fluxes on these surfaces must be equal. In conducting wall Hall thrusters, magnetic shielding isolates the walls sufficiently from the plasma to eliminate this requirement for equal ion and electron fluxes, and the walls tend to float near the anode potential because they are in contact with magnetic field lines that contain cold electrons ($T_e \rightarrow 0$) from the anode region. Ions reaching the wall of a TAL can attain nearly the entire applied discharge voltage, which results in erosion rates that are of the same order as unshielded Hall thrusters. In a magnetically shielded thruster, the ion energy is only a few volts, which results in orders of magnitude lower erosion rates than either an unshielded Hall thruster or a TAL.

Magnetic shielding is also markedly different from other techniques that have been pursued to protect surfaces from erosion, like in the Highly Efficient Multistage Plasma Thruster (HEMP-T) [61] and the Diverging Cusped Field Thruster (DCFT) [62]. These configurations exploit the magnetic mirror effect on electrons by employing multi-cusped magnetic fields to reduce plasma bombardment of the walls at the cusps. Such cusped arrangements also provide the magnetic field direction needed to induce the azimuthal electron motion and, in turn, the accelerating electric field at the cusped regions. In Hall thrusters with magnetic shielding, there is no magnetic mirror effect because there are no cusps.

Finally, in the first proof-of-principle laboratory experiments of magnetic shielding with the H6 (described in Section 8.3.2), no channel erosion was observed but the surface of the front poles facing the near-plume plasma of the thruster appeared eroded after only 150 h of wear testing [63, 64]. Such erosion of the front poles had not been observed in unshielded Hall thrusters. To achieve magnetic shielding, the magnetic field must be applied in a manner in which the location of the peak field flux density is slightly downstream of that in an equivalent unshielded thruster, forcing the acceleration region to move further downstream. The potential implication on thruster wear is therefore that such displacement could lead to more radially divergent ions with sufficient energy to sputter the front pole surfaces. In fact, some of the misconstrued early arguments were that magnetic shielding simply traded wear out processes by protecting the channel walls at the expense of the front poles. This is of course not true. Numerical simulations of both unshielded and magnetically shielded versions of the H6 showed that the current density of the radially divergent high-energy ions was high enough to cause observable sputtering of the poles in the H6MS. But it was also found that this current density was three orders of magnitude lower than that of axially accelerated ions [65]. Consequently, the sputtering of these surfaces in magnetically shielded thrusters is considerably lower than those along the channel walls of unshielded thrusters. In fact, it is low enough that front pole erosion can be easily mitigated for long duration missions using thin protective pole covers made of low-sputtering material such as graphite. This approach was successfully demonstrated in the H6MS thruster [12, 66], and subsequently in both HERMeS [26] and MaSMi [58]. The ion energy impinging on the pole covers was found to be reduced and regulated by electrically connecting the cathode common to the thruster body [67, 68]. The subsequent front pole cover

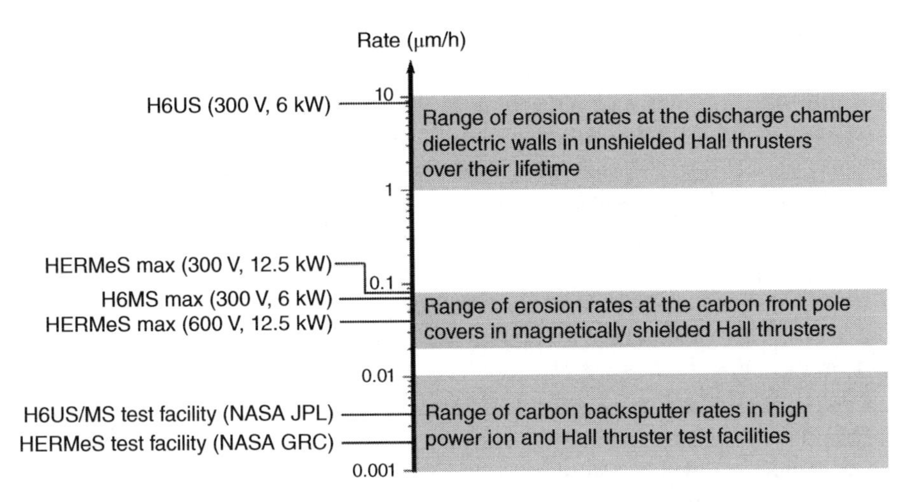

Figure 8-15 Typical erosion rates along the channel walls and graphite front pole covers of unshielded and magnetically shielded Hall thrusters. Also shown is a typical range of carbon backsputtered rates in vacuum facilities used for testing high power ion and Hall thrusters. The data for the specific thrusters and facilities shown on the left are taken from [12, 26, 66].

erosion rates were measured to be at least one order of magnitude lower than those along the channel walls of unshielded Hall thrusters. A comparison of typical erosion rates along these surfaces is provided in Fig. 8-15. Also included on the plot for comparison is a typical range of backsputtered carbon rates in some existing vacuum facilities used for testing high power ion and Hall thrusters.

References

1 A. I. Morozov and V. V. Savelyev, "Fundamentals of Stationary Plasma Thruster Theory," in *Reviews of Plasma Physics*, vol. 21, edited by B. B. Kadomtsev and V. D. Shafranov, Springer, pp. 203–391, 2000.

2 Manzella, D., Jankovsky, R., and Hofer, R. "Laboratory Model 50 kW Hall Thruster," AIAA-2002-3676, Presented at the 38th AIAA Joint Propulsion Conference, Indianapolis, July 7–10, 2002.

3 R. R. Hofer and A. D. Gallimore, "High-Specific Impulse Hall Thrusters, Part 2: Efficiency Analysis," *Journal of Propulsion and Power*, vol. 22, no. 4, pp. 732–740, 2006.

4 B. Welander, C. Carpenter, K. de Grys, R. Hofer, T. Randolph, and D. Manzella, "Life and Operating Range Extension of the BPT-4000 Qualification Model Hall Thruster," AIAA-2006-5263, Presented at the 42nd AIAA Joint Propulsion Conference, Sacramento, July 9–12, 2006.

5 H. Kamhawi, D. Manzella, L. Pinero, T. Haag, A. Mathers, and H. Liles, "In-Space Propulsion High Voltage Hall Accelerator Development Project Overview," AIAA-2009-5282, Presented at the 45th AIAA Joint Propulsion Conference, Denver, August 2–5, 2009.

6 A. I. Morozov, Y. V. Esinchuk, G. N. Tilinin, A. V. Trofimov, Y. A. Sharov, and G. Y. Shchepkin, "Plasma Accelerator with Closed Electron Drift and Extended Acceleration Zone," *Soviet Phys.-Tech. Phys.*, vol. 17, no. 1, pp. 38–45, 1972.

7 Y. Raitses, M. Keidar, D. Staack, and N. J. Fisch, "Effects of Segmented Electrode in Hall Current Plasma Thrusters," *Journal of Applied Physics*, vol. 92, no. 9, pp. 4906–4911, 2002; doi:10.1063/1.1510556.

8 L. Garrigues, G. J. M. Hagelaar, J. Bareilles, C. Boniface, and J. P. Boeuf, "Model Study of the Influence of the Magnetic Field Configuration on the Performance and Lifetime of a Hall Thruster," *Physics of Plasmas*, vol. 10, no. 12, pp. 4886–4892, 2003.

9 I. G. Mikellides, I. Katz, R. R. Hofer, D. M. Goebel, K. de Grys, and A. Mathers, "Magnetic Shielding of the Channel Walls in a Hall Plasma Accelerator," *Physics of Plasmas*, vol. 18, no. 3, p. 033501, 2011.

10 I. G. Mikellides, I. Katz, R. R. Hofer, and D. M. Goebel, "Magnetic Shielding of Walls from the Unmagnetized Ion Beam in a Hall Thruster," *Applied Physics Letters*, vol. 102, no. 2, 2013; doi:10.1063/1.4776192.

11 K. H. De Grys, A. Mathers, B. Welander, and V. Khayms, "Demonstration of >10,400 Hours of Operation on 4.5 kW Qualification Model Hall Thruster," AIAA-2010-6698, Presented at the 46th AIAA Joint Propulsion Conference, Nashville, July 2010.

12 R. R. Hofer, D. M. Goebel, and I. G. Mikellides, "Magnetic Shielding of a Laboratory Hall Thruster, II. Experiments," *Journal of Applied Physics*, vol. 15, no. 043304, 2014, doi:10.1063/1.4862314.

13 L. Grimaud and S. Mazouffre, "Performance Comparison Between Standard and Magnetically Shielded 200 W Hall Thrusters with BN-SiO2 and Graphite Channel Walls," *Vacuum*, vol. 155, pp. 514–523, 2018, doi:10.1016/j.vacuum.2018.06.056.

14 L. Grimaud and S. Mazouffre, "Ion Behavior in Low-Power Magnetically Shielded and Unshielded Hall Thrusters," *Plasma Sources Science & Technology*, vol. 26, no. 5, 2017; doi:10.1088/1361-6595/aa660d.

15 A. Piragino, E. Ferrato, F. Faraji, M. Reza, T. Andreussi, A. Rossodivita, and M. Andrenucci, "Experimental Characterization of a 5 kW Magnetically-Shielded Hall Thruster," SP2018-427, Presented at the Space Propulsion 2018, Seville, May 14–18, 2018.

16 A. Piragino, E. Ferrato, F. Faraji, M. Reza, A. Kitaeva, D. Pedrini, M. Andrenucci, and T. Andreussi, "SITAEL's Magnetically Shielded 20 kW Hall Thruster Tests," IEPC-2019-879, Presented at the 36th International Electric Propulsion Conference, Vienna, September 15–20, 2019.

17 Y. J. Ding, H. Li, L. Q. Wei, Y. L. Hu, Y. Shen, H. Liu, Z. X. Ning, W. Mao, and D. R. Yu, "Overview of Hall Electric Propulsion in China," *IEEE Transactions on Plasma Science*, vol. 46, no. 2, pp. 263–282, 2018, doi:10.1109/Tps.2017.2776257.

18 X. Duan, M. Cheng, X. Yang, N. Guo, X. Li, M. Wang, and D. Guo, "Investigation on Ion Behavior in Magnetically Shielded and Unshielded Hall Thrusters by Laser-Induced Fluorescence," *Journal of Applied Physics*, vol. 127, no. 9, p. 093301, 2020; doi:10.1063/1.5140514.

19 A. Mishra, E. S. Rupesh, and H. Gole, "Development of a 1.5 KW High Specific Impulse Magnetic Shielded Hall Thruster," IEPC-2019-140, Presented at the 36th International Electric Propulsion Conference, Vienna, September 15–20, 2019.

20 H. Watanabe, S. Cho, K. Kubota, G. Ito, K. Fuchigami, K. Uematsu, Y. Tashiro, S. Iihara, and I. Funaki, "Performance Evaluation of a Two-Kilowatt Magnetically Shielded Hall Thruster," *Journal of Propulsion and Power*, vol. 36, no. 1, pp. 14–24, 2020; doi:10.2514/1.B37550.

21 H. Kamhawi, W. Huang, T. Haag, R. Shastry, G. C. Soulas, T. B. Smith, I. G. Mikellides, and R. R. Hofer, "Performance and Thermal Characterization of the NASA-300MS 20 kW Hall Effect Thruster," IEPC-2013-444, Presented at the 33rd International Electric Propulsion Conference, Washington, October 6–10, 2013.

22 R. W. Conversano, D. M. Goebel, R. R. Hofer, I. G. Mikellides, and R. E. Wirz, "Performance Analysis of a Low-Power Magnetically Shielded Hall Thruster: Experiments," *Journal of Propulsion and Power*, pp. 1–9, 2017, doi:10.2514/1.b36230.

23 R. W. Conversano, D. M. Goebel, I. G. Mikellides, R. R. Hofer, and R. E. Wirz, "Performance Analysis of a Low-Power Magnetically Shielded Hall Thruster: Computational Modeling," *Journal of Propulsion and Power*, pp. 1–10, 2017, doi:10.2514/1.b36231.

24 R. R. Hofer, R. Lobbia, V. Chaplin, A. Lopez Ortega, I. G. Mikellides, J. E. Polk, H. Kamhawi, J. Frieman, W. Huang, and P. Peterson, "Completing the Development of the 12.5 kW Hall Effect Rocket with Magnetic Shielding (HERMeS)," IEPC-2019-193, Presented at the 36th International Electric Propulsion Conference, Vienna, September 15–20, 2019.

25 D. A. Herman, T. Tofil, W. Santiago, H. Kamhawi, J. E. Polk, J. S. Snyder, R. R. Hofer, and F. Picha, "Overview of the Development and Mission Application of the Advanced Electric Propulsion System (AEPS)," IEPC-2017-284, Presented at the 35th International Electric Propulsion Conference, Atlanta, October 8–12, 2017.

26 J. D. Frieman, J. H. Gilland, H. Kamhawi, J. Mackey, G. J. Williams., R. R. Hofer, and P. Y. Peterson, "Wear Trends of the 12.5 kW HERMeS Hall Thruster," *Journal of Applied Physics*, vol. 130, no. 14, p. 143303, 2021, doi:10.1063/5.0062579.

27 W. Huang and H. Kamhawi, "Counterstreaming Ions at the Inner Pole of a Magnetically Shielded Hall Thruster," *Journal of Applied Physics*, vol. 129, no. 043305, 2021, doi:10.1063/5.0029428.

28 V. H. Chaplin, B. A. Jorns, A. L. Ortega, I. G. Mikellides, R. W. Conversano, R. B. Lobbia, and R. R. Hofer, "Laser-Induced Fluorescence Measurements of Acceleration Zone Scaling in the 12.5 kW HERMeS Hall Thruster," *Journal of Applied Physics*, vol. 124, no. 18, 2018, doi:10.1063/1.5040388.

29 I. G. Mikellides, P. Guerrero, A. Lopez Ortega, and J. E. Polk, "Spot-to-Plume Mode Transition Investigations in the HERMeS Hollow Cathode Discharge Using Coupled 2-D Axisymmetric Plasma-Thermal Simulations," AIAA-2018-4722, Presented at the AIAA Propulsion and Energy Forum, Cincinnati, July 9–11, 2018.

30 I. G. Mikellides, A. Lopez Ortega, R. Hofer, J. Polk, H. Kamhawi, J. Yim, and J. Myers, "Hall2De Simulations of a 12.5-kW Magnetically Shielded Hall Thruster for the NASA Solar Electric Propulsion Technology Demonstration Mission," IEPC-2015-254, Presented at the 34th International Electric Propulsion Conference, Hyogo-Kobe, July 2015.

31 A. Lopez Ortega, I. G. Mikellides, V. Chaplin, H. W. Huang, and J. D. Frieman, "Anomalous Ion Heating and Pole Erosion in the 12.5-kW Hall Effect Rocket with Magnetic Shielding (HERMeS)," AIAA-2020-3620,m Presented at the AIAA Propulsion and Energy Forum, New Orleans, August 24–28, 2020.

32 I. G. Mikellides and A. L. Ortega, "Growth of the Modified Two-Streaminstability in the Plume of a Magnetically Shielded Hall Thruster," *Physics of Plasmas*, vol. 27, no. 10, p. 100701, 2020, doi:10.1063/5.0020075.

33 I. G. Mikellides and A. L. Ortega, "Growth of the Lower Hybrid Drift Instability in the Plume of a Magnetically Shielded Hall Thruster," *Journal of Applied Physics*, vol. 129, no. 19, 2021; doi:10.1063/5.0048706.

34 R. L. Ticker, M. Gates, D. Manzella, A. Biaggi-Labiosa, and T. Lee, "The Gateway Power and Propulsion Element: Setting the Foundation for Exploration and Commerce," AIAA-2019-3811, Presented at the AIAA Propulsion and Energy 2019 Forum, Indianapolis, August 19–22, 2019.

35 G. D. Hobbs and J. A. Wesson, "Heat Flow Through a Langmuir Sheath in Presence of Electron Emission," *Plasma Physics*, vol. 9, no. 1, pp. 85–87, 1967.

36 I. G. Mikellides, I. Katz, R. R. Hofer, and D. M. Goebel, "Magnetic Shielding of a Laboratory Hall Thruster. I. Theory and Validation," *Journal of Applied Physics*, vol. 115, no. 4, 2014, doi:10.1063/1.4862313.

37 I. G. Mikellides and I. Katz, "Numerical Simulations of Hall-Effect Plasma Accelerators on a Magnetic-Field-Aligned Mesh," *Physical Review E*, vol. 86, no. 4, p. 046703, 2012, doi:10.1103/Physreve.86.046703.

38 I. Katz and I. G. Mikellides, "Neutral Gas Free Molecular Flow Algorithm Including Ionization and Walls for Use in Plasma Simulations," *Journal of Computational Physics*, vol. 230, no. 4, pp. 1454–1464, 2011, doi:10.1016/j.jcp.2010.11.013.

39 B. A. Jorns, C. A. Dodson, J. R. Anderson, D. M. Goebel, R. R. Hofer, M. J. Sekerak, A. Lopez Ortega, and I. G. Mikellides, "Mechanisms for Pole Piece Erosion in a 6-kW Magnetically-Shielded Hall Thruster," AIAA-2016-4839, Presented at the 52nd AIAA Joint Propulsion Conference, Salt Lake City, July 25–27, 2016.

40 B. A. Jorns, D. M. Goebel, and R. R. Hofer, "Plasma Perturbations in High-Speed Probing of Hall Thruster Discharge Chambers: Quantification and Mitigation," AIAA 2015-4006, Presented at the 51th AIAA Joint Propulsion Conference, Orlando, July 27–29, 2015.

41 J. D. Frieman, H. Kamhawi, J. Mackey, T. W. Haag, P. Y. Peterson, and R. R. Hofer, "Expanded Performance Characterization of the NASA HERMeS Hall Thruster," AIAA-2022-1353, Presented at the AIAA SciTech 2022 Forum, San Diego, January 3–7, 2022.

42 H. Kamhawi, W. Huang, J. Gilland, T. Haag, J. A. Mackey, J. Yim, L. Pinero, G. Williams, P. Peterson, and D. Herman, "Performance, Stability, and Plume Characterization of the HERMeS Thruster with Boron Nitride Silica Composite Discharge Channel," IEPC-2017-392, Presented at the 35th International Electric Propulsion Conference, Atlanta, October 8–12, 2017.

43 R. W. Conversano, R. B. Lobbia, T. V. Kerber, K. C. Tilley, D. M. Goebel, and S. W. Reilly, "Performance Characterization of a Low-Power Magnetically Shielded Hall Thruster with an Internally-Mounted Hollow Cathode," *Plasma Sources Science & Technology*, vol. 28, no. 10, 2019; doi:10.1088/1361-6595/ab47de.

44 D. M. Goebel, R. R. Hofer, I. G. Mikellides, I. Katz, J. E. Polk, and B. Dotson, "Conducting Wall Hall Thrusters," *IEEE TPS Special Issue on Plasma Propulsion*, vol. 43, no. 1, pp. 118–126, 2015.

45 Y. Raitses, J. Ashkenazy, G. Appelbaum, and M. Guelman, "Experimental Investigation of the Effect of Channel Material on Hall Thruster Characteristics," IEPC-1997-056, Presented at the 25th International Electric Propulsion Conference, Cleveland, August 1977.

46 N. Gascon, M. Dudeck, and S. Barral, "Wall Material Effects in Stationary Plasma Thrusters I: Parametric Studies of an SPT-100," *Physics of Plasmas*, vol. 10, no. 10, pp. 4123–4136, 2003.

47 S. Barral, K. Makowski, Z. Peradzynski, N. Gascon, and M. Dudeck, "Wall Material Effects in Stationary Plasma Thrusters II. Near-Wall and In-Wall Conductivity," *Physics of Plasmas*, vol. 10, no. 10, pp. 4137–4152, 2003.

48 D. Staack, Y. Raitses, and N. J. Fisch, "Control of Acceleration Region in Hall Thrusters," IEPC-03-0273, Presented at the 28th International Electric Propulsion Conference, Toulouse, March 17–21, 2003.

49 A. Dunaevsky, Y. Raitses, and N. J. Fisch, "Secondary Electron Emission from Dielectric Materials of a Hall Thruster with Segmented Electrodes," *Physics of Plasmas*, vol. 13, no. 1, pp. 2574–2577, 2003.

50 Y. Raitses, A. Smirnov, D. Staack, and N. J. Fisch, "Measurements of Secondary Electron Emission Effects in Hall Thrusters," *Physics of Plasmas*, vol. 13, no. 1, 014502-1–014502-4, 2006.

51 Y. Raitses, I. Kaganovich, A. Khrabrov, D. Sydorenko, N. Fisch, and A. Smolyakov, "Effect of Secondary Electron Emission on Electron Cross-Field Current in ExB Discharges," *IEEE Transactions on Plasma Science*, vol. 39, pp. 995–10062011.

52 R. R. Hofer, R. Lobbia, and S. Arestie, "Performance of a Conducting Wall, Magnetically Shielded Hall Thruster at 3000-S Specific Impulse," IEPC-2022-401, Presented at the 37th International Electric Propulsion Conference, Cambridge, 19–23 June 2022.

53 L. Grimaud and S. Mazouffre, "Conducting Wall Hall Thrusters in Magnetic Shielding and Standard Configurations," *Journal of Applied Physics*, vol. 122, no. 033305, 2017; doi:10.1063/1.4995285.

54 Y. Ding, H. Li, S. Hezhi, L. Wei, B. Jia, H. Su, W. Peng, P. Li, and D. Yu, "A 200-W Permanent Magnet Hall Thruster Discharge with Graphite Channel Wall," *Physics Letters A*, vol. 382, no. 42–43, pp. 3079–3082, 2018; doi:10.1016/j.physleta.2018.08.017.

55 R. R. Hofer, L. Johnson, D. M. Goebel, and R. Wirz, "Effect of Internally-Mounted Cathodes on a Hall Thruster Plume Properties," *IEEE Transactions on Plasma Science*, vol. 36, pp. 2004–2014, 2008.

56 R. R. Hofer and J. R. Anderson, "Finite Pressure Effects in Magnetically Shielded Hall Thrusters," AIAA-2014-3709, Presented at the 50th AIAA Joint Propulsion Conference, Cleveland, July 28–30, 2014.

57 R. W. Conversano, D. M. Goebel, R. R. Hofer, and R. Wirz, "Development and Initial Testing of a Magnetically Shielded Miniature Hall Thruster," *IEEE Transactions on Plasma Science* vol. 43, no. 1, pp. 103–117, 2015; doi:10.2514/1.B36230.

58 R. W. Conversano, R. B. Lobbia, S. M. Arestie, A. Lopez Ortega, V. H. Chaplin, S. W. Reilly, and D. M. Goebel, "Demonstration of One Hundred Kilogram Xenon Throughput by a Low-Power Hall Thruster," *Journal of Propulsion and Power*, 2022; doi:10.2514/1.B38733.

59 V. Kim, G. Popov, B. Arkhipov, V. Murashko, O. Gorshkov, A. Koroteyev, V. Garkusha, A. Semenkin, and S. Rverdokhlebov, "Electric Propulsion Activity in Russia," IEPC-01-05, Presented at the 27th Interntional Electric Propulsion Conference, Pasadena, October 15–19, 2001.

60 V. V. Zhurin, H. R. Kaufman, and R. S. Robinson, "Physics of Closed Drift Thrusters," *Plasma Sources Science & Technology*, vol. 8, no. 1, pp. R1–R20, 1999.

61 G. Kornfeld, N. Koch, and H. Harmnann, "Physics and Evolution of HEMP-Thrusters," IEPC-2007-108, Presented at the 30th International Electric Propulsion Conference, Florence, September 17–20, 2007.

62 D. Courtney, P. Lozano, and M. Martinez-Sanchez, "Continued Investigation of Diverging Cusped Field Thruster," AIAA-2008-4631, Presented at the 44th AIAA Joint Propulsion Conference, Hartford, CT, July 21–23, 2008.

63 D. M. Goebel, B. A. Jorns, R. R. Hofer, I. G. Mikellides, and I. Katz, "Pole-piece interactions with the plasma in a magnetically shielded Hall thruster," AIAA-2014-3899, Presented at the 50th AIAA Joint Propulsion Conference, Cleveland, July 28–30, 2014.

64 I. G. Mikellides, A. Lopez Ortega, and B. A. Jorns, "Assessment of Pole Erosion in a Magnetically Shielded Hall Thruster," AIAA-2014-3897, Presented at the 50th AIAA Joint Propulsion Conference, Cleveland, July 28–30, 2014.

65 A. Lopez Ortega, I. G. Mikellides, M. J. Sekerak, and B. A. Jorns, "Plasma Simulations in 2-D (r-z) Geometry for the Assessment of Pole Erosion in a Magnetically Shielded Hall Thruster," *Journal of Applied Physics*, vol. 125, no. 3, 2019; doi:10.1063/1.5077097.

66 M. J. Sekerak, R. R. Hofer, J. E. Polk, B. A. Jorns, and I. G. Mikellides, "Wear Testing of a Magnetically Shielded Hall Thruster at 2000-s Specific Impulse," IEPC 2015-155, Presented at the 34th International Electric Propulsion Conference, Hyogo-Kobe, July 6–10, 2015.

67 R. R. Hofer, J. E. Polk, M. Sekerak, I. G. Mikellides, H. Kamhawi, T. Verhey, D. Herman, and G. Williams, "The 12.5 kW Hall Effect rocket with Magnetic Shielding (HERMeS) for the Asteroid Redirect Robotic Mission," AIAA-2016-4825, Presented at the 52nd AIAA Joint Propulsion Conference, Salt Lake City, July 25–27, 2016.

68 I. Katz, A. L. Ortega, D. M. Goebel, M. J. Sekerak, R. R. Hofer, B. A. Jorns, and J. R. Brohpy, "Effect of Solar Array Plume Interactions on Hall Thruster Cathode Common Potentials," Presented at the 14th Spacecraft Charging Technology Conference, Noordwijk, April 4, 2016. http://hdl.handle.net/2014/46085.

Chapter 9

Electromagnetic Thrusters

9.1 Introduction

Electromagnetic thrusters produce thrust by accelerating the plasma with the Lorentz force, which is generated when an electrical current flows transverse to an applied and/or self-induced magnetic field. There are several different thruster configurations that can generate the Lorentz force for propulsion. This chapter describes three of the more technologically mature examples of these types of thrusters.

9.2 Magnetoplasmadynamic Thrusters

Magnetoplasmadynamic (MPD) thrusters are high-power electric propulsion devices [1–3] that are classified as electromagnetic (EM) thrusters because they utilize electromagnetic forces to accelerate the plasma. An MPD thruster basically consists of a central rod-shaped cathode insulated from a cylindrical anode that surrounds the cathode, as seen in Fig. 9-1. A high-current electron discharge is driven between cathode and anode that ionizes the propellant gas injected from the backplane to create plasma. The azimuthal component of the magnetic field interacts with the current flowing radially from cathode to anode to produce an electromagnetic Lorentz force that accelerates the plasma out of the engine generating thrust. MPD thrusters will be classified here either as *self-field* where an azimuthal magnetic field is generated only by the internal plasma current flowing from cathode to anode, or *applied-field* where an external solenoid (shown in the figure) is used to provide an additional magnetic field that can help stabilize discharge instabilities and, if large enough, produce additional acceleration of the plasma.

Electromagnetic acceleration was first recognized [4] in a hydrogen arcjet thruster during experiments at low propellant gas flow rates and high discharge current aimed at increasing the level of ionization. High Isp was reported with efficiencies over 50% and power levels over 250 kW. The authors concluded that the high current discharges were generating self-magnetic fields that "helped to contain and accelerate the plasma" [4].

In MPD thrusters there are three main effects that generate electromagnetic thrust. First, the plasma is accelerated in the axial direction because of the Lorentz force from the radial electron current flowing from the cathode to the anode crossed with the self-generated azimuthal magnetic field from the current running axially on the cathode electrode. This contribution has been historically called electromagnetic "blowing". Second, there is radially inward acceleration of the plasma

Fundamentals of Electric Propulsion, Second Edition. Dan M. Goebel, Ira Katz, and Ioannis G. Mikellides.

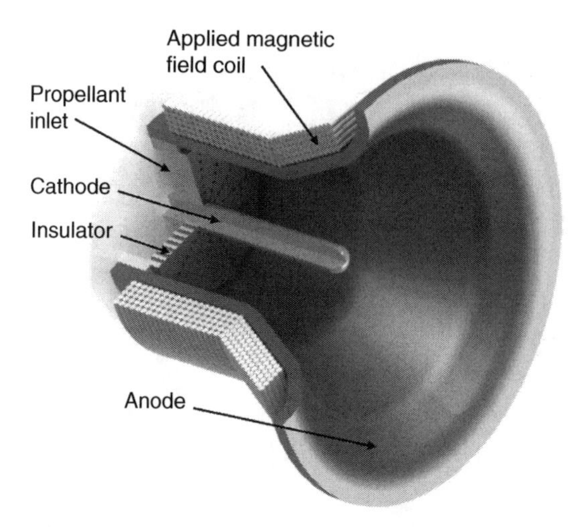

Figure 9-1 Illustration of an MPD thruster that includes an example of applied magnetic field coils.

because of an axial component of the electron current flowing from cathode to anode crossed with the azimuthal magnetic field. This second term is historically called electromagnetic "pinching", and represents the force in the thruster that compresses the plasma radially toward the axis. Finally, if there is an applied magnetic field in the thruster that has components in the axial and radial directions, the plasma will "swirl" azimuthally about the axis because of the $j_r B_z$ and $j_z B_r$ forces generated in the thruster. Cold gas contributions to the thrust are usually significantly smaller than those from the electromagnetic force.

MPD thrusters utilize electron discharges at power levels from kilowatts up to several megawatts to produce tens to hundreds of newtons of thrust at Isps of 1,000–10,000 s and higher [5–9]. They process more power and produce higher thrust than ion and Hall thrusters, but equivalent thruster lifetimes and efficiencies have not yet been demonstrated. MPD thrusters have demonstrated operation with a wide variety of propellants including H_2, He, Li, N_2, NH_3, Ne, Na, Ar, K, Kr, Xe, and Bi. They typically run at low discharge voltages (<250 V) to minimize erosion of the discharge electrodes, and light propellants are used to obtain higher Isp. Self-field MPDs typically have efficiencies between 30–50%, and obtaining thruster lifetimes beyond 1000 h has to date been challenging because of component erosion. High-power MPD thrusters that use lithium as the propellant have demonstrated efficiencies above 60% [7, 10]. MPD thrusters also have significantly higher thrust density than ion and Hall thrusters, making high-power thrusters more compact.

As a result of these promising demonstrated performance parameters, there is continuing research and development in MPD thrusters to achieve improved efficiencies and lifetimes in the higher power and thrust regime not achievable by other electric thrusters. This work will increase in interest as power levels on the order of megawatts or more becomes available in space. The research and development goals are to obtain higher Isp at the increased power levels, enhancing the efficiency, and significantly improving thruster lifetime, perhaps up to an order of magnitude or more.

9.2.1 Self-Field MPD Thrusters

A self-field thruster has the same basic coaxial configuration of two electrodes shown in Fig. 9-1 but without the external solenoid coils. The magnetic field that interacts with the plasma and produces the Lorentz force acceleration is generated purely by the discharge current from cathode to anode. The cathode is one of the main life-limiting components in the thruster. MPD cathodes have been made of refractory metals such as tungsten [11], thoriated tungsten [12], or barium-supplied or impregnated tungsten [13, 14], hollow cathodes [15] and multi-channel hollow cathodes [16, 17]. Anode erosion can also limit the thruster life, and anodes have been made of refractory metals such as molybdenum and tungsten [12] in self-radiating designs, or water-cooled metals such as copper in higher power, continuous operation designs.

The mechanical configuration of various self-field thrusters has been described in the literature [7, 12]. Self-field MPD thrusters require high-discharge power to efficiently produce significant levels of thrust because the self-induced magnetic field is relatively weak unless very high discharge currents are applied, typically of several kiloamperes and above. This has prompted much of the testing of these thrusters to be pulsed on millisecond time scales to avoid the high-power electronics, the high heat loads on the thruster components, and the high gas load on the vacuum facility that causes high background pressures and spurious thrust measurements. Pulsed high-power MPD thrusters can achieve quasi-steady electromagnetic acceleration of the ionized plasma, but not thermal equilibrium, and the results can be difficult to scale directly to steady-state performance [2]. Models that can predict the thrust and performance are then critically important in guiding the development of this technology for flight applications.

9.2.1.1 Idealized Model of the Self-Field MPD Thrust

Thrust generation in self-field MPD thrusters was first described in an analytic model of the acceleration mechanisms by Maecker [18]. The axial force (the "blowing" mechanism) was analyzed by breaking the cathode into a straight cylindrical section of radius r_c and a conical section representing the end of the rod, as shown in the schematic drawing in Fig. 9-2. A purely radial electron current flow between the cathode and the anode at a radius r_a that is uniform along the length of the cathode rod in the region from $z = 0$ to $z = z_1$ is assumed. From Ampere's Law given by Eq. 3.2-4, neglecting the displacement current and assuming vacuum, the self-induced magnetic field in this region is purely azimuthal and proportional to $1/r$ in the gap. It is assumed that the radial discharge current is uniform along the cathode and goes to zero at the end of the cylindrical section. The magnet field can then be described by:

$$B_\theta = \frac{\mu_o I}{2\pi r}\left(1 - \frac{z}{z_1}\right), \tag{9.2-1}$$

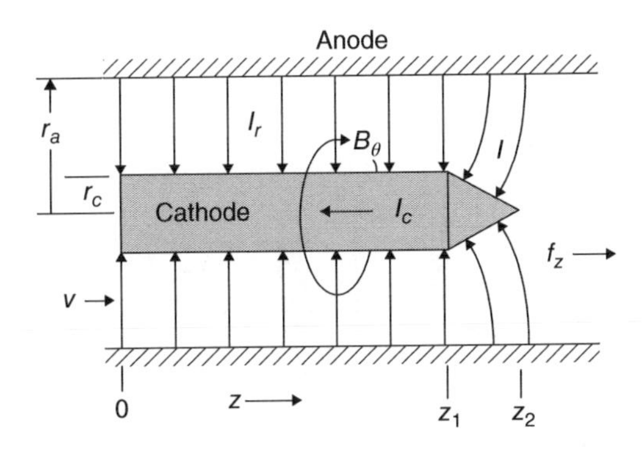

Figure 9-2 Schematic drawing of an idealized MPD thruster showing current paths and induced magnetic field.

where $I = 2\pi r z_1 J_r$ is the total discharge current in this region. The Lorentz force density is then purely axial:

$$f_z(r,z) = J_r B_\theta = \frac{\mu_o I^2}{4\pi^2 r^2 z_1^2}(z_1 - z). \quad (9.2\text{-}2)$$

The total axial force applied to the particle stream is the integral of the force density in Eq. 9.2-2 over the volume of the gap:

$$F_z = \frac{\mu_o I^2}{4\pi^2 z_1^2} \int_0^{z_1} \int_0^{2\pi} \int_{r_c}^{r_a} \frac{z_1 - z}{r^2} r dr\, d\theta\, dz = \frac{\mu_o I^2}{4\pi} \ln \frac{r_a}{r_c} \quad (9.2\text{-}3)$$

This expression is valid for any distribution of the discharge current along the cathode cylinder as long as azimuthal symmetry of the generated magnetic field is maintained.

Now we need to consider the force associated with the current from the tip of the cathode. If all of the current flows into the face of the cathode tip with uniform radial current density, the total axial force is

$$F_z = \frac{\mu_o I^2}{4\pi^2 (z_2 - z_1)^2} \int_{z_1}^{z_2} \int_0^{2\pi} \int_{r_c(1-(z-z_1)/(z_2-z_1))}^{r_a} \frac{z_2 - z}{r^2} r dr\, d\theta\, dz = \frac{\mu_o I^2}{4\pi}\left[\ln \frac{r_a}{r_c} + \frac{1}{2}\right] \quad (9.2\text{-}4)$$

If the current is not uniform over the cathode surface, the integral in Eq. 9.2-4 must include this spatial dependence, but the thrust produced will generally be a lower contribution to the total blowing force from the tip current [1].

The "pinching" contribution to the electromagnetic force can be estimated by assuming all the discharge current enters the cathode end axially along the thruster axis. The magnetic field produced is azimuthal, but is now proportional to the cathode radius:

$$B_\theta(r) = \frac{\mu_o I\, r}{2\pi r_c^2} \quad (9.2\text{-}5)$$

The Lorentz force generated from Eq. 9.2-5 is now radial, and must be balanced by a radial pressure gradient:

$$f_r = J_z B_\theta = \frac{\mu_o I^2\, r}{2\pi^2 r_c^4} = -\frac{dp}{dr}. \quad (9.2\text{-}6)$$

Assuming the end of the cathode is flat and integrating Eq. 9.2-6, the profile of the fluid pressure over the cathode face is parabolic:

$$p(r) = p_o + \frac{\mu_o I^2}{4\pi^2 r_c^2}\left[1 - \left(\frac{r}{r_c}\right)^2\right], \quad (9.2\text{-}7)$$

where p_o is the pressure in the gap outside the radius of the cathode. Since this pressure is not balanced at the anode, the integral of $(p - p_o)$ over the surface of the cathode tip generates additional thrust:

$$F_c = 2\pi \int_0^{r_c} (p - p_o) r dr = \frac{\mu_o I^2}{8\pi} \quad (9.2\text{-}8)$$

This result does not depend on the radial distribution of the current over the cathode face provided the azimuthal symmetry of the magnetic field is again maintained.

The total thrust T is the sum of the blowing and pinching terms, which for a uniform current density over the cathode surface area gives

$$T = F_z + F_c = \frac{\mu_o I^2}{4\pi}\left(\ln\frac{r_a}{r_c} + \delta\right) \tag{9.2-9}$$

with $0 < \delta \leq 3/4$. This expression is referred to as the Maecker formula [18], although the constant δ does not appear in his original derivation. Jahn showed that constants on the order of unity should be included depending on variations in the cathode geometry [1], and 3/4 tends to provide the best agreement with data at high current levels [19].

It is possible to show that Eq. 9.2-9 does not depend on the specific path of the discharge current between the anode and cathode [1]. The forces that need to be considered are on the thruster surfaces inside a control volume contoured to the inside of the thruster. The downstream boundaries of the volume are sufficiently far from the thruster that it can be assumed that the magnetic field is zero and pressure is the ambient value. The total thrust is described by

$$T = T_{GD} + \int_V J_r B_\theta \, dV - \int_S p(\hat{\mathbf{z}} \cdot d\mathbf{S}) \tag{9.2-10}$$

where V and S are the volume and surface of the control volume, p is the pressure, $\hat{\mathbf{z}}$ is the unit vector. The first term on the right, T_{GD}, is the gasdynamic contribution to the thrust that results from the cold gas propellant injected into the control volume. The second term on the right is the blowing contribution to the thrust, and the third term on the right is the pinching contribution.

The evaluation of the blowing term through volume integration requires detailed knowledge of the current distribution inside the thruster. Using the divergence theorem, the volume integral can be replaced by a surface integral and the Lorentz force written in terms of a Maxwell magnetic stress sensor, $\widetilde{\boldsymbol{\beta}}$, that satisfies the following equation:

$$\mathbf{J} \times \mathbf{B} = \frac{\mathbf{1}}{\mu_o}(\nabla \times \mathbf{B}) \times \mathbf{B} = \nabla \cdot \widetilde{\boldsymbol{\beta}} - \frac{\mathbf{1}}{\mu_o}\mathbf{B}\nabla \cdot \mathbf{B} \tag{9.2-11}$$

Since the divergence of $\mathbf{B}$ is zero, the integral of the axial component of the Lorentz force over the thruster volume gives the blowing contribution to the thrust:

$$\int_V J_r B_\theta dV = \int_V (\nabla \cdot \tilde{\boldsymbol{\beta}})_z dV = \int_S (\tilde{\boldsymbol{\beta}} \cdot d\mathbf{S})_z \tag{9.2-12}$$

For a coaxial self-field MPD thruster with a symmetric discharge, the magnetic field has only azimuthal components and the magnetic stress tensor is given by

$$\widetilde{\boldsymbol{\beta}} = \frac{1}{\mu_o}\begin{bmatrix} -\frac{B_\theta^2}{2} & 0 & 0 \\ 0 & \frac{B_\theta^2}{2r^2} & 0 \\ 0 & 0 & -\frac{B_\theta^2}{2} \end{bmatrix}. \tag{9.2-13}$$

The control volume surfaces that contribute to the surface integral are then only those perpendicular to the thrust axis over which the magnetic field is finite. Therefore,

$$\widetilde{\boldsymbol{\beta}} \cdot \hat{\mathbf{z}} = 0, 0, \frac{B_\theta^2}{\mu_o}. \tag{9.2-14}$$

Assuming uniform current density over the cathode surface, this gives the profile of the azimuthal magnetic field as follows:

$$B_\theta = \begin{cases} \dfrac{\mu_o I}{2\pi r_c^2} r & r < r_c \\ \dfrac{\mu_o I}{2\pi r} & r > r_c \end{cases}. \tag{9.2-15}$$

Using this profile of the magnetic field, the surface integral becomes:

$$\int_S \left(\widetilde{\boldsymbol{\beta}} \cdot d\mathbf{S}\right)_z = \frac{\mu_o I^2}{4\pi}\left(\ln \frac{r_a}{r_c} + \frac{1}{4}\right) \tag{9.2-16}$$

which is the blowing contribution to the Maecker formula in Eq. 9.2-9. This shows that the blowing contribution in self-field MPD thrusters does not depend on the details of the current path from cathode to anode, and so the Maecker formula is applicable to a large range of thruster geometries [1].

The thrust from self-field MPD thrusters is, therefore, often described as

$$T = bI^2, \tag{9.2-17}$$

where b is a parameter largely dependent on the geometry. Note that the thrust predicted by this equation is independent of the propellant type and the propellant mass flow rate, and scales simply as the discharge current squared and the radial size of the thruster.

The validity of the Maecker formula was investigated by Choueiri [5, 20] using data from the Princeton Benchmark Thruster (PBT) [21, 22]. The data on the thrust versus discharge current from these papers for two argon mass flow rates is shown in Fig. 9-3, with the Maecker prediction. The overall agreement between the Maecker formula and the data seems reasonable, especially for such a simple model.

However, the discrepancy of these simple predictions with discharge mass flow rate and current is exposed by defining a dimensionless thrust coefficient C_T:

$$C_T = \frac{4\pi\,\mathrm{T}}{\mu_o I^2} \tag{9.2-18}$$

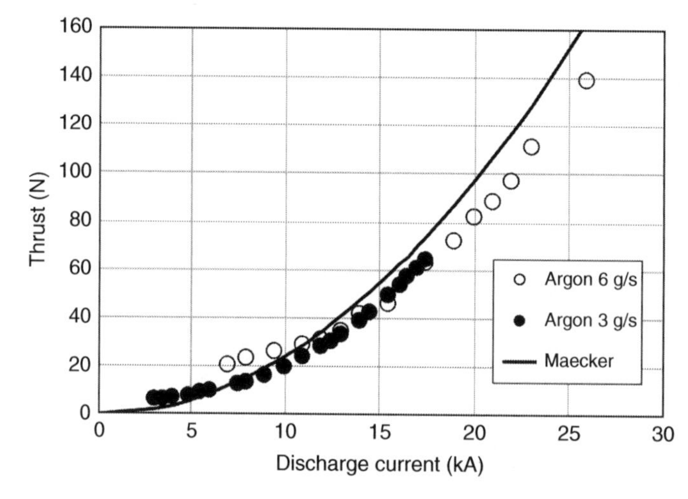

Figure 9-3 Thrust versus discharge current for the Princeton Baseline Thruster at two different flow rates of argon propellant (*Source:* Data From [5]).

which for the case of the Maecker formula is equal to

$$C_T = \ln \frac{r_a}{r_c} + \delta. \tag{9.2-19}$$

Figure 9-4 shows the thrust coefficient C_T calculated from the measured thrust by the PBT [5, 20, 21] versus discharge current for $\delta = 3/4$. This depiction of the data from Fig. 9-3 shows that at low discharge currents the thrust is dependent on the mass flow rate, and that it differs from the current squared behavior in the Maecker formula at low discharge currents. At higher discharge currents where the electromagnetic force is expected to dominate, the Maecker formula overpredicts the magnitude of C_T by a significant amount. Although discrepancy can be obviated by using a different constant for δ in the Maecker formula than 3/4, this is generally arbitrary and will vary from thruster to thruster.

To better evaluate the surface integral of the third term in Eq. 9.2-10 (the pinching thrust contribution), Choueiri [5] evaluated the radial and axial pressures in the control volume and integrated over the PBT surfaces assuming a nonuniform profile of the pressure over the backplate. He found that at low current the pinching pressure effects on the backplate of the thruster caused an increase in C_T above the Maecker value, in better agreement with the measured data. Using empirical data for the unknown current distributions along the electrodes, he showed that at high current the main contribution to the thrust is from the blowing component at the backplate with a small contribution from the pinching component. As the current is decreased the pressure profile on the backplate changes and results in the rise in C_T above the Maecker value. The rise in C_T at lower current is not from changes in the thrust caused by the expansion of an ohmically heated gas, but rather by the scaling of pressure distributions induced by the pinching effect of the volumetric Lorentz forces.

Tikhonov and Semenihin [23] derived an analytic model that assumes the magnetic and thermal pressures are equal at the downstream end of the channel, and set the upstream end of the

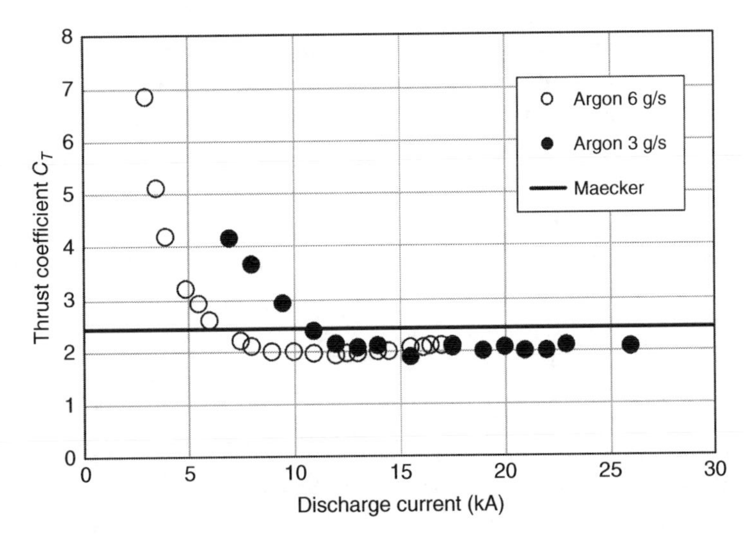

Figure 9-4 Dimensionless thrust coefficient C_T versus discharge current for two argon propellant flow rates (*Source:* [5]).

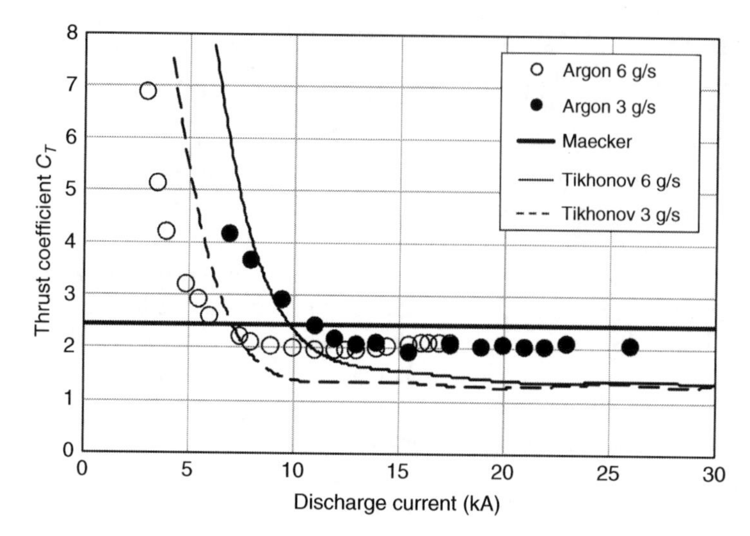

Figure 9-5 Comparison of the Tikhonov thrust coefficient with the measured PBT thrust coefficient versus discharge current (*Source:* [5]).

thruster to be immediately behind all the enclosed current. They derived another expression for the thrust coefficient:

$$C_T = \frac{\gamma + 1}{2} + \frac{\alpha_o^{-2}}{2}, \tag{9.2-20}$$

where α_o is a dimensionless parameter evaluated at the upstream end of the thruster channel,

$$\alpha_o = \frac{\gamma \mu_o I^2}{8\pi a_o \dot{m}}, \tag{9.2-21}$$

γ is the ratio of the specific heats of the propellants, and a_o is the ion acoustic speed at the upstream end of the channel. The Tikhonov formula plotted in Fig. 9-5 predicts the correct shape of the thrust coefficient at low current, scaling as I^{-4} and the constant C_T at high current. However, the quantitative agreement with the data remains poor.

9.2.1.2 Semi-empirical Model of the Self-Field MPD Thrust

Choueiri developed a semiempirical scaling model [5] for the thrust coefficient in self-field MPD thrusters:

$$C_T = \frac{\dot{m}}{\dot{m}^*}\frac{1}{\xi^4} + \ln\left[\frac{r_a}{r_c} + \xi^2\right], \tag{9.2-22}$$

where $\dot{m}$ is the mass flow rate, and $\dot{m}^*$ is a reference mass flow rate that must be determined empirically from the thruster data. The nondimensional thruster current parameter ξ is defined as

$$\xi \equiv \frac{I}{I_{ci}}, \tag{9.2-23}$$

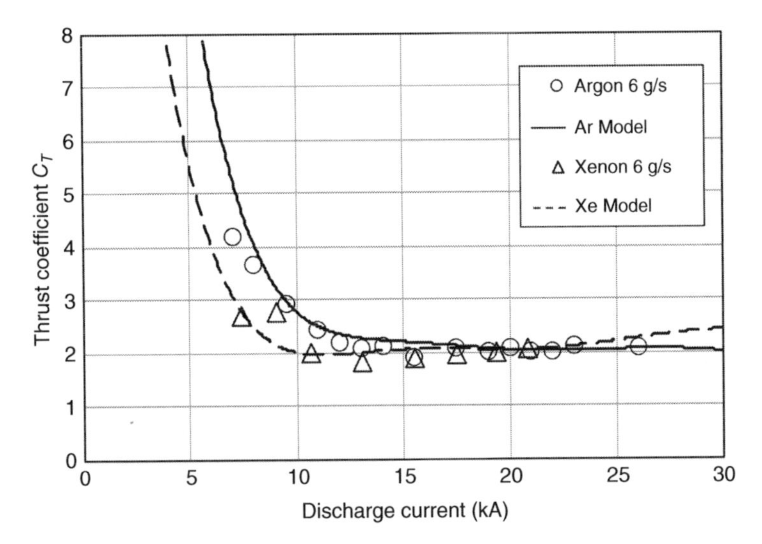

Figure 9-6 Semiempirical self-field MPD thrust coefficient model predictions for argon and xenon at fixed flow rate in the PBT (*Source:* [5]).

where I_{ci} is the thruster current that produces an exhaust velocity equal to the critical ionization velocity:

$$I_{ci} = \left[\frac{4\pi \dot{m} u_{ci}}{\mu_o C_T}\right]^{1/2}. \tag{9.2-24}$$

The critical ionization velocity, originally derived by Alfven [24], is given by

$$u_{ci} = \left(\frac{2\varepsilon_i}{M}\right)^{1/2}, \tag{9.2-25}$$

where ε_i is the first ionization potential of the neutral propellant atom. Combining Eqs. 9.2-23 and 9.2-24 gives

$$\xi = \sqrt{\frac{\mu_o C_T}{4\pi u_{ci}}}\frac{I}{\sqrt{\dot{m}}}. \tag{9.2-26}$$

For self-field MPD thrusters, similar thrust and Isp can be expected if the value of ξ is similar. In addition, operation at values of $\xi > 1$ is related to the onset of current-driven instabilities and large voltage fluctuations termed "onset", which will be discussed further in Section 9.2.3.

Figure 9-6 shows the thrust coefficient calculated from Eq. 9.2-22 versus discharge current for the PBT operating in argon and xenon. The semiempirical model provides quantitative agreement with the thruster data, and the model predictions are consistent with the data trends at both low and high current. This scaling model can, therefore, be useful in self-field MPD thruster design and performance optimizations.

9.2.2 Applied-Field MPD Thrusters

In the applied-field MPD thruster, an external solenoid is added around the thruster body to produce a magnetic field that diverges toward the thruster exit, as shown in Figs. 1–5. Different mechanical configurations of applied-field MPD thrusters have been described in several review papers [2, 7, 8, 12, 25]. The self-induced field from the discharge current is usually significantly less than the applied-field at low discharge currents, but as the current is increased the self-fields become comparable to or higher than the applied-induced fields, and the net magnetic field lines become twisted in a helical fashion. The strong axial component of the applied magnetic field between the cathode and anode inhibits the electron flow to the anode and causes the electrons to circulate around the cathode rod reminiscent of the Hall current in Hall thrusters. This also forces the discharge current and the plasma plume to move downstream of the cathode end to reach the anode. The fraction of the thrust generated within the channel can be relatively small and the Lorentz force due to the applied field in this region results in a swirling of the plasma around the axis of symmetry. Downstream of the cathode tip where the discharge current bends toward the anode, the axial magnetic field produces an azimuthal component of the force that contributes to the swirl. The total axial and radial force components then produce the blowing and pinching contributions to the thrust.

The thrust from an applied field MPD can be produced by a combination of several acceleration mechanisms, with both the electromagnetic and gas dynamic processes contributing. Four potential acceleration mechanisms have been identified [2, 8, 26]:

- **Self-Field Acceleration** Just as in MPD thrusters without an applied magnetic field, the axial component of the cathode current in applied-field thrusters produces an azimuthal magnetic field inside the thruster. The radial component of the discharge current between cathode and anode induces a Lorentz $J_r B_\theta$ force density that produces the axial blowing force, and the axial component of the discharge current induces a Lorentz $J_z B_\theta$ force density that produces the radial pinching force. The axial blowing force contributes directly to the total self-field thrust T_{SF}, and the radial pinching force contributes indirectly to the thrust because of a pressure imbalance on the center cathode electrode.
- **Hall Acceleration** If the applied magnetic field is high enough, and the internal pressure is low enough such that the Hall parameter (defined in Chapter 3) is significantly greater than 1, an azimuthal current will be induced reminiscent of that found in Hall thrusters. The induced azimuthal current and the applied magnetic field produce the blowing $J_\theta B_r$ force density and pinching $J_\theta B_z$ force density components in a similar manner to the self-field case. This contribution of the Hall acceleration thrust T_H to the total thrust likely only becomes significant for large applied magnetic fields.
- **Swirl Acceleration** The Lorentz force interaction of the radial and axial components of the applied magnetic field and the discharge current ($J_r B_z$ and $J_z B_r$) causes the plasma to rotate (swirl) in the azimuthal direction. A magnetic nozzle formed by the diverging magnetic field in the exit region of the thruster can convert this rotational energy to axial thrust T_{SW} because of the expansion of the rotating plasma through the magnetic field, at which point the accelerated plasma detaches from the magnetic field lines [27].
- **Gasdynamic Acceleration** A gasdynamic component of the thrust T_{GD} is also produced from joule heating of the propellant gas similar to that found in electrothermal arcjets. This term becomes significant when the mass flow rate is high or collisional heating of the gas is substantial.

Based on these descriptions, the total thrust produced by an applied-field MPD thruster has been traditionally expressed as the sum of these contributions:

$$T = T_{\mathrm{SF}} + T_H + T_{\mathrm{SW}} + T_{\mathrm{GD}}, \tag{9.2-27}$$

The self-field contribution T_{SF} was described in Section 9.2.2 above and still applies to applied-field MPD thrusters. The Hall and swirl components in Eq. 9.2-27 are discussed in the next Section. The gasdynamic contribution to the thrust, T_{GD}, is the result of the conversion of internal energy into directed kinetic energy in the nozzle formed by the anode. This term is often neglected at high powers where the other electromagnetic terms dominate, but at low values of the discharge current and applied magnetic field, or at high mass flow rates, T_{GD} can be substantial [26]. This component depends on the mass flow rate, the velocity at the injection site, the gas dynamic pressure inside the nozzle, and the nozzle area over which that pressure is applied.

9.2.2.1 Empirical and Semi-empirical Thrust Models

Self-field MPD thrusters operating below the critical current for onset show a relatively straightforward thrust scaling proportional to bI^2, as described above. However, the numerous possible acceleration mechanisms present in an applied-field MPD represent a much more complicated endeavor in producing a thrust model. Most models published to date [23, 26, 28–32] are empirical or semi-empirical; thus, partially obscuring a comprehensive explanation of the dominant acceleration mechanisms. Moreover, in most cases these models have not been validated by comparisons to a wide and diverse range of experimental data, but rather applied to one or two specific thrusters from the original authors. A notable exception is the recent work by Coogan [26] who assembled an extensive database of relevant measurements from applied-field thruster experiments, reviewed almost all existing thrust models, provided comparisons and, ultimately, proposed a new empirical model.

With the exception of Coletti [33] and Sasoh and Arakawa [32], most of the cited thrust models above provide simple analytic expressions, which is a sought-after feature by the MPD designer because they are straightforward to apply. However, they also suffer from notable drawbacks. Specifically, most of them [23, 26, 28–31] scale the applied-field thrust T_{AF} linearly with the product of current and applied magnetic field, $T_{AF}\,\tilde{}\,IB_A$. For example, in the empirical formula developed by Coogan [26] based on fits to measurements in their extensive database, this linear dependence on IB_A is clear:

$$T_{\text{Coogan}} = 1.14\, IB_A r_a \overline{\Phi}^{\,-0.13} \left({}^{r_a}\!/\!{}_{r_c}\right)^{-0.3} \left(10 + {}^{l_c}\!/\!{}_{l_a}\right)^{-0.67}. \tag{9.2-28}$$

In Eq. 9.2-28, $\overline{\Phi}$ is the magnetic flux at the anode exit plane normalized to the magnetic flux at the anode throat (minimum anode inner diameter), l_c is the cathode length, and l_a is the anode length. Physically $\overline{\Phi} = 1$ represents an anode contoured to the magnetic field, and $\overline{\Phi}>1$ represents an anode that diverges rapidly compared to the magnetic field. The authors emphasized the significance of the parameter $\overline{\Phi}$, which was absent in previous models, and led to an improved fit to the thrust measurements. They argued that the improvement was largely due to a better representation of the effective anode radius and the volume over which the Lorentz force acts. Nevertheless, the linear dependence of thrust on IB_A remains intact in Eq. 9.2-28 and consistent with previous empirical models that are widely reported in the literature. The drawback with such scaling is that the best linear fits to the data do not yield the expected limit that the thrust must approach zero, or more accurately approach the cold thrust value, as the current goes to zero at finite applied field B_A. By "cold thrust" we mean the force produced by injecting propellant through the thruster in the absence of a discharge, which is typically negligible compared to the self-field and applied-field components (in the presence of the discharge) except when the flow is very high. This is clearly illustrated by the dashed lines in the representative example of the Moscow Aviation Institute's (MAI) 30-kW Lithium thruster in Fig. 9-7. Moreover, empirical models akin to those in Eq. 9.2-28 do not account for all controllable parameters, like the mass flow rate and propellant mass, which are both known to affect thrust as shown below in Figs. 9-8 and 9-9.

9.2.2.2 First-principles Thrust Model

The complexity of the physics in the applied-field MPD made first-principles *ab initio* thrust models challenging to develop and validate. A notable exception is the theoretical work of P. Mikellides and Turchi [34] who employed extensive 2-D axisymmetric resistive magnetohydrodynamic simulations to guide the development of their thrust model. They recognized that the (viscous) Reynolds number associated with the plasma swirl can be in the tens to hundreds [36] and argued that applied-field MPD operation is dominated and limited by viscous effects. Specifically, they proposed that the rotational speed produced by $J_r B_z$ is opposed by viscous forces, which gives rise to a critical maximum rotational speed v_*:

$$v_* = \frac{IB_A}{\mu} \tag{9.2-29}$$

where μ is the ion dynamic viscosity. Thus, as IB_A is increased the swirling flow becomes fully developed and thus the rotational speed reaches the maximum critical speed beyond which point it is independent of IB_A. Further increase of IB_A does not result in increasing azimuthal speed, rather viscous dissipation converts such rotational kinetic energy to internal energy [36]. This is similar to the Alfven critical speed [24] that limits self-field thrusters; once the plasma reaches such critical speed, deposition to internal modes dominates as opposed to exhaust kinetic energy conversion, the so-called onset phenomenon.

For applied-field thrusters, this critical speed arises from viscous dissipation, thus the main acceleration mechanism is conversion of enthalpy – increased by viscous heating – to exhaust kinetic

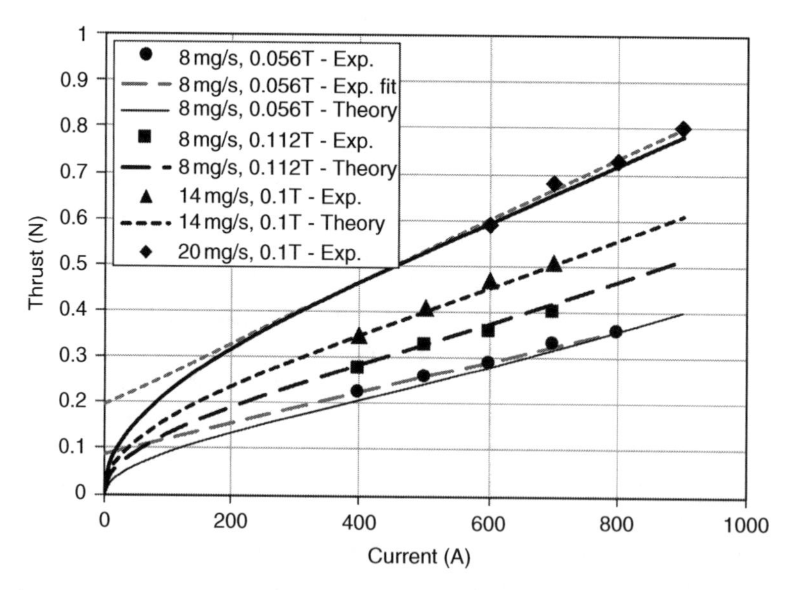

Figure 9-7 Comparison of the predicted thrust from Eq. 9.2-30, using the theoretical model by P. Mikellides for the applied-field contribution T_{AF} Eq. 9.2-31 [34], to experimental data obtained from the MAI (30-kW, Lithium) AF MPD thruster [26, 35]. The dotted lines represent the best linear fit to the data.

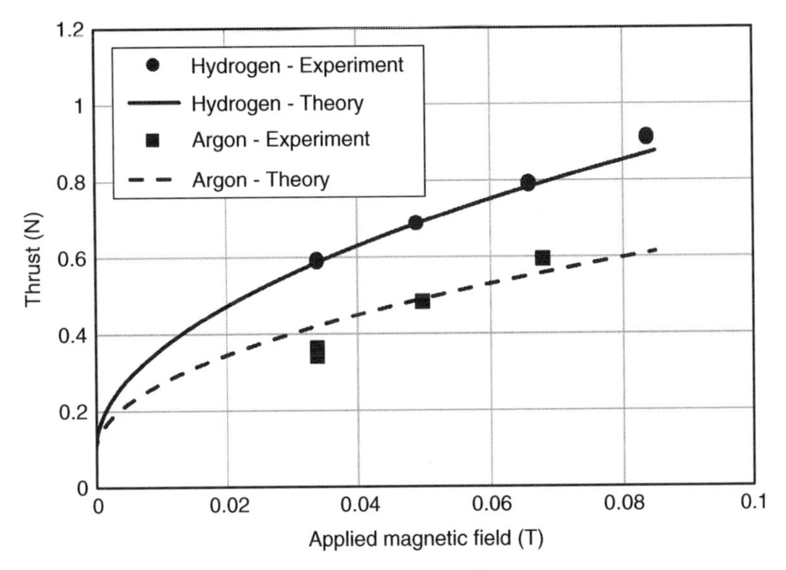

Figure 9-8 Comparison of the predicted thrust from Eq. 9.2-30, using the theoretical model by P. Mikellides for the applied field contribution T_{AF} Eq. 9.2-31 [34], to experimental data obtained from the LeRC 100-kW AF MPD thruster [31] for different propellants at 750 A and 25 mg/s.

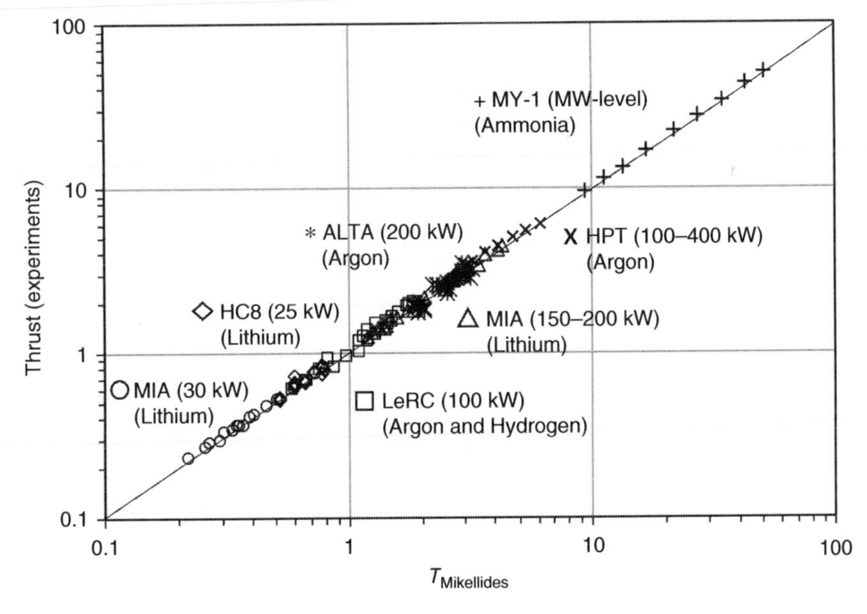

Figure 9-9 Comparison of the predicted total thrust from the theoretical model by P. Mikellides Eq. 9.2-30 [34] to a wide range of applied-field MPD experimental thrust data.

energy. In the simulations performed by P. Mikellides, the Reynolds number was found to be ~10 implying that viscous forces are indeed quite significant when compared to inertial forces [36]. Furthermore, the simulations revealed that Hall acceleration is almost non-existent due to anomalous resistivity effects with minimal contributions by the magnetic nozzle. Based on these insights, the authors developed a first-principles analytical model, which in conjunction with the self-field thrust contribution T_{SF} from Eq. 9.2-9, yields the total thrust in the applied-field MPD thruster:

$$T_{\text{Mikellides}} = T_{\text{AF}} + T_{\text{SF}} \tag{9.2-30}$$

where the applied-field thrust T_{AF} is given by [34]

$$T_{\text{AF}} = \frac{C}{M^{1/4}} \sqrt{\frac{a}{\zeta}} \frac{R(1+R)\sqrt{R-1}}{\sqrt{R^{3.8}-1}} \sqrt{\dot{m} I B_A} \tag{9.2-31}$$

and $C = 25\ ms^{1/2}$, M is the dimensionless numerical value of the propellant's atomic/molecular mass, $a = r_c/l$, ζ is the average degree of ionization, $R = r_a/r_c$. For the self-field thrust T_{SF}, the parameter δ is in the range $0 \le \delta \le 3/4$. The dimensional variables are in SI units. The parameter l represents the electrode length along which current conduction occurs, in most cases the anode length suffices. For flared anodes an average radius will produce more accurate results. For calculation of the average degree of ionization, the Saha equation can be used even though for most cases assuming fully singly ionized plasma, ζ=1, will produce acceptable estimates.

The model is distinct from all others as it scales $T_{\text{AF}} \sim \sqrt{\dot{m} I B_A}$ and includes dependence on the propellant mass. The square-root dependence with the product IB_A is also consistent with the conclusions from the experimental findings of Coogan [35] in two lithium MPD thrusters, who found that the applied-field thrust component is not proportional to current or applied field strength, as has been assumed in much of the literature, but it is more closely proportionate to $\sqrt{IB_A}$. Coogan also found the square-root dependence on the mass flow rate, $\sqrt{\dot{m}}$, as P. Mikellides predicted, but incorrectly contended that viscous effects were not significant.

A comparison of the thrust predicted by Eqs. 9.2-30 and 9.2-31 with measurements from two different thrusters is presented in Figs. 9-7 and 9-8. The model by P. Mikellides predicts the correct limit for the thrust as the current approaches zero at finite applied field. Furthermore, the data at 0.1 T showed the significant effect of the different mass flow rates on the thrust. It is worthwhile to note that the MAI thruster utilized a recessed hollow cathode with a flared anode. Figure 9-8 compares the model to the LeRC 100-kW AF MPD [31] operating with two different propellants, hydrogen and argon, which have substantially different atomic/molecular masses. The dependence on propellant type is clear, which implies that any high-fidelity thrust model must include such dependence.

To fully validate any first-principles model, a comparison to a wide and diverse range of experimental data is required. Figure 9-9 shows such comparison of the model by P. Mikellides Eq. 9.2-30, which includes both the self-field and applied-field effects, to selected applied-field MPD thrust data (MAI [35], LeRC [31], Alta [37], HC8 [38], HPT [39] and MY-I [40], which include thrusters with recessed cathodes, hollow cathodes, flared anodes, pre-ionization, and propellant injection schemes as well as a variety of propellants, geometries, and operating conditions. The excellent agreement establishes the high fidelity of the model and, perhaps more importantly, considering that Eqs. 9.2-30 and 9.2-31 are the derivations from a purely analytical and computational exercise, it offers the opportunity to gain deeper insight into the driving processes of applied-field MPD operation compared to empirical or semi-empirical approaches. It should be recognized that research in this area is still ongoing.

9.2.2.3 Lithium Applied-Field MPD Thrusters

Applied-field MPD thrusters that use lithium are sometimes called Lorentz Force Accelerators (LFA), and are characterized by the use of lithium propellant and the inclusion of a multichannel hollow cathode as the electron emitter. Lithium has several advantages over conventional gaseous propellants like xenon and argon, and other alkali metals [15]. Lithium has a lower first ionization potential (5.4 eV) than the other propellants (e.g. 12.1 eV for xenon), which means less power is spent on ionization. In addition, lithium has a very high second ionization potential (75.6 eV compared to 21.0 for Xenon), reducing the production of multiply charged ions in the discharge. Using lithium as the propellant, therefore, helps to reduce these frozen flow losses, and may reduce the work function of the cathode surface.

The performance using lithium as the propellant has significantly exceeded those of other propellants. Lithium MPD thrusters developed in Russia have reported efficiencies above 50% in steady state operation for hundreds of hours at power levels of 400–500 kW [10]. Operation at lower power levels is improved because the applied field offsets the lower self-field occurring from lower discharge currents. For example, a 100 kW MPD thruster, shown in Fig. 9-10, operating with <0.1 T applied fields achieved up to 40% efficiency at a discharge current of 1–2 kA and a voltage of only 50–70 V [41]. Note in the design of this steady-state thruster the use of a central lithium vaporizer inside the multichannel hollow cathode, the external magnetic coil, and the conical anode. Operating a modified version of this thruster at higher powers up to 200 kW led to increases in efficiency over 50% and Isp to over 5000 s [42].

Research on lithium-fed applied-field MPD thrusters continues in the US, primarily at Princeton University [43], and a 1-MW steady-state test facility has been established but not yet operated at the Jet Propulsion Laboratory. Significant research on various MPD thruster configurations also continues in Europe, Russia, and Asia [8, 44]. Lithium MPD thrusters are attractive from an efficiency viewpoint, and using a solid, storable propellant allows more compact tankage for long duration operation in space. However, lithium is a condensable propellant with the potential to coat sensitive spacecraft and payload surfaces, and such plume effect must be carefully considered. The use of non-condensable gas propellants mitigates possible ground handling and spacecraft contamination issues associated with lithium/metallic propellants, but significant further development is required to provide comparable efficient thruster performance.

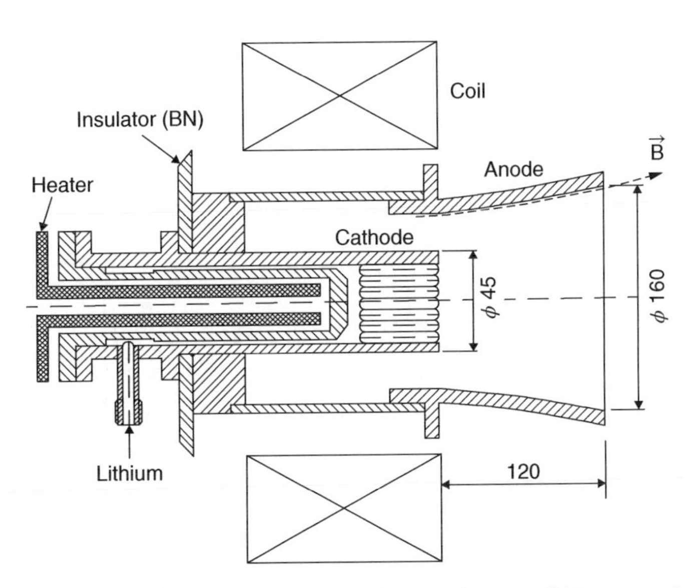

Figure 9-10 Schematic of a 130 kW applied-field MPD thruster that uses lithium propellant and a multichannel hollow cathode (*Source:* [41]/IEPC).

9.2.3 Onset Phenomenon

It has been observed in all MPD thrusters that there is a critical discharge current above which the discharge voltage increases nonlinearly and becomes noisy, and electrode erosion (primarily the anode) is observed to increase. An example of the increase in the discharge voltage magnitude and noise level measured in an MPD thruster as the discharge current is raised [45] is shown in Fig. 9-11. This behavior, first reported in the United States by Malliaris in 1972 [46], became known as the *onset phenomenon*, or simply *onset*, and has been confirmed by a number of researchers in many laboratories worldwide. Onset in MPD thrusters is a significant problem that limits the Isp attainable by constraining the discharge current, and, therefore, limiting the thrust, at a given flow rate [47]. Application of a weak magnetic field in the downstream region of the anode in self-field thrusters was shown to delay onset to higher currents [10, 48]. These thrusters then enter the class of applied-field thrusters, and optimizing the shape and strength of the magnetic field has been shown to significantly reduce onset-related behaviors [48].

To describe onset, Malliaris identified a critical value k, which is essentially a constant for a given thruster, above which onset occurs:

$$k = \frac{I^2}{\dot{m}}. \tag{9.2-32}$$

The onset condition is usually defined as the value of k at which high-frequency voltage fluctuation levels exceed 10% peak-to-peak of the discharge voltage. An example of the discharge voltage trend in an MPD thruster as the current is increased for three increasing propellant flow rates is shown in Fig. 9-12. The voltage first increases linearly with discharge current, indicative of resistivity in the plasma associated with the current crossing the magnetic field lines to the anode. At a higher current, the voltage scales as I^3 associated with electromagnetic effects dominating the discharge voltage [12]. At a critical current that depends on the gas flow rate, the discharge voltage increases significantly associated with onset.

An onset criterion that includes a dependence on propellant atomic weight was later proposed by Hügel [49]:

$$k^* = \left(\frac{I^2 M_a^{1/2}}{\dot{m}}\right). \tag{9.2-33}$$

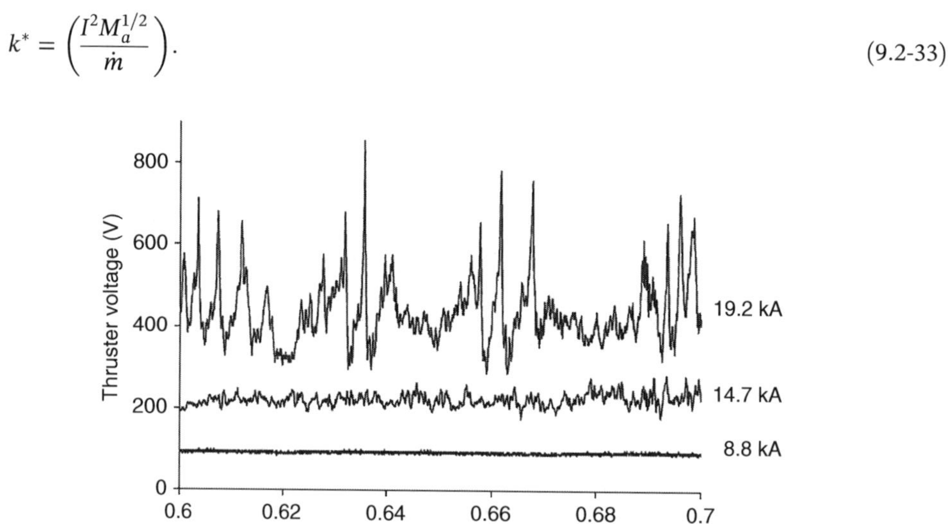

Figure 9-11 MPD discharge voltage versus time traces for three different currents showing increase in magnitude and noise with current (*Source:* [45]).

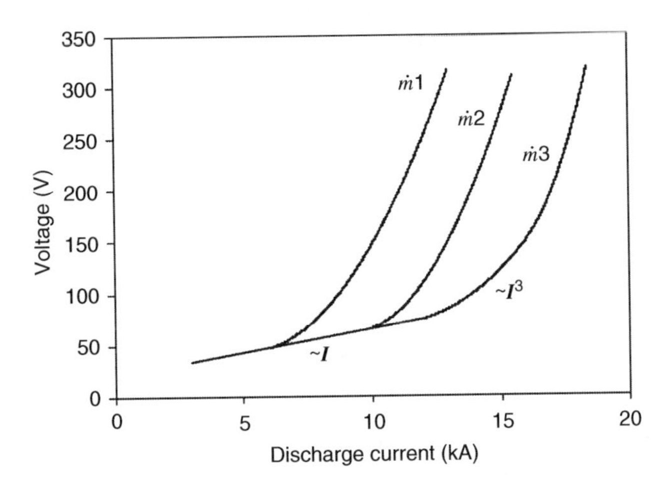

Figure 9-12 Trend of discharge voltage versus current for three propellant flow rates showing onset producing increasing voltage in MPD thrusters.

Since k^* is thruster-dependent constant, using lower atomic mass propellants (e.g. lithium, hydrogen) in a given thruster design allows stable operation at higher discharge currents before the transition to onset-related behavior occurs. Also, thruster geometries and flow conditions that increase the plasma density near the anode have been found to increase the transitional value of k^*. For example, propellant injection close to the anode surface was found to increase the value of k^* at which onset occurs [50].

A detailed review of the large body of literature published over the years on onset can be found in Appendix D of the Ph.D. thesis of Uribarri [51]. Once the onset threshold current is exceeded, the magnitude of the voltage noise (hash) increases significantly and the erosion of all thruster components, in particular that of the anode, rises steadily with increasing current [52]. In addition, localized *anode spots* (sometimes erroneously called anode arcs) associated with current concentration and local melting appear on the anode surface as the discharge current is increased above onset [45, 53]. The development of anode spots is indicative of large anode sheath voltages that increase the power into the anode surface and evolve material into the gap that can then be ionized and create more charge carriers. The concentration of the discharge and localization of the heating to erode more material to fuel the spot is similar to the process in cathode arc formation, except that it is electron bombardment heating and sublimating the anode surface instead of ions bombarding a cathode surface. The voltage fluctuations ("hash") are then explained by the formation, extinction, and movement of anode spots [53]. While not triggering onset, anode spots appear to be generated in onset and are unacceptable for long-duration thruster operation in space due to the extreme erosion of the anode surface.

The critical current at which onset occurs was discussed in Section 9.2.2.2. In his derivation of the thrust coefficient C_T described there, Choueiri [5] defined the critical current I_{ci} (given in Eq. 9.2-22). This led to the derivation of a nondimensional thruster current $\xi = I/I_{ci}$ given in Eq. 9.2-24 that was used to estimate the thrust in self-field MPDs using Eq. 9.2-20, but where $\xi > 1$ also results in onset.

Another approach toward an analytic description of the onset current in self-field MPDs was developed by Tikhonov and Semenihin [23] and described by Kodys [8]. Tikhonov defined a term

A_o as the ratio of the electromagnetic to gasdynamic pressure at the location of the transition from subsonic to supersonic flow:

$$A_o = \frac{\mu_o \gamma I^2}{8\pi a_o \dot{m}}, \tag{9.2-34}$$

where a_o is the plasma sound speed given by

$$a_o = \sqrt{\frac{\gamma k T_e}{M}}. \tag{9.2-35}$$

He then defined the critical current I^* empirically by plotting A_o versus r_a/r_c for several thrusters operating just below onset. From this, Tikhonov then found for self-field MPD thrusters:

$$I^* = \sqrt{\frac{28.8\pi a_o \dot{m}}{\mu_o \gamma (r_a/r_c - 1/2)}}. \tag{9.2-36}$$

The nondimensional thruster current for this case is

$$\xi_T \equiv \frac{I}{I^*} = \sqrt{\frac{\mu_o \gamma (r_a/r_c - 1/2)}{28.8\pi a_o}} \frac{I}{\sqrt{\dot{m}}}. \tag{9.2-37}$$

Kodys asserted [8] that over a range of operating conditions the values of ξ and ξ_T agree well, despite the different derivation approaches.

An empirical relationship for the critical current for onset in applied-field MPD thrusters was also developed by Tikhonov and explained by Kodys [8]. Tikhonov defined a term A_o^B, which is the ratio of the electromagnetic to gasdynamic pressure at the location of the transition from subsonic to sonic flow in applied-field thrusters:

$$A_o^B \equiv \frac{(r_a - r_c) I B_A}{4\pi a_o \dot{m}}, \tag{9.2-38}$$

where B_A is the applied magnetic field and a_o is the sound speed. He again empirically determined the applied-field critical current I_B^* by plotting A_o^B versus r_a/r_c for several thrusters operating at the edge of stability [8]:

$$I_B^* = \frac{14.4\pi a_o \dot{m}}{B_A r_a (1 - r_c/r_a)(r_a/r_c - 1/2)}, \tag{9.2-39}$$

Similar to the self-field case, the ratio of the discharge current to the applied-field critical current I_B^* is given by [8]:

$$\xi_B \equiv \frac{I}{I_B^*} = \frac{I B_A r_a (1 - r_c/r_a)(r_a/r_c - 1/2)}{14.4\pi a_o \dot{m}}. \tag{9.2-40}$$

The values of ξ_T and ξ_B predict the stable operating limits in self-field and applied-field thruster. Operation with these terms greater than 1 results in reduced performance [8], higher voltage fluctuations, and increased erosion.

As discussed below, there are two primary theories for onset; *anode starvation* and *plasma instabilities*, although, additional effects that may contribute to onset have also been discussed in the literature [2, 51].

9.2.3.1 Anode Starvation

The discharge current is carried to the anode by the plasma electrons that have a flux to anode sheath edge determined by the plasma density and electron temperature. In the MPD thruster, the axial component of the electron current interacting with the azimuthal component of the self-induced magnetic field produces a radial inward Lorentz force that results in "pinching" of the plasma toward the axis described in Section 9.2.2.1. This motion reduces the plasma density at the anode, which lowers the flux of electrons arriving at the anode sheath edge and potentially "starving" the discharge current collection.

Anode starvation can be understood by examining the electron current collected by the anode from the local plasma. For plasma potentials positive relative to the anode (an "electron-repelling" sheath), the electron current that passes through the sheath, I_a, and collected at the anode is the surface integral of the random electron flux from the local plasma density over the anode area A times the Boltzmann factor for the effect of the sheath on the transmitted current:

$$I_a = \left[\iint \frac{1}{4} n(r,z) e \sqrt{\frac{8kT_e}{\pi m}} dA\right] e^{\left(\frac{e(V_a - \phi_p)}{kT_e}\right)}, \tag{9.2-41}$$

where $n(r,z)$ is the plasma density at the anode sheath, k is Boltzmann's constant, V_a is the anode potential, and ϕ_p is the plasma potential relative to the anode. If the system is trying to drive more electron current to the anode than the random electron flux from the plasma integrated over the anode area, then the plasma potential goes negative relative to the anode potential [54] (called a "positive going" or "electron accelerating" sheath that produces an "anode fall" voltage). This increases the electron current collection by the anode by pulling in electrons in the Maxwellian distribution that are not initially headed toward the anode. The plasma electron current collected by the anode then becomes [55]

$$I_a = \left[\iint \frac{1}{4} n(r,z) e \sqrt{\frac{8kT_e}{\pi m}} dA\right] e^{\left(\frac{e(V_a - \phi_p)}{kT_e}\right)} \left[1 - \text{erf}\left(\frac{-e\left(V_a - \phi_p\right)}{\text{kT}_e}\right)^{1/2}\right]^{-1} \tag{9.2-42}$$

where $(V_a - \phi_p)$ is now negative and the current integral is still over the entire anode surface. The increased anode sheath voltage acts to accelerate the electrons into the anode, which increases the power deposition and can lead to anode spots and plasma instabilities that show up as voltage fluctuations or "hash" on the discharge voltage signals.

Anode starvation is a direct result of the pinch force in MPD thrusters reducing the plasma density near the anode. Applied-field MPDs provide a counter to this mechanism because the axial component of the applied field that inhibits the radial plasma motion and results in rotation or swirl. Optimizing the shape and magnitude of the applied magnetic field relative to the anode has been shown to significantly decrease the start of onset [48]. Over a broad range of currents, optimally applied magnetic fields reduce the amplitude and frequency of the voltage fluctuations, decrease the anode fall voltages, and lower the mean discharge voltages. These results imply substantial improvements in efficiency and lifetime are likely to be obtained through the use of appropriately designed and tailored applied magnetic fields to locally influence near-anode phenomena that drive onset [56]. This also suggests that increases in the anode-adjacent particle density, through local propellant injection [57] or geometry changes [47], could delay starvation and onset.

9.2.3.2 Plasma Instabilities

Another explanation for onset phenomena is related to plasma instabilities. Various theories have been developed that suggest that operation at or above the critical onset current creates conditions in the thruster channel that results in the growth of unstable oscillation modes. Choueiri [58] investigated current driven instabilities such as ion cyclotron waves in MPDs. Tilley et al. [59] investigated lower hybrid and electron-cyclotron drift instabilities in a 10 kW self-field MPD thruster. Wagner et al. [60] investigated space-charge gradient-driven instabilities, and Maurer et al. [61] and Wagner et al. [62] examined drift waves in MPD thrusters. However, as pointed out by Andrenucci [2], these results and others in the literature contribute to understanding the fluctuation levels observed in the thruster voltage and in the plasma parameters [63], and anomalous transport and energy transfer effects in MPD thrusters, but don't appear to be directly related to the fundamental occurrence of onset phenomena.

9.2.3.3 Other Onset Effects

In addition to the theories described above, a number of additional theories can be found in the literature. For example, Lawless et al. [64] attribute onset to the generation of a back EMF that reduces the electric field and flow in the plasma. They examined a fully ionized, compositionally-frozen flow and both equilibrium and non-equilibrium flow from cathode to anode in the discharge and found, under conditions related to onset, that a back-EMF was generated that reduced the total electric field to levels insufficient to drive the desired current. Several macroscopic instabilities have also been proposed for explaining onset. Zuin et al. [65, 66] investigated MHD instabilities that produced a rotating kink instability that they related on onset occurrence. Schrade et al. [67, 68] examined a macroscopic helical instability in a discharge column between cathode and anode, and claimed that the concentration of the discharge into the helical channel was indicative of onset behavior and also anode damage. Additional papers describing other potential theories for onset are listed in Uribarri's Ph.D. thesis Appendix [51].

9.2.4 MPD Thruster Performance Parameters

There have been surveys of MPD thruster performance published periodically over the years [7, 8, 12, 22, 44, 69, 70]. These papers show the significant improvements in performance and life due to better understanding of the thruster physics combined with the addition of applied magnetic fields and the use of propellants such as lithium. Some trends, summarized by Kodys [8] are:

- At low powers (<50 kW), applied-field thrusters can obtain the same levels of efficiency and Isp achieved by megawatt-class self-field thrusters, of course at much lower thrust. This is primarily because the Isp is increased by the addition of the applied magnetic field at a given power.
- An optimal applied field strength and shape exists that depends on the thruster power, propellant, mass flow rate, and geometry.
- Injection of some fraction of the propellant through the anode or near the anode surface has been shown to improve performance.
- Lithium and hydrogen propellants have demonstrated the highest efficiency and Isp.
- A hollow cathode or multi-channel hollow cathode can sustain higher currents with less erosion than a solid rod cathode.

Several papers have also described scaling laws for the thruster parameters [6, 12, 22]. The thrust in self-field MPDs scales (below onset) as bI^2. The corresponding discharge voltage was described by Sovey [12] as given by

$$V = \int \frac{\mathbf{J}}{\sigma} d\mathbf{l} + \int (\mathbf{v} \times \mathbf{B}) \cdot d\mathbf{l} + V_F \tag{9.2-43}$$

where **J** is the current density, σ is the plasma conductivity, **v** is the plasma velocity, **B** is the magnetic field strength, ***dl*** is the path length from cathode to anode, and V_F is a voltage "fall" accounting for the drop in the sheath potentials. Integrating the first term yields an Ohm's law dependence that produces a linear dependence of the voltage on the discharge current. Integrating the second term for the electromagnetic effects, assuming thrust proportional to bI^2 as the dominant component in self-field MPDs, gives a voltage term proportional to the current cubed:

$$V \propto \frac{b^2 I^3}{\dot{m}}. \tag{9.2-44}$$

As the discharge current is increased, the electromagnetic voltage term will dominate over the linear ohmic term, as illustrated in Fig. 9-10. For the voltage at high discharge currents described by Eq. 9.2-47, the power into the thruster (I times V) scales as the discharge current to the fourth power. These ideal self-field MPD thruster scaling laws were summarized by Andrenucci and Paganucci [6] as:

$$T_{\mathrm{SF}} \propto I^2, V \propto I^3, P \propto I^4 . \tag{9.2-45}$$

Applied-field thrusters are much more complicated and simple scaling laws like these are not generally available. The thrust from the empirical and semi-empirical applied-field thruster models described in Section 9.2.2.1 scale as

$$T_{\mathrm{AF}} \propto k\, I B_A \tag{9.2-46}$$

where k is a constant that depends on the anode radius and other geometric terms. The Isp then scales as

$$\mathrm{Isp}_{\mathrm{AF}} \propto \frac{k I B_A}{\dot{m}}. \tag{9.2-47}$$

However, the thruster discharge voltage and power are now complicated parameters depending on the magnitude of the various thrust mechanisms given in Eq. 9.2-25 that vary as the geometry and operating conditions of the thruster are established and the discharge current is increased.

The first-principles theoretical model by P. Mikellides and Turchi [36] differs from all the others in that it proposes that the applied-field thrust scales as

$$T_{\mathrm{AF}} \propto k \sqrt{\dot{m} I B_A} \tag{9.2-48}$$

where k depends on geometry and propellant type and suggests additional and specific avenues for improvements. Likewise, the Isp then scales as

$$\mathrm{Isp}_{\mathrm{AF}} \propto k \sqrt{\frac{I B_A}{\dot{m}}}. \tag{9.2-49}$$

The P. Mikellides model for the applied-field component Eq. 9.2-31 also contains a propellant mass dependence that is consistent with observations that low atomic/molecular mass propellants, e.g. hydrogen, improve performance, and provides a physical explanation for this effect. Furthermore, the effort produced analytic expressions for plasma voltage and flow efficiency that suggests an optimum thruster geometry. Specifically, the model proposes that narrow flow channels ($1.5 \le R \le 2$) in combination with short electrodes will maximize efficiency.

Finally, the life of MPD thrusters for proposed mission applications is a serious concern and active area of research. The power density and the total power of MPD thrusters are generally so high that materials issues and cooling become serious issues. Onset and anode spots must be avoided at all cost because of the severe erosion of the anode observed. The use of large area hollow cathodes, anode propellant injection, flared-anode geometries with optimized applied magnetic fields, and light propellants have all been observed to reduce the erosion rates and extend the life of MPD thrusters. More theoretical and experimental efforts are needed to increase the life and total impulse capabilities of MPD thruster by thoroughly examining the effects of electrode design, applied fields, and operating conditions on the electrode lifetime and performance of the thruster.

9.3 Ablative Pulsed Plasma Thrusters

Pulsed Plasma Thrusters (PPT) were employed on a microsatellite very early in the history of electric propulsion due to their simplicity, robustness and the ability to operate in pulses at low average power. In fact, one of the first application of electric propulsion was on board the Russian spacecraft Zond-2 during a mission to Mars in late 1964 where six, ablation-fed, coaxial PPTs were used to provide attitude control. Only four years later, the Lincoln Experimental Satellite 6 (LES-6) successfully used PPTs in a rectangular configuration for east-west station keeping. The four PPTs were developed and flight-qualified by the Massachusetts Institute of Technology (MIT) Lincoln Laboratory and, combined with cold gas microthrusters, became the first completely automatic, self-contained station keeping system to be demonstrated in space [71].

Missions successfully employed ablative PPTs in the 1970s and 1980s mainly for drag make-up and precision spacecraft position control [72, 73]. Flight qualification of PPTs at various laboratories continued [74, 75] during this period. For a more in-depth review of the early history and laboratory investigations of the PPT, the reader may refer to Burton and Turchi [76]. The first PPT flight for a NASA mission was on the Earth Observer – 1 (EO-1) spacecraft launched in 2000. The onboard PPT experiment, shown in Figures 1–8, was aimed at demonstrating precision pointing accuracy, response and stability, and to confirm that the thruster plume and EMI effects on the spacecraft and instruments were inconsequential. The single PPT on EO-1 was used successfully for pitch attitude control and accumulated over 26 h of operation with over 96,000 pulses [77]. More recently, PPTs have been flown and operated successfully on the HuskySat-1 [78] and 1-U Cubsats [79]. Finally, in 2019, Champaign-Urbana Aerospace, LLC, was selected by NASA to develop and demonstrate in-space operation of a fiber-fed PPT on the 6U CubeSat DUPLEX [80].

Of all the existing electric thruster concepts, the PPT was the first to gain acceptance for space flight applications mainly because of its system simplicity, use of a solid propellant, simple electrode geometry and a common form of energy storage (capacitors). Owing to its state as a solid, the propellant, which is usually some form of polytetrafluoroethylene (Teflon®), requires no special tankage to contain it in space and thereby bypasses the complexities and risks associated with gas valves. As an alternative to reaction wheels, PPTs are especially well suited for attitude precision control on small satellites considering they are also low-mass and with power requirements in the range of <1 W to only a few hundred watts. In addition to its simplicity, robustness, low mass and power requirements, the PPT is also a low-cost device that can provide relatively high Isps, in some cases exceeding 1000 s, as will be shown later. In the selection process for space applications, these

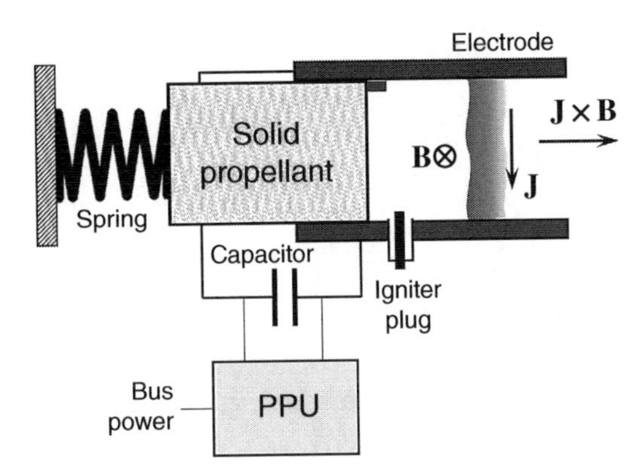

Figure 9-13 Schematic of an ablative PPT in rectangular configuration showing the propellant feed and the basic discharge acceleration mechanism in slug mode. Also shown are the rail electrodes, igniter plug and energy source.

advantages overshadowed the historically poor performance of this thruster; thrust efficiency generally ranged only 2–12%. Interest in PPTs continues today and is now largely driven by the rapidly growing microsatellite industry.

The basic set-up and operation of the PPT are illustrated in Fig. 9-13. The more traditional breech-fed rectangular geometry is shown here for ease of representation. Its main components are two parallel electrodes that are connected to an energy storage source, a solid insulator used as the propellant (usually Teflon®) and an igniter plug, which is mounted on one of the two electrodes. After charging the energy storage capacitor to a given voltage, the igniter plug is fired producing a minuscule amount of initial plasma that has sufficient electrical conductivity to trigger a current discharge across the exposed propellant surface. The discharge contains enough energy to cause ablation of material from the surface, which is subsequently ionized and accelerated by electromagnetic and gas dynamic forces.

9.3.1 Thruster Configurations and Performance

Traditionally, these thrusters have been operated under pulsed conditions. The term "pulsed" refers to operation with transient current waveforms during which all relevant parameters are (in general) strong functions of time. In most such cases, the discharge travels along the electrodes away from the ablating surface in a "slug" mode (Fig. 9-13) since the flow from the propellant does not provide electrically conducting material fast enough to sustain the arc position along the propellant surface. Other ablative devices were designed to operate in a "quasi-steady" mode in which all pertinent discharge and flow parameters approached their steady state values. The current waveforms in these cases are typically more prolonged and non-reversing and the discharge usually operates in an "ablation-arc" mode, adjacent to the propellant surface.

In the traditional slug mode, the first PPTs were driven by an inductance-resistance-capacitance (LRC) electrical circuit, which typically exhibits at least a few current oscillations due to the external impedance. These current reversals are unfavorable since the corresponding voltage oscillations can reduce capacitor life: for high-voltage capacitors, lifetime tends to scale inversely with charging voltage (to a high power) and decreases rapidly with the number of peak-to-peak voltage reversals.

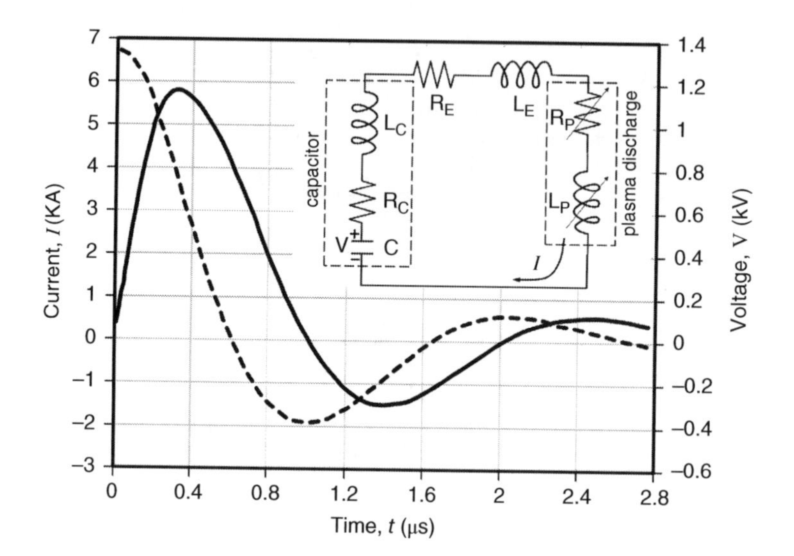

Figure 9-14 Typical voltage and current waveforms produced by numerical simulations of the LES-6 PPT, with $C = 2\ \mu\text{F}$, $L_E + L_C = 34\ \text{nH}$ and $R_E + R_C = 30\ \text{m}\Omega$ and $V = 1360\ \text{V}$. A schematic of the LRC circuit is also shown for reference (*Source:* [81]).

A typical LRC waveform produced during laboratory testing of a LES-6 PPT is shown in Fig. 9-14. Also shown for reference on the top right of the plot is a diagram of the electrical circuit depicting the various circuit elements, namely, the capacitor (subscript "*C*"), the external transmission line (subscript "*E*") and the plasma discharge (subscript "*P*"). A more favorable waveform would be one that does not undergo reversals but such pulses involve more complicated circuitry and usually require pulse forming networks (PFN).

The pulsed nature of its operation requires a different approach to characterizing the performance of the PPT compared to that of steady-state electromagnetic thrusters like the MPD. The thrust (T) is produced from the acceleration of many discrete increments of mass released from the propellant, or mass bits m_b. Each mass bit is generated during a single pulse that discharges the energy E_0 from the energy-storage element. Since the thrust generated during the pulse is a function of time, it is convenient to define the impulse per pulse, more commonly referred to as the impulse-bit I_b, as

$$I_b = \int T dt. \tag{9.3-1}$$

If the frequency of the discharge pulses is f then the corresponding average thrust is given by

$$T = f I_b. \tag{9.3-2}$$

Along the same lines, the equivalent average mass flow rate $\dot{m}$ is

$$\dot{m} = f m_b. \tag{9.3-3}$$

By combining the two equations above we can define the thrust per unit flow rate or specific impulse Isp as

$$\text{Isp} = \frac{T}{\dot{m} g_0} = \frac{I_b}{m_b g_0} \tag{9.3-4}$$

where g_0 is the gravitational acceleration. Finally, the thrust efficiency η_T is defined as

$$\eta_T = \frac{T^2}{2\dot{m}P} = \frac{I_b^2}{2m_b E_0} \tag{9.3-5}$$

where $P = fE_0$ is the corresponding electrical power delivered to the thruster from the energy source.

In the laboratory, performance is determined simply by direct measurement of the impulse and mass bits. The former measurement is typically performed by operating the thruster for thousands of discharge pulses, measuring the accumulated thrust, and then dividing that value by the pulse frequency. The mass bit is determined by measuring the mass of the propellant before and after operation of the thruster and then dividing by the total number of pulses.

The theoretical prediction of the performance is not as straightforward. That is because there are two contributions to the propulsive force from the acceleration of the ablated material – electromagnetic and gas dynamic – neither of which can be easily and accurately determined using rudimentary analyses. This topic is discussed in more detail in Section 9.3.2.1. Nevertheless, we shall attempt such analyses here for the purpose of exposing some of the driving processes behind the performance of this device. Later, we will expand further based on insights from numerical and more in-depth theoretical modeling.

Following the same lines for the determination of thrust in the MPD in Section 9.2.1.1, the divergence of Maxwell's magnetic stress tensor, $\beta_{ij} = (B_iB_j - \delta_{ij}B^2/2)/\mu_0$ with B, δ_{ij} and μ_0 being the magnetic flux density, Kronecker delta and vacuum permeability, respectively, yields the Lorentz force density $(\mathbf{J} \times \mathbf{B})_i$ acting on a fluid volume V_F. In β_{ij}, the second term in the parenthesis is known as the magnetic pressure and the second is the magnetic tension, which becomes important in highly curved magnetic field topologies. Conversely, the divergence of the particle stress tensor, $\sigma_{ij} = nm\langle v_iv_j\rangle$ yields the gas dynamic force density, where n, m, and v are the number density, mass, and velocity of the particles, respectively. The integral of the divergence of the two stresses over the volume V_F yields the total force and equals the rate of change of momentum of the ablated material. Using Gauss' theorem, the volume integral may then be expressed in terms of a surface integral over a control surface S enclosing V_F.

The direct integration of the momentum conservation as described here is not trivial and usually requires numerical simulation, as we will discuss later in Section 9.3.2.1. With a few simplifying assumptions, which come at the expense of accuracy, a more idealized yet still insightful approach can be formulated by considering the thruster as an electrical circuit element with an inductance L. For ease of illustration, we consider here a PPT with rectangular geometry in which the arc discharge has a height h and width w. The two lengths define the area over which the magnetic and gas dynamic stresses act. Then the electromagnetic (EM) contribution to the impulse bit requires only knowledge of the inductance gradient $L' = \nabla L$ and the current waveform:

$$I_{b,\mathrm{EM}} = \frac{L'}{2}\int_0^\infty I^2 dt \approx \frac{\mu_0}{2}\frac{h}{w}\int_0^\infty I^2 dt \tag{9.3-6}$$

where we have approximated $L' \approx \mu_0 h/w$. For more complicated geometries, L' will take a different form as shown for example in Burton and Turchi [76]. The gas dynamic (GD) contribution can be expressed in most general terms as follows [82]:

$$I_{b,\mathrm{GD}} = \iint dS \int_0^\infty \sigma_{ij} dt \approx \sum_a m_a c_a \tag{9.3-7}$$

where m_a and c_a are the mass and thermal velocities of each species a. The total impulse bit then is just the sum of the two contributions:

$$I_b = I_{b,\mathrm{EM}} + I_{b,\mathrm{GD}}. \tag{9.3-8}$$

Although highly idealized, Eqs. 9.3-6 and 9.3-7 illustrate the significance of the current pulse waveform, the discharge and propellant geometry and the gas dynamic velocity of the ablated species. It is largely these few elements that have driven PPT research efforts to improve performance since the late 1960s.

Over several decades now, a wide range of thruster geometries, current waveforms, and solid propellants have been investigated at various institutions worldwide. To review some of the different approaches and findings from these efforts, it is helpful to employ a loosely defined taxonomy of PPTs as follows. Depending on electrode geometry and orientation of the propellant feed, there are basically four types of ablative PPT configurations, two in rectangular and two in coaxial arrangements. These are described briefly in the next two sections. As pointed out by Molina-Cabrera [83], a broader classification is necessary when gas- and liquid-fed pulsed thrusters are considered and when different ignition approaches are implemented. Here, because we are focusing only on ablative PPTs which are typically ignited in a similar way, the narrower geometry-based classification is sufficiently useful.

9.3.1.1 Rectangular Configurations

The rectangular or parallel-rail configuration utilizes two parallel rectangular electrodes as shown in Fig. 9-15 and can be fed with the propellant either in a breech-fed (BF) or a side-fed (SF) manner, each of which may influence flow behavior. The electrodes are usually parallel to each other but some investigations explored the effects of angled electrodes on the flow expansion and overall thruster performance (e.g. see [82]).

The simplicity of the device and the success of the LES-6 PPTs in the late 1960s, inspired the start of several research programs in the United States during the 1970s to improve thruster performance and to broaden its applicability. During this time, similar efforts were launched in Japan and China that continued into the 1980s [84, 85]. Table 9-1 summarizes performance characteristics of selected PPTs in rectangular geometry with either BF or SF propellant configurations, developed in the late

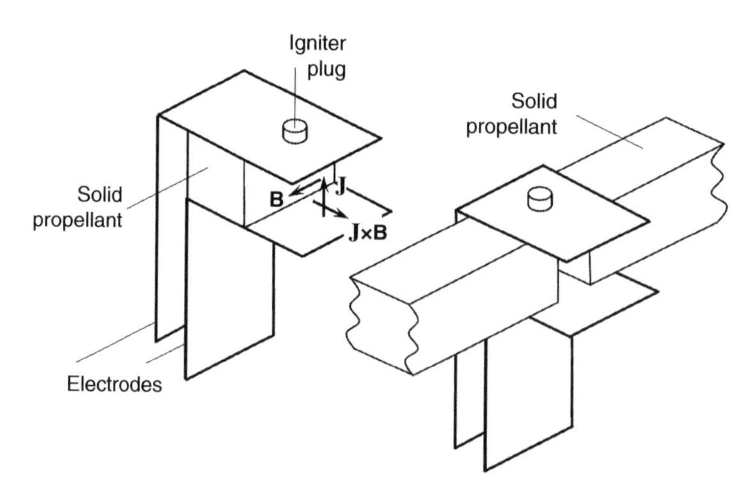

Figure 9-15 Basic schematic of rectangular PPT geometries with breech-fed (left) and side-fed (right) propellant configurations.

Table 9-1 Performance characteristics of selected PPTs in rectangular geometry with either breech-fed (BF) or side-fed (SF) configurations, taken in the late 1960s through the 1980s. (*Source:* Data taken from [76] where citations to the different thrusters listed in the table can also be found).

PPT	Config.	E_0 (J)	Isp (s)	I_b (μN-s)	m_b/E_0 (μg/J)	η_T (%)
LES-6	BF	1.85	300	26	4.8	2.1
SMS	BF	8.4	450	133	3.4	3.7
LES-8/9	BF	20	1000	297	1.5	7.4
TIP-II(NOVA)	BF	20	850	375	2.3	7.6
MIT Lab	SF	20	600	454	2.8	9.2
MIPD-3	SF	100	1130	2250	2	12.7
Millipound	SF	750	1210	22,300	2.5	17.7
Primex-NASA	BF	43	1136	737	1.5	9.8
Japan lab	BF	30.4	423	469	10	18
China lab	BF	23.9	990	448	3.7	3.2

1960s through the 1980s (data has been taken from [76]). These early efforts were largely laboratory investigations that provided integrated measurements, such as mass loss and impulse bit, over many discharges. They eventually subsided in the late 1980s, but a renewed interest in the mid-90s saw the restart of some of those earlier empirical studies. For the interested reader, a much more extensive PPT database than the one provided in Table 9-1 was compiled in 2011 by Molina-Cabrera et al. [83].

It was recognized early that poor propellant mass utilization was one of the major contributors to the low thrust efficiency, which rarely exceeded 10% at energies $E_0 < 50$ J. After a series of spectroscopic diagnostics and Faraday Cup measurements in a 20-J BF PPT, Thomassen and Vondra [86] reported that production of heavy molecules and neutral atoms from the propellant surface persisted long after the capacitor energy was expended, presumably as a result of thermal bombardment by "hot plasma particles." In an effort to optimize performance by better controlling the mass ablated per shot, a SF mechanism was attempted and notable performance improvements over the BF configuration were claimed, when combined with an arrangement that utilized flared electrodes [86]. A later study, however, by Palumbo and Guman [87] using a 450-J thruster, demonstrated higher thrust efficiency (as high as 53.4%) with a BF geometry. In a similar effort to discover the effects of propellant geometry on performance in a 20-J, BF PPT, Yuan-Zhu [85] determined that higher ratios of propellant height-to-width (for the same exposed area) improved performance as impulse bit, Isp, and thrust efficiency all increased. A survey of different thermoplastics and Teflon® seeding yielded no performance improvements, as shown by Palumbo and Guman [87]. Later, attempts were made to improve Isp by using laminated propellant configurations with alternating layers of Teflon® and polyethylene, possibly to take advantage of the lower molecular weight of polyerthylene. These efforts, however, were hampered by surface carbonization and ultimately yielded lower thrust efficiencies [88].

9.3.1.2 Coaxial Configurations

Most initial designs of ablation-fed, coaxial devices closely resembled the gas-fed MPD thruster prototypes. Typically, they were made of a central, cylindrical electrode surrounded by an outer ring-conductor and, as in the rectangular device, could be supplied with the solid propellant from the

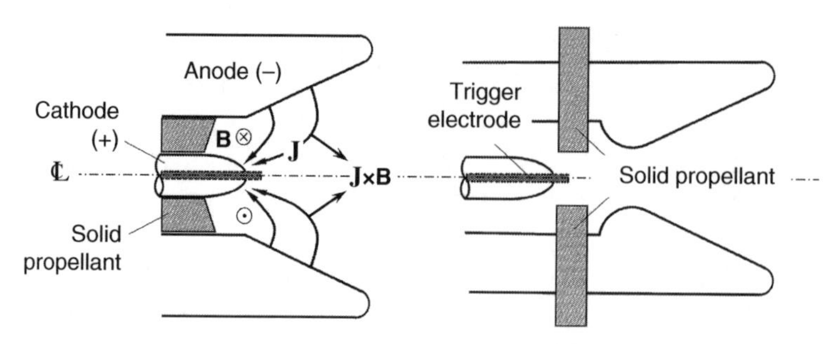

Figure 9-16 Breech-fed (left) and side-fed (right) coaxial PPTs based on the configurations tested in [89, 90].

breech or the sides. A BF coaxial thruster with frontal ablation surface is shown in Fig. 9-16 (left). Figure 9-16 (right) also shows a coaxial device but with a radial-feed and frontal ablation surface. The nozzle in these devices is usually conical and serves as the anode.

After the Soviets developed the first ablation-fed, coaxial, PPT (later integrated on the Zond-2 spacecraft) most ensuing PPT programs refocused on rectangular PPTs, largely because of the mission success of the simpler LES-6 thruster. By the late 1970s, activities in electric propulsion had shifted to other higher power devices, most of which involved gaseous-fed mechanisms. However, investigations of ablative-fed thrusters continued in Italy using pulsed coaxial accelerators similar in geometry to their steady-state MPD gas-fed counterparts [89, 90]. A considerable amount of empirical work was performed on these devices during this time, but their kJ-level operation and coaxial set-up presented a major difficulty in correlating such work with that from their much lower energy, rectangular counterparts.

Other coaxial configurations at much lower energy levels ($E_0 \lesssim 10$ J) were also pursued by Guman [91], Spanjers [92] and Bushman and Burton [93]. The thruster by Bushman and Burton [93] was designed to induce an arc-discharge into a cylindrical cavity (resembling a linear pinch device) that was fed with the solid propellant from two sides. Owing to the large ratio of cavity length to diameter, this device was characterized by a predominantly electrothermal acceleration mechanism with a relatively high-impulse bit per joule and low Isp. Based on the values provided from the Illinois PPT-3 Lab in [76], at $E_0 = 7.5$ J the thruster produced Isp = 600 s, $I_b = 45$ μN-s and $m_b = 75$ μg yielding an efficiency of $\eta_T = 18\%$. The coaxial version developed by Spanjers [92] was particularly simple and miniaturized: the current pulse was delivered to the exposed end of a coaxial cable where the ablation occurred, and, therefore, required no spark plug and associated circuitry. This micro-PPT produced $I_b = 2$ N-s at $E_0 = 0.8$ J in a package with a mass of less than 0.5 kg. The thruster by Guman [91] was more akin to an MPD thruster geometry but with a conical nozzle section serving as the cathode, whereas the interior electrode was the anode. That thruster produced Isp > 2000 s at $E_0 =$ but impulse bit measurements and thrust efficiencies were not reported.

More coaxial variants have been pursued over the last two decades. For example, Markusic et al. [94] investigated a thruster in a z-pinch configuration and reported Isp and η_T spanning 240–760 s and 2–9%, respectively, in the energy range $25 < E_0 < 76$ J. The Fiber-Fed Pulsed Power Thruster (FPPT) by Burton [80] uses coaxial electrodes in a motor-driven system that feeds a polytetrafluoroethylene (PTFE) fiber, a few mm in diameter, through a hollow central anode. During the discharge, the fiber tip ablates to a conical shape. The coaxial cathode has an inner diameter that is 3–6 times the anode diameter. The operational principle is different from all previous PPTs in that the magnetic field topology in the traditional PPT is inverted in the FPPT, which also reverses the direction of the Lorentz force at the ablating surface. The radially inward magnetic pinch forces produce a high gas-dynamic pressure on the axis of the thruster and a repeatable pulse-to-pulse m_b with no surface charring. The unfavorable process of propellant charring was observed experimentally in micro-PPTs like that by Spanjers [92], and it was postulated that it is formed by the released carbon flux returning from the plasma rather than an incomplete decomposition of the Teflon® [95]. The discharge energy in the FPPT is typically in the range $8 < E_0 < 32$ J with a measured performance of 800 < Isp < 2450 s and $\eta_T < 10\%$.

9.3.2 Physics and Modeling

Even though it is structurally and operationally a simple device, the PPT incorporates a variety of complex physical processes. The most challenging physics are those associated with the transient ablation of the solid propellant and subsequent breakdown and acceleration of the gaseous products. Heat from the arc discharge diffuses into the solid propellant, causing ablation of the material, only a portion of which is ionized and accelerated electromagnetically to speeds exceeding 10 km/s. The rest of the ablated mass is released at much lower speeds (<1 km/s). During the early investigations in the 1970s and 1980s, the lack of comprehensive theoretical and numerical models in programs largely devoted to empirical analyses and development of these thrusters limited their modeling to simplified expressions of the performance. These formulations depended on measurements that were conducted over many discharges and were driven primarily by an effort to separate the electromagnetic and gas dynamic contributions to the thrust, and/or to correlate mass loss to external parameters such as capacitor energy and initial circuit impedance by assuming a fixed LRC circuit [96]. All these formulations provided little insight into the strongly coupled physics of the ablation and acceleration processes during a single pulse.

9.3.2.1 Numerical Simulations

The failure of the early empirical investigations and idealized models to improve thruster performance pointed strongly to the need for a more comprehensive analysis approach that employed (at least) 2-D numerical simulation. The anticipation was that by self-consistently capturing the coupling between all three critical phases of thruster operation – ablation, breakdown, and acceleration – critical (yet elusive at the time) insight could be gained to inform not only how to improve performance, but also to allow for the advancement of more comprehensive and predictive analytical models.

The first successful attempt to employ advanced numerical simulation of PPTs was made in the 1990s with the time-dependent, 2½-D code MACH2 [97]. The non-ideal magnetohydrodynamics (MHD) code was augmented with ablative boundary conditions at the solid propellant surface, which allowed for the determination of the propellant's temperature not only at its interface with the vapor/plasma but also deeper into the propellant [81]. The latter was achieved by solving the transient 2-D heat diffusion equation in the solid simultaneously with the evolution of the plasma during the current pulse. An example of the computed temperatures inside the Teflon® propellant from the first simulations of the LES-6 PPT is depicted in Fig. 9-17.

This capability and subsequent simulations provided several new insights. For example, the LES-6 PPT simulations captured well the measured impulse bit and showed that the computed mass of ablated propellant that was electromagnetically accelerated during a single pulse was much less than the measured mass consumed after thruster operation through many pulses. The simulation results and comparisons with the measurements are plotted in Fig. 9-18. The mass measurement was an average value that was determined by dividing the total mass released (over many current pulses) by the number of pulses. Therefore, it also included mass that was released from the

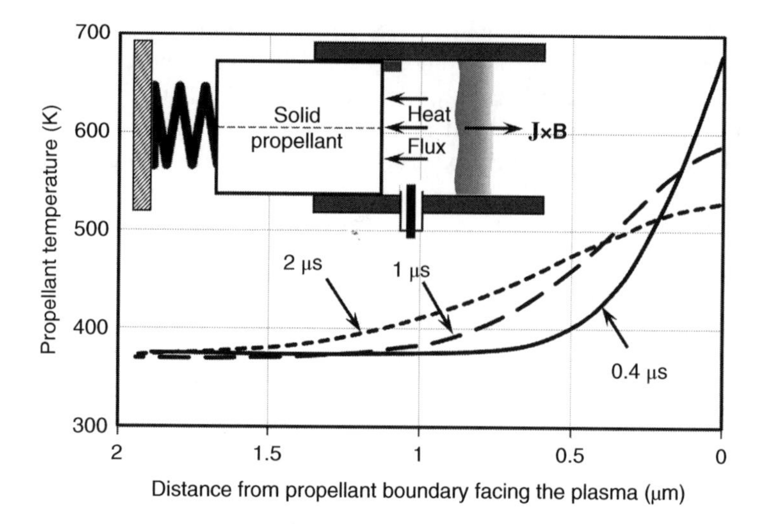

Figure 9-17 Computed temperature in the solid propellant of the LES-6 PPT from the first fully coupled ablation-MHD time-dependent numerical simulations in two dimensions (*Source:* [81]).

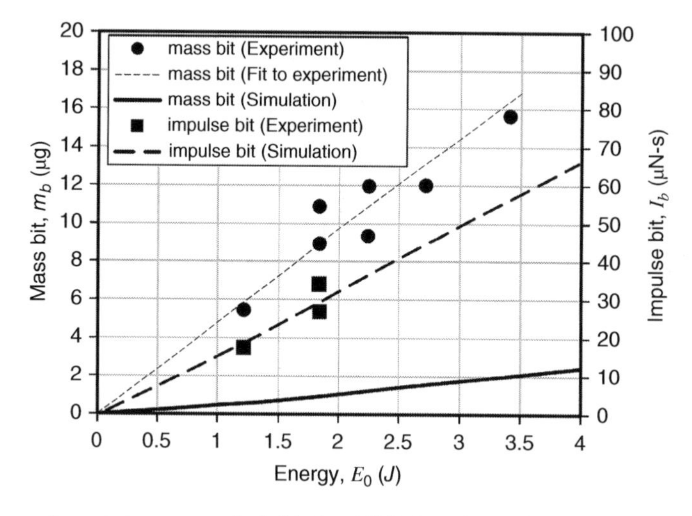

Figure 9-18 Comparisons between coupled ablation-MHD time-dependent numerical simulations and measurements of the mass and impulse bits (*Source:* [81]).

solid propellant between pulses, and/or released during each pulse but not accelerated electromagnetically.

Although the presence of this "late-time" mass was long conjectured from laboratory investigations to be the driver of the poor propellant utilization [81, 98], this was the first time that 2-D numerical simulations coupled a transient temperature-dependent ablation model with the thruster magnetohydrodynamics, and quantified the fraction of the total ablated mass that is accelerated by the Lorentz force. This improved the understanding of the processes associated with the release of the late-time mass and enabled more in-depth investigations into methods that could reduce it. For example, in subsequent investigations with MACH2, the authors argued that because the thickness of propellant containing the mass needed to sustain the discharge increases linearly with time (in steady state), whereas heat from the plasma diffuses into the solid to a depth that increases with the square root of time, the fraction of mass within this heated depth that is above the temperature at which material can decompose can be released from the solid surface without the benefit of electromagnetic acceleration. They proposed then that electromagnetic utilization of the ablated mass could be improved using current pulses with shorter rising times and longer decay times than the traditional LRC oscillatory waveforms [99]. Specifically, their numerical simulations showed that an optimized pulse in a typical rectangular configuration, similar to that in the LES-6 PPT, would require current pulse times in the order of milliseconds at several kA or a steep rise of the current to tens of kA followed by a long (critically damped) decline in the tens of microseconds time scale [99, 100].

Some of the first attempts to produce fast-rising currents with non-reversing long declines using inductively-driven circuits failed to improve performance at low capacitor energies in the tens of joules [101]. However, later, a combination of high-current, long pulses and high capacitor energies achieved using a PFN, improved thrust efficiency considerably. For example, using a PPT with PTFE propellant, Kamhawi [102] designed and demonstrated thrust efficiency of 36.4% and a Isp of 3940 s. The measured current waveforms and thrust efficiencies for the three configurations tested – 1a, 1b and 2 – are depicted in Figs. 9-19 and 9-20, respectively. The main differences between the two configurations 1 and 2 were in the electrical transmission line layout; initial testing with configuration 1 indicated that the transmission line inductance was limiting the peak discharge current magnitudes, thereby affecting thruster performance. Therefore, the second iteration, configuration 2, was designed to reduce the external inductance. The differences between 1a and 1b were in the electrode geometry and material, with the former having twice the width of the latter. Also, 1a and 1b used Al and Cu electrodes, respectively. The electrical circuit produced a peak current of 70 kA at 700 J, with rise times in the order of 5 μs and decay times of about 20 μs, as shown in Fig. 9-19. Kamhawi argued that a more improved transmission line would increase further efficiency, to the levels observed by Palumbo [87] (with maximum $\eta_T = 53\%$) from his testing of high-energy PPTs, because it would achieve an even faster rise and longer decay times of the current as suggested by the theory. The mass and impulse bits

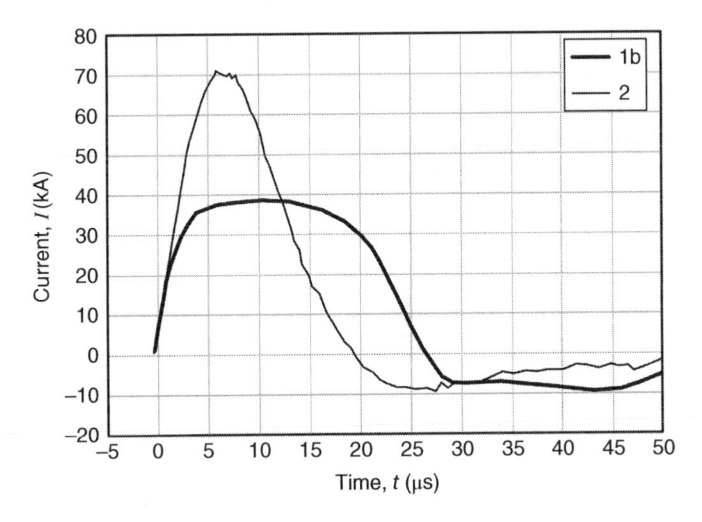

Figure 9-19 Current waveforms produced in laboratory investigations of high-energy PPTs at discharge energy of 700 J and capacitance of 260 μF (*Source:* [102]).

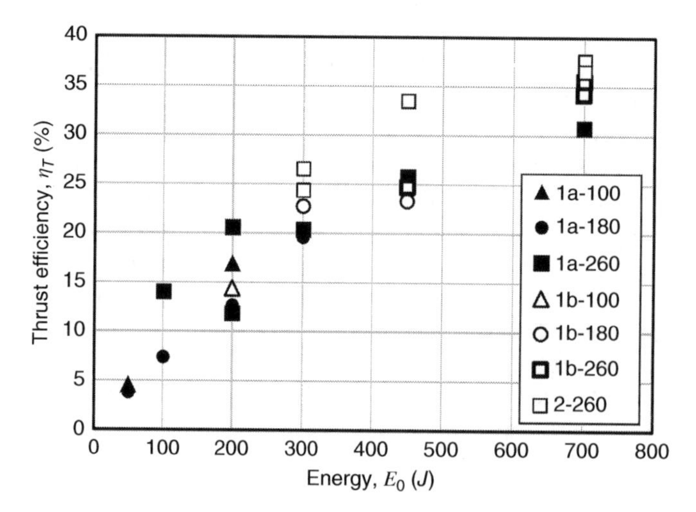

Figure 9-20 Thrust efficiency as a function of discharge energy for three high-energy PPT configurations tested in the laboratory – 1a, 1b and 2 – at three different capacitances, 100, 180 and 260 μF (*Source:* [102]).

of the PTFE PPT were 342 μg and 13,200 μN-s, respectively. Similar efficiencies, 25–30%, were achieved later at discharge energies >100 J by Andropov [103].

9.3.2.2 First-principles Idealized Models

Since the early intensive computational efforts described above that used large-scale coupled ablation-MHD numerical simulations, there have been several attempts to obtain engineering models that are high-fidelity yet simple enough for practical use in the engineering design cycle. Some researchers have, in fact, been quite innovative in their approach. For example, Zeng et al. [104] used statistical methods to model the random degradation of molecular chains in a PTFE propellant and developed expressions for the mass bit as a function of the ablation energy, the energy required to break the carbon-carbon bonds and the ratio of the mass that breaks away from the propellant surface to the total ablated mass. Although the authors recognized that their model may only be suitable for the propellant and structural parameters of their specific thruster experiment, within the very limited range of comparisons they reported, the agreement with the measured mass bit was encouraging. Hossain et al. [105] used machine learning techniques to train a relatively simple electromechanical model of a LES-6-like PPT, also with some encouraging success. Such models, though potentially useful for engineering purposes, provided limited insight into the driving physics of the device and their scaling with controllable parameters.

One of the few analytical investigations of breech-fed, rectangular PPTs that relied on insights gained from previous experiments and simulations and agreed very well with a wide range of PPT data was that by P. Mikellides et al. [106]. The theoretical model was based on the hypothesis that the total ablated mass during a single pulse is released early in the waveform and consists of a variety of species that expand at different speeds, some of which persist in the chamber after discharge cessation. Hence, as also implied in earlier sections, there is a strong contribution from both electromagnetic and gas dynamic acceleration. He then solved the magnetohydrodynamic conservation equations imposing quasi-steady, 1-D flow assumptions for a control volume that extends to the magnetosonic point. This latter condition associated with the magnetosonic point was based on earlier theoretical and numerical work by Mikellides and Turchi who argued [107] and showed by numerical simulation [99, 100] that the speed υ of the flow that is electromagnetically accelerated ultimately reaches the Alfven wave speed:

$$\upsilon_A = \frac{|\boldsymbol{B}|}{\sqrt{\mu_0 \rho}} \tag{9.3-9}$$

where ρ is the mass density of the plasma. This magnetosonic condition, $\upsilon_* = \upsilon_A$, is a special case of the plasma flow speed approaching the phase velocity of a magnetoacoustic wave when the ratio of the plasma pressure to the magnetic pressure approaches zero. In the limit of high magnetic Reynolds number, $R_M = \mu_0 \upsilon l / \eta$, with υ, l and η being the flow speed, characteristic length and resistivity, respectively, and when the required ionization energy dominates the change in flow enthalpy, this flow speed tends to remain fixed and proportional to the Alfven critical speed, that is, $\upsilon_* \sim u_A \propto \sqrt{\Delta H}$ where ΔH is the change in the flow enthalpy.

Under these assumptions, P. Mikellides et al. [106] produced the following expression for the total ablated mass during the pulse, the ionized portion of which is dominated by internal energy deposition:

$$m_b = \frac{\mu_0}{\beta u_A} \frac{h}{w} \int_0^\infty I^2(t) dt \tag{9.3-10}$$

where w is the electrode (or solid propellant) width, h is the distance between the two electrodes, $\beta = 4.404$ and u_A is the Alfven critical speed. For Teflon® PPTs driven by LRC circuitry, $u_A = 13.3$ km/s and $I(t)$ is the underdamped waveform solution for the current based on constant electrical elements. A comparison of the theoretical model to a range of experimental data is shown in Fig. 9-21, depicting agreement within experimental error for almost all thrusters addressed. For Teflon® PPTs it shows that ablated mass scales with electrode aspect ratio, h/w, and stored energy, E_0 and it is inversely proportional to the total resistance, R.

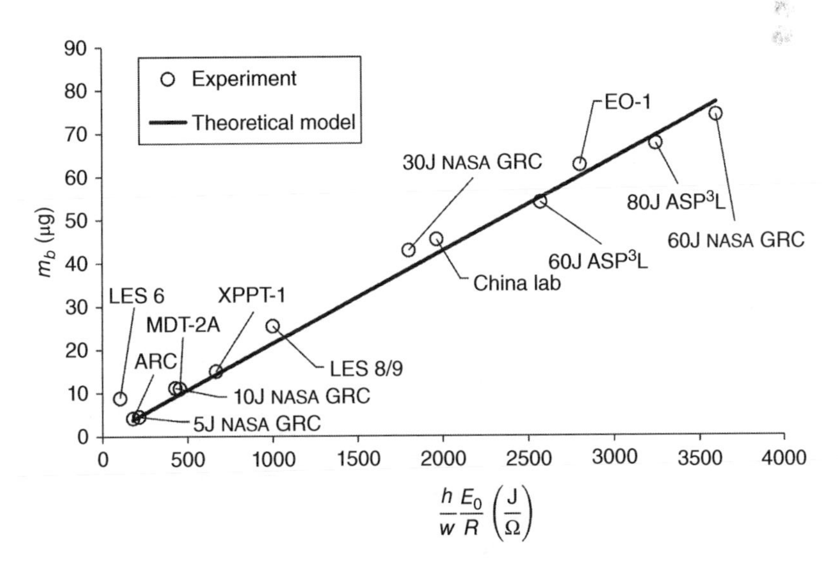

Figure 9-21 Comparison of the theoretical model for total ablated mass Eq. 9.3-10 to diverse range of PPT experimental data (*Source:* [106]/American Institute of Aeronautics and Astronautics).

Development of the theory for the total impulse-bit proceeded by modeling the electromagnetic and gas dynamic body forces F separately, as with Eq. 9.3-8:

$$F = F_{EM} + F_{GD} \tag{9.3-11}$$

where

$$F_{EM} = \frac{L'}{2} I^2 \tag{9.3-12}$$

and

$$F_{GD} = c_f p_0 A. \tag{9.3-13}$$

In Eq. 9.3-13 p_0 is the stagnation pressure, A is the cross-sectional area, and c_f is the thrust coefficient which can be estimated based on a constant-area isentropic expansion to vacuum as follows [106]: $c_f = (1+\gamma)(2/1+\gamma)^{\gamma/\gamma-1} = 1.255$ for $\gamma = 1.3$. The stagnation pressure was related to the propellant's recession rate by Saint Robert's Law. Integration of the body forces (Eqs. 9.3-12 and 9.3-13) over the volume produced the final expression for the total impulse bit,

$$I_b = \frac{\mu_0}{2\pi}\left[\frac{3}{2} + ln\left(\frac{h}{w+d}\right)\right]\int_0^\infty I^2(t)dt + \left(\frac{\mu_0 c_f^n}{\beta a \rho_s v_A}\right)^{1/n} \frac{h}{w^{(2/n)-1}}\int_0^\infty I^{2/n}(t)dt \tag{9.3-14}$$

where d is the electrode thickness, ρ_s is the solid propellant's density, and a and n are propellant-dependent constant parameters and are empirically determined. For Teflon®, $a = 0.207$ mm/kPa^{n}s and $n = 0.8$ [106]. A comparison to experimental data is depicted by Fig. 9-22 and shows reasonable agreement, which implies the dominant acceleration mechanisms are well captured. The dashed line on Fig. 9-22 is the best linear fit to the theoretical model values and is included to better illustrate the relevant scaling: for Teflon®-fed, LRC-driven PPTs the impulse-bit scales with the

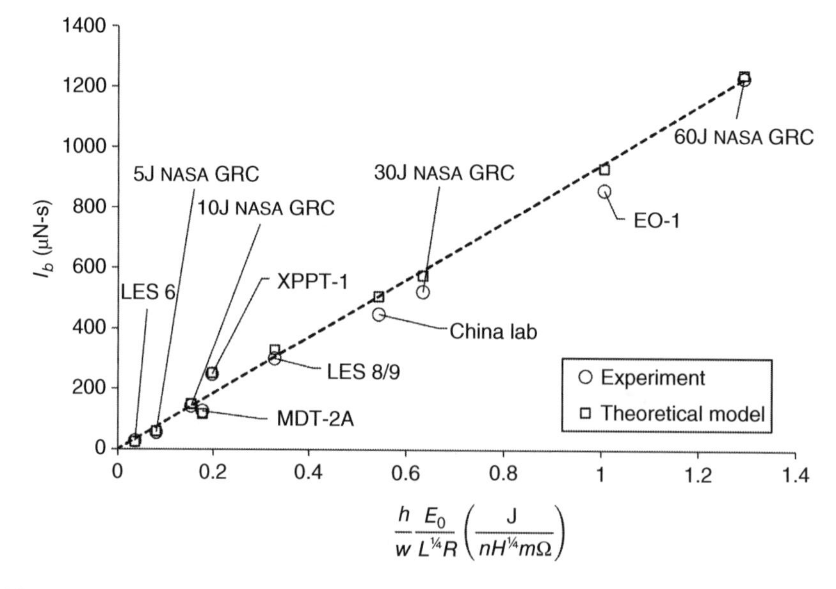

Figure 9-22 Comparison of the theoretical model for total impulse bit given by Eq. 9.3-14 to a diverse range of Teflon®, breech-fed, rectangular PPTs (*Source:* [106]/American Institute of Aeronautics and Astronautics).

electrode aspect ratio, h/w and stored energy, E_0 and it is inversely proportional to the total resistance, R and the fourth root of inductance, L.

The development of expressions for mass ablated and impulse-bit allows for the determination of the PPT thrust efficiency, which for Teflon® PPTs driven by an underdamped LRC circuit, becomes:

$$\eta_T = \frac{\left[\frac{L'}{2} + \frac{c_2}{\sqrt{w}} \frac{h}{w}\left(\frac{E_0}{L}\right)^{1/4}\right]^2}{2c_1 \frac{h}{w} R} \tag{9.3-15}$$

where $c_1 = 0.0214\,\mu\text{g/sA}^2$ and $c_2 = 0.134\,\mu\text{g m}^{3/2}/\text{s}^2\text{A}^{5/2}$. This expression reveals that efficiency increases with stored energy and it is inversely proportional to the total resistance R; trends that have been experimentally established, as shown for the case of the former for example in Fig. 9-20. Furthermore, the model implies that for a given underdamped LRC circuit, an optimum electrode aspect ratio and electrode width exist that maximize PPT efficiency.

9.4 Pulsed Inductive Thrusters (PIT)

One way to evade the poor propellant utilization associated with the challenges of controlling the ablation of a solid propellant in pulsed thrusters is to provide the propellant in gaseous form, through controlled injections. In this section, we discuss a different class of pulsed electromagnetic accelerators in which the plasma is created by inductive breakdown of a layer of gas that is transiently "puffed" onto the surface of a flat spiral induction coil, as shown in Fig. 9-23 (left). At the instant of optimum placement of the propellant along the inductor cover-glass ring, energy stored in a bank of capacitors is released into the induction coil. The induced azimuthal electric field

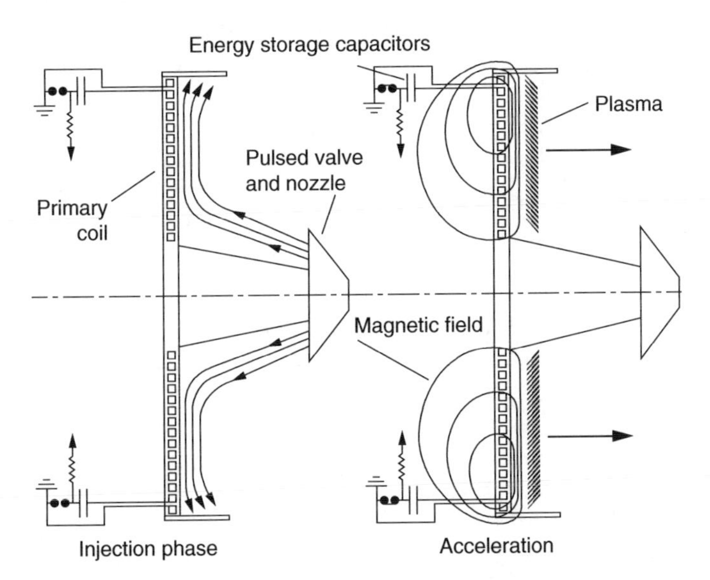

Figure 9-23 Fundamental operation sequence of the Pulsed Inductive Thruster (PIT). Left: Gas Injection. Right: Breakdown and electromagnetic acceleration. (*Source:* [108]).

produces rapid ionization of the gas, and establishes a flat ring of current that provides a piston against which the rising magnetic field acts, entraining and ionizing the balance of the propellant, and ejecting it along the thruster axis in Fig. 9-23 (right). That is, the resulting plasma current ring is repelled away from the coil by the **J** × **B** force that arises from the interaction between the radial magnetic field and the azimuthal plasma current. Such inductive acceleration circumvents the need for conventional electrodes and, in turn, the known lifetime limitations associated with their erosion. Also, since the plasma is lifted away from the coil surface early in the current pulse, thermal loads to the accelerator structure are manageable. This electrodeless operation, along with its potential to throttle over a range of thrust and Isp at constant power and almost constant efficiency (as it will be shown later in the chapter), and its use of plentiful, cheap and potentially in situ- replenishable propellants, have served as the impetus behind that decades-long research and development of this type of thruster.

The first laboratory application to embody this type of inductive acceleration was invented at Thompson Ramo Wooldridge Inc. (TRW) in the early 1960s. The laboratory investigations of the PIT at TRW were led by Dailey and Lovberg in the late 1960s and 1970s [109–113], and continued intermittently at Northrop Grumman Space Technology in the 1980s and 1990s [114–120]. For the interested reader, a comprehensive summary of the PIT's early development history is also provided in a review article by Polzin [121]. Through several iterations of the PIT, these efforts produced the Mark (Mk) V, the best-performing version of this thruster thus far [118]. In the early 2000s, a renewed interest in high-power electric propulsion for lunar and Mars cargo missions [122, 123] allowed further development that focused mostly on thruster components [124, 125]. These efforts were also accompanied by new theoretical investigations that now included more rigorous analytical modeling and the first comprehensive MHD 2-D numerical simulations of this type of thruster [126–129].

The PIT is relatively heavy and necessarily large due to the nature of its propellant puff technology and the electrical circuitry that enables pulsed operation at high voltages. Early idealized modeling also suggested that efficiency would improve substantially with thruster size [130]. For example, the electrical coil used in the PIT MkV was configured as parallel sets of windings forming a Marx-generator coil topology that was approximately 1 m in diameter. As a consequence of its size and weight, the thruster is best suited for high-power interplanetary missions. For reference, in early mission application studies, a 1-m diameter MkV thruster was estimated to carry a mass in excess of 100 kg and, if operated with a 10^5-h lifetime at an average power of 20 kW, its specific mass would be 8 kg/kW at a pulse repetition rate of approximately 6 Hz [118]. A wider range of thruster specifications and mission applications can be found in [119, 122, 123].

In addition to its relatively high mass and size, this type of accelerator is also challenged by the complexities associated with achieving efficient conversion of the input electric energy to propulsive work. In particular, motion of the plasma current ring (or sheet) away from the coil ring decouples them rapidly, so the design of the accelerator must be such that most of the stored energy is deposited to the plasma before this decoupling occurs. Also, the timing of the propellant breakdown relative to the applied current pulse is critical since, if such breakdown is delayed, energy will be lost irreversibly to the external circuit. Using a separate mechanism to ionize the gas before the application of the pulse may offer some mitigation of this effect and was a primary motivation for the design of a more recent inductive accelerator called the Faraday Accelerator with Radiofrequency Assisted Discharge (FARAD) [131, 132]. In this configuration the plasma acceleration is produced inductively as in the PIT. But unlike the PIT, FARAD also employed a separate radiofrequency (RF)-assisted pre-ionization stage that allowed for the formation of the inductive current sheet at much lower discharge energies and voltages. This advantage, however, came at the expense of the added complexity associated with incorporating an RF plasma source and applied magnetic field to produce and guide the plasma onto the face of the inductive coil [133]. The coil for this accelerator was similar to the PIT's Marx-type configuration but smaller, with an outer diameter of 0.2 m. Performance optimization however remained beyond the scope of that first proof-of-concept effort.

Finally, even though inductive acceleration eliminates concerns associated with conventional electrode erosion, there are other practical challenges that can limit the life of the thruster. These are mainly associated with its pulsed operation, namely, the life of the electrical switches and the propellant injection valves. For example, the MkV pulsed configuration employed spark gap switches with lifetime in the order of 10^5 pulses, which would only last less than 5 h at 6 Hz of repetition rate. Newer switch and valve technologies, however, have been proposed as potential mitigations such as solid-state switches (thyristors), long-life poppet valve designs [124, 125] and/or the use of long-life materials for the valves like polycrystalline thermoplastics (e.g. Polyetheretherketone). The solid-state switches have also been proposed as potential candidates for improving thruster efficiency, specifically, by using them to turn off the current pulse before it reverses sign thereby reducing energy losses from the LRC ringdown. Finally, since the current sheet is accelerated away from the coils it seems implausible that a sufficient flux of energetic ions can be created and directed towards the coils to cause any damage. Hence, lifetime risks associated with conventional erosion of the coils appears to be marginally low. Nevertheless, such low risk must be demonstrated in the laboratory through sufficiently long wear tests. It is worth mentioning here that the above-mentioned variants of the PIT were never tested continuously for hundreds of thousands of pulses or more to demonstrate marginal to no damage of the inductive coils.

9.4.1 Thruster Performance

The PIT MkV remains today the state of the art in performance for this type of electromagnetic thruster. We, therefore, summarize some of the most salient results from the early laboratory investigations of this version of the thruster. The pulsed nature of its operation requires the same approach in defining performance as that described in Section 9.3 for the ablative PPT. Therefore, Eqs. 9.3-2–9.3-5 apply here as well so we will not repeat them.

As in the ablative PPT, a typical laboratory test of the PIT will produce a measurement of the impulse bit (I_b) when the thruster is operated at a given total energy E_0 associated with the electrical circuit's capacitance and charging voltage. Unlike the ablative PPT, the mass bit (m_b) is well controlled through the injection mechanism of the (gaseous) propellant and is therefore known. The MkV employed a fast-opening solenoid valve, located on the axis of the coil and just above its plane. The valve releases the contents of a small, high-pressure plenum chamber into a radial supersonic nozzle, having a cylindrical aperture of 0.25 m radius and 0.025 m height. It is worth pointing out that the proper initial deposition of the propellant gas over the face of the coil is one of the most difficult technical challenges in achieving efficient operation with this thruster. Any gap between the initial plasma breakdown and the coil will produce a parasitic inductance which will degrade efficiency. If the gas layer is too thick, then snowplow losses will occur through most of the stroke, which will also reduce efficiency. With knowledge of I_b, m_b, and E_0, the specific impulse (Isp) and thruster efficiency (η_T) are determined using Eqs. 9.3-4 and 9.3-5. respectively.

Two versions of the PIT MkV were in fact tested and only in single-pulse operation. A first version called the MkV was designed and tested with a 2000-J capacitor bank. Its upgraded successor, the PIT MkVa (see Figs. 1.1–1.9 in Chapter 1), used a 4000-J bank and was the more successful design. Both designs were guided by the investigations of earlier devices in this series, particularly the MkI and MkIV [114]. Rather than using charging voltages in excess of 30 kV, the MkV employed a Marx connection of its capacitors to allow the total voltage around the coil to be an integer multiple of the capacitor charging voltage. It used a two-segment Marx connection and was operated up to 32 kV coil voltage. The MkV capacitors had a total effective capacitance of 4.5 μF whereas the MkVa had twice that value. Each of the 18 capacitors in the MkVa had a capacitance of 21 μF, and were charged in the range of 12–16 kV. Thus, the bank at an effective capacitance of 9 μF and total voltage of up to 32 kV produced a maximum energy of 4600 J. The total inductance of the circuit was 740 nH.

A first series of tests with the MkV explored the performance of the thruster with several different propellants, specifically, He, Ar, CO_2, NH_3 and simulated hydrazine ($N_2 + 4NH_3$). Because of its availability, storability ease of handling and better performance, NH_3 was given a greater emphasis than the other propellants. Performance measurements (from [126, 127] and private communication with R. Lovberg)] showing the specific impulse Isp and total efficiency η_T as functions of the specific energy, E_0/m_b are plotted in Fig. 9-24. For all propellants it was found that Isp increased with E_0/m_b. The efficiency also increased with E_0/m_b when the thruster was operated with NH_3 but remained relatively constant, around 20%, for all other propellants.

The low performance of the MkV was thought to be because of the challenges in consistently forming a magnetically impermeable current sheet at low discharge energies (as reported in [121]). Mainly for this reason, the Marx circuit in the MkVa was upgraded to allow operation at larger discharge energies. This upgrade improved the performance considerably, as shown in Fig. 9-25. The thruster operating with NH_3 achieved the best performance over all previous versions and propellants tested, with Isp ranging approximately 2000–9000 s and increasing with E_0/m_b. Thrust efficiency exceeded 50% and remained fairly constant at energies $E_0 \gtrsim 3500$ J and $E_0/m_b \gtrsim 2000$ J/mg, producing impulse bits in the range of 0.05–0.12 N-s [128].

The $\eta_T -$ Isp trend in the improved PIT MkVa was, generally, that efficiency increased with increasing Isp until a certain critical (low) value of the propellant mass was injected. For even lower mass bits the efficiency degraded with increasing Isp. The existence of a so-called "critical mass" will be discussed further in Section 9.4.2. The maximum value of η_T in the measured $\eta_T -$ Isp characteristics of the thruster was conjectured to represent an optimum match between the transit time of the plasma to its point of decoupling from the coil ring and the electrical period of the current waveform. Hydrazine was also tested with the MkVa but the performance was generally found to be lower compared to NH_3 [118]. Before leaving this section, we should note that the PIT tests described here were performed in a vacuum chamber that was not much larger than the thruster itself. Therefore, even though the measured trends associated with the microseconds-long single-pulse operation of this thruster were unlikely to have been affected significantly, facility effects on thruster performance were never interrogated extensively. Clearly, close attention must be given to such effects in any future development programs that will consider multi-pulse operation of this thruster.

9.4.2 Physics and Modeling

9.4.2.1 Numerical Simulations

Because the laboratory work on this type of accelerator was exclusively performed by the TRW group for the few decades following its invention in the 1960s, it is no surprise that some of the earliest models and physics insights were produced by that team as well. Indeed, the first attempts to simulate numerically the acceleration of the plasma sheet in the PIT were reported by Lovberg and Dailey in the early 1980s [130, 134, 135]. In their first attempt, the electrical circuit components

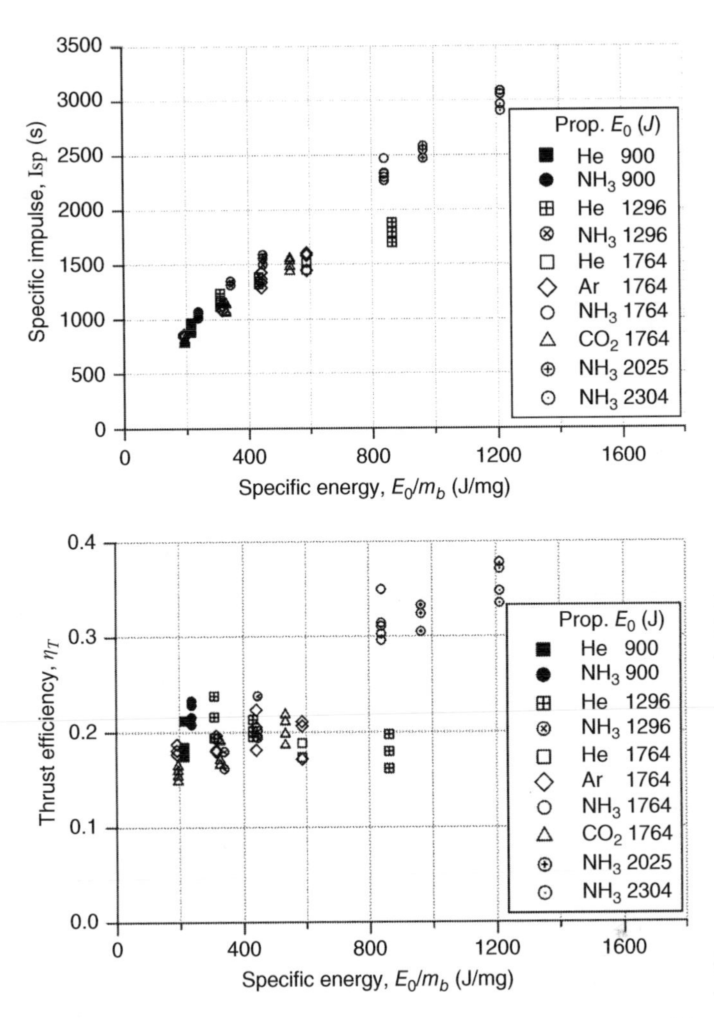

Figure 9-24 Measured performance of the PIT MkV with different propellants (*Source:* [126, 127] and private communication with R. Lovberg; the data is also reported in [121]). Top: specific impulse. Bottom: thrust efficiency.

were represented quite accurately but the plasma was modeled simply as a 1-D slab having uniform density and constant resistivity. The single-current layer formed at the rear of the plasma was taken to be a thinner uniform slab whose thickness was given by classical diffusion scaling [130]. Despite its simplicity, the model provided several useful insights regarding performance and agreed quite well with the results of earlier experiments. Of significance was the finding that thruster efficiency should improve substantially with scale size which led to the experiments with a 1-m diameter thruster soon thereafter.

The computer model was then upgraded to a more comprehensive 1-D resistive MHD formulation that was also solved numerically to provide the spatiotemporal distributions of plasma kinetic and magnetic field energies as well as the resistive losses [135]. There were, nevertheless, several simplifications to the model. Specifically, the plasma was assumed to be isothermal and an empirical relation was used to model the dependence of the resistivity on the current density. Also, the electrical circuit was not explicitly included in the model; a semiempirical analytical form was used

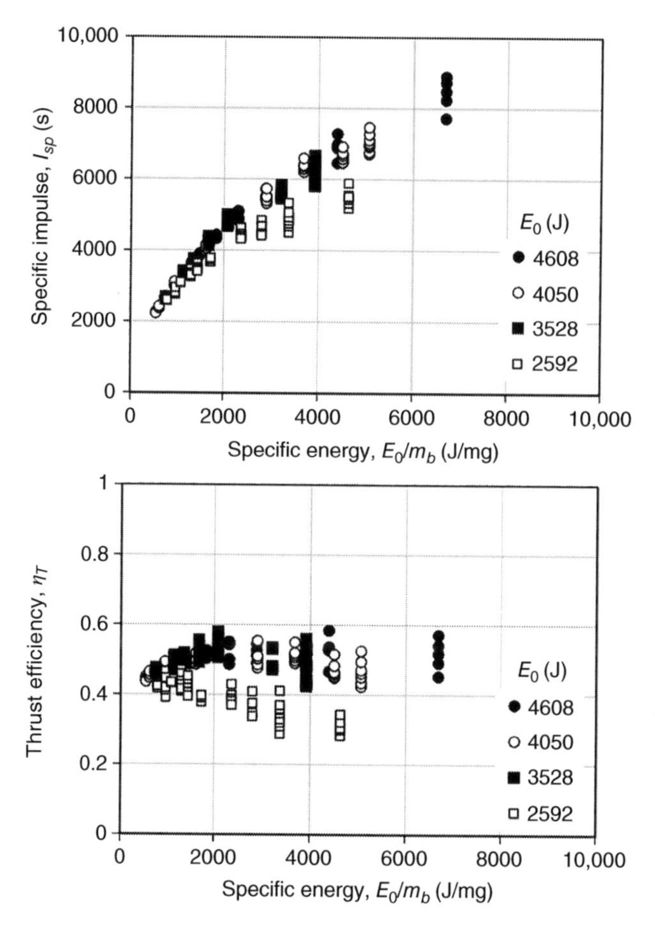

Figure 9-25 Measured performance of the PIT MkVa with NH3 (*Source:* [118]/Public domain /US Government). Top: Specific impulse. Bottom: thrust efficiency.

to specify the magnetic field as a function of time at the insulator. A main finding from this effort was on the effects of the finite plasma resistivity. Specifically, it was argued that an optimum value of the resistivity exists for which efficiency is higher than that for ideal MHD acceleration (i.e. zero resistivity). These early efforts underscored the need for an even more comprehensive resistive-MHD model of the PIT. Such models came quite a bit later during the renewed interest in high-power electric propulsion in the early 2000s, at a time when such capabilities had become more readily available.

The first successful resistive-MHD simulations of the PIT, in 2-D axisymmetric geometry, were performed by P. Mikellides [127–129] using the MACH2 code [136, 137]. The code is a highly advanced MHD simulation tool with a long record of successful simulations mostly for pulsed power and, of course, electromagnetic plasma propulsion applications as already discussed in Sections 9.2 and 9.3. As we will discuss shortly, of particular significance to the PIT are MACH2's radiation modeling capabilities. The electrons, ions, and radiation field in MACH2 are treated separately so the code actually solves up to three energy equations. There are three options for the radiation field: non-equilibrium radiation diffusion, simple radiation cooling in the optically thin limit, and equilibrium radiation diffusion in the optically thick limit. Also of importance to the PIT physics is that the evolution of the magnetic field is prescribed by Maxwell's induction equation that includes resistive diffusion. Various models for the plasma resistivity are available, including classical resistivity due to coulomb electron-ion collisions as well as electron-neutral collisions. Finally, in many engineering applications such as the ablative PPT and PIT, the source of magnetic flux is currents produced from externally applied voltage differentials. For this, MACH2 includes a variety of circuit models such as LRC, PFN, sine-waveforms, and several others. MACH2, therefore, offered a unique and most fitting simulation capability for the PIT.

The computed evolution of the mass density in the PIT MkVa operating with NH_3 at $E_0 = 4050$ J and $E_0/m_b = 942$ J/mg is depicted in Figs. 9-26 and 9-27. The left plot on Fig. 9-26 shows the

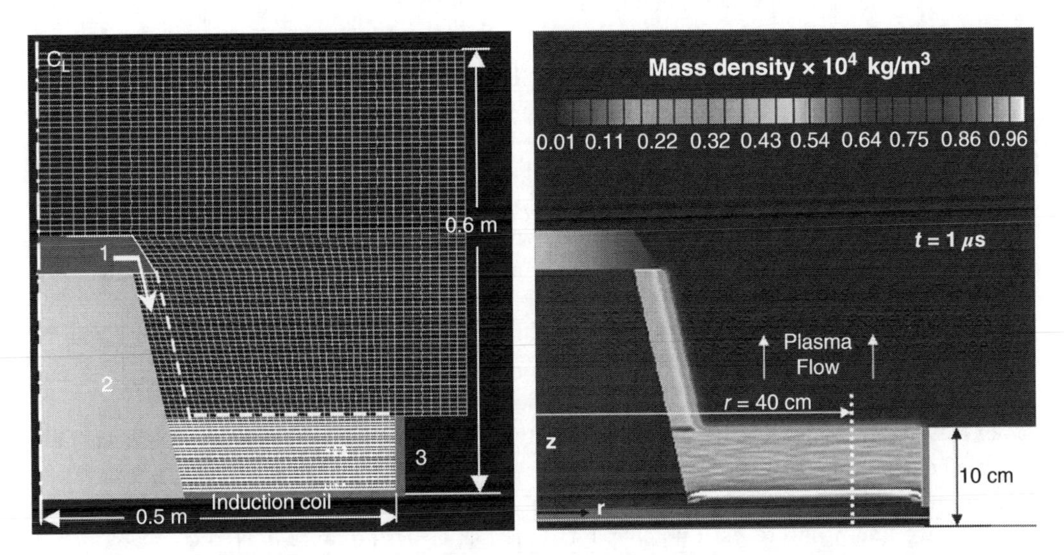

Figure 9-26 Left: Half plane and computational grid used in the numerical simulations of the PIT MkV with the 2-D½ resistive-MHD code MACH2; (1) nozzle with pulsed mass valve showing propellant mass injection, (2) conical pylon and (3) confining cuff. Right: Mass density from the simulations with NH_3 propellant at $t = 1$ ms; $E_0 = 4050$ J and $m_b = 4.3$ mg (*Source:* [127]).

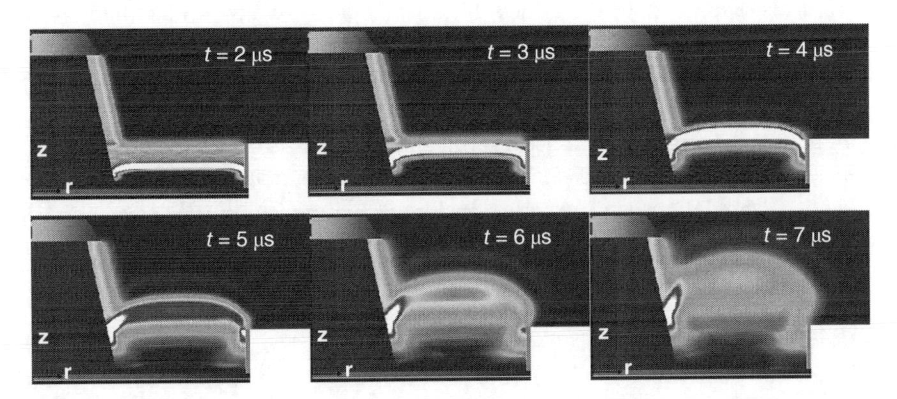

Figure 9-27 Mass density from the simulations with NH_3 propellant for times during the pulse in the range 2–7 μs; $E_0 = 4050$ J and $m_b = 4.3$ mg. The density contours correspond to the same legend as that in Figure 9-26-right (*Source:* [127]).

computational domain and mesh with the nozzle and location of propellant mass injection, the conical pylon and the confining cuff. The right plot shows the computed mass density at time $t = 1\ \mu s$ during the pulse and few relevant dimensions. The evolution of the mass density for several microseconds during the pulse is shown in Fig. 9-27. A close look at the distribution at $t = 1\ \mu s$ suggests a relatively efficient acceleration. Specifically, it is observed that the sheet exhibits an approximately uniform plasma distribution over the radius, which in turn implies that the axial distributions are approximately representative of the entire propellant. The neutral gas is entrained by the accelerating plasma in the expected snowplow fashion.

The distributions at $t = 2\ \mu s$ (Fig. 9-27) show that the radial uniformity of the mass density is still largely retained as is the efficient coupling between the magnetic flux and the accelerating plasma. At $t = 3\ \mu s$ the radial uniformity is maintained even though the majority of the plasma sheet has reached the end of the confining region. By this point, the majority of the propellant has been entrained by the accelerating sheet and compressed to the highest density. This marks the end of the "snowplow" phase and the start of the a "slug" phase characterized by acceleration of the entire propellant mass. At $t = 4\ \mu s$ the plasma is now expanding beyond the confining region. At this point the simulations predict some decoupling of the magnetic flux and plasma. The results at later times show the formations of a secondary conduction zone involving a reduced amount of plasma that is accelerated in a similar snowplow fashion. The calculations also show that the plasma continues to accelerate beyond the confining region because of expansion and conversion of enthalpy to kinetic energy [127].

The abovementioned MACH2 simulations have produced the most accurate representation of the plasma dynamics in the PIT since its invention now over six decades ago. This accuracy has been demonstrated for a variety of specific energies and propellants, not just for NH_3 [126, 127]. Figure 9-28 depicts representative comparisons between the MACH2 simulations and performance measurements for He, Ar and NH_3.

Such model fidelity permits unique insights into processes that drive the performance of this thruster. Two findings were most notable. First, the higher performance of NH_3 compared to all other propellants tested appeared to be because of negligible radiation losses during operation with this propellant [127]. The conclusion was reached after simulations with Ar and He that accounted for radiation losses showed excellent agreement with performance measurements, whereas the simulations with NH_3 neglected such losses but, nevertheless, also produced excellent agreement with the measurements as shown in Fig. 9-28. The argument about reduced radiation losses in NH_3 is also supported by the values of its mean opacity, which P. Mikellides estimated to be almost three times lower than that of He [127].

Second, a critical value of the specific energy, ε^* was identified below which thruster operation remains highly efficient. This implied that energy losses to modes other than kinetic energy are independent of the propellant mass above a "critical mass" value. In the range of $E_0/m_b < \varepsilon^*$ high thruster efficiencies can be achieved, 45–50% for NH_3, and are independent of the input energy. When $E_0/m_b > \varepsilon^*$ simulations and experiments showed diminishing efficiency at a rate that becomes more dramatic at lower discharge energies. The origin and significance of ε^* are discussed in more detail in the next section.

9.4.2.2 First-principles Idealized Modeling

Collectively, the early simplified models of the PIT [130, 135], the numerous evolutionary laboratory investigations ([114, 118, 121] and references therein), and the 2-D numerical simulations that followed them [126–128], have provided sufficient insight to permit the formulation of an idealized performance model for this thruster.

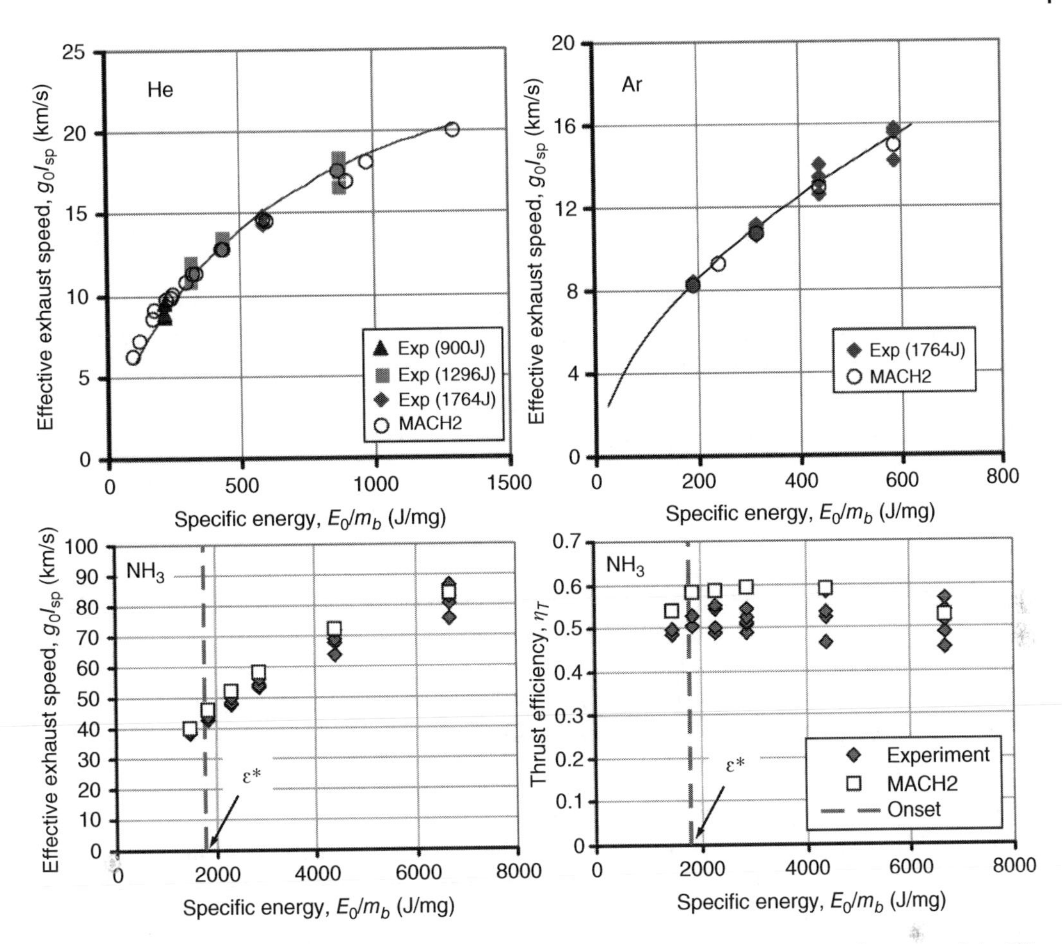

Figure 9-28 Comparison between numerical simulations with MACH2 and measured performance of the PIT MkV and MkVa. Top: He and Ar propellants for $900 < E_0 < 1764$ J in the MkV. Bottom: NH_3 propellant at E_0=4050 J in the MkVa. Also shown is the computed critical value of the specific energy, ε^* (*Source:* [127, 128]).

We begin with the simple Eq. 9.4-1 for an LRC circuit, assuming that the values of the circuit elements associated with the plasma remain constant:

$$V(t) - L\frac{\partial I(t)}{\partial t} - RI(t) = 0 \tag{9.4-1}$$

where the total inductance and resistance are given by:

$$L = L_E + L_P = L_E + L'\delta \quad R = R_E + R_P + \frac{\partial L_P}{\partial t}. \tag{9.4-2}$$

In Eq. 9.4-2, all circuit elements retain the same definitions as those provided in Section 9.3 (see Fig. 9-14). The length δ represents the thickness of the current sheet and will be discussed further shortly. The solution to the differential Eq. 9.4-1 for the current $I(t)$ yields the familiar waveform of a decaying oscillation for an underdamped ($\alpha/\omega_0 < 1$) LRC circuit:

$$I(t) = \frac{V_0}{\omega L}\sin(\omega t)e^{-\alpha t} \tag{9.4-3}$$

where the angular frequency of the system is $\omega = \sqrt{\omega_0^2 - \alpha^2}$ and the natural frequency and damping factor are given by $\omega_0 = 1/\sqrt{LC}$ and $\alpha = R/2L$, respectively. As in Section 9.3, the impulse bit associated with electromagnetic acceleration of the current sheet can then be determined as follows:

$$I_b = \frac{L'}{2}\int_0^{\infty} I^2(t)dt = L'\frac{E_0}{2R}. \tag{9.4-4}$$

Since there is no significant gas dynamic contribution to the impulse bit in this thruster, we have dropped subscript "EM" from $I_{b,EM}$ in Eq. 9.3-6 for simplicity. P. Mikellides proposed the following model to determine L' [126]. Referring to the schematic in Fig. 9-29, the induced radial magnetic field B_r is assumed to be uniform in both radial and azimuthal directions, and to diffuse into the induced current sheet of thickness delta, δ, according to an arbitrary function f that depends only on the axial direction:

$$B_r(z,t) = \mu_0 I(t)\frac{f(z)}{r_0 - r_i}. \tag{9.4-5}$$

Although arbitrary in z, function $f(z)$ is subject to the following boundary conditions: $f(\bar{z}=0)=1$, $f(\bar{z}=1) = 0$ and $f'(\bar{z}=1)=0$ where $\bar{z} \equiv z/\delta$. The assumed variation of the magnetic field in Eq. 9.4-5 allows for a simplification in the determination of the axial electromagnetic force F (and ultimately L'), namely, that such force is independent of the spatial details associated with the diffusion of the magnetic field in the sheet. That is,

$$F(t) = \frac{L'}{2}I^2(t) = \iiint j_\theta B_r dU \tag{9.4-6}$$

which upon integration of the $\mathbf{J} \times \mathbf{B}$ force over the plasma control volume U yields

$$L' = \pi\mu_0\frac{r_0 + r_i}{r_0 - r_i}. \tag{9.4-7}$$

In Eq. 9.4-6 the azimuthal current density has been determined using Ampere's law: $J_\theta(z,t) = \mu_0 dB_r(z,t)/dz$.

To complete the performance model the resistance R must also be determined. For that, P. Mikellides invoked a relatively simple statement of energy conservation as follows [126]:

$$E_R + E_K = (1-\xi)E_0 \tag{9.4-8}$$

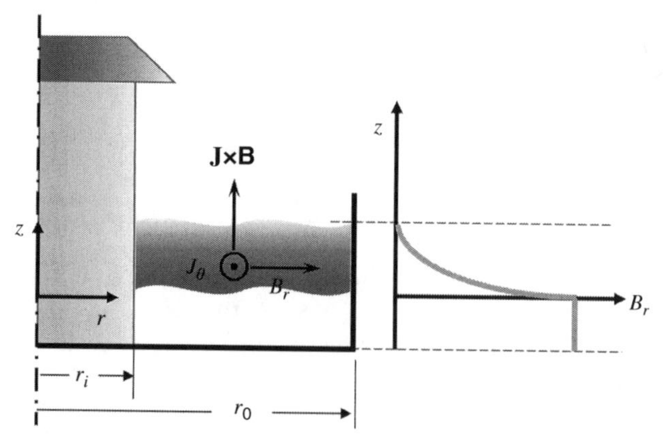

Figure 9-29 Left: Simplified representation of the current sheet acceleration used in the development of the idealized PIT performance model. Right: The assumed variation of $B_r(z)$ plotted schematically on z-B_r axes.

where the kinetic energy of the plasma, E_K, and ohmic losses associated with the total resistance of the system (plasma + external), E_R, are given by

$$E_K = \frac{I_b^2}{2m_b} = \frac{1}{m_b}\left(\frac{L'E_0}{2R}\right)^2 \quad E_R = R_E\int_0^\infty I^2(t)dt = \frac{R_E}{R}E_0. \tag{9.4-9}$$

The parameter ξ in Eq. 9.4-8 is defined as the ratio of all the energy sinks over the input energy E_0:

$$\xi = \frac{E_{\text{Loss}}}{E_0} \tag{9.4-10}$$

where E_{Loss} represents losses to internal modes of the propellant (ion/neutral, electron heating and ionization), nonutilized magnetic energy, and radiation dissipation.

In general, the determination of ξ is unamenable to simplified analysis. However, the numerical simulations described in the previous section provided invaluable insight into its determination. It was found that above a certain critical mass m^*, the fraction ξ was fairly constant and independent of the input energy and propellant mass bit. On the other hand, when $m_b < m^*$, E_{Loss} increased exponentially leading to rapid degradation of the efficiency. The processes in which a significant fraction of the input energy is deposited to internal energy modes (most notably to ionization) once the plasma reaches a certain critical speed may be familiar to the reader because it was invoked previously in reference to related processes in other electromagnetic thrusters such as the MPD and ablative PPT. P. Mikellides argued a similar limiting process occurs in the PIT at low mass bits whereby a large fraction of the input energy is deposited to internal modes when that energy becomes comparable to the propellant ionization energy, leaving negligible amount for conversion to kinetic energy of the plasma [126]. This then allowed the following model for ξ:

$$\xi = \xi_0 + (1-\xi_0)e^{-m_b/m^*} \tag{9.4-11}$$

where the critical mass m^* associated with the ionization of the propellant has been defined as

$$m^* = \frac{ME_0}{\sum_i Q_i}. \tag{9.4-12}$$

The denominator on the right-hand side of Eq. 9.4-12 denotes the sum of the ionization energy Q [J/mol] over all ionization energy levels i, and M [kg/mol] is the molecular weight of the propellant. A corresponding critical specific energy can also be defined as $\varepsilon^* = E_0/m^*$. In Eq. 9.4-11, ξ_0 is the asymptotic value of the energy fraction ξ as $m_b \to \infty$ (or as $E_0/m_b \to 0$), and is a function of the propellant. Importantly, as it is also implied by Eqs. 9.4-11 and 9.4-12, ξ_0 represents all energy losses except that associated with ionization. It is, therefore, a challenging quantity to determine without measurement or simulation. MACH2 simulations provided guidance on how this value varies with propellant mass. An example is shown for He at $E_0 = 1296$ J in Fig. 9-30, which compares the idealized model for the energy fraction ξ Eq. 9.4-11 with numerical simulations that account for all contributions in E_{Loss}, not just ionization. The comparison confirms the dominance of ionization losses below the critical mass m^* and quantifies the value of ξ_0 to be approximately 0.75 [126]. Further insight on ξ_0 is provided below as the idealized performance model is completed and the results are combined with those from the numerical simulations.

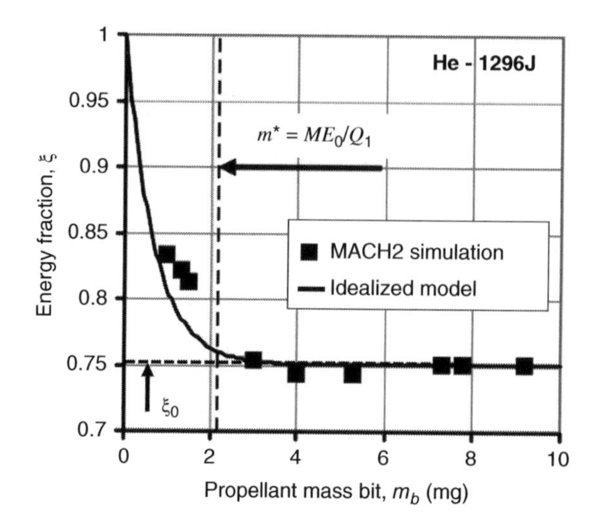

Figure 9-30 Comparison between the idealized model, Eq. 9.4-11, and MACH2 numerical simulations for the energy loss fraction ξ as a function of injected mass of He propellant (*Source:* [126]).

With the introduction of ξ in Eq. 9.4-11 we can now use Eqs. 9.4-8 and 9.4-9 to express the total resistance R as follows:

$$R = \frac{R_E + \sqrt{R_E + \frac{1-\xi}{2m_b} E_0 L'^2}}{2(1-\xi)} \tag{9.4-13}$$

which, in turn, allows for the determination of the impulse bit from Eq. 9.4-4. The thrust efficiency may then also be calculated using Eq. 9.3-5.

The idealized model results are compared with performance measurements in Fig. 9-31 for a variety of propellants and specific energies. For all propellants investigated in this analysis, the effective exhaust speed g_0Isp is predicted quite well when a constant value for ξ_0 is used in the model. This implies that at the lower energies the authors investigated in [126] ($E_0 < 1764$ J), the fraction of the total available energy expended in sinks other than ionization is independent of the specific energy when $m_b > m^*$. Based on Fig. 9-31 (and additional results reported in [126]), ξ_0 was also found to be independent of the propellant type since good agreement with the measurements was attained using the same value ($\xi_0 = 0.77$) for all three propellants (He, Ar, and CO_2). Also, the computed efficiency from the idealized model did not exceed 23% for these propellants. The exception was found to be NH_3 which required a lower value, $\xi_0 = 0.69$, to capture the measured trend for g_0Isp in the energy range 900–1764 J [126]. With this value of ξ_0, the maximum efficiency was predicted by the model to be 31% for NH_3, which is consistent with the measurements, as shown in Fig. 9-24 (bottom). Considering the results collectively, the model predictions are also consistent with the observed trends on the effect of different propellants on thrust efficiency; they showed that efficiency is lower for He, Ar, and CO_2 than for NH_3, regardless of propellant mass and energy. Further analysis of the results suggested that the lower value of ξ_0 responsible for the superior performance of NH_3 was because of lower radiation losses in this propellant compared to the other propellants.

The analysis discussed above was performed in the energy range of the PIT MkV (namely E_0<2000 J), which tested the performance of several propellants. At the higher energies of the MkVa, which focused mostly on NH_3, the electrical circuit model in the MACH2 simulations had to be advanced to include the effects of the time-dependent plasma resistance and inductance in the calculation of the current waveform. The results for a range of energies and propellant masses are reported in [128] but a representative set of comparisons has already been shown in Fig. 9-24 (bottom).

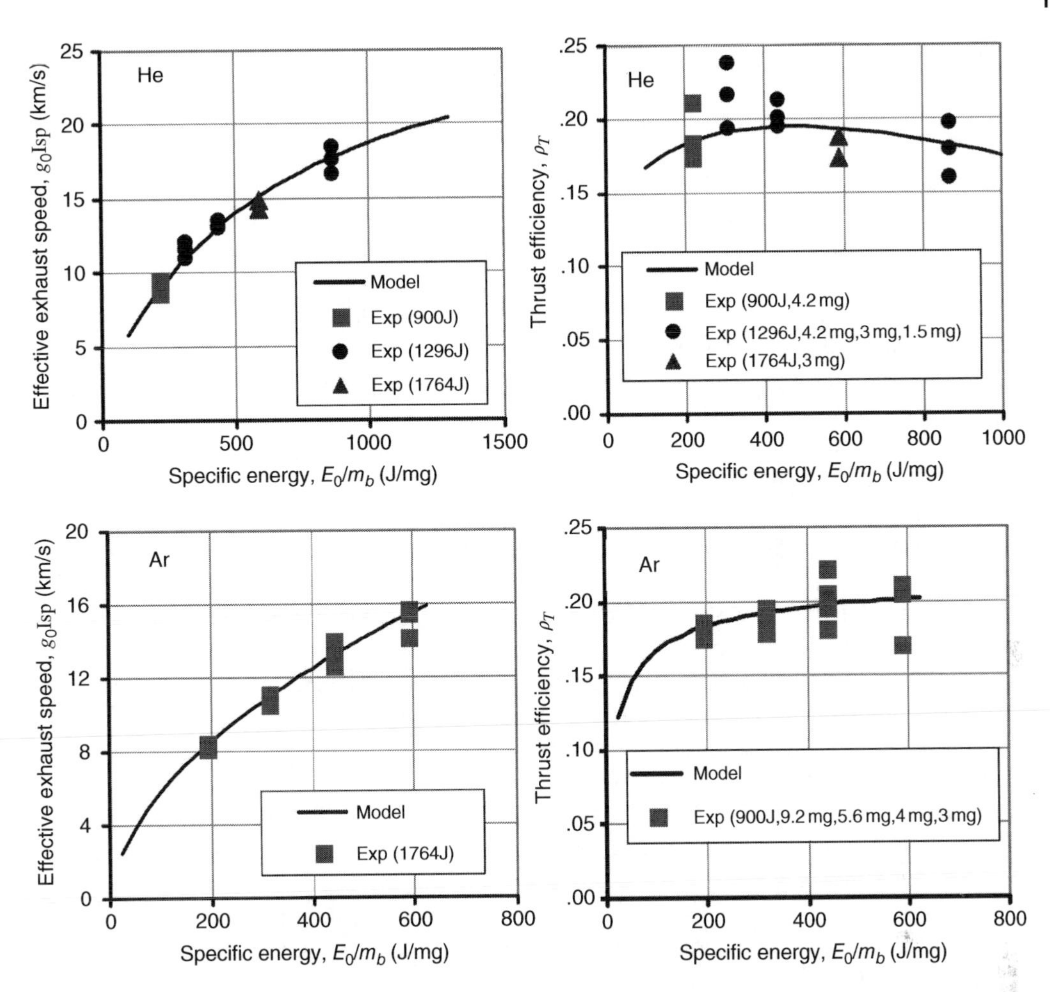

Figure 9-31 Comparisons between the idealized model and performance measurements for He and Ar propellants in the energy range $900<E_0<1764$ J, and at different values of m_b. The value of ξ_0 was 0.77 for all cases (*Source:* [126]).

These high-energy PIT simulations confirmed the existence of a critical mass below which efficiency begins to degrade because of increased deposition of the available energy to ionization. They also revealed that at higher energies, ξ is less amenable to the idealized modeling discussed above (per Eq. 9.4-11) as additional processes, such as incomplete gas breakdown and azimuthal and radial nonuniformities, become important when the thruster is operated beyond the critical specific energy value ε^*.

In fact, the preservation of the sheets' uniformity was a main motivation behind a variant of the FARAD thruster mentioned earlier [131, 132] in which the planar coil was replaced by a conical arrangement. The variant appropriately called Conical Theta Pinch (CTP) FARAD [133, 138] aimed at producing a coil geometry that more closely aligned with the natural path of the injected propellant thereby, potentially, better preserving its uniformity while also enabling easier pre-ionization. A development and testing program at NASA MSFC continued investigations of both the planar and conical variants of this thruster that culminated in 2012 in the continuous operation of the CTP at a repetition rate of 5 Hz. At a charging voltage of a 40-μF bank to 5 kV, the thruster was operated at a power of 2.5 kW which, to the authors' knowledge [139], is over an order of magnitude greater than any previous operation of a pulsed plasma thruster.

References

1 R. G. Jahn, *Physics of Electric Propulsion*. New York: McGraw-Hill, 1968.
2 M. Andrenucci, "Magnetoplasmadynamic Thrusters," in *Encyclopedia of Aerospace Engineering*, edited by R. Blockley and W. Shyy, Wiley, 2010.
3 P. J. Turchi, "Electromagnetic Propulsion - Magnetoplasmadynamic Thrusters," in *Space Propulsion Analysis and Design*, edited by R. W. Humble, G. N. Henry, and W. J. Larson, first edition, Learning Solutions, pp. 563–573, 1995.
4 A. C. Ducati, G. M. Giannini, and E. Muehlberger, "Experimental Results in High Specific Impulse Thermo-Ionic Acceleration," *AIAA Journal*, vol. 2, no. 8, pp. 1452–1454, 1964.
5 E. Y. Choueiri, "Scaling of Thrust in Self-Field Magnetoplasmadynamic Thrusters," *AIAA Journal of Propulsion and Power*, vol. 15, no. 5, pp. 744–753, 1998.
6 M. Andrenucci and F. Paganucci, "Fundamental Scaling Laws for Electric Propulsion Concepts Part 2: MPD Thrusters," AIAA-2004-3468, Presented at the 40th AIAA Joint Propulsion Conference, Fort Lauderdale, July 11–14, 2004.
7 O. A. Gorshkov, V. N. Shutov, K. N. Kozubsky, V. G. Ostrovsky, and V. A. Obukhov "Development of High Power Magnetoplasmadynamic Thrusters in the USSR," IEPC-2007-136, Presented at the 30th International Electric Propulsion Conference, Florence, September 1–20, 2005.
8 A. D. Kodys and E. Y. Choueiri, "A Critical Review of the State-of-the-Art in the Performance of Applied-Field Magnetoplasmadynamic Thrusters," AIAA-2005-4247, Presented at the 41st AIAA Joint Propulsion Conference, Tucson, July 10–13, 2005.
9 M. Auweter-Kurtz, "Plasma Thruster Development Program at the IRS," *Acta Astronautica*, vol. 32, no. 5, pp. 377–391, 1994.
10 V. P. Ageyev, V. G. Ostrovsky, and V. A. Petrosov, "High Current Stationary Plasma Accelerator of High Power," IEPC-93-117, Presented at the 23rd International Electric Propulsion conference, Seattle, September 1993.
11 J. E. Polk, A. J. Kelly, and R. G. Jahn, "Characterization of Cold Cathode Erosion Processes," IEPC-88-075, Presented at the 20th International Electric Propulsion Conference, Garmisch-Partenkirchen, October 3–6, 1988.
12 J. S. Sovey and M. A. Mantenieks, "Performance and Lifetime Assessment of Magnetoplasmadynamic Arc Thruster Technology," *AIAA Journal of Propulsion*, vol. 7, no. 1, pp. 71–83, 1991.
13 F. R. Chamberlain, A. J. Kelly, and R. G. Jahn, "Electropositive Surface Layer MPD Thruster Cathodes," AIAA-89-2706, Presented at the 25th AIAA Joint Propulsion Conference, Monterey, July 12–16, 1989.
14 V. V. Zhurin, A. A. Porotnikov, A. A. Porotniko, Jr., and V. P. Shadov, "MPD Thruster Opportunities and Perspectives," AIAA-81-0689, Presented at the 15th International Electric Propulsion Conference, Las Vegas, April 21–23, 1981.
15 J. E. Polk and T. J. Pivirotto, "Alkali Metal Propellants for MPD Thrusters," AIAA-91-3572, Presented at the AIAA Conference on Advanced SEI TEchnologies, Cleveland, September 4–6, 1991.
16 J. L. Delcroix, H. Minoo, and A. R. Trindade, "Gas Fed Multichannel Hollow Cathode Arcs," *Review of Scientific Instruments*, 40, pp. 1555–1562, 1969, https://doi.org/10.1063/1.1683861.
17 R. Albertoni, M. DeTata, R. Rossetti, F. Paganucci, M. Andrenucci, M. Cherkasova, and V. Obukhov, "Experimental Study of a Multichannel Hollow Cathode for High Power MPD Thrusters," AA-2011-6075, Presented at the 47th AIAA Joint Propulsion Conference, San Diego, July 31–August 3, 2011.
18 H. Maecker, Plasma Flows in Arcs Due to Self-Magnetic Compression," (in German)*Zeitschrift für Physik* vol. 141, pp. 198–216, 1955; doi:10.1007/BF01327300.

19 J. Gilland and G. Johnston, "MPD Thruster Performance Analytic Models," *AIP Conference Proceedings*, vol. 654, no. 1, 2003; doi:10.1063/1.1541334.

20 E. Y. Choueiri, "On the Thrust of Self-Field MPD Thrusters," IEPC-97-121, Presented at the 25th International Electric Propulsion Conference, Cleveland, October 14–17, 1997.

21 J. H. Gilland, "The Effect of Geometrical Scale Upon MPD Thruster Behavior," Master's thesis. Princeton University.

22 J. H. Gilland, A. J. Kelly, and R. G. Jahn, "MPD Thruster Scaling," AIAA-87-0997, Presented at the 19th AIAA Joint Propulsion Conference, Colorado Springs, May 11–13, 198.

23 V. B. Tikhonov and S. A. Semenihin, "Research of Plasma Acceleration Processes in Self-Field and Applied Magnetic Field Thrusters," IEPC-93-076, Presented at the 23rd International Electric Propulsion Conference, Seattle, September 1993.

24 H. Alfven, "Collision Between a Nonionized Gas and a Magnetized Plasma," *Reviews of Modern Physics*, vol. 32, no. 4, pp. 710–713, 1960; doi:10.1103/RevModPhys.32.710.

25 G. Krülle, M. Auweter-Kurtz, and A. Sasoh, "Technology and Application Aspects of Applied Field Magnetoplasmadynamic Propulsion," *AIAA Journal of Propulsion and Power*, vol. 14, no. 5, 1998; doi:10.2514/2.5338.

26 W. J. Coogan and E. Y. Choueiri, "A Critical Review of Thrust Models for Applied-Field Magnetoplasmadynamic Thrusters," AIAA-2017-4723, Presented at the 53rd AIAA Joint Propulsion Conference, Atlanta, July 10–12, 2017.

27 E. Ahedo and M. Merino, "On Plasma Detachment in Propulsive Magnetic Nozzles," *Physics of Plasmas*, vol. 18, no. 053504, 2011, doi:10.1063/1.3589268.

28 R. Albertoni, F. Paganucci, and M. Andrenucci, "A Phemenological Model for Applied-field MPD Thrusters," *Acta Astronautica*, vol. 107, pp. 177–186, 2015; doi:10.1016/j.actaastro.2014.11.017.

29 D. B. Fradkin, A. W. Blackstock, D. J. Roehling, F. F. Stratton, M. Williams, and K. W. Liewer, "Experiments Using a 25 kW Hollow Cathode Lithium Vapor MPD Arcjet," AIAA-1969-241, Presented at the 7th AIAA Electric Propulsion Conference, Williamsburg, 1969.

30 G. Herdrich, A. Boxberger, D. Petkow, R. A. Gabrielli, S. Fasoulas, M. Andrenucci, R. Albertoni, F. Paganucci, and P. Rosetti, "Advanced Scaling Model for Simplified Thrust and Power Scaling of an Applied-Field Magnetoplasmadynamic Thruster," AIAA-2010-6531, presented at the 46th AIAA Joint Propulsion Conference, Nashville, July 25–28, 2010.

31 R. M. Myers, "Scaling of 100 kW Class Applied-Field MPD Thrusters," AIAA-92-3462, Presented at the 28th AIAA Joint Propulsion Conference, Nashville, July 6–8, 1992.

32 A. Sasoh and Y. Arakawa, "Thrust Formula for Applied-Field Magnetoplasmadynamic Thrusters Derived from Energy Conservation Equation," *AIAA Journal of Propulsion and Power*, vol. 11, no. 2, pp. 351–356, 1995.

33 M. Coletti, "A Thrust Formula for an MPD Thruster with Applied-Magnetic Field," *Acta Astronautica*, vol. 81, pp. 667–674, 2012; doi:10.1016/j.actaastro.2012.08.014.

34 P. G. Mikellides and P. J. Turchi, "Applied-Field Magnetoplasmadynamic Thrusters, Part 2: Analytic Expressions for Thrust and Voltage," *AIAA Journal of Propulsion and Power*, vol. 16, no. 5, pp. 894–901, 2000; doi:10.2514/2.5657.

35 W. J. Coogan, "Thrust Scaling in Applied-Field Magnetoplasmadynamic Thrusters," PhD. Princeton University, 2018.

36 P. G. Mikellides and P. J. Turchi, "Applied-Field Magnetoplasmadynamic Thrusters, Part 1: Numerical Simulations Using the MACH2 Code," *Journal of Propulsion and Power*, vol. 16, no. 6, pp. 887–893, 2000; doi:10.2514/2.5656.

37 R. Albertoni, P. Rossetti, F. Paganucci, M. Andrenucci, M. Zuin, E. Martinez, and R. Cavazzana, "Experimental Study of a 100-kW Class Applied-Field MPD Thruster," EPC-2011-110, Presented at the 32nd International Electric Propulsion Conference, Wiesbaden, September 11–15, 2011.

38 D. B. Fradkin, A. W. Blackstock, D. J. Roehling, T. F. Stratton, M. Williams, and K. W. Liewer, "Experiments Using a 25 kW Hollow Cathode Lithium Vapor MPD Arcjet," *AIAA Journal*, vol. 8, no. 5, pp. 886–894, 1969; doi:10.2514/6.1969-241.
39 F. Paganucci, P. Rossetti, M. Andrenucci, V. B. Tikhonov, and V. A. Obukhov, "Performance of an Applied Field MPD Thruster," IEPC-01-132, Presented at the 27th International Electric Propulsion Conference. Pasadena, October 15–19, 2001.
40 H. Tahara, Y. Kagaya, and T. Yoshikawa, "Hybrid MPD Thruster with Axial and Cusp Magnetic Fields," IEPC-88-058, Presented at the 26th International Electric Propulsion Conference, Garmisch-Partenkirchen, October 3–6, 1988.
41 V. Tikhonov, S. Semenikhin, J. R. Brophy, and J. E. Polk, "The Experimental Performance of the 100 kW Li MPD Thruster with External Magnetic Field," IEPC-95-105, Presented at the 24th International Electric Propulsion Conference, Moscow, September 19–23, 1995.
42 V. B. Tikhonov, S. A. Semenikhin, J. R. Brophy, and J. E. Polk, "Performance of the 130 kW MPD Thruster with an External Magnetic Field and Li as Propellant," IEPC-97-117, Presented at the 25th International Electric Propulsion Conference, Cleveland, October 14–17, 1997.
43 E. Y. Choueiri, "Electric Propulsion and Plasma Dynamics Laboratory," https://alfven.princeton.edu, 2023.
44 E. Bögel, M. Collier-Wright, K. Aggarwal, and M. L. Betancourt, "State of the Art Review in Superconductor-Based Applied-Field Magnetoplasmadynamic Thruster Technology," IEPC-2022-476, Presented at the 37th International Electric Propulsion Conference, Cambridge, June 19–23, 2022.
45 L. Uribarri and E. Y. Choueiri, "Relationship Between Anode Spots and Onset Voltage Hash in Magnetoplasmadynamic Thrusters," *AIAA Journal of Propulsion and Power*, vol. 24, no. 3, pp. 571–577, 2008; doi:10.2514/1.34525.
46 A. Malliaris, R. John, R. Garrison, and D. Libby, "Performance of Quasi-Steady MPD Thrusters at High Powers," *AIAA Journal*, vol. 10, no. 2, pp. 121–122, 1972; doi:10.2514/3.50074.
47 L. K. Rudolph, "The MPD Thruster Onset Current Performance Limitation," PhD thesis. Princeton University, 1980.
48 R. C. Moeller and J. E. Polk, "Influence of Tailored Applied Magnetic Fields on High-Power MPD Thruster Current Transport and Onset-Related Phenomena," IEPC-2013-379, Presented at the 33rd International Electric Propulsion Conference, Washington, October 6–10, 2013.
49 H. Hügel, "Flow Rate Limitations in the Self-FIELD ACcelerator," AIAA-73-1094, Presented at the 10th AIAA Electric Propulsion Conference, Lake Tahoe, October 1973.
50 D. J. Merfeld, A. J. Kelly, and R. G. Jahn, "MPD Thruster Performance: Propellant Distribution and Species Effects," *AIAA Journal of Propulsion and Power*, vol. 2, no. 4, pp. 317–322, 1986; doi:10.2514/3.22889.
51 L. Uribarri, "Onset Voltage Hash and Anode Spots in Quasi-Steady Magnetoplasmadynamic Thrusters," PhD thesis. Princeton University, 2008.
52 M. Andrenucci, V. Antoni, M. Bagatin, C. A. Brghi, M. R. Carraro, A. Cristofolini, R. Ghidini, F. Paganucci, P. Rossetti, and G. Serianni, "Magneto-Plasma-Dynamic Thrusters for Space Applications," AIAA-2002-2185, Presented at the 33rd Plasmadynamics and Lasers Conference, Maui, May 20–23, 2002.
53 K. D. Diamant, E. Y. Choueiri, and R. G. Jahn, "Spot Mode Transition and the Anode Fall of Pulsed Magnetoplasmadynamic Thrusters," *AIAA Journal of Propulsion and Power*, vol. 14, no. 6, pp. 1036–1042, 1998.
54 D. M. Goebel, "Ion Source Discharge Performance and Stability," *Physics of Fluids*, vol. 25, no. 6, pp. 1093–1102, 1982.

55 I. Hutchinson, *Principles of Plasma Diagnostics*, Second edition, Cambridge: Cambridge University Press, 2002.

56 R. C. Moeller, "Current Transport and Onset-Related Phenomena in an MPD Thruster Modified by Applied Magnetic Fields," PhD thesis. California Institute of Technology, 2013.

57 M. J. Boyle, K. E. Clark, and R. G. Jahn, "Flowfield Characteristics and Performance Limitations of Quasi-Steady Magnetoplasmadynamic Accelerators," *AIAA Journal*, vol. 14, no. 7, pp. 955–962, 1976.

58 E. Y. Choueiri, "MPD Thruster Plasma Instability Studies," AIAA-87-1067, Presented at the 19th International Electric Propulsion Conference, Colorado Springs, May 11–13, 1987.

59 D. L. Tilley, E. Y. Choueiri, A. J. Kelly, and R. G. Jahn, "Microinstabilities in a 10-kilowatt Self-Field Magnetoplasmadynamic Thruster," *AIAA Journal of Propulsion and Power*, vol. 12, no. 2, pp. 381–389, 1996; doi:10.2514/3.24040.

60 H. P. Wagner, H. J. Kaeppeler, and M. Auweter-Kurtz, "Instabilities in MPD Thruster Flows: 1. Space Charge Instabilities in Unbounded and Inhomogeneous Plasmas," *Journal of Physics D: Applied Physics*, vol. 31, pp. 519–528, 1998.

61 M. Maurer, H. J. Kaeppeler, and W. Richert, "Calculation of Nonlinear Drift Instabilities in Magnetoplasmadynamic Thruster Flows," *Journal of Physics D: Applied Physics*, vol. 28, pp. 2269–2278, 1995.

62 H. P. Wagner, J. J. Kaeppeler, and M. Auweter-Kurtz, "Instabilities in MPD Thruster Flows: 2. Investigation of Drift and Gradient Driven Instabilities Using Multi-Fluid Plasma Models. *Journal of Physics D: Applied Physics*, vol. 31, pp. 529–541, 1998.

63 M. Zuin, V. Anton, F. Paganucci, R. Cavazzana, G. Serianni, M. Spolaore, E. Martines, N. Vianello, M. Bagatin, P. Rosetti, and M. Andrenucci, "Plasma Fluctuations in an MPD Thruster with and Without the Application of an External Magnetic Field," IEPC-2003-299, Presented at the 28th International Electric Propulsion Conference, Toulouse, March 17–21, 2003.

64 J. Lawless and V. V. Subramaniam, "Theory of Onset in Magnetoplasmadynamic Thrusters," *AIAA Journal of Propulsion and Power*, vol. 3, no. 2, pp. 121–127, 1987.

65 M. Zuin, R. Cavazzana, E. Martines, G. Serianni, V. Antoni, M. Bagatin, M. Andrenucci, F. Paganucci, and P. Rosetti, "Kink Instability in Applied Field MPD Thrusters," *Physical Review Letters*, vol. 92, no. 225003, 2004.

66 M. Zuin, R. Cavazzana, E. Martines, G. Serianni, V. Antoni, M. Bagatin, M. Andrenucci, F. Paganucci, and P. Rosetti, "Critical Regimes and Magnetohydrodynamic Instabilities in a Magnetoplasmadynamic Thruster," *Physics of Plasmas*, vol. 11, no. 10, pp. 4761–4770, 2004.

67 H. O. Schrade, M. Auweter-Kurtz, and H. Kurtz, "Stability Problems in Magneto Plasmadynamik Arc Thrusters," AIAA-85-1633, Presented at the 18th AIAA Fluid Dynamics, Plasma Dynamics and Lasers Conference, Cincinnati, July 1985.

68 H. O. Schrade, T. Wegmann, and T. Rosgen, "The Onset Phenomena Explained by Run-Away Joule Heating," IEPC-91-022, Presented at the 22nd International Electric Propulsion Conference, Viareggio, October 14–17, 1991.

69 E. Y. Choueiri and J. K. Ziemer, "Quasi-Steady Magnetoplasmadynamic Thruster Performance Database," AIAA-1998-3472, Presented at the 34th AIAA Joint Pro7pulsion Conference. Cleveland, July 13–16, 1998.

70 M. R. LaPointe, "High Power MPD Thruster Performance Measurements," AIAA-2004-3467, Presented at the 40th AIAA Joint Propulsion Conference, Fort Lauderdale, July 11–14, 2004.

71 W. J. Guman and D. M. Nathanso "Pulsed Plasma Microthruster Propulsion System for Synchronous Orbit Satellite," *Journal of Spacecraft and Rockets*, vol. 7, no. 4, pp. 409, 1970; doi:10.2514/3.29955.

72 W. J. Guman and T. E. Williams, "Pulsed Plasma Microthruster for Synchronous Meteorological Satellite," *Journal of Spacecrafi and Rockets*, vol. 11, no. 10, pp. 729–731, 1974.

73 W. Ebert, S. Kowal, and R. Sloan, "Operational NOVA Spacecraft Teflon Pulsed Plasma Thruster System," AIAA-1989-2497, Presented at the 25th AIAA Joint Propulsion Conference, Monterey, July 12–16, 1989.

74 S. M. An, H. J. Wu, X. Z. Feng, and W. X. Liu, "Space-Flight Test of Electric Thruster System Mdt-2a," (in English), *Journal of Spacecraft and Rockets*, vol. 21, no. 6, pp. 593–594, 1984; doi:10.2514/3.25701.

75 R. J. Vondra and K. I. Thomassen, "Flight Qualified Pulsed Electric Thruster for Satellite Control," *Journal of Spacecraft and Rockets*, vol. 11, no. 9, pp. 613–617, 1974; doi:10.2514/3.62141.

76 R. L. Burton and P. J. Turchi, "Pulsed Plasma Thruster," *Journal of Propulsion and Power*, vol. 14, no. 5, pp. 716–735, 1998; doi:10.2514/2.5334.

77 C. Zakrzwski, S. Benson, P. Sanneman, and A. Hoskins, "On-Orbit Testing of the EO-1 Pulsed Plasma Thruster," AIAA-2002-3973, 38th AIAA Joint Propulsion Conference, Indianpolis, July 7–10, 2002; doi:10.2514/6.2002-3973

78 P. Northway, C. Aubuchon, H. Mellema, R. Winglee, and I. Johnson, "Pulsed Plasma Thruster Gains in Specific Thrust for CubeSat Propulsion," AIAA-2017-5040, Presented at the 53rd AIAA Joint Propulsion Conference, Atlanta, July 10–12, 2017.

79 N. Buldrin, D. Jelem, A. Reissner, C. Scharlmann, B. Seifert, and R. Sypniewski, "Small Sat Propulsion Developments at FHWN and FOTEC," 1st International Conference on Micropropulsion and CubeSats, MPCS-2017-Cs03, Bari, Italy, June 26-28, 2017.

80 R. L. Burton, C. A. Woodruff, D. M. King, and D. L. Carroll, "Analysis of Fiber-Fed Pulsed Plasma Thruster Performance," *Journal of Propulsion and Power*, vol. 37, no. 1, pp. 176–178, 2021.

81 P. Mikellides and P. Turchi, "Modeling of Late-Time Ablation in Teflon Pulsed Plasma Thrusters," AIAA-1996-2733, 32nd AIAA Joint Propulsio Conference, Lake Buena Vista, July 1–3, 1996; doi:10.2514/6.1996-2733.

82 R. J. Vondra and K. I. Thomassen, "Performance Improvements in Solid Fuel Microthrusters," *Journal of Spacecraft and Rockets*, vol. 9, no. 10, pp. 738–742, 1972.

83 P. Molina-Cabrera, G. Herdrich, M. Lau, S. Fausolas, T. Schoenherr, and K. Komurasaki, "Pulsed Plasma Thrusters: A Worldwide Review and Long Yearned Classification," IEPC-2011-340, 32nd International Electric Propulsion Conference, Wiesbaden, September 11–15, 2011.

84 I. Kimura, K. Ogiwara, and Y. Suzuki, "Effect of Applied Magnetic Fields on a Solid-Propellant Pulsed Plasma Thruster," IEPC-79-2098, 14th International Electric Propulsion Conference, Princeton, October 30–November 1, 1979.

85 K. Yuan-Zhu, "Effects of Propellant Geometry on PPT Performance," IEPC-1984-94, 17th International Electric Propulsion Conference, Tokyo, May 28–31, 1984.

86 K. I. Thomassen and R. J. Vondra, "Exhaust Velocity Studies of a Solid Teflon Pulsed Plasma Thruster," *Journal of Spacecraft and Rockets*, vol. 9, no. 1, pp. 61–64, 1972.

87 D. J. Palumbo and W. J. Guman, "Effects of Propellant and Electrode Geometry on Pulsed Ablative Plasma Thruster Performance," *Journal of Spacecraft and Rockets*, vol. 13, no. 3, pp. 163–167, 1976.

88 R. Leiweke, P. Turchi, H. Kamhawi, and R. Myers, "Experiments with Multi-material Propellants in Ablation-Fed Pulsed Plasma Thrusters," AIAA-95-2916, 31st AIAA Joint Propulsion Conference, San Diego, July 10–12, 1995.

89 G. Paccani, L. Petrucci, and W. D. Deininger, "Scale Effects on Solid-Propellant Coaxial Magnetoplasmadynamic Thruster Terformance," *Journal of Propulsion and Power*, vol. 19, no. 3, pp. 431–437, 2003.

90 G. Paccani, U. Chiarotti, and W. D. Deininger, "Quasisteady Ablative Magnetoplasmadynamic Thruster Performance with Different Propellants," *Journal of Propulsion and Power*, vol. 14, no. 2, pp. 254–260, 1998.

91 W. J. Guman and P. E. Peko, "Solid-Propellant Pulsed Plasma Microthruster Studies," *Journal of Spacecraft and Rockets*, vol. 5, no. 6, pp. 732–733, 1968.

92 G. Spanjers, D. Bromaghim, J. Lake, M. Dulligan, D. White, J. Schilling, and S. Bushman, "AFRL MicroPPT Development for Small Spacecraft Propulsion," AIAA-2002-3974, 38th AIAA Joint Propulsion Conference, Indianapolis, July 7–10, 2002.

93 S. Bushman, R. Burton, and E. Antonsen, "Arc Measurements and Performance Characteristics of a Coaxial Pulsed Plasma Thruster," AIAA-1998-3660, 34th AIAA Joint Propulsion Conference, Cleveland, July 13–15, 1998.

94 T. E. Markusic, K. A. Polzin, E. Y. Choueiri, M. Keidar, I. D. Boyd, and N. Lepsetz, "Ablative Z-pinch Pulsed Plasma Thruster," *Journal of Propulsion and Power*, vol. 21, no. 3, pp. 392–400, 2005.

95 M. Keidar, I. D. Boyd, F. S. Gulczinski, III, E. L. Antonsen, and G. G. Spanjers, "Analyses of Teflon Surface Charring and Near Field Plume of a Micro-Pulsed Plasma Thruster," IEPC-2001-155, 27th International Electric Propulsion Conference, Pasadena, October 14–19, 2001.

96 A. Solbes and R. J. Vondra, "Performance Study of a Solid Fuel-Pulsed Electric Microthruster," *Journal of Spacecraft and Rockets*, vol. 10, no. 6, pp. 406–410, 1973.

97 M. H. Frese, "MACH2: A Two-Dimensional Magnetohydrodynamic Simulation Code for Complex Experimental Configurations," Interim Report AD-A-192285/5/XAB, 1987.

98 P. Turchi, I. Mikellides, P. Mikellides, and C. Schmahl, "Theoretical Investigation of Pulsed Plasma Thrusters," AIAA-1998-3807, 34th AIAA Joing Propulsion Conference, Cleveland, July 13–15, 1998.

99 I. Mikellides and P. Turchi, "Optimization of Pulsed Plasma Thrusters in Rectangular and Coaxial Geometries," IEPC-99-211, 26th International Electric Propulsion Conference, Kitakyushu, October 17–21, 1999.

100 I. G. Mikellides, "Theoretical Modeling and Optimization of Ablation-Fed Pulsed Plasma Thrusters," PhD thesis. Ohio State University, 2000.

101 H. Kamhawi and P. Turchi, "Design, Operation, and Investigation of an Inductively-Driven Pulsed Plasma Thruster," AIAA-1998-3804, 34th AIAA Joint Propulsion Conference, Cleveland, July 13–15, 1998.

102 H. Kamhawi, L. Arrington, E. Pencil, and T. Haag, "Performance Evaluation of a High Energy Pulsed Plasma Thruster," 41st AIAA Joint Propulsion Conference, Tucson, July 10–13, 2005.

103 N. N. Antropov, A. V. Bogatyy, G. A. Dyakonov, N. V. Lyubinskaya, G. A. Popov, S. A. Semenikhin, V. K. Tyutin, M. M. Khrustalev, and V. N. Yakovlev, "A New Stage in the Development of Ablative Pulsed Plasma Thrusters at the RIAME," *Solar System Research*, vol. 46, no. 7, pp. 531–541, 2012.

104 L. H. Zeng, Z. W. Wu, G. R. Sun, T. K. Huang, K. Xie, and N. F. Wang, "A New Ablation Model for Ablative Pulsed Plasma Thrusters," *Acta Astronautica*, vol. 160, pp. 317–322, 2019.

105 N. Hossain, N. F. Wang, G. R. Sun, H. Li, and Z. W. Wu, "A Reliable Data-Driven Model for Ablative Pulsed Plasma Thruster," *Aerospace Science and Technology*, vol. 105, 105953, October 2020.

106 P. G. Mikellides, E. M. Henrikson, and S. S. Rajagopalan, "Theoretical Formulation for the Performance of Rectangular, Breech-Fed Pulsed Plasma Thrusters," *Journal of Propulsion and Power*, vol. 35, no. 4, pp. 811–818, 2019.

107 P. J. Turchi, I. G. Mikellides, P. G. Mikellides, and H. Kamhawi, "Pulsed Plasma Thrusters for Microsatellite Propulsion: Techniques for Optimization," *Micropropulsion for Small Spacecraft*, pp. 353–368, 2000; doi:10.2514/5.9781600866586.0353.0368

108 R. H. Lovberg and C. L. Dailey, "A PIT Primer," Techical Report 005. RLD Associates, Encino, 1994.

109 C. L. Dailey, "Investigation of Plasma Rotation in a Pulsed Inductive Accelerator," *AIAA Journal*, vol. 7, no. 1, pp. 13–19, 1969.

110 C. L. Dailey, "Pulsed Electromagnetic Thruster," AFRPL-TR-71-107. TRW Systems Group, Redondo Beach, 1971 December.

111 C. L. Dailey and R. H. Lovberg, "Current Sheet Structure in an Inductive-Impulsive Plasma Accelerator," *AIAA Journal*, vol. 10, no. 2, pp. 125–129, 1972.

112 C. L. Dailey, "Pulsed Plasma Propulsion Technology," AFRPL-TR-73-81.TRW Systems Group, Redondo Beach, 1973 July.

113 C. L. Dailey, "Plasma Properties in an Inductive Pulsed Plasma Accelerator," AIAA-1965-637, 6th Biennial Gas Dynamics Symposium, Evanston, 1965 August.

114 C. L. Dailey and R. H. Lovberg, "Pulsed Inductive Thruster Component Technology," AFALTR-87-012. TRWSpace and Technology Group, Redondo Beach, April 1987.

115 C. L. Dailey and R. H. Lovberg, "PIT Clamped Discharge Evolution," AFOSR-TR-89-0130. TRW Space and Technology Group, Redondo Beach, 1988 December.

116 R. H. Lovberg and C. L. Dailey, "Current Sheet Development in a Pulsed Inductive Thruster," 25th AIAA Joint Propulsion Conference, Monterey, July 10–12, 1989.

117 R. H. Lovberg and C. L. Dailey, "PIT Mark V Design," Conference on Advanced SEI Technologies, Cleveland, September 4–6, 1991.

118 C. L. Dailey and R. H. Lovberg, "The PIT MkV Pulsed Inductive Thruster," NASA-CR-191155, TRW Systems Group, Redondo Beach, 1993 July.

119 C. L. Dailey, J. Hieatt, and R. H. Lovberg, "Nuclear Propulsion for Mars Exploration - Electric Versus Thermal," AIAA-1992-3871, 28th AIAA Joint Propulsion Conference, Nashville, 1992 July.

120 R. H. Lovberg and C. L. Dailey, "A Lightweight Efficient Argon Electric Thruster," AIAA-82-1921, 16th International Electric Propulsion Conference, New Orleans, November 17–19, 1982.

121 K. A. Polzin, "Comprehensive Review of Planar Pulsed Inductive Plasma Thruster Research and Technology," *Journal of Propulsion and Power*, vol. 27, no. 3, pp. 513–531, 2011.

122 R. H. Frisbee, "Evaluation of High-Power Solar Electric Propulsion Using Advanced Ion, Hall, MPD, and PIT Thrusters for Lunar and Mars Cargo Missions," AIAA-2006-4465, 42nd AIAA Joint Propulsion Conference, Sacramento, July 9–12, 2006.

123 R. H. Frisbee and I. G. Mikellides, "The Nuclear Electric Pulsed Inductive Thruster (NuPIT): Mission Analysis for Prometheus," AIAA-2005-3892, 41st AIAA Joint Propulsion Conference, Tuscon, July 2005.

124 J. Poylio, R. H. Lovberg, C. L. Dailey, W. Goldstein, D. Russell, and B. Jackson, "Pulsed Inductive Thruster: Flight-Scale Proof of Concept Demonstrator," AIAA-2004-3640, 40th AIAA Joint Propulsion Conference, Ford Lauderdale, July 2005.

125 D. Russell, C. L. Dailey, W. Goldstein, R. H. Lovberg, J. Poylio, and B. Jackson, "The PIT Mark VI Pulsed Inductive Thruster," AIAA-2004-6054, Space 2004 Conference and Exhibit, San Diego, September 2004.

126 P. G. Mikellides and C. Neilly, "Modeling and Performance Analysis of the Pulsed Inductive Thruster," *Journal of Propulsion and Power*, vol. 23, no. 1, pp. 51–58, 2007.

127 P. G. Mikellides and N. Ratnayake, "Modeling of the Pulsed Inductive Thruster Operating with Ammonia Propellant," *Journal of Propulsion and Power*, vol. 23, no. 4, pp. 854–862, 2007.

128 P. G. Mikellides and J. K. Villarreal, "High Energy Pulsed Inductive Thruster Modeling Operating with Ammonia Propellant," *Journal of Applied Physics*, vol. 102, no. 10, 103301, 2007.

129 P. G. Mikellides and J. K. Villarreal, "Numerical Modeling of a Low Energy Pulsed Inductive Thruster," AIAA-2008-4726, 44th AIAA Joint Propulsion Conference, Hartford (July 2008).

130 R. H. Lovberg and C. L. Dailey, "Large Inductive Thruster Performance Measurement," *AIAA Journal*, vol. 20, no. 7, pp. 971–977, 1982.

131 E. Y. Choueiri and K. A. Polzin, "Faraday Acceleration with Radio-Frequency Assisted Discharge," *Journal of Propulsion and Power*, vol. 22, no. 3, pp. 611–619, 2006.

132 K. A. Polzin, "Faraday Accelerator with Radio-Frequency Assisted Discharge (FARAD)," PhD. Mechanical and Aerospace Engineering, Princeton University, 2006.

133 A. K. Hallock, E. Y. Choueiri, and K. A. Polzin, "Current Sheet Formation in a Conical Theta Pinch Faraday Accelerator with Radio-Frequency Assisted Discharge," IEPC-2007-165, 30th International Electric Propulsion Conference, Florence, September 2007.

134 C. L. Dailey and R. H. Lovberg, "Large Diameter Inductive Plasma Thrusters," in *Electric Propulsion and Its Applications to Space Missions*, edited by R. Finke, pp. 529–542, American Institute of Aeronautics and Astronautics, Inc, 1981.

135 R. H. Lovberg and C. L. Dailey, "Numerical Simulation of Pulsed Inductive Thruster Plasma," AIAA-83-1396, 19th AIAA Joint Propulsion Conference, Seattle, June 27–29, 1983.

136 M. H. Frese, "MACH2: A Two-Dimensional Magnetohydrodynamic Simulation Code for Complex Experimental Configurations," National Technical Information Service Document No. ADA 192285, Mission Research Corporation Report AMRC-R-874, 1987.

137 R. E. Peterkin, M. H. Frese, and C. R. Sovinec, "Transport of Magnetic Flux in an Arbitrary Coordinate ALE Code," *Journal of Computational Physics*, vol. 140, no. 1, pp. 148–171, 1998.

138 A. K. Hallock, E. Y. Choueiri, and K. A. Polzin, "Current Sheet Formation in a Conical Theta Pinch Faraday Accelerator with Radio-Frequency Assisted Discharge," AIAA-2008-5201, 44th AIAA AIAA Joint Propulsion Conference, Hartford, July 2008.

139 K. A. Polzin, A. K. Martin, R. H. Eskridge, A. C. Kimberlin, B. M. Addona, A. P. Devineni, N. R. Dugal-Whitehead, and A. K. Hallock, "Summary of the 2012 Inductive Pulsed Plasma Thruster Development and Testing Program," NASA/TP-2013-217488, NASA Marshall Space Flight Center, Huntsville, 2013.

Chapter 10

Future Directions in Electric Propulsion

Hall thrusters and ion thrusters are now routinely used to provide in-space propulsion in Earth-orbiting satellites and deep space science missions. These two types of thrusters are considered mature flight technologies. However, research and development efforts on them continue worldwide primarily aimed at scaling to higher power levels (>10 kW) and lower power levels (<1 kW), extending life and throughput, understanding stability and oscillations, and improving modeling and simulation capabilities. Examples of some additional investigation activities on these types of thrusters are provided below. Electromagnetic thrusters also have an extensive history of research and development and some emerging flight experience. Several examples of the ongoing activities in electromagnetic thruster development are also given below.

Other electric thruster concepts that do not have flight heritage to date are of interest to discuss briefly to illustrate research and development directions in the field. Reviews of new concepts and prospects for electric thrusters have been recently published by Levchenko et al. [1–3]. A review of thruster technologies and paths toward maturation of existing thrusters and new concepts was also recently published by Dale et al. [4]. Examples of some of these thruster concepts are provided in this chapter and the others can be found in the references in these reviews. Again, we will ignore electrothermal thrusters as mature technology with relatively low performance for future space propulsion applications, and also not discuss micropropulsion as it is too extensive a field to be covered sufficiently in this book.

10.1 Hall Thruster Developments

Hall thrusters have emerged as the most commonly flown electric thrusters worldwide due to the recent emphasis on the new communications constellations and Earth observing spacecraft. Conventional Hall thrusters are now flying routinely in the power range of 0.1–5 kW using both xenon and krypton propellants. Magnetically shielded Hall thrusters are now in development for use in the next generation of EP flight systems. Investigations on simple improvements and novel innovations to Hall thrusters continue at universities and laboratories worldwide.

10.1.1 Alternative Propellants

Although xenon has historically been the most commonly used propellant in Hall thrusters, the increasing cost and concerns about the availability of xenon has spurred use and development of alternative propellants. Fortunately, Hall thrusters will run on nearly any propellant if properly

Fundamentals of Electric Propulsion, Second Edition. Dan M. Goebel, Ira Katz, and Ioannis G. Mikellides.

configured and operated at the right temperature consistent with the propellant properties. Krypton has emerged as the most commonly used alternative propellant in now thousands of Hall thrusters on communications satellites. A recent review [5] of the performance changes with krypton compared to xenon in a conventional Hall thruster showed both the performance benefits of krypton (primarily higher Isp associated with the lower atom mass) with the detriments of lower thrust and efficiency (primarily because of the lower mass utilization efficiency that results from higher atom velocity in the channel, higher ionization potential and lower ionization cross section). Issues with higher oscillation levels and modified stability limits have also been identified when using krypton. These differences and potential mitigation techniques are areas of active research at this time.

Iodine has long been considered an attractive alternative propellant for Hall thrusters compared to xenon [6] because of its similar mass and ionization potential, and its liquid state at room temperature for storage and launch considerations. Experiments have shown good performance for iodine propellants in Hall thrusters, but also difficulties associated with the corrosive nature of iodine when in contact with thruster or spacecraft components. Compatible cathode technology is also an issue with iodine propellant. Other metal propellants such as magnesium and zinc have been investigated [7–9] in Hall thrusters, and bismuth has been tested in Hall thrusters in the United States [10, 11] and reported to be used Russian [12] TAL thrusters. In addition, other gaseous propellants such as carbon dioxide, methane, ammonia, hydrogen, nitrogen, oxygen, helium, air, and vaporized water/ice associated with potential use from in-situ resource utilization from planets, comets and asteroids have been investigated [13]. All these alternative propellants can have both advantages in ionization characteristics and storage capabilities in space, and disadvantages in ionization rates, sputtering, surface chemistry, and mass utilization that require specialized design of the Hall thruster.

10.1.2 Nested Channel Hall Thrusters for Higher Power

As described in the previous chapters on Hall thrusters, scaling to higher power levels is constrained by the requirements on the magnetic field strength and shape achievable in a given discharge channel width, and on the maximum ion current density achievable in the Hall thruster design. Discharge channels cannot be made arbitrarily wide to produce higher total power capability without serious compromises in the magnetic field magnitude and the coil power required to produce that field. Increasing the power level requires increasing the channel diameter to produce more plasma area, but eventually the thruster becomes just a large ring with unutilized area toward the center. Therefore, an attractive path toward higher total power is to use multiple annular discharge channels in a nested configuration. This not only provides the capability to operate at a higher total power level, but to also operate the channels independently and so provides a very wide throttling range for the thruster.

The first reported development and testing at total power levels up to 6 kW of a two-channel Hall thruster called the X2 was by Liang and Gallimore [14]. The operation of this conventional discharge channel magnetic field geometry was explored in early simulations at Michigan [15], and interactions between the nested discharge channels was later investigated to understand the coupling and performance [16]. The two-channel nested design was improved to provide total power levels in excess of 30 kW and include magnetic shielding for long discharge channel life in the N30 thruster shown in Figure 10-1 by Cusson [17].

Three channel nested Hall thrusters have also been pursued to demonstrate significantly higher power capability. The three-ring X3 Hall thruster [18, 19] shown in Figure 10-2 demonstrated operation at 100 kW of total discharge power in xenon [20] in testing at the University of Michigan and at NASA GRC. The large size of this thruster made fabrication of boron nitride walls problematic and produced several breakdown and cracking failures in test. Fortunately, the next generation of this thruster is planned to use magnetic shielding to extend the thruster life, and so can also use non-ceramic discharge chamber walls (such as graphite) [21] that are easier to fabricate and robust to the thermal and plasma bombardment environment in the Hall thruster at very high discharge powers.

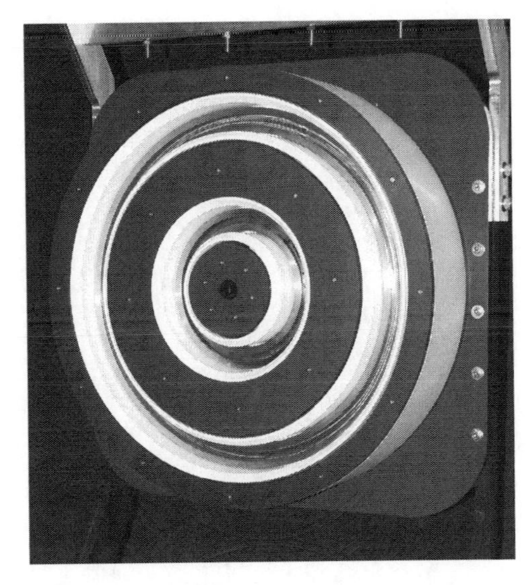

Figure 10-1 Two-channel 30 kW N30 nested Hall thruster photo (*Source:* Courtesy of University of Michigan).

The X3 thruster also utilized a LaB_6 hollow cathode specifically designed and fabricated at JPL [22] to operate in the X3 thruster at discharge currents in excess of 250 A. The cathode also incorporated gas injectors in the near-cathode plume [23] to damp discharge oscillations and reduce energetic ion generation that are a significant problem at discharge currents in excess of 100 A. An added benefit of the cathode-plume gas injectors is to reduce the total cathode flow required to stably operate the thruster [24]. Cathode flow fractions below 5% of the anode flow were achieved, increasing the overall thruster efficiency by several percent. Higher power Hall thrusters will require hollow cathodes capable of even higher discharge currents and must develop additional techniques to minimize discharge oscillations in the cathode plume that generate energetic ions.

Figure 10-2 Three-channel 100 kW X3 nested Hall thruster photo (*Source:* Courtesy of University of Michigan).

Nested discharge channel Hall thruster research and development aimed at higher powers and improved life and efficiency will continue. This includes developments on TAL thrusters. Nakajima reported [25] on the design and testing of a two-channel TAL thruster at discharge powers up to 1 kW. Additional research on configurations of the magnetic field and anode are planned.

10.1.3 Double Stage Ionization and Acceleration Regions

Hall thrusters normally use an $\mathbf{E} \times \mathbf{B}$ discharge to perform both ionization and acceleration of the propellant gas in the discharge channel. These functions can be separated to potentially improve the performance by reducing the discharge cost of the ion production and decoupling the thrust and

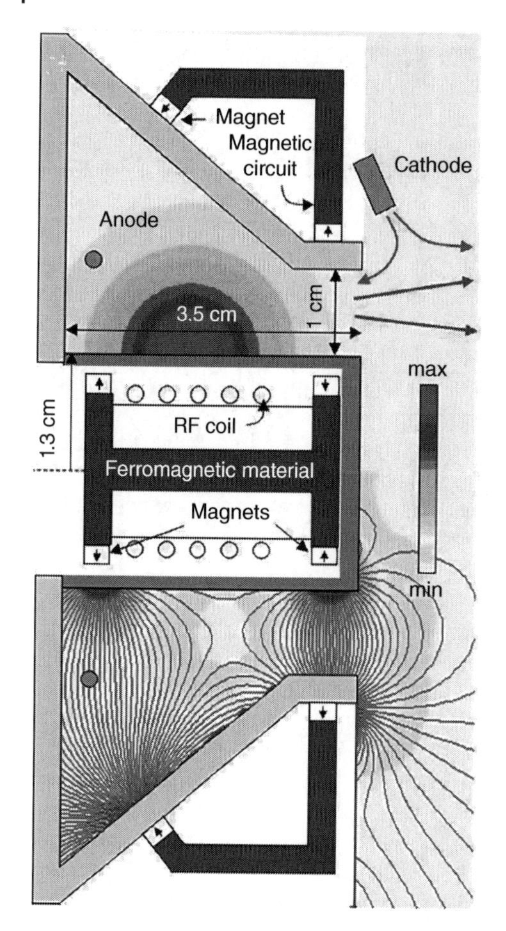

Figure 10-3 Double layer Hall thruster showing the rf ionization section and the transverse magnetic field acceleration region (*Source:* [27]/American Institute of Physics).

Isp relationship typically found in Hall thrusters. First attempts utilized biased electrodes placed on the discharge chamber wall [26] to modify the local electric field and change the ion losses to the wall in the ionization region. Another concept, shown in Figure 10-3, is the Double Stage Hall thruster [27] where the ionization region is physically separated from the transverse magnetic field acceleration region at the thruster exit. The plasma is produced near the anode region using an rf coil configured either as an inductive or helicon plasma source. This concept permits operation at either high thrust or high specific impulse (Isp) for a given power into the discharge region by changing the ionization rate independent of the acceleration potential, providing a more versatile design that can be used for a variety of maneuvers in space. Modeling of the thruster performance has been published [27], and initial experiments on hardware using a helicon plasma source in the Hall thruster have shown some propellant utilization improvements associated with the helicon plasma generation, but slight performance decrease associated with the additional rf power [28]. Other configurations and ionization techniques are being pursued at various laboratories to reduce the ionization cost and improve the overall thruster performance.

10.1.4 Multipole Magnetic Fields in Hall Thrusters

Another technique to make the ionization region separate from the acceleration region is to change the magnetic field topology. Traditional Hall thrusters use a single magnetic pole that produces a peak in the local transverse magnetic field near the exit plane. Ding reports [29] studies of a Hall thruster configuration with multiple permanent magnetic rings to form two magnetic peaks in the discharge channel. Figure 10-4 shows a schematic representation of one version of this concept with two permanent magnet rings and two electrodes for the anode and an intermediate electrode near the exit plane. This configuration is intended to reduce the power deposition on the thruster wall and decrease the wall erosion rate, and increases the ionization rate upstream of the exit plane. Experimental investigations of this novel geometry using different anode configurations [30] in a 200-W class Hall thruster and with different magnetic configurations are continuing [31].

Other magnetic field configurations that use a number of cusp fields or additional magnetic poles are also possible. For example, the HEMP (High Efficiency Multi-Stage) thruster [32] uses multiple permanent magnet rings around a coaxial channel to separate the anode from the external hollow cathode. A variation on this design is the Divergent Cusped-Field thruster [33] that uses a conical discharge chamber lined by permanent magnet rings. Likewise, the Magnetic Octupole Plasma

Thruster (MOPT) [34] replaces the conventional magnetic circuit of a Hall thruster with an octupole magnetic field generated by eight permanent magnets and a magnet yoke. All these configurations attempt to vary the magnetic field configuration to improve the performance and life of the thruster, and additional geometries are conceivable for investigations going forward.

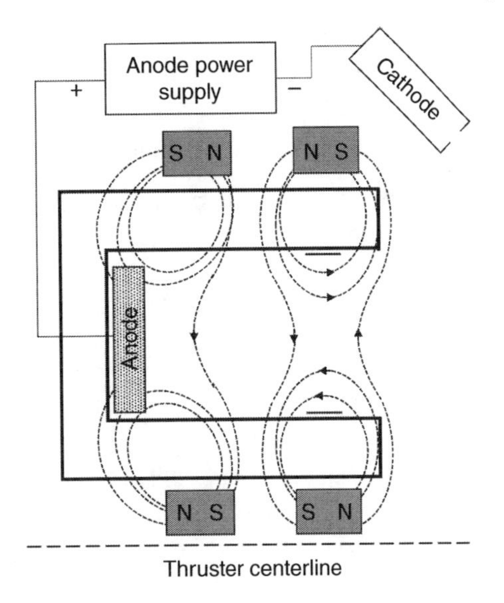

Figure 10-4 Simplified schematic of the two-peak magnetic field Hall thruster concept.

10.2 Ion Thruster Developments

Ion thrusters are a very mature electric propulsion technology with significant flight heritage using DC discharge, Kaufman and microwave plasma generator configurations. Development is ongoing for rf ion thrusters and larger scale microwave ion thrusters. Work also continues on improving the performance and life of the accelerator grids and the utilization of alternative propellants. These will be described briefly below.

10.2.1 Alternative Propellants

While xenon is the standard propellant used in flight ion thrusters as of now, ion thrusters will run on nearly any gaseous propellant that the thruster materials are compatible with. Alternative propellants, specifically liquid metals, were used in the original development and flights of ion thrusters. The very first ion thruster utilized cesium as the propellant [35], and the first electron-bombardment ion thruster used mercury as the propellant [36]. The first flight of ion thrusters on SERT-1 [37] carried both these cesium and mercury ion thrusters. Xenon emerged in the early 1980s because of its high mass, low ionization potential, inert gas properties that minimize spacecraft interactions, and high-density storage capability on the spacecraft.

Surveys of potential propellants describing their physical properties, (ionization potential, vaporization temperature, etc.) performance in the thruster, storage in the spacecraft, impacts on flow controller and PPU, cathode operation, plume-spacecraft interactions, toxicity, and thruster lifetime have been previously published [38, 39]. The most viable alternative propellant to xenon is likely krypton. Iodine and mercury were found to have high performance, but were discounted in that study because of their compatibility issues, especially in terms of spacecraft contamination, corrosion and toxicity. In addition, a 2021 United Nations provision banned the use of mercury in spacecraft propellant [40] over health hazard concerns, and environmental cleanup of mercury contaminated test facilities on the ground has been a significant issue for further use of this material. Nevertheless, iodine is still being pursued as an ion thruster propellant, especially for Earth-orbiting CubeSat and SmallSat thrusters [41]. A small rf-ion thruster using iodine propellant and operating at up to 65 W of total power was recently launched into a 480 km orbit and demonstrated operation in space [42]. Iodine thrusters are being scaled to higher power and performance levels for other applications. Other propellants described in these surveys are being pursued in many laboratories, and investigations of the modifications needed to the thrusters to accommodate those propellants and optimize the thruster performance is an active area of research and development.

10.2.2 Grid Systems for High Isp

The accelerator grids for ion thrusters have historically been made of molybdenum because of its machineability, high strength to survive launch loads and high voltage standoff capability. A large effort was made about 20 years ago to develop alternative grid materials such as titanium, graphite, carbon-carbon composite and pyrolytic graphite that have longer life because of lower sputtering rates from ion bombardment that occurs within the grid structure. This is necessary to operate the grids at higher voltages (over 2 kV) required to obtain Isp of over 4000 s. The carbon-based grid materials all have low sputtering yields compared to molybdenum, but poorer high voltage standoff and robustness to arc formation [43]. Nevertheless, carbon-carbon composite grids were successfully developed for the NEXIS ion thruster [44] that operated at beam voltages up to 6.5 kV producing a Isp as high as 8500 s and a total power up to 28 kW.

As described in Chapter 6, flight ion accelerators have historically used two and three grid systems. These consist of the "screen" grid that is near the discharge chamber plasma potential and forms the beamlets of ions to be accelerated, the negatively biased "accel" grid that provides the high electric field for accelerating the ions, and sometimes the "decel" grid biased at near the ion beam potential that protects the accel grid from backstreaming ion bombardment and sputtering. Accelerator systems with combined grid materials of a relatively thick graphite for the accel grid for strength and low sputtering yield, and a thin molybdenum grid for the screen grid for high strength and transparency have been developed to produce high Isp with long life. The T6 Kaufman thruster [45] used in the BepiColombo mission [46] has a 2 kV accelerator system with the combined moly-graphite grid materials produce a Isp of about 4000 s at a power of 4.5 kW. However, scaling of pure graphite grids to larger diameters is problematic structurally to survive launch loads, and the thin molybdenum screen grids limit the thruster life and throughput capabilities from sputtering of the grid by low energy ions from the discharge chamber. Alternative grid materials such as carbon-based electrodes or others, combined with the introduction of additive manufacturing techniques [47], are needed in two-and three-grid systems to achieve higher total power at high Isp.

Another technique to obtain high Isp (and higher power levels) is to increase the number of accelerator grids. Dual-stage four-grid ion thrusters were first proposed [48] based on the original development of four-grid accelerator systems for neutral beam injectors used in fusion research [49]. Initial experiments [50] demonstrated total acceleration potentials of up to 30 kV producing a Isp up to 15,000 s. This work also demonstrated an order of magnitude increase in the beam power density compared to conventional two- to three-grid system thrusters, enablilqng ion thrusters with total power levels over 100 kW. Work is continuing on the development of four-grid ion thrusters [51] with recent experiments investigating the decoupling of the ion extraction from the discharge plasma from the ion acceleration to high voltage of the dual-stage ion four-grid optical system.

10.3 Helicon Thruster Development

Helicon thrusters use rf fields from a specially designed antenna around a cylindrical ceramic chamber to launch helicon waves in the resultant plasma that ionize and heat propellant gas injected from one end. The device requires a strong axial magnetic field, which is produced by

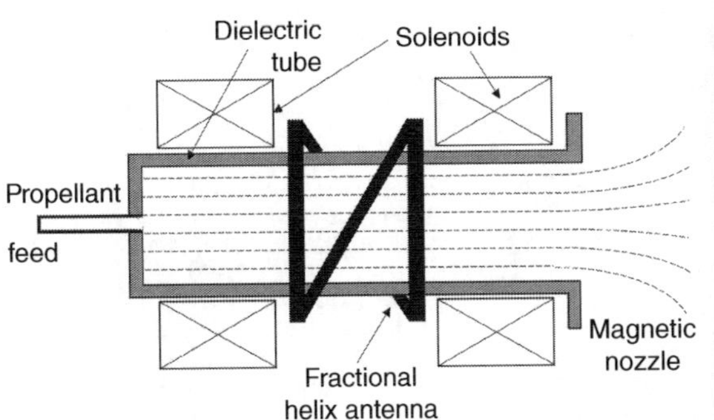

Figure 10-5 Helicon thruster schematic (left) and photograph of a helicon plasma thruster (right) operating in the laboratory (*Source:* [55]/Springer Nature).

either solenoids, superconducting magnets, or permanent magnets. The plasma generated in the source region flows along the magnetic field lines out of the thruster, and the ions are accelerated by axial ambipolar fields [52], the generation of free-standing double layers [53], and the diverging field magnetic nozzle [54] at the exit. Helicon thrusters are considered electroless (no grids or anode/cathode electrodes in contact with the plasma), and do not require a neutralizer cathode.

Figure 10-5 shows a simplified schematic of a helicon thruster (left) and a photograph of laboratory helicon thruster [55] (right) with the solenoid coils, antenna structure and plasma plume. A comprehensive review of helicon thruster technology and physics was recently published by Takahashi [55], and the plume structure and ion acceleration mechanisms described by Williams [56]. Helicon thrusters have been characterized by lower Isp and efficiency than ion and Hall thrusters, but recent work optimizing the magnetic field configuration in the helicon source region and the magnetic nozzle at the exit have increased the efficiency to levels approaching 30% [57]. Research and development continue on the helicon plasma source design, the ion acceleration physics for higher Isp, reducing the losses in the solenoids and rf generator, improving the magnetic nozzle design, and understanding the mechanisms for detachment of the plasma from the magnetic field.

The problem with lower Isp from helicon plasma thrusters has been solved by the development of the Variable Specific Impulse Magnetoplasma Rocket (VASIMR®) [58]. Figure 10-6 shows a schematic layout of the VASIMR® thruster that has a high-density helicon plasma source and an ion cyclotron resonance heating (ICRH) section. The plasma produced by the helicon source is guided by the 1-to-2 T magnetic field from an external superconducting magnet to the ICRH section where the ions are heated in the direction perpendicular to the magnetic field lines. The increased perpendicular energy of the ions is converted into the axial energy as the magnetic field expands in the magnetic nozzle [59]. The separate helicon and ICRH sections enable the thrust and Isp to be varied independently, which can be useful in the mission trajectory planning. The VASMIR® system has no physical material electrodes in contact with the plasma, enabling longer life because the ion erosion is minimized. A thruster efficiency of 72% has been demonstrated [60] using argon propellant with a specific impulse of 4880 s when operated at a total coupled RF power of 200 kW. Development of the VX-200 thruster continues including using krypton propellant, which is anticipated to yield improved thrust-to-power ratios and a higher thruster efficiency at lower Isp values.

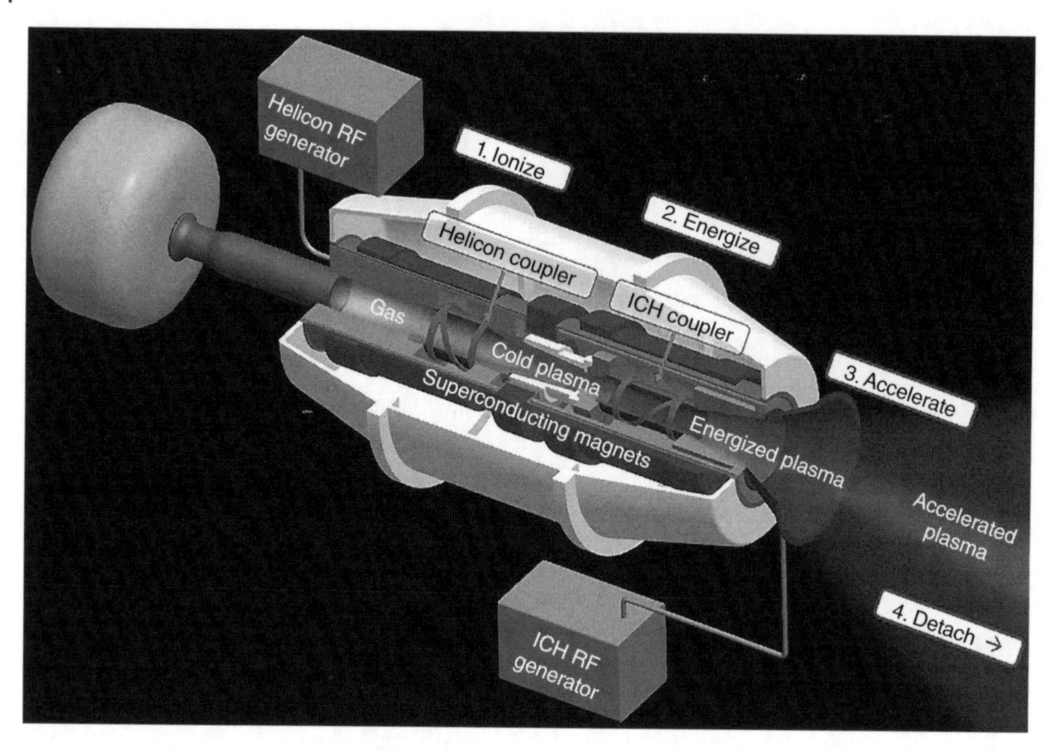

Figure 10-6 Schematic of the VASIMR® thruster showing the helicon, ICRF and magnetic nozzle sections (*Source:* Courtesy of Ad Astra Rocket Company).

10.4 Magnetic Field Dependent Thrusters

There is significant interest in thruster concepts that manipulate magnetic fields to produce ion acceleration. This includes rotating magnetic fields originating from reversed field configurations developed in the nuclear fusion community where similar physics is used to magnetically confine plasma for fusion purposes. There is also research and development on continuous induction-acceleration thrusters and thrusters that use ion acceleration from magnet reconnection. These will be described briefly in this section.

10.4.1 Rotating Magnetic Field (RMF) Thrusters

Rotating magnetic field (RMF) thrusters are a propulsion application of field-reversed configurations (Rotamak-FRC) and spherical tokamaks (Rotamak-ST) developed in the fusion community [61]. They work by inducing an azimuthal current by means of a rotating magnetic field in a seed plasma produced by rf coils or helicon sources [62–65]. The rotating magnetic field is usually generated using sets of saddle coil antennas around the thruster body, as illustrated in Figure 10-7. Each of these forms a Helmholtz pair surrounding the thruster and oriented in either the x or y direction, with the z direction oriented along the thruster's axis. By injecting sinewave currents into each antenna 90° out of phase, a magnetic field is generated which effectively rotates in the plane transverse to the thruster's central axis [62]. Given a sufficiently strong RMF magnitude, this causes

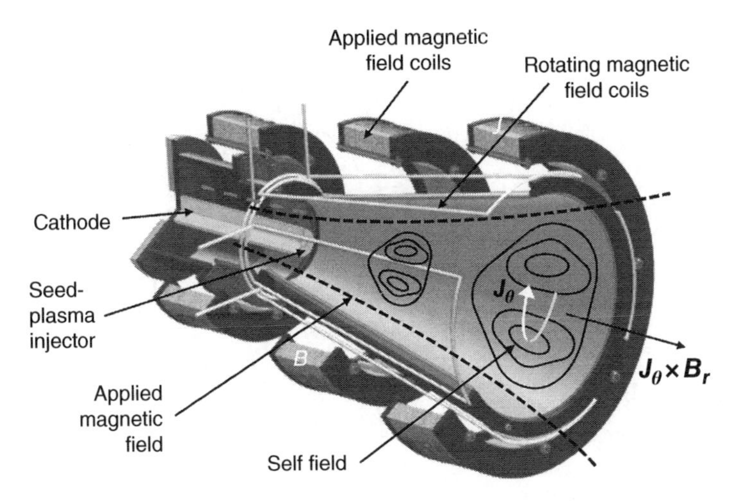

Figure 10-7 Rotating Magnetic Field (RMF) Thruster operating sequence: (1) seed plasma fills thruster in an applied magnetic field, (2) RMF is formed by antennas while ionization occurs, (3) azimuthal current is generated and ions accelerated by the Lorentz force.

ionization by electrons in the seed plasma and generates the azimuthal current [66]. This current can interact via the Lorentz force with the radial component of an applied bias field generated by external electromagnets. Additionally, if conductive structural elements such as flux conservers or the bias magnets themselves are present, the rapid rise of plasma current will induce secondary transient currents in those structures via mutual inductance, and the plasma slug (called a "plasmoid") will be repelled from those structures via the Lorentz force. The force from direct interaction with the bias magnet scales linearly with plasma current, and force owing to interaction with the secondary induced currents in nearby structures goes quadratically with plasma current. However, these forces together do not account for directly measured thrust, indicating that thermal effects may be at play, generating force in a similar manner as a magnetic nozzle using the applied magnetic field [67].

At present, the RMF thruster remains an inefficient device [62] likely due to losses in the plasma at the high plasma density ($>10^{19}$ m^{-3}) throughout the plasma volume. There is development is aimed at producing steady state operation at lower plasma densities, which will reduce ionization and radiation losses to boost the efficiency. This will also allow a proportionally lower RMF current amplitude and make operation possible at similar power levels. In this mode, it is expected that the thruster will behave more like an applied-field MPD thruster with the benefit of no plasma-wetted electrodes.

10.4.2 Magnetic Induction Plasma Thrusters

The Magnetic Induction Plasma Engine (MIPE) is a 15-kW class RF-based magnetic induction thruster [68] that in contrast to the pulsed inductive thruster described in Chapter 9 is capable of steady-state operation. In the thruster, multiple phased-driven coils, similar in concept to a linear induction motor, are wound around an insulating cylindrical tube, as shown in Figure 10-8. The coils are excited with delayed phases to create an axially propagating magnetic field. The plasma breaks down because of the electrostatic fields from the coils, and a plasma current is then driven by

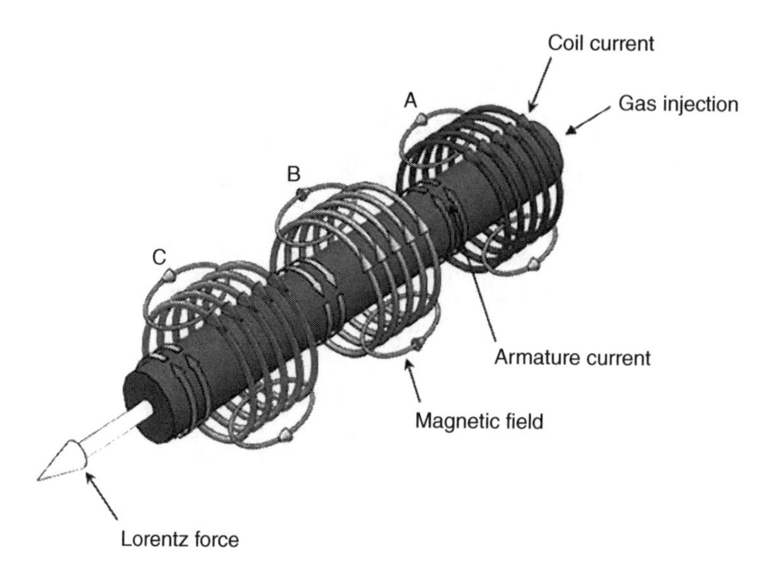

Figure 10-8 Magnetic Induction Plasma Engine (MIPE) showing phased coils inducing a traveling magnetic field and Lorentz force acceleration (*Source:* [68]).

magnetic induction transformer action. The Lorentz force is exerted on the plasma current by the traveling magnetic field wave. The magnetic force on the current is exerted on the electrons, which drag the ions with them electrostatically to maintain charge neutrality and generate thrust. Thruster efficiencies up to 37%, Isp up to 2300 s, and thrust up to 100 mN were measured in laboratory testing, although not simultaneously. Research continues on improving the simultaneous thruster performance and expanding the capabilities of this novel approach.

10.4.3 Magnetic Reconnection Thrusters

Magnetic reconnection is one of the primary drivers of particle acceleration processes in space and astrophysical plasmas by conversion of magnetic energy into charged particle thermal and kinetic energy [69–71]. Magnetic reconnection has been studied with respect to solar flares [72], fusion-research tokamaks [73, 74], and in laboratory plasmas [75, 76].

Magnetic reconnection occurs in the sudden change in the topology of a magnetic field that converts the energy stored into the magnetic field into energetic charged particles. This mechanism [71] is responsible for the plasma acceleration of interest for thruster applications. The energy release can occur from annihilation and creation of magnetic flux, expansion and compression of the magnetic field, and contraction or stretching of curved magnetic field lines. One of the issues for propulsion is that the energy release in terms of plasma acceleration is nearly symmetric from the event location [76], which eliminates net thrust. An asymmetric field geometry or plasma density profile is then needed to generate net thrust. Magnetic reconnection thruster concepts for use in space propulsion systems have been proposed based on results from fully kinetic modeling [77] and MHD simulations [78]. Initial experimental results of a magnetic reconnection thruster [79] demonstrated ion acceleration from an argon plasma with a calculated velocity corresponding to an Isp of 860 s. Although magnetic reconnection plasma acceleration has been recognized for decades, research on how to use this mechanism in a thruster is relatively new and expanding worldwide.

10.5 Laser-Based Propulsion

Electromagnetic radiation can provide thrust for spacecraft for interplanetary and interstellar missions [3]. Laser-based propulsion concepts are an active area of research because of the potential for both high power and high Isp. There are basically three viable laser-based propulsion concepts:

- Laser radiation pressure acceleration
- Laser ablation
- Laser driven collisionless shock acceleration of ions

Laser radiation pressure propulsion uses photon pressure beamed from an Earth-based or near-Earth orbiting laser to accelerate the spacecraft. An example of this system is the Starshot Initiative that set a goal of sending a spacecraft beyond our Solar System to a nearby star within the next half-century using laser-based propulsion. Starshot proposed an ultralight spacecraft design that is accelerated by laser radiation pressure from an Earth-based source on a lightsail [80]. A related concept is to use ultrasmall spacecraft called "Chipsats" equipped with a solar sail that would escape the Solar System using photon pressure from a laser, travel to the interstellar target, and send back some telemetry [81]. Lightsails represent a materials challenge to achieve optimum reflectance in a multilayer structure [82] for high momentum transfer over a very large area while maintaining low mass. The laser technology is also challenging to achieve the power and pointing accuracy from the ground or from space required for beaming power to spacecraft for deep space travel.

Propulsion using laser ablation of materials has been investigated since the 1970s [83] when it was first proposed for Earth-based lasers to propel satellites into orbit. A detailed review was published in 2010 [84]. A high intensity laser pulse hitting a material surface generates several processes [85] such as production of a surface plasma that expands and ejects ionized particles, expulsion of jet of vapor from evaporation or sublimation of the surface, and imparting momentum to the surface by reflection of the laser photons or emission of secondary electrons. The propellant can be with liquid or solid, both complicated by the delivery requirements of the material to the surface interacting with the laser beam. However, the potential for low power average consumption using pulsed lasers, and the possibility of high Isp (>1000 s) and variable thrust controlled by the pulse repetition rate, is driving continued investigations [86] into new laser-ablation propulsion system.

Laser-driven collisionless shockwaves in a plasma produce significant acceleration of ions [87, 88]. Measurements of the acceleration of protons and multiply-carbon ions in shockwave experiments using ultra-short, high peak-power laser pulses are consistent with reflection of particles off the moving potential of a shock front in the plasma at near the critical plasma density. Gaseous hydrogen targets driven by intense 10 μm laser pulses have generated ~MeV energy protons accelerated by the radiation pressure driven shock wave [89]. Experiments using 1 μm and 10 μm lasers with $>10^{21}$ W/cm^2 produced ion energies well in excess of 1 MeV [87, 90]. Although not configured for thruster applications, these results are promising for the future development of high power, high Isp (>10,000 s) propulsion systems.

10.6 Solar Sails

Solar sails are a well-known concept [91–93] that uses radiation from the sun to provide propulsion without the need for propellant. Owing to the low solar radiation pressure, a solar sail must have a large area to produce significant amounts of thrust, which mandates a light-weight surface material and support structure. Sails that use photo pressure are made up of reflecting surfaces used to redirect sunlight for the purpose of creating forces for thrust and torques for attitude control, and separate structures or mechanisms for deploying the sail. Many complex sail shapes and different rigid and non-rigid support structures have been proposed [91] to satisfy these criteria.

The first spacecraft to successfully demonstrate interplanetary solar sail technology was IKAROS (Interplanetary Kite-craft Accelerated by Radiation Of the Sun) launched by the Japan Aerospace Exploration Agency (JAXA) in 2010 [94]. IKAROS used a diagonal spinning square sail with an area of 196 m^2 made of a 7.5-μm thick polyimide with a mass of only 77 g/m^2 [95]. IKAROS spent six months traveling to Venus, and then performed a three-year journey around the Sun [96]. Another successful solar sail mission was LightSail [97], which was a project to demonstrate controlled solar sailing within low Earth orbit using a CubeSat. The program consisted of two spacecraft — LightSail 1 with a mission duration of 25 days and LightSail 2 with a mission duration over 3 years [98, 99]. A recently launched (2022) solar sail mission is the NASA Near-Earth Asteroid (NEA) Scout [100] intended to demonstrate a controllable CubeSat solar sail spacecraft capable of encountering near-Earth asteroids. As of this writing, radio contact with the NEA Scout spacecraft has not been achieved and the mission is feared lost. The next major planned mission is Solar Cruiser, a NASA mission intended to launch in 2028 and planned to test a very large $>1000\ m^2$ solar sail in an artificial orbit between the Earth and Sun [101].

Challenges for solar sails include developing extremely lightweight deployable materials for the reflector/membrane surface [102], understanding the structural dynamics both for support of the membrane and compatibility with the attitude control system of the spacecraft [97], solar sail geometry changes with temperature during the mission affecting the thrust profile, and modeling of the sail performance [91].

Electric solar-wind sails were invented in 2004 by Janhunen [103] and provide thrust by reflecting charged solar wind particles using electric fields, thus converting momentum from particles in the solar wind into propulsion [104]. The pressure associated with charge particles in the solar wind is over a thousand times lower than the photon pressure, but this difference can be made up by increasing the size of the much lighter electric sail and taking advantage of the electric fields to cover more area [105].

The electric sail concept uses conducting wires placed radially around the spacecraft and straightened by rotation of the spacecraft [105]. The wires are electrically charged to create an electric field, which extends some tens of meters into the plasma of the surrounding solar wind. Charged particles are reflected by the electric field much like the photons on a traditional solar sail. The sail is much lighter than a traditional solar sail because the radius of the sail is determined by the electric field rather than the wire. An example of an electric sail conceptional design may have 50–100 straightened wires with a length of about 20 km each. The fundamental mechanism of electric sails, momentum transfer from steaming ion deflection in a positive sheath, was measured in tests performed at NASA/MSFC (K. H. Wright Jr., personal Communication) and agreed with the analytical thrust estimates of Jahunen [103]. Research is continuing on sail performance and mission applications.

10.7 Hollow Cathode Discharge Thrusters

Plasma discharges between a hollow cathode and various configurations of the anode, both with and without applied magnetic fields, have been investigated for nearly 50 years for hollow cathode development work and plasma source applications. For example, Figure 10-9 shows a hollow

Figure 10-9 Hollow cathode test setups at JPL showing a hollow cathode with a solenoid magnetic field coil and a segmented anode (a) or cylindrical anode (b) for investigating discharge plasma properties.

cathode test setup at JPL with a solenoid coil positioned around it, and coupled to a segmented anode (left) and a cylindrical anode (right). Various configurations of this setup, with solenoids around the cathode and/or the anode, different anode geometries and dimensions, and including plasma diagnostics and ion energy analyzers, were found to generate promising parameters for thruster applications such as high ionization fractions, high discharge currents and energetic ion production. Investigations of hollow cathode-based thrusters illuminated the thrust production mechanisms [106], and experimental work using a T6 hollow cathode with a cylindrical anode produced thrust levels of over 1 mN at an Isp of 200–1000 s (depending on the flow rate and discharge voltage) using discharge currents of up to 30 A.

Recently, experiments with various combinations of solenoids and anodes positioned around a central hollow cathode have been made to investigate the performance as a thruster [107]. The thrust characteristics of what is termed "electrostatic-magnetic hybrid acceleration" have been measured on this "central-cathode electrostatic thruster (CC-EST)" shown in Figure 10-10. This device injects the majority of the propellant though an annular slit in the ring anode, and features a diverging magnetic field that produces a Lorentz force on the ions and also acts as a magnetic nozzle. Both electrostatic and electromagnetic contributions to the measured thrust were found [107]. Various combinations of anode electrodes and magnetic field shapes from permanent magnets and solenoids continue to be investigated to increase the performance of this thruster concept.

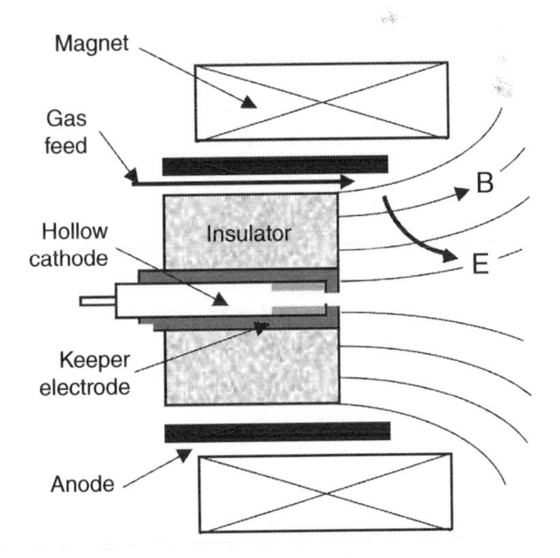

Figure 10-10 Simplified schematic diagram of a hollow cathode plasma discharge thruster showing cathode, anode and magnetic field (*Source:* [107]).

References

1 I. Levchenko, S. Xu, S. Mazouffre, D. Lev, D. Pedrini, D. M. Goebel, L. Garrigues, F. Taccogna, D. Pavarin, and K. Bazaka, "Perspectives, Frontiers and New Horizons for Plasma Based Space Electric Propulsion," *Physics of Plasmas*, vol. 27, no. 02061, 2020; doi:10.1063/1.5109141.

2 I. Levchenko, O. Baranov, D. Pedrini, C. Riccardi, H. E. Roman, S. Xu, D. Lev, and K. Bazaka, "Diversity of Physical Processes: Challenges and Opportunities for Space Electric Propulsion," *Applied Sciences*, vol. 12, no. 11143, 2022; doi:10.3390/app122111143.

3 I. Levchenko, L. Bazaka, S. Mozouffre, and S. Xu, "Prospects and Physical Mechanisms for Photonic Space Propulsion," *Nature Photonics*, vol. 12, pp. 649–657, 2018; doi:10.1038/s41566-018-0280-7.

4 E. Dale, B. A. Jorns, and A. D. Gallimore, "Future Directions for Electric Propulsion Research," *Aerospace*, vol. 7, no. 9, 2020; doi:10.3390/aerospace7090120.

5 T. Andreussi, M. M. Saravia, E. Ferrato, A. Piragino, A. Rossodivita, M. Andrenucci, and D. Estublier, "Identification, Evaluation and Testing of Alternative Propellants for Hall Effect Thrusters," IEPC-2017-380, Presented at the 35th International Electric Propulsion Conference, Atlanta, October 8–12, 2017.

6 R. A. Dressler, Y. H. Chiu, and D. J. Levandier, "Propellant Alternatives for Ion and Hall Effect Thrusters," AIAA-2000-0602, Presented at the 38th Aerospace Sciences Meeting and Exhibit, Reno, January 10–13, 2000).

7 J. Szabo, M. Robin, J. Douglas, and R. R. Hofer, "Light Metal Propellant Hall Thrusters," IEPC-2009-138, Presented at the 31st International Electric Propulsion Conference, Ann Arbor, September 20–24, 2009.

8 M. Hopkins and L. B. King "Performance Comparison Between a Magnesium and Xenon Fueled 2 kW Hall Thruster," AIAA-2014-3818, Presented at the 50th AIAA Joint Propulsion Conference, Cleveland.

9 L. B. King and M. Hopkins, "Magnesium Hall Thruster with Active Thermal Mass Flow Control," *AIAA Journal of Propulsion and Power*, vol. 30, no. 3, pp. 637–644, 2014; doi:10.2514/1.B34888.

10 J. Szabo, M. Robin, and V. Hruby, "Bismuth Vapor Hall Effect Thruster Performance and Plume Experiments," IEPC-2017-25, Presented at the 35th International Electric Propulsion Conference, Atlanta, October 8–12, 2017.

11 D. Massey, L. B. King, and J. Makela, "Development of a Direct Evaporation Bismuth Hall thruster," AIAA 2008-4520, Presented at the 44th AIAA Joint Propulsion Conference, Hartford, July 21–23, 2008.

12 A. Sengupta, C. Marrese-Reading, A. Semenkin, L. Zakharenkov, S. Tverdokhlebov, and S. Tverdokhlebov, "Summary of the VHITAL thruster technology demonstration program: a two-stage bismuth-fed very high specific impulse TAL," IEPC-2007-005, Presented at the 30th International Electric Propulsion Conference, Florence.

13 T. Itsuki, T. Nagayoshi, H. Tahara, T. Ikeda, and Y. Takao. "Research and Development of Hall Thrusters for Transportation in the Solar System - Use of Carbon Dioxide, Methane, Ammonia, Hydrogen, Helium, Air and Ice/Water etc. in the Planets and Satellites." IEPC-2022-282, Presented at the 37th International Electric Propulsion Conference, Cambridge, 19–23 June 2022.

14 Liang, R. and Gallimore, A. D. "Constant-Power Performance and Plume Properties of a Nested-Channel Hall-Effect Thruster," IEPC-2011-049, Presented at the 32nd International Electric Propulsion Conference, Wiesbaden, September 11–15, 2011.

15 H. C. Dragnea and I. D. Boyd, "Simulation of a Nested Channel Hall Thruster," IEPC-2015-250, Presented at the 34th International Electric Propulsion Conference, Hyogo-Kobe, July 4–10, 2015.

16 S. E. Cusson, M. P. Georgin, H. C. Dragnea, E. T. Dale, V. Dhaliwal, I. D. Boyd, and A. D. Gallimore, "On Channel Interactions in Nested Hall Thrusters," *Journal of Applied Physics*, vol. 123, no. 133303, 2018; doi:10.1063/1.5028271.

17 S. E. Cusson, R. R. Hofer, D. M. Goebel, M. P. Georgin, A. R. Vazsonyi, B. A. Jorns, and A. D. Gallimore "Development of a 30-kW Class Magnetically Shielded Nested Hall Thruster," IEPC-2019-266, Presented at the 36th International Electric Propulsion Conference, Vienna, September 11–15, 2019.

18 S. J. Hall, R. E. Florenz, A. D. Gallimore, H. Kamhawi, D. L. Brown, J. E. Pol, D. M. Goebel, and R. R. Hofer, "Implementation and Initial Validation of a 100-kW Class Nested-Channel Hall Thruster," AIAA-2014-3815, Presented at the 50th AIAA Joint Propulsion Conference, Cleveland, July 28–30, 2014.

19 S. J. Hall, B. A. Jorns, A. D. Gallimore, H. Kamhawi, T. W. Haag, J. A. Mackey, J. H. Gilland, P. Y. Peterson, and M. J. Baird, "High-Power Performance of a 100-kW Class Nested Hall Thruster," IEPC-2017-228, Presented at the 35th International Electric Propulsion Conference, Atlanta, October 8–12, 2019.

20 S. W. Shark, S. J. Hall, B. A. Jorns, R. R. Hofer, and D. M. Goebel, "High Power Demonstration of a 100 kW Nested Hall Thruster System," AIAA-2019-3809, Presented at the AIAA Propulsion and Energy Forum, Indianapolis, August 19–22, 2019.

21 D. M. Goebel, R. R. Hofer, I. G. Mikellides, I. Katz, J. E. Polk, and B. Dotson, "Conducting Wall Hall Thrusters," *IEEE Transactions on Plasma Science*, vol. 43, no. 1, pp. 118–126, 2015.

22 D. M. Goebel and E. Chu, "High Current Lanthanum Hexaboride Hollow Cathode for High Power Hall Thrusters," *AIAA Journal of Propulsion and Power*, vol. 30, no. 1, pp. 35–40, 2014; doi:10.2514/1.B34870.

23 E. Chu, D. M. Goebel, and R. E. Wirz, "Reduction of Energetic Ion Production in Hollow Cathodes by External Gas Injection," *AIAA Journal of Propulsion and Power*, vol. 30, pp. 1155–1162, 2013; doi:10.2514/1.B34799.

24 S. J. Hall, B. A. Jorns, A. D. Gallimore, and D. M. Goebel, "Operation of a High-Power Nested Hall Thruster with Reduced Cathode Flow Fraction," *AIAA Journal of Propulsion and Power*, vol. 36, no. 6, 2020; doi:10.2514/1.B37929.

25 T. Nakajima and A. Kakami, "Performance Evaluation of a Double-Channel TAL-Type Hall Thruster," IEPC-2022-354, Presented at the 37th International Electric Propulsion Conference, Cambridge, June 19–23, 2022.

26 S. Langendorf, K. Xu, and M. Walker, "Effects of Wall Electrodes on Hall Effect Thruster Plasma," *Physics of Plasmas*, vol. 22, no. 023508, 2015; doi:10.1063/1.4908273.

27 L. Dubois, F. Gaboriau, L. Liard, D. Harribey, C. Henaux, L. Garrigues, G. J. H. Hagelaar, S. Mazouffre, C. Boniface, and J. P. Beouf, "ID-HALL, a New Double Stage Hall Thruster Design. I. Principle and Hybrid Model of ID-HALL," *Physics of Plasmas*, vol. 25, no. 093503, 2018; doi:10.1063/1.5043354.

28 A. Shabshelowitz, A. D. Galimore, and P. Y. Peterson, "Performance of a Helicon Hall Thruster Operating with Xenon, Argon, and Nitrogen," *AIAA Journal of Propulsion and Power*, vol. 30, no. 3, 2014; doi:10.2514/1.B35041.

29 Y. Ding, P. Li, H. Sun, L. Wei, Y. Xu, W. Peng, H. Su, H. Li, and D. Yu, "Simulation of Double Stage Hall Thruster with Double-Peaked Magnetic Field," *European Physics Journal D*, vol. 71, no. 192, 2017; doi:10.1140/epjd/e2017-80124-8.

30 Y. Ding, H. Sun, P. Li, L. Wei, H. Su, W. Peng, H. Li, and D. Yu, "Application of Hollow Anodes in a Hall Thruster with Double-Peak Magnetic Fields," *Journal of Physics D*, vol. 50, no. 335201, 2017; doi:10.1088/1361-6463/aa7bbf.

31 H. Li, Y. Ding, L. Wang, H. Fan, P. Li, L. Wei, and D. Yu, "Effect of Magnetic-Field Intensity Near an Intermediate Electrode on the Discharge Characteristics of a Hall Thruster with a Double-Peaked Magneticfield," *European Physics Journal D*, vol. 73, 2019; doi10.1140/epjd/e2019-100162-6.

32 N. Koch, H. P. Harmann, and G. Kornfeld, "First Test Results of the 1 to 15 kW Coaxial HEMP 30250 Thruster," AIAA-2005-4224, Presented at the 41st AIAA Joint Propulsion Conference, Tucon, July 10–13, 2005.

33 S. R. Gildea, M. Martinez-Sanchez, M. R. Nakles, and W. A. Hargus, "Experimentally Characterizing the Plume of a Divergent Cusped-Field Thruster," IEPC-2009-259, Presented at the 31st International Electric Propulsion Conference, Ann Arbor, September 20–24, 2009.

34 J. Hsieh, Y. H. Li, M. M. Shen, and B. H. Huang, LaB6 Hollow Cathode Design and Development for Magnetic Octupole Plasma Thruster," IEPC-2022-137, Presented at the 37th International Electric Propulsion Conference, Cambridge.

35 A. T. Forrester and R. C. Speiser, "Cesium-Ion Propulsion," *Astronautics*, vol. 4, pp. 34–44, 1959.

36 H. R. Kaufman and P. D. Reader, *Progress in Astronautics and Rocketry, Vol. 5, Electrostatic Propulsion*. New York: Academic Press.

37 J. S. Sovey, V. K. Rawlin, and M. J. Patterson, "Ion Propulsion Development Projects in U.S.: Space Electric Rocket Test I to Deep Space I," *AIAA Journal of Propulsion and Power*, vol. 17, no. 3, pp. 517–526, 2001.

38 Fazio, N., Gabriel, S. B., and Golosnoy, I. O. (2018). "Alternative Propellants for Gridded Ion Engines," SP2018-00102, Presented at the Space Propulsion 2018, Seville, May 18, 2018.

39 K. Holste, W. Gartner, P. Kohler, P. Dietz, J. Konrad, S. Schippers, P. J. Klar, A. Muller, and P. R. Schreiner, "In Search of Alternative Propellants for Ion Thrusters," IEPC-2015-320, Presented at the 34th International Electric Propulsion Conference, Hyogo-Kobe, 2015.

40 M. Koziol. (April 19, 2022) "U.N. Kills Any Plans to Use Mercury as a Rocket Propellant," *IEEE Spectrum*, vol. 29, no. 4.

41 M. Tsay, R. Terhaar, K. Emmi, and C. Barcroft, "Volume PRODUCTION of Gen-2 Iodine BIT-3 Ion Propulsion System," IEPC-2022-267, Presented at the 37th International Electric Propulsion Conference, Cambridge, 2022.

42 D. Rafalskyi, J. M. Martinez, L. Habl, E. Z. Rossi, P. Proynov, A. Bore, T. Baret. In-Orbit Demonstration of an Iodine Electric Propulsions System," *Nature*, vol. 599, pp. 411–415, 2021; doi:10.1038/s41586-021-04015-y.

43 D. M. Goebel and A. Schneider, "High Voltage Breakdown and Conditioning of Carbon and Molybdenum Electrodes," *IEEE Transactions on Plasma Science*, vol. 33, pp. 1136–1148, 2005.

44 J. E. Polk, D. M. Goebel, J. S. Snyder, A. C. Schneider, J. R. Anderson, and A. Sengupta, "A High Power Ion Thruster for Deep Space Missions," *Review of Scientific Instruments*, vol. 83, no. 073306, 2012.

45 N. Wallace, D. Mundy, D. Fearn, and C. Edwards, "Evaluation of the Performance of the T6 Ion Thruster," AIAA-1999-2442, Presented at the 35th AIAA Joint Propulsion Conference, Los Angeles, 1999.

46 A. N. Grubisic, S. Clark, N. Wallace, C. Collingwood, and F. Guarducci, "Qualification of the T6 Ion Thruster for the BepiColombo Mission to the Planet Mercury," IEPC-2011-234, Presented at the 32nd International Electric Propulsion Conference, Wiesbaden, 11–15 September 2011.

47 C. C. Farnell, S. J. Thompson, and J. D. Williams, "Additive manufacturing for ion optics," IEPC-2022-156, Presented at the 37th International Electric Propulsion Conference, Cambridge, 2002.

48 D. G. Fearn, "The Application of Gridded Ion Thrusters to High Thrust, High Specificimpulse Nuclear-Electric Missions," *Journal of the British Interplanetary Society*, vol. 58, no. 7/8, pp. 257–267, 2005.

49 M. M. Menon, C. C. Tsai, D. E. Schechter, P. M. Ryan, G. C. Barber, R. C. Davis, W. L. Gardner, J. Kim, H. H. Haselton, N. S. Ponte, J. H. Whealton, and R. E. Wright, "Power Transmission Characteristics of a Two-Stage Multiaperture Neutral Beam Source," *Review of Scientific Instruments*, vol. 51, pp. 1163–1167, 1980.

50 R. Walker, C. Bramanti, O. Sutherland, R. Boswell, C. Charles, D. Fearn, and J. G. D. Amo, "Initial Experiments on a Dual-Stage 4-Grid Ion Thruster for Very High Specific Impulse and Power," AIAA-2006-4669, Presented at the 42nd AIAA Joint Propulsion Conference, Sacramento, July 6–12, 2006.

51 L. Jia, L. Yang, Y. Zhao, M. Liu, Y. Jia, T. Zhang, J. Zhuang, F. Yang, Y. Zhao, and X. Wu, "An Experimental Study of a Dual-Stage 4-Grid Ion Thruster," *Plasma Sources Science and Technology*, vol. 28, no. 105003, 2019; doi:10.1088/1361-6595/ab424c.

52 C. Charles, R. Boswell, and K. Takahashi, "Boltzmann Expansion in a Radiofrequency Conical Helicon Thruster Operating in Xenon and Argon," *Applied Physics Letters*, vol. 102, no. 223510, 2013.

53 C. Charles and R. W. Boswell, "Current-Free Double-Layer Formation in a High-Density Helicon Discharge," *Applied Physics Letters*, vol. 82, no. 9, pp. 1356–1358, 2003.

54 E. Ahedo and M. Merino, "Two-Dimensional Supersonic Plasma Acceleration in a Magnetic Nozzle," *Physics of Plasmas*, vol. 17, no. 7, 2010; doi:10.1063/1.3442736.

55 K. Takahashi, Helicon-Type Radiofrequency Plasma Thrusters and Magnetic Plasma Nozzles *Reviews of Modern Plasma Physics*, vol. 3, no. 3, 2019; doi:10.1007/s41614-019-0024-2.

56 L. T. Williams and M. L. R. Walker, "Plume Structure and Ion Acceleration of a Helicon Plasma Source," *IEEE Transactions on Plasma Science*, vol. 43, no. 5, pp. 1694–1705, 2015; doi:10.1109/TPS.2015.2419211.

57 K. Takahashi, "Thirty Percent Conversion Efficiency from Radiofrequency Power to Thrust Energy in a Magnetic Nozzle Plasma Thruster," *Scientific Reports*, vol. 12, no. 18618, 2022; doi:10.1038/s41598-022-22789-7.

58 F. R. Chang-Diaz, "The VASIMR Rocket," *Scientific American*, vol. 283, no. 5, pp. 90–97, 2000; doi:10.1038/scientificamerican1100-90.

59 B. W. Longmier, E. A. Bering, M. D. Carter, L. D. Cassady, W. J. Chancery, F. R. Chang-Diaz, T. W. Glover, N. Hershkowity, A. V. Llin, G. E. McCaskill, C. S. Olsen, and J. P. Squire, "Ambipolar Ion Acceleration in An Expanding Magnetic Nozzle," *Plasma Sources Science and Technology*, vol. 20, no. 015007, pp. 1–9, 2011; doi:10.1088/0963-0252/20/1/015007.

60 B. W. Longmier, J. P. Squire, C. S. Olsen, L. D. Cassady, M. G. Ballenger, M. D. Carter, A. V. Ilin, T. W. Glover, G. E. McCaskill, F. R. C. Díaz, and E. A. Bering, "Improved Efficiency and Throttling Range of the VX-200 Magnetoplasma Thruster," *AIAA Journal of Propulsion and Power*, vol. 30, no. 1, pp. 123–132, 2014; doi:10.2514/1.B34801.

61 I. R. Jones, "A Review of Rotating Magnetic Field Current Drive and the Operation of the Rotamak as a Field-Reversed Configuration (Rotamak-FRC) and a Spherical Tokamak (Rotamak-ST)," *Physics of Plasmas*, vol. 6, pp. 1950–1957, 1999; doi:10.1063/1.873452.

62 C. L. Sercel, T. M. Gill, J. M. Woods, and B. A. Jorns, "Performance Measurements of a 5 kW-Class Rotating Magnetic Field Thruster," AIAA-2021-3384, Presented at the AIAA Propulsion and Energy 2021 Forum, Virtual Event, August 9–11, 2021.

63 T. Furukawa, K. Takizawa, D. Kuwahara, and S. Sinohara, "Study on Electromagnetic Plasma Propulsion Using Rotating Magnetic Field Acceleration Scheme," *Physics of Plasmas*, vol. 24, no. 043505, 2021; doi:10.1063/1.4979677.

64 T. Furukawa, K. Takizawa, K. Kuwahara, and S. Shinohara, "Electrodeless Plasma Acceleration System Using Rotating Magnetic Field Method," *AIP Advances*, vol. 7, no. 115204, pp. 1–13, 2017; doi:10.1063/1.4998248.

65 S. Shinohara, D. Kuwahara, T. Furukawa, S. Nishimura, T. Yamase, Y. Ishigami, H. Horita, A. Igarashi, and S. Nishimoto, "Development of Featured High-Density Helicon Sources and Their Application to Electrodeless Plasma Thruster," *Plasma Physics and Controlled Fusion*, vol. 61, no. 014017, 2019; doi:10.1088/1361-6587/aadd67.

66 T. M. Gill, C. L. Sercel, J. M. Woods, and B. A. Jorns, "Experimental Characterization of Efficiency Modes in a Rotating Magnetic Field Thruster," AIAA-2022-2191, Presented at the AIAA SCITECH 2022 Forum, San Diego, January 8–12, 2022.

67 C. L. Sercel, T. M. Gill, and B. A. Jorns, "Inductive probe measurements during plasmoid acceleration in an RMF thruster," IEPC-2022-554, Presented at the 37th International Electric Propulsion Conference, Cambridge, June 19–23, 2022.

68 K. Quinlan and D. B. Cope, "Magnetic Induction Plasma Engine, IEPC-2022-412, Presented at the 37th International Electric Propulsion Conference, Cambridge, June 19–23, 2022.

69 J. Latham, E. V. Belova, and M. Yamada, "Numerical Study of Coronal Plasma Jet Formation," *Physics of Plasmas*, vol. 28, no. 012901, 2021; doi:10.1063/5.0025136.

70 M. Hesse and P. A. Cassak, "Magnetic Reconnection in the Space Sciences: Past, Present, and Future," *Journal of Geophysical Research: Space Physics*, vol. 125, no. 2; doi:10.1029/2018JA025935.

71 X. Li, F. Guo, and Y. H. Liu, "The Acceleration of Charged Particles and Formation of Power-Law Energy Spectra in Nonrelativistic Magnetic Reconnection," *Physics of Plasmas*, vol. 28, no. 052905, 2021; doi:10.1063/5.0047644.

72 J. F. Drake, P. A. Cassak, M. A. Shay, M. Swisdak, and E. Quataert, "A Magnetic Reconnction Mechanism for Ion Acceleration and Abundance Enhancements in Impulsive flares," *The Astrophysical Journal*, vol. 700, pp. 16–20, 2009; doi:10.1088/0004-637X/700/1/L16.

73 A. H. Boozer, "Flattening of the Tokamak Current Profile by a Fast Magnetic Reconnection with Implications for the Solar Corona," *Physics of Plasmas*, vol. 27, no. 102305, 2020; doi:10.1063/5.0014107.

74 A. H. Boozer, "Magnetic Reconnection and Thermal Equilibration.," *Physics of Plasmas*, vol. 28, no. 032102, 2021; doi:10.1063/5.0031413.

75 J. Yoo, M. Yamada, H. Ji, J. Jara-Almonte, and C. E. Myers, "Bulk Ion Acceleration and Particle Heating During Magnetic Reconnection in a Laboratory Plasma," *Physics of Plasmas*, vol. 21, no. 055706, 2014; doi:10.1063/1.4874331.

76 M. Yamada, J. Yoo, and C. E. Myers, "Understanding the Dynamics and Energetics of Magnetic Reconnection in a Laboratory Plasma: Review of Recent Progress on Selected Fronts," *Physics of Plasmas*, vol. 23, no. 055402, 2016; doi:10.1063/1.4948721.

77 E. Cazzola, D. Curreli, and G. Lapenta, "On Magnetic Reconnection as Promising Driver for Future Plasma Propulsion Systems," *Physics of Plasmas*, vol. 25, no. 073512, 2018; doi:10.1063/1.5036820.

78 F. Ebrahimi, "An Alfenic Reconnecting Plasmoid Thruster," *Journal of Plasma Physics*, vol. 86, no. 905860614, 2020; doi:10.1017/S0022377820001476.

79 S. N. Bathgate, M. M. Bilek, I. H. Cairns, and D. R. McKenzie, A Thruster Using Magnetic Reconnection to Create a High-Speed Plasma Jet *European Physics J Applied Physics*, vol. 84, no. 20801, 2018; doi:10.1051/epjap/2018170421.

80 H. A. Atwater, A. R. Davoyan, O. Ilic, D. Jariwala, M. C. Sherrorr, C. M. Went, W. S. Shitney, and J. Wong, "Materials Challenges for the Starshot Lightsail," *Nature Materials*, vol. 17, pp. 861–897, 2018; doi:10.1038/s41563-018-0075-8.

81 W. Hu, C. Welch, and E. Ancona, "A Minimal Chipsat Interstellar Mission: Technology and Mission Architecture," IAC-18-B4.8.16.46958, Presented at the 69th International Astronautical Congress (IAC), Bremen, October 1–5, 2018.

82 G. Santi, G. Favaro, A. J. Corso, M. Bazzan, R. Ragazzoni, D. Garoli, and M. G. Pelizzo, "Multilayers for Directed Energy Accelerated Lightsails," *Communications Materials*, vol. 3, no. 16, 2022; doi:10.1038/s43246-022-00240-8.

83 A. Kantrowitz, "Propulsion to Orbit by Ground-Based Lasers," *Astronautics and Aeronautics*, vol. 10, no. 5, pp. 74–76, 1972.

84 C. Phipps, M. Birkan, W. Bohn, H. A. Eckel, T. Lippert, M. Michaelis, Y. Rezunkoc, A. Sasoh, W. Schall, S. Scharring, and J. Sinko, "Review: Laser Ablation Propulsion," *AIAA Journal of Propulsioin and Power*, vol. 26, no. 4, pp. 609–637, 2010; doi:10.2514/1.43733.

85 A. V. Pakhomov and D. A. Gregory, "Ablative Laser Propulsion: An Old Concept Revisited," *AIAA Journal*, vol. 38, no. 4, pp. 725–727, 2012; doi:10.2514/2.1021.

86 B. Duan, H. Zhang, L. Wu, Z. Hua, Z. Bao, N. Guo, Y. Ye, and R. Shen, "Acceleration Characteristics of Laser Ablation Cu Plasma in the Electrostatic Field," *European Physics J Applied Physics*, vol. 93, no. 20802, 2021; doi:10.1051/epjap/2021200349

87 S. Tochitsky, A. Pak, F. Fiuza, D. Haberberger, N. Lemos, A. Link, D. H. Froula, and C. Joshi, "Laser-Driven Collisionless Shock Acceleration Oions from Near-Critical Plasmas," *Physics of Plasmas*, vol. 27, no. 083102, 2020; doi:10.1063/1.5144446.

88 P. Singh, V. B. Pathak, J. H. Shin, I. W. Choi, K. Nakajima, S. K. Lee, J. H. Sung, H. W. Lee, Y. J. Rhee, C. Anichlaesei, C. M. Kim, K. H. Pae, M. H. Cho, C. Hojbota, S. G. Lee, F. Mollica, V. Malka, C. M. Ryu, H. T. Kim, and C. H. Nam, "Electrostatic Shock Acceleration of Ions in Near-Critical-Density Plasma Driven by a Femtosecond Petawatt Laser," *Scientific Reports*, vol. 10, no. 18452, 2010; doi:10.1038/s41598-020-75455-1.

89 C. A. J. Palmer, N. P. Dover, I. Pogorelsky, M. Babzien, G. I. Dudnikov, M. Ispiriyan, M. N. Polyanskiy, J. Schreiber, P. Scholnikov, V. Yakimenko, and Z. Najmudin, "Monoenergetic Proton Beams Accelerated by a Radiation Pressure Driven Shock," *Physics Review Letters*, vol. 106, no. 014801, 2011. doi:10.1103/PhysRevLett.106.014801.

90 A. Pak, S. Kerr, N. Lemos, A. Link, P. Patel, F. Albert, L. Divol, B. B. Pollock, D. Habergerger, D. Froula, M. Gauthier, S. H. Genzer, A. Longman, L. Manzoor, F. Fedosejevs, S. Tochitsky, C. Joshi, and F. Fiuza, "Collisionless Shock Acceleration of Narrow Energy Spread Ion Beams from Mixed Species Plasmas Using 1 μm Lasers," *Physical Review Accelerator and Beams*, vol. 21, no. 103401, 2018; doi:10.1103/PhysRevAccelBeams.21.103401.

91 B. Fu, E. Sperber, and F. Eke, "Solar Sail Technology—A State of the Art Review," *Progress in Aerospace Science*, vol. 86, pp. 1–19, 2016; doi:10.1016/j.paerosci.2016.07.001.

92 C. R. McInnes, *Solar Sailing Technology, Dynamics and Mission Applications*. Springer.

93 G. Vulpetti, L. Johnson, and G. Matloff, *Solar Sails: A Novel Approach to Interplanetary Travel*, 2nd ed., Springer Praxis Books.

94 Y. Tsuda, O. Mori, R. Funase, H. Sawada, R. Yamamoto, T. Saiki, T. Endo, K. Yonekura, H. Hoshino, and J. Kawaguchi, "Achievement of IKAROS—Japanese Deep Space Solar Sail Demonstration Mission," *Acta Astronautica*, vol. 82, pp. 183–188, 2013. doi:doi:10.1016/j.actaastro.2012.03.032.

95 Y. Shirasawa, O. Mori, H. Sawada, Y. Chishki, K. Kitamura, and J. Kawaguchi, "A Study on Membrane Dynamics and Deformation of Solar Power Sail Demonstrator "IKAROS," AIAA 2012-1747, Presented at the 53rd AIAA/ASME Structures, Structural Dynamics and Materials Conference, Honoluu, April 23–26, 2012.

96 H. Sawada, O. Mori, Y. Shirasawa, Y. Miyazaki, M. Natori, S. Matunaga, H. Furuya, and H. Sakamoto, "Mission Report on the Solar Power Sail Deployment Demonstration of IKAROS," AIAA 2011-1887, Presented at the 52nd AIAA/ASME Structures, Structural Dynamics and Materials Conference, Denver, April 4–7,2011.

97 R. Ridenoure, R. Munakata, A. Wong, S. Spencer, D. Setson, B. Betts, B. Plante, J. Bellard, and J. Foley, "Testing the Lightsail Program: Demonstrating Solar Sailing Technology Using a CubeSat Platform," *Journal of Small Satellites*, vol. 5, no. 3, pp. 531–550, 2016.

98 B. Betts, D. Spencer, J. Bellardo, B. Nye, A. Diaz, B. Plante, K. Mansell, J. Fernandez, C. Gillespie, and D. Garber, "LightSail 2: controlled solar sail propulsion using a CubeSat," IAC-19,C4,8-B4.5A,2, x51593, Presented at the International Astronautical Congress 2019, Washington, October 21–25, 2019.

99 D. Spencer, B. Betts, J. Bellardo, A. Diaz, B. Plante, and J. Mansell, "The LightSail 2 Solar Sailing Technology Demonstration," *Advances in Space Research (ASR)*, vol. 67, no. 9, pp. 2878–2889, 2021.

100 L. McNutt, L. Johnson, P. Kahn, J. Castillo-Rogez, and A. Frick, "Near Earth Asteroid (NEA) Scout," AIAA-2014-4435, Presented at the AIAA SPACE 2014 Conference and Exposition, San Diego, August 4–7, 2014.

101 L. Johnson, J. Everett, D. McKenzie, D. Tyler, D. Wallace, J. Wilson, J. Newmark, D. Turse, M. Cannella, and M. Feldman, "The NASA Solar Cruiser Mission – Solar Sail Propulsion Enabling Heliophysics Missions," SSC22-II-03, Presented at the 36th Annual Small Satellite Conference, Utah State University, Loga, August 2022.

102 A. R. Davoyan, J. N. Munday, N. Tabiryan, G. A. Swartzlander, and L. Johnson, "Photonic Materials for Interstellar Solar Sailing," *Optica*, vol. 8, no. 5, pp. 722–734; doi:10.1364/OPTICA.417007.

103 P. Janhunen, "Electric Sail for Spacecraft Propulsion," *AIAA Journal of Propulsion*, vol. 20, pp. 763764, 2004.

104 P. Janhunen and A. Sandroos, "Simulation Study of Solar Wind Push on a Charged Wire: Basis of Solar Wind Electric Sail Propulsion," *Annales Geophysicae*, vol. 25, no. 3, pp. 755–767, 2007; doi:10.5194/angeo-25-755-2007.

105 G. Mengali, A. A. Quarta, and P. Janhunen, "Electric Sail Performance Analysis," *Journal of Spacecraft and Rockets*, vol. 45, no. 1, pp. 122–129, 2008; doi:10.2514/1.31769.

106 A. N. Grubisic and S. B. Gabriel, "Hollow Cathode Thrust Production Mechanisms," AIAA-2009-4823, Presented at the 45th AIAA Joint Propulsion Confernece, Denver, August 2–5, 2009.

107 A. Sasoh, H. Kasuga, Y. Nagagawa, T. Matsuba, D. Ichihara, and A. Iwakawa, "Electrostatic-Magnetic-Hybrid Thrust Generation in Central–Cathode Electrostatic Thruster (CC–EST)," *Acta Astronautica*, vol. 152, pp. 137–145, 2018; doi10.1016/j.actaastro.2018.07.052.

Chapter 11

Electric Thruster Plumes and Spacecraft Interactions

11.1 Introduction

Electric propulsion offers advantages for many missions and applications, but like many spacecraft systems, integration of electric thrusters on spacecraft can present significant systems engineering challenges. Assessing thruster plume interactions with the spacecraft is key in determining thruster location and other spacecraft configuration issues, often requiring trades between thrust efficiency and the life of other subsystems, such as the solar arrays. In this chapter we discuss these topics, concentrating almost exclusively on ion engines and Hall thrusters since they comprise the largest fraction of electric propulsion with flight heritage.

Electric thruster plumes consist of energetic ions, un-ionized propellant neutral gas, low energy ions and electrons, and sputtered thruster material. Spacecraft systems engineers must account for the interaction between each of the plume components and other spacecraft systems. For example, in North-South station keeping by electric thrusters on geosynchronous communications satellites, plume impingement on solar arrays can be a significant issue. As shown in Fig. 11-1, geosynchronous satellites are in a circular orbit coplanar with the Earth's equator, with an orbital period of exactly one day. The satellite appears stationary to an observer on the earth; however, the earth's equator is tilted by 23.5° with respect to the plane of the Earth's orbit around the sun. The plane of the Earth's orbit is called the ecliptic plane. The sun's gravity pulls on a geosynchronous satellite to change the satellite's plane toward the ecliptic. If the orbital plane were allowed to change, the satellite would appear from the ground to move North and South in the sky. Optimal communication would then require the ground-based antennas to constantly scan North and South to track the satellite, defeating the big advantage of geosynchronous satellites. Electric thrusters are used on satellites to counter the sun's pull and prevent the orbital plane from changing. This application is referred to as "North-South station keeping", and Fig. 11-1 shows the Hughes/Boeing patented [1] strategy for this function.

Most modern satellites are three-axis stabilized with solar arrays that rotate to keep the cells pointed toward the sun. From a thrust perspective, North-South station keeping is accomplished most efficiently if the thrusters point in the North and South directions. In geosynchronous orbit, the solar array axis of rotation points North and South, directly in the path of plumes from the North-South station keeping thrusters. The thruster energetic ion beam would impinge on the solar arrays and quickly damage them, dramatically shortening satellite life.

The usual solution is to mount thrusters such that the resultant force is in the North-South direction, but each plume is then at some finite angle with respect to the solar array axis. The larger the

Fundamentals of Electric Propulsion, Second Edition. Dan M. Goebel, Ira Katz, and Ioannis G. Mikellides.

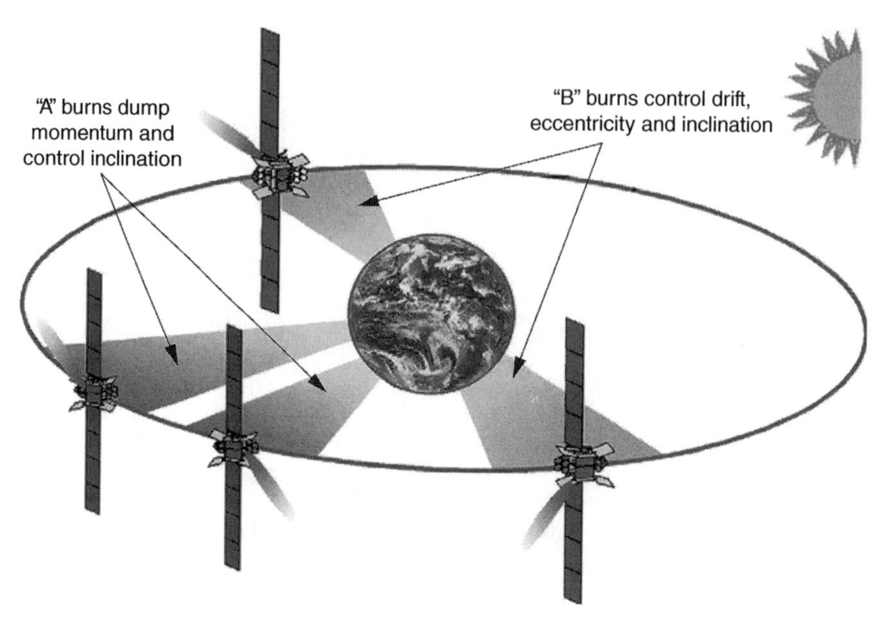

Figure 11-1 Illustration of the burn arcs of the ion thrusters used for electric propulsion station keeping on Boeing satellites [1, 2].

angle, the greater the thrust loss for station keeping, which leads to requirements for larger thrusters and more propellant mass; the smaller the angle, the greater the array damage, which reduces satellite life. This trade between North-South thrust efficiency and solar array life requires detailed knowledge of thruster plumes and their interactions.

Electric thrusters used for primary propulsion, such as those on NASA's Dawn spacecraft [3] and the Psyche Mission [4] flying to the asteroid belt, can also create issues associated with plume impact on the spacecraft solar arrays, exposed components, and scientific instruments. Thruster plumes and their interactions with the spacecraft must be understood and accommodated for the spacecraft to perform to specification for the required mission life.

11.2 Plume Physics in Ion and Hall Thrusters

The thruster plume is composed of ions and electrons of various energies and some neutral gas. The energetic beam ions accelerated by the thruster fields are the dominant ion species and the major source of thrust. The velocity and angular distributions of these ions can be measured in the laboratory and calculated by the thruster computer models discussed in previous chapters. For ion thrusters, where the accelerating voltages are typically a thousand volts or more, the weak plume electric fields have little influence on energetic ion trajectories. In this case, the challenge is usually determining the ion trajectories from the shaped-grid accelerator structure. However, for Hall thrusters, where the accelerating voltages may be only a few hundred volts, the plume electric fields can significantly broaden the energetic ion plume.

The second source of ions is due to charge exchange reactions between beam ions and neutral propellant gas. The neutral gas is due to un-ionized particles leaving both the thruster and the neutralizer (hollow cathode) and, in the case of laboratory measurements, background neutrals

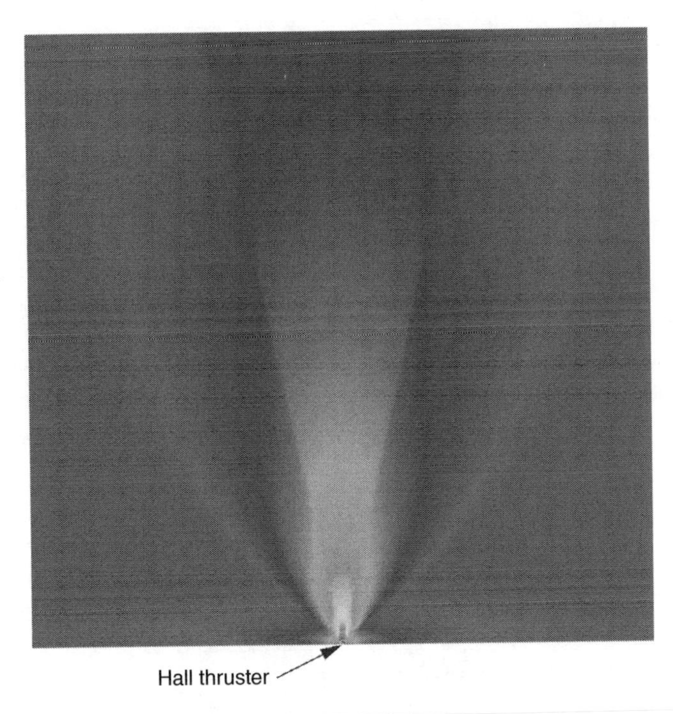

Figure 11-2 Total ion density in the plume of a 4 kW Hall thruster.

present in the vacuum chamber. Charge exchange reactions have usually been associated with inelastic collisions processes yielding low-energy ions at large angles with respect to the main beam direction. However, as thruster voltages increase to provide higher Isp, the energy of these scattered ions can become significant. The total plume plasma density, including all three ion components, is shown schematically in Fig. 11-2 for a 4-kW Hall thruster.

11.2.1 Plume Measurements

Thruster plume characteristics have been measured extensively in the laboratory and on only a few spacecraft in space. In the laboratory, most measurements have been of the ion velocities and densities, and some thruster erosion products, but not of the unionized neutral gas, which is in most cases dominated by background gas in the test chambers. The balance of the thruster gas flow and the speed of test facility's vacuum pumps determine the background gas pressure. The maximum facility pressure during high power testing is usually limited to less than 10^{-4} Torr. Therefore, the density of un-ionized propellant from Hall and ion thrusters is greater than the background only within a few centimeters of the thruster.

The dominance of test facility background neutral gases makes it difficult to directly measure in a laboratory the secondary plasma environment, which consists of the ions generated by charge exchange and/or elastic scattering with neutrals, that would be seen on a spacecraft. Spacecraft system engineers, therefore, use detailed models [5–9] of the plume and secondary ion generation to predict the in-flight plasma environment. Several of these models have been validated with flight data from a few electric propulsion spacecraft.

11.2.2 Flight Data

The first in-flight measurements of the plasma environment generated by an ion thruster were made on NASA's Deep Space 1 (DS1) spacecraft [10]. The NSTAR Diagnostics Package that flew on DS1 included contamination monitors, plasma sensors, magnetometers, and a plasma wave antenna. The plasma sensors and contamination monitors were mounted on the Remote Sensor Unit (RSU) [10] as shown in Fig. 11-3. The measured plasma density was an order of magnitude lower than that measured during ground tests, but it was in good agreement with model predictions. Figure 11-4 shows a comparison of the ion fluxes measured during the DS1 mission by the remote sensing unit and the computed values [11]. The ion fluxes at the sensor location are primarily the result of charge exchange between beam ions and un-ionized propellant in the beam.

Measurements of the plume and secondary ions from Hall thrusters were carried out on a Russian communications satellite, Express-A3 [7]. The satellite had instruments to measure ion fluxes both

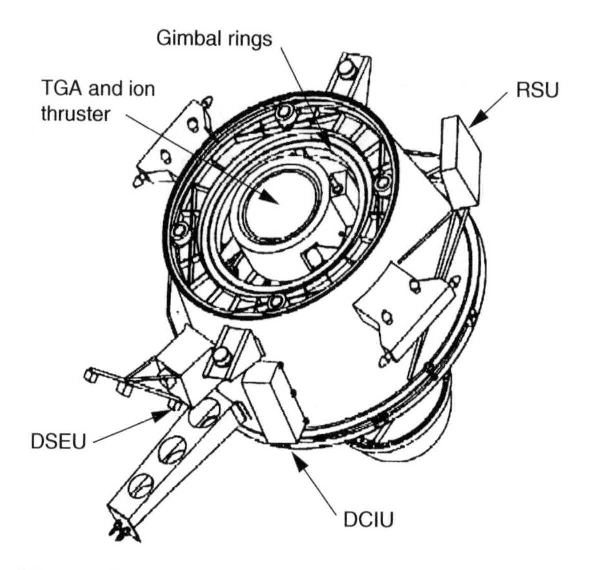

Figure 11-3 Location of Remote Sensor Unit on DS1 with respect to the ion thruster.

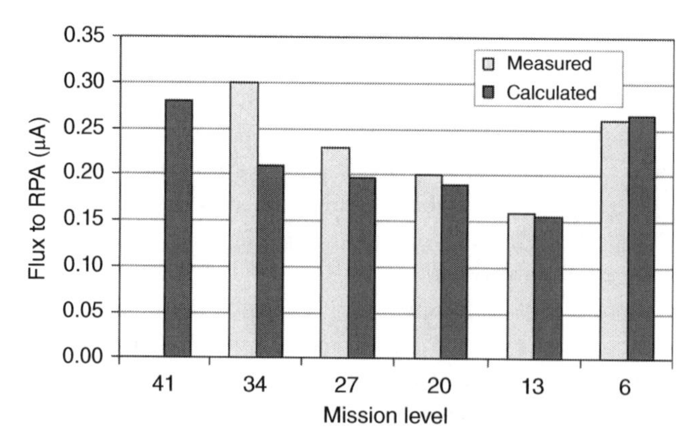

Figure 11-4 Calculated and measured charge-exchange ion fluxes in the plume of NSTAR at various operating points (*Source:* [11]).

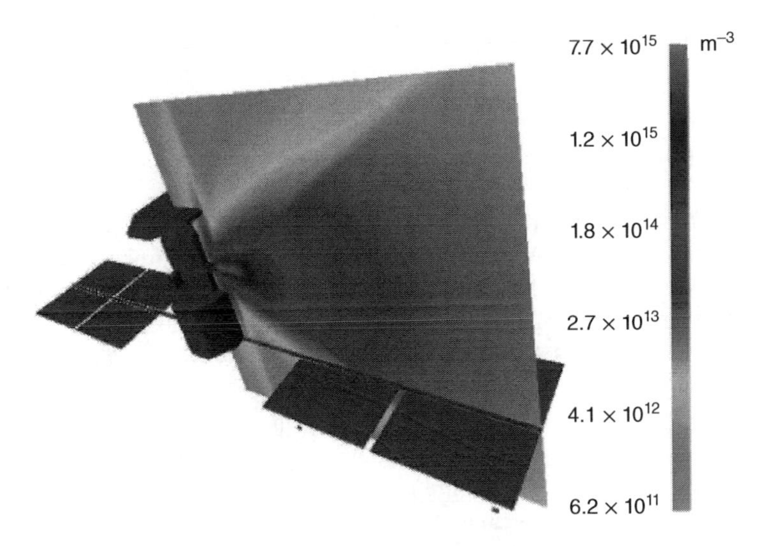

Figure 11-5 EPIC model of the Express-A spacecraft showing the plume ion density profile during operation of RT4 SPT-100 thruster (*Source:* [7]/American Institute of Aeronautics and Astronautics, Inc.).

on the spacecraft body, 90° from the thrust direction, and on the solar arrays. These diagnostics monitored effects from the central beam over a cone with a half angle of about 40°. The SPT-100 Hall thruster plume calculated using the Electric Propulsion Interactions Code (EPIC) [8, 11] is shown in Fig. 11-5. As was the case for ion thrusters, the measured secondary ion fluxes were an order of magnitude less than fluxes measured in ground based chambers, but, again, in good agreement with plume models. The accuracy of the models is illustrated in Fig. 11-6, where the current density measurements on the Express-A spacecraft are compared to the computed values.

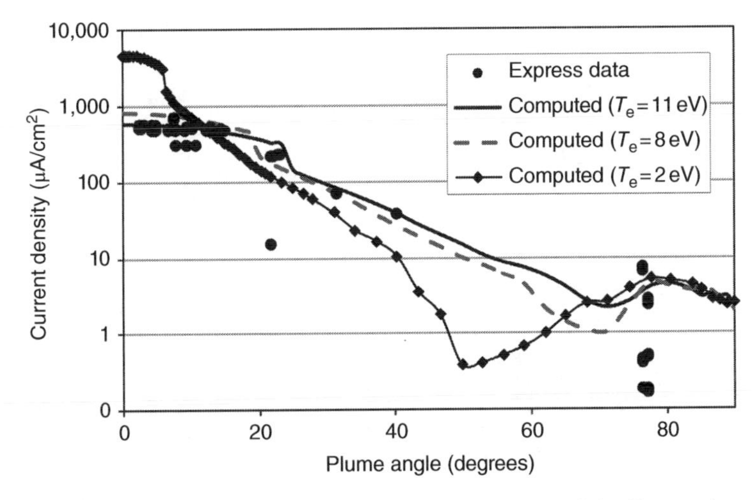

Figure 11-6 Comparison between current density measurements on board the Express-A spacecraft and computed values (*Source:* [7]/American Institute of Aeronautics and Astronautics, Inc.).

11.2.3 Laboratory Plume Measurements

While the flight measurements show the validity of the models to predict thruster generated plasma environments, tests in ground-based chambers provide much more detailed measurements than those made in space.

Experiments conducted by The Aerospace Corporation for Lockheed Martin Space Systems Company on the Busek-Primex Hall Thruster (BPT-4000) provided plume data [12] for comparison with computer models. Measurements were taken using fully exposed flux probes ("uncollimated") for assessing the non-directional ion flux and probes inside graphite collimators ("collimated").

Figure 11-7 shows experimental data [12] taken from the BPT-4000 Hall thruster at discharge power of 3 kW and voltage of 300 V using a collimator for energy spectra at different angles with respect to the thruster axis. The angle-independent, high-energy peak at E/q ~ 280 V associated with the main beam is clearly evident. Also apparent is a small-amplitude peak at the lowest energy values of the collimated spectra from the background chamber plasma. This peak was dominant in the uncollimated spectra. Figure 11-7 reveals the existence of secondary current density peaks with relatively high energies compared to the primary resonant charge exchange peak. For example, at an angle of 40°, the energy associated with the second maximum is approximately 150 eV. These observed ion-flux crests show a marked energy dependence on angle. In an ideal elastic collision between a moving sphere and an identical stationary sphere, the magnitude of the final velocity for each sphere is proportional to the cosine of the angle between its final velocity and the initial velocity of the moving sphere, and the sphere's kinetic energy varies with the square of the cosine. Because the RPA data in Fig. 11-7 shows a peak with energy dependence given roughly by $E_b\cos^2\theta_{\text{lab}}$, where E_b is the main ion beam energy and θ_{lab} is the angle with respect to the thruster axis, these peaks have been attributed to simple elastic scattering (momentum transfer) between beam ions and neutral atoms. Numerical simulations using calculated differential scattering cross sections confirm that elastic scattering is the cause of the observed mid-energy peak [13, 14].

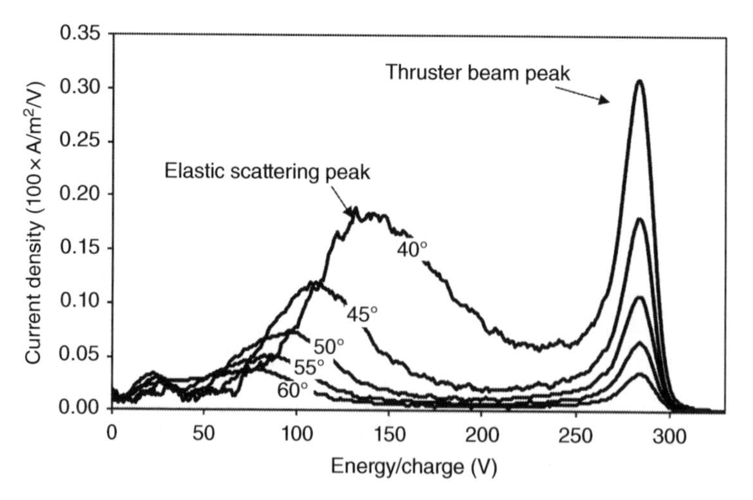

Figure 11-7 Collimated RPA data for the BPT-4000 showing the angle-independent, high-energy main beam peaks and the angle-dependent, elastic scattering peaks (*Source:* [12]).

11.3 Plume Models for Ion and Hall Thrusters

11.3.1 Primary Beam Expansion

Before the advent of multidimensional computer models of thruster plumes, empirical models of the primary beam expansion were used. These models reproduce the general features of the ion beam angular distribution. Because they are very simple, they are invaluable for initial trades when planning electric propulsion system accommodation on spacecraft.

Parks and Katz [15] derived an analytical model of the expansion of an ion beam with Gaussian profile in its self-consistent, quasi-neutral electric field with or without an initial distribution of radial velocities. This model is very useful for calculating thruster ion-beam plume characteristics analytically. The steady state ion continuity and momentum equations in the absence of ionization are

$$\nabla \cdot (\rho_m \mathbf{v}) = 0 \tag{11.3-1}$$

$$\nabla \cdot (\rho_m \mathbf{v}\mathbf{v}) = -\nabla p \tag{11.3-2}$$

where the mass density, ρ_m, is the ion mass density.

Assuming the beam has cylindrical symmetry, the axial beam velocity remains constant everywhere and the axial derivative of the pressure can be neglected compared with its radial derivative. The ion continuity and momentum equations can then be rewritten as

$$\frac{1}{r}\frac{\partial}{\partial r}(r\rho_m \upsilon_r) + \frac{\partial(\rho_m \upsilon_z)}{\partial z} = 0 \tag{11.3-3}$$

$$\upsilon_r \frac{\partial \upsilon_r}{\partial r} + \upsilon_z \frac{\partial \upsilon_r}{\partial z} = -\frac{1}{\rho_m}\frac{\partial p}{\partial r} \tag{11.3-4}$$

The second equation was obtained from the momentum equation by using the continuity equation to eliminate derivatives of the density. The pressure is given by

$$p = nkT_e. \tag{11.3-5}$$

Using the assumption of constant axial velocity, the axial distance, z, can be replaced by the product of the beam velocity, υ_z, and t, the time since the beam left the thruster

$$z = \upsilon_z t. \tag{11.3-6}$$

The axial derivative can be replaced with a time derivative as follows:

$$\frac{\partial}{\partial z} = \frac{1}{\upsilon_z}\frac{\partial}{\partial t} \tag{11.3-7}$$

Equations 11.3-3 and 11.3-4 can then be rewritten as

$$\frac{1}{r}\frac{\partial}{\partial r}(r\rho_m \upsilon_r) + \frac{\partial \rho_m}{\partial t} = 0 \tag{11.3-8}$$

$$\upsilon_r \frac{\partial \upsilon_r}{\partial r} + \frac{\partial \upsilon_r}{\partial t} = -\frac{1}{\rho_m}\frac{\partial p}{\partial r} \tag{11.3-9}$$

These approximations are quite good if the axial velocity is much greater than both the initial radial velocities and the ion sound speed.

With the assumption that the ion beam profile starts out and remains a Gaussian profile, the set of equations can be solved analytically. The beam profile is written as

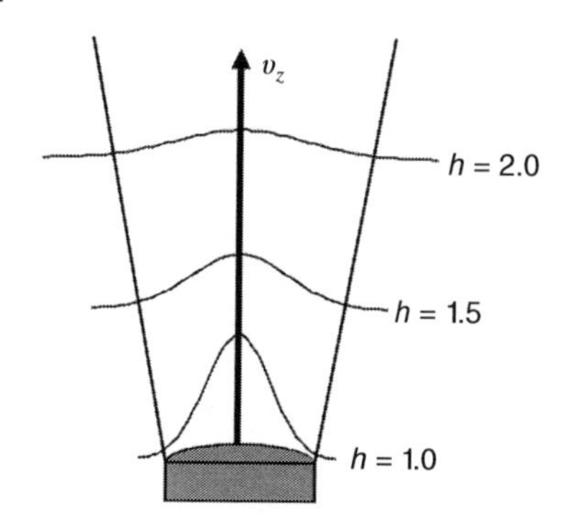

Figure 11-8 The Gaussian beam density profile broadens as ions move downstream from the thruster exit plane.

$$\rho_m(r) = \frac{\rho_o}{h(t)^2} \exp\left(-\frac{r^2}{2R^2 h(t)^2}\right) \tag{11.3-10}$$

where the initial ion beam mass density, ρ_o, is

$$\rho_o = \frac{MI_b}{2\pi e v_b R^2} \tag{11.3-11}$$

and the non-dimensional function *h(t)* describes how the beam expands radially. The parameter *R* is chosen to best represent initial beam width and the initial value of the expansion parameter *h*(0) is unity. The density spreads out as the beam moves axially, but the beam profile remains Gaussian, as shown in Fig. 11-8.

An ion that starts out at a radial position r_o will move radially outward proportionally to *h*(*t*):

$$r(r_o, t) = r_o h(t). \tag{11.3-12}$$

This implies that the radial velocity, v_r, is proportional to the time derivative $\dot{h}(t)$.

$$v_r(r,t) = r_o \dot{h}(t) \tag{11.3-13}$$

Equations 11.3-12 and 11.3-13 can be combined to obtain an expression for the local velocity in which the initial radial position is eliminated:

$$v_r(r,t) = r\frac{\dot{h}(t)}{h(t)} \tag{11.3-14}$$

The solution obtained below is valid for a beam with no initial radial velocity, or for an initial radial velocity distribution that is proportional to radius:

$$v_r(r,0) = v_r^0\, r \tag{11.3-15}$$

The density, defined by Eq. 11.3-10, and the radial velocity defined in Eq. 11.3-15, both satisfy the ion continuity equation (Eq. 3.5-10) for any function *h*(*t*). The first term in Eq. 11.3-8 then becomes

$$\frac{1}{r}\frac{\partial(r\rho_m v_r)}{\partial r} = \frac{1}{r}\frac{\partial}{\partial r}\left(r^2\frac{\dot{h}}{h}\rho_m\right) = \left(2\frac{\dot{h}}{h} - \frac{r^2\dot{h}}{R^2h^2}\right)\rho_m \tag{11.3-16}$$

and the second term in Eq. 11.3-8 becomes

$$\frac{\partial\rho_m}{\partial t} = \left(-2\frac{\dot{h}}{h} + \frac{r^2\dot{h}}{R^2h^2}\right)\rho_m \tag{11.3-17}$$

Making the same substitutions into the momentum equation (Eq. 11.3-9), an equation for *h*(*t*) is obtained that is independent of the radius:

$$v_r\frac{\partial v_r}{\partial r} + \frac{\partial v_r}{\partial t} = -\frac{1}{\rho_m}\frac{\partial p}{\partial r}$$

$$r\frac{\dot{h}}{h}\frac{\partial}{\partial r}r\frac{\dot{h}}{h}+\frac{\partial}{\partial t}r\frac{\dot{h}}{h}=-\frac{eT_e}{M\rho_m}\frac{\partial\rho_m}{\partial r}$$

$$-r\frac{\dot{h}^2}{h^2}+r\frac{\dot{h}^2}{h^2}+r\frac{\ddot{h}}{h}=\frac{eT_e}{M\rho_m}\frac{r}{R^2h}\rho_m$$

$$h\ddot{h}=\frac{kT_e}{MR^2}=\frac{\upsilon_B^2}{R^2} \tag{11.3-18}$$

where υ_B is the Bohm velocity. In Eq. 11.3-18, the right-hand side is a constant. This equation can be integrated by the usual substitution of a new function $w = dh/dt$ for the time derivative of $h(t)$:

$$\ddot{h}=\frac{\mathrm{d}}{\mathrm{d}t}\dot{h}=\frac{dw}{dt}=\frac{dw}{dh}\frac{dh}{dt}=w\frac{dw}{dh} \tag{11.3-19}$$

Using this, Eq. 11.3-18 can be rewritten as

$$w\frac{dw}{dh}=\frac{\upsilon_B^2}{R^2h} \tag{11.3-20}$$

Integrating once yields

$$\int_{\dot{h}(0)}^{\dot{h}(t)}wdw=\int_1^{h(t)}\frac{\upsilon_B^2}{R^2h}dh \tag{11.3-21}$$

Writing this expression in terms of h and its time derivative gives

$$\frac{1}{2}\dot{h}^2=\frac{\upsilon_B^2}{R^2}\ln h+\frac{1}{2}\dot{h}^2(0) \tag{11.3-22}$$

Taking the square root and integrating again, an equation relating h to time is obtained. For the case of no initial radial velocity, $\dot{h}(0) = 0$, the time derivative of h is

$$\dot{h}=\frac{\upsilon_B}{R}\sqrt{2\ln h} \tag{11.3-23}$$

Equation 11.3-23 can be rewritten and integrated to give

$$\frac{dh}{\sqrt{\ln h}}=\frac{\upsilon_B}{R}\sqrt{2}\,dt \tag{11.3-24}$$

$$\int_1^h\frac{dx}{\sqrt{\ln x}}=\int_0^t\sqrt{2}\frac{\upsilon_B}{R}\mathrm{d}t=\sqrt{2}\frac{\upsilon_B}{R}t \tag{11.3-25}$$

An approximate numerical solution of Eq. 11.3-25 for the expansion parameter, h, is given by

$$h\approx 1.0+0.6524\tau+0.0552\tau^2-0.0008\tau^3, \tag{11.3-26}$$

where τ is given by

$$\tau\equiv\sqrt{2}\frac{\upsilon_B}{R}t \tag{11.3-27}$$

These expressions describe the beam expansion for the case of no initial radial velocity or for an initial radial velocity distribution that is proportional to the radius. Examples of schematic beam profiles as a function of distance from the thruster were given in Fig. 11-8. For the case of an initial radial velocity profile, the integral in Eq. 11.3-25 is

$$\dot{h}_0 t = \int_1^h \left(1 + 2\frac{v_B^2}{R^2 \dot{h}_0^2} \ln x\right)^{-1/2} dx, \tag{11.3-28}$$

where the integral has to be calculated numerically. Park's model has been extended by Ashkenazy and Fruchtman [16] to include thermal gradient and two-dimensional effects.

The Park's formula is very similar to an empirical formula developed earlier by Randolph and Pencil for Hall thrusters [17]. Randolph's formula has two Gaussians, but does not have the curved trajectories of the Park's formula. The four parameters, k_0 through k_3, in Randolph's formula are chosen to fit plume measurements:

$$j = \frac{R^2}{r^2}\left\{k_0 \exp\left(-\frac{(\sin\theta)^2}{k_1^2}\right) + k_2 \exp\left(-\frac{\theta^2}{k_3^2}\right)\right\} \tag{11.3-29}$$

While the analytical expressions above are invaluable for estimating plume interactions, multidimensional computer models are normally used for detailed calculations. There is general agreement on the physics that control the expansion of the main ion beam from ion and Hall thruster, but there are differences in the numerical algorithms used to calculate the expansion. Some researchers [5, 6] employ Particle-In-Cell (PIC) algorithms, where the beam is modeled as a collection of macro-particles with each particle representing a large number of ions. The velocity and acceleration of each particle is followed in the self-consistently calculated electric field.

Another approach, which is much less computationally intensive, is to model the thruster beam as a drifting fluid of cold ions and warm electrons. In this method, the expansion of the fluid-like ion beam is calculated using a Lagrangian algorithm [7, 8]. The ion beam profile for the Nuclear Electric Xenon Ion System (NEXIS) ion thruster [18] calculated using this algorithm is shown in Fig. 11-9. The primary beam is assumed to be comprised of a collisionless, singly ionized, quasi-neutral plasma expanding in a density-gradient electric field. The electron drift velocity is small compared to the electron thermal speeds, so momentum balance for electrons can be written as:

$$m_e \frac{d\mathbf{v}_e}{dt} = e\nabla\phi - \nabla p_e = 0 \tag{11.3-30}$$

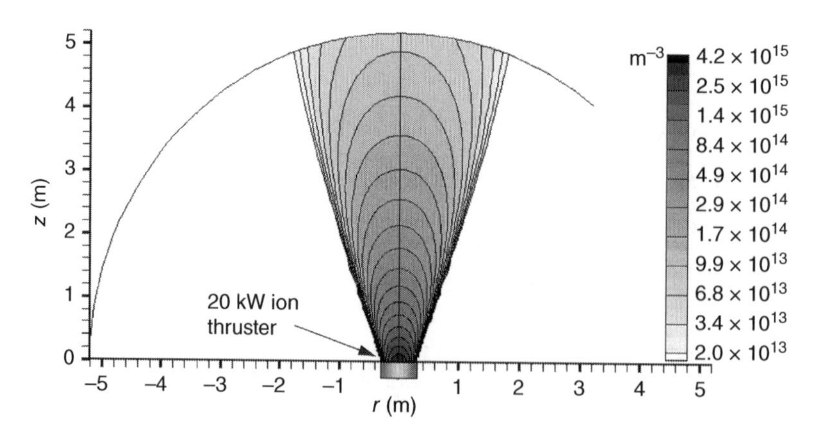

Figure 11-9 Calculated primary ion beam density profile for the 20 kW NEXIS ion thruster *Source:* [18].

where $\mathbf{v}_e$, ϕ, and p_e are the electron velocity, electric potential, and electron pressure, respectively. Assuming an ideal gas electron pressure, the potential follows the Boltzmann relation,

$$\phi = \frac{kT_e}{e} \ln\left(\frac{n_e}{n_\infty}\right) \tag{11.3-31}$$

with T_e being the electron temperature, n_e is the plasma density, and n_∞ is a reference plasma density. The plume is also assumed to be isothermal. This is a better approximation for space conditions than in the laboratory where inelastic collisions with background neutrals will tend to cool the electrons.

In this model, ions are assumed to be cold compared with the electrons ($p_i \approx 0$), and their acceleration dominated by the electric field:

$$M\frac{D\mathbf{v}_i}{Dt} = -e\nabla\phi \tag{11.3-32}$$

Since the drift velocity of the ions is much greater than their thermal velocity, the high velocity ions are modeled as a fluid, with a velocity of $\mathbf{v_i}$. The governing equations, solved in 2-D (R-Z) geometry, are conservation of mass and momentum:

$$\begin{aligned} &\nabla \cdot n\mathbf{v}_i = 0 \\ &M\mathbf{v}_i \cdot \nabla n\mathbf{v}_i = -en\nabla\phi \end{aligned} \tag{11.3-33}$$

where $n = n_e = n_i$. The accuracy of the algorithm has been confirmed by comparisons of analytical solutions with model problems in one and two dimensions [11]. The Lagrangian modeling approach leads to reduced numerical noise compared with PIC algorithms. However, unlike PIC algorithms, the fluid technique ignores spreading of the beam because of ion temperature and, in the case of ion thrusters, the angular distribution coming out of each grid aperture.

11.3.2 Neutral Gas Plumes

The neutral gas density in a laboratory vacuum chamber has three components: gas from the thruster, gas from the neutralizer hollow cathode, and the background chamber density. To model the neutral gas density, the gas from ion thrusters can be approximated by isotropic emission from a disk with the diameter of the grid:

$$n_o \sim \frac{\cos\theta}{r^2} \tag{11.3-34}$$

For Hall thrusters, the neutral gas density can be approximated using an annular anode gas flow model with isotropic emission from the channel. This is done by calculating emissions from two disks, one large and one smaller, and subtracting the smaller from the larger. The neutral density drop-off with r and z from a disk emitting a Maxwellian distribution is calculated using an approximate view factor. Energetic CEX neutrals are negligible compared to the total neutral density and are therefore not included when modeling the neutral gas density.

For plume models, the neutral gas from the neutralizer hollow cathode is usually assumed to be from isotropic emission at a constant temperature equal to the neutralizer cathode orifice temperature. Although the neutralizer is offset from the thruster axis of symmetry, in cylindrical 2-D codes there are an equal number of points from the thruster axis closer to and farther from the neutralizer. The cylindrically averaged neutral density for any point at a distance z downstream is estimated as if the point was along the thruster centerline. The vacuum chamber background neutral density is

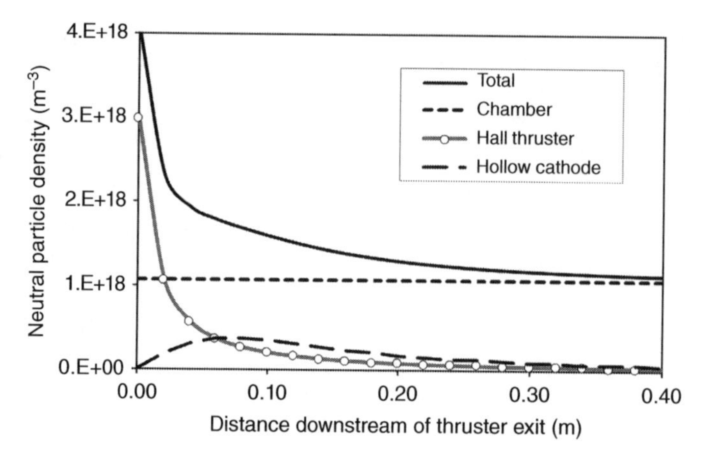

Figure 11-10 Neutral gas density downstream of the BPT-4000 exit plane (*Source:* [14]).

usually assumed to be constant. Based on values of the ambient temperature and pressure the background density can be determined assuming an ideal gas law. No background density is assumed for calculations in space conditions. Figure 11-10 shows each of the three components and the total calculated neutral density [14] for the BPT-4000 Hall thruster.

11.3.3 Secondary Ion Generation

Low energy ions are created near a thruster exit plane by charge exchange collisions between the main ion beam and the neutral gas. The mechanism is the same for both gridded ion and Hall thrusters. Charge exchange (CEX) ion density can be computed using a two-dimensional, R, Z-geometry PIC code, while using the main beam ion densities computed by the Lagrangian calculations and the neutral gas profile as inputs. The charge exchange ion production rate, $\dot{n}_{\text{CEX}}$, is calculated assuming that the beam ions have a velocity, υ_b, much greater than the neutral gas velocity:

$$\dot{n}_{\text{CEX}} = n_i n_0 \upsilon_b \sigma_{\text{CEX}} \tag{11.3-35}$$

Resonant charge exchange cross sections between singly charged xenon ions and neutral xenon atoms range from 30 Å^2 to 100 Å^2 for typical ion and Hall thruster energies [19].

The charge exchange ion density is calculated by tracking particle trajectories in density gradient electric fields using a finite current barometric law for the electron density (electron current equals ion current). Poisson's Equation is solved on a finite element grid and iterated until steady state CEX densities and density gradient potentials are self-consistent. Comparisons of the CEX plume model with flight data from the NSTAR's ion engine exhibited good agreement [11].

Figure 11-11 shows plume maps at 1 m calculated using this method for the BPT-4000 under both lab and space conditions. The CEX density in the lab is found to be more than one order of magnitude greater than it is in space because of the dominance of the background neutral gas in the chamber. With the exception of the neutral gas density, all the terms in the expression for charge exchange ion generation (Eq. 11.3-35 above) are identical for the laboratory and space. Figure 11-10 showed that at distances greater than about a tenth of a meter downstream of the thruster exit plane, the chamber gas density is much greater than the gas coming directly from the thruster, resulting in greater charge exchange ion generation. The computed total ion current in the laboratory case (5.3 A) is in approximate agreement with measurements of the integrated ion current (5–6 A for collector potential of 20 V) [9]. The calculations assumed a charge exchange cross-section for 300 V ions of 55 Å^2 based on the calculations and measurements by Miller [19].

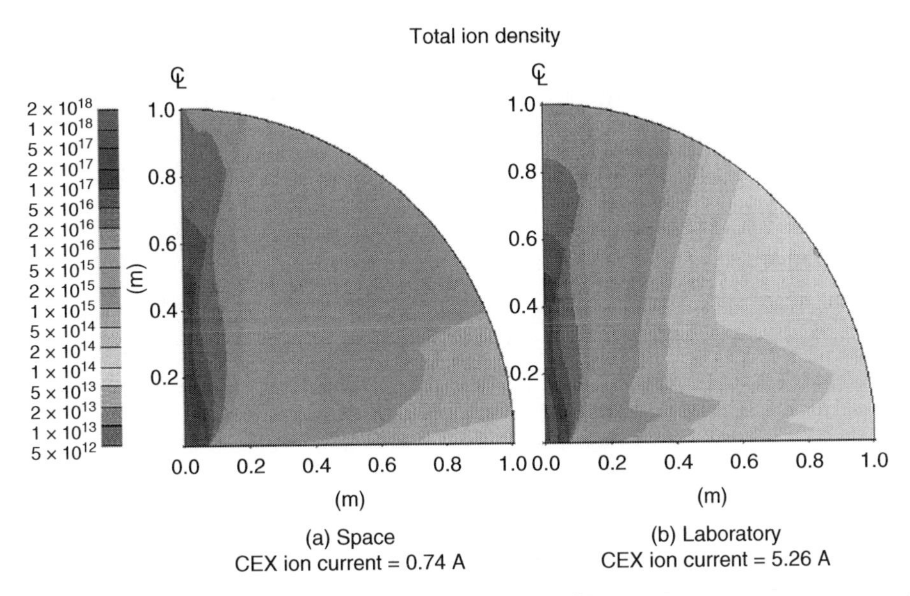

Figure 11-11 Hall thruster plume maps of the ion density (in m^{-3}) for (a) space and (b) laboratory and conditions showing dominance of background density in the charge-exchange plume production (*Source:* [14]).

The distinctive second peak in the energy spectra captured by the collimated retarding potential analyzer (RPA) data shown in Fig. 11-7 is from elastic scattering of xenon ions by neutral xenon atoms. Mikellides et al. [14] have calculated differential cross section data for elastic $Xe^+ - Xe$ scattering in a center-of-mass frame of reference. The calculations involve averaging over the pertinent $Xe_2{}^+$ potentials, without inclusion of charge exchange. The results are then subsequently corrected for charge exchange.

The derived, center-of-mass differential cross sections were converted to values in a fixed frame of reference relative to the laboratory and implemented in the plume model. For comparisons with RPA measurements, the flux of scattered ions Γ_{is},

$$\Gamma_{\text{is}} = \int_0^x \frac{I_b n_o}{d^2} \frac{d\sigma}{d\Omega} d\xi$$

$$\approx \frac{I_b n_o x_c}{d^2} \frac{d\sigma}{d\Omega} \tag{11.3-36}$$

was computed at a radius of 1 m (the RPA location). In Eq. 11.3-36, I_b and n_o are the main beam ion current and neutral density, respectively. The dimension x_c is the characteristic length of the beam column and d is the radial distance between the thruster and the RPA. The differential contribution due to the column element along the beam is denoted by $d\xi$, and $d\sigma/d\Omega$ is the differential cross section.

The results from the complete calculation compared with data are shown in Fig. 11-12. Plotted in the figures are the results of the calculations of the expanding beam ions, and the beam and scattered ions combined. Also plotted is the ion current probe data for two voltage (V) bias levels of 50 V and 100 V. The probe bias potential prevents lower energy ions from being collected. As expected, the beam-only values compare best with the ion probe biased to 100 V since, at this value, most of the scattered and charge exchanged ions are excluded. The calculation combining beam and elastic scattering compares well with 50 V-biased probe data since this data includes most of the elastically scattered ions.

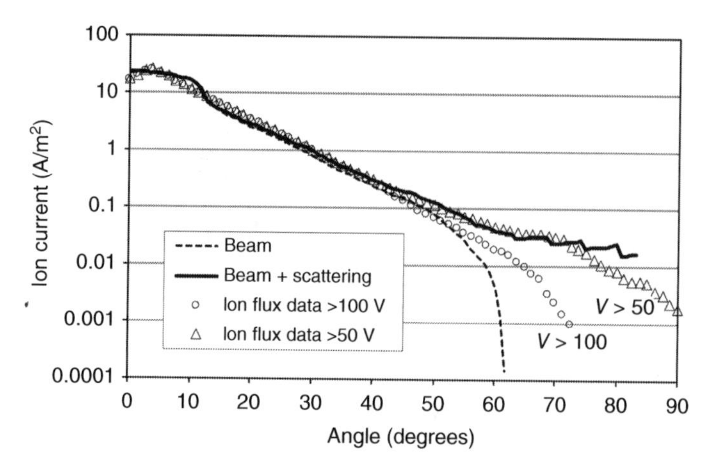

Figure 11-12 Comparison of high-energy ion current between the calculations and measurements for the BPT-4000 (*Source:* [14]).

11.3.4 Combined Models and Numerical Simulations

The previous sections outlined some of the different approaches being used to model individual species and plasma properties in the plumes of ion engines and Hall thrusters. Ultimately, of course, since the main objective of these models is to support the assessment of plume interactions with spacecraft, numerical simulations that account for all these effects must be performed, self-consistently and at least in two dimensions. In cases where thruster clusters are used on the spacecraft and the plumes are in close enough proximity to affect the expansion of each plume, three-dimensional simulations are needed (e.g. [20]). Due largely to the relative simplicity of the plume physics compared to those in the interior of many electric thrusters, plume models and simulations matured early and to sufficient levels of fidelity to support the thruster-spacecraft integration cycle.

Nonetheless, improvements to the plume physics, their mathematical models, and the computational approaches used to solve them have continued. One challenge in plume modeling, for example, has been related to the wide range of length scales that must be resolved in a typical plume calculation. To be more specific, for typical spacecraft interaction assessments the domain of the simulation must span many times the thruster diameter, which can be several to tens of meters away from the engine. At the core of this challenge is the sensitivity of the plume solution to the details of the conditions in the vicinity of the thruster exit. Fortunately, in ion engines the very near plume is quite amenable to idealized analytical modeling (as also discussed in Section 11.3.1), and the accuracy of these models can be considerably improved with only a few basic plasma measurements. These models can then be incorporated easily in a plume simulation to define thruster boundary conditions, without degrading accuracy and computational speed.

The Hall thruster near-plume plasma, however, is much more complicated and less amenable to such idealized modeling. To meet the challenge in these thrusters, the most accurate and computationally manageable approach is to use directly the results from a thruster simulation. This approach formed the basis of a plume code called HallPlume2D that was developed at JPL in the mid-2010s [21]. We discuss it briefly here to provide an example of a relatively recent plume code in which such a strategy was successfully implemented. It is a fitting example also because HallPlume2D has integrated many of the lessons learned in plume modeling over several decades, some of which have already been discussed in the previous sections.

HallPlume2D employs a radial-axial (r-z) domain and uses the simulation results from a Hall thruster code called Hall2De [22] to define boundary conditions at the plume inlet, as illustrated by the schematic in Fig. 11-13. Referring to Fig. 11-13 (top), Hall2De solves numerically, also in r-z geometry, the governing equations for the plasma inside the Hall thruster discharge channel and in the near-plume region. Typically, the latter spans a distance that can be several times the channel length. Of particular relevance to the discussion here is that in Hall2De ions can be modeled as

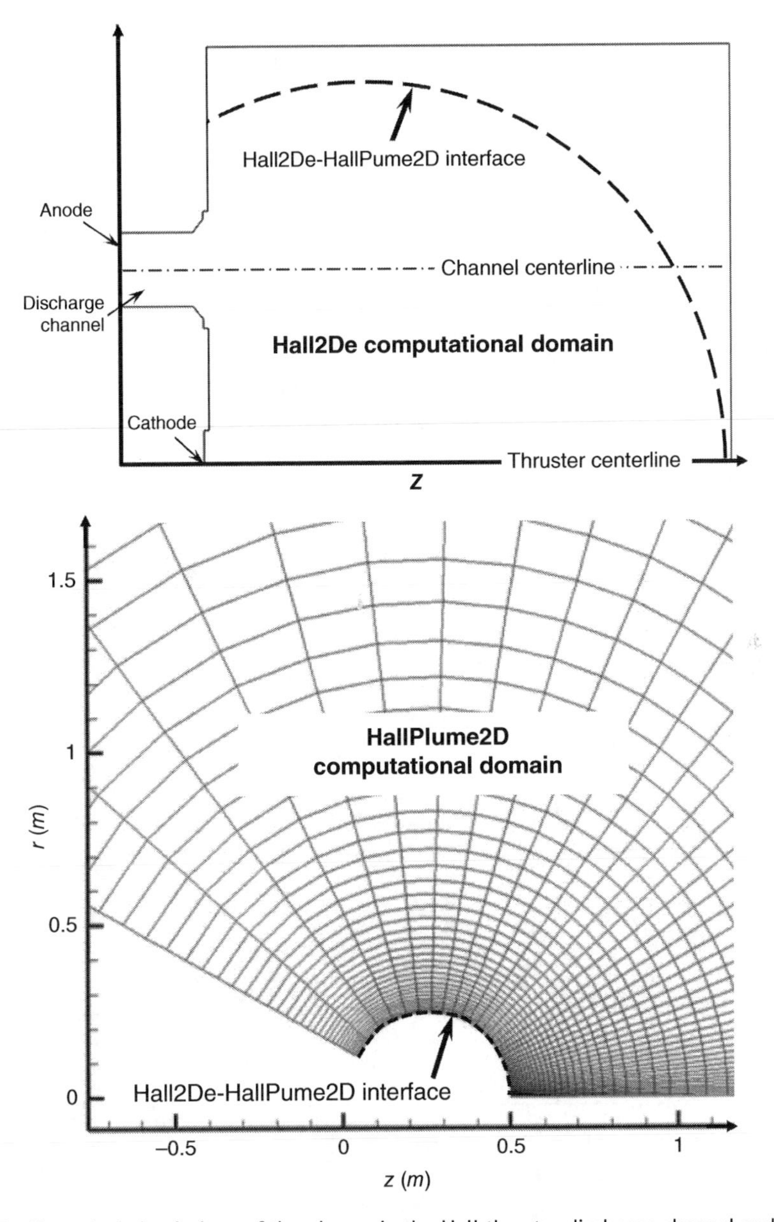

Figure 11-13 Numerical simulations of the plasma in the Hall thruster discharge channel and near plume regions (top) using the Hall2De code [22] provide the boundary conditions for simulations of the thruster plume in a much larger computational domain (bottom) using the HallPlume2D code [21].

distinct fluids or tracked individually using discrete macroparticles. This means that the information passed to the HallPlume2D simulation must be consistent with the equations *it* solves to avoid interpolations that can lead to inaccuracies and numerical difficulties. Therefore, the ion solver in HallPlume2D follows the same strategy as that in Hall2De, allowing for both multi-fluid and discrete particle methods. In both strategies, ionization and charge exchange collisions are accounted for. The neutrals gas also is solved in the same manner as that in Hall2De [23]. For the electrons, the temperature is computed by solving an energy conservation equation that is subject to Dirichlet boundary conditions at the Hall2De-HallPlume2D interface. This interface is depicted by the dashed line in Fig. 11-13 (bottom). Because the thruster boundary in the plume simulation is deliberately chosen to be far from the thruster exit, the applied magnetic field is neglected and the electric field and plasma potential are computed using Boltzmann's law for the electrons.

The intentional similarity of the approaches implemented to solve the mathematical models in the two codes allows for the unambiguous definition of boundary conditions at the inlet of the plume simulation with relative ease, and leads to higher-fidelity solutions. Some representative results are shown in Fig. 11-14. The simulations were of a magnetically shielded Hall thruster developed for the Advanced Electric Propulsion System that will propel the Power and Propulsion Element on Gateway [24]. Gateway is a planned outpost that will orbit the Moon and serves as a critical piece of infrastructure for NASA's Artemis program. Before using the results for thruster–spacecraft integration assessments, the simulations were first validated using measurements taken in a vacuum facility during thruster testing, when the background pressure was on the order of 5 μTorr (6.7×10^{-4} Pa). The comparison shown in Fig. 11-14 is between simulation and measurement of the total ion current density, 0.8 m away from thruster as a function of plume angle. The simulation was performed by imposing the same background pressure as that in the ground facility during thruster testing, at a discharge voltage and power of 600 V and 12 kW, respectively. Also shown for comparison is the simulation result in vacuum. As with the results discussed in the preceding section, the two simulations expose the effect of the facility background on the ion current density

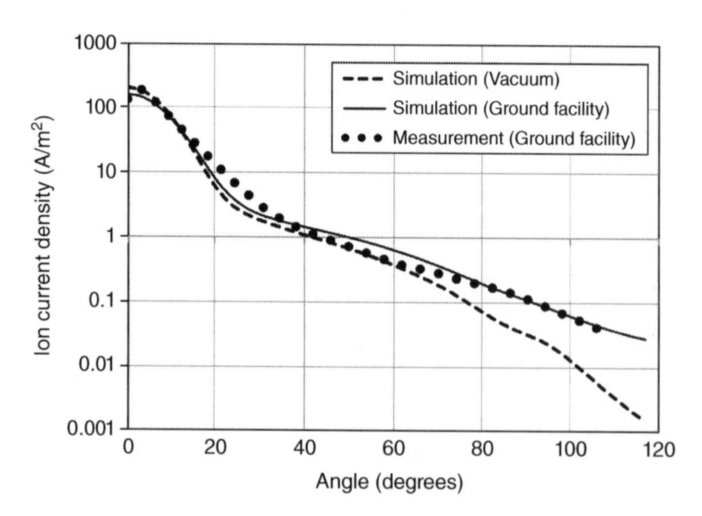

Figure 11-14 Ion current density as a function of plume angle, 0.8 m from a magnetically shielded Hall thruster operating at a discharge voltage and power of 600 V and 12 kW, respectively. The plot compares measurements performed in a ground facility with simulations from the HallPlume2D plume code (*Source:* [21]).

at large angles and re-emphasize the value of accurate plume models in the integration of electric propulsion on spacecraft.

11.4 Spacecraft Interactions

To design a spacecraft to accommodate electric thrusters, it is necessary to understand how the thruster plumes interact with the spacecraft and its payloads. Thruster plumes affect the spacecraft immediately during their operation, for example, by momentum transfer from plume impingement or optical emissions, and by slow, cumulative processes, such as ion erosion of spacecraft surfaces or contamination of surfaces by materials generated by thruster wear. The immediate interactions may affect spacecraft operations; the longer term interactions may affect spacecraft life.

Unique to electric propulsion is the interaction between the thruster plumes and the spacecraft electrical system, in particular the solar arrays. Electric thruster plumes are composed of charged particles and can carry currents between the thruster electrical power system and exposed electrical conductors such as solar array cell edges and interconnects. Although the currents that flow through the thruster plumes are in general quite small, they may cause changes in subsystem potentials. These potential changes, if not anticipated, may be mistaken for system anomalies by spacecraft operators.

As described in previous sections, while most of the plume is in the thrust direction, a small fraction of the thruster exhaust is emitted at large angles. The large angle component is mostly composed of low velocity particles. Some high-energy ions in Hall thruster plumes can be found at angles greater than 45°, but at such a low flux density that they will have little impact on spacecraft life. Techniques for quantitatively calculating the effects of thruster plumes on spacecraft are presented in the following sections.

11.4.1 Momentum of the Plume Particles

Just as with chemical thrusters, when electric thruster plumes impact spacecraft surfaces, they exert a force which causes a torque on the spacecraft. The force is easily calculated as the difference in momentum between the plume particles that impact the surface and the momentum of particles that leave the surface. The momentum of the plume particles is the sum of the ion and neutral atom fluxes. Since the plume consists primarily of ions, and the velocity of the ionized particles is much greater than the neutral atoms, the neutral component can usually be neglected. The ion momentum is

$$\mathbf{p}_i = n_i M_{\mathrm{Xe}} \mathbf{v}_i, \tag{11.4-1}$$

and the neutral momentum is

$$\mathbf{p}_o = n_o M_{\mathrm{Xe}} \mathbf{v}_o, \tag{11.4-2}$$

so that the total plume momentum is

$$\mathbf{p}_{\text{plume}} = \mathbf{p}_i + \mathbf{p}_o \approx \mathbf{p}_i. \tag{11.4-3}$$

In one extreme, an ion that impacts a surface may scatter elastically and leave the surface with its kinetic energy unchanged, but its velocity component normal to the surface is reversed:

$$\mathbf{p}_{\text{surface}}^{\text{elastic}} \cdot \mathbf{n} = -2\mathbf{p}_{\text{plume}} \cdot \mathbf{n}, \tag{11.4-4}$$

where **n** is the unit vector normal to the impacted surface. In the other extreme, the incident xenon ion resides on the surface long enough to transfer its momentum and energy to the surface, and the particle leaves the surface with a velocity distribution corresponding to the surface temperature. This process is called *accommodation*, and the fraction of particles that undergo this process is called the *accommodation coefficient.*

Since spacecraft surfaces are typically less than a few hundred degrees Kelvin, the velocities of accommodated atoms are orders of magnitude less than energetic thruster ions. For example, the speed of a xenon atom leaving a 300 K surface is

$$v_i(300K) = \sqrt{\frac{kT}{M}} = 137\ [\text{m/s}], \tag{11.4-5}$$

while the speed of a beam ion from a 300-V Hall thruster is approximately

$$v_b(300\text{eV}) = \sqrt{\frac{2eV}{M}} = 22,000\ [\text{m/s}]. \tag{11.4-6}$$

Because the thermal speeds are so small compared with the beam speeds, the momentum of re-emitted *surface-accommodated* ions can be ignored when calculating surface torques. The momentum transfer per unit area is approximated by

$$\mathbf{F} = (2 - A_c)\mathbf{p}_{\text{plume}}, \tag{11.4-7}$$

where A_c is the surface accommodation coefficient, which has a range of values from 0 to 1. Flight data from the Express-A satellite shows that accommodation coefficients for Hall thruster ions on the solar arrays were close to unity [7].

11.4.2 Sputtering and Contamination

A major concern for implementing electric thrusters on earth orbiting satellites is that energetic ions from the thruster beam will erode spacecraft surfaces. As discussed above, North-South station-keeping with body-mounted thrusters invariably leads to high-energy ions bombarding some part of the solar arrays. When these high-energy ions impact the solar arrays or other spacecraft surfaces, they can cause erosion by sputtering atoms. However, with proper placement and orientation of the thrusters, and the use of stay-out "zones" during which the thrusters are not operated because the plume would impinge on the array, the ion flux can be small enough to keep electric thrusters from limiting satellite life. Whether a given surface erodes or accumulates material depends on the relative rates of sputtering and the deposition of sputter deposits. The deposits result from erosion products from the thruster itself, as well as material sputtered from other spacecraft surfaces.

Sputtering affects spacecraft in two ways. First, spacecraft surfaces can erode by sputtering, or be contaminated by the buildup of sputtering products. Primary thruster beam ions are the principal source of sputtering, and spacecraft surfaces within a narrow cone angle of the thrust direction will erode significantly because of ion sputtering. The cone angle where sputtering is important depends on the specific thruster and is usually narrower for ion thrusters than for Hall effect thrusters. For example, the NEXIS ion thruster primary beam plume, shown in Fig. 11-9, has a half angle for all particles of only about 20° and 95% of the particles are within a 10° half angle.

Second, while ion and Hall thrusters typically usually use an inert gas propellant, any type of thruster can potentially contaminate spacecraft surfaces. The sources of contamination are thruster material sputtered by energetic ions and spacecraft material sputtered by the main thruster beam. In ion thrusters, sputter erosion of grid material not only limits thruster life, but the sputtered grid material may be a significant source of contamination to spacecraft surfaces. This was recognized early in the development of commercial ion thrusters [25], and as a result, a third grid was added to reduce the amount of sputtered grid material coming from the thruster and to shield the spacecraft from grid sputter products. The third grid has the added benefit of dramatically reducing the grid sputter rate by preventing charge exchange ions made downstream of the third grid from hitting the accelerator grid [26]. For ion thrusters with metal grids, the problem of contamination in the absence of a third grid can be quite important. Only a few monolayers of a metallic contaminant can make large changes to the optical, thermal, and electrical properties of spacecraft surfaces.

For Hall thrusters, the situation can be quite different. The plume from Hall thrusters normally has about twice the angular divergence of an ion thruster, and so sputtered thruster material comes out at large angles. The plume also changes significantly between conventional Hall thruster designs with exterior hollow cathodes, and Hall thrusters that utilize internal centrally-mounted cathodes [27]. Early in life most of the contamination comes from sputter erosion of the ceramic channel wall in conventional Hall thruster designs. Although this can produce a substantial flux of sputter products, the products are mainly insulating molecules. Deposition of sputtered insulators, such as Hall thruster channel ceramic or solar cell cover glass materials, has little effect on the spacecraft surface optical and thermal properties. More problematic is the sputtered graphite material from the pole face covers in magnetically shielded thrusters, and the metallic material from the late life erosion of Hall thruster magnetic pole pieces. In the same manner as with ion thrusters, very thin layers of the deposited metal can radically change the properties of spacecraft surfaces.

One effect found with Hall thrusters, but common to both ion and Hall thrusters, is that surfaces can experience net deposition of sputter products or can be eroded away by energetic beam ions, depending on their location with respect to the thruster ion beam [17]. As shown in Fig. 11-15, the plume of sputtered products coming from the thruster is normally much broader than the main ion beam. For surfaces at small angles with respect to the thrust vector, sputtering from the beam ions is greater than the deposition of thruster erosion molecules. These surfaces will erode over time. However, surfaces located at large angles to the thruster vector are contaminated by thruster erosion products faster than they can be sputter away by energetic beam ions. Over time, sputtered thruster material will accumulate on these surfaces. For the SPT-100 Hall thruster, the dividing line between erosion and deposition is about 65° [17].

Besides thruster erosion products, the other source of contamination is spacecraft surface material sputtered by thruster beam ions. Computer codes, such as the Electric Propulsion Interactions Code (EPIC) [8], are used to calculate the erosion and redeposition over the entire spacecraft. EPIC

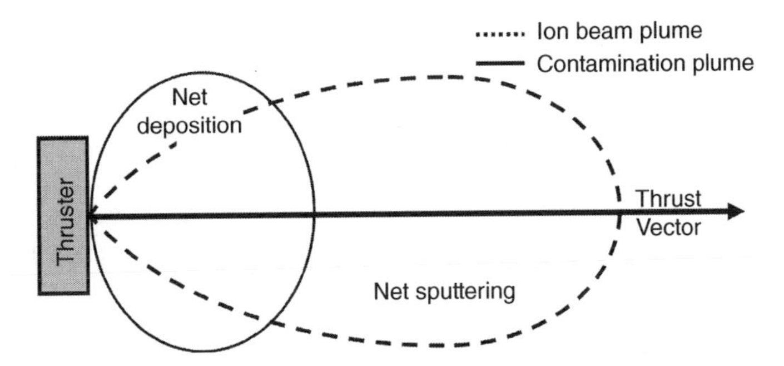

Figure 11-15 Sputtering by main beam ions dominates at angles close to thrust vector direction; deposition of thruster erosion products occurs at angles far from the thrust direction.

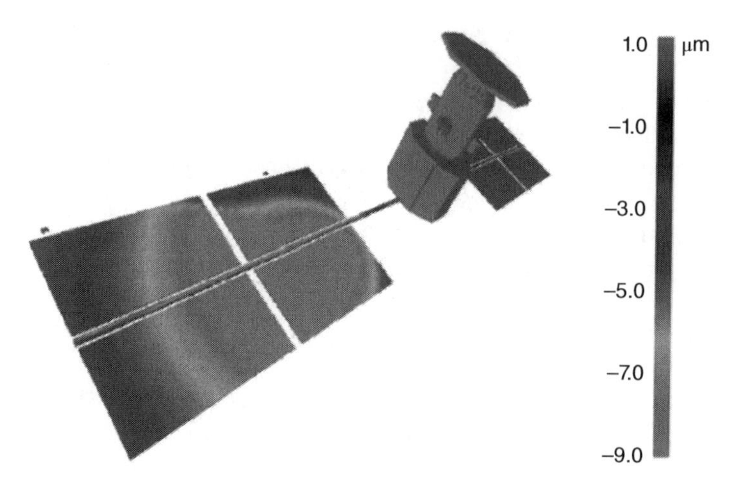

Figure 11-16 Contours of the erosion (negative numbers) and deposition depths (positive numbers) due to sputtering during operation of the SPT100 Hall Thruster onboard the Express-A spacecraft. The calculation was performed with EPIC (*Source:* [8]).

is an integrated package that models the interactions between a spacecraft and its electric propulsion system. The user provides EPIC with spacecraft geometry, surface materials, thruster locations, and plume parameters, along case study parameters such as orbit and hours of thruster operation. EPIC outputs thruster plume maps, surface interactions on the 3-D spacecraft, 1-D plots along surfaces (e.g. erosion depth on a solar array as a function of distance from the thruster), and integrated results over duration of mission (e.g. total induced torque in a given direction, total deposition of eroded material at a specific location on the spacecraft). Figure 11-16 shows results of a sample EPIC calculation for the Express-A spacecraft during firing of one its four Stationary Plasma Thrusters. The calculation shows both sputter erosion and deposition depths. The thruster erodes the solar array surface that is along the thruster direction. Some of the eroded material deposits on other spacecraft surfaces.

11.4.3 Plasma Interactions with Solar Arrays

Ion and Hall thruster plasma plumes connect thrusters electrically to the exposed spacecraft conducting surfaces. It is important to account for current paths through the plasma to prevent current loops or unintended propulsion system floating potentials.

To understand the plasma currents and floating potentials between the electric propulsion system and the rest of the spacecraft, first consider the thruster external cathode as the source of the plasma. As discussed in Chapter 4, the sheath voltage drop internal to the hollow cathode and resistive heating in the orifice region produce energetic electrons that ionize the propellant gas and generate plasma. The combined insert and orifice potential drops are typically between 10 and 15 V, causing the external plasma to be about the same value above cathode common, as illustrated in Fig. 11-17. The hollow cathode generated plasma has an electron temperature of about 2–3 eV, typical of many laboratory plasmas.

The spacecraft acts as a Langmuir probe in the thruster plume plasma and will float to a potential where the ion and electron currents from the plasma to the spacecraft body are equal.

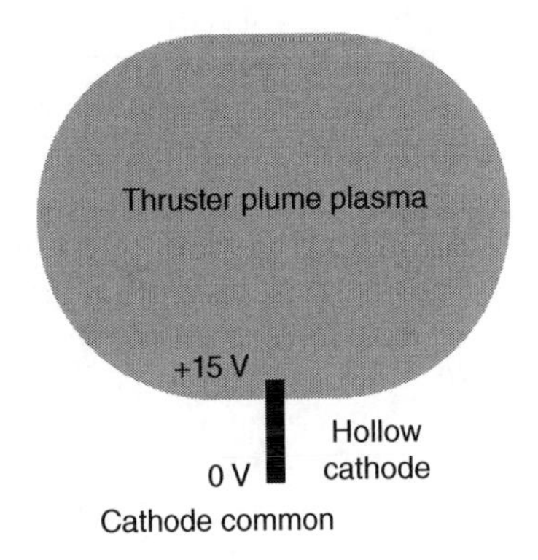

Figure 11-17 The thruster neutralizer hollow cathode generates a plasma typically 10–20 V above cathode common.

As discussed in Chapter 3, plasma electron velocities are much higher than ion velocities, so current balance is achieved by repelling most of the plasma electrons. This balance occurs when the surface is a few times the electron temperature negative of the local plasma potential. If the electric propulsion system were isolated from spacecraft ground by a very high impedance, cathode common would float around 10 V negative with respect to spacecraft ground, as illustrated in Fig. 11-18.

When the spacecraft has exposed surfaces at different voltages, predicting the cathode common floating potential is more difficult. An extreme case would be if the spacecraft solar arrays had a large area at high positive voltage immersed in the thruster plume. Then, to achieve current balance, the high voltage area would be close in potential to the thruster plume plasma. For example, assume that the spacecraft had 100 V solar arrays. Since the cathode common is only about 10 V negative with respect to the thruster plume plasma, cathode common would be 90 V positive compared to spacecraft ground as illustrated in Fig. 11-19.

On operational spacecraft, cathode common will float somewhere between the two extremes, −15 to +90 V for a 100 V bus, depending on the array construction, and may vary with orientation and season. Cathode common potential can be held at a fixed potential with respect to spacecraft chassis ground by tying the electric propulsion system circuit ground to spacecraft ground with a resistor. Plasma currents collected by exposed spacecraft surfaces will flow through the resistor. These currents can be limited by reducing the exposed conducting area in the thruster plumes. The plasma currents are usually quite small. For example, if the charge exchange plasma plume density one meter from the thruster axis is $\sim 10^{14}$ m^{-3}, the conducting area on a square meter of solar array would collect only a few milli-amperes of electron current. A 1 kilo-ohm resistor could then clamp cathode common within a few volts of spacecraft ground.

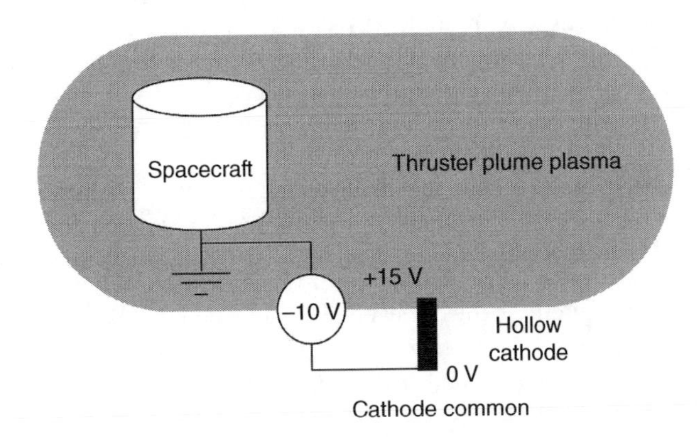

Figure 11-18 Cathode common would float on the order of 10 V negative on a spacecraft with a conducting surface.

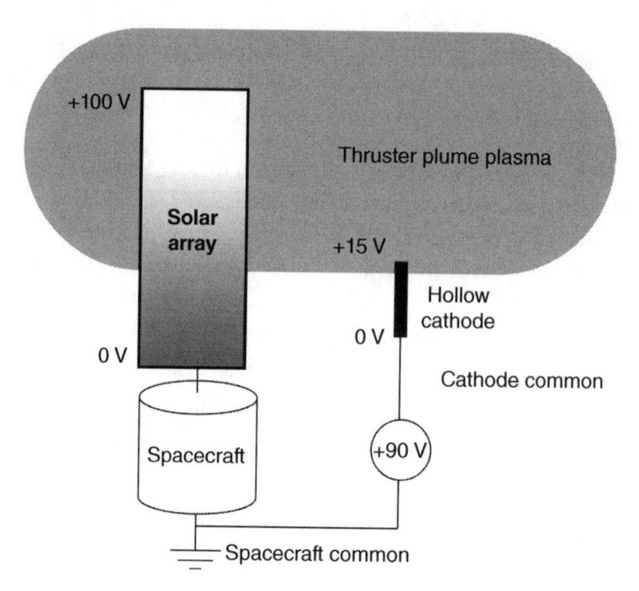

Figure 11-19 A large area of high voltage solar array exposed to the thruster plume causes the cathode common to float the order of the array voltage positive of spacecraft chassis ground.

11.5 Interactions with Payloads

11.5.1 Microwave Phase Shift

Electromagnetic waves interact with plasmas, particularly if the wave frequency is on the order of or lower than the plasma frequency along its path of propagation. In most spacecraft applications, the communications and payload frequencies are so high (>1 GHz) that there is little effect. For a typical thruster, the main beam plume density drops below 10^{15} m^{-3} less than a meter from the thruster, and then it drops even more rapidly at greater distances. The plasma frequency at this density about 1 m from the thruster, from Eq. 3.5-25, is 285 MHz.

As a result, microwave signals with frequencies below a few hundred megahertz could be affected by the thruster plasma-plume. However, even at higher frequencies, highly directional antenna patterns should be analyzed for possible distortion by small phase shifts caused by the plasma. A plane wave with frequency f passing through a plasma with density n_e will undergo a phase shift $\nabla\varphi$ according to the following formula

$$\nabla\varphi \approx \frac{\pi f}{c}\int_0^L \frac{n_e}{n_c}\,dl, \tag{11.5-1}$$

where c is the speed of light and n_c is the critical density at which the plasma density has a plasma frequency equal to the microwave frequency. Since the plasma density drops rapidly with distance from the thruster, the scale length over which the plasma frequency is comparable to the wave frequency is usually small.

11.5.2 Plume Plasma Optical Emission

The optical emissions from ion and Hall thrusters are very weak, but can be measured by sensitive instruments. The only in-space measurement of the optical emissions from a xenon plasma plume generated by an electric propulsion device is from a shuttle flight that had a "plasma contactor" as part of the Space Experiments with Particle Accelerators (SEPAC) [28] flown on the NASA Space Shuttle Mission STS-45. The "plasma contactor" was actually a XIPS 25-cm thruster without the accelerator grids or neutralizer hollow cathode. Plasma and electron current from the discharge chamber were allowed to escape into space, unimpeded by an ion accelerator grid set.

The absolute intensity optical emission spectrum measured in space of the xenon plasma plume of the operating plasma source is shown in Fig. 11-20. The spectrum was measured by the Atmospheric Emissions Photometric Imaging (AEPI) spectrographic cameras. The source was the SEPAC plasma contactor [28, 29] that generated about 2A of singly charged xenon ions from a ring-cusp discharge chamber. The plasma density was about 10^{17} m^{-3} and its temperature was about 5 eV. On leaving the discharge chamber, the quasi-neutral plasma expanded into the much less dense surrounding ionosphere. The spectrum was taken about 15 m from the contactor plume, focusing on the plume about 1.5 m downstream of the contactor exit plane. The apparent broadness of the lines is because of the spectrograph's relatively wide slit [30].

Optical emissions from the SEPAC plasma contactor are higher than the emissions expected from a similarly sized ion thruster for two reasons. First, the plasma contactor ion density is higher because the contactor ions are traveling about a quarter as fast as thruster beam ions. Second, the electrons in the SEPAC plasma contactor plume originate in the discharge chamber and are much hotter than the neutralizer cathode electrons in an ion thruster plume, 5 eV versus 2 eV. As a result, the absolute magnitude of the spectrum in Fig. 11-20 is about two orders of magnitude more intense than one would expect from an operating ion thruster.

The source of the strong visible lines in the xenon spectrum is interesting. Visible emissions from states with allowed transitions to ground contribute very little to the total observed visible spectra. Most of the visible emissions originate from states that do not have allowed transitions to ground. The reason for this is that an optically allowed transition to ground is typically a thousand times more probable than a transition to another excited state. Thus, if allowed, almost every excitation will lead to an ultraviolet (UV) photon. Indeed, most of the radiation from xenon plasmas is in the UV, with only a small part in the visible. Line emissions in the visible are dominated by radiative decay from states where the radiative transitions to ground are forbidden. When an electron collision excites one of these states, it decays though a multi-step process to ground, since the direct

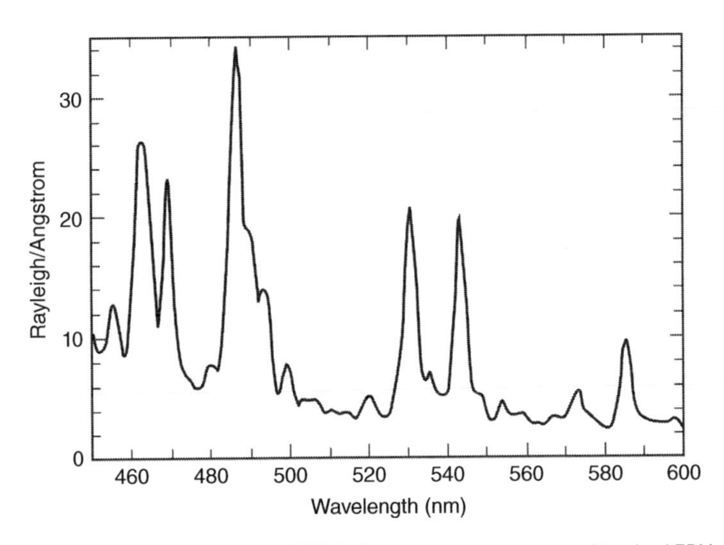

Figure 11-20 Visible xenon spectra from the SEPAC plasma contactor observed by the AEPI hand held camera during Shuttle Mission STS-45 (*Source:* [30]).

radiative decay to ground is forbidden and the collisional decay rate is orders of magnitude slower than the allowed radiative transitions. Although the excitation cross sections from ground to these states are smaller than those to states with optically allowed transitions, the absence of a competing single step decay path to ground allows these states to dominate the visible emissions.

The total power radiated by a thruster plume, both in the visible and UV, can be estimated by assuming that both the ion beam and the neutral gas expand with the same effective cone angle θ. The radius of the beam and neutral plumes as a function of the distance z from the thruster is then

$$R = R_o + z \tan \theta, \tag{11.5-2}$$

where R_o is the initial radius. Assuming a quasi-neutral beam, the ion and electron densities are

$$n_e = n_i = \frac{I_b}{e v_i \pi R^2} \tag{11.5-3}$$

The neutral density is given by

$$n_o = \frac{I_b}{e v_n \pi R^2} \left(\frac{1 - \eta_m}{\eta_m} \right) \tag{11.5-4}$$

where v_n is the neutral velocity and η_m is the thruster mass utilization efficiency. Emission from the neutral gas is proportional to the product of the electron density, the neutral gas density, the electron velocity, and the Maxwellian-averaged excitation cross section:

$$\begin{aligned} P_{\text{emission}} &= \int n_e n_o v_e \langle \sigma_{\text{excite}} \rangle \, e E^* dV \\ &= \int_0^\infty n_e n_o v_e \langle \sigma_{\text{excite}} \rangle \, e \, E^* \, 2\pi R^2 dz \\ &= \int_{R_0}^\infty n_e n_o v_e \langle \sigma_{\text{excite}} \rangle \, e \, E^* \frac{2\pi R^2}{\tan \theta} dR \end{aligned} \tag{11.5-5}$$

where E^* is the energy of the lowest excited state of the neutral gas and the temperature averaged excitation cross section, $\langle \sigma_{\text{excite}} \rangle$ is from Ref. [31], which is contained in Appendix D.

$$\sigma_{\text{excite}}(T_{\text{eV}}) = \frac{19.3 \exp(-11.6/T_{\text{eV}})}{\sqrt{T_{\text{eV}}}} \times 10^{-20} \, m^2 \tag{11.5-6}$$

For example, at 2 eV, the value of $\langle \sigma_{\text{excite}} \rangle$ is about 0.8×10^{-20} m^2. Integrating over the plume volume, assuming that E^* is 10 eV, (approximately the energy of the lowest lying excited state of xenon), and that the neutral temperature is 500 °C, the total radiated power in the NSTAR thruster plume at the full power point ($I_b = 1.76$ A, beam voltage $V_b = 1100$ V) is

$$\begin{aligned} P_{\text{emission}} &= \frac{2 I_b^2}{e v_i v_n \pi R \tan \theta} \left(\frac{1 - \eta_m}{\eta_m} \right) v_e \langle \sigma_{\text{excite}} \rangle E^* \\ &\approx 0.04 \, W \end{aligned} \tag{11.5-7}$$

which is much less than a tenth of a Watt. Emissions in the visible range are usually only about one percent of the total radiated power.

Problems

11.1 An ion thruster 20 cm in diameter produces a Xe^{+} ion beam at 2000 V.

a) If there is no electron neutralization of the beam, what is the maximum current in the beam if the beam diameter doubles in a distance of 1 m?

b) What is the effective angular divergence of this beam?

c) At what current density is electron neutralization required to keep the angular divergence less than 10°?

d) (Hint: Find the radial acceleration using Gauss's Law for the radial electric field in the beam)

11.2 You have just been hired as a propulsion engineer by a spacecraft manufacturer who plans to launch a commercial satellite that uses a 30-cm xenon ion engine operating for station-keeping. The manufacturer plans to perform a costly test to assess whether a 1-mil-thick Kapton coating over a critical spacecraft surface located near the engine will survive 1500 h of thruster operation. You immediately recall that your coursework may allow you to determine the sputtering erosion of the Kapton layer by analysis, and thus possibly save your employer the high cost of performing the test. The spacecraft surface in question is a flat panel located perpendicular to the thruster's r-z plane as shown in Fig. 11-21. The panel length exceeds 6 m.

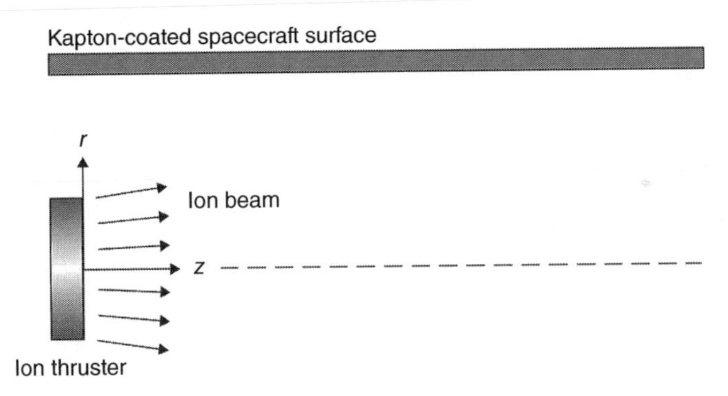

Figure 11-21 Flat panel positioned over an ion thruster plume.

a) Assuming that the ion beam consists of singly charged ions only:

b) Use the equations in your textbook to express the ion beam density, n, as a function of spatial coordinates (r, z). Produce contour plots of the beam density within a radius of $r = 0.5$ m from the center of the thruster exit ($r = 0$, $z = 0$). Assume that the ion density n_0 at ($r = 0$, $z = 0$) is 4×10^{15} m^{-3} and that the ion beam velocity V_0 is 40 km/s. Also, assume that $u_{Bohm}/V_0 = 0.03$.

c) Derive an expression for the radial component of the ion beam flux, $\Gamma_r = nu_r$, as a function of spatial coordinates (r, z). Plot the radial ion beam flux as a function of z for $r = 0.3$, 0.4 and 0.5 m.

d) Perform a literature search to find the sputtering yield Y of Kapton as a function of ion energy/ion charge, E, and incidence angle β, and then plot Y for 300-V and 1000-V ions between 0–90° of incidence angle. (Hint: The sputtering yield for many materials is usually expressed as $Y(E, \beta) = (a + bE) f(\beta)$, where $f(\beta)$ is a polynomial function and a,b are constants.)

e) Compute the erosion rate in (Å/sec) caused by the main ion beam along the Kapton plate (in the r-z plane), as a function of z, for $r = 0.3$, 0.4 and 0.5 m. Assume that the molecular weight AW of Kapton is 382 g/mol and that its mass density is 1.42 g/cm^3. (Å $= 10^{-10}$ m)

f) If the panel was placed at $r = 0.5$ m from the thruster how long would it take for the main ion beam to erode completely the Kapton layer?

g) For partial credit, choose one answer to the following question: how would you advise your boss based on your results?
   i) The Kapton coating will be just fine; no need to perform a test. Build the spacecraft as is (panel radial location = 0.5 m).
   ii) The Kapton coating will not survive; we must consider changing the location of the panel relative to the thruster.
   iii) The Kapton coating will not survive; why don't we just use chemical propulsion?
   iv) I must perform more calculations.
   v) ii and iv
   vi) The Kapton coating will not survive, the mission cannot be launched.

11.3 In Section 11.3.3, the differential scattering cross section was introduced.

a) What is its physical meaning and what are its units?

b) Figure 11-22 represents the basic picture of a classical scattering trajectory, viewed from the frame of reference of the target particle. In the sketch, R is the distance of closest approach, b is the impact parameter, and θ is the defection angle. For elastic scattering the conservation equations of angular momentum and energy allow us to predict the deflection angle as follows:

$$\theta = \pi - 2b \int_R^{\infty} \frac{dr}{r^2 \left[1 - \left(b/r\right)^2 - \frac{\Phi(r)}{E}\right]}$$

where $\Phi(r)$ is the interaction potential, which is related to the force field between the colliding particles. E is the (relative) energy of the incident particles. The differential $(d\sigma/d\Omega)$ and total σ cross-sections are given by,

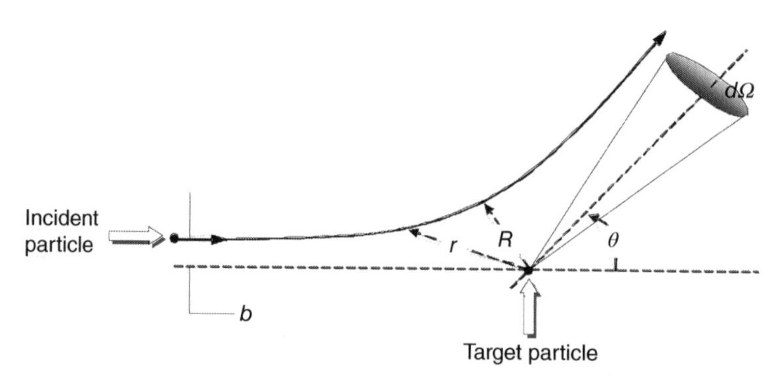

Figure 11-22 Classic scattering diagram for an incident particle on a target particle.

$$\frac{d\sigma}{d\Omega} = \frac{b}{\sin\theta}\left|\frac{db}{d\theta}\right|$$

$$\sigma = \int \frac{d\sigma}{d\Omega} d\Omega$$

Compute the differential and total cross sections for (i) collisions between hard spheres of diameter d, and (ii) a repulsive force field between particles that varies as k/r^2 (k is a constant).

11.4 The most general elastic collision process between two particles of unequal masses m_1 and m_2, velocity vectors before the collision, $\mathbf{u}_1$ and $\mathbf{u}_2$, and after the collision, $\mathbf{u}_1'$ and $\mathbf{u}_2'$, can be represented by the geometrical construction in Fig. 11-23 using the following definitions:

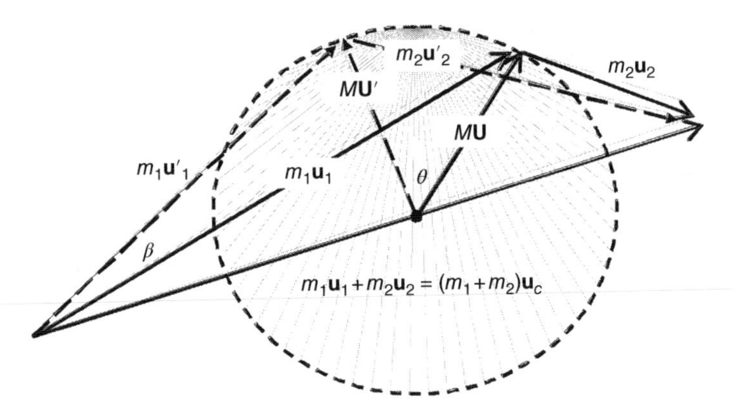

Figure 11-23 Center of mass depiction of an elastic scattering event.

Relative velocities: $\mathbf{U} = \mathbf{u}_1 - \mathbf{u}_2$, $\mathbf{U}' = \mathbf{u}_1' - \mathbf{u}_2'$
Center-of-mass (CM) velocity: $\mathbf{u}_c = (m_1\mathbf{u}_1 + m_2\mathbf{u}_2)/(m_1 + m_2)$
Reduced mass: $M = m_1 m_2/(m_1 + m_2)$

a) In the case of equal masses $m_1 = m_2 = m$ and one stationary particle, $\mathbf{u}_2 = 0$, draw the new geometrical construction. What is the relationship between the scattering angle in the center-of-mass (CM) frame, θ, and the scattering angle in laboratory frame, β?
b) Convert the center-of-mass differential cross section $d\sigma/d\Omega_{CM}$ into the laboratory frame of reference, $d\sigma/d\Omega_L$.

11.5 Derive Eq. 11.5-1 for the phase shift of electromagnetic radiation passing through a plasma. (Hint: Assume the phase shift is small)

11.6 A spacecraft has a 32 GHz communications system that passes into the diverging plume of an ion propulsion system 1 m from the thruster. If the NSTAR thruster beam has an initial radius of 15 cm and produces 1.76 A of xenon ions at 1100 V with a 10° half-angle divergence from the initial area, what is the total phase shift in degrees produced when the thruster is turned on or off?

References

1 B. M. Anzel, "Method and Apparatus for a Satellite Station Keeping," US Patent #5,443,231, August 22, 1995.

2 D. M. Goebel, M. Martinez-Lavin, T. A. Bond, and A. M. King, "Performance of XIPS Electric Propulsion in Station Keeping of the Boeing 702 Spacecraft," AIAA-2002-4348, 38th Joint Propulsion Conference, Indianapolis, July 7–10, 2002.

3 J. R. Brophy, D. E. Garner, and S. Mikes, "Dawn Ion Propulsion System: Initial Checkout After Launch," *Journal of Propulsion and Power*, vol. 25, no. 6, pp. 1189–1202, 2009; doi:10.2514/1.40480

4 J. S. Snyder, V. Chaplin, D. M. Goebel, R. R Hofer, A. Lopez Ortega, I. G. Mikllides, T. Kerl, G. Lenguito, F. Aghazadeh, and I. Johnson, "Electric Propulsion for the Psyche Mission: Development Activities and Status," AIAA-2020-3607, AIAA Propulsion and Energy 2020 Forum, Virtual Event, August 24–28, 2020; doi:10.2514/6.2020-3607.

5 I. D. Boyd and A. Ketsdever, "Interactions Between Spacecraft and Thruster Plumes," *Journal of Spacecraft and Rockets*, vol. 38, no.3, p. 380, 2001.

6 J. Wang, D. Brinza, and M. Young, "Three-Dimensional Particle Simulations of Ion Propulsion Plasma Environment for Deep Space 1," *Journal of Spacecraft and Rockets*, vol. 38, no.3, pp. 433–440, 2001.

7 I. G. Mikellides; G. A. Jongeward; I. Katz, and D. H. Manzella, "Plume Modeling of Stationary Plasma Thrusters and Interactions with the Express-A Spacecraft," *Journal of Spacecraft and Rockets*, vol. 39, no.6, pp. 894–903, 2002.

8 I. G. Mikellides, M. J. Mandell, R. A. Kuharski, D. A. Davis, B. M. Gardner, and J. Minor, "The Electric Propulsion Interactions Code (EPIC)," AIAA 2003-4871, 39th AIAA Joint Propulsion Conference, Huntsville, July 20–23, 2003.

9 A. Lopez-Ortega, I. Katz, I. G. Mikellides, and D. M. Goebel, "Self-Consistent Model of a High Power Hall Thruster Plume," *IEEE Transactions on Plasma Science*, vol. 43, no. 9, pp. 2875–2886, 2015.

10 D. E. Brinza, M. D. Henry, A. T. Mactutis, K. P. McCarty, J. D. Rademacher, T. R. van Zandt, P. Narvaez, J. J. Wang, B. T. Tsurutani, I. Katz, V.A. Davis, S. Moses, G. Musmann, F. Kuhnke, I. Richter, C. Othmer, and K.H. Glassmeier, "An Overview of Results from the Ion Diagnostics Sensors Flown on DS1," AIAA-2001-0966, presented at the 39th AIAA Aerospace Sciences Meeting & Exhibit, Reno, NV, January 8–11, 2001.

11 V. A. Davis, I. Katz, M. J. Mandell, D. E. Brinza, M. D. Henry, J. J. Wang, and D. T. Young, "Ion Engine Generated Charge Exchange Environment - Comparison Between NSTAR Flight Data and Numerical Simulations," AIAA-2000-3529, 36th AIAA Joint Propulsion Conference, Huntsville, AL, July 16–19, 2000.

12 J. Pollard, K. D. Diamant, V. Khayms, L. Werthman, D.Q. King, and K.H. DeGrys, "Ion Flux, Energy, Charge-State Measurements for the BPT-4000 Hall Thruster,"AIAA-2001-3351, 37th AIAA Joint Propulsion Conference, Salt Lake City, UT, July 8–11, 2001.

13 I. Katz, G. Jongeward, V. Davis, M. Mandell, I. Mikellides, R. Dressler, I. Boyd, K. Kannenberg, J. Pollard, and D. King, "A Hall Effect Thruster Plume Model Including Large-Angle Elastic Scattering," AIAA-2001-3355, 37th AIAA Joint Propulsion Conference, Salt-Lake City, UT, July 8–11, 2001.

14 I. G. Mikellides, I. Katz, R. A. Kuharski, and M. J. Mandell, "Elastic Scattering of Ions in Electrostatic Thruster Plumes," *Journal of Propulsion and Power*, vol. 21, no. 1, pp. 111–118, 2005.

15 D. E. Parks and I. Katz, "A Preliminary Model of Ion Beam Neutralization," *Electric Propulsion and Its Applications to Space Missions*", edited by R. C. Finke, Progress in Astronautics and Aeronautics, vol. 79, pp. 1–10, 1981.

16 J. Ashkenazy and A. Fruchtman, "A Far-Field Analysis of Hall Thruster Plume," Division of Plasma Physics (DPP) and International Congress on Plasma Physics Meeting (ICPP), Paper BMI-004, Quebec City, October 2000.

17 E. J. Pencil, T. Randolph, and D. H. Manzella, "End-of-Life Stationary Plasma Thruster Far-Field Plume Characterization," AIAA-1996-2709, 32nd AIAA Joint Propulsion Conference, Lake Buena Vista, July 1–3, 1996.

18 J. E. Polk, D. M. Goebel, J. S. Snyder, and A. C. Schneider, "Performance and Wear Test Results for a 20-kW Class Ion Engine with Carbon-Carbon Grids," AIAA_2005-4393, 41th AIAA Joint Propulsion Conference, Tucson, July 11–14, 2005.

19 J. S. Miller, S. H. Pullins, D. J. Levandier, Y. Chiu, and R. A. Dressler, "Xenon Charge Exchange Cross Sections for Electrostatic Thruster Models," *Journal of Applied Physics*, vol. 91, no. 3, pp. 984–991, 2002.

20 F. Cichocki, A. Dominguez-Vazquez, M. Merino, and E. Ahedo, "Hybrid 3D Model for the Interaction of Plasma Thruster Plumes with Nearby Objects," *Plasma Sources Science & Technology*, vol. 26, no. 12, 2017.

21 A. Lopez Ortega, I. Katz, I. G. Mikellides, and D. M. Goebel, "Self-Consistent Model of a High-Power Hall Thruster Plume," *IEEE Transactions on Plasma Science*, vol. 43, no. 9, pp. 2875–2886, 2015; doi:10.1109/Tps.2015.2446411.

22 I. G. Mikellides and I. Katz, "Numerical Simulations of Hall-Effect Plasma Accelerators on a Magnetic-Field-Aligned Mesh," *Physical Review E*, vol. 86, no. 4, p. 046703, 2012; doi:10.1103/Physreve.86.046703

23 I. Katz and I. G. Mikellides, "Neutral Gas Free Molecular Flow Algorithm Including Ionization and Walls for Use in Plasma Simulations," (in English), *Journal of Computational Physics*, vol. 230, no. 4, pp. 1454–1464, 2011.

24 D. Herman, T. Gray, I. Johnson, S. Hussein, and T. Winkelmann, "Development and Qualification Status of the Electric Propulsion Systems for the NASA PPE Mission and Gateway," IEPC-2022-465, 37th International Electric Propulsion Conference, Boston, MA, June 19–23, 2022.

25 J. R. Beattie and S. Kami, "Advanced Technology 30-cm Diameter Mercury Ion Thruster," AIAA-82-1910, 16th International Electric Propulsion Conference, New Orleans, LA, November 17–19, 1982.

26 R. Wirz and I. Katz, "XIPS Ion Thruster Grid Erosion Predictions for Deep Space Missions," 30th International Electric Propulsion Conference, Florence, Italy, September 17–20, 2007.

27 R. Hofer, L. Johnson, D. M. Goebel, and R. Wirz, "Effect of Internally-Mounted Cathodes on a Hall Thruster Plume Properties," *IEEE Transactions on Plasma Science*, vol. 36, no. 5, pp. 2004–2014, 2008.

28 J. R. Beattie, J. A. Marshall, J. L. Burch, and W. C. Gibson, "Design, Qualification, and on-Orbit Performance of the ATLAS Plasma Contactor," IEPC-93-010, International Electric Propulsion Conference, Seattle, September 13–15, 1993.

29 J. R. Beattie, W. S. Williamson, J. N. Matossian, E. J. Vourgourakis, and J. L. Burch, "High-Current Plasma Contactor Neutralizer System," AIAA-1989-1603, 3rd International Conference on Tethers in Space – Toward Flight, San Francisco, May 17–19, 1989.

30 I. Katz, D. E. Parks, B. M. Gardner, S. B. Mende, H. I. Collin, D. H. Manzella and R. M. Myers, "Spectral Line Emission by the SEPAC Plasma Contactor: Comparison Between Measurement & Theory," AIAA-95-0369, 33rd Aerospace Sciences Meeting and Exhibit, Reno, NV, January 9–12, 1995.

31 M. Hayashi, "Determination of Electron-Xenon Total Excitation Cross-Sections, from Threshold to 100-eV, from Experimental Values of Townsend's a," *Journal of Physics D: Applied Physics*, vol. 16, no. 4, pp. 581–589, 1983.

Chapter 12

Flight Electric Thrusters

12.1 Introduction

Electric thruster technology development programs continue to improve the performance of these engines. It is worthwhile to survey the state-of-the-art thrusters that have flown to date. Table 12-1 shows a list of the significant electric thruster first flights [1, 2]. Thousands of ion and Hall thrusters are presently in use in space in communications satellites and deep space exploration missions. In this chapter, thrusters that have flown in the last 25 years in satellite station-keeping and spacecraft prime-propulsion applications are described. These thrusters are ion thruster and Hall thruster systems that use xenon and krypton as the propellant. SpaceX has launched thousands of Earth-orbiting spacecraft with a krypton Hall thruster on board each, but the performance parameters are not public so these are not included in Table 12-1. The parameters given for the thrusters include the neutralizer cathode flow rates, since the total mass utilization efficiency is important at the system level for flight operation on satellites and spacecraft.

12.2 Ion Thrusters

The first modern inert-gas ion thrusters to fly were intended for station-keeping applications on geosynchronous satellites and were developed by Mitsubishi Electric Corporation (MELCO) for use on the Japanese "Engineering Test Satellite (ETS-6)" in 1994 [3, 4]. These 13-cm Kaufman thrusters produced nominally 20 mN of thrust at an Isp of about 2400 s. Despite launch vehicle problems that caused the satellite to fail to reach its planned orbit, the thrusters were successfully operated in orbit. The same electric propulsion subsystem was launched on the COMETS satellite in 1996, which also failed to reach its planned orbit. Development of ion thrusters for communications satellite station-keeping applications is continuing at JAXA in Japan. The first successful use of ion thrusters in a commercial station-keeping application was the Hughes 13-cm Xenon Ion Propulsion System (XIPS©) [5, 6], which was launched into orbit in 1997 on the Hughes PAS-5 satellite. The XIPS system utilizes two fully redundant subsystems, each consisting of two thrusters, a power supply, and a xenon gas controller. The performance parameters for the 13 cm XIPS thruster are shown in Table 12-2. The thrusters produce nominally 18 mN of thrust at an Isp of 2800 s and a total efficiency of about 51%. A schematic of the 13 cm XIPS thruster is shown in Fig. 12-1, and a photograph of the thruster is shown in Fig. 12-2. Over 56 of these thrusters were launched into orbit and successfully used for North–South station keeping on Hughes and Boeing communications satellites.

Fundamentals of Electric Propulsion, Second Edition. Dan M. Goebel, Ira Katz, and Ioannis G. Mikellides.

Table 12-1 List of selected electric propulsion flights.

Mission	Purpose	Source	Launch Date	Thruster	Propellant	Thrust (mN)	Isp (s)
SERT-1	Technology test	US	1964	Ion	Hg Cs	28 5.6	4900 8050
Zond-2	Exploration	USSR	1964	PPT	Teflon	2	410
SERT-II	Technology test	US	1970	Ion Kaufman	Hg	28	4200
Meteor 1	Meteorology	USSR	1971	Hall SPT-50	Xe	20	1100
Intelsat V2	Communication	US	1980	Resistojet	Hydrazine	0.45	300
Plasma	Communication	USSR	1987	Hall SPT-70	Xe	40	1500
Telstar 401	Communication	US	1993	Arcjet	Hydrazine	250	500
GALS	Communication	USSR	1995	Hall SPT-100	Xe	80	1600
Hughes	Communication	US	1997	Ion XIPS-13	Xe	17.2	2507
Deep Space 1	Technology test	US	1998	Ion NSTAR	Xe	20–90	3100
Hughes	Communication	US	1999	Ion XIPS-25	Xe	80–165	3500
Artemis	Communication	ESA	2001	Ion Kaufman rf-RIT	Xe Xe	18 15	3200 3400
SMART-1	Technology test	ESA	2003	Hall	Xe	67	1540
Hayabusa-1	Exploration	Japan	2003	Ion μ10	Xe	8	3000
Maxar	Communication	US	2004	Hall SPT-100	Xe	80	1600
ETS-VIII	Communication	Japan	2006	Kaufman	Xe	21–23	2600
Dawn	Exploration	US	2007	Ion NSTAR	Xe	20–90	3100
GOCE	Earth science	ESA	2009	Ion T5	Xe	1–20	3000
Lockheed M.	Communication	US	2010	Hall XR-5	Xe	290	2020
Hayabusa-2	Exploration	Japan	2014	Ion μ10	Xe	10	3000
BepiColombo	Exploration	ESA	2018	Ion T6	Xe	145	3900
Maxar	Communication	US	2018	Hall SPT-140	Xe	170–270	1750
Safran	Communication	France	2021	PPS5000	Xe	267–308	1973
DART	Technology test	US	2021	NEXT-C	Xe	235	4190

Table 12-2 13-cm XIPS©Performance.

Parameter	Station keeping
Active grid diameter (cm)	13
Thruster input power (W)	421
Average Isp (seconds)	2507
Thrust (mN)	17.2
Total efficiency (%)	50.0
Mass utilization efficiency (%)	77.7
Beam voltage (V)	750
Beam current (A)	0.4

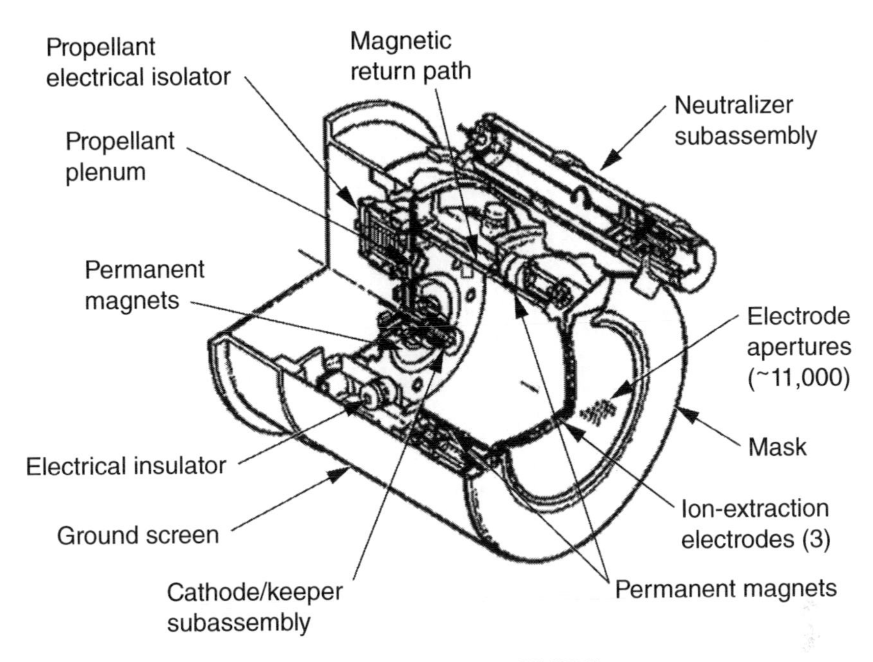

Figure 12-1 Schematic of the 13 cm XIPS© thruster (*Source:* [5]/IEPC).

The next ion thruster to fly was NASA's NSTAR ion engine [7, 8], which is a ring-cusp, DC electron-bombardment discharge thruster with an active grid diameter of 28.6 cm. NSTAR was developed and manufactured by a team of NASA GRC, JPL and Hughes/Boeing EDD, and launched in 1998 on the Deep Space 1 spacecraft. This ion engine has arguably been the most analyzed and tested ion thruster in history, with over 16,000 h of operation in space, over 40,000 h of life testing, and hundreds of papers published on its design and performance. NSTAR was operated over a wide throttle range in the DS1 application from a minimum input power to the PPU of 580 W to a maximum power of over 2550 W. The Extended Life Test of this thruster at JPL demonstrated 30,252 h of operation distributed across several of the throttle levels, and was terminated with the engine still running to provide life status and data for the subsequent DAWN mission [9]. The throttle table used on DS1, with parameters for the NSTAR thruster from a review by Brophy [8], is shown in Table 12-3. A photograph of the NSTAR engine, which was manufactured by Hughes Electron Devices Division (now Stellant Systems, Torrance, CA), is shown in Fig. 12-3.

Figure 12-2 Photograph of the 13 cm XIPS© ion thruster (*Source:* Courtesy of Stellant Systems).

The next ion thruster technology launched was designed for both orbit-raising and station-keeping applications on a commercial communications satellite. The 25-cm XIPS© thruster was

Table 12-3 NSTAR throttle table.

NSTAR throttle level	PPU input power (W)	Engine input power (W)	Calculated thrust (mN)	Specific impulse (s)	Total efficiency (%)
15	2567	2325	92.7	3127	61.8
14	2416	2200	87.9	3164	62.4
13	2272	2077	83.1	3192	63.0
12	2137	1960	78.4	3181	62.8
11	2006	1845	73.6	3196	63.1
10	1842	1717	68.4	3184	62.6
9	1712	1579	63.2	3142	61.8
8	1579	1456	57.9	3115	61.1
7	1458	1344	52.7	3074	59.6
6	1345	1238	47.9	3065	59.0
5	1222	1123	42.6	3009	57.4
4	1111	1018	37.4	2942	55.4
3	994	908	32.1	2843	52.7
2	825	749	27.5	2678	48.7
1	729	659	24.6	2382	47.2
0	577	518	20.7	1979	42.0

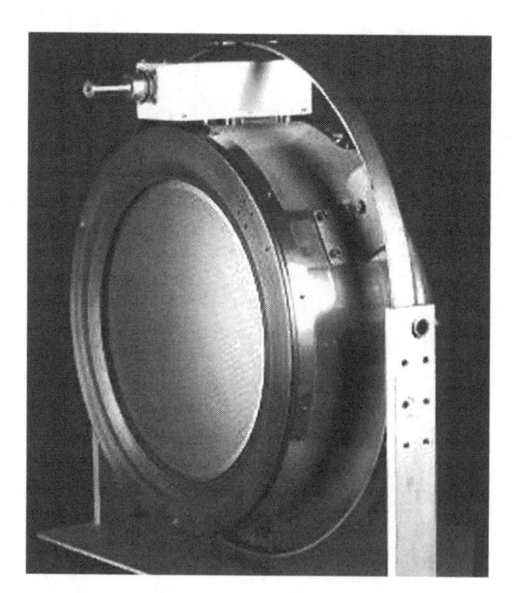

Figure 12-3 Photograph of the NASA NSTAR ion thruster (*Source:* Courtesy of Stellant Systems).

first launched in 1999 on a Hughes/Boeing 702 satellite. Although the 25-cm XIPS ion thruster was developed [10] at Hughes Research Laboratories in the same time frame as the NSTAR engine and has a similar basic design as the 13-cm XIPS thruster, the 25-cm thruster entered production after the 13-cm version, and incorporated sufficient improvements to be considered a second-generation device. A photograph of the 25-cm XIPS thruster, which was manufactured by Hughes Electron Devices Division (now Stellant Systems, Torrance, CA), is shown in Fig. 12-4. There are presently well over a hundred XIPS-25 cm thrusters in orbit that have been successfully used with over 99% reliability for orbit raising and station keeping, and momentum control.

The initial operation of the 25-cm thrusters in space on the 702 satellites was described in 2002 [11]. After launch, these thrusters are first used for orbit raising and then provide all the propulsion requirements for orbit control, including North–South and East–West station keeping, attitude control, and momentum dumping. The ion thrusters are also used for any optional station change strategies and will ultimately be used for de-orbit at the end of the satellite's lifetime. The "high power" orbit insertion mode requires nearly continuous operation by two of the thrusters for times of 500–1000 h, depending on the launch vehicle and satellite weight. This mode utilizes about 4.5 kW of bus power to each thruster generate a 1.2 kV, 3 A ion beam, which produces 165 mN thrust at a Isp of about 3500 s. Once orbit insertion is completed, each of the four thrusters is fired once daily for an average of about 45 min in a "low power", 2 kW mode for station keeping. In this mode, the beam voltage is kept the same, and the discharge current and gas flow are reduced to generate a 1.2 kV, 1.5 A beam that produces nominally 79 mN of thrust at an Isp of 3400 s. The thruster performance parameters are shown in Table 12-4. Recently, tests by the manufacturer and JPL have demonstrated that the XIPS engine and PPU can be throttled from a PPU input power level of 400 W to over 5 kW. Over this range, the performance of the 25-cm thruster significantly exceeds the NSTAR thruster performance [12].

Figure 12-4 Photograph of the 25-cm XIPS© ion thruster (*Source:* Courtesy of Stellant Systems).

The next flight of ion thrusters was on the European Space Agency Artemis spacecraft launched in 2001. Artemis carried four ion thruster assemblies, two EITA (Electron-bombardment Ion Thruster Assembly) systems manufactured by Astrium UK, and two RITA (Radio-frequency Ion Thruster Assembly) systems developed by Astrium Germany. The EITA system, also called the UK-10 system, used copies of the T5 thruster [13, 14], and the RITA system used RIT–10 ion thrusters [15, 16]. Artemis was intended to be launched into a geosynchronous orbit, but a malfunction of the launcher's upper stage placed the satellite into a lower orbit. The ion thrusters were used in an unplanned orbit-raising role to rescue the spacecraft from the lower 31,000 km parking orbit and raise the spacecraft to the proper geosynchronous orbit. The thrusters then successfully performed standard EP station-keeping activities.

Table 12-4 25-cm XIPS©performance parameters.

Parameter	Low power station keeping	High power orbit raising
Active grid diameter (cm)	25	25
Thruster input power (kW)	2.0	4.2
Average Isp (seconds)	3420	3550
Thrust (mN)	80	165
Total efficiency (%)	67	68.8
Mass utilization efficiency (%)	80	82.5
Electrical efficiency (%)	87.1	87.5
Beam voltage (V)	1215	1215
Beam current (A)	1.45	3.05

Table 12-5 T5 Kaufman thruster performance parameters.

Parameter	Station keeping
Active grid diameter (cm)	10
Thruster input power (W)	476
Nominal Isp (seconds)	3200
Thrust (mN)	18
Total efficiency (%)	55
Mass utilization efficiency (%)	76.5
Electrical efficiency (%)	76.6
Beam voltage (V)	1100
Beam current (A)	0.329

The EITA/UK-10/T5 thruster is a 10-cm Kaufman thruster [13] that was manufactured by QinetiQ in England. The performance of the T5 Kaufman thruster in station-keeping applications [13] is given in Table 12-5. A schematic of a generic Kaufman thruster was shown in Chapter 4, and a photograph of the T5 thruster is shown in Fig. 12-5. The T5 thruster generates an 1100 V, 0.329 A xenon ion beam that produces about 18 mN of thrust at a nominal Isp of 3200 s with a total efficiency of about 55%.

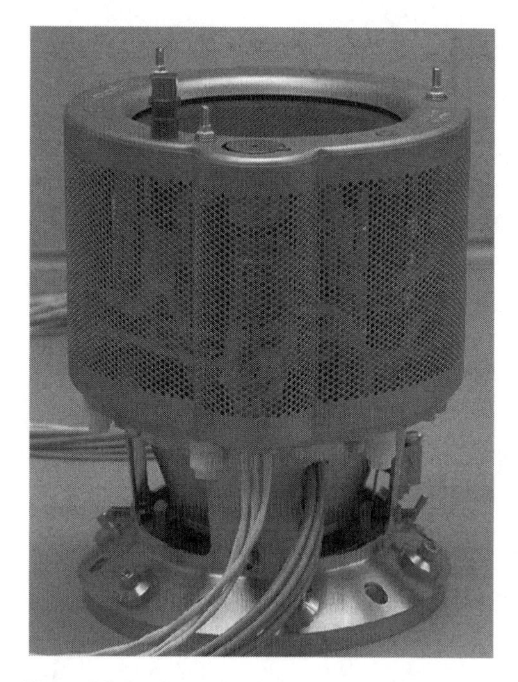

Figure 12-5 Photograph of the T5 Kaufman ion thruster (*Source:* Photo courtesy of Qinetiq Limited [14]).

The RITA system used a RIT-10 rf ion thruster originally developed [15] at the University of Giessen in Germany and manufactured [16] for Artemis by Astrium in Germany. The performance of the RIT-10 thruster in station-keeping applications [16] is shown in Table 12-6. A schematic of a generic rf thruster was shown in Chapter 4, and a photograph of the RIT-10 rf ion thruster from [16] is shown in Fig. 12-6. The RIT-10 thruster generates a 1500 V, 0.234 A xenon ion beam that produces 15 mN of thrust at an Isp of 3400 s and a total efficiency in excess of 51%.

The Institute of Space and Astronautical Science (ISAS) of the Japan Aerospace Exploration Agency (JAXA) launched four μ10 electron cyclotron resonance (ECR) ion thrusters on the Hayabusa (formerly Muses-C) spacecraft [18] in 2003. These 10-cm grid-diameter thrusters [19, 20] successfully provided primary propulsion for this asteroid sample return mission. An upgraded version of the μ10 with higher thrust and performance [21] was then used the Hayabusa-2 mission that launched in 2014. The

Table 12-6 RIT-10 rf thruster performance parameters.

Parameter	Station keeping
Active grid diameter (cm)	10
Thruster input power (W)	459
Nominal Isp (seconds)	3400
Thrust (mN)	15
Total efficiency (%)	52
Mass utilization efficiency (%)	69.3
Beam voltage (V)	1500
Beam current (A)	0.234

thruster uses 4.2 GHz microwaves to produce the main plasma in the thruster and drive the electron neutralizer. A schematic drawing of the thruster was shown in Chapter 4. The performance of the upgraded μ10 thruster is shown in Table 12-7, and a photograph of the μ10 thruster from [22] is shown in Fig. 12-7. The 10-cm ECR ion thruster generates a 1500 V, 0.178 A xenon ion beam that produces 12 mN of thrust at an Isp of 3122 s and a peak total efficiency of 39.6%.

Figure 12-6 Photograph of the RIT-10 rf ion thruster.

The Japan Aerospace Exploration Agency (JAXA) launched four 20 mN-class Kaufman ion thrusters developed by Mitsubishi Electric Corporation on the Engineering Test Satellite VIII (ETS-VIII) [23] in 2006. The 12-cm grid-diameter Kaufman thrusters provide North–South station keeping for this large geosynchronous communications satellite. The performance of the 12-cm Kaufman thruster [24] is shown in Table 12-8, and a photograph of the thruster from [25] is shown in Fig. 12-8. At its nominal

Table 12-7 μ10 ECR Microwave ion thruster performance.

Parameter	Primary propulsion
Active grid diameter (cm)	10
Thruster input power (W)	400
Average Isp (seconds)	3100
Thrust (mN)	12
Total efficiency (%)	39.6
Discharge loss (eV/ion)	162
Beam voltage (V)	1500
Beam current (A)	0.178

Figure 12-7 Photograph of the μ10 ECR microwave discharge ion thruster and microwave neutralizer (*Source:* [22]. Courtesy of Institute of Space and Astronautical Science: ISAS).

operating condition, the thruster generates a 996-V, 0.432–0.480-A xenon ion beam that produces 20.9–23.2 mN of thrust at an Isp of 2402 to 2665 s and a total efficiency of about 46–50%.

The next flight of ion thrusters were T6 Kaufman ion thrusters [26–28] developed by QinetiQ and launched on the BepiColombo mission in 2018 [29]. The 20-cm grid-diameter Kaufman thrusters operating singly and in pairs to provide prime propulsion for this mission. The performance of the T6 Kaufman ion thruster [27, 28] is shown in Table 12-9, and a photograph of the thruster is shown in Fig. 12-9.

A flight demonstration of the NASA Evolutionary Xenon Thruster (NEXT) [30], developed by NASA Glenn Research Center (GRC), was performed by the DART Mission [31] that launched in 2021. The NEXT-C thruster, manufactured by Aerojet, successfully operated in space at about 3 kW of power [32], limited by the power available from the solar arrays. The NEXT thruster is capable of throttling from 0.5 kW to 6.9 kW, producing 25–236 mN of thrust at an Isp ranging from 1400 to 4190 s. The performance capabilities of the NEXT thruster are listed in Table 12-10, and a photograph of the NEXT ion thruster is shown in Fig. 12-10.

There are a significant number of ion thrusters in development world-wide for prime propulsion and satellite station-keeping applications. Since these thrusters have not flown as of this date, they will not be covered in detail and only mentioned here. NASA's Jet Propulsion Laboratory (JPL) led the development of the 25-kW Nuclear Electric Xenon Ion thruster System (NEXIS) [33], which produced the highest efficiency (>81%) xenon ion thruster developed to date. NASA's GRC also led the development of the 30 kW High Power Electric Propulsion (HiPEP) thruster [34], which featured a rectangular geometry with both rf and DC hollow cathode plasma-production versions. In Japan, the Institute of Space and Astronautical Science is developing a 20-cm diameter, 30-mN class microwave

Table 12-8 ETS-8 Kaufman thruster performance parameters.

Parameter	NS-station keeping
Active grid diameter (cm)	12
Thruster input power (W)	541–611
Nominal Isp (seconds)	2402–2665
Thrust (mN)	20.9–23.2
Total efficiency (%)	45.6–49.7
Mass utilization efficiency (%)	66.2–73.5
Beam voltage (V)	996
Beam current (A)	0.43–0.48

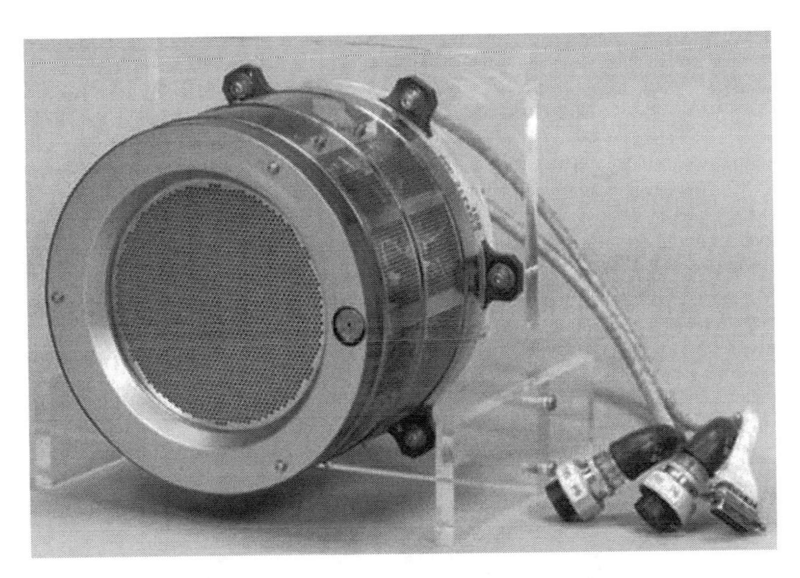

Figure 12-8 Photograph of the ETS-8 Kaufman ion thruster (*Source:* [25]).

Table 12-9 T6 Kaufman ion thruster performance parameters.

	Nominal throttle levels			
Parameter	**75 mN**	**100 mN**	**125 mN**	**145 mN**
Active grid diameter (cm)	20	20	20	20
Thruster input power (kW)	2.52	3.28	4.04	4.62
Specific impulse (seconds)	3720	3880	3990	3980
Thrust (mN)	76	102	128	148
Total efficiency (%)	55.2	59.3	61.8	62.5
Mass utilization efficiency (%)	69.3	70.7	71.0	71.2
Discharge loss (eV/ion)	314	280	263	245
Beam voltage (V)	1850	1850	1850	1849
Beam current (A)	1.13	1.50	1.88	2.17

Figure 12-9 Photograph of the T6 Kaufman ion thruster (*Source:* Courtesy of QinetiQ Ltd.).

Table 12-10 NEXT thruster performance.

Parameter	Lowest power	High power
Active grid diameter (cm)	36	36
Thruster input power (kW)	0.55	6.9
Specific impulse (seconds)	1400	4190
Thrust (mN)	25.5	235
Total efficiency (%)	32.0	70.4
Discharge loss (eV/ion)	224.8	138.1
Beam voltage (V)	275	1800
Beam current (A)	1.0	3.52

Figure 12-10 Photograph of the NEXT ion thruster (*Source:* NASA/GRC, Public Domain).

discharge ion thruster [35]. In Germany, Astrium is developing higher power rf ion thrusters for station-keeping and orbit-raising applications [36]. Finally, ring-cusp and rf ion thrusters are being miniaturized for applications that require thrust levels of the order of 1 mN or less. The 3-cm Miniature Xenon Ion thruster (MiXI) [37] uses a DC discharge, ring-cusp geometry with closely spaced ion optics to produce up to 3 mN of thrust at beam voltages of up to 1200 V. The micro-Newton rf Ion Thruster (μN-RIT) [38] use a low frequency (≈1 MHz) rf discharge scaled down to 2–4 cm in diameter to produce precision thrust levels as low as 20 μN at beam voltages in excess of 1 kV. There are many additional small research and development programs at universities and in small businesses, but these are too numerous to be covered here.

12.3 Hall Thrusters

The most successful and extensive electric propulsion development and application has been by the Russians flying Hall thrusters for station keeping on satellites [39, 40]. Over 140 Hall thrusters have been operated in space since 1971 when the Soviets first flew a pair of Hall thrusters called Stationary Plasma Thrusters (SPT) on the Meteor satellite [40]. This name is translated from the Russian literature, but refers to the continuous operation ("stationary") of the Hall thruster in comparison to the pulsed-plasma thrusters (PPT) that the Russians had previously tested and flown in the 1960s [40]. SPT thrusters for satellite applications have been developed with different sizes characterized by the outside diameter of the plasma discharge slot of 50 to over 140 mm [40].

Table 12-11 STP Hall thruster performance.

Parameter	SPT-50	SPT-70	SPT-100	SPT-140
Channel outside diameter (cm)	5	7	10	14
Thruster input power (W)	350	700	1350	5000
Average Isp (seconds)	1100	1500	1600	1750
Thrust (mN)	20	40	80	280
Total Efficiency (%)	35	45	50	55

The performance of four sizes of the SPT thruster manufactured by Fakel in Russia is shown in Table 12-11. These are all "conventional" Hall thruster designs and do not incorporate magnetic shielding to date. The SPT-100 is the most widely used of this family, operating nominally at a discharge voltage of 300 V and current of 4.5 A to produce 82 mN of thrust at an Isp of 1600 s and a total efficiency of 50% averaged over the life of the thruster. The different SPT thrusters shown in the table have been tested at discharge voltages of 200 to 500 V and have all operated in space at power levels of a few hundred watts up to 4.5 kW. These Hall thrusters have also been tested on a variety of gases such as argon and krypton, but xenon is the present standard for space applications. A schematic of Hall thrusters was shown in Chapter 7.

The first flight of a Hall thruster on a US spacecraft was the 1998 launch of a D-55 TAL (Thruster with Anode Layer) Hall thruster [41, 42] manufactured by TsNIIMASH in Russia on the National Reconnaissance Office's Space Technology Experiment Satellite (STEX). The STEX mission was intended to develop and demonstrate advanced spacecraft technologies in space, including Hall thrusters. The D-55 TAL thruster nominally operates in xenon at 1.35 kW with an Isp of about 1500 s, but due to power limitations on the spacecraft was required to run at a discharge of 300 V and 2.2 A (660 W).

The European Space Agency (ESA) demonstrated the use of commercial Hall thruster technology on the SMART-1 (Small Mission for Advanced Research in Technology) spacecraft in a lunar orbiting mission [43]. A PPS-1350-G Hall thruster [44], manufactured by what is now SAFRAN in France, was launched on SMART-1 in 2003 and provided primary propulsion for this mission. This thruster is based on the SPT-100 design and is similar in size and power level. The thruster was operated over a throttleable power range of 462–1190 W for this lunar mission, producing a maximum thrust of 70 mN at an Isp of 1600 s. The PPS-1350 Hall thruster accumulated about 5000 h of operation in space, and processed 82 kg of xenon in a very successful mission that featured several extensions of the mission life due to the thruster capabilities. The nominal performance of this thruster [45] at 1.35 kW is shown in Table 12-12. The thruster schematic was shown in Chapter 7, and a photograph of the PPS-1350 Hall thruster is shown in Fig. 12-11.

The first commercial use of Hall thrusters by a U.S. spacecraft manufacturer was in 2004 by what is now Maxar Space on the MBSAT satellite [46], which used Fakel SPT-100 s provided by International Space Technologies Inc.d (ISTI). Maxar has launched dozens of communications satellites to date that each use two pairs of SPT–100 Hall thrusters to provide North–South station keeping in Earth orbit. A photograph of a Fakel SPT-100 thruster from [46] is shown in Fig. 12-12.

In 2018, Maxar started flying the SPT-140 Hall thruster on their high-power communications satellites [47, 48]. Like their lower power SPT-100 propulsion system, the SPT-140 thruster is manufactured by Fakel in Russia. This thruster has a "conventional" design and was optimized to operate on the Maxar 1300 series communications satellites at high power for orbit raising at

Table 12-12 PPS-1350 Hall thruster performance.

Parameter	Primary Propulsion
Channel outside diameter (cm)	10
Thruster input power (W)	1500
Average Isp (seconds)	1650
Thrust (mN)	88
Total efficiency (%)	55
Discharge voltage (V)	350
Discharge current (A)	4.28

Figure 12-11 Photograph of the PPS 1350 Hall thruster (*Source:* Safran Electronics & Defense, France).

Figure 12-12 Photograph of the newest version of the SPT-100 Hall thruster (*Source:* Photo from Maxar [46]/American Institute of Aeronautics and Astronautics, Inc.).

4.5 kW, and lower power at 3-kW for station keeping. The thruster performance is listed in Table 12-11 and produces a maximum thrust of about 270 mN at an Isp of 1750 s. The STP-140 will used by the NASA Psyche Mission [49] that is scheduled to be launched in 2023. For this deep space prime propulsion and momentum control application, the EP system from Maxar (thruster, power processing unit and xenon flow controller) was modified and extensively tested over a throttle range of 0.9–4.5 kW [50]. A photograph of the SPT-140 Hall thruster is shown in Fig. 12-13.

The first flight of the PPS®5000 Hall thruster [51, 52] manufactured by Safran in France was made 2021 on a French military communications satellite supplied by Thales Alenia Space. The thruster performed orbit raising for six months and then initiated station-keeping activities. Nearly a hundred of these thrusters have been ordered to date for use in European and U.S. communications satellites. The PPS®5000 Hall thruster was originally based on the design of the Russian SPT-140, but was developed at Safran to have improved performance and life. The extended qualification life test [53] of this thruster achieved 19,370 h, 17.2 MNs of total impulse, 9539 ON/OFF cycles, and 972 kg of xenon processed when the test was voluntarily terminated with the thruster still operational. A photograph of the PPS®5000 Hall thruster is shown in Fig. 12-14, and the thruster performance is given in Table 12-13.

The first US company to provide flight Hall thrusters for a spacecraft was Busek, Inc. with the 200-W BHT-200 that flew on board the U.S. Air Force TacSat-2 spacecraft that was launched in 2006 [54, 55]. The Air Force continues to fly the BHT-200 on various FalconSAT spacecraft. Busek is continuing qualification of their BHT-6000 Hall thrusters for the Power and Propulsion Element (PPE) as part of NASA's Artemis program [56]. The PPE is a high-power, 60-kW solar electric propulsion spacecraft built by Maxar Technologies and operated by NASA and planned to launch in 2024.

Lockheed Martin Space Systems began flying the XR-5 Hall thruster (previously named the BPT-4000 Hall thruster) developed in the US by Aerojet [57] on Air Force Advanced-EHF defense communications satellites in 2010. Aerojet and JPL have jointly investigated the applicability of the XR-5 thruster to NASA deep space missions [58] where throttle range and efficiency are important. The ability of the XR-5 to throttle from power levels of 1 to 4.5 kW was demonstrated with very high efficiency [59]. The XR-5 Hall thruster is presently also being used for orbit raising and station keeping by several commercial communications satellites in the US. The performance of the XR-5 at two power levels and two discharge voltages is listed in Table 12-14, and a photograph of the XR-5 is shown in Fig. 12-15.

The recent emergence of LEO communications constellations has resulted in unprecedented numbers of electric thrusters to be launched into orbit. Since 2018, SpaceX has launched over 3000 Starlink spacecraft [60], with plans for over 10,000 in the constellation, each with a Hall thruster on board utilizing krypton propellant for orbit insertion and station keeping. A number of other constellations are in work, many of which use electric propulsion. These constellations will drive significant advances in the commercial use of electric propulsion technology.

Figure 12-13 Photograph of the SPT-140 Hall thruster (*Source:* Photo courtesy of Maxar).

Figure 12-14 Photograph of the PPS®5000 Hall thruster (*Source:* Photo ©Bertrand Lecuyer, used with permission, Safran Electronics & Defense).

Table 12-13 **PPS5000 thruster performance from [53].**

Parameter	High thrust-to-power	High Isp
Thruster input power (kW)	5.2	5.2
Specific impulse (seconds)	1741	1973
Thrust (mN)	308	267
Total efficiency (%)	51	50
Discharge voltage (V)	300	400
Discharge current (A)	16.7	12.5

Table 12-14 XR-5 Hall thruster performance.

Parameter	Throttle range			
Thruster input power (kW)	2.0	2.0	4.5	4.5
Specific impulse (seconds)	1676	1858	1790	2020
Thrust (mN)	132	117	290	254
Total efficiency (%)	–	49	–	55
Discharge voltage (V)	300	400	300	400
Discharge current (A)	6.7	5	15	11.3

Figure 12-15 Photograph of the XR-5 Hall thruster (*Source:* Courtesy of Aerojet).

12.4 Electromagnetic Thrusters

The first electromagnetic thruster to fly in space was a Pulsed Plasma Thruster (PPT) on the "Zond-2" Russian satellite in 1964 [61], which was used for attitude control. In the U.S., four PPTs were launched into orbit in 1968 on the Lincoln Experimental Satellite 6 (LES 6) experimental military UHF communications satellite [62]. This PPT system operated for over a year without issue and demonstrated no interference with the telemetry or the communications system of the satellite. Follow-on flights on the LES 8/9 missions were originally planned to be equipped with pulsed plasma thrusters that were fully flight qualified [63], but the spacecraft were launched with gas thrusters. Figure 1-6 shows a photograph of the PPT that was used for attitude control on the Earth Observing (EO-1) Mission in 2000 [64]. PPTs are now a commercial product in the US and are finding applications in CubeSat and other microsatellites.

Experimental prototype MPD thrusters were first flown on Soviet spacecraft [65], and there have been two additional flights of MPD thrusters by Japan to date. These flight units have all been pulsed technology demonstration thrusters due to the high power needed to operate continuously. An MPD thruster was launched onboard the MS-T4 test satellite by the Institute of Space and Aeronautical Science (ISAS) and the University of Tokyo, in 1980 [66]. This thruster operated on ammonia at up to 700 A of discharge current for 1.5 ms pulses, and demonstrated up to 22% efficiency at and Isp of about 2500 s. A second tech-demo MPD was launched as part of Electric Propulsion Experiment (EPEX) in 1995 by ISAS [67] and used liquid hydrazine as the propellant. This thruster demonstrated operation at 6-kA (peak) currents and 300 μs FWHM pulse widths with an Isp over 1000 s. MPD thruster development is continuing at universities and national laboratories worldwide.

References

1 K. Holste, P. Dietz, S. Scharmann, K. Keil, T. Henning, D. Zschätzsch, M. Reitemeyer, B. Nauschütt, F. Kiefer, F. Kunze, J. Zorn, C. Heiliger, N. Joshi, U. Probst, R. Thüringer, C. Volkmar, D. Packan, S. Peterschmitt, K.-T. Brinkmann, H.-G. Zaunick, M. H. Thoma, M. Kretschmer, H. J. Leiter, S. Schippers, K. Hannemann and P. J. Klar, "Ion Thrusters for Electric Propulsion: Scientific Issues Developing a Niche Technology into a Game Changer Featured", *Review of Scientific Instruments*, vol. 91, 061101 (2020); doi:10.1063/5.0010134

2 S. Mazouffre, "Electric Propulsion for Satellites and Spacecraft: Established Technologies and Novel Approaches", *Plasma Sources Science and Technology*, vol. 25, no. 3, pp.033002 (2016); doi:10.1088/0963-0252/25/3/033002

3 S. Shimada, K. Satho, Y. Gotoh, E. Nistida, I. Terukina, T. Noro, H. Takegahara, K. Nakamaru, and H. Nagano, "Development of an Ion Engine System for ETS-6," IEPC-93-009, 23rd International Electric Propulsion Conference, Seattle, WA, September 13–16, 1993.

4 T. Ozaki, E. Nishida, and Y. Gotoh, "Development Status of 20mN Xenon Ion Thruster," AIAA-2000-3277, 36th AIAA Joint Propulsion Conference, Huntsville, Al, July 16–19, 2000.

5 J. R. Beattie, J. D. Williams, and R. R. Robson, "Flight Qualification of an 18-mN Xenon Ion Thruster," IEPC 93-106, 23rd International Electric Propulsion Conference, Seattle, September 13–16, 1993.

6 J. R. Beattie, "XIPS Keeps Satellites on Track," *The Industrial Physicist*, pp. 24–26, 1998.

7 M. J. Patterson, T. W. Haag, V. K. Rawlin, and M. T. Kussmaul, "NASA 30-cm IonThruster Development Status," AIAA-1994-2849, 30th AIAA Joint Propulsion Conference, Indianapolis, IN, June 27–29, 1994.

8 J. R. Brophy, "NASA's Deep Space 1 Ion Engine," *Review of Scientific Instruments*, vol. 73, no. 2, pp. 1071–1078, 2002; doi:10.1063/1.1432470.

9 J. R. Brophy, M. Marcucci, G. Ganapathi, C. Garner, M. Henry, B. Nakazono, and D. Noon, "The Ion Propulsion System for Dawn," AIAA 2003-4542, 39th AIAA Joint Propulsion Conference, Huntsville, AL, July 20–23, 2003.

10 J. R. Beattie, J. N. Matossian, and R. R. Robson, "Status of Xenon Ion Propulsion Technology," *Journal of Propulsion and Power*, vol. 6, no. 2, p.145–150, 1990.

11 D. M. Goebel, M. Martinez-Lavin, T. A. Bond, and A. M. King, "Performance of XIPS Electric Propulsion in Station Keeping of the Boeing 702 Spacecraft", AIAA-2002-5117, 38th AIAA Joint Propulsion Conference, Indianapolis, IN, July 7–10, 2002.

12 W. Tighe, K. Chien, E. Solis, P. Robello, D. M. Goebel, and J.S. Snyder, "Performance Evaluation of the XIPS 25-cm Thruster for Application to NASA Missions," AIAA-2006-4999, 42nd AIAA Joint Propulsion Conference, Sacramento, California, July 9–12, 2006.

13 H. Gray, P. Smith, and D. G. Fern, "Design and Development of the UK-10 Ion Propulsion System," AIAA-96-3084, 32nd AIAA Joint Propulsion Conference, Lake Buena Vista, FL, July 1–3, 1996.

14 Courtesy of Neil Wallace, QinetiQ Limited, United Kingdom.

15 K. H. Groh, O. Blum, H. Rado, and H.W. Loeb, "Inert Gas Radio-Frequency Thruster RIT 10," IEPC-79-2100, 14th International Electric Propulsion Conference, October 30–November 1, 1979.

16 R. Killinger, H. Bassner, H. Leiter, and R. Kukies, "RITA Ion Propulsion for Artemis", AIAA-2001-3490, 37th AIAA Joint Propulsion Conference, Salt Lake City, UT, July 8–11, 2001.

17 http://cs.space.eads.net/sp/SpacecraftPropulsion/Rita/RIT-10.html.

18 H. Kuninaka, K. Nishiyama, I. Funakai, Tetsuya, and Y. Shimizu, "Asteroid Rendezvous of Hayabusa Explorer Using Microwave Discharge Ion Engines," IEPC Paper 2005-010, 29th International Electric Propulsion Conference, Princeton, NJ, October 31–November 4, 2005.

19 S. Tamaya, I. Funaki, and M. Murakami, "Plasma Production Process in an ECR Ion Thruster", AIAA-2002-2196, 33rd AIAA Plasma Dynamics and Lasers Conference, Maui, HI, May 20–23, 2002.

20 H. Kuninaka, I. Funaki, K. Nishiyama, Y. Shimizu, and K. Toki, "Results of 18,000 Hour Endurance Test of Microwave Discharge Ion Thruster Engineering Model," AIAA-2000-3276, 36th AIAA Joint Propulsion Conference, Huntsville, Alabama, July 16–19, 2000.

21 Y. Tani, R. Tsukizaki, D. Kodab, K. Nishiyama, and H. Kuninaka, "Performance Improvement of the μ10 Microwave Discharge Ion Thruster by Expansion of the Plasma Production Volume," *Acta Astronautica*, vol. 157, pp. 425–434 (2019); doi:10.1016/j.actaastro.2018.12.023

22 Courtesy of Prof Kuninaka, Institute of Space and Astronautical Science, Japan Aerospace Exploration Agency.

23 T. Ozaki, Y. Kasai, T. Nakagawa, T. Itoh, K. Kajiwara, and M. Ikeda, "In-Orbit Operation of 20 mN Class Xenon Ion Engine for ETS-VIII," IEPC-2007-084, 28th International Electric Propulsion Conference, Florence, Italy, September 17–20, 2007.

24 T. Ozaki, E. Nishida, Y. Kasai, Y. Gotoh, T. Itoh, and K. Kajiwara, "Development Status of Xenon Ion Engine Subsystem for ETS-VIII," AIAA-2003-2215, 21st International Communications Satellite Systems Conference, Yokohama, Japan, April 15–19, 2003.

25 T. Ozaki, Y. Kasai, and E. Nishida, "Improvement of 20mN Xenon Ion Thruster," IEPC-99-153, 26th International Electric Propulsion Conference, Kitakyushu, Japan, October 17–21, 1999.

26 N. Wallace, D. Mundy, D. Fearn, and C. Edwards, "Evaluation of the Performance of the T6 Ion Thruster," AIAA-1999-2442, 35th AIAA Joint Propulsion Conference, Los Angeles, CA, June 20–24, 1999.

27 S. Clark, P. Randall, R. Lewis, D. Marangone, D. M. Goebel, V. Chaplin, H. Gray, K. Kempkens, and N. Wallace, "BepiColombo – Solar Electric Propulsion System Test and Qualification Approach," IEPC-2019-586, 36th International Electric Propulsion Conference, Vienna, September 15–20, 2019.

28 V. Chaplin, D. M. Goebel, R. A. Lewis, F. Lockwood Estrin, and P. N. Randall, "Accelerator Grid Life Modeling of the T6 Ion Thruster for BepiColombo," *Journal of Propulsion and Power*, vol. 37, no. 3, pp. 436–449, 2021; doi:10.2514/1.B37938.

29 A.N. Grubisic, S. Clark, and N. Wallace, C. Collingwood and F. Guarducci, "Qualification of the T6 Ion Thruster for the BepiColombo Mission to the Planet Mercury, IEPC-2011-234, 32nd International Electric Propulsion Conference, Wiesbaden, Germany, Sept. 11-15, 2011). IEPC-2011-234.

30 M. Patterson, J. Foster, T. Haag, V. Rawlin, and G. Soulas, "NEXT: NASA's Evolutionary Xenon Thruster," AIAA-2002-3832, 38th AIAA Joint Propulsion Conference, Indianapolis, Indiana, July 7–10, 2002.

31 A. S. Rivkin, N. L. Chabot, A. M. Stickle, C. A. Thomas, D. C. Richardson, O. Barnouin, E. G. Fahnestock, C. M. Ernst, A. F. Cheng, S. Chesley, S. Naidu, T. S. Statler, B. Barbee, H. Agrusa, N. Moskovitz, R. T. Daly, P. Pravec, P. Scheirich, E. Dotto, V. Della Corte, P. Michell, M. Küppers, J. Atchison, and M. Hirabayashi, "The Double Asteroid Redirection Test (DART): Planetary Defense Investigations and Requirements," *The Planetary Science Journal*, vol. 2, 173, 2021; doi:10.3847/PSJ/ac063e

32 J. John, R. Thomas, A. Hoskin, J. Fisher, J. Bontempo, and A. Birchenough, "NEXT-C Ion Propulsion System Operations on the DART Mission," IEPC-2022-281, 37th International Electric Propulsion Conference, Boston, MA, June 19–23, 2022.

33 J. E. Polk, D. M. Goebel, I. Katz, J. Snyder, A. Schneider, L. Johnson, and A. Sengupta, "Performance and Wear Test Results for a 20-kW Class Ion Engine with Carbon-Carbon Grids," 41st AIAA Joint Propulsion Conference, Tucson, July 2005.

34 J. Foster, T. Haag, H. Kamhawi, M. Patterson, S. Malone, and F. Elliot, "The High Power Electric Propulsion (HiPEP) Ion Thruster," AIAA-2004-3812, 40th AIAA Joint Propulsion Conference, Fort Lauderdale, FL, July 11–14, 2004.

35 K. Nishiyama, H. Kuninaka, Y. Shimizu, and K. Toki, "30-mN Class Microwave Discharge Ion Thruster," IEPC-2003-62, 28th International Electric Propulsion Conference, Bordeau, France, March 17–21, 2003.

36 H. J. Leiter, D. Lauer, P. Bauer, M. Berger, and M. Rath, "The Ariane Group Electric Propulsion Program 2019–2020," 36th International Electric Propulsion Conference, Vienna, September 15–20, 2019. IEPC-2019-592.

37 R. Wirz, J. E. Polk, C. Marrese, and J. Mueller, "Experimental and Computational Investigation of the Performance of a Micro-Ion Thruster," AIAA-2002-3835, 38th AIAA Joint Propulsion Conference, Indianapolis, July 7–10, 2002.

38 D. Feili, H. W. Loeb, K. H. Schartner, St. Weis, D. Kirmse, B. K. Meyer, R. Kilinger, and H. Mueller, "Testing of New μN-RITs at Giessen University," AIAA 2005-4263, 41st AIAA Joint Propulsion Conference, Tucson, Arizona, July 10–13, 2005.

39 A. J. Morozov, "Stationary Plasma (SPT) Development Steps and Future Perspectives," IEPC-1993-101, 23rd International Electric Propulsion Conference, Seattle, WA, September 13–16, 1993.

40 V. Kim, "Electric Propulsion Activity in Russia," IEPC-2001-005, 27th International Electric Propulsion Conference, Pasadena, CA, October 14–19, 2001.

41 M. T. Domonkos, C. M. Marrese, J. M. Haas, and A. D. Gallimore, "Very Near-Field Plume Investigation of the D55," AIAA-1997-3062, 33rd AIAA Joint Propulsion Conference, Seattle, WA, July 6–9, 1997.

42 S. O. Tverdokhlebov, A. V. Semenkin, and A. E. Solodukhin, "Current Status of Multi-Mode TAL Development and Areas of Potential Application," AIAA-2001-3779, 37th AIAA Joint Propulsion Conference, Salt Lake City, UT, July 8–11, 2001.

43 C. R. Koppel and D. Estublier, "The SMART-1 Hall Effect Thruster Around the Moon: in Flight Experience," IEPC-2005-119, 29th International Electric Propulsion Conference, Princeton, NJ, October 31–November 4, 2005.

44 M. Lyszyk, E. Klinger, J. Bugeat, and D. Valentian, "Development Status of the PPS-1350 Plasma Thruster," AIAA-1998-3333, 34th AIAA Joint Propulsion Conference, Cleveland, OH, July 13–15, 1998.

45 C.R. Koppel and D. Estublier, "The SMART-1 Electric Propulsion Subsystem," AIAA-2003-4545, 39th AIAA Joint Propulsion Conference, Huntsville, AL, July 20–23, 2003.

46 D. L. Pidgeon, R. L. Corey, B. Sauer, and M. L. Day, "Two Years on-Orbit Performance of SPT-100 Electric Propulsion," AIAA 2006-5353, 42nd AIAA Joint Propulsion Conference, Sacramento, CA, July 9–12, 2006.

47 I. K. Johnson, G. Santiago, J. Li, and J. Baldwin, "100,000 hrs of on-Orbit Electric propulsion and MAXAR's First ELECTRIC Orbit raising," AIAA 2020-0189, AIAA SCITECH 2020 Forum, Orlando, FL, January 6–10, 2020; doi:10.2514/6.2020-0189.

48 A. Komarov, S. Pridannikov, and G. Lenguito, "Typical Transient Phenomena of Hall Effect Thrusters," IEPC-2019-304, 36th International Electric Propulsion Conference, Vienna, 15–20 September 2019.

49 D. Y. Oh, S. Collins, D.M. Goebel, W. Hart, T. Imken, C. Lawler, J. Maxwell, M. Martin, C. Polanskey, J. S. Snyder, and T. Weiss, "Preparation for Launch of the Psyche Spacecraft for NASA's Discovery Program," IEPC-2022-452, 37th International Electric Propulsion Conference, Cambridge, MA, June 19–23, 2022.

50 J. S. Snyder, V. H. Chaplin, D. M. Goebel, R. R. Hofer, A. Lopez Ortega, I. G. Mikellides, T. Kerl, G. Lenguito, F. Aghazadeh, and I. K. Johnson, "Electric Propulsion for the Psyche Mission: Development Activities and Status," AIAA 2020-3607, AIAA Propulsion and Energy 2020 Forum, Virtual Event, August 24–28, 2020; doi:10.2514/6.2020-3607

51 O. Duchemin, J. Rabin, L. Balika, M. Diome, V. Guyon, D. Vuglec, X. Cavelan, and V. Lero, "Development and Qualification Status of the PPS® 5000 Hall Thruster Unit," AIAA-2018-4420, 2018 AIAA Joint Propulsion Conference, Cincinnati, OH, July 9–11, 2018: doi:10.2514/6.2018-4420.

52 G. Coduti, O. Duchemin, J. Rabin, J. Pasquiet, D. Pagano, A. Arde, F. Scortecci, V. Leroi, and P. Le Meur, "Plume Characterization and Influence of Background Pressure on the PPS®5000 Hall Thruster," IEPC-2022-367, 37th International Electric Propulsion Conference, Cambridge, June 19–23, 2022.

53 O. Duchemin, J. Rabin, G. Coduti, M. Diome, X. Cavelan, V. Leroi, P. Le Meur, C. Edwards, and B. Fallis, "Extended Qualification Life Test of the PPS®5000 Hall Thruster Unit," SP2022-364, Space Propulsion 2022 Conference, Estoril, Portugal, May 9–13, 2022.

54 T. Yee, "Roadrunner, a High-Performance Responsive Space Mission," Proceedings of the 18th AIAA/USU Conference on Small Satellites, SSC04-I-5, Logan, August 2004.

55 D. R. Bromaghim, J. T. Singleton, R. Gorecki, F. Dong Tan, and H. Choy, "200 W Hall Thruster Propulsion Subsystem Development for Microsatellite Missions," Proceedings of the 53rd JANNAF Propulsion Meeting, Monterey, CA, December 5–8, 2005.

56 D.A. Herman, T. Gray, and I. Johnson, T.Kerl, T. Lee and T. Silva, "The Application of Advanced Electric Propulsion on the NASA Power and Propulsion Element (PPE)," IEPC-2019-651, 36th International Electric Propulsion Conference, Vienna, Austria, September 15–20, 2019.

57 K. H. de Grys, B. Welander, J. Dimicco, S. Wenzel, B. Kay, V. Khayms, and J. Paisley, "4.5 kW Hall Thruster System Qualification Status," AIAA 2005-3682, 41st AIAA Joint Propulsion Conference, Tucson, AZ, July 10–13, 2005.

58 R. R. Hofer, T. M. Randolph, D. Y. Oh, and J. S. Snyder, "Evaluation of a 4.5 kW Commercial Hall Thrusters System for NASA Science Missions," AIAA-2006-4469, 42nd AIAA Joint Propulsion Conference, Sacramento, CA, July 9–12, 2006.

59 B. Welander, C. Carpenter, K. H. de Grys, R. R. Hofer, T. M. Randolph, and D. H. Manzella, "Life and Operating Range Extension of the BPT-4000 Qualification Model Hall Thruster," AIAA-2006-5263, 42nd AIAA Joint Propulsion Conference, Sacramento, 9–12 July 2006.

60 www.starlink.com.

61 A. S. Bober, V. Kim, A. S. Koroteyev, et al., "State of Works on Electrical Thrusters in the USSR," IEPC-91-03, 22nd International Electric Propulsion Conference, Viareggio, October 14–17, 1991.

62 W. J. Guman and D. M. Nathanson, "Pulsed Plasma Microthruster Propulsion System for Synchronous Orbit Satellite," *Journal of Spacecraft*, vol. 7, no. 4, pp. 409–415, 1970.
63 R. J. Vondra and K. I. Thomassen, "Flight Qualified Pulsed Electric Thruster for Satellite Control", *Journal of Spacecraft*, vol. 11, no. 9, pp. 613–617, 1974.
64 S. Benson, L. Arrington, W. Hoskins, and N. Meckel, "Development of a PPT for the EO-1 Spacecraft," AIAA-99-2276, 35th AIAA Joint Propulsion Conference, Los Angeles, June 20–24, 1999; doi:10.2514/6.1999–2276
65 O. A. Gorshkov, V. N. Shutov, K. N. Kozubsky, V. G. Ostrovsy, and V. A. Obukhov, "Development of High Power Magnetoplasmadynamic Thrusters in the USSR," IEPC-2007-136, 30th International Electric Propulsion Conference, Florence, Italy, September 17–20, 2007.
66 K. Kuriki, S. Morimoto, and K. Nakamaru, "Flight Performance Test of MPD Thruster System," AIAA-81-0664, 15th International Electric Propulsion Conference, Las Vegas, NV, April 21–23, 1981; doi:10.2514/6.1981-664
67 K. Toki, Y. Shimizu, and K. Kuriki, "Electric Propulsion Experiment (EPEX) of a Repetitively Pulsed MPD Thruster System on Board Space Flyer," IEPC-97-120, 25th International Electric Propulsion Conference, Cleveland, Ohio, August 24–28, 1997.

Appendix A

Nomenclature

A.1 Constants

A_v	Avogadro's number (atoms/mole)	$6.02214179 \times 10^{23}$
AMU	Atomic Mass Unit	$1.6602176487 \times 10^{-27}$ kg
c	velocity of light	2.9979×10^{8} m/s^2
e	electron charge	$1.602176487 \times 10^{-19}$ C
g	gravitational acceleration	9.80665 m/s^2
k	Boltzmann's constant	1.3807×10^{-23} J/K
m	electron mass	$9.1093822 \times 10^{-31}$ kg
M	proton mass	$1.67262164 \times 10^{-27}$ kg
e/m	electron charge to mass ratio	1.75882×10^{11} C/kg
M/m	proton to electron mass ratio	1836.153
M_{Xe}	mass of a xenon atom	131.293 AMU 2.17975×10^{-25} kg
M_{Kr}	mass of a krypton atom	83.798 AMU 1.3915×10^{-25} kg
ε_o	permittivity of free space	8.8542×10^{-12} F/m
μ_o	permeability of free space	$4\pi \times 10^{-7}$ H/m
πa_0^2	atomic cross section	8.7974×10^{-21} m^2
e/k	temperature associated with 1 electron volt	11,604.5 °K
eV	energy associated with 1 electron volt	$1.602176487 \times 10^{-19}$ J
T_o	standard temperature (0°C)	273.15 °K
p_o	standard pressure (760 Torr = 1 atm)	1.0133×10^{5} Pa
n_o	Loschmidt's number (gas density at STP)	2.6868×10^{25} m^{-3}

A.2 Acronyms

BaO	Barium Oxide
CC–EST	Central–Cathode Electrostatic Thruster
DC	Direct Current (steady-state)
DS1	Deep Space 1 mission
ECR	Electron Cyclotron Resonance (microwave)
ELT	Extended Duration Life Test (NSTAR thruster life test)

Fundamentals of Electric Propulsion, Second Edition. Dan M. Goebel, Ira Katz, and Ioannis G. Mikellides.

EP	Electric Propulsion
EPIC	Electric Propulsion Interactions Code
ESA	European Space Agency
ETS	Engineering Test Satellite
FEEP	Field Emission Electric Propulsion
GRC	NASA Glenn Research Center
HET	Hall Effect Thruster
HRL	Hughes Research Laboratories
IKAROS	Interplanetary Kite-craft Accelerated by Radiation Of the Sun
ICRH	Ion Cyclotron Resonance Heating
JAXA	Japan Aerospace Exploration Agency
JPL	Jet Propulsion Laboratory
LaB_6	Lanthanum Hexaboride
LDT	Life Demonstration Test (8200 h NSTAR thruster wear test)
LEO	Low Earth Orbit
LHDI	Lower Hybrid Drift Instability
MHD	Magnetohydrodynamic
MIPE	Magnetic Induction Plasma Engine
MTSI	Modified Two-Stream Instability
MPD	Magneto-Plasma-Dynamic thruster
MSFC	Marshall Space Flight Center
NASA	National Aeronautics and Space Administration
NEA	Near Earth Asteroid
NSTAR	Solar electric propulsion Technology Applications Readiness
NEXIS	Nuclear Electric Xenon Ion thruster System
NEXT	NASA's Evolutionary Xenon Thruster
PIC	Particle In Cell
PIT	Pulsed Inductive Thruster
PPT	Pulsed Plasma Thruster
rf	Radio Frequency
RIT	Radio frequency Ion Thruster
RITA	Radio frequency Ion Thruster Assembly
RMF	Rotating Magnetic Field
TAL	Thruster with Anode Layer (a type of Hall thruster)
SERT	Space Electric Rocket Test
SMART	Small Mission for Advanced Research in Technology
SPT	Stationary Plasma Thruster (a type of Hall thruster)
US	United States
VASIMR®	Variable Specific Impulse Magnetoplasma Rocket
XIPS	Xenon Ion Propulsion System (manufactured by Stellant Systems)

A.3 Defined Terms

Isp	specific impulse
F_t	correction to thrust force due to beam divergence

T_e	electron temperature in °K
T_{eV}	electron temperature in electron volts
$\ln \Lambda$	Coulomb logarithm
$Q_{injested}$	gas flow recycled into thruster from vacuum system

A.4 Variables

A	area, coefficient in Richardson-Dushman Equation
A_a	electron loss area at anode
A_{as}	total surface area of anode exposed to plasma
A_c	surface accommodation coefficient
A_g	area of grids
A_p	primary electron loss area at anode
A_s	area of screen grid
A_w	discharge chamber wall area
b	constant in self field MPD thrust scaling
B	magnetic field
B_o	magnetic field a the surface of the magnet
B_r	radial magnetic field
B_θ	poloidal magnetic field
$\bar{c}$	neutral gas thermal velocity
C	constant, conductance of grids for neutral gas flow
C_1	experimental fitting coefficient in barium depletion model
C_T	dimensionless thrust coefficient
d	gap distance (between electrodes or magnets), distance, electrode thickness
d_a	accel grid aperture diameter
d_b	beamlet diameter
d_s	screen grid aperture diameter
D	diffusion coefficient, Richardson-Dushman coefficient
D_a	ambipolar diffusion coefficient
D_B	Bohm diffusion coefficient
D_e	electron diffusion coefficient
D_E	thermionic insert inner diameter
D_i	ion diffusion coefficient
$D_\perp$	perpendicular diffusion coefficient
E	electric field
E_{accel}	electric field at the accel grid
E_{screen}	electric field at the screen grid
E	energy
E'	Fitting parameter in the Longo barium surface coverage model
E_o	energy per pulse in PPTs
E_{eff}	effective atom activation energy
E^*	energy of the lowest excited state of the neutral gas
ε	ion energy at the sheath edge
f	frequency, fraction of ions with a radial velocity, ion distribution function, force density, arbitrary function

f_a	open area fraction of accel grid
f_b	beam flatness parameter
f_c	ion confinement factor for fraction of Bohm current lost
f_i	current fraction of the ith species, frequency of ion oscillations
f_n	edge to average plasma density ratio in cathode plasma
f_p	electron plasma frequency
F	force
F_{accel}	force on the accel grid
F_e	force on the electrons
F_i	force on the ions
F_{is}	flux of scattered ions
F_c	force due to collisions causing momentum transfer
F_{EM}	electromagnetic force
F_{GD}	gas dynamic force
F_L	Lorentz force
F_p	pressure gradient force
F_{screen}	force the screen grid
F_t	thrust vector correction factor
F_z	total axial Lorentz force
g	the acceleration by gravity = 9.8067 m/s^2
h	plume expansion parameter, height
$H(T)$	total heat lost by hollow cathode (a function of the temperature)
I_a	electron current leaving plasma to anode
I_A	accel grid current
I	current
I_b	beam current, impulse bit (impulse per pulse)
I_B	Bohm current
I_{ck}	current to the discharge cathode keeper
I_d	discharge current
I_{DE}	decel grid current
I_e	electron current, emission current from hollow cathodes
I_{ea}	electron current to anode
I_{eb}	electron backstreaming current
I_{ec}	electron current flowing backwards in a Hall thruster
I_{ew}	electron current to the wall
I_H	Hall current
I_i	ion current
I_{ia}	ion current lost to anode
I_{ib}	ion current in the beam
I_{ic}	ion current lost to cathode
I_{iw}	ion current to the wall
I_k	ion current back to the hollow cathode
I_L	primary electron current lost directly to anode
I_{nk}	current to the neutralizer cathode keeper
I_p	ion production rate in the plasma
I_r	random electron flux

I_s	ion current to the screen grid
I_t	thermionic emission current
I_w	current to the walls
I^+	singly charged ion current
I^{++}	doubly charged ion current
I^*	excited neutral production rate in the plasma
j_o	equilibrium current density
j_1	perturbed current density
J	current density
J_e	electron current density
J_i	ion current density
J_{Hall}	Hall current density = $-qn_e v_e$
J_{max}	maximum Child-Langmuir current density
$J_{0,1}$	zero and first order Bessel functions
k	Boltzman's constant, wavenumber $=2\pi/\lambda$
$k_{0,1,2,3}$	fit parameters for Randolph's plume divergence formula
K	proportionality constant
K_o	areal density in adatom model for surface coverage
l	length for radial ion diffusion between cusps
l_d	distance to merged beamlets in plume
l_e	sheath thickness length
l_g	grid gap length
L	primary electron path length, plasma length, microwave interaction length, length of the ionization region in Hall thrusters, inductance
L_c	total length of magnetic cusps
L_g	path length for electron gyration
L_j	penetration depth of the electron current density into the insert region
L_n	penetration depth of neutral gas into the insert plasma
L_T	total path length for helical electron motion
m	mass, electron mass
m_a	mass flow injected into the anode region
m_c	mass flow injected through the cathode
m_d	delivered spacecraft mass
m_i	propellant mass due to ions
m_p	propellant mass
m_s	mass of species "s"
m_t	total mass flow
$\dot{m}_a$	Hall thruster anode mass flow rate
$\dot{m}_c$	Hall thruster cathode mass flow rate
$\dot{m}_i$	ion mass flow rate
$\dot{m}_p$	total propellant mass flow rate
M	ion mass, total spacecraft mass, dipole strength per unit length
M_a	ion mass in AMU
M_d	delivered mass
M_f	final mass
M_i	initial mass

M_p	propellant mass
N	total number of particles, number of magnet coil turns
n	particle density
n_a	neutral atom density
n_b	beam plasma density
n_c	critical density at which the plasma frequency equals the microwave frequency
n_e	electron density
n_f	neutral density flowing from cathode
n_i	ion density
n_o	neutral density, plasma density at center of symmetry
n_p	primary electron density
n_R	reference electron density along a magnetic field line
n_s	source or sink density term, secondary electron density, density of species "s"
n^+	singly ionized particle density
n^{++}	doubly ionized particle density
p	plasma pressure
p_e	electron pressure
p_i	ion pressure
p_o	neutral pressure
P	neutral gas pressure, probability of a collision, power, perveance
P_a	power to the anode
P_{abs}	absorbed rf power
P_b	beam electrical power
P_d	discharge electrical power
P_{ei}	electron-ion momentum change from collisions
P_f	final neutral pressure
P_{ie}	ion-electron momentum change from collisions
P_{in}	power into the plasma discharge
P_{jet}	jet power (defined in Eq. 2.3-3)
P_k	keeper discharge electrical power
P_{max}	maximum perveance
P_o	initial neutral pressure, "other" electrical power in the thruster
P_{out}	power out of the plasma
P_T	total electrical power into thruster, pressure in Torr
P_w	power into the wall
q	charge, number of magnetic dipoles
q_s	charge of species "s"
Q	total charge $=qn$, propellant flow rate or throughput, ionization energy
Q_{injested}	equivalent flow due to backstreaming facility gas
Q_s	heat exchange terms in the energy equations
r	radius
r_a	aperture radius
r_e	electron Larmor radius
r_h	hybrid Larmor radius
r_i	ion Larmor radius
r_L	Larmor radius

r_o	internal radius of the cathode orifice
r_p	primary electron Larmor radius
R	ratio of beam voltage to total voltage in ion thrusters, outside channel radius in Hall thrusters, radiation losses in the electron energy equation
R_e	change in electron momentum due to collisions with ions and neutrals
R	resistance
R_m	magnetic mirror ratio
R_o	initial beam radius
R_s	mean change in the momentum of particles "s" due to collisions
R^{++}	rate of double ion production
$\dot{\mathfrak{R}}$	erosion rate of the walls
S	pumping speed, ionization losses in the electron energy equation
t	time
t_a	accel grid thickness
t_s	screen grid thickness
S	ionization energy loss, pumping speed
T	thrust, temperature [°K]
T_a	optical transparency of the grid
T_c	cold propellant thrust
T_e	electron temperature [K]
T_{eV}	electron temperature [eV]
T_g	grid transparency
T_i	ion temperature [K]
T_{iV}	ion temperature [eV]
T_m	sum of thrust from multiple species
T_o	temperature of the neutral gas
T_n	temperature of nth species
T_s	effective transparency of the screen grid, temperature of secondary electrons from wall, temperature of species "s"
T_w	wall temperature
u_{ci}	critical velocity
U^+	first ionization potential
U^*	average excitation potential
v,υ	velocity
υ_a	ion acoustic velocity
υ_b	beam velocity
υ_A	Alfven wave speed
υ_B	Bohm velocity
υ_D	diamagnetic drift velocity
υ_e	electron velocity
υ_{ex}	exhaust velocity
υ_E	ExB drift velocity
υ_f	final velocity
υ_i	ion velocity, initial velocity
υ_n	velocity of the neutral species, velocity of the nth species
υ_o	neutral velocity, initial ion velocity

v_p	primary electron velocity
v_{th}	thermal electron drift velocity
v_{wall}	particle velocity radial boundary wall
$v_\perp$	perpendicular velocity
$v_\parallel$	parallel velocity
V	volume, voltage
V_a	accel grid voltage, activation energy
V_{arc}	arc discharge voltage
V_b	net beam voltage
V_{bp}	potential of beam plasma
V_{ck}	potential of discharge cathode keeper
V_c	voltage drop inside the hollow cathode, coupling voltage from neutralizer common potential to beam potential
V_{cg}	cathode to ground potential
V_d	discharge voltage
V_f	floating potential
V_G	coupling voltage relative to ground in ion thrusters
V_k	net voltage of electrons (primaries) from the cathode
V_m	magnet volume, minimum potential in grids
V_{nk}	potential of neutralizer cathode keeper
V_o	initial voltage
V_p	voltage drop in plasma, plasma generator potential
V_s	screen power supply voltage
V_T	total voltage across accelerator gap $= V_s + V_a$
w	width
x	distance, characteristic length of beam column
y	insert thickness
Y	sputtering yield
Y_{ad}	adatom production yield on cathode surface
Y_{ps}	sputtered particle yield from cathode surface
Z	ion atomic number

A.5 Symbols

α	thrust correction factor for doubly charged ions, work function correction constant, e-folding distance for plasma density decrease, constant in Bessel's function argument, fraction of ion density
α_m	mass utilization correction factor
α_o	dimensionless parameter evaluated at the upstream end of the MPD thruster channel
β	coefficient in Bohm collision frequency, coefficient in anomalous frequency ν_α
$\widetilde{\beta}$	Maxwell stress tensor
γ	total thrust correction factor $= \alpha F_t$, secondary electron yield
γ_ι	ratio of the ion specific heats
γ_o	secondary electron yield at the space charge limit
Γ	flux of particles

Γ_{is}	flux of scattered ions
Γ_o	initial flux of particles
$\Gamma(x)$	Gamma function
Δv	change in velocity
ΔV	potential modification in grids due to space charge
δ	magnet half-height
ε	electron energy density
ε_b	electrical cost of a beam ion
ε_e	energy that an electron removes from the plasma
ε_i	energy that an ion removes from the plasma
ζ	viscosity
ζ_i	density fraction of the ith species
η	total plasma resistivity
η_a	anode efficiency of a Hall thruster
η_b	beam current fraction of discharge current
η_c	Clausing factor (neutral flow conductance reduction factor)
η_d	discharge loss
η_e	electrical efficiency
η_{ei}	plasma resistivity due to electron-ion collisions
η_{en}	plasma resistivity due to electron-neutral collisions
η_m	mass utilization efficiency
η_{m^*}	mass utilization efficiency for multiply charged particles
η_{md}	mass utilization efficiency of the discharge chamber
η_o	electrical efficiency for other power in a Hall thruster
η_T	total thruster efficiency
$\eta_\perp$	perpendicular resistivity
η_v	beam voltage fraction of discharge voltage
θ	angle, surface coverage fraction
θ_s	heat transported by conduction
κ	parameter in double sheath equation $\approx 1/2$, thermal conductivity
λ	mean free path, wavelength
λ_D	Debye length
λ_{01}	first zero of the Bessel function
μ	mobility
μ_B	Bohm mobility
μ_e	electron mobility
μ_{ei}	electron mobility due only to electron-ion collisions
μ_i	ion mobility
$\eta_\perp$	perpendicular electron mobility
v	velocity
ν	collision frequency
ν_{ab}	collision frequency between species a and b
ν_{ee}	electron-electron collision frequency
ν_{ei}	electron-ion collision frequency
ν_{en}	electron-neutral collision frequency
ν_{ii}	ion-ion collision frequency

ν_{in}	ion-neutral collision frequency
ν_m	total momentum transferring collision frequency
ν_{sn}	collision frequency between species "s" and the nth species
ν_{scat}	scattering frequency
ν_α	anomalous collision frequency
ν_w	electron-wall collision frequency
ξ	normalized dimension = x/λ_D, nondimensional current parameter, ratio of the energy sinks over the input energy
ρ	charge density = qn
ρ_m	ion mass density = qM
ρ_o	initial ion mass density
σ	cross section, surface charge density
σ_{CEX}	charge-exchange cross section
σ_e	excitation cross section
σ_i	ionization cross section
τ	collision time, mean electron or ion confinement time
τ_c	time for electron-neutral collision
τ_e	thermal equilibration time
τ_m	total collision time for momentum transferring collisions
τ_p	primary electron confinement time
τ_s	Spitzer primary electron thermalization time with plasma electrons
τ_t	total thermalization time
ϕ	potential, work function
ϕ_o	potential at sheath edge
ϕ_R	reference potential along a magnetic field line
ϕ_s	sheath potential
ϕ_{wf}	work function of a material or surface
ϕ^*	thermalized potential
χ	normalized potential = $e\phi/kT$
ω	cyclic frequency ($=2\pi f$)
ω_c	electron cyclotron frequency
ω_{ci}	ion cyclotron frequency
ω_p	electron plasma cyclotron frequency
Ψ_s	energy loss by species "s" due to inelastic collisions
Ω_p	ion plasma frequency
Ω_e	electron Hall parameter

Appendix B

Gas Flow Units Conversions and Cathode Pressure Estimates

Conversion between the different systems of flow units is necessary to calculate various parameters used in evaluating thruster performance. Owing to the precision required in calculating the thruster performance, it is necessary to carry several significant digits in the constants used to calculate the conversion coefficients, which are obtained from the National Institute of Standards and Technology (NIST) database that can be found in the NIST website.

Converting flow in Standard Cubic Centimeters per Minute (SCCM) to other flow units for an ideal gas is achieved as follows. A mole of gas at standard pressure and temperature is Avogadro's number ($6.02214179 \times 10^{23}$) of particles at one atmosphere pressure and 0 °C (273.15 °K), which occupies 22.413996 liters. The conversions are:

$$\begin{aligned} 1\ \text{sccm} &= \frac{6.02214179 \times 10^{23}[\text{atoms/mole}]}{22.413996\ [\text{liters/mole at STP}]*10^{3}[\text{cc/liter}]*60\ [\text{s/min}]} \\ &= 4.477962 \times 10^{17}\left[\frac{\text{atoms}}{\text{s}}\right] \end{aligned} \tag{B-1}$$

$$\begin{aligned} 1\ \text{sccm} &= 4.477962 \times 10^{17}\left[\frac{\text{atoms}}{\text{s}}\right]*1.6021765 \times 10^{-19}[\text{Coulombs/charge}] \\ &= 7.174486 \times 10^{-2}[\text{equilvalent amperes}] \end{aligned} \tag{B-2}$$

$$1\ \text{sccm} = \frac{10^{-3}[\text{liters}]*760[\text{Torr}]}{60[\text{s/min}]} = 0.01267\left[\frac{\text{Torr}-\text{l}}{\text{s}}\right] \tag{B-3}$$

$$\begin{aligned} 1\ \text{sccm} &= 4.47796 \times 10^{17}\left[\frac{\text{atoms}}{\text{s}}\right]*1.660539 \times 10^{-27}*\text{M}_a*10^{6}, \\ &= 7.43583 \times 10^{-4} M_a\left[\frac{\text{mg}}{\text{s}}\right] \end{aligned} \tag{B-4}$$

where M_a is the propellant mass in atomic mass units, AMU.

For xenon, $M_a = 131.293$ AMU, and a correction must be made for its compressibility at STP which changes the mass flow rate by 0.9931468. Therefore, using Eq. B-4, for xenon:

$$\begin{aligned} 1\ \text{sccm (Xe)} &= \frac{7.17448 \times 10^{-2}}{0.9931468} \\ &= 0.0722399\ [\text{equilvalent amperes}] \\ &= 0.0983009\ [\text{mg/s}] \end{aligned} \tag{B-5}$$

Fundamentals of Electric Propulsion, Second Edition. Dan M. Goebel, Ira Katz, and Ioannis G. Mikellides.

For krypton, $M_a = 83.798$ AMU, so the conversion is

$$1\ \text{sccm (Kr)} = 7.17448 \times 10^{-2}[\text{equilvalent amperes}] \\ = 0.062311[\text{mg/s}] \tag{B-6}$$

It is possible to estimate the neutral gas pressure inside of a hollow cathode insert region and in the orifice as a function of the propellant flow rate and cathode temperature using analytic gas flow equations. While these equations may not be strictly valid in some locations, especially the relatively short orifices found in discharge cathodes, they can still provide an estimate that is usually within 10 to 20% of the actual measured pressures.

In the viscous flow regime, where the transport is because of gas atoms or molecules primarily making collisions with each other rather than walls, the pressure through a cylindrical tube is governed by the Poiseuille law [1, 2] modified for compressible gas [3]. The rate at which compressible gas flows through a tube of length l and radius a (in moles per second) is given [2] from this law by:

$$N_m = \frac{\pi}{8\zeta}\frac{a^4}{l}\frac{P_a(P_1 - P_2)}{R_o T} = \frac{\pi}{16\zeta}\frac{a^4}{l}\frac{P_1^2 - P_2^2}{R_o T}, \tag{B-7}$$

where a is the tube radius, l is the tube length, P_a is the average pressure in the tube given by $(P_1 + P_2)/2$, ζ is the viscosity, P_2 is the downstream pressure at the end of the tube, P_1 is the upstream pressure of the tube, R_o is the universal gas constant, and T is the temperature of the gas. The measured gas flow rate, or the gas throughput, is given by the ideal gas law:

$$Q = P_m V_m = N_m R_o T_m, \tag{B-8}$$

where P_m is the pressure and V_m is the volume where the flow is measured for gas at a temperature T_m, and N_m is the mole flow rate. The mole flow rate is then $N_m = P_m V_m / R_o T_m$. Defining $T_r = T/T_m$ and substituting the mole flow rate into Eq. B-1 gives the measured flow to be:

$$Q = \frac{\pi}{16\zeta}\frac{a^4}{l}\frac{P_1^2 - P_2^2}{T}T_m = \frac{\pi}{16\zeta}\frac{a^4}{l}\frac{P_1^2 - P_2^2}{T_r}. \tag{B-9}$$

Putting this in useful units and writing it in terms of a conductance of the tube, which is defined as the gas flow divided by the pressure drop, gives

$$Q = \frac{1.28\, d^4}{\zeta T_r l}\left(P_1^2 - P_2^2\right), \tag{B-10}$$

where Q is the flow in sccm, ζ is the viscosity in poises, d is the orifice diameter, and l the orifice length in cm, and the pressures are in Torr. The pressure upstream of the cathode orifice is then

$$P_1 = \left(P_2^2 + \frac{0.78\, Q \zeta T_r l}{d^4}\right)^{1/2} \tag{B-11}$$

While Eq. B-11 requires knowledge of the downstream pressure, for this rough estimate it is acceptable to assume $P_2 << P_1$ and neglect this term. For xenon, the viscosity in poises [4] is

$$\zeta = 2.3x10^{-4} T_r^{(0.71 + 0.29/T_r)} \quad \text{for } T_r > 1, \tag{B-12}$$

where $T_r = T(°\text{K})/289.7$. The viscosity in Eq. B-12 is different than Eq. 4.5-9 because $1\ \text{Ns/m}^2 = 10$ poise. It should be noted that the temperature of the gas in the hollow cathode can exceed the temperature of the cathode by factors of 2 to 4 because of charge-exchange heating with the ions, which then affects the viscosity.

As an example, take the NSTAR discharge cathode operating at a nominal flow of 3.7 sccm, with an orifice diameter of 1 mm and the length of the cylindrical section of the orifice as 0.75 mm. Assuming the gas in the orifice is 4000°K because of charge-exchange heating and $P_2 = 0$, the upstream pressure is found from Eq. B-11 to be 6.7 Torr. The pressure measured upstream of the cathode tube for this TH15 case is about 8 Torr [5]. Correcting for the pressure drop in the insert region (also because of Poiseuille flow), the actual pressure upstream of the orifice plate is about 7.2 Torr. The pressure calculated from Eq. B-11 is low because the downstream pressure is finite (about 2 Torr where the barrel section ends) and the bevel region at the output of the orifice has a finite molecular conductance in collisionless flow regime. In general, it can be assumed that the results of Eq. B-11 are about 10% low due to these effects. Similar agreement has been found for neutralizer cathodes with straight bore orifices, suggesting that this technique provides reasonable estimates of the pressure in the cathodes.

Finally, once the pressure inside the cathode or in the orifice region entrance is estimated, it is straightforward to calculate the local neutral density from Eq. 2.7-2:

$$n_o = 9.65 \times 10^{24} * \frac{P}{T} \left[\frac{\text{particles}}{\text{m}^3}\right], \tag{B-13}$$

where P is the pressure in Torr and T is the gas temperature in °K.

References

1 S. Dushman and J. Lafferty, *Scientific Foundations of Vacuum Techniques*, New York: Wiley and Sons, 1962.

2 K. F. Herzfeld and H. M. Smallwood, *Taylor's Treatise on Physical Chemistry*, 2nd edition, vol. 1, New York: D. VanNostrand Co., p. 175, 1931.

3 A. Roth, *Vacuum Technology*, New York: North-Holland, 1990.

4 R. C. Reid, *The Properties of Gases and Liquids*, New York: McGraw-Hill, p. 403, 1977.

5 K.K. Jameson, D.M. Goebel, and R.M. Watkins, "Hollow cathode and keeper-region plasma measurements," AIAA-2005-3667, 41st Joint Propulsion Conference, Tucson, AZ, July 11–13, 2005.

Appendix C

Energy Loss by Electrons

The energy lost from the plasma because of electrons being lost to an anode that is more negative than the plasma potential is derived. Figure C-1 shows the plasma potential distribution in the negative-going sheath toward the anode wall. The Maxwellian electrons are decelerated and repelled by the sheath potential. To determine the average energy removed from the plasma by each electron, moments of the Maxwellian distribution are taken. The electron current density reaching the wall is given by

$$J_e = en\int_{-\infty}^{\infty} dv_x \int_{-\infty}^{\infty} dv_y \int_{\sqrt{\frac{2e\phi}{m}}}^{\infty} v_z \left(\frac{m}{2\pi\, kT_e}\right)^{3/2} e^{\left(\frac{-m\left(v_x^2+v_y^2+v_z^2\right)}{2\,kT_e}\right)} dv_z$$
$$= \frac{1}{4} en\sqrt{\frac{8\,kT_e}{\pi m}}\, e^{\left(-\frac{e\phi}{kT_e}\right)} \tag{C-1}$$

The electrons must overcome the sheath potential to reach the wall so the minimum electron speed toward the wall (assumed to be in the z-direction) is $\sqrt{2e\phi/m}$. The plasma electrons lose kinetic energy as they traverse the sheath, so the power flux from plasma is

$$P_e = en\int_{-\infty}^{\infty} dv_x \int_{-\infty}^{\infty} dv_y \int_{\sqrt{\frac{2e\phi}{m}}}^{\infty} v_z \left(\frac{m_e\left(v_x^2+v_y^2+v_z^2\right)}{2}\right)\left(\frac{m}{2\pi kT_e}\right)^{3/2}$$
$$e^{\left(\frac{-m\left(v_x^2+v_y^2+v_z^2\right)}{2kT_e}\right)} dv_z \tag{C-2}$$
$$= \frac{1}{4} en\sqrt{\frac{8\,kT_e}{\pi m}}\left(2\frac{kT_e}{e}+\phi\right) e^{\left(-\frac{e\phi}{kT_e}\right)}$$

where ϕ is expressed in electron volts (eV). The average energy that an electron removes from the plasma (in eV) is then the ratio of the power per electron to the flux of electrons:

$$E_{\text{ave}} = \frac{P_e}{J_e} = 2\frac{kT_e}{e} + \phi = 2T_{\text{eV}} + \phi, \tag{C-3}$$

where T_{eV} is in electron volts (eV). This is the energy removed from the plasma per electron striking the wall through a negative-going sheath.

It should be noted that this energy loss from the plasma per electron is different than the average energy that each electron has when it hits the wall. The flux of electrons hitting the anode wall is the same as analyzed above. The plasma electrons lose kinetic energy as they traverse the sheath, hence

Fundamentals of Electric Propulsion, Second Edition. Dan M. Goebel, Ira Katz, and Ioannis G. Mikellides.

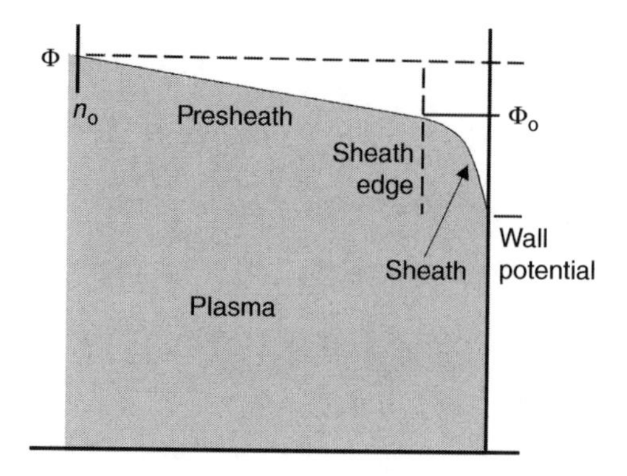

Figure C-1 Schematic of plasma in contact with the anode wall.

a $-e\phi$ term must be included in the particle energy expression for each electron. The power flux to the insert from plasma electrons is then

$$P_e = en\int_{-\infty}^{\infty} dv_x \int_{-\infty}^{\infty} dv_y \int_{\sqrt{\frac{2e\phi}{m}}}^{\infty} v_z \left(\frac{m_e\left(v_x^2 + v_y^2 + v_z^2\right)}{2} - e\phi\right)\left(\frac{m}{2\pi kT_e}\right)^{3/2} e^{\left(\frac{-m\left(v_x^2 + v_y^2 + v_z^2\right)}{2kT_e}\right)} dv_z \tag{C-4}$$

$$= \frac{1}{4}en\sqrt{\frac{8\,kT_e}{\pi m}}\left(2\frac{kT_e}{e}\right)e^{\left(-\frac{e\phi}{kT_e}\right)}$$

The average energy of each electron is then the ratio of the power to the flux

$$E_{ave} = \frac{P_e}{J_e} = 2\frac{kT_e}{e} = 2T_{eV} \text{ [energy per electron that strikes the wall]} \tag{C-5}$$

Appendix D

Ionization and Excitation Cross Sections for Xenon and Krypton

Ionization and excitation cross sections for xenon and krypton by monoenergetic electrons are available from the following references:

[1] D. Rapp and P. Englander, "Total Cross Sections for Ionization and Attachment in Gases by Electron Impact. I. Positive Ionization," *The Journal of Chemical Physics*, vol. 43, pp. 1464–1479, 1965. https://doi.org/10.1063/1.1696957.

[2] M. Hayashi, "Determination of Electron-Xenon Total Excitation Cross-Sections, from Threshold to 100-eV, from Experimental Values of Townsend's a," *Journal of Physics D: Applied Physics*, vol. 16, pp. 581–589, 1983. https://doi.org/10.1088/0022-3727/16/4/018.

[3] K. Stephen and T. D. Mark, "Absolute Partial Electron Impact Ionization Cross Sections of Xe from Threshold up to 180 eV," *The Journal of Chemical Physics*, vol. 81, 3116–3117, 1984. https://doi.org/10.1063/1.448013.

[4] J. A. Syage, "Electron Impact Cross Sections for Multiple Ionization of Kr and Xe," *Physical Review A*, vol. 46, pp. 5666–5680, 1992. https://doi.org/10.1103/PhysRevA.46.5666.

[5] S. Trajmar, S. K. Srivastava, H. Tanaka, and H. Nishimura, "Excitation Cross Sections for Krypton by Electrons in the 15–100-eV Impact-Energy Range," *Physical Review A*, vol. 23, pp. 2167–2177, 1981. https://doi.org/10.1103/PhysRevA.23.2167.

[6] J. E. Chilton, M. D. Stewart, Jr., and C. C. Lin, "Cross Sections for Electron-Impact Excitation of Krypton," *Physics Review A*, vol. 62, 032714, 2000. https://doi.org/10.1103/PhysRevA.62.032714.

The ionization and excitation cross sections for xenon from threshold to 100 eV electron temperatures from the above references are plotted in Fig. D-1 and tabulated in Table D-1 below. The ionization cross sections for krypton for 1–10 eV electron temperatures are tabulated in Table D-2 and plotted in Fig. D-2.

Fundamentals of Electric Propulsion, Second Edition. Dan M. Goebel, Ira Katz, and Ioannis G. Mikellides.

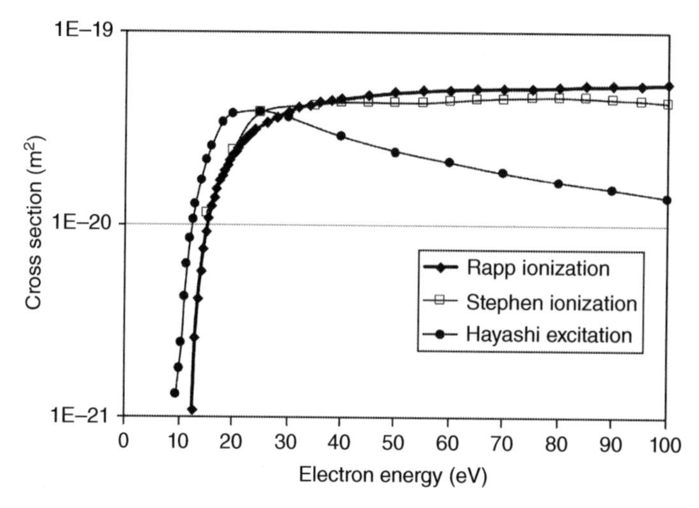

Figure D-1 Ionization and excitation cross sections for xenon.

Table D-1 Ionization and excitation cross sections for xenon.

Electron energy (eV)	Rapp and Englander [1] Ionization (m^2)	Stephen and Mark [3] Ionization (m^2)	Hayashi [2] Total Excitation (m^2)
			2.6E−22
9.0			1.26E−21
9.5			1.31E−21
10.0			1.8E−21
10.5			2.4E−21
11			4.2E−21
11.5			6.2E−21
12			8.4E−21
12.5	1.099E−21		1.05E−20
13.0	2.558E−21		1.28E−20
13.5	4.123E−21		
14.0	5.714E−21		1.7E−20
14.5	7.420E−21		
15.0	9.055E−21	1.15E−20	2.14E−20
15.5	1.073E−20		
16.0	1.231E−20		2.55E−20
16.5	1.380E−20		
17.0	1.529E−20		
17.5	1.670E−20		
18.0	1.802E−20		3.35E−20

Table D-1 (Continued)

Electron energy (eV)	Rapp and Englander [1] Ionization (m^2)	Stephen and Mark [3] Ionization (m^2)	Hayashi [2] Total Excitation (m^2)
18.5	1.925E−20		
19.0	2.048E−20		
19.5	2.163E−20		
20.0	2.277E−20	2.42E−20	3.73E−20
20.5	2.382E−20		
21.0	2.488E−20		
21.5	2.619E−20		
22.0	2.734E−20		
22.5	2.831E−20		
23.0	2.928E−20		
24.0	3.095E−20		
25.0		3.81E−20	3.85E−20
26.0	3.367E−20		
28.0	3.613E−20		
30.0	3.851E−20		3.57E−20
32.0	4.044E−20		
34.0	4.185E−20		
35.0		4.17E−20	
36.0	4.290E−20		
38.0	4.387E−20		
40.0	4.475E−20	4.30E−20	2.85E−20
45.0	4.677E−20	4.31E−20	
50.0	4.835E−20	4.29E−20	2.4E−20
55.0	4.941E−20	4.27E−20	
60.0	5.029E−20	4.37E−20	2.1E−20
65.0	5.081E−20	4.47E−20	
70.0	5.117E−20	4.54E−20	1.85E−20
75.0	5.134E−20	4.57E−20	
80.0	5.178E−20	4.59E−20	1.66E−20
85.0	5.249E−20	4.55E−20	
90.0	5.266E−20	4.48E−20	1.52E−20
95.0	5.328E−20	4.42E−20	
100.0	5.380E−20	4.31E−20	1.38E−20

Table D-2 Example Ionization ion cross sections for krypton.

Electron energy (eV)	Kr^{+} Syage [1] (m^2)	Kr^{++} Syage [1] (m^2)	Kr^{++} Syage [1] (m^2)
18	5.69E−23		
20	1.09E−22		
22	1.47E−22		
24	1.8E−22		
26	2.15E−22		
28	2.41E−22		
30	2.68E−22		
32	2.89E−22		
34	3.08E−22	1.6E−26	
36	3.18E−22	4.8E−26	
38	3.3E−22	1.77E−25	
40	3.41E−22	5.07E−25	
42	3.43E−22	1.11E−24	
44	3.52E−22	2.09E−24	
46	3.61E−22	3.52E−24	
48	3.64E−22	5.35E−24	
50	3.67E−22	7.45E−24	
52	3.68E−22	9.71E−24	
56	3.7E−22	1.41E−23	
60	3.67E−22	1.78E−23	
64	3.69E−22	2.12E−23	
68	3.71E−22	2.38E−23	
72	3.7E−22	2.58E−23	
76	3.7E−22	2.75E−23	4E−27
80	3.71E−22	2.89E−23	2.8E−26
84	3.71E−22	3.01E−23	8.9E−26
88	3.67E−22	3.07E−23	1.8E−25
92	3.64E−22	3.11E−23	2.92E−25
96	3.62E−22	3.15E−23	4.35E−25
100	3.63E−22	3.18E−23	5.81E−25

Source: Adapted from [1].

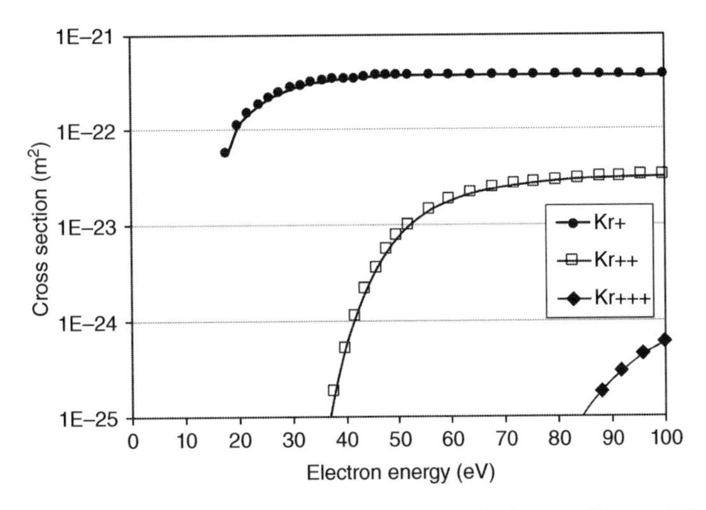

Figure D-2 Ionization cross sections for monoenergetic electrons for krypton (*Source:* [1]).

Ionization and excitation cross sections for other gases such as argon and others are available from the following references:

[1] J. A. Syage, "Electron Impact Cross Sections for Multiple Ionization of Kr and Xe," *Physical Review A*, vol. 46, pp. 5666–5680, 1992. https://doi.org/10.1103/PhysRevA.46.5666.

[2] M. Hayashi, "Bibliography of Electron and Photon Cross Sections with Atoms and Molecules Published in the 20th Century: Argon," NIFS-DATA-72. National Institute for Fusion Science (Japan), ISSN 0915-6364, 2003.

[3] R. Rejoub, B. G. Lindsay, and R. F. Stebbings, "Determination of the Absolute Partial and Total Cross Sections for Electron-Impact Ionization of Rare Gases," *Physical Review A*, vol. 65, 042713, 2002.

[4] A. Yanguas-Gil, J., Cotrino, and L. L. Alves, "An Update of Argon Inelastic Cross Sections for Plasma Discharges," *Journal of Physics D*, vol. 38, pp. 1588–1598, 2005.

[5] G. G. Raju, "Electron-Atom Collision Cross Sections in Argon: An Analysis and Comments," *IEEE Transactions on Dielectrics and Electrical Insulation*, vol. 11, pp. 649–673, 2004.

[6] A. A. Sorokin, L. A. Shmaenok, S. V. Bobashey, et al., "Measurements of Electron-Impact Ionization Cross Sections of Argon, Krypton, and Xenon by Comparison with Photoionization," *Physical Review A*, vol. 61, 022723, 2000.

Appendix E

Ionization and Excitation Reaction Rates in Maxwellian plasmas

Ionization and excitation reaction rate coefficients $<\sigma v>$ for xenon calculated from the data in Appendix D averaged over a Maxwellian electron distribution are given in Table E-1 below. Over the ranges indicated, the data can be well fit to the cross section averaged over a Maxwellian

Table E-1 Total ionization and excitation cross sections for xenon by Maxwellian electrons.

Electron energy (eV)	Ionization (m^3/s)	Excitation (m^3/s)
0.5	4.51E−25	1.99E−22
0.6	3.02E−23	4.01E−21
0.7	6.20E−22	3.61E−20
0.8	6.04E−21	1.95E−19
0.9	3.58E−20	7.44E−19
1.0	1.50E−19	2.21E−18
1.5	1.16E−17	6.64E−17
2.0	1.08E−16	4.02E−16
2.5	4.24E−16	1.23E−15
3.0	1.08E−15	2.66E−15
3.5	2.13E−15	4.66E−15
4.0	3.59E−15	7.12E−15
4.5	5.43E−15	9.93E−15
5.0	7.61E−15	1.30E−14
5.5	1.01E−14	1.61E−14
6.0	1.28E−14	1.94E−14
6.5	1.57E−14	2.26E−14
7.0	1.88E−14	2.57E−14
7.5	2.20E−14	2.87E−14
8.0	2.53E−14	3.14E−14
8.5	2.86E−14	3.34E−14
9.0	3.20E−14	3.41E−14
9.5	3.55E−14	3.21E−14
10.0	3.90E−14	2.48E−14

Fundamentals of Electric Propulsion, Second Edition. Dan M. Goebel, Ira Katz, and Ioannis G. Mikellides.

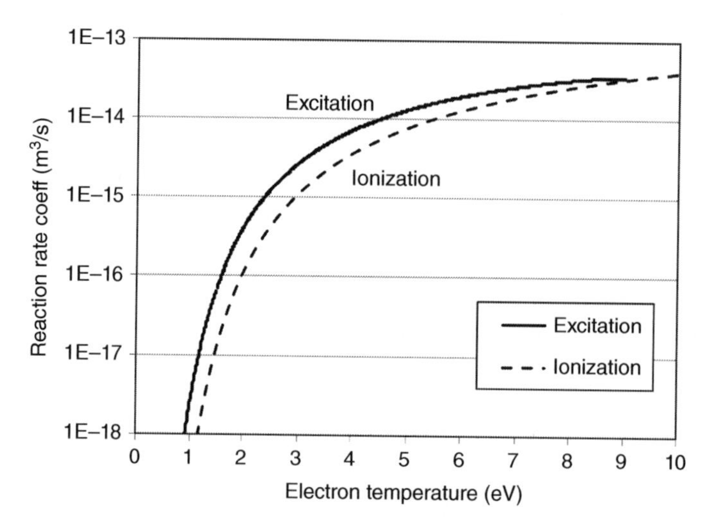

Figure E-1 Ionization and excitation reaction rate coefficients for xenon in a plasma with Maxwellian electrons.

distribution times the electron thermal velocity [1], where T_{eV} is in electron volts (eV). The fits to the calculated values in SI units are:

Ionization ($T_{\text{eV}} < 5$ eV):

$$\langle \sigma_i v_e \rangle \approx \langle \sigma_i \rangle \bar{v}_e = 10^{-20} \left[\left(3.97 + 0.643 T_{\text{eV}} - 0.0368 T_{\text{eV}}^2 \right) e^{-12.127/T_{\text{eV}}} \right] \left(\frac{8 T_{\text{eV}}}{\pi m} \right)^{1/2}$$

Ionization ($T_{\text{eV}} > 5$ eV):

$$\langle \sigma_i v_e \rangle \approx \langle \sigma_i \rangle \bar{v}_e = 10^{-20} \left[-\left(1.031 \times 10^{-4} \right) T_{\text{eV}}^2 + 6.386 e^{-12.127/T_{\text{eV}}} \right] \left(\frac{8 T_{\text{eV}}}{\pi m} \right)^{1/2}$$

Excitation:

$$\langle \sigma^* v_e \rangle \approx \langle \sigma^* \rangle \bar{v}_e = 1.93 \times 10^{-19} \frac{e^{-11.6/T_{\text{eV}}}}{\sqrt{T_{\text{eV}}}} \left(\frac{8 T_{\text{eV}}}{\pi m} \right)^{1/2}$$

The ionization and excitation reaction rate coefficients for xenon found from integrating the cross sections over a Maxwellian distribution of electrons are given in Fig. E-1.

The ionization and excitation reaction rate for krypton can be found from an integration of the cross sections in the literature cited in Appendix D over a Maxwellian distribution of electrons.

Reference

1 I.G. Mikellides, I. Katz, and M. Mandell, "A 1-D Model of the Hall-Effect Thruster with an Exhaust Region," AIAA-2001-3505, 37th Joint Propulsion Conference, Salt Lake City, NV, July 8–11, 2001.

Appendix F

Electron Relaxation and Thermalization Times

Spitzer [1] derived an expression for the slowing down time of test particles (primary electrons in our case) with a velocity $v = \sqrt{2V_p/m}$, where eV_p is the test particle energy in electron volts, in a population of Maxwellian electrons at a temperature T_e. Spitzer defined the inverse mean velocity of the Maxwellian electron "field particles" in one dimension as $l_f = \sqrt{m/2kT_e}$. The slowing down time is then given by

$$\tau_s = \frac{v}{\left(1 + {}^{m}/_{m_f}\right) A_D l_f^2 G(l_f v)}, \tag{F-1}$$

where m is the mass of the test particles, m_f is the mass of the field particles, A_D is a diffusion constant given by

$$A_D = \frac{8\pi e^4 n_f Z^2 Z_f^2 \ln\Lambda}{m^2}, \tag{F-2}$$

where Z is the charge, and $\ln\Lambda$ is the collisionality parameter [2] equal to 23-$\ln(n_f^{1/2}/T_e^{3/2})$. The function $G(l_f\, v)$ is defined as

$$G(x) = \frac{\Phi(x) - x\Phi'(x)}{2x^2}, \tag{F-3}$$

and $\Phi(x)$ is the erf function:

$$\Phi(x) = \frac{2}{\pi^{1/2}} \int_0^x e^{-y^2} dy. \tag{F-4}$$

Spitzer gave the values of $G(x)$ in a table, which is plotted in Fig. F-1 and fitted. For $x = l^2 v$ greater than 1.8, a power function fits best with the relation $G(x) = 0.4638x^{-1.957}$.

In our case, the field particles and the test particles have the same mass, which is the electron mass, and charge $Z = e$. The slowing down time is plotted in Fig. F-2 as a function of the primary particle energy for three representative plasma densities found in the discharge chamber and near the grids of ion thrusters.

For 15 eV primaries in the discharge chamber plasma with an average temperature of 4 eV and a density approaching 10^{18} m^{-3}, the slowing down time is about 10^{-6} s. The slowing down time is also plotted in Fig. F-3 as a function of the plasma density for several values of the primary electron

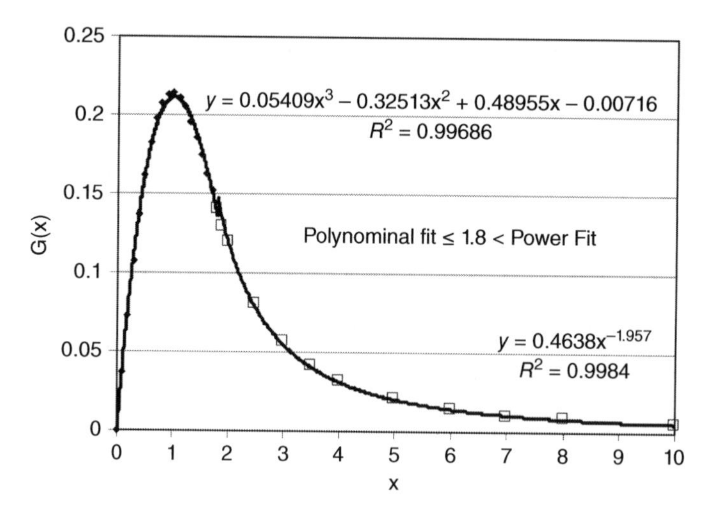

Figure F-1 Spitzer's $G(x)$ with curve fits.

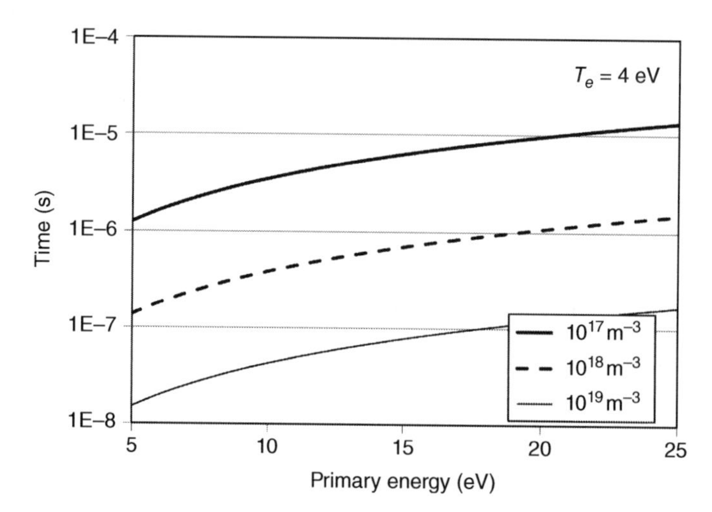

Figure F-2 Spitzer's slowing down time as a function of the primary electron energy for three densities of electrons at 4 eV.

energy, again assuming the plasma has an electron temperature of about 4 eV. As the plasma density increases, the slowing down time becomes very small ($<10^{-6}$ s). This will lead to rapid thermalization of the primary electrons.

For the case of primary electron with some spread in energy, we can examine the time for the equilibration between that population and the plasma electrons. Assuming that the primaries have a temperature T_1 and the plasma electrons have a temperature T_2, the time for the two populations to equilibrate is

$$\tau_{\text{eq}} = \frac{3m^{1/2}(kT_1 + kT_2)^{3/2}}{8(2\pi)^{1/2} n e^4 \ln\Lambda}. \tag{F-5}$$

As an example, the slowing time for monoenergetic primaries and primaries with a Maxwellian distribution of energies injected into a 4 eV plasma is shown in Fig. F-4. The slowing time is significantly faster than the equilibration time.

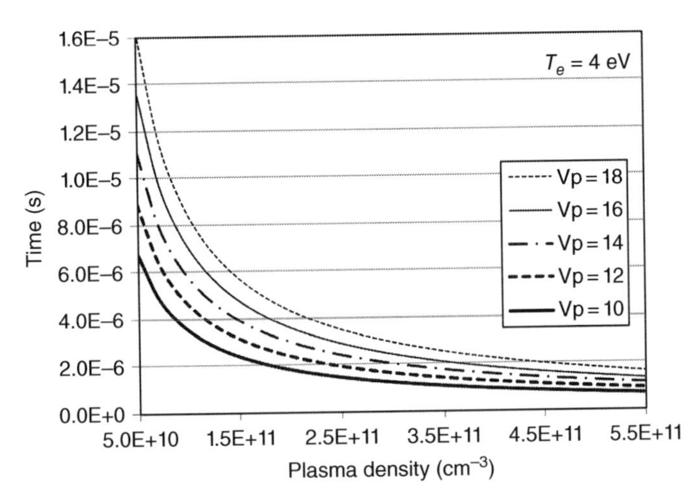

Figure F-3 Spitzer's slowing down time as a function of the plasma density with an electron temperature of 4 eV for several primary electron energies in eV.

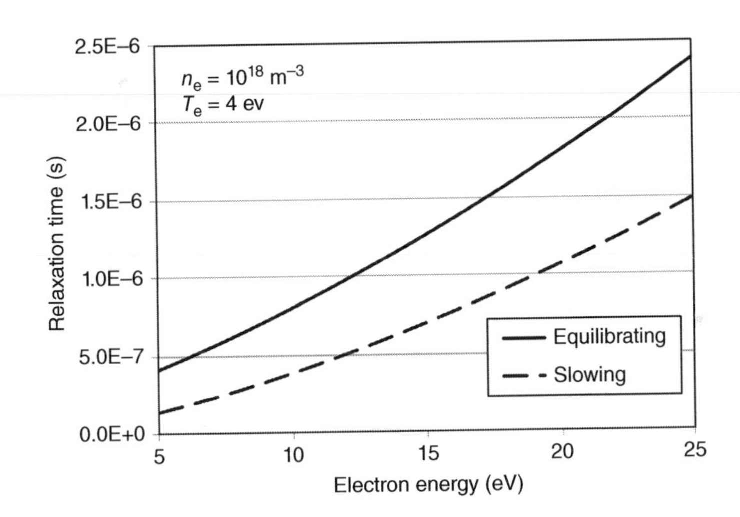

Figure F-4 Relaxation times of monoenergetic primaries and a Maxwellian primary population in a 4 eV, 10^{18} m^{-3} plasma.

References

1 L. Spitzer, Jr., *Physics of Fully Ionized Gases*, New York: Interscience, pp. 127–135, 1962.

2 D. L. Book, *NRL Plasma Formulary*, Washington: Naval Research Laboratory, pp. 33–34, 38, 1987.

Appendix G

Clausing Factor Monte Carlo Calculation

Visual Basic Monte-Carlo calculation of Clausing Factor for thruster grids.

Inputs:

Clausing Factor Calculator		
Inputs	**Radius mm**	**Diameter**
thickScreen	0.381	
thickAccel	0.5	
rScreen	0.9525	1.905
rAccel	0.5715	1.143
gridSpace	0.5	
npart	1.00E+05	

Result: Clausing factor = 0.314

Code:

```
Sub Clausing()
    thickScreen = Range("C4")
    thickAccel = Range("C5")
    rScreen = Range("C6")
    rAccel = Range("C7")
    gridSpace = Range("C8")
    npart = Range("C9")
'   Monte Carlo Routine that calculates Clausing factor for CEX
'   returns Clausing Factor and Downstream Correction factor
   Dim gone As Boolean
        Pi = 3.14159265358979
'assumes rTop = 1
    rBottom = rScreen / rAccel
    lenBottom = (thickScreen + gridSpace) / rAccel
    lenTop = thickAccel / rAccel
    Length = lenTop + lenBottom
        iescape = 0
        maxcount = 0
```

Fundamentals of Electric Propulsion, Second Edition. Dan M. Goebel, Ira Katz, and Ioannis G. Mikellides.

```
        icount = 0
        nlost = 0
            vztot = 0#
            vz0tot = 0#
    For ipart = 1 To npart
     ' launch from bottom
            notgone = True
            r0 = rBottom * Sqr(Rnd)
            z0 = 0#
            costheta = Sqr(1# - Rnd)
            If (costheta > 0.99999) Then costheta = 0.99999
            phi = 2 * Pi * Rnd
            sintheta = Sqr(1# - costheta ^ 2)
            vx = Cos(phi) * sintheta
                vy = Sin(phi) * sintheta
                vz = costheta
                rf = rBottom
                t = (vx*r0 + Sqr((vx^2 + vy^2) * rf^2 -(vy* r0) ^2)) /
(vx^2 + vy^2)
                z = z0 + vz * t
                vz0tot = vz0tot + vz
                    icount = 0
            Do While notgone
                    icount = icount + 1
                If (z < lenBottom) Then
                ' hit wall of bottom cylinder and is re-emitted
                        r0 = rBottom
                        z0 = z
                        costheta = Sqr(1# - Rnd)
                        If (costheta > 0.99999) Then costheta = 0.99999
                        phi = 2 * Pi * Rnd
                        sintheta = Sqr(1# - costheta ^ 2)
                        vz = Cos(phi) * sintheta
                        vy = Sin(phi) * sintheta
                        vx = costheta
                        rf = rBottom
                        t = (vx*r0 + Sqr((vx^2+vy^2)*rf^2-(vy*r0)^2))/
(vx^2 +vy^2)
                        z = z0 + t * vz
                End If ' bottom cylinder re-emission
                If ((z >= lenBottom) And (z0 < lenBottom)) Then
                ' emitted below but going up
                    ' find radius at lenBottom
                        t = (lenBottom - z0) / vz
                        r = Sqr((r0 - vx * t) ^ 2 + (vy * t) ^ 2)
                    If (r <= 1) Then
                    '  continuing upward
```

```
                        rf = 1#
                        t = (vx*r0 + Sqr((vx^2+vy^2)*rf^2-(vy*r0)^2))/
(vx^2 +vy^2)
                        z = z0 + vz * t
                    Else
                    '  hit the upstream side of the accel grid and is
re-emitted downward
                        r0 = r
                        z0 = lenBottom
                        costheta = Sqr(1# - Rnd)
                        If (costheta > 0.99999) Then costheta = 0.99999
                        phi = 2 * Pi * Rnd
                        sintheta = Sqr(1# - costheta ^ 2)
                            vx = Cos(phi) * sintheta
                            vy = Sin(phi) * sintheta
                            vz = -costheta
                        rf = rBottom
                        t = (vx*r0 + Sqr((vx^2+vy^2)*rf^2-(vy*r0)^2))/
(vx^2 +vy^2)
                        z = z0 + vz * t
                    End If
                End If ' end upward
                If ((z >= lenBottom) And (z <= Length)) Then
                    '  hit the upper cylinder wall and is re-emitted
                        r0 = 1#
                        z0 = z
                        costheta = Sqr(1# - Rnd)
                        If (costheta > 0.99999) Then costheta = 0.99999
                        phi = 2 * Pi * Rnd
                        sintheta = Sqr(1# - costheta ^ 2)
                        vz = Cos(phi) * sintheta
                        vy = Sin(phi) * sintheta
                        vx = costheta
                        rf = 1#
                        t = (vx*r0 + Sqr((vx^2+vy^2)*rf^2-(vy*r0)^2))/
(vx^2 +vy^2)
                        z = z0 + t * vz
                        If (z < lenBottom) Then
                        ' find z when particle hits the bottom cylinder
                            rf = rBottom
                        If ((vx ^ 2 + vy ^ 2) * rf ^ 2 - (vy * r0) ^ 2 <
0#) Then
                             t = (vx * r0) / (vx ^ 2 + vy ^ 2) 'if sqr
arguement is less than 0 then set sqr term to 0 12 May 2004
                            Else
                            t=(vx*r0+Sqr((vx^2+vy^2)*rf^2(vy*r0)^2))/
(vx^2+vy^2)
```

```
                        End If
                        z = z0 + vz * t
                    End If
                End If ' end upper cylinder emission
                If (z < 0#) Then
                    notgone = False
                End If
                If (z > Length) Then
                    iescape = iescape + 1
                    vztot = vztot + vz
                    notgone = False
                End If
                If (icount > 1000) Then
                    notgone = False
                    icount = 0
                    nlost = nlost + 1
                End If
            Loop  ' while
            If (maxcount < icount) Then maxcount = icount
        Next ipart ' particles
        Range("C11") = (rBottom^2) * iescape / npart ' Clausing factor result
            vz0av = vz0tot / npart
            vzav = vztot / iescape
        DenCor = vz0av / vzav ' Downstream correction factor
    End Sub ' Clausing
```

Index

a

Fundamentals of Electric Propulsion, Second Edition. Dan M. Goebel, Ira Katz, and Ioannis G. Mikellides.

d

i

j

k

l

m

n

O

p

q

t

U

Printed and bound by CPI Group (UK) Ltd, Croydon, CR0 4YY

22/04/2025

14659799-0001